Neurophysiological Monitoring during Intensive Care and Surgery

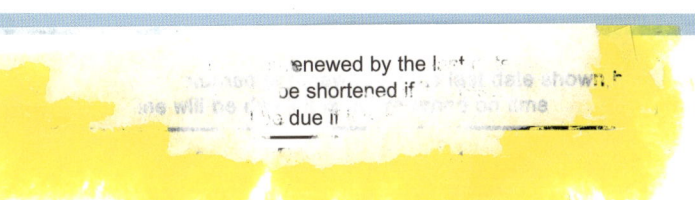

For Elsevier Mosby:

Commissioning Editor: Alison Taylor
Development Editor: Kim Benson
Production Manager: Yolanta Motylinska
Project Manager: Helius
Cover Design: Kneath Associates

Neurophysiological Monitoring during Intensive Care and Surgery

EDITED BY

N Jollyon Smith MA FRCP
Consultant Clinical Neurophysiologist, University Hospital, Nottingham, UK

Mark van Gils PhD
Senior Research Scientist, VTT Technical Research Centre of Finland, Tampere, Finland

Pamela Prior MD FRCP
Consultant Clinical Neurophysiologist, St. Bartholomew's and the Royal London Hospitals, West Smithfield, London, UK

FOREWORD BY: Professor Arvi Yli-Hankala
Research Professor and Anaesthetist, Department of Anaesthesia, University of Tampere, Finland

MOSBY
ELSEVIER

EDINBURGH   LONDON   NEW YORK   OXFORD   PHILADELPHIA   ST LOUIS   SYDNEY   TORONTO   2006

ELSEVIER | MOSBY

© 2003 Elsevier B.V., 2004 Elsevier B.V., Amsterdam, The Netherlands.

© 2006 this compilation, Elsevier Ltd. All rights reserved.

First published 2006

No part of this publication may be reproduced, stored in a retrieval system, or transmitted in any form or by any means, electronic, mechanical, photocopying, recording or otherwise, without either the prior permission of the publishers or a licence permitting restricted copying in the United Kingdom issued by the Copyright Licensing Agency, 90 Tottenham Court Road, London W1T 4LP. Permissions may be sought directly from Elsevier's Health Sciences Rights Department in Philadelphia, USA: phone: (+1) 215 239 3804, fax: (+1) 215 239 3805, e-mail: healthpermissions@elsevier.com. You may also complete your request on-line via the Elsevier homepage (http://www.elsevier.com), by selecting 'Customer Support' and then 'Obtaining Permissions'.

ISBN-13: 978 0 7234 3381 1
ISBN-10: 0 7234 3381 X

British Library Cataloguing in Publication Data
A catalogue record for this book is available from the British Library.

Library of Congress Cataloging in Publication Data
A catalog record for this book is available from the Library of Congress.

Note
Knowledge and best practice in this field are constantly changing. As new research and experience broaden our knowledge, changes in practice, treatment and drug therapy may become necessary or appropriate. Readers are advised to check the most current information provided (i) on procedures featured or (ii) by the manufacturer of each product to be administered, to verify the recommended dose or formula, the method and duration of administration, and contraindications. It is the responsibility of the practitioner, relying on their own experience and knowledge of the patient, to make diagnoses, to determine dosages and the best treatment for each individual patient, and to take all appropriate safety precautions. To the fullest extent of the law, neither the Publisher nor the Editors assumes any liability for any injury and/or damage to persons or property arising out of or related to any use of the material contained in this book.
The Publisher

Working together to grow
libraries in developing countries

www.elsevier.com | www.bookaid.org | www.sabre.org

ELSEVIER   BOOK AID   Sabre Foundation
             International

ELSEVIER your source for books,
          journals and multimedia
          in the health sciences
www.elsevierhealth.com

The publisher's policy is to use paper manufactured from sustainable forests

Printed in China

Contributors

Colin D Binnie MD FRCP
Emeritus Professor of Clinical Neurophysiology
Department of Clinical Neurophysiology
King's College Hospital
London, UK

Borka Ceranic MD ENTspec PhD
Consultant in Audiological Medicine
St. George's Hospital
London, UK

Ray Cooper BSc PhD
Formerly Scientific Director
Burden Neurological Institute
Bristol, UK

Clare J Fowler FRCP
Professor of Uro-Neurology
Institute of Neurology
The National Hospital for Neurology and Neurosurgery
London, UK

Luis García-Larrea MD PhD
Research Director
Department of Clinical Neurophysiology
Neurological Hospital Pierre Wertheimer
Lyon, France

Graham E Holder PhD
Director of Electrophysiology
Moorfields Eye Hospital
London, UK

Linda M Luxon BSc(Hons) MBBS (Hons) FRCP
Professor of Audiological Medicine
University of London
Institute of Child Health
University College London
Great Ormond Street Hospital
London, UK
and
Department of Neuro-otology
National Hospital for Neurology and Neurosurgery
London, UK

Bruce B MacGillivray FRCP
Consulting Clinical Neurophysiologist
Department of Clinical Neurophysiology
The Royal Free Hospital
London, UK
and
Consultant in Clinical Neurophysiology
The National Hospital for Neurology and Neurosurgery
London, UK

François Mauguière MD PhD
Professor of Neurology
Department of Functional Neurology and Epileptology
Neurological Hospital Pierre Wertheimer
Lyon, France

Kulandavelu Nagendran MB FRCP
Consultant Clinical Neurophysiologist
Department of Clinical Neurophysiology
St. Bartholomew's and the Royal London Hospitals
London, UK

John W Osselton BSc
Formerly Senior Lecturer in EEG
University of Newcastle on Tyne
Newcastle on Tyne, UK

John C Shaw PhD BSc CEng MIEE *(deceased)*
Formerly Medical Research Council
Clinical Psychiatry Research Unit
Chichester, UK

Brian M Tedman BSc BMedSci BM BS PhD FRCP
Consultant Clinical Neurophysiologist
The Walton Centre for Neurology and Neurosurgery
Liverpool, UK

Tom Wisman
Product Manager
Tefa-Portanje
Woerden, The Netherlands

Foreword

The principal aim of neurophysiological monitoring is to prevent damage to the nervous system. Achieving this aim places great demands on both personnel and equipment. Traditionally the need for monitoring has been determined by clinicians (surgeons, anaesthetists, intensivists), and it has been provided by clinical neurophysiologists acting as visiting consultants in the operating theatre and intensive care unit. While there can be no doubt that neurophysiological monitoring is vital for the patient at risk of permanent neurological damage, monitoring presents challenging practical problems for anaesthetist and neurophysiologist. On the one hand, the neurophysiologist finds that the level of electrical interference from patient-connected anaesthetic and surgical equipment is an obstacle to recording low amplitude signals from the nervous system; on the other hand, the anaesthetist may be asked to avoid neuromuscular blocking drugs and powerful anaesthetic agents, so that the task of keeping the patient asleep and immobile is more difficult. These competing requirements should be approached in a spirit of cooperative teamwork and compromise.

In the intensive care unit neurophysiological recordings can give diagnostic and prognostic information, and sometimes influence treatment. Prolonged monitoring of brain function allows clinicians to recognize spontaneous fluctuations of activity in comatose patients, the arousal responses of inadequately sedated patients, and the previously unrecognized incidence of epileptiform activity, particularly after head injury or neurosurgical procedures. For these reasons neurophysiological monitoring is being used more often in general as well as neuro-intensive care units.

During surgical operations monitoring has been used at the request of the surgeon to detect impending damage to neural pathways, the onset of ischaemia, and the development of epileptiform discharges. In addition, the possibility of unintentional awareness has now become an important legal issue for anaesthetists, and any means of avoiding it is warmly welcomed. As a result of the great increase in power of readily-available computers in the 1990s, monitoring devices began to appear which extracted certain characteristics of the EEG or an EP, and in some cases compressed them into a single number thought to indicate clinically important states of the nervous system. Enthusiastic marketing ensured that anaesthetists and surgeons became interested in brain waves; now they had a box to tell the anaesthetist how close the patient was to awareness, or at least that was what the makers claimed it could do.

As often happens following the introduction of new technology, clinicians were divided into the enthusiasts, the sceptics, and those waiting for compelling evidence before deciding one way or the other. According to the enthusiasts, EEG-monitored anaesthesia reduces the incidence of unintentional awareness, shortens recovery time, and makes anaesthesia a pleasanter experience for the patient, who is more alert and less nauseous postoperatively. The enthusiasts therefore advocate monitoring every anaesthetized patient with an EEG or EP index. Sceptics, on the other hand, point out that some anaesthetic agents do not affect the EEG or EPs in the expected manner, and so render the indices unreliable. Poor electrode application, external electrical interference, and muscle and movement artefacts produce misleading results. Because the indices are validated in normal subjects, they are less reliable, and in some cases useless, in the presence of epileptiform discharges or the EEG patterns produced by encephalopathies. The reluctance of some manufacturers to publish the algorithms underlying their indices has only served to heighten scepticism. The situation is not helped by the fact that few anaesthetists or surgeons are expert in neurophysiology, and merely follow the behaviour of an index number, without understanding the EEG or EP signal from which it is derived; some have even attempted to detect cerebral ischaemia with monitoring devices designed for prevention of awareness. Editors of journals therefore have a particularly difficult task in ensuring that published reports on surgery and anaesthesia monitored by EEG or EPs are of the same scientific standard as those on other subjects.

While EEG and EP monitors undoubtedly have their place in the operating theatre, their precise contribution needs to be clarified. They must not be regarded as replacing conventional EEG or EPs, but merely as indicating the effect upon the brain of clearly defined and rigidly specified anaesthetic and surgical techniques. The inappropriate perioperative use of neurophysiological methods will only be avoided by educating anaesthetists and surgeons. They should not be expected to become neurophysiologists, but they should understand the physiology of neuronal activity, and the basics of signal acquisition and interpretation. In particular, anaesthetists should be encouraged to demand a high quality display of the raw EEG or EP along with a depth-of-anaesthesia index, just as they would expect to see the ECG while monitoring heart rate. They must learn to distinguish the EEG from artefacts, because analysing garbage will not help patients. Finally, anaesthetists and surgeons who understand neurophysiology will take a critical attitude to what is published, and thereby prevent forests of paper and barrels of ink from being wasted in reporting studies of questionable value.

Perioperative neurophysiological monitoring is important. Properly performed, it can save lives and vital tissue, and diminish human suffering. This book is an important contribution in making neurophysiology understandable for the clinician.

Arvi Yli-Hankala MD PhD
Research Professor and Anaesthetist
Department of Anaesthesia
University of Tampere
Finland

Preface

This book is derived from the two volumes entitled *Clinical Neurophysiology* (Binnie et al, 2003, 2004). We have compiled it using material extracted, rearranged and expanded from these parent volumes to cater for the particular needs of those working in neurophysiological monitoring. Such monitoring in the intensive care unit and the operating theatre has developed into a rapidly expanding subspecialty of clinical neurophysiology, to which many neurophysiologists and their technologists now devote a substantial part of their professional lives. At the same time, it was appreciated that colleagues in other specialties needed to understand what could – and, more important, what could not – be expected from monitoring. Also, in many hospitals, monitoring was being carried out by anaesthetists, surgeons, intensivists and others who lacked a firm grounding in the basics of clinical neurophysiology.

There are several important differences between monitoring and other branches of clinical neurophysiology. Firstly, the principal aim of monitoring is to warn of deterioration, and thereby to avoid catastrophe, rather than to provide diagnostic information. Secondly, the population norm is of much less importance because the patient usually serves as his or her own control. Thirdly, monitoring has found applications in many apparently disparate branches of surgery and intensive care, and no one can claim to be expert, or even reasonably competent, in all of them. We have tried to keep these points in mind in preparing this book, and to provide sufficient explanation of the neurophysiological background for those entering the field from other specialties, including biomedical engineers and medico-legal experts as well as clinicians and clinical scientists. Nevertheless, in many cases the reader is referred to one or other of the parent volumes for an account of the basics; more detailed information on neurophysiological technology is given in Cooper, Binnie and Billings (2005), which provides a convenient companion to this present volume.

We are acutely aware that neurophysiological monitoring is still developing rapidly, so that parts of the present volume may soon be out of date, if they are not already. Faced with such an actively developing subject, a printed book is perhaps not the best way of presenting information; an Internet version would provide greater possibilities of regular updating if the demand arises.

Finally, we must reiterate the injunction with which the Editors of the parent volumes concluded their Preface. Although the information in this book is correct to the best of our knowledge, no one should attempt to carry out any of the techniques described without having undertaken adequate, recognized training, and without consulting up-to-date local, national and international safety regulations, and local, national and international recommendations on good practice.

N Jollyon Smith
Mark van Gils
Pamela Prior

REFERENCES

Binnie CD, Cooper R, Mauguière F, Osselton JW, Prior PF, Tedman BM 2003 *Clinical Neurophysiology*, Volume 2: *EEG, Paediatric Neurophysiology, Special Techniques and Applications*. Churchill Livingstone, Amsterdam.

Binnie CD, Cooper R, Mauguière F, Osselton JW, Prior PF, Tedman BM 2004 *Clinical Neurophysiology*, Volume 1: *EMG, Nerve Conduction and Evoked Potentials*, revised and enlarged edition. Elsevier, Amsterdam.

Cooper R, Binnie CD, Billings R 2005 *Techniques in Clinical Neurophysiology: A Practical Manual*. Churchill Livingstone, Edinburgh.

Dedication

We were encouraged to prepare the present work by our good friend and colleague, the distinguished biomedical engineer, Professor Annelise Rosenfalck who died just as we were nearing its completion.

Annelise Rosenfalck will probably be remembered principally for her pioneering work on EMG and nerve conduction with Fritz Buchthal and her husband Poul Rosenfalck in Copenhagen from the 1950s (Trojaborg et al 2005). However, in the 1990s she had an important role in two multidisciplinary projects co-funded by the European Union; these were 'Improving control of patient status in critical care' (Biomed-I project IMPROVE; Carson and Saranummi 1996, Cerutti and Saranummi 1997) and 'Improved monitoring for brain dysfunction in intensive care and surgery' (Biomed-II project IBIS; Carson et al 2000). Annelise Rosenfalck was one of the greatest driving forces in 'bridging the communication gap' between clinicians and engineers on numerous occasions. Her warm personal interest in people and talent to create the most unexpected synergic results will be sorely missed.

REFERENCES

Carson E, Saranummi N (eds) 1996 Improving control of patient status in critical care: the IMPROVE project. Comput Method Program Biomed 51(1,2):1–130.

Carson E, Van Gils M, Saranummi N (eds) 2000 Special Issue. IBIS: Improved Monitoring for Brain Dysfunction in Intensive Care and Surgery. Comput Method Program Biomed 63(3):157–235.

Cerutti S, Saranummi N (eds) 1997 Improving control of patient status in critical care. IEEE Eng Med Biol Mag 16:19–79.

Trojaborg W, Arendt-Nielsen L, Krarup C 2005 Obituary: Annelise Rosenfalck 1922–2004. Clin Neurophysiol 116:991–992.

Acknowledgements

We have received a considerable amount of good advice and practical help from many people and, in particular, we would like to thank Jessica Clark, Ray Cooper, John Gade, Ilkka Korhonen, Tarmo Lipping, Gerlinde Mandersloot, Douglas Maynard, Niilo Saranummi, Carsten Thomsen, Maarten van de Velde, Peter Weller and Arvi Yli-Hankala for their generosity. Our publishers and production team have given us great support and we are grateful to Nello Spiteri, Elly Tjoa, Caroline Makepeace and Peter Bakker for the original planning and to Kim Benson, Julie Gorman and Alison Taylor for its implementation. As always our patients and clinical colleagues have provided inspiration and wise words and our families have given us the strength to complete the task.

Abbreviations

AAEM	American Association of Electrodiagnostic Medicine (formerly AAEE)	CD ROM	computer disc read-only memory
		CDSA	colour density spectral array
		CE	Conformité Européenne
AAI	auditory evoked potential index	CEN	European Committee for Standardization
AC	alternating current		
ACAS	Asymptomatic Carotid Artherosclerosis Study Group	CEPOD	Confidential Enquiry into Perioperative Deaths
acf	autocorrelation function	CFAM	cerebral function analysing monitor
AD	analogue to digital	CFM	cerebral function monitor
ADAM	Advanced Depth of Anaesthesia Monitor	CJD	Creutzfeldt–Jakob disease
		CMAP	compound muscle action potential
AEP	auditory evoked potential	CMRR	common-mode rejection ratio
AI	artificial intelligence	CNE	concentric needle electrode
AICA	anteroinferior cerebellar artery	CN-EMG	concentric needle electrode EMG
AIDP	acute inflammatory demyelinating polyneuropathy	CNS	central nervous system
		CNV	contingent negative variation
AIDS	acquired immune deficiency syndrome	CPAP	continuous positive airway pressure
AMAN	acute motor axonal neuropathy	CPB	cardiopulmonary bypass
AMSAN	acute motor sensory axonal neuropathy	CPP	cerebral perfusion pressure
ANN	artificial neural network	cpd	cycles per degree (spatial frequency)
AP, APG IEC	marks relating to equipment explosion risks with flammable vapours	CSA	compressed spectral array
		CSF	cerebrospinal fluid
ApEn	approximate entropy	CT	computed tomography
AR	autoregressive	CUSA	Cavitron ultrasonic surgical aspirator
ARMA	autoregressive moving average	CVA	cerebral vascular accident
ASTM	American Society for Testing and Materials	CVP	central venous pressure
		CWT	continuous wavelet transform
AWF	airway flow		
AWP	airway pressure	dB	decibel
		DC	direct current, directly coupled (amplifier)
BAEP	brainstem auditory evoked potential		
BIS	Bispectral Index	DFT	discrete Fourier transform
BRAC	basic rest and activity cycle	DSA	density spectral array
BSI	biosignal interpretation	DSA	density-modulated spectral array
BSI	British Standards Institute	DSP	digital signal processor
BSR	burst suppression ratio	DWT	discrete wavelet transform
BUN	black up negative		
		ECG	electrocardiogram, electrocardiograph, electrocardiographic
C	central (electrode site)		
C	capacitance		
CABG	coronary artery bypass graft (surgery)	ECochG	electrocochleogram, electrocochleographic
CAP	compound action potential		
CAP	cyclic alternating pattern	ECS	electrocerebral silence
CBF	cerebral blood flow	ECST	European Carotid Surgery Trialists' Collaborative Group
CBF_{isi}	initial slope index of CBF		
CCT	central conduction time	EDF	European data format

Abbreviations

EEG	electroencephalogram, electroencephalograph, electroencephalography	kΩ	kilo-ohm
		kHz	kilohertz
ELAE	episodic low amplitude event	kPa	kilopascal
EMG	electromyogram, electromyograph, electromyographic	LAN	local area network
EOG	electro-oculogram	LED	light-emitting diode
EP	evoked potential	LF	low frequency
EPSP	excitatory postsynaptic potential	LL	lateral lemniscus
ERG	electroretinogram, electroretinography, electroretinographic	LLAEP	long latency auditory evoked potential
		LMS	least mean squares
ERP	event-related potential	LTI	linear time-invariant
F	frontal (electrode site)	MAC	minimum alveolar concentration
FFT	fast Fourier transform	MAP	mean arterial pressure
FIR	finite impulse response	MCV	motor conduction velocity
FIRDA	frontal intermittent rhythmic delta activity	MFS	Miller–Fisher syndrome
		MLAEP	middle latency auditory evoked potential
fMRI	functional MRI	MN-SEP	median nerve stimulation SEP
FN	false negatives	MPF	median power frequency
Fp	frontal pole (electrode site)	MRI	magnetic resonance imaging
FP	false positives	MRSA	methicillin-resistant *Staphylococcus aureus*
FT	Fourier transform	ms	millisecond
		mV	millivolt
GBS	Guillain–Barré syndrome	MΩ	mega-ohm
GCS	Glasgow Coma Scale, Glasgow Coma Scale score	μA	microampere
		μF	microfarad
		μV	microvolt
HF	high frequency		
HIV	human immunodeficiency virus	NAP	nerve action potential
HL	hearing level	8NAP	eighth cranial nerve action potential
Hz	hertz, cycles per second (frequency)	NASCET	North American Symptomatic Carotid Endarterectomy Trial Collaborators
IBI	interburst intervals	NCEPOD	National Confidential Enquiry into Perioperative Deaths
IC	inferior colliculus		
ICA	independent component analysis	NCS	nerve conduction studies
ICP	intracranial pressure	nHL	normal hearing level
ICU	intensive care unit	NICE	National Institute for Clinical Excellence
ID	identity		
IEC	International Electrochemical Commission	NLEO	non-linear energy operator
IEEE	Institute of Electrical and Electronic Engineers	O	occipital (electrode site)
		OAA/S	Observer's Assessment of Alertness/Sedation scale
IFCN	International Federation for Clinical Neurophysiology (previously IFSECN)	OIRDA	occipital intermittent rhythmic delta activity
IFSECN	International Federation of Societies for Electroencephalography and Clinical Neurophysiology (now IFCN)	OR	operating room
		OSET	International Organisation of Societies for Electrophysiological Technology
IFMBE	International Federation for Medical and Biological Engineering		
IIR	infinite impulse response	P	parietal (electrode site)
IPI	interpeak interval	Pa	pascal
IPSP	inhibitory postsynaptic potential	PaCO$_2$	arterial blood carbon dioxide tension
ISO	International Organization for Standardization	PaO$_2$	arterial blood oxygen tension
		PAP	pulmonary artery pressure
IT	information technology	PC	personal computer
i.v., i/v	intravenous	PCA	principal component analysis

PCO$_2$	carbon dioxide tension	SAP	systemic arterial pressure
PFA	principal factor analysis	SAP	systolic arterial pressure
PLED	periodic lateralized epileptiform discharge	SAS	sedation–agitation scale
		SD	standard deviation
PPF	peak power frequency	SE	state entropy
ppm	parts per million	SEF	spectral edge frequency
PSD	power spectral density	SEF90	SEF marker at 90% of spectral range
PSI	Patient State Index	SEF95	SEF marker at 95% of spectral range
PSP	post-synaptic potential	SEP	somatosensory evoked potential
PVS	persistent vegetative state, permanent vegetative state	SFEMG	single fibre electromyography
		SI	Système Internationale
		SL	sensation level
QALY	quality of life measurement unit	SMA	supplementary motor area
qEEG	quantitative EEG	SNAP	sensory nerve action potential
QEEG	quantitative EEG	SNR	signal to noise ratio
QUOL, QOL	quality of life	SOM	self-organizing map
		SPL	sound pressure level
RAM	random access memory	SSPE	subacute sclerosing panencephalitis
RASS	Richmond agitation–sedation scale	STFT	short-time Fourier transform
R&D	research and development		
rCBF	regional cerebral blood flow	T	temporal (electrode site)
RE	response entropy	T	tesla
RF	radiofrequency	TC	time constant
rINN	recommended international non-proprietary name (of drugs)	TCI	target controlled infusion
		TN	true negative
r.m.s., RMS	root mean square	*TP*	true positive
ROC	receiver operating characteristic		
ROM	read-only memory	V	vertex (electrode site)
		VDU	video display unit, visual display unit
SAH	subarachnoid haemorrhage	VEP	visual evoked potential
SaO$_2$	arterial oxygen saturation		
SAP	sensory action potential	WAN	wide area network

Contents

Contributors   *v*
Foreword   *vii*
Preface   *ix*
Dedication   *x*
Acknowledgements   *x*
Abbreviations   *xi*

1 The why and the how of neurophysiological monitoring in the ICU and surgery   *1*

Introduction   *1*
Applications of clinical neurophysiology in the ICU   *1*
Monitoring during surgical operations   *3*

Assessment of results   *7*
Statistical significance of changes   *7*
Clinical significance of changes   *8*
Confidence intervals   *10*
Determining the optimal cut-off point   *11*
Comparing two methods of monitoring   *11*
The effect of corrective action   *12*
Is the risk eliminated by monitoring?   *13*

References   *14*

2 Neurophysiological instrumentation, connection with patients and recording methods   *17*

Introduction   *17*

Recording the electrical activities of the nervous system   *18*
Electrodes for neurophysiological recording   *18*
Electrode placement systems   *24*
Connecting electrodes to amplifiers and the recording convention   *29*
Amplifiers   *32*
Signal bandwidth and filters   *35*
Digital systems   *37*

EEG recording equipment   *38*
General features of EEG machines   *38*
Input circuits for EEG recording   *39*
Amplifiers for EEG recording   *39*
Filters for EEG recording   *40*
Analogue EEG systems   *40*
Digital EEG systems   *42*

Evoked potential recording systems   *48*
Electrodes for EP recording   *48*
Amplifiers for EP recording   *48*
Filters for EP recording   *48*
Averaging   *48*
EP display   *49*
Calibration of EP recorders   *49*
EP stimulators   *49*

EMG and nerve conduction studies   *53*
Electrodes for EMG and nerve conduction studies   *53*
Amplifiers for EMG and nerve conduction   *54*
Displaying EMG and nerve conduction data   *55*
Stimulation for nerve conduction studies   *55*

Practical aspects of intraoperative and ICU neurophysiological recording   *55*
Safety during neurophysiological recording in the ICU and operating theatres   *55*
Intraoperative and ICU EEG recording   *59*
Intraoperative and ICU EP recording   *65*
EMG and nerve conduction studies   *68*

References   *68*

3 Introduction to methods for continuous EEG and evoked potential monitoring   *73*

Introduction   *73*

EEG monitoring   *73*
Historical development of EEG monitors   *73*
Technical requirements for EEG monitoring in the ICU and during surgical operations   *85*

Evoked potential monitoring   *94*
Technical requirements for EP monitoring   *94*

Comparisons between currently used methods   *100*

Conclusions   *102*

References   *103*

4 Normal and pathological phenomena in EEG, evoked potential, EMG and nerve conduction studies 109

General introduction 109

EEG WAVEFORMS AND INTERPRETATION 109
Introduction 109
Describing EEG phenomena 109
Wave shape (morphology) 110
Rhythmicity 110
Frequency of repetition 110
Amplitude 110
Transients 112
Spatial distribution 113
Spatiotemporal patterns 113
Symmetry and synchrony 113
Inherent variability of the EEG and other biological rhythms 113
Reactivity 114

General categories of abnormality in the EEG of importance for ICU and intraoperative monitoring 114
Change in frequency content 114
Amplitude reduction 115
Localized and lateralized abnormalities 116
Rhythms at a distance (projected rhythms) 117
Altered reactivity 118
Epileptiform activity 119
Periodicity 120
Burst suppression pattern 123
Electrocerebral inactivity – electrocerebral silence (ECS) – the isoelectric EEG 124

EVOKED POTENTIAL WAVEFORMS AND INTERPRETATION 125
General definition – limitations – clinical utility of evoked potentials 125
Responses and EP components 126
The electroretinogram and electrocochleogram 126
Relation between neuronal responses and surface evoked potentials 127
Action potentials 127
Postsynaptic potentials 128
Near-field versus far-field evoked potentials 128
How to localize evoked potential sources from surface recordings 130
Normal findings by modality 131
Flash VEPs 131
Auditory evoked potentials: BAEPs and MLAEPs 132
Somatosensory evoked potentials 134

EMG FINDINGS AND INTERPRETATION 145
Features of motor units recorded by needle electrodes 145
Other normal EEG phenomena 145
Insertion activity 145
End-plate noise 145
Fibrillations at single sites 146
Fasciculations 146
Nerve conduction 147
Effects of limb temperature on nerve conduction 147
References 148

5 Neurophysiological monitoring during sedation and anaesthesia 155

Introduction 155
Sedation: assessment with EEG and evoked potentials 155
Effects of sedative drugs on the EEG 155
Effects of sedative drugs on EPs 157
Assessment of sedation 159
Anaesthesia: assessment with EEG and evoked potentials 161
Effect of anaesthetic agents on the EEG 161
Effect of anaesthetic agents on EPs 172
Combined EEG and evoked potential measures 176
Awareness during anaesthesia 176
Neurophysiological features useful in prediction of possible awareness during anaesthesia 177
Medico-legal aspects of awareness during anaesthesia 180
References 182

6 Neurophysiological work in the ICU 189

Introduction 189
Neurophysiological parameters to be monitored and procedures 189
Sleep in the ICU 189

CLINICAL CONDITIONS AFFECTING THE CENTRAL NERVOUS SYSTEM ENCOUNTERED IN THE ICU 189
Coma and related states 189
EEG recording and interpretation in coma and related states 189
EPs in comatose patients 199
EPs combined with other variables 215
Cardiac arrest and hypoxic–ischaemic encephalopathies 216
Other metabolic and toxic encephalopathies and multiple organ failure 218
Encephalitis 218
Epileptiform discharges and status epilepticus 218
Head injury and other neurosurgical applications 221

Vegetative states and brainstem death (brain death) 225
Vegetative states 225
Brainstem death (brain death) 228

CLINICAL CONDITIONS AFFECTING THE PERIPHERAL NERVOUS SYSTEM AND MUSCLES ENCOUNTERED IN THE ICU 231
Neuromuscular syndromes of critical illness (critical illness neuropathy) 231
Acute onset neuropathies 234
Guillain–Barré syndrome (GBS) 234
Acute intermittent porphyria 237
Other causes of acute or subacute peripheral neuropathy 242

Acute weakness due to disorders of neuromuscular transmission 242
Botulism 242
Myasthenia gravis 242
Familial periodic paralysis and the channelopathies 242
Non-peripheral causes of acute onset generalized weakness 243

Conclusions on present status of neurophysiological work in the ICU 243
References 243

7 Neurophysiological monitoring during surgical operations 253

Introduction 253
Intracranial surgery 253
EEG monitoring in intracranial surgery 253
Visual evoked potential monitoring 256
SEP monitoring in intracranial surgery 257
Posterior fossa surgery 260

Spinal cord function monitoring 268
Experimental studies and mechanisms of spinal cord damage 268
Methods of monitoring cord function 269
Practical aspects of spinal cord monitoring 272
Clinical applications of spinal cord monitoring 274

Spinal root surgery and peripheral nerve surgery 278

Cerebral ischaemia during non-intracranial surgery 279
Cardiac surgery 279
Carotid endarterectomy 285

References 292

8 Further signal analysis 309

Introduction 309
Data acquisition 311
Technology 311
Safety issues 315
Improving signal quality 317
Filtering methods 318
Artefact detection and rejection methods 327
Signal processing and interpretation 327
Common processing tasks 329
Processing methods 332
Integration of features into the clinical context – pattern classification 346
Statistical process control 354
Use of the tools 358
Decision assistance 358
Development and uptake of methods 359
Performance assessment 359
Display of results 360
Use of IT facilities 363
References 365

9 Legal implications of neurophysiological monitoring 371

Safety of patients and staff 371
Electromedical equipment 372
Infection control 373
Drugs 373
Identification of patients and consent to procedures 373
Inherent risk of procedures 374
Responsibilities in training and supervising staff 374
Responsibilities in respect of reports on investigations 375
References 375

Index 377

Chapter 1

The Why and the How of Neurophysiological Monitoring in the ICU and Surgery

INTRODUCTION

Applications of clinical neurophysiology in the ICU

Neurophysiological work in the intensive care unit (ICU) can add a new dimension to that in general clinical neurophysiology. The tasks are often much more demanding in practical, and even emotional, terms and have to be adapted to rapidly changing clinical circumstances. Even the 'routine' ICU management activities and use of pharmacological agents will manipulate the situation – but this can be an advantage in that variations in the level of arousal and even associated biological 'artefacts' allow a broader view of the functional state of the nervous system. In stuporous or comatose patients, documentation of the effects of systematic auditory, tactile and painful stimulation forms a routine part of the assessment, unless clinically contraindicated. If any emergencies arise, their neurophysiological consequences can be immediately seen and the efficacy of treatments evaluated. Considerable flexibility is generally used in utilizing electroencephalography (EEG), evoked potentials (EPs), nerve conduction studies (NCS) and electromyography (EMG) in an overlapping manner to improve clinically relevant diagnostic information. Indeed, there is something of the philosophical approach of an interactive specialist consultation rather than just 'doing a test'.

Both short-term conventional diagnostic recordings and prolonged surveillance by purpose-built monitoring systems are usually required. Monitoring modules are increasingly integrated so that the behaviour of cardiac, respiratory, intracranial pressure (ICP) and central nervous system (CNS) functional states may be viewed simultaneously on one one screen. Modern, flexible, monitoring systems increasingly provide analysis of trends, review of events preceding any sudden deterioration, and methods to help optimize control of various forms of pharmacological and other treatments.

Neurophysiological recordings are generally demonstrated to ICU staff on-line and interpreted in the context of discussion of the present clinical state of the patient, medication, results of other investigations and the overall care plan for the patient. This is very much a team effort with technologist, neurophysiologist and ICU nursing and medical staff each contributing to high quality, appropriate recordings and a clinically relevant interpretation.

Reports are promptly documented in writing, preferably stored in the ICU computerized system, together with illustrative samples of signals. Such an interactive approach inevitably contains a 'teaching' component and can increase mutual understanding of the clinical contribution of neurophysiology in critical care medicine.

In this section, we give an overview of EEG, EP and EMG applications particular to the general and the neuroscience ICU, and discuss the methodology and clinical utility of EEG and EP monitoring techniques. The important topics of safety, professional standards, quality control and medico-legal responsibilities are set out in *Safety during neurophysiological recording in the ICU and operating theatres* (p. 55) and in Chapter 9; the principles of electrical and physical safety and infection control requirements are reiterated below. The reader is urged to study these carefully and to seek expert advice on national regulations and local safety procedures before undertaking any neurophysiological work in the ICU. We endorse the specific standards for safety in the ICU (International Task Force on Safety in the Intensive Care Unit 1993) and the IFCN recommended standards for electrophysiological monitoring in comatose and other unresponsive states (Chatrian et al 1996, Guérit et al 1999).

Common clinical questions

Although the reader might expect a list of specific medical conditions in which neurophysiology is useful in patients undergoing intensive care, in reality it is a range of more general clinical issues that occasion a request for help. Specific problems in neurological examination of ICU patients arise when sedative and muscle relaxant drugs impair signs of reactivity or when assessment of cranial, tendon or superficial reflexes is limited by local wounds, swelling or limb immobilization. A few of the questions frequently asked by ICU doctors, nurses and physiotherapists, and indeed by patients' relatives, may serve to exemplify the unique contribution that an interactive, problem-orientated, neurophysiological approach can make:

About good management of the patient:
- Is this 'unresponsive' patient feeling distress or pain from routine ICU procedures?
- Is this patient, who is receiving muscle relaxants and some sedation, aware of the words or actions of relatives or ICU staff?

About prognosis:
- Are the movements we see purposeful responses, reflexes or seizures?
- Is the brain or spinal cord functionally intact or irreversibly damaged?
- Is the patient 'locked-in' or 'brain dead'?

About the nature of common practical problems:
- Why is it so 'difficult to wean' the patient from the ventilator?
- Why is the patient 'slow to wake' and so unresponsive following withdrawal of sedation?
- Why do the patient's muscles seem to have become so weak?

Such everyday dilemmas highlight the role of neurophysiology in providing information that may be otherwise inaccessible. They emphasize the value of a step-wise, multimodality approach achieved by combined use of EEG, EP, nerve conduction and EMG techniques. Answers can be obtained in more than one way: thus cortical responses to painful stimuli can be confirmed either with specific somatosensory evoked potentials (SEPs) or by means of reproducible, temporally related, EEG arousal responses. Differentiation between peripheral and central causes of difficulties in 'weaning' or 'waking' a patient may require a sophisticated battery of techniques in the hands of an experienced neurophysiologist eliminating common causes until a diagnosis is reached by a logical process.

Diagnostic versus monitoring applications

It will already be evident that there are considerable differences between classical diagnostic neurophysiology and the on-going, interactively supportive role of well-planned and well-interpreted neurophysiological monitoring techniques in the ICU. Both are essential, indeed complementary, in that monitoring usually requires preliminary 'calibration' in an individual patient with a conventional EEG or EP recording, and these may need to be repeated at intervals for validation of prolonged monitoring.

Diagnostic work is often disadvantaged by the need to work at the bedside in the ICU. Nonetheless, examples such as the documentation of subclinical seizure discharges in the EEG or of reactivity to external stimulation by EPs or EEG arousal phenomena in an apparently comatose patient, show the unique contribution that can be made. For EMG and nerve conduction studies, diagnostic procedures form the main contribution in the ICU; serial studies are also common.

Monitoring work, the main topic of this section, is well suited to the nature of the problems arising in the ICU. It shares many of the methodological and technological developments with intraoperative monitoring; these are discussed together in Chapter 3 with advanced signal processing described in detail in Chapter 8. Experienced workers emphasize the value of neurophysiological monitoring in the ICU to provide continuous information which will allow accurate definition of the severity of brain injury, especially during regimens with heavy sedation, predict the patient's likely progress, permit early detection of secondary intracranial changes allowing more precise indications for treatment and prognosis, help in control of status epilepticus and in identifying its non-convulsive forms (Prior and Maynard 1986, Bricolo et al 1987, Chatrian 1990, Crippen 1994, Jordan 1994 1999, Chatrian et al 1996, Young et al 1996, Chiappa and Hill 1998, Vespa et al 1999a,b, Huang et al 2002a,b).

These advantages have to be balanced against the difficulties of recording in a busy clinical environment where actions of staff may lead to disturbance of electrodes or leads, or of the patient, with consequent artefacts. It can be argued that some biological artefacts appearing in the EEG may provide information that is useful when interpreting the recordings. This is because they may indicate changes in the state of the patient, such as arousal responses, or may help document the timing of movements with epileptic seizures.

The pitfalls and caveats associated with EEG monitoring in the ICU have been outlined by Young and Campbell (1999) who report problems arising from faulty electrodes, connections of electronic equipment, induced artefacts from electronic devices and non-electrical equipment, issues of electrode placement, and biological, including movement-related, artefacts. They rightly emphasize the importance of continuous quality improvement strategies and of prompt trouble-shooting and regular review sessions. Such considerations also apply to EEG monitoring in other locations.

Advantages of coordinated EEG, EP and EMG assessment

Our objective in this book has been to consider clinical neurophysiology as a speciality offering a spectrum of overlapping techniques. Rigid compartmentalization of investigations into 'electroencephalography', 'evoked potentials', 'nerve conduction studies' and 'electromyography' makes nonsense of attempts to assess the 'integrated nervous system' of a critically ill patient. This is reflected in the development of modern multipurpose equipment which is increasingly capable of recording, analysing and displaying signals from the brain, central sensory and motor pathways and peripheral nerves and muscles.

Examples of this approach are evident in the protocols used to address two of the common clinical questions: whether an unconscious victim of a road traffic accident has evidence of primary brain and/or spinal cord injury; and whether problems in weaning a patient from ventilatory support are due to sedative medication, neuromuscular or muscular malfunction. In the first example, a combination of EEG and EP techniques would be appropriate, whereas in the second, EEG, EPs, nerve conduction studies and EMG might all be contributory in elucidating the problem.

Two more general advantages result from combining EEG with multimodality EPs in comatose or sedated patients in the ICU. First, more refined assessment of the location and extent of areas of dysfunction in the CNS becomes possible, including information not otherwise obtainable about the integrity of spinal cord, brainstem and subcortical pathways. This may account for the increased sensitivity obtained by use of both serial EEG and serial SEP recordings for prediction of outcome after cardiac arrest (Chen et al 1996). Secondly, when neurological assessment and EEG activity are significantly impaired by sedative regimens or previous drug overdose, the SEPs and brainstem auditory evoked potentials (BAEPs) are usually intact (Sutton et al 1982, Ganes and Lundar 1983, Marsh et al 1984), although some exceptions such as transient BAEP abolition by lidocaine–thiopental combination are reported (García-Larrea et al 1988). This important resistance of short latency EPs to regimens such as deep and sustained thiopentone narcosis sufficient to produce pronounced burst suppression or even electrocerebral inactivity, makes their utilization an important aid to the management of severely brain injured patients.

In the light of the range of common clinical questions in the ICU, it is helpful to take this holistic approach one stage further. It is essential to view all the neurophysiological findings in the context of the functional state of the patient's other physiological systems and the effects of any pharmacological or physical interventions during intensive care. This means that we must understand not only the clinicopathological disturbances due to the presenting disease, but also the effects of drugs, exhaustion, pain, disturbed sleep–wake cycles, multisystem failure, and a continuous background of 'activity' (noises, bright lights, frightening procedures) which can affect the patient. Relevant information is maximized by combining recordings from the CNS with those from other physiological systems into a single display. Signal processing techniques can assist in detection of significant trends and interrelationships.

The IFCN recommendations for clinical practice of EEG and EPs in comatose and other unresponsive states (Guérit et al 1999) consider continuous neuromonitoring as 'a neurophysiological tool to be used by non-neurophysiologists' and emphasize that the neurophysiological information must be 'coded in a language that is readily interpretable by the user'. To this end they suggest a three-step approach to its implementation: (i) quality screening, followed by automatic feature extraction, (ii) selecting and combining parameters to define patterns, and (iii) integration of neurophysiological features into the clinical context (see *Table 3.3*, p. 102). They emphasize that 'because the neurophysiologist cannot be present throughout the entire period of neuromonitoring, it is essential that intensivists and the nursing team be trained in some basic principles of EEG interpretation and the way to react to the most commonly encountered technical problems'. They add that 'a trained neurophysiologist and an EEG technologist should be reachable for the remaining problems'.

Comparisons with coma scores and other outcome predictors

Clinical systems for scoring the depth of coma, such as the Glasgow Coma Scale (GCS) (see p. 189 *et seq.*), have contributed substantially to patient triage after trauma, communication regarding patient state and prognostication. These systems were designed to be widely applicable and repeatable by trained personnel without the need for special tools. They are used around the world in the routine nursing and medical monitoring of the level of consciousness of patients at the site of accidents, during transport and in hospitals.

There are two main disadvantages when using such scales either as a monitor of patient state or for prognostication in the ICU. First, scoring is generally dependent on fairly simple, reproducible, but necessarily subjective, grading of responses to external stimuli. Typically, these involve eye opening, vocalization and movement – with pupillary responses and other signs incorporating cranial nerve function also used in some schemes. All of these may be masked as a result of pathological processes, drugs, or restrictions due to physical treatments, reducing the efficiency or applicability of the techniques in the ICU. Secondly, there are limitations to what can be evaluated by such simple methods; valuable additional information about responsiveness in coma can be gained by recording neurophysiological signs of reactivity in parallel with the clinical scoring. Indeed, neurophysiological monitoring and clinical scoring systems provide prognostic information from quite different backgrounds and should be considered complementary rather than competitive (Häkkinen et al 1988, Matousek et al 1996). Overall, in the neuro-ICU, neurophysiological findings have become influential in clinical decision-making in over 85% of patients and are considered contributory factors to improvement in outcome combined with significant cost reduction (Chiappa and Hill 1998, Jordan 1999, Vespa et al 1999a).

Monitoring during surgical operations

In the last decade, intraoperative monitoring has undergone remarkable changes consequent upon improved computation and technological developments which have led to the more advanced forms of monitoring equipment that are now becoming available. In parallel, there has been an intensive clinical appraisal of the urgent need for, and the cost-effectiveness of, monitoring to improve the quality and outcome of anaesthetic and certain surgical procedures. This has applied to neurophysiological monitoring as much as to that in other fields and has meant a considerable change in the work patterns, training and outlook in most clinical neurophysiology departments. Working interactively in a team in the operating suite with clinical colleagues managing the patients has become common practice, rather than the more traditional relationship of service provider issuing reports from a distant laboratory. Such developments have proved effective in increasing scientific standards and all-round quality of service to patients undergoing surgical procedures but necessitate an increase in neurophysiological staffing and specialized training.

Specific clinical reasons for intraoperative neurophysiological monitoring

Operative neurophysiological recording has three purposes: to prevent permanent neurological disability by detecting early damage which can be corrected; to locate specific neural structures when anatomy is distorted; to demonstrate functional continuity, particularly in cases of peripheral nerve injury. Operative monitoring to prevent disability should only be undertaken if the following conditions are satisfied (Grundy 1982, 1995, Guérit 1998, Burke et al 1999):

1. The operation carries a risk of producing a specific disability.
2. The neural structure at risk conveys or produces electrical activity (either spontaneous or evoked) which can be recorded during operations, preferably without adding excessively to the duration or complexity of the procedure. In some cases a neighbouring structure can be monitored, if pathological changes affect both equally. This is the assumption underlying sensory EP monitoring to prevent paraplegia during spinal surgery, but such monitoring always carries the risk of a false negative result arising from a lesion which spares the dorsal columns.
3. Significant electrophysiological changes occur at the stage of early reversible damage, i.e. the recordings are *sensitive*.
4. Criteria are well established for identifying significant changes in the recordings. This implies that changes having a high probability of predicting a postoperative deficit have been identified during operations in which the recordings have not been allowed to influence the surgical procedure. There is, therefore, an important distinction between intraoperative recording carried out with a view to defining significant changes, and recording to detect the onset of such changes in the hope of altering the outcome. Only the latter constitutes clinical monitoring; the former is research.
5. Extraneous factors, such as anaesthesia and hypotension, have no effect on the recordings, or if there is an effect then it can be distinguished from the effects of surgery, i.e. the recordings are *specific*.
6. Having been warned that a significant change has occurred, the surgeon can correct or minimize the damage.

There is therefore no justification for monitoring operations in which there is no predictable risk, or if the structure at risk is not amenable to electrophysiological recording of any kind, or where the nature of the operation precludes the surgeon from correcting adverse effects. It is only fair to point out that insufficient information is yet available to satisfy each of the above conditions for all types of operation in which monitoring is currently used. The literature consists of many case reports intended to demonstrate the usefulness of monitoring, but detailed analyses of the sensitivity, specificity and contribution of monitoring to the reduction of surgical morbidity are only beginning to appear. In many operations the decision to use monitoring is based on deductions from animal experiments, retrospective comparison with operations having a poor outcome and on the assumption that the surgeon should be helped by knowing that he or she has (or has not) produced a significant alteration in neural function.

Unfortunately, few controlled trials have been performed with patients randomized to have operations either with or without monitoring (Fletcher et al 1988). Because of the low risk of a poor outcome from most operations, large numbers of patients would have to be randomized into two groups; in one the surgeon would be informed of changes identified by monitoring, and would take the appropriate action, while in the other group the surgeon would have no information about monitoring (Jones 1993). Apart from the difficulty of recruiting sufficient patients to give a statistically significant result, there would be serious ethical objections to withholding potentially beneficial data in half the procedures. For many types of surgery information is simply not available as to whether neurophysiological recordings can detect reversible damage or even predict postoperative disability if an abnormality is not corrected. The opinion is therefore firmly held in some quarters that, except for certain clearly defined applications, operative neurophysiological monitoring is not sufficiently reliable to justify widespread clinical use (Michenfelder 1987, Aminoff 1989, Gelber 1999). On the other hand, it must be noted, firstly, that each of these papers warning against reliance upon monitoring was accompanied by a clear statement of the benefits of monitoring (Friedman and Grundy 1987, Daube 1989, 1999), and, secondly, that careful review of the available evidence has led others to the conclusion that certain applications of monitoring – spinal cord monitoring by SEPs, EEG or SEP monitoring for cerebral ischaemia, BAEP, SEP and cranial nerve EMG monitoring in posterior fossa surgery, identification of sensory cortex by SEP, functional localization by direct stimulation, and EMG and nerve conduction measurements in peripheral nerve surgery – are efficacious if performed by experienced personnel (American Academy of Neurology 1990, Fisher et al 1995, Bejjani et al 1998, Wiedemayer et al 2002).

A further possible justification for monitoring, particularly during removal of cerebral and spinal tumours, is that a more complete tumour removal may be possible if the surgeon can be sure that he has not damaged an important structure, such as the spinal cord or an eloquent area of cerebral cortex. This aspect has received only limited attention so far, but may become more important in future. Finally, monitoring may assume increasing significance in surgical training, by demonstrating the adverse effects of excessive retractor pressure and clumsy manipulation of sensitive neural structures (Andrews et al 1993). Quite independently of any consideration of efficacy, medico-legal pressures have rendered monitoring mandatory in many branches of surgery, often because failure to monitor is regarded as a reason for awarding larger damages should a mishap occur. Furthermore, at least in the USA, insufficient expenditure on prevention constitutes negligence (Schwartz and Komesar 1978).

Many reports of monitoring could be summarized as 'We did this in *N* operations with a good result in every case, so the probability of a bad result must be less than one in *N*'. This statement ignores the fact that one in twenty series of *N* consecutive operations would consist entirely of good results even if the risk of a mishap was as high as 3/*N* (Hanley and Lippman-Hand 1983). Therefore, for example, if we wish to be 95% certain that the risk of a mishap is less than 1%, we need a series of 300 consecutive cases with good results (see *Assessment of results*, p. 7).

Intraoperative monitoring has taken the first four steps on a path well trodden by medical innovations (McKinlay 1981). After a new technique is introduced, *promising reports* lead to *professional adoption*, after which *public acceptance and state endorsement* follow automatically, and the new technique is firmly established as a standard procedure. This is the position with respect to intraoperative monitoring today. Frequently, however, as McKinlay goes on to point out, when a *randomized controlled trial* is belatedly performed, the results are so unfavourable that *professional denunciation* is the only possible reaction, after which *erosion and discreditation* are inevitable. It remains to be seen whether this will apply to intraoperative monitoring, but for many operations it would now be considered unethical to randomize some patients to have the procedure without monitoring, so randomized controlled trials are unlikely to be performed. The lesson to be drawn from this is that there should be no 'pilot studies' or 'preliminary reports'; all patients should be randomized as soon as a new technique is introduced, and before possibly erroneous opinions are formed (Chalmers 1981).

A theme throughout this volume is the integrated use of complementary tools for clinical neurophysiology, as opposed to rigid compartmentalization into EEG, EPs, nerve conduction and EMG. Nowhere is this more important than in modern intraoperative neurophysiological monitoring and in recent developments of multifunctional digital neurophysiological equipment. For monitoring in the ICU and operating theatre, we strongly believe in the present trend towards 'fusion' of data extracted from various neurophysiological, cardiovascular and other biological signals with anaesthetic and patient state data into composite displays and as a basis for intelligent interpretative algorithms to guide better management of patients during high-risk procedures.

Medico-legal issues in intraoperative monitoring, cost-effectiveness and staffing implications

Advocates of neurophysiological monitoring who seek to increase the safety and quality of anaesthetic and surgical procedures, generally cite scientifically-based evidence in justification of their position. They follow the steps of Popper (1959) who propounded the view that sound scientific knowledge is based on the development of testable hypotheses that may then be falsified by the new evidence that emerges from the process of testing. In discussing some limitations of evidence-based medicine, Black (1998) suggests that we need greater emphasis on the prior intellectual analysis of the problems, be they clinical or organizational, and that we must carefully define the *type of evidence* that is really relevant to the issue under consideration. Whilst randomized controlled trials and other forms of group evidence may provide clarity in structured evidence-based decision-making protocols, many 'grey areas' still exist in the daily work of a good clinician. Black endorses the view that, in spite of scientific advances in medicine and surgery, the art of medicine still lies in identifying and treating the individual patient's problem rather than his disease.

These introductory observations provide some background to medico-legal views on 'good practice' and the need to consider the consequences of our actions or lack of action in our daily clinical work. By extrapolation, 'good practice' requires implementation of monitoring whenever there is evidence (i) of a calculable risk of (iatrogenic) 'harm' or 'injury' to a patient, (ii) that early warning systems such as monitoring reliably indicate when that risk is impending, and (iii) that effective action can realistically be taken to avoid damage. This implies that neurophysiological monitoring during high risk procedures, like any other warning system, must provide warning of adverse changes that are potentially reversible by prompt action and not merely evidence that damage or disaster has already occurred.

Various categories of evidence exist as to the causes, nature, consequences and possibilities for prevention of injury to patients relating to surgical or anaesthetic procedures. These are reports of clinical outcome studies, national reports of confidential enquiries into perioperative mortality, reports of medico-legal organizations concerning claims for death or injury, objective assessments of the effectiveness of monitoring, and international or national guidelines to good practice and technical standards.

Clinical outcome studies will be discussed in the context of the various clinical applications of EEG monitoring. This is an area of rapid development and those considering whether they should institute particular monitoring procedures will want to check the latest publications available. However, in more general terms of the well-being of patients undergoing surgery, some comments from two earlier reviews are of interest. The safety and efficacy of general anaesthesia were studied in 17,201 patients who were followed-up for 7 days after anaesthesia and surgery in 15 university hospitals in the USA and Canada (Forrest et al 1990a,b). The rates of death, myocardial infarction, and stroke in the study population were reassuringly low (less than 0.15%) and adverse outcomes were generally minor. From the standpoint of neurological injury sustained during hospital procedures, a study of 1,500 neurological consultations over 6 years in one large general hospital revealed that iatrogenic neurological conditions were present in 14% of the patients, the most common causes being angiography, cardiac surgery and immunosuppression (Moses and Kaden 1986). There is a continuing need for better surveillance of both the incidence and of the implementation of routine use of methods for prevention of neurological injury in hospitals.

National enquiries into perioperative mortality have been undertaken in various forms in many countries over very many years. We may cite the report of the UK National Confidential Enquiry into Perioperative Deaths (National CEPOD 1987) in which the assessors found that one of the commonest reasons for suboptimal care, occurring in about 45% of cases, was the failure to use appropriate monitoring. At that time this mainly implied monitoring of the cardiovascular system, an area that attracted less attention a decade later (National CEPOD 1998).

Medico-legal organization reports reveal much about the sequelae of perioperative morbidity and mortality, in particular the social and financial implications for individual patients and their families, the consequences for the medical practitioners involved and for the community at large. Over the last decade in the UK, some of the largest settlements were for claims relating to the practice of anaesthesia which, in one report, represented approximately double the sums for the next highest risk specialities (Medical Defence Union 1995). Of interest in the context of neurophysiological monitoring is the report that of 150 anaesthesia-related claims in 1989–1990, brain and/or spinal cord damage accounted for 24%, awareness during general anaesthesia for 12% and peripheral nerve damage for 4% of claims (Aitkenhead 1991). Guidelines on risk management in anaesthesia were issued by the Medical Defence Union in 1997 (Green et al 1997); the background documentation noted the balance between incidence of various causes and their financial consequences in 114 claims relating to anaesthetic practice over a 6-year period. This showed that 6% of claims were for death (44.3% of the indemnity payments), 4% for brain damage (21.1% of payments), 5% for neurological complications including paraplegia (18.4% of payments), 3% for severe vascular complications such as cardiac arrest (8.1% of payments) and 5% for anaesthetic awareness (2.7% of payments). By contrast, dental damage during anaesthesia accounted for 63% of claims but for only 1.4% of payments (Green et al 1997). Significant risk factors included failure to use essential equipment, failure to use pre-, peri- or postoperative monitoring and the incorrect setting of alarm parameters or failure to use or respond to alarms.

Objective studies of the effectiveness of monitoring have only begun to appear relatively recently. The American Academy of Neurology provided an assessment of intraoperative neurophysiology in 1990 concluding that 'several electrophysiology tests and monitoring techniques are safe and efficacious, to a variable degree, as commonly applied in the operating room' and listed EEG or SEP to monitor for cerebral ischaemia amongst applications of other modalities (American Academy of Neurology 1990). They added the caveat that the techniques 'need to be applied by a well-trained, knowledgeable physician–neurophysiologist or personnel directly under his or her supervision'. Two other reviews of American practice appeared 5 years later: the multi-centre study of SEPs in spinal monitoring (Nuwer et al 1995) discussed in *False positives, false negatives and the predictive value of spinal monitoring* (p. 269) and that of Fisher et al (1995) on the efficacy of EEG and EP monitoring for carotid artery, intracranial and spinal surgery. Fisher and his colleagues concluded that the following applications of neurophysiological monitoring were beneficial to the patient:

1. EEG (conventional visual assessment or quantitative analysis) and SEPs for carotid and brain surgery potentially compromising cerebral blood flow;
2. BAEPs and cranial nerve monitoring for brainstem or inner ear surgery; and
3. SEPs for surgery potentially involving ischaemia or mechanical trauma to the spinal cord.

Like the American Academy of Neurology, they too, added a rider that intraoperative monitoring must be performed by an experienced team of surgeons, anaesthetists, neurophysiologists and technologists. They also noted that 'Board subspecialty licensing' now exists for physician neurophysiologists and technologists to document a minimum level of knowledge and training. With sophisticated modern monitoring equipment, recording and interpretation are no longer quite as simple as when Nuwer (1993) suggested that 'The well-trained electroencephalographer should find recording and interpreting EEG in the surgical setting to be a natural extension of routine EEG.'

This question of appropriate personnel and training for EEG monitoring has exercised most groups involved in this work and led to considerable debate, not least on roles, responsibilities and competence of various categories of neurophysiology staff (Borresen et al 1993, Nuwer and Nuwer 1997). Many authorities now require proper advanced training courses and accreditation systems for physician–neurophysiologists and their technological teams. Nursing staff working in the surgical teams should also receive training about the background to and aims of neurophysiological monitoring (Pearlman and Schneider 1994). The main argument at present centres on the requirement for expert neurophysiologists and neurophysiological technologists, as opposed to personnel from other specialities, to undertake all neurophysiological monitoring. Unfortunately, inadequate establishments for these new roles, and indeed a perception of lack of interest, mean that there may not be sufficient expert manpower available in neurophysiology departments. It is then that other clinicians begin to seek guidance about undertaking the work themselves rather than neglect a valuable source of help during high risk procedures. Although there have been notable exceptions, in general it is better that there are sufficient fully trained professionals available to provide these demanding services.

Studies of cost-effectiveness have provided arguments for devoting the necessary finance to provide expert staff and appropriate equipment for EEG monitoring. These have demonstrated benefits, both to the patient and, potentially, in overall costs, from reducing the depth and duration of anaesthesia and hence recovery time; this means taking care to use no more than a dose sufficient to cover the requirements of the particular patient and the type and stage of surgery (Gan et al 1997, Yli-Hankala et al 1999, Vakkuri et al 2005). However, the cost of electrodes may sway the balance of costs for short procedures. Reduction of morbidity, thought to be a consequence of EEG monitoring in both adult myocardial revascularization procedures and paediatric cardiac surgery, has allowed reduction in postoperative hospital stays with benefits to both patients and hospitals (Edmonds et al 1992, Austin et al 1997). In the 'managed health era', protocols are also being proposed for standardization of procedures such as carotid endarterectomy 'without adversely affecting well-established low morbidity and mortality rates and with significant hospital cost savings' (Syrek et al 1999) – a phenomenon that seems likely to increase! On the other hand the appropriateness of carotid endarterectomy, along with other procedures in patients over 60 years old, has itself been questioned by health scientists concerned about wide disparities of care (Brook et al 1990).

International and national guidelines to good practice and technical standards for monitoring in intensive care and surgery have been promulgated by the International Federation for Clinical Neurophysiology (IFCN) (Nuwer et al 1993) and are complementary to those for coma and intensive care mentioned above (pp. 1–2) (Chatrian et al 1996). Detailed updates of their earlier guidelines for intraoperative monitoring of sensory EPs have been published by the American Electroencephalographic Society (1994). Some recommendations for technology and equipment for intraoperative monitoring have also been proposed (Anon. 1993). Recently updated IFCN recommendations on intraoperative monitoring are given by Burke et al (1999) in a volume collating all the Federation's current *Recommendations for the Practice of Neurophysiology* (Deuschl and Eisen 1999). Similar guidelines for technologists are produced by the International Organization of Societies for Electrophysiological Technology (OSET) (1999).

ASSESSMENT OF RESULTS

As explained above (pp. 4–5), the primary purpose of operative monitoring is to detect changes which, if uncorrected, are likely to be followed by neurological disability. Monitoring followed by appropriate corrective action should therefore prevent disability. However, it is not justified to assume that monitoring plus corrective action will *always* be successful in preventing disability; the risk can only be reduced, not eliminated. Unfortunately it has to be admitted that most of the monitoring currently performed has been adopted without any systematic assessment of its contribution to reducing risk. The present section indicates how a newly developed monitoring method could be assessed rationally, and emphasizes that while monitoring should reduce risk it is unrealistic to expect risk to be eliminated completely, a conclusion with important medico-legal implications.

Statistical significance of changes

When a neurophysiological recording is performed for diagnostic purposes the result is interpreted by comparison with the range of values to be expected in healthy subjects usually expressed as the *95% reference interval*: the interval within which values from 95% of the population will lie, a concept which has replaced the 'normal range' (Smith 2000). But when recordings are performed to monitor a patient's condition over a period of hours or days the reference interval for healthy subjects is of much less importance, because the patient's value may already lie outside the reference interval as a result of disease, metabolic disturbance or drugs such as anaesthetics or sedatives, and what we are principally concerned with is whether there is a significant change in the patient's value over a period of time. Each patient therefore acts as his or her own control for monitoring purposes. However, even in healthy subjects repeated values show some variation, so whether a change is significant can only be assessed if we know the range of variation to be expected.

Let us consider a hypothetical situation in which only two values are available, one obtained before and one after an intervention of some sort, and we need to know whether the intervention has produced a significant change in the patient's condition. Provided that variability is independent of the level, in other words repeated values are within $\pm X$ units of the original value, rather than being within $\pm X\%$ (Bland and Altman 1996a), we can answer this question if we know the overall standard deviation (s) of repeated values produced by the technique. A second value is significantly different at the 95% probability level if it is more than $2.8s$ from the first value (Bland and Altman 1996b, Fraser and Fogarty 1989), the quantity $2.8s$ being the *repeatability limit* as defined in the relevant International Standard (ISO 5725-6 1994). However, in patients undergoing operations or receiving intensive care, the variability of repeated values is likely to be different from the variability of repeated values in other patients or healthy subjects. In some situations the variability will be higher because of the greater number and variety of influences to which the patient is exposed, while in other circumstances one influence, often anaesthetic drug administration, may predominate, so that variability is reduced. Therefore, for monitoring we need a unique measure of repeatability for each patient.

Where a number of pre-intervention values are available, the method of statistical process control can establish control limits for a single patient. This is done by estimating the standard deviation from the mean difference between consecutive values. The estimated standard deviation (σ) is the mean consecutive difference divided by 1.128. 'Control limits' are set at 3σ above and below the mean (Wheeler and Chambers 1992). If 12 or more values are available, all lying within the control limits, then we can say that the process is 'in control', and these control limits can be used to monitor the patient's condition subsequently. During some operations sufficient values can be obtained to establish control limits in this way after induction of anaesthesia, but before any surgical intervention that could damage the nervous system (*Fig. 1.1(a)*). Early warning of an emerging problem is provided by two further criteria: four out of five values more than 1σ from the mean on the same side, or two out of three values more than 2σ from the mean on the same side (Nelson 1984) (*Fig. 1.1(b)*). These and other methods of statistical control are described more fully in Chapter 8. In some circumstances in which monitoring is used, either an EP amplitude reduction of 50%, or a latency increase of 10%, or both, have been found to produce satisfactory results (e.g. Brown et al 1984) and the 50% amplitude criterion in particular has been applied arbitrarily in other situations (Banoub et al 2003). The few workers who have attempted to establish a cut-off value for a particular operation have used clinical criteria rather than relying solely upon statistical calculations.

Clinical significance of changes

Apparently significant changes identified by the methods described above may not be clinically significant. Whether a change during monitoring is clinically significant depends upon the risk of a poor outcome, and on the likelihood ratio associated with a change in monitoring. To understand why this is so, we need to consider the result of monitoring a number of procedures without allowing changes in monitoring to influence what is done. After some procedures, patients will be found to have suffered the complication monitoring is intended to prevent, and patients may be divided into four groups according to their clinical state and whether any change in monitoring was identified:

- change identified by monitoring, complication develops (true positives: *TP*);
- *no* change identified by monitoring, *no* complication (true negatives: *TN*);
- change identified by monitoring, *no* complication (false positives: *FP*); and
- *no* change identified by monitoring, complication develops (false negatives: *FN*).

The numbers in the four groups can be arranged in a two-by-two table as in *Table 1.1*. The probability of a complication arising, regardless of the monitoring findings, is $(TP + FN)/(TP + FN + TN + FP)$; this is the *pre-test probability*, or *prevalence*, of the complication among patients who have undergone the particular procedure. Among patients with the complication, the proportion in whom monitoring detected a change is the sensitivity of the monitoring method, $TP/(TP + FN)$; among those without the complication the proportion who showed no change in monitoring is the *specificity*, $TN/(TN + FP)$. From the values in *Table 1.2* the probability of a deficit developing as a result of the operation is 3/100 or 3%, the sensitivity of monitoring is 3/3 or 100%, and the specificity is 92/97 or 94.9%. Few diagnostic tests

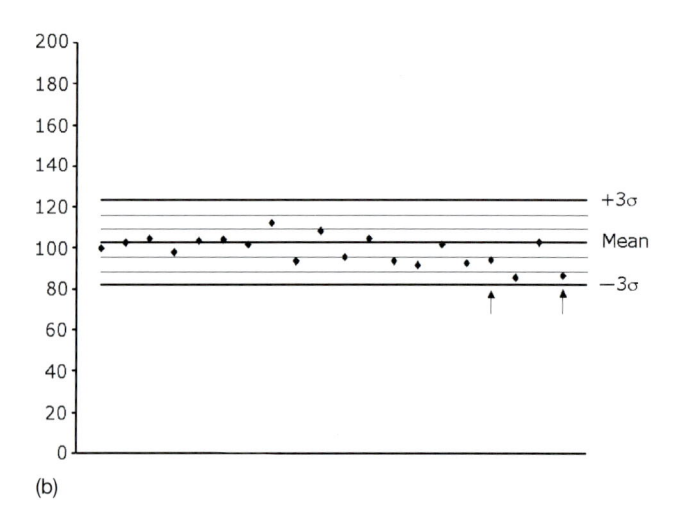

Fig. 1.1 Control charts showing the 3σ limits established from the first 12 data points. (a) All 12 points are within the limits, so the measurements are 'in control'. (b) The limits established in (a) are applied to subsequent points. At point 17 (first arrow) four out of five values have been more than 1σ below the mean, and at point 20 (second arrow) two out of three values have been more than 2σ below the mean.

Assessment of results

Table 1.1 The results of monitoring 100 operations, with the sensitivity, specificity, predictive values, likelihood ratios and their 95% confidence intervals

	New deficit	No new deficit	Total
Change in monitoring	3 (TP)	5 (FP)	8
No change	0 (FN)	92 (TN)	92
Total	3	97	100

	Proportion	Value	95% confidence interval
Prevalence	(TP + FN)/(TP + FN + TN + FP)	3.0%	0.623% to 8.52%
Sensitivity	TP/(TP + FN)	100%	29.2% to 100%
Specificity	TN/(TN + FP)	94.8%	88.4% to 98.3%
Positive predictive value	TP/(TP + FP)	37.5%	8.52% to 75.5%
Negative predictive value	TN/(TN + FN)	100%	96.1% to 100%
Likelihood ratio +	TP(FP + TN)/FP(TP + FN)	19.4	8.3 to 45.6
Likelihood ratio −	TN(TP + FN)/FN(FP + TN)	Undefined (FN = 0)	–

Table 1.2 The results of monitoring 1000 operations as in *Table 1.1*. The risk of a new deficit following operation is still 3%, but in this larger series one patient developed a new deficit without any change in monitoring being identified: a false negative result. The confidence intervals are narrower than those in *Table 1.1* because the sample is much larger. The difference in sensitivity between *Table 1.1* (3/3 or 100%) and Table 1.2 (29/30 or 96.7%) is 3.33%, 95% confidence interval −3.09% to 9.76%. The confidence interval includes zero, so the difference is not statistically significant, but clearly the occurrence of one false negative is highly significant clinically

	New deficit	No new deficit	Total
Change in monitoring	29 (TP)	50 (FP)	8
No change	1 (FN)	920 (TN)	92
Total	30	970	1000

	Proportion	Value	95% confidence interval
Prevalence	(TP + FN)/(TP + FN + TN + FP)	3.0%	2.03% to 4.25%
Sensitivity	TP/(TP + FN)	96.7%	82.8% to 99.9%
Specificity	TN/(TN + FP)	94.8%	93.3% to 96.2%
Positive predictive value	TP/(TP + FP)	36.7%	26.1% to 48.3%
Negative predictive value	TN/(TN + FN)	99.9%	99.4% to 100%
Likelihood ratio +	TP(FP+TN)/FP(TP+FN)	18.8	14.2 to 24.8
Likelihood ratio −	TN(TP+FN)/FN(FP+TN)	28.5	4.14 to 195

achieve sensitivity or specificity approaching these values, but a sensitivity of 100% for operative monitoring is generally regarded as a realistic target, if not a rigid requirement, and failure to achieve it, leading to the appearance of a false negative result, is commonly followed by publication of a case report, and in some instances by allegations of negligence.

Following the identification of a change in monitoring, the proportion of patients with the complication who had a change in monitoring is $TP/(TP + FP)$; this is the *post-test probability* or *positive predictive value* of the identified change. Conversely, the proportion of patients *without* a deficit postoperatively who had *no change* in monitoring during the operation is $TN/(TN + FN)$; this is the *negative predictive value*. In *Table 1.1* the positive predictive value is 3/8 or 37.5%, and the negative predictive value is 92/92 or 100%. This illustrates the important point that the predictive value of a diagnostic test such as monitoring is critically dependent upon the prevalence or pre-test probability of the disease or complication in which we are interested, as well as the sensitivity and specificity of the test. In this instance, the positive predictive value is only 37.5%, in spite of 100% sensitivity and 94.9% specificity, because the prevalence or pre-test probability is only 3%. With the same sensitivity and specificity, if the prevalence rose to 30%, the positive predictive value would be nearly 90%.

An alternative method of determining the predictive values is to consider the *likelihood ratio* of a change in monitoring. This is the ratio of the probability of a change in monitoring among patients with the complication to the probability of a change in monitoring among patients without the complication, which can be calculated from *Table 1.1* as $TP(TP \times FN)/FP(FP + TN)$, or more simply as sensitivity/(1 – specificity). The likelihood ratio is an indication of how a change in monitoring affects the probability of the complication arising. If the pre-test probability and likelihood ratio for a change in monitoring are known, the positive predictive value can be determined from a nomogram (Sackett et al 2000), or calculated directly by considering odds rather than probabilities. Odds are simply the ratio of the probability of an event happening to the probability of its not happening, so odds = probability/(1 – probability), and probability = odds/(1 + odds). In *Table 1.1* the pre-test probability is 3%, so the odds are 3/97. The post-test odds are given by multiplying the pre-test odds by the likelihood ratio, so the post-test odds will be 19.4 × 3/97, or 0.6. From these post-test odds the post-test probability is 0.6/(1 + 0.6) = 0.375, or 37.5%. Calculating the post-test probability in this way emphasizes its dependence upon the pre-test probability. In this case the surprisingly low value arises because the probability of a poor outcome, the pre-test probability, is low (3%), so that even with the relatively high likelihood ratio, the post-test probability is less than 50%.

On the other hand, if there is no change in monitoring, the probability of a good outcome (no complication) is $TN/(FP + TN)$, and the likelihood ratio of no change in monitoring is $TN (TP + FN)/FN(FP + TN)$ or specificity/(1 – sensitivity). In the example in *Table 1.1*, there are no false negatives (*FN* is zero), so the probability of a good outcome following no change in monitoring is 1, or 100%, which is the maximum possible value. No likelihood ratio can be calculated because the formulae lead to division by zero. It must be emphasized that it is the 100% sensitivity, *not* the specificity, that ensures the 100% probability of no complication following uneventful monitoring.

Confidence intervals

A quantity such as the sensitivity of a diagnostic test is usually estimated in a sample of the population, and it is important to know how close the sample value might be to the value for the whole population. This is done by calculating the confidence interval, which is the interval within which the population value will lie with a particular level of probability. For example, if a test is positive in 90 patients out of a group of 100 with a certain disease, then the estimated sensitivity is 90%, but the sensitivity of the same test in all patients with the disease could lie between 82.4% and 95.1% with 95% probability. The sensitivity is therefore quoted as 90%, 95% confidence interval from 82.4% to 95.1%.

For large samples, confidence intervals are easily calculated from the standard error of the sample estimate, because sample estimates are normally distributed. However, this method only produces reliable results if the number of positive results and the number of negative results in the sample are both greater than five (Bland 2000, p. 128). A further problem arises if the proportions are close to either zero or one, because then the method based on the normal distribution may give a confidence interval including values less than zero or greater than one. This is clearly nonsense, because proportions such as sensitivity and specificity always lie *between* zero and one (Deeks and Altman 1999). Fortunately an exact method is available for small samples or where the proportions are close to zero or one. This method depends upon the binomial distribution and makes no assumption that the values follow a normal distribution (Altman et al 2000, Bland 2000). The appropriate confidence intervals can be obtained from tables (Lentner 1982, pp. 89–102) or calculated by the computer programs which accompany the books by Altman et al (2000) and Bland (2000); this was the method used to produce the confidence intervals in *Tables 1.1, 1.2* and *1.3*.

In *Table 1.1* the confidence interval for sensitivity is from 29.2% to 100%, because the number with a new deficit is small. If we could study more operations, it is possible that a false negative result might be encountered, that is a patient who developed the complication in spite of no change being identified in monitoring, as shown in *Table 1.2*. Here the sensitivity, although less than 100%, lies within the wide confidence interval established for a small number of operations in *Table 1.1*, but the new confidence interval is much narrower, from 82.8% to 99.9%.

If we are comparing the difference in prevalence of new deficits following two different operations, or the sensitivity or specificity of two different methods of monitoring the same operation, then the correct method is to calculate the confidence interval of the difference in proportions, rather than considering whether the confidence intervals of the two proportions overlap or not (Altman et al 2000). In *Table 1.1* the sensitivity is 3/3 or 100%, and in *Table 1.2* it is 29/30 or 96.7%. The difference in sensitivity is 3.3%, 95% confidence interval from –3.09% to

9.76%. The confidence interval includes zero, so the difference is not statistically significant, but clearly the occurrence of one false negative is highly significant clinically.

Determining the optimal cut-off point

In the above example it was assumed that there was no uncertainty about the definition of a significant change in monitoring. If we are monitoring the amplitude of an EP, as is done in many situations, we may adopt the widely used criterion of a 50% amplitude reduction as indicating a significant change; this is called the *cut-off value*. However, it is perfectly possible that a 60% reduction in amplitude would still detect all patients who develop a complication, while giving fewer false positives. The optimum amplitude reduction can be determined by plotting sensitivity against (1 – specificity) for all possible cut-off values. The resulting graph is called a *receiver operating characteristic* (ROC) curve, a method originally developed in the 1940s to assess the detection of target radar signals in the presence of noise, and applied to interpretation of radiographs in the 1970s (Henderson 1993, Zweig and Campbell 1993, Obuchowski 2003) (*Fig. 1.2*). Few workers have so far applied the ROC curve to monitoring (Cao et al 1997, de Letter et al 1998, Ackerstaff et al 2000, Belardi and Lucertini 2003, Crossman et al 2003).

In order to use this graph to determine the optimum cut-off point we need to know the risk of a poor outcome (prevalence), and the costs, both human and financial, associated with false positive and false negative results, the aim being to minimize the costs of misclassification (Lindley 1985, Strike 1996). If good and bad outcomes are equally likely, and the costs of the two different types of error (false positive and false negative) are identical, the optimum cut-off value will be the one producing the point on the ROC curve farthest from the diagonal (A in *Fig. 1.2*). This point gives the maximum proportion of correct results, or accuracy, defined as $(TP + TN)/(TP + FN + TN + FP)$, and the slope of the line OA is the positive likelihood ratio, if all values greater than or equal to the cut off value are taken as indicating a positive result (Choi 1998).

If there is a cost associated with a false negative, but no cost associated with a false positive, then the cut-off value must be selected to give 100% sensitivity, and any associated specificity will be acceptable. If there is a range of cut-off values producing 100% sensitivity, and the cost associated with a false positive is not zero but very small compared with the cost of a false negative, then misclassification costs are minimized when the cut-off value gives the maximum specificity associated with 100% sensitivity, the point B in *Fig. 1.2*. The slope of the line OB is the positive likelihood ratio if the cut-off and all higher values indicate a positive result; the corresponding negative likelihood ratio is zero because the cut-off value has been chosen to give 100% sensitivity. The combination of 100% sensitivity with maximum specificity seems to be the aim in selecting the cut-off value for many monitoring applications, although this is rarely stated explicitly. For example, in spinal deformity surgery monitoring is required to provide 100% detection of developing paraplegia, while minimizing the number of operations in which the surgeon has to terminate or modify the procedure without achieving full correction of deformity.

For many monitoring applications the probabilities of good and bad outcomes are not equal, and appreciable but different costs are associated with false negative and false positive results. An example is the prevention of stroke during carotid endarterectomy when the corrective action is to induce hypertension. Here a false negative result leads to a stroke, while a false positive exposes the patient to a risk of myocardial ischaemia or even infarction (Smith et al 1988). The misclassification costs associated with a particular cut-off value depend upon the risk of a poor outcome, the relative costs of false positive and false negative results, and the sensitivity and specificity resulting from the cut-off value. The optimum cut-off value is the one associated with minimum misclassification costs for the known prevalence and cost ratio (Lindley 1985, Strike 1996, Remaley et al 1999). Further difficulty arises if corrective action is not always effective, so that some patients incur both costs, one because corrective action was applied, and the other because it was ineffective.

Comparing two methods of monitoring

In addition to determining the optimal cut-off value, the ROC curve can be used to compare two different methods of monitoring. The area under an ROC curve is the

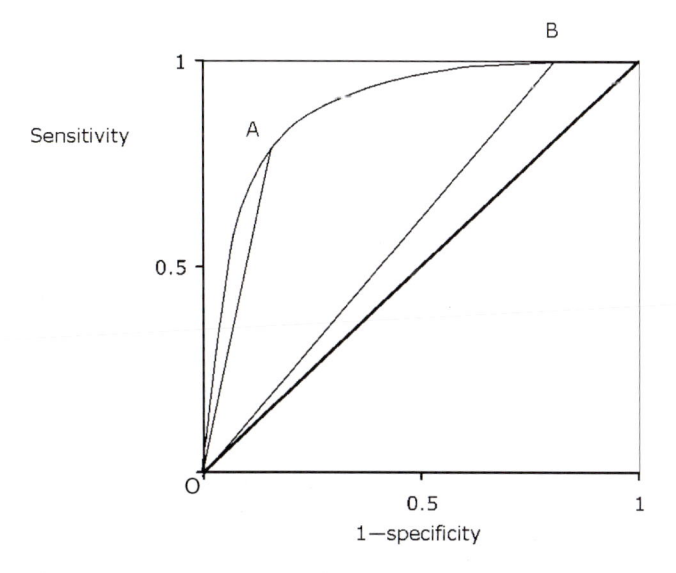

Fig. 1.2 Receiver operating characteristic (ROC) curve. A is the point furthest from the diagonal, and would represent the optimal cut-off value if the costs of false positives and false negatives are equal. The cut-off value associated with B produces 100% sensitivity and maximum specificity; this is the optimum cut-off value if the cost of a false negative is very much greater than the cost of a false positive. The slopes of the lines OA and OB are the positive likelihood ratios if each cut-off value and all greater values are taken as positive results.

probability of correctly classifying a pair of patients, one with and one without the disease in question, in this case a complication arising during the procedure. A monitoring method with a greater area under the ROC curve will in general have higher likelihood ratios for all possible cut-off values than a method with a lower area under the ROC curve. However, if the ROC curves cross it is possible for a particular cut-off value with one method to give a higher likelihood ratio than any cut-off value with a second method, but the area under the second ROC curve is greater than that under the first. The areas under ROC curves, and the difference between the areas of two curves, should always be quoted with confidence intervals; calculating the appropriate interval is not straightforward for small samples (Obuchowski and Lieber 1998).

The effect of corrective action

In previous paragraphs we have considered the detection of incipient neurological deficits by monitoring, and the costs associated with false negative and false positive results, and we now turn to the benefit arising from appropriate corrective action. Ideally the old and new methods (no monitoring, and monitoring plus corrective action) would be compared in a randomized controlled trial. Let us assume that corrective action prevents the appearance of a new deficit in a proportion of patients in whom monitoring detects a change. If this is the case, the number of *TP*s will fall, and the number of *FP*s will rise by the same number. In *Table 1.3* monitoring plus corrective action avoids complications in two patients, so the new risk of a complication is 1/100 or 1%, compared with the 3% risk without corrective action. There are several different ways of expressing the change from old risk to new risk: *absolute risk reduction*, *relative risk*, *relative risk reduction*, *odds ratio* and *number needed to treat* (Cook and Sackett 1995). The *absolute risk reduction* achieved by corrective action resulting from monitoring is simply the difference between the old risk and the new risk, and the *relative risk reduction* is the absolute risk reduction divided by the old risk, or (old risk − new risk)/(old risk). The *relative risk* is the new risk divided by the old risk. The *odds ratio* is the new odds divided by the old odds. The odds ratio has a particular application in case–control studies (Sandercock 1989), and when the results of different trials are combined

Table 1.3 An example of the results of trials in which patients were randomly allocated to undergo an operation either without monitoring (control group), or with monitoring plus corrective action (experimental group). In one trial there were 100 patients in each group, and in the other each group included 1000 patients. The risks were identical in the two trials, but in the first trial (100 per group) the confidence intervals indicate that the risk reduction is not statistically significant, whereas with 1000 patients per group the risk reduction is statistically significant. This emphasizes the importance of including adequate numbers in trials. (No confidence interval is given for the number needed to treat in the 100 per group trial because the absolute risk reduction includes zero, so the confidence interval for the number needed to treat includes a discontinuity)

	New deficit	No new deficit	Total		
Experimental group: monitoring + corrective action (New treatment)	1(0) *a*	99(0) *b*	100(0)		
Control group: no corrective action (Old treatment)	3(0) *c*	97(0) *d*	100(0)		

	Proportion	Value	95% confidence interval: 100 patients per group	95% confidence interval: 1000 patients per group
Old risk	c/(c + d)	3%	0.623% to 8.52%	2.03% to 4.26%
New risk	a/(a + b)	1%	0.0253 to 5.45%	0.48% to 1.83%
Absolute risk reduction	Old risk − new risk	2%	−1.87% to 5.87%	0.776% to 3.22%
Relative risk	New risk / Old risk	33.3%	3.53% to 315%	16.4% to 67.8%
Relative risk reduction	(Old risk − new risk) / Old risk	66.7%	−215% to 94.7%	32.2% to 83.6%
Number needed to treat	1 / (Old risk − new risk)	50	*	32 to 129
Odds ratio	(a/b)/(c/d) = ad/bc	0.327	0.0334 to 3.19	0.159 to 0.672

in the procedure known as meta-analysis (Egger et al 1997), although reliance upon the odds ratio has been criticized (Sackett et al 1996). The *number needed to treat* is the number of patients needed to undergo the operation with the new procedure (i.e. monitoring plus corrective action) in order to prevent one postoperative deficit; this is 1/(old risk – new risk), or 1/(absolute risk reduction), rounded up to the nearest whole number (Cook and Sackett 1995). The values in *Table 1.3* give an absolute risk reduction of 2%, a relative risk reduction of 2/3 or 66%, a relative risk of 1/3 or 33.3%, an odds ratio of 0.327 and the number needed to treat is 50. However, with only 100 patients in each group the confidence intervals are wide; the intervals for absolute and relative risk reduction include zero, and the intervals for relative risk and odds ratio include 1. The apparent risk reduction in this small sample could therefore have occurred by chance, and could have arisen if corrective action was not really beneficial, and might even have been harmful. (No confidence interval is given for the number needed to treat, because the confidence interval for the absolute risk reduction includes zero, so there is a discontinuity in the confidence interval for the number needed to treat. If a confidence interval is required in this situation it should include a *number needed to treat to harm*, as well as a *number needed to treat to benefit* (Altman 1998).) If each group had included 1000 patients, with the same proportions (old risk, new risk, etc.), the confidence intervals would be narrower, intervals for absolute and relative risk would not include zero, and intervals for relative risk and odds ratio would not include 1, so the reduction in risk would have been significant (see *Table 1.3*).

This illustrates the important point that adequate numbers of patients must be studied to achieve a statistically significant result. The number of patients needed depends on the existing risk, the expected risk reduction, the significance level and the power of the study. 'Power' is the probability of obtaining a result which is significant at the required level if the expected effect is present. A desirable power level is between 0.8 and 0.9 (80% and 90%). Sample size calculation is complicated, but values can be obtained from tables (Machin and Campbell 1987), a nomogram (Altman 1982, 1991, p. 456), or from computer programs (Thomas and Krebs 1997), including one that accompanies the book by Bland (2000).

If we know the cost of monitoring one operation, we can calculate the cost of preventing one unsatisfactory outcome. This is the cost of monitoring one procedure multiplied by the number needed to treat, because on average only one unsatisfactory outcome is prevented for every number needed to treat. For example, if the cost of monitoring one operation is £1000 and the number needed to treat is 50, then the cost of preventing one unsatisfactory outcome will be £50 000 on average, because 50 operations must be monitored to avoid one poor outcome. Provided that there is no cost associated with a false positive result, monitoring will be cost-effective if the cost of prevention is less than the cost incurred following one unsatisfactory outcome. In most cases there is a cost associated with a false positive result, which will increase the cost of preventing one unsatisfactory outcome.

Is the risk eliminated by monitoring?

It is tempting to think that monitoring plus corrective action eliminates risk entirely, and this view is reinforced when a number of operations have been performed with monitoring plus corrective action without any patient developing a complication. However, if 100 procedures have been completed without the complication appearing, although the estimated prevalence of the complication as a result of the procedure is zero, the 95% confidence interval extends from zero to 3.62%. This means that there is a 95% probability that a consecutive series of 100 procedures might be completed without mishap if the prevalence of the complication was in fact as high as 3.62%, and the confidence intervals for 99% and 99.9% probability are even wider. As more procedures are performed the confidence intervals decline, but never reach zero (*Fig. 1.3*). An approximate value for the 95% confidence limit is $3/N$, where N is the number of operations performed without a new deficit (Hanley and Lippman-Hand 1983).

Clearly the above reasoning has profound medico-legal implications. Patients and their lawyers need to accept that failure of monitoring plus corrective action to prevent a complication may not be anyone's fault; it may simply arise from a small but non-zero residual risk. This argument is entirely compatible with the approach taken in other safety-critical activities such as aviation and nuclear power. Here it is axiomatic that risk can only be reduced, not eliminated. The acceptable risk in civil aviation is less than 1 in 10^7 hours (Mitchell and Evans 2004). The corresponding surgical risk for an operation lasting 1 hour would be 0.00001%, rising to 0.0001% for an operation lasting 10 hours. The best estimate of the risk for many surgical operations is considerably higher than this, and a surgeon would have to perform approximately 3 million

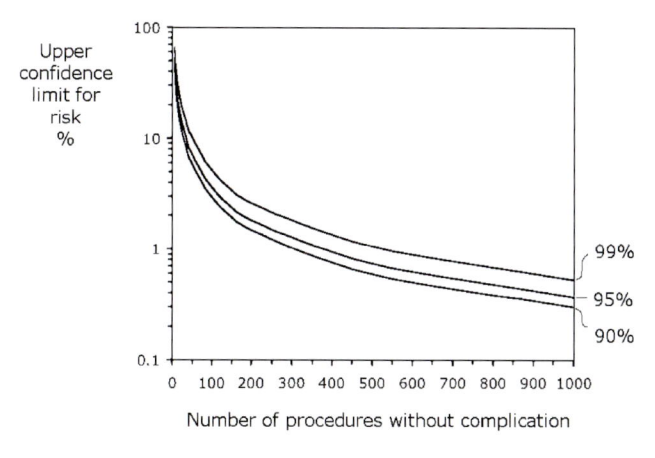

Fig. 1.3 The upper 90%, 95% and 99% confidence intervals for overall risk following a consecutive series of procedures without a complication appearing. (Note that risk is plotted on a logarithmic scale).

operations without a complication before he could be 95% certain that the risk for a 10-hour operation had been reduced to one in a million, or 0.0001%.

REFERENCES

Ackerstaff RG, Moons KG, van de Vlasakker CJ, et al 2000 Association of intraoperative transcranial Doppler monitoring variables with stroke from carotid endarterectomy. Stroke 31:1817–1823.

Aitkenhead AR 1991 Risk management in anaesthesia. J Med Defence Union 4:86–90.

Altman DG 1982 How large a sample? In: *Statistics in Practice* (eds SM Gore and DG Altman). British Medical Association, London, pp. 6–8.

Altman DG 1991 *Practical Statistics for Medical Research*. Chapman & Hall, London.

Altman DG 1998 Confidence intervals for the number needed to treat. BMJ 317:1309–1312.

Altman DG, Machin D, Bryant TN, et al 2000 *Statistics with Confidence*, 2nd edition. BMJ Books, London.

American Academy of Neurology 1990 Assessment: intraoperative neurophysiology. Report of the Therapeutics and Technology Assessment Subcommittee of the American Academy of Neurology. Neurology 40:1644–1646.

American Electroencephalographic Society 1994 Guideline eleven: guidelines for intraoperative monitoring of sensory evoked potentials. J Clin Neurophysiol 11:77–87.

Aminoff MJ 1989 Intraoperative monitoring by evoked potentials for spinal surgery: the cons. Electroencephalogr Clin Neurophysiol 73:378–380.

Andrews RJ, Bringas JR, Muizelaar JP, et al 1993 A review of brain retraction and recommendations for minimizing intraoperative brain injury. Neurosurgery 33:1052–1063 (discussion 1063–1064).

Anon. 1993 Technology and equipment review. Intraoperative monitoring equipment. J Clin Neurophysiol 10:526–533.

Austin EH 3rd, Edmonds HL Jr, Auden SM, et al 1997 Benefit of neurophysiologic monitoring for pediatric cardiac surgery. J Thorac Cardiovasc Surg 114:707–715, 717.

Banoub M, Tetzlaff JE, Schubert A 2003 Pharmacologic and physiologic influences affecting sensory evoked potentials. Anesthesia 99:716–737.

Bejjani GK, Nora PC, Vera, PL, et al 1998 The predictive value of intraoperative somatosensory evoked potential monitoring: Review of 244 procedures. *Neurosurgery* 43:491–498.

Belardi P, Lucertini G 2003 Cerebral vasoreactivity does not predict cerebral ischaemia during carotid endarterectomy. J Cardiovasc Surg 44:731–735.

Black D 1998 The limitations of evidence. J R Coll Physicians London 32:23–26.

Bland MJ 2000 *An Introduction to Medical Statistics*, 3rd edition. Oxford University Press, Oxford.

Bland MJ, Altman DG 1996a Measurement error proportional to the mean. BMJ 313:106.

Bland MJ, Altman DG 1996b Measurement error. BMJ 313:744.

Borresen TE, Overstreet JE, Truax WD, et al 1993 EEG monitoring during carotid surgery: who should do the monitoring? Clin Electroencephalogr 24:70–77.

Bricolo A, Faccioli F, Grosslercher JC, et al 1987 Electrophysiological monitoring in the intensive care unit. In: *The London Symposia. Electro-encephalography and Clinical Neurophysiology* Suppl 39 (eds RJ Ellingson, NMF Murray, AM Halliday). Elsevier, Amsterdam, pp. 255–263.

Brook RH, Park RE, Chassin MR, et al 1990 Predicting the appropriate use of carotid endarterectomy, upper gastrointestinal endoscopy, and coronary angiography. N Engl J Med 323:1173–1177.

Brown RH, Nash CL, Berilla JA, et al 1984 Cortical evoked potential monitoring. A system for intraoperative monitoring of spinal cord function. Spine 9:256–261.

Burke D, Nuwer MR, Daube J, et al 1999 Intraoperative monitoring. In: *Recommendations for the Practice of Clinical Neurophysiology. Guidelines of the International Federation of Clinical Neurophysiology* (EEG Suppl 52) (eds G Deuschl, A Eisen). Elsevier Science, Amsterdam, pp. 133–148.

Cao P, Giordano G, Zannetti S, et al 1997 Transcranial Doppler monitoring during carotid endarterectomy: is it appropriate for selecting patients in need of a shunt? J Vasc Surg 26:973–979.

Chalmers TC 1981 The clinical trial. Milbank Mem. Fund Q 59:324–339.

Chatrian GE 1990 Coma, other states of altered consciousness and brain death. In: *Current Practice of Clinical Electroencephalography* (eds DD Daly, TA Pedley). Raven, New York, pp. 425–187.

Chatrian GE, Bergamasco B, Bricolo A, et al 1996 IFCN recommended standards for electrophysiologic monitoring in comatose and other unresponsive states. Report of an IFCN committee. Electroencephalogr Clin Neurophysiol 99:103–122.

Chen R, Bolton CF, Young B 1996 Prediction of outcome in patients with anoxic coma: a clinical and electrophysiologic study. Crit Care Med 24:672–678.

Chiappa KH, Hill RA 1998 Evaluation and prognostication in coma. Electroencephalogr Clin Neurophysiol 106:149–155.

Choi BCK 1998 Slopes of receiver operating characteristic curve and likelihood ratios for a diagnostic test. Am J Epidemiol 148:1127–1132.

Cook RJ, Sackett DL 1995 The number needed to treat: a clinically useful measure of treatment effect. BMJ 310(6977):452–454.

Crippen DW 1994 Neurologic monitoring in the intensive care unit. New Horizons 2:107–120.

Crossman J, Banister K, Bythell V, et al 2003 Predicting clinical ischaemia during awake carotid endarterectomy: use of the SJVO2 probe as a guide to selective shunting. Physiol Meas 24:347–354.

Daube JR 1989 Intraoperative monitoring for spinal cord surgery: the pros. Electroencephalogr Clin Neurophysiol 73:374–377.

Daube JR 1999 Intraoperative monitoring reduces complications and is therefore useful. Muscle Nerve 22:1151–1153.

Deeks JJ, Altman DG 1999 Sensitivity and specificity and their confidence intervals cannot exceed 100%. BMJ 318:193–194.

de Letter JA, Sie HT, Thomas BM, et al 1998 Near-infrared reflected spectroscopy and electro-encephalography during carotid endarterectomy – in search of a new shunt criterion. Neurol Res 20(Suppl 1):S23–S27.

Deuschl G, Eisen A (eds) (1999) *Recommendations for the Practice of Clinical Neurophysiology. Guidelines of the International Federation of Clinical Neurophysiology*, 2nd revised and enlarged edition. *Electroencephalography Clinical Neurophysiology* Suppl 52. Elsevier Science, Amsterdam.

Edmonds HL Jr, Griffiths LK, van der Laken J, et al 1992 Quantitative electroencephalographic monitoring during myocardial revascularization predicts postoperative disorientation and improves outcome. J Thorac Cardiovasc Surg 103:555–563.

Egger M, Smith GD, Phillips AN 1997 Meta-analysis. Principles and procedures. BMJ 315:1533–1537.

References

Fisher RS, Raudzens P, Nunemacher M 1995 Efficacy of intraoperative neurophysiological monitoring. J Clin Neurophysiol 12:97–109.

Fletcher JP, Morris JGL, Little JM, et al 1988 EEG monitoring during carotid endarterectomy. ANZ J. Surg 58:285–288.

Forrest JB, Rehder K, Goldsmith CH, et al 1990a Multicenter study of general anesthesia. I. Design and patient demography. Anesthesiology 72:252–261.

Forrest JB, Cahalan MK, Rehder K, et al 1990b Multicenter study of general anesthesia. II. Results. Anesthesiology 72:262–268.

Fraser CG, Fogarty Y 1989 Interpreting laboratory results. BMJ 298:1659–1660.

Friedman WA, Grundy BL 1987 Monitoring of sensory evoked potentials is highly reliable and helpful in the operating room. J Clin Monit 3:38–44.

Gan TJ, Glass PS, Windsor A, et al 1997 Bispectral index monitoring allows faster emergence and improved recovery from propofol, alfentanil, and nitrous oxide anesthesia. BIS Utility Study Group. Anesthesiology 87:808–815.

Ganes T, Lundar T 1983 The effect of thiopentone on somatosensory evoked responses and EEGs in comatose patients. J Neurol Neurosurg Psychiatry 46:509–514.

García-Larrea L, Artru F, Bertrand O, et al 1988 Transient drug-induced BAEP abolition in coma. Neurology 38:1487–1489.

Gelber DA 1999 Intraoperative monitoring is of limited use in routine practice. Muscle Nerve 22:1154–1156.

Green S, Goodwin H, Moss J 1997 *Risk Management in Anaesthesia*. Medical Defence Union, London.

Grundy BL 1982 Monitoring of sensory evoked potentials during neurosurgical operations: methods and applications. Neurosurgery 11:556–575.

Grundy BL 1995 The electroencephalogram and evoked potential monitoring. In: *Monitoring in Anesthesia and Critical Care Medicine*, 3rd edition (eds CD Blitt, RL Hines). Churchill Livingstone, New York, pp. 423–489.

Guérit JM 1998 Neuromonitoring in the operating room: why, when and how to monitor. Electroencephalogr Clin Neurophysiol 106:1–21.

Guérit JM, Fischer C, Facco E, et al 1999 Standards of clinical practice of EEGs and EPs in comatose and other unresponsive states. In: *Recommendations for the Practice of Clinical Neurophysiology. Guidelines of the International Federation of Clinical Neurophysiology*, 2nd revised and enlarged edition (eds G Deuschl, A Eisen). *Electro-encephalography and Clinical Neurophysiology* Suppl 52. Elsevier Science, Amsterdam, pp. 117–131.

Häkkinen VK, Kaukinen S, Heikkilä H 1988 The correlation of EEG compressed spectral array to Glasgow Coma Scale in traumatic coma patients. Int J Clin Monit Comput 5:97–101.

Hanley JA, Lippman-Hand A 1983 If nothing goes wrong, is everything alright? Interpreting zero numerators. JAMA 249:1743–1745.

Henderson AR 1993 Assessing test accuracy and its clinical consequences: a primer for receiver operating characteristic curve analysis. Ann Clin Biochem 30:521–539.

Huang Z, Dong W, Yan Y, et al 2002a Effects of intravenous mannitol on EEG recordings in stroke patients. Clin Neurophysiol 113:446–453.

Huang Z, Dong W, Yan Y, et al 2002b Effects of intravenous human albumin and furosemide on EEG recordings in patients with intra-cerebral hemorrhage. Clin Neurophysiol 113:454–458.

International Organization of Societies for Electrophysiological Technology (OSET) 1999 Guidelines for performing EEG and evoked potential monitoring during surgery. Am J END Technol 39:257–277.

International Task Force on Safety in the Intensive Care Unit 1993 International standards for safety in the intensive care unit. Intensive Care Med 19:178–181.

ISO 5725-6 1994 *Accuracy (trueness and precision) of measurement methods and results – Part 6: Use in practice of accuracy values*. International Organization for Standardization, Geneva.

Jones SJ 1993 Evoked potentials in intra-operative monitoring. In: *Evoked Potentials in Clinical Testing*, 2nd edition (ed. AM Halliday). Churchill Livingstone, Edinburgh, pp. 565–588.

Jordan KG 1994 Status epilepticus. A perspective from the neuroscience intensive care unit. Neurosurg Clin North Am 5:671–686.

Jordan KG 1999 Continuous EEG monitoring in the neuroscience intensive care unit and emergency department. J Clin Neurophysiol 16:14–39.

Lentner C 1982 *Geigy Scientific Tables*. Volume 2: *Introduction to Statistics, Statistical Tables, Mathematical Formulae*, 8th edition. Ciba-Geigy, Basle.

Lindley DV 1985 *Making Decisions*, 2nd edition. John Wiley, London.

Machin D, Campbell MJ 1987 *Statistical Tables for the Design of Clinical Trials*. Blackwell, Oxford.

Marsh RR, Frewent TC, Sutton LN, et al 1984 Resistance of the auditory brainstem responses to high barbiturate levels. Otolaryngol Head Neck Surg 92:685–688.

Matousek M, Takeuchi K, Starmark JE, et al 1996 Quantitative EEG analysis as a supplement to the clinical coma scale RLS85. Acta Anaesthesiol Scand 40:824–831.

McKinlay JB 1981 From 'promising report' to 'standard procedure': seven stages in the career of a medical innovation. Milbank Memorial Fund Q 59:374–411.

Medical Defence Union 1995 Top ten settled cases. J Med Defence Union 11:13.

Michenfelder JD 1987 Intraoperative monitoring of sensory evoked potentials may be neither a proven nor an indicated technique. J Clin Monit 3:45–47.

Mitchell SJ, Evans AD 2004 Flight safety and medical incapacitation risk of airline pilots. Aviation, Space Environ Med 75:260–269.

Moses H 3rd, Kaden I 1986 Neurologic consultations in a general hospital. Spectrum of iatrogenic disease. Am J Med 81:955–958.

National CEPOD 1987 *The Report of a Confidential Enquiry into Perioperative Deaths*. Nuffield Provincial Hospitals Trust, King's Fund Publishing Office, London.

National CEPOD 1998 *The National Confidential Enquiry into Perioperative Deaths*. NCEPOD, London.

Nelson LS 1984 The Shewhart Control Chart – tests for special causes. J Quality Technol 16:237–239.

Nuwer JM, Nuwer MR 1997 Neurophysiologic surgical monitoring staffing patterns in the USA. Electroencephalogr Clin Neurophysiol 103:616–620.

Nuwer MR 1993 Intraoperative electroencephalography. J Clin Neurophysiol 10:437–444.

Nuwer MR, Daube J, Fischer C, et al 1993 Neuromonitoring during surgery. Report of an IFCN committee. Electroencephalogr Clin Neurophysiol 87:263–276.

Nuwer MR, Dawson EG, Carlson LG, et al 1995 Somatosensory evoked potential spinal cord monitoring reduces neurologic deficits after scoliosis surgery: results of a large multicenter survey. Electroencephalogr Clin Neurophysiol 96:6–11.

Obuchowski NA 2003 Receiver operating characteristic curves and their use in radiology. Radiology 229:3–8.

Obuchowski NA, Lieber ML 1998 Confidence intervals for the receiver operating characteristic area in studies with small samples. Academic Radiol 5:561–571.

Pearlman RC, Schneider PL 1994 Intraoperative neural monitoring. An introduction for perioperative nurses. Assoc Operating Room Nurses J 59:841–849.

Popper KR 1959 *The Logic of Scientific Discovery*. Routledge and Kegan Paul, London [Popper's own translation of his *Logik der Forschung* of 1934].

Prior PF, Maynard DE 1986 *Monitoring Cerebral Function. Long-term Monitoring of EEG and Evoked Potentials*. Elsevier, Amsterdam.

Remaley AT, Sampson ML, DeLeo JM, et al 1999 Prevalence-value–accuracy plots: a new method for comparing diagnostic tests based on misclassification costs. Clin Chem 45:934–941.

Sackett DL, Deeks JJ, Altman DG 1996 Down with odds ratios! Evidence-based Med 1:164–166.

Sackett DL, Straus SE, Richardson WS, et al 2000 *Evidence-based Medicine. How to Practice and Teach EBM*, 2nd edition. Churchill Livingstone, Edinburgh.

Sandercock P 1989 The odds ratio: a useful tool in the neurosciences. J Neurol Neurosurg Psychiatry 52:817–820.

Schwartz WB, Komesar NK 1978 Doctors, damages and deterrence. An economic view of medical malpractice. N Engl J Med 298:1282–1289.

Smith JS, Roizen MF, Cahalan MK, et al 1988 Does anesthetic technique make a difference? Augmentation of systolic blood pressure during carotid endarterectomy: effects of phenylephrine versus light anesthesia and of isoflurance versus halothane on the incidence of myocardial ischemia. Anesthesiology 69:846–853.

Smith NJ 2000 What is normal? Am J END Technol 40:196–214.

Strike PW 1996 *Measurement in Laboratory Medicine*. Butterworth-Heinemann, Oxford.

Sutton LN, Frewen T, Marsh R, et al 1982 The effects of deep barbiturate coma on multimodality evoked potentials. J Neurosurg 57:178–185.

Syrek JR, Calligaro KD, Dougherty MJ, et al 1999 Five-step protocol for carotid endarterectomy in the managed health care era. Surgery 125:96–101.

Thomas L, Krebs CJ 1997 A review of statistical power analysis software. Bull Ecol Soc Am 78:126–139.

Vakkuri A, Yli-Hankala A, Sandin R, et al 2005 Spectral entropy monitoring is associated with reduced Propofol use and faster emergence in Propofol–nitrous oxide–Alfentanyl anesthesia. Anesthesiology 103:274–279.

Vespa PM, Nenov V, Nuwer MR 1999a Continuous EEG monitoring in the intensive care unit: early findings and clinical efficacy. J Clin Neurophysiol 16:1–13.

Vespa PM, Nuwer MR, Nenov V, et al 1999b Increased incidence and impact of nonconvulsive and convulsive seizures after traumatic brain injury as detected by continuous electroencephalographic monitoring. J Neurosurg 91:750–760.

Wheeler DJ, Chambers DS 1992 *Understanding Statistical Process Control*. SPC Press, Knoxville.

Wiedemayer H, Fauser B, Sandalcioglu IE, et al 2002 The impact of neurophysiological intraoperative monitoring on surgical decisions: a critical analysis of 423 cases. J Neurosurg 96:255–262.

Yli-Hankala A, Vakkuri A, Annila P, et al 1999 EEG Bispectral Index monitoring in sevoflurane or propofol anaesthesia: analysis of direct costs and immediate recovery. Acta Anaesthesiol Scand 43:545–549.

Young GB, Campbell VC 1999 EEG monitoring in the intensive care unit: pitfalls and caveats. J Clin Neurophysiol 16:40–45.

Young GB, Jordan KG, Doig GS 1996 An assessment of nonconvulsive seizures in the intensive care unit using continuous EEG monitoring: an investigation of variables associated with mortality. Neurology 47:83–89.

Zweig MH, Campbell G 1993 Receiver–operation characteristic (ROC) plots: a fundamental evaluation tool in clinical medicine. Clin Chem 39:561–577.

Chapter 2

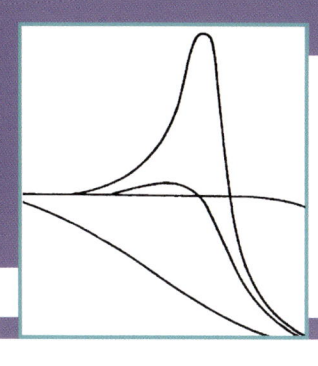

Neurophysiological Instrumentation, Connection with Patients and Recording Methods

INTRODUCTION

In this chapter we shall set out the essential basic principles of recording techniques in clinical neurophysiology utilized in diagnostic and monitoring applications for patients in the intensive care unit (ICU) and during anaesthesia and surgery in the operating room. Safety considerations for this rather demanding type of work in the ICU and operating room will be covered in *Practical aspects of intraoperative and ICU neurophysiological recording*, p. 55, whilst legal responsibilities are covered in Chapter 9.

The range of techniques and equipment used is given in *Table 2.1*. The reader will find more advanced information on the design and technology relating to such monitoring in Chapter 8 of this volume and may also wish to refer to the two volumes of *Clinical Neurophysiology* (Binnie et al 2003, 2004) on which this work is based and a related volume by Cooper et al (2005).

The growth in use of neurophysiological monitoring techniques in recent years has undoubtedly contributed to increased understanding of the effects of anaesthesia and the critical limits for ischaemic damage to neural tissues in the individual patient. As Professor Yli-Hankala indicates in his Foreword to this book, these benefits have been accompanied by various methodological controversies, some relating to recording techniques (mainly patient connections) and others to choice of signal processing algorithms. In this present chapter, we address the former

Table 2.1 Equipment for clinical neurophysiology

	Type of recording	Purpose	Duration (approx.)	Equipment
EEG	Multichannel	Diagnosis	1 h	Multichannel EEG recorder (analog or digital)
	Multichannel	(a) Diagnosis of intermittent events (b) Monitoring	Hours or days	Multichannel EEG recorder. (May be same as for ambulatory recording)
	Limited number of channels	Monitoring	Hours or days	Special EEG monitoring device, often with data reduction and analysis
EP	Multichannel	Diagnosis	1–2 h	Multichannel EP recorder
	Limited number of channels	Monitoring	Hours or days	Multichannel EP recorder
Nerve conduction	Measurement of action potential amplitude and velocity in several nerves	Diagnosis	30 min	Electrical stimulator and recorder (part of EMG machine)
Neuromuscular transmission	Simple assessment by repetitive stimulation at a single rate	(a) Diagnosis (b) Monitoring	5 min	Anaesthetist's peripheral nerve stimulator ('train of four')
	Detailed assessment including single fibre EMG	Diagnosis	30–60 min	EMG machine
EMG	Detection of abnormal spontaneous activity	Diagnosis	10 min	EMG machine
	Assessment of motor unit features and recruitment	Diagnosis	20–30 min	EMG machine

with detailed accounts of the basic principles and good practice concerning electrodes and neurophysiological recording techniques. As Professor Yli-Hankala has also pointed out, the extremely small size of the signals concerned, particularly the electroencephalogram (EEG) and evoked potentials (EPs), makes it essential to give careful attention to such technological matters to ensure that valid signals are being monitored. No one takes seriously a heart rate monitor displaying a rate of 225/min while the raw electrocardiogram (ECG) is badly corrupted, whereas if a monitor of anaesthetic depth displays an unexpected value while surgical equipment is producing an EEG artefact, the occurrence is greeted with surprise and deemed worthy of publication. Those using EEG monitoring devices should adopt the same critical attitude with respect to validity of the raw signal that they already apply to the ECG.

Our own experience of the benefits of working in multidisciplinary teams leads us to urge those without formal training in clinical neurophysiology to consult their local experts to plan a joint approach to the requirements of the surgical and intensive care clinicians.

RECORDING THE ELECTRICAL ACTIVITIES OF THE NERVOUS SYSTEM

Electrodes for neurophysiological recording

General features of electrodes

Electrodes are used to make connections between the conducting fluid of the tissue in which the electrical activity is generated and the input circuit of the amplifier. They can take many forms but there is always, somewhere, a liquid–metal junction. The signal should not be distorted at this interface, nor should artefactual activity be added. This is not always possible and, as most neurophysiological recording systems are not designed to test the electrodes with calibration signals, hidden and important modifications to the electrical activity can occur.

Any metal, when in contact with an ionized liquid, exhibits a steady electrical potential – the *electrode potential*. These potentials can be useful in other fields – batteries and accumulators are designed to exploit them – or they can be a nuisance, for example, by setting up local currents, which promote metallic corrosion. The potentials *per se* do not distort the fluctuating electrical activity recorded by the electrode, but only add a fixed value to it. If the steady potential arising at each of the two electrodes connected to an amplifier is of the same value – even if it is 1 V or more – the design of the differential amplifier will ensure that the microvolt neurophysiological signals will still be recorded accurately. If the electrode potentials are unequal, however, they will generate a constant potential across the amplifier input and, worse, if the are variable they will generate spurious signals possibly of much greater amplitude than the bioelectrical activity under study. Moreover, distortion of the neurophysiological signal can occur at the electrode surface because some metals act like a high-pass filter or time constant circuit, attenuating the low-frequency activity. This occurs when an electrode is *polarized*. To understand this, we need to know what is happening at the electrode surface.

When an electrode is placed in an electrolyte, some metallic atoms give up electrons and pass into solution as positively charged ions. Conversely, positive ions in solution can pick up electrons from the electrode, are converted to neutral atoms, and are deposited on its surface. An equilibrium is established between these two processes; the net shift of positive ions into solution causes the electrode to become negatively charged with respect to the electrolyte. The ions do not go far from the electrode's surface and the electrode is surrounded by a layer of ions forming an *electrical double layer* (Fig. 2.1). The electrode potential can be measured only against another electrode (which will have its own potential) and values are quoted as measured against a standard electrode, usually made of hydrogen adsorbed on platinum black. This is an inconvenient electrode and substandards such as calomel or chlorided

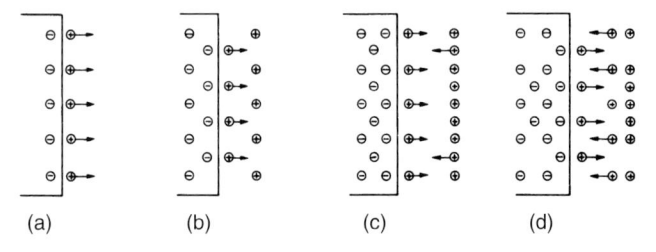

Fig. 2.1 Simplified diagram showing formation of electrical double layer. (a) Ionic flow into solution immediately after immersion. (b) Accumulation of ions in solution. (c) Ionic flow into and out of solution at different rates. (d) Equilibrium when rates are equal. Note excess of positive ions in solution giving rise to double layer. (From Cooper (1962), by permission.)

Table 2.2 Electrode potentials of various metals referred to the hydrogen electrode

Ion and metal in equilibrium	Electrode potential
Aluminium(3+) ↔ aluminium	−1.66
Titanium(2+) ↔ titanium	−1.63
Chromium(3+) ↔ chromium	−0.70
Iron(2+) ↔ iron	−0.44
Nickel(2+) ↔ nickel	−0.14
Lead(2+) ↔ lead	−0.13
Hydrogen(+) ↔ hydrogen	Arbitrary zero
Copper(2+) ↔ copper	+0.34
Copper(+) ↔ copper	+0.52
Silver(+) ↔ silver	+0.80
Platinum(2+) ↔ platinum	+1.20
Gold(+) ↔ gold	+1.70

silver are used (their potentials referred to the hydrogen electrode are –0.280 and 0.224 V, respectively, at 25°C). *Table 2.2* shows some electrode potentials measured or calculated at 25°C referred to the hydrogen electrode.

Clearly, the potential differences between unlike metals can be very large compared with neurophysiological signals. It might be hoped that if two electrodes were made from the same metal their electrode potentials would be exactly the same (and the potential difference zero), but slight impurities and surface contamination can cause considerable differences of potential between electrode pairs, often of many millivolts.

When a pair of electrodes is connected across the input of an alternating current (AC) amplifier, this difference in electrode potential is 'blocked' by the coupling capacitors and does not appear with the recorded fluctuating signal once the potentials on each side of the capacitors have equalized. However, if the electrode montage is changed, the new combination of electrode potentials can cause the amplifiers to 'block' until the capacitors have equalized; when using long time constants this can take tens of seconds. Most amplifiers used for low-frequency recording are fitted with antiblock circuits. In direct current (DC) amplifiers there are no coupling capacitors and the difference of steady potential between electrodes has to be 'backed off' by connecting an adjustable potential source (usually a battery and potentiometer) in series with each channel. The matter is made worse by the potentials slowly changing in time because of changing conditions at the electrode–skin interface. This gives rise to drift of DC recordings.

Although large differences of electrode potentials are blocked by the coupling capacitors in AC amplifiers and cause little trouble, they can exacerbate movement artefact. This arises because the potential differences between electrodes cause currents to flow in the input circuit of the amplifier and any change of electrode impedance (e.g. by movement) alters the current, causing artefact in the recording. Movement of the electrode also disturbs the electrical double layer, changing the electrode potential itself, again causing artefact. This can be reduced in surface recording by not placing the metallic surface of the electrode in direct contact with the skin, connection being made with saline jelly.

Electrode potentials are unavoidable, but the potential differences between electrodes can be minimized by making all electrodes of the same metal of high purity, and by avoiding contamination of the surfaces. Silver coated with silver chloride provides a stable electrode and is usually used for low-frequency and DC recordings.

Despite the input impedances of amplifiers being large (many mega-ohms), the small neurophysiological potentials cause tiny currents (10^{-10} to 10^{-13} amps) to flow through the input circuit and electrodes, thus crossing the electrical double layer. A metal interface that allows this to happen unhindered is said to be *reversible*. The separation of electrical charge at the electrode surface (electrical double layer) is very similar to the action of a capacitor (indeed there are electrolytic condensers): the steady biological potentials can be blocked and the low-frequency components attenuated. When this happens the electrode is said to be *polarized*, and an electrode subject to polarization is described as *non-reversible*. Clearly, if this significantly distorts the signal, the electrode is not suitable for neurophysiological recordings. Fortunately this occurs only when the electrodes are small and of particular metal, such as stainless steel (as sometimes used in intracerebral recordings), and only affects the low frequencies of the EEG. Whether an electrode is polarized or not depends upon the surface conditions.

One type of reversible or non-polarized electrode consists of a metal electrode in contact with one of its insoluble salts and immersed in a solution of a soluble salt of the same anion, for example, silver in contact with silver chloride immersed in saline (sodium chloride). In this case, when an external current passes such that the silver electrode is positive, silver atoms become ionized (Ag^+), but immediately combine with chloride ions (Cl^-) from the solution, producing neutral molecules of insoluble silver chloride which are deposited on the electrode. Chlorine ions are thus removed from solution:

$$NaCl \rightarrow Na^+ + Cl^-$$

$$Cl^- + Ag^+ \rightarrow AgCl$$

If the passage of current across the electrical double layer is in the opposite direction, silver ions from the solution regain electrons and are deposited as metallic silver on the electrode surface, the solution becomes unsaturated of silver and silver chloride dissociates into silver and chlorine ions:

$$Ag^+ + electron \rightarrow Ag$$

$$AgCl \rightarrow Ag^+ + Cl^-$$

Thus, chlorine ions pass into solution. The system behaves like a reversible chlorine electrode, with the silver chloride acting as a bridge helping the current to flow equally well either way with little impairment. Passing small amounts of charge through the electrode in either direction will make the chloride layer thicker or thinner but will not change its properties. Unless a substantial net charge is passed in one direction (for instance, by injudicious testing of electrode resistance using a DC source), the electrochemical state of the electrode surface remains stable and the electrode potential constant.

The equivalent circuit of the electrode tissue junction

It is sometimes helpful to consider the equivalent circuit diagram of an electrode in solution, although it should be noted that there is no *a priori* reason why such an

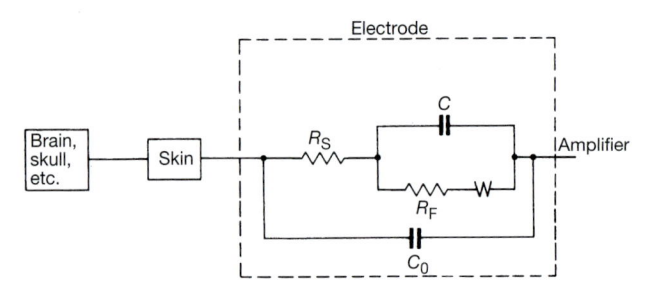

Chapter 2 Neurophysiological Instrumentation, Connection with Patients and Recording Methods

Fig. 2.2 Equivalent circuit diagram of an electrode. (From Cooper (1962), by permission.)

equivalent circuit can be drawn using conventional components (resistors, capacitors and inductors). *Figure 2.2* is an equivalent circuit suggested by Grahame (1952). R_s is the resistance of the electrode jelly or saline in contact with the electrode. Its value depends upon the concentration of sodium chloride and is of the order of hundreds of ohms. C is the capacitance of the electrical double layer and is of the order of $\mu F/mm^2$. R_F is the electrical resistance of the chemical change taking place when the current flows. —W— is called the 'Warburg impedance' and is equivalent to a resistor and capacitor whose values are proportional to frequency. C_0 is a small capacitance made up of strays and can be neglected in most practical cases.

The most important elements are C shunted by R_F and —W—. In a reversible electrode, R_F and —W— are resistive and small, and bypass the capacitance C. In a polarized or non-reversible electrode, R_F is large (many mega-ohms) and the effective equivalent circuit becomes a series capacitor of value C. The effect on the neurophysiological signal will depend on the value of C since, with the input impedance of the amplifier, it acts as a filter attenuating low frequencies and is equivalent to the introduction of a time constant (cf. *Fig. 2.16* below).

The electrical double layer, which gives rise to the capacitance C, is similar to a parallel plate capacitor; the capacitance is proportional to the surface area. For conventional surface EEG or EMG electrodes, this capacitance is large, perhaps 10 µF or more, and even if R_F is also large, as in a polarized electrode, this input capacitance will not distort recordings to any great extent, provided that the input impedance of the amplifier is greater than, say, 5 MΩ. However, electrodes with very small contact areas will have correspondingly small capacitance, which may affect recording of low frequencies. Moreover, it happens that those very types of electrodes which have the smallest surface areas are generally made of polarizable materials. There are practical reasons for this: needle electrodes used for EMG and sometimes for scalp EEG recording have to be made of hard metals, such as stainless steel, which can be sharpened; the fine wires used as intracerebral electrodes cannot be made from non-polarizable materials such as chlorided silver, because silver chloride is toxic. A stainless steel intracerebral electrode also has a chromic oxide layer on the surface (which

keeps it stainless) and this acts as another capacitor in series with the double layer, thus giving a smaller net value of capacitance. Typical values of C and R_F for 38-gauge (0.006 in., 150 µm) stainless steel, 2 mm bared, are 0.05 µF and 10 MΩ. EMG needle electrodes have a very small surface area and are usually made of stainless steel; fortunately they are required only to record high-frequency activity and do not affect the data significantly.

The effect of the capacitance of the electrical double layer can be seen by measuring the impedance of the electrode using ACs at less than 10 Hz. *Figure 2.3* shows the steeply rising values of needle and sphenoidal electrode impedances at frequencies less than 1 Hz. There could be attenuation of low-frequency signals using these electrodes, especially if the input impedance were 1 MΩ or less. This may not affect signals in the conventional EEG or EMG ranges, but would be a major drawback for recording some types of slow EPs, and for various kinds of polygraphy, monitoring eye position, skin potential, etc.

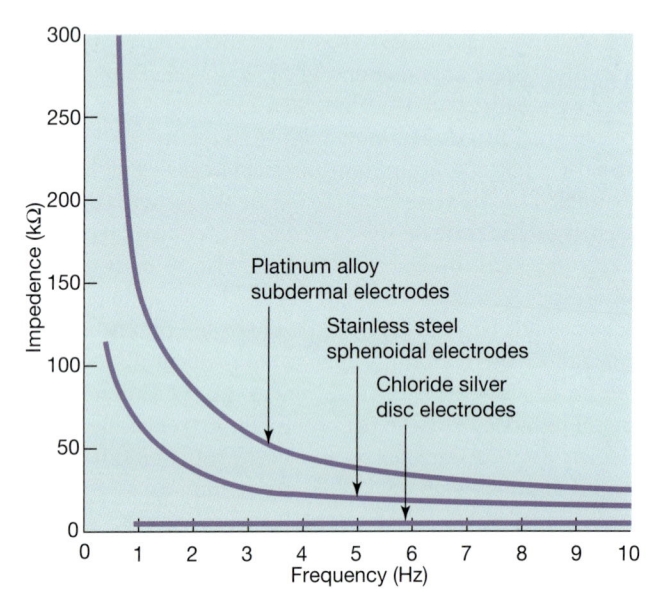

Fig. 2.3 Change of electrode impedance with frequency. Note the high impedance of subdermal and sphenoidal needle electrodes at frequencies less than 1 Hz. (From Cooper et al (1980), by permission.)

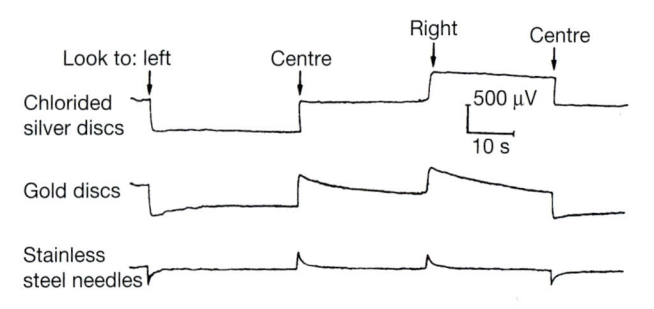

Fig. 2.4 Slow speed recording of eye movements using various types of electrodes. Note slow paper speed. (From Cooper et al (1980), by permission.)

If DC amplifiers are available, the electrical characteristics of electrodes can be seen by using them to record the positions of the eyes (*Fig. 2.4*). Electrodes for DC recording are best made from chlorided silver. With care, differences of electrode potential of less than 1 mV can be achieved. Care should also be taken in choosing the best type of amplifier and in preparation of the skin (Bauer et al 1989).

Although the requirements of electrodes for routine EEG recording using time constants of 1 s or less are not so stringent as those for DC, time spent in preparation and care of electrodes is not wasted. We have already seen how differences of electrode potential can occur and how the artefacts caused by them can be minimized by using the same metal for all electrodes. In practice, this is not always possible as stainless steel sphenoidal needles may have to be connected in montages with chlorided silver scalp electrodes.

Another source of artefact not previously mentioned is that caused by the attachment of the electrode to a lead of different metal. Often this is insulated copper wire soldered to the electrode metal and covered with insulation. If this bimetallic junction comes in contact with electrolyte (jelly, etc.) because of defects in the insulation, it can act as a battery and produce large fluctuating potentials which appear as artefacts. Similarly, clips that connect the pad-type EEG electrodes to the input leads must be kept dry.

Reversible electrodes are generally to be preferred because of their reduced susceptibility to artefact. They are most often made of pure silver coated with a layer of silver chloride. An electrolytic method is usually used to coat the electrodes, although dipping them in a strong solution of bleaching agent will both coat and sterilize them. Before old electrodes are re-chlorided they should be stripped of existing silver chloride electrolytically by placing them in saline and making them electronegative (by about 9 V) with respect to another electrode, for a few minutes, until the surface is free from any purple chloride. The electrodes are then placed in a glass dish containing a solution of 2–5% sodium chloride in water (2–5 g/100 ml) and each connected to the positive terminal of a 1.5 V battery. The negative terminal is connected to another electrode immersed in the saline. After a few seconds the electrodes connected to the positive pole will be covered with a dark brown or purple coating of silver chloride. The currents, which will be about 2.5 mA/cm^2 of electrode surface, should be allowed to flow for about a minute. The chemical changes are shown by the following equations:

$$NaCl \rightarrow Na^+ + Cl^-$$

$$Cl^- + Ag^+ \rightarrow AgCl$$

The positive sodium ions react at the cathode surface to produce hydrogen:

$$2Na^+ + 2H_2O + 2\text{ electrons} \rightarrow 2NaOH + H_2$$

Electrodes are a vulnerable but vital part of the recording system, especially in recording the slow potentials of the EEG: they should be prepared and treated with care.

Desirable characteristics of electrodes

Electrodes are described here in the context of EEG recording; however, most would also be suitable for recording EPs, and some indeed may also be appropriate for peripheral studies such as surface recording of nerve action potentials. The important electrical properties of EEG electrodes can be summarized as follows.

Associated with each electrode in contact with tissue is a standing potential difference of some tens of millivolts and an impedance of a few kilo-ohms (as measured at frequencies from DC to 10 Hz). If the potentials are constant and nearly equal between pairs of electrodes, they present little problem. Inequalities in potential between pairs of electrodes will, like batteries, cause currents to flow in the input circuits of the amplifiers and produce potential differences across the input connections. Provided the currents are unchanging, they will not be recorded by AC amplifiers, although they may produce temporary blocking when amplifiers are switched between different pairs of electrodes. However, they will be modulated by any changes in electrode impedance (e.g. because of scalp movement), thus generating out-of-phase signals which will be amplified and contaminate the tracing (movement artefact). Changes in the standing potential of only a few per cent may be far larger than the EEG itself.

It follows that 'electrode impedances' (or more strictly the impedances at each electrode/scalp interface) should so far as possible be low, equal and stable. They should be low because they form a potential-divider circuit with the input impedance of the amplifiers, and if electrode impedances approach the values of input impedance (which is rarely the case with modern machines), a substantial part of the signal will be lost across the electrodes, and not be available for amplification. Electrode impedances should be stable, because any fluctuations will modulate currents from standing potentials and produce artefacts as described earlier. They should be equal to ensure rejection of common mode signals, as for instance those produced by interference sources causing the subject to change in potential with respect to the chassis of the EEG machine. If the electrode impedances are unequal, such potential changes will produce unequal currents in the two input leads, will not be rejected by the amplifier as common mode signals and will therefore appear in the output.

There is a trade-off between the desirable characteristics of low, equal and stable electrode potentials and impedances. Low potentials may compensate for unstable impedance (a feature exploited in some special electrodes which were developed for recording under unfavourable conditions during space missions) and if the potential differences are high and/or unequal between electrodes, the situation may yet be saved by skilled application, giving stable impedances.

A further consideration in electrode design must be practicality. They must be easy to attach and retain at the appropriate location; considerations of cost and time have hitherto limited the use of complex procedures to position scalp electrodes, but careful measurement from bony landmarks remains essential to make standardized and valid recordings. Simple linear measurements using callipers may suffice (Le et al 1998). Obviously electrodes and their means of fixation should not be harmful and should not damage the skin or cause pain; indeed the current awareness of health and safety requirements has given rise to stringent guidelines regarding their preparation and application (see *Electrical safety*, p. 56). Furthermore, any associated head movement or discomfort should not be such as to disturb the subject unduly, nor jeopardize the cerebral circulation in very ill patients.

A range of electrode types has been used over the years in the EEG laboratory. Many of these are unsuitable for work in the ICU and the operating theatre, particularly for long duration studies lasting hours or days (*Table 2.3*). Basic types of electrode are described here; further information about those appropriate for ICU and intraoperative monitoring is discussed on page 59.

Self-retaining electrodes

The self-retaining electrodes in general use are either discs retained by some form of adhesive, or needles inserted into the scalp (*Fig. 2.5*).

Disc and cup electrodes Disc or cup electrodes may be fabricated from chlorided silver, tin, solder, platinum or gold. Some electrical characteristics of these materials are given in *Figs 2.3* and *2.4*. Chlorided silver is the material most widely used, because the chloride layer provides a reversible low impedance bridge between metal and tissue. They are non-polarizable and will record steady (DC) and very low frequency activity arising in the tissue. Small polarized electrodes made of gold, platinum or stainless steel block steady potentials and attenuate low frequency potentials. The electrodes may be applied in either of two ways (*Fig. 2.6(a–b)*).

For maximum stability, they may be glued by a thin layer of collodion spread over the outer surface of the electrode and extending over the adjacent scalp. In this case some form of conducting jelly must be placed between the undersurface of the electrode and the skin in order to establish contact. Electrodes to be attached in this way are

Table 2.3 Characteristics of EEG electrodes used for ICU and intraoperative monitoring)

Type	Application/removal methods	Advantages	Disadvantages	Comments
Self-retaining electrodes				
Cup electrodes (chlorided silver)	(i) Collodion + conductive jelly (ii) Skin preparation with ether mixture or special paste (iii) Acetone to dissolve and help remove collodion	(i) Secure for long periods (hours to days) (ii) Low contact impedance	(i) Application and removal time consuming (ii) Danger of transmission of viral diseases if scarification used to lower impedance (iii) Difficult to remove all collodion from hair	The 'gold standard' for diagnostic and monitoring applications
Disk electrodes (chlorided silver)	(i) Adhesive–conductive pastes (ii) Skin preparation with ether mixture or special paste (iii) Remove with damp swab	(i) Quick and easy to	(i) Less secure than collodion over long periods or in restless patients	Increasingly used for diagnostic and shorter duration monitoring applications
Needle electrodes (usually platinum–iridium alloy, disposable type)	(i) Cleanse skin with alcohol (ii) Insert intradermally, parallel to the skin surface, using sterile technique (iii) Firm pressure with swab on withdrawal (iv) Safe disposal	(i) Quick and easy to apply and remove	(i) Transmission of viral infections (ii) Bleeding in certain patients	No longer justified except in a few, limited, circumstances
Pre-gelled ECG electrodes	(i) Good skin preparation essential (ii) Typically placed on forehead	(i) Ease of application (ii) Standard interelectrode distance	(i) Cost relatively high (ii) EEG contaminated by muscle and eye movement potentials	Use only for work where performance is demonstrably optimal for the purpose

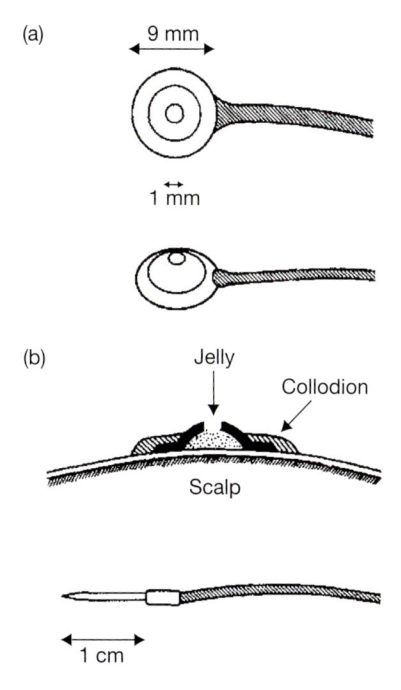

Fig. 2.5 Electrodes for EEG and EP monitoring. (a) Silver–silver chloride disc electrodes and method of application. (b) Platinum–iridium alloy intradermal needle electrodes. (From Prior and Maynard (1986), by permission.)

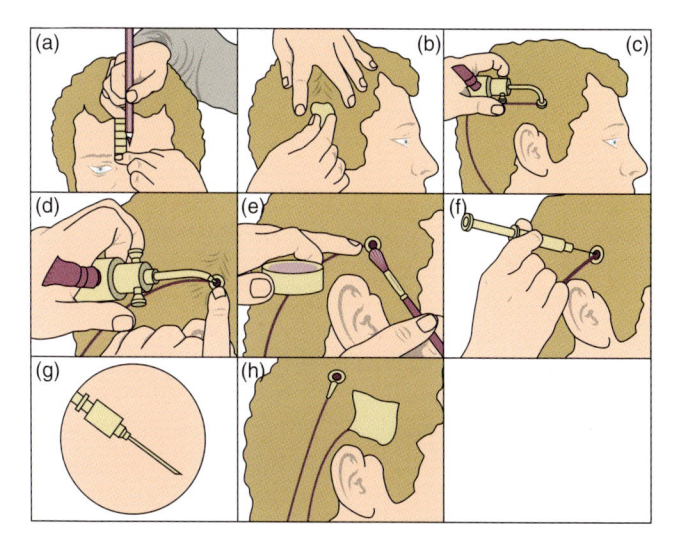

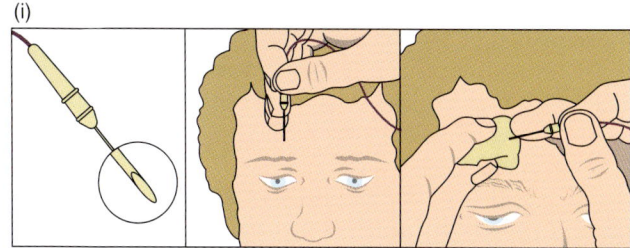

Fig. 2.6 Application of stick-on cup electrode. (a) Marking electrode site. (b) Cleansing scalp. (c) Use of compressed air applicator for holding electrode and drying collodion. (d) Spreading collodion with finger tip while drying with compressed air. (e) Alternative method of collodion application using brush. (f) Injection of contact medium and gentle abrasion of skin. (g) Needle with blunt end for injection of electrode paste. (h) Collodion-impregnated gauze may be placed over electrode for greater stability. (i) Insertion and removal of subdermal needle electrode. (From Binnie CD, Rowan AJ, Gutter Th 1982b *A Manual of Electroencephalographic Technology*, Cambridge University Press, by permission.)

shaped like a shallow cup with a flattened rim (see *Fig. 2.5*). There may be a hole through which the electrode jelly can be inserted, after the disc has been stuck in place. First the skin is prepared by parting the hair at the point where the electrode is to be placed, then cleansing the scalp with vigorous rubbing with gauze soaked in an ether mixture, or preferably a special skin preparation paste (e.g. Omniprep or NuPrep, DO Weaver & Co., Aurora, CO, USA, http://www.doweaver.com) applied with a cotton bud; then jelly is injected into the space between the cup and the skin through a hollow blunt needle. Formerly, it was usual to abrade the skin under the electrode by means of the needle. This practice reduces skin resistance but can no longer be recommended because of the risk of transmission of viruses in blood or exudate from the abraded skin. Jellying causes some discomfort to the patient as the electrolyte in the jelly comes into contact with the prepared skin surface. Where no hole is provided in the cup, it is necessary for the jelly to be inserted before the electrode is applied. Particularly when disc electrodes without a hole are used, and also to provide extra security for ICU or intraoperative recording, a 2 cm gauze square soaked in collodion may be placed over the electrode after it is in place. Application and drying of collodion may be carried out in various ways. The collodion (5% solution in ether) may be squeezed from a tube, or applied with a paint brush from a pot. It is usually dried by means of compressed air, or with a hand-held electric hairdryer; if these are not available, the technologist can dry the collodion by blowing on it. An applicator is available, which holds the electrode in place and also directs the stream of compressed air (see *Fig. 2.6*). A skilled operator can apply disc electrodes faster than pads, particularly in a restless or unconscious subject.

An interesting variant of the cup electrode is constructed of plastic, the actual electrode interface being represented by a coil of chlorided silver wire recessed within the cup. Electrode jelly forms a bridge between the chlorided surface and the skin. The advantage of this recessed type of electrode is that it avoids changes in the physical state of the electrode/electrolyte interface due to mechanical contact with the skin. Such electrodes are very stable and are particularly suitable for DC recording.

To remove the electrodes the collodion is dissolved with acetone applied by means of a swab. The remaining fragments can be loosened with further acetone and removed first with a coarse steel comb and then with a fine-toothed nit comb. It is not acceptable for the patient to be left with collodion fragments in the hair but, as the acetone dries, fine deposits of collodion will reappear; this is unavoidable and the subject should be warned that a fine powder of collodion, resembling dandruff, may persist for a few days after the investigation. Acetone is not importantly toxic under the conditions of use described above, but smells unpleasant and is a fire hazard. The area which is used for electrode removal should be well ventilated, preferably with an extractor fan. The vapour is heavier than air and flows down the face so the patient's eyes should be covered with gauze during electrode removal.

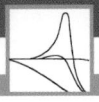

The second method of applying disc electrodes is to use the same medium as the adhesive and as a conducting bridge between the electrode and the skin. Various commercial products are now available, which are non-toxic and provide good adhesion. Cup electrodes as described above may be used, but electrodes intended specifically for this method of fixation are discs with a flat undersurface. After preparation of the skin, a pea-sized pellet of paste is placed on the electrode, which is then pressed into place. This method of fixation is less secure than collodion, but is the most convenient choice for routine recording for electrodes near the face which are not subject to disturbance during monitoring.

Needle electrodes Needle electrodes, generally made of platinum or stainless steel with an iridium tip, are inserted tangentially just below the skin (see *Fig. 2.6(i)*). Their insertion is surprisingly painless and needle electrodes can even be applied to a sleeping subject without causing waking. These electrodes can be inserted very swiftly and therefore were formerly the preferred choice where speed is essential, for instance to obtain a post-ictal EEG, or in intensive care. An obvious drawback of needle electrodes is the risk of infection. Increasing concern with the possible transmission of viruses in blood or exudate by needle-stick injury should cause these electrodes, once very popular in some countries, to be used very selectively. In addition, it should be noted that many patients undergoing intensive care may have abnormal bleeding tendencies, or be receiving anticoagulants. These factors mean that even single-use, disposable needle electrodes (re-useable needles, even with careful sterilization procedures of the type described in *Particular physical hazards and safety considerations* (p. 57), are no longer acceptable) can no longer be recommended. Indeed, there is a strong case for considering that needle electrodes should no longer be used at all.

In exceptional circumstances, however, the use of single-use disposable needle electrodes may be considered justifiable providing no abnormal bleeding tendency is present. These include patients after head injury or recent neurosurgery when the scalp is traumatized or oedematous making it inappropriate to use disc electrodes. In such patients, the skin should be stretched taut between two gloved fingers after skin cleaning with alcohol to reduce the risk of local infection and the needle then inserted. All the electrodes should preferably be inserted in the same direction (e.g. pointing forwards), and the leads gathered together. It may be noted that the midpoint of the volume of tissue from which the electrode records does not correspond to the point of entry through the skin and this must be taken into account in determining placements. The technologist must take particular care to avoid self-infecting by needle-stick. Gloves must be worn whilst handling needle electrodes, for preparation, application and for subsequent cleaning. Subcutaneous bleeding may occur, particularly in a restless subject, but it is uncommon and generally easily controlled by pressure. To remove a needle electrode after use, the technologist takes a small pledglet of cotton wool, presses firmly over the site of entry of the electrode and then quickly withdraws it. The puncture wounds are extremely small and no special aftercare is necessary, unless there has been significant bleeding.

Electrodes for use in the magnetic resonance imaging (MRI) unit Simultaneous recording of the EEG during functional MRI (fMRI) investigation has created a need for special electrode systems, on account of the intense magnetic fields employed. Special electrode caps have been developed (Baumann and Noll 1999) providing a high resolution array of 64 electrodes and, of course, free of ferromagnetic materials. Goldman et al (2000) describe the use of silver chloride coated plastic cup electrodes with carbon fibre leads. The electrodes were made up into a prewired bipolar montage with the lead from each pair twisted together to minimize electromagnetic induction of currents in the input circuits. Mirsattari et al (2004) produced electrodes without ferromagnetic materials that were consistent with artefact free MRI images when tested in volunteers and then in 100 patients undergoing long-term monitoring for epilepsy.

Electrode placement systems

It is necessary that the locations of electrodes should be known in relation to the topography of the scalp and so far as possible that of the underlying brain, and should be standardized so that comparisons can be made within and between subjects. After an initial period of confusion during which many workers developed their own systems, the International Federation of Societies for Electroencephalography and Clinical Neurophysiology (IFSECN) proposed an international standard, the so-called 10–20 system (Jasper 1958) which is now widely if not universally accepted throughout the world. Despite some obvious deficiencies which are discussed below, no serious attempt has been made to modify or improve the 10–20 system in the four decades since its introduction and the individual user must therefore choose between accepting the system with its undoubted deficiencies or choosing not to comply with the only international standard.

General considerations in the design of placement systems

In a preamble to the specification of the 10–20 system, the IFSECN's Committee on Methods of Clinical Examination in EEG set out a number of general principles; these are listed below (in italics) together with some comments and additions by the present authors.

1. *'Position of electrodes should be determined by measurement from the standard landmarks on the scalp. Measurements should be proportional to skull size and shape, in so far as possible.'* Proportional measurements are suggested, as absolute measurements could not take account of variations in cranial size and shape.

2. *'Adequate coverage of all parts of the head should be provided with standard designated positions even though all would not be used in a given examination.'* Unfortunately, the coverage provided by the 10–20 system cannot reasonably be regarded as adequate, as the lowest standard placements are well above the lower limit of the cerebral convexity. Coverage of the temporal lobes is particularly poor and has led to various proposals for additional anterior temporal placements.

3. *'Electrode placements should so far as possible be symmetrical about the sagittal plane.'* The normal EEG is essentially symmetrical and asymmetries may be important evidence of abnormality. Unfortunately, most heads are asymmetrical. If the two halves of the cranium were of similar shape and differed only in size, a suitable proportional method of electrode placement could take account of this. However, the four quadrants bounded by the inion, nasion and the external auditory meati are unequal not only about the sagittal but also about the coronal plane. Thus, in most right-handed normal subjects the posterior quadrant is larger on the left, and the anterior on the right (*Fig. 2.7*), a

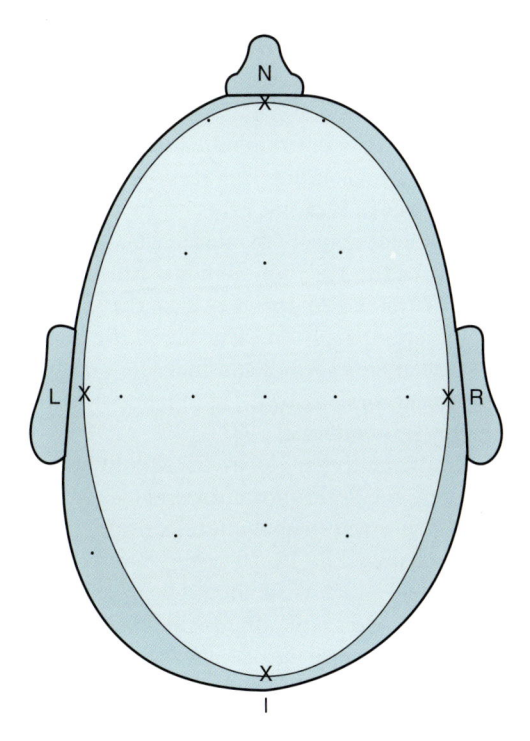

Fig. 2.7 Less than perfect symmetry of the typical average adult head. The quadrants between nasion (N), right preauricular point (R), inion (I), and left preauricular point (L) are not equal, as the 10–20 system (see *Fig. 2.8*) assumes. (From Binnie et al (2003), by permission.)

situation far removed from that represented in the standardized schema of the 10–20 system (*Fig. 2.8*) (Binnie et al 1982a). It is possible to compensate for

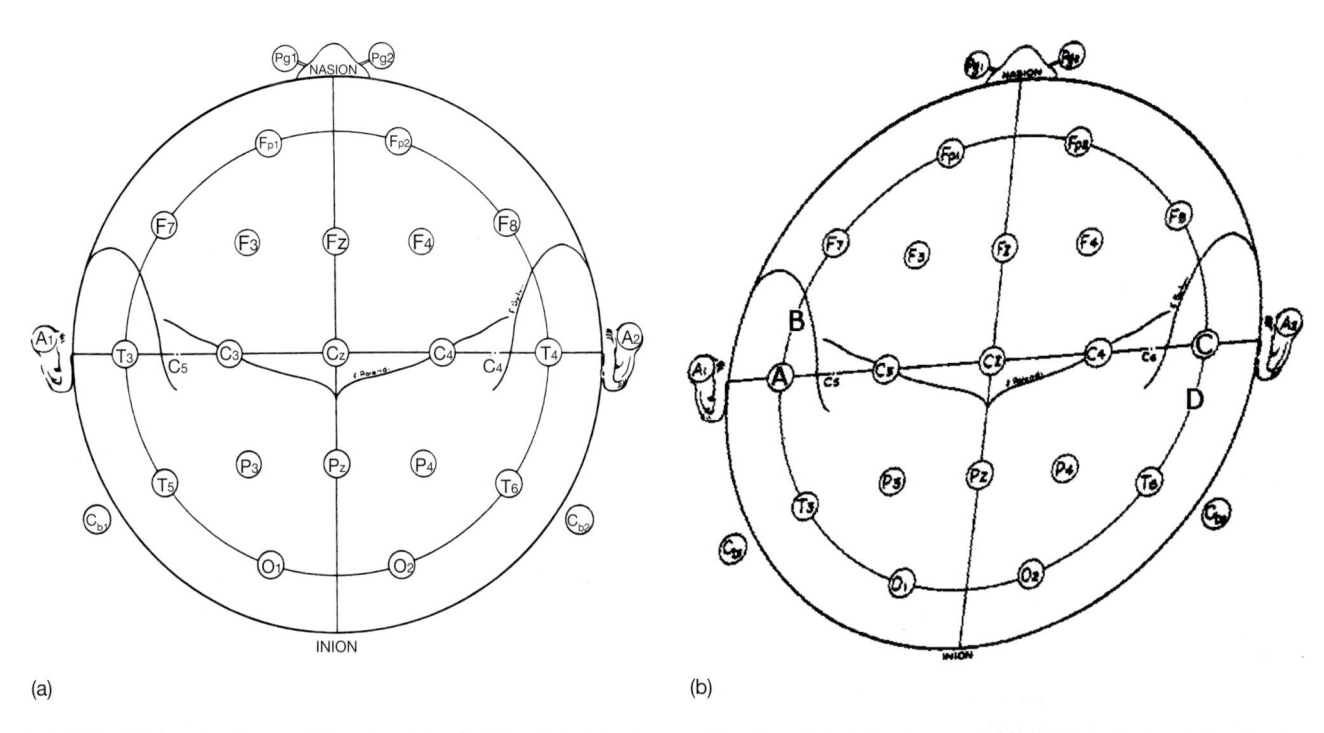

Fig. 2.8 (a) Formal schema of the international 10–20 electrode placement system. (b) The locations of T3/T4 (at B and D) as determined by measurements along the anteroposterior lines and those based on transverse measurements (A and C) are unlikely to coincide on an asymmetrical head.

this but only by a system of placement which involves separate proportional measurements within each quadrant of the scalp. On an asymmetric head, symmetrical placement might be taken to mean location of corresponding right- and left-sided electrodes over homotopic areas of the brain. Unfortunately it appears that, even on a symmetrical head, some of the 10–20 placements (specifically F7/8, T3/4 and T5/6) do not overlie the same cytoarchitectural areas on each side of the head, and that all placements show a considerable variability in relation to cerebral anatomy, independent of cranial asymmetry (Homan et al 1987).

4. It is generally held that electrodes should be equally spaced along the anteroposterior and transverse axes of the head to ensure equal interelectrode distances in bipolar chains. The 10–20 system attempts to meet this ideal which is, as should be apparent from the preceding paragraph, impossible to attain on most heads. As interelectrode distance influences the amplitude of the recording, it is certainly desirable to employ a system of placement which avoids excessively large or disproportionately small intervals. Nevertheless, no particular importance is known to attach to relative amplitudes of different activities along the anteroposterior axis, nor between the midline and the lateral extremities of the scalp. Equality of interelectrode distance should not, therefore, be a major consideration in the design of placement systems, when this is irreconcilable with ensuring constant relationships between electrode sites and underlying cerebral anatomy.

5. It should be easy to determine the sites and to retain electrodes in position. For some research applications, elaborate arrays of very large numbers of electrodes have been used, sometimes requiring special locating devices or the construction of helmets specially shaped to fit the head of each individual. Such systems are clearly not suitable for routine use. It is difficult to keep pad electrodes in place towards the lower margins of the scalp, and arrangements providing more extensive coverage than the 10–20 system may require the use of self-retaining (stick-on or needle) electrodes rather than pads.

6. *'Designations of positions should be in terms of brain areas (frontal, parietal, etc.), rather than only in numbers so that communication would become more meaningful to the non-specialist.'* It should however be pointed out that EEG machines on which the headbox sockets and corresponding selector controls are designated only by anatomical sites are extremely confusing to use with non-standard placements, notably during electrocorticography. Some digital computer-controlled machines allow the operator to select either anatomical or numeric designation of input leads, or indeed complete freedom to name the input sockets.

7. *'Anatomical studies should be carried out to determine the cortical areas most likely to be found beneath each of the standard electrode positions in the average subject.'* Such studies were rarely performed until the development of computed tomography and MRI provided convenient technologies, and increasing interest in detailed topography of evoked responses has provided the motivation for them to be carried out. Thus far they have served largely to highlight the variability of the relationships between standard 10–20 placements and cerebral anatomy (Homan et al 1987).

It is clear that, incredibly, electroencephalographers have largely ignored the problems arising from the complex shape and asymmetries of the skull, and of the brain and its cytoarchitecture, and have not yet properly addressed the issue of what they are trying to achieve in devising systems of electrode placement The international 10–20 system, described below, while far from perfect, represents the only method to have achieved general recognition and wide acceptance.

The international 10–20 system of electrode placement

The following specification of the international 10–20 system is reprinted, by permission, from Jasper (1958) (*Figs 2.9* and *2.10*).

Method of measurement The anterior–posterior measurements are based upon the distance between the nasion and the inion over the vertex in the midline. Five points are then marked along this line, designated frontal pole (Fp), frontal (F), central (C), parietal (P) and occipital (O). The first point (Fp) is 10% of the nasion–inion distance above the nasion; the second point (F) is 20% of this distance back from the point Fp, and so on in 20% steps back for the central, parietal and occipital midline points, the last being 10% above the inion – hence the name 10–20 system. These divisions are illustrated in *Fig. 2.8(a)*. It will be noted that this places the central line of electrodes one-half the distance between nasion and inion.

For example, if the nasion–inion distance is 30 cm for a given subject, the Fp line will be 3 cm above the nasion, the F line 6 cm back of the Fp line (or 9 cm from the nasion), the C line 6 cm back of the F line (or 15 cm from the nasion). There will also be 6 cm between the C and P lines, and 6 cm between P and O. The occipital line is then 3 cm above the inion.

Lateral measurements are based upon the central coronal plane. The distance is first measured from left to right preauricular points (felt as depressions at the root of the zygoma just anterior to the tragus). These points were selected because they seemed easier to determine with accuracy than the external auditory meati. Be sure the tape is passing through the predetermined central point at the vertex when making this measurement. Ten per cent of this distance is then taken for the temporal point up from the preauricular point on each side. The central points are

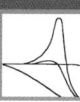

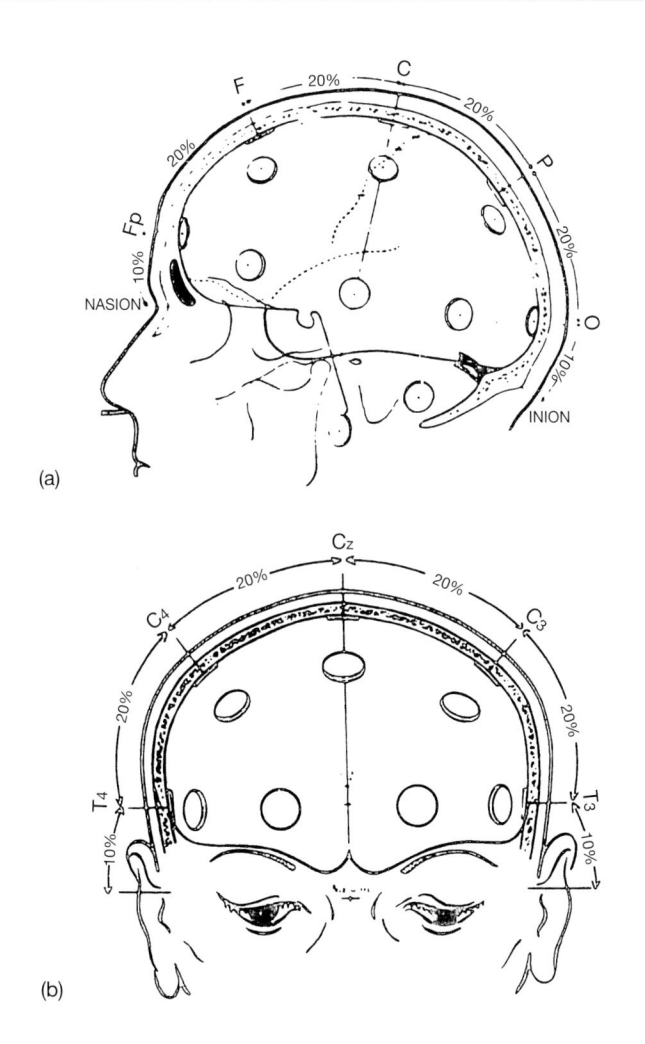

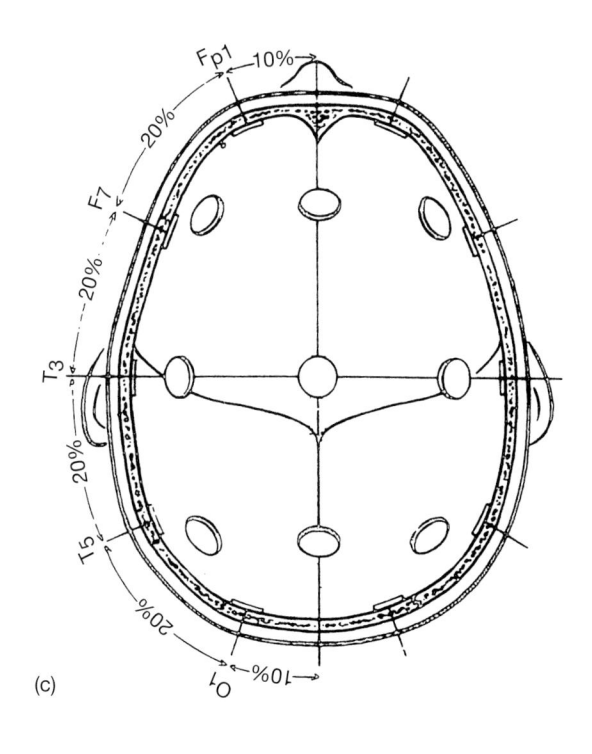

Recording the electrical activities of the nervous system

Fig. 2.9 10–20 system: (a) measurements from nasion to inion along the midline (see text); (b) measurements between the preauricular points in the central–coronal plane (see text); (c) measurements between the midfrontal and midoccipital points through the midtemporal point (see text).

then marked 20% of the distance above the temporal points, as shown in *Fig. 2.9(b)*.

The A–P line of electrodes over the temporal lobe, frontal to occipital, is determined by measuring the distance between the Fp midline point (as determined above), through the T position of the central line, and back to the midoccipital point. The Fp electrode position is then marked 10% of this distance from the midline in front, and the occipital electrode position 10% of the distance from the midline at the back. The inferior frontal and posterior temporal positions then fall 20% of the distance from the Fp and O electrodes respectively along this line, as shown in *Fig. 2.9(c)*.

The remaining midfrontal (F3 and F4) and midparietal (P3 and P4) electrodes are then placed along the frontal and parietal coronal lines respectively, equidistant between the midline and temporal line of electrodes on either side, as shown in *Fig. 2.10(a)*.

This provides a total of 21 standard electrode positions, including midline electrodes in frontal, central and parietal regions, and the two auricular electrodes. Electrode separations are approximately the same for all pairs in the A–P direction. Coronal lines of electrodes are also approximately equally spaced, with the exception of the shorter distance between the auricular and midtemporal points.

Additional electrodes may be placed between any of these principal standard positions for especially refined localization studies (with numbers provided for these special positions as well, as indicated below).

Designation of electrode positions Traditional anatomical terms have been employed to designate electrode positions over the various lobes of the brain, with the exception of the central region which is, strictly speaking, partly frontal and partly parietal. It represents the cortex in the vicinity of the central sulcus, both pre- and postcentral. It is sometimes called the sensorimotor area.

In order to differentiate between homologous positions over the left and right hemispheres, it was decided to use even numbers as subscripts for the right hemisphere, and odd numbers for the left hemisphere. Fp2, F4, F8, C4, P4, T4, T6 and O2 became standard positions on the lateral aspect of the right hemisphere, while Fp1, F3, F7, C3, P3, T3, T5 and O1 became standard lateral positions over the left hemisphere. These numbers were selected to allow for intermediate positions (e.g. F2, C2, C6, etc.) for special localization studies.

Electrodes at the midline in frontal, central and parietal regions were originally designated Fo, Co and Po but this led to some confusion since Po, for example, might be

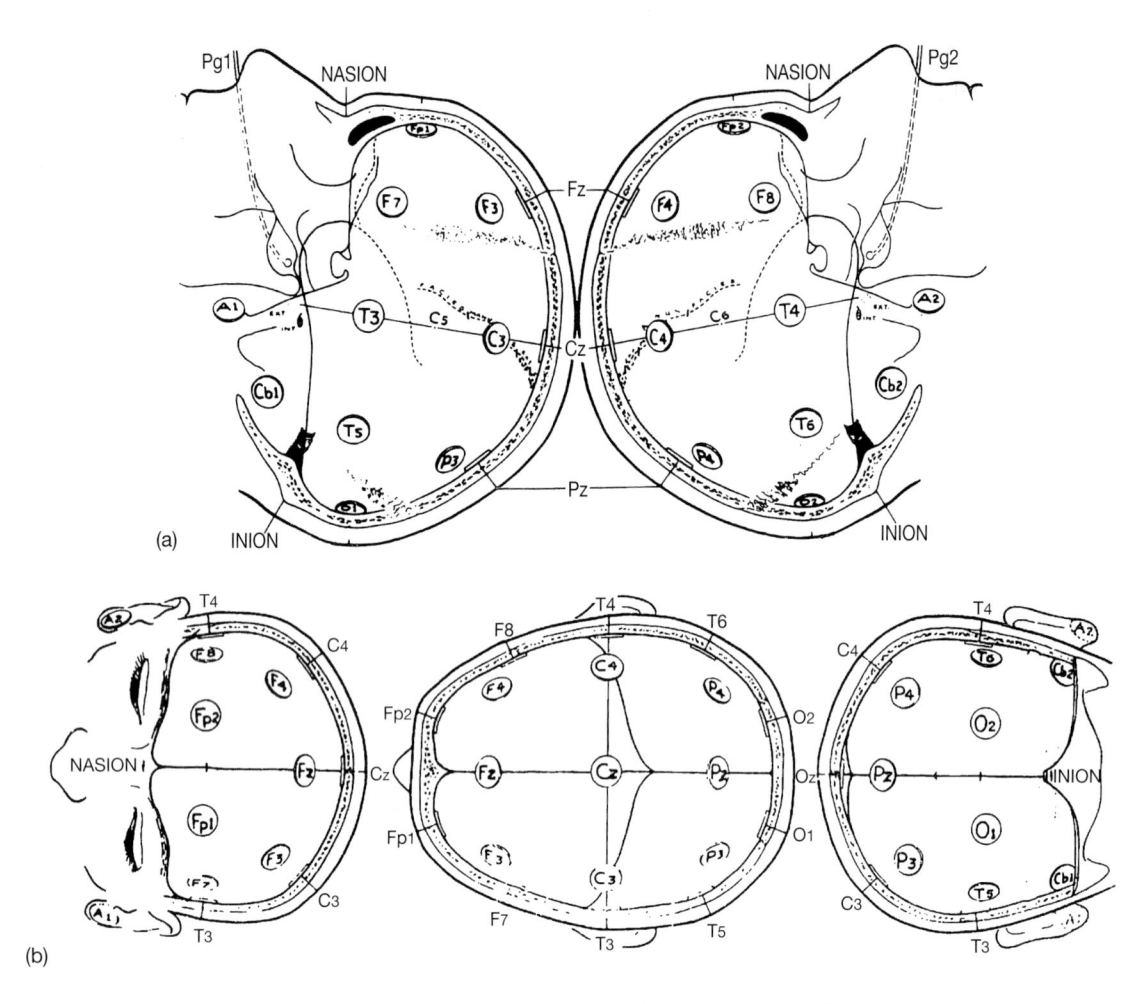

Fig. 2.10 (a) Lateral views showing distribution of 10–20 system electrodes over the left and right hemispheres. (b) Frontal, superior and posterior views showing all the electrodes of the 10–20 system.

interpreted as parieto-occipital. Consequently, the midline positions have been changed to Fz, Cz and Pz (z for zero). The complete system of placements with designations is shown in *Figs 2.8* and *2.10*.

In addition to the positions described above, pharyngeal electrodes are designated Pg1 or Pg2 for the left and right side respectively. Additional electrode positions over the posterior fossa are also shown, designated Cb1 and Cb2 (Cb for cerebellar) for the left and right sides, respectively.

Anatomical studies After these electrode positions were agreed upon, anatomical studies were carried out with the help of Dr Penfield, Dr McRae and Dr Caveness to determine the cortical areas over which each position would lie in the average brain. Two methods were employed: (1) metal clips placed along the central and Sylvian fissures at operation were then used to identify these fissures in x-ray studies of the skull after the EEG electrodes had been applied; and (2) electrode positions were carefully marked in the head of cadavers, drill holes placed through the skull and the cortex marked with India ink in each position before removing the brain for examination. Brains with gross lesions or local atrophy were excluded.

Although some variability was found, and is to be expected, the position of the two principal fissures should be within plus or minus about 1 cm of that indicated on the drawings, provided the head measurements are carefully made and the brain is free of gross distortion due to expanding or contracting lesions. Due to the obliquity of the central fissure, the upper central electrodes will usually lie precentral while the lower ones will be postcentral in most cases. The spatial relations between the 10–20 system placement and underlying brain anatomy have been investigated more recently by Towle et al (1993).

Some deficiencies of the 10–20 system should be apparent from the discussion in *General considerations in the design of placement systems* (p. 24). A further problem, which is not widely acknowledged, is that the specification as published and never revised is also inherently ambiguous. It assumes, for instance, that the T3 and T4 positions as measured in the central coronal plane are equidistant along a horizontal circumference from the nasion and inion. In a majority of normal subjects there will be discrepancies on one or both sides of at least 1 cm in these measurements, so anteroposterior and transverse measurements commonly indicate different locations for these electrodes (see *Fig. 2.8(a)*). The system as published offers no guidance as to what action should then be taken. In practice, most experienced technologists appear to have devised their own individual solutions, which are usually not known to their medical colleagues.

Several workers have suggested augmenting the 10–20 system by means of additional placements. Radical enhancements of the 10–20 placement are described by Rémond and Torres (1964) (57 electrodes), the American Electroencephalographic Society (1994c) (75 electrodes) and Morris and Lüders (1985), all of which include additional rows of electrodes for detailed topographic studies of EEG and EPs. Variations in the positions of a 33-electrode system relative to a best-fit sphere were measured by Lagerlund et al (1993) by MRI scanning of Vaseline-filled capsules attached to the scalp. The lines of equally spaced electrodes are extended by extrapolation to provide fuller coverage of the periphery of the scalp including particularly the anterior temporal region. Comparisons between the original 10–20 system and the 75-electrode system of the American Electroencephalographic Society (1994c) as given by Klem et al (1999) in the volume *Recommendations for the Practice of Clinical Neurophysiology: Guidelines of the International Federation of Clinical Neurophysiology* (Deuschl and Eisen 1999) are shown in Fig. 2.11.

Connecting electrodes to amplifiers and the recording convention

It cannot be emphasized too often or too strongly that only differences of potential can be measured. There is no absolute zero of potential against which potentials arising in the nervous system can be measured. Nevertheless we shall use the concept of, and write about, 'the potential' or 'the activity' at a point or at an electrode, even though we cannot measure its absolute value, by assuming that the second recording point is far away from the process that generates the potential.

As we have just seen, neurophysiological amplifiers have two 'active' inputs and an earth. There are many ways of connecting amplifiers to electrodes especially for recording the EEG, but all methods result in a potential difference being measured between two electrodes or between one electrode and a combination of electrodes.

The basic method of recording is to connect the (two) input terminals of an amplifier to two electrodes and the earth terminal to another electrode. For example, when using a concentric needle electrode placed in muscle, the recording is the difference between the electrical activity occurring at the central electrode and that occurring in the tissue in contact with (any part) of the concentric metal sheath. If identical electrical activity occurs simultaneously at both electrodes, it will not be amplified (in-phase signal). Any activity recorded in these circumstances means that there are steep potential gradients in the tissue surrounding the needle, so that even when the two electrodes are very close together there are significant differences of potential.

Similarly, the inputs can be connected to two scalp electrodes: one, say, on the vertex and the other on the occiput with the earth electrode on the temporal region (the earth electrode should be on the head but its position is not critical). In this case the activity recorded is the difference between the electrical signals being picked up

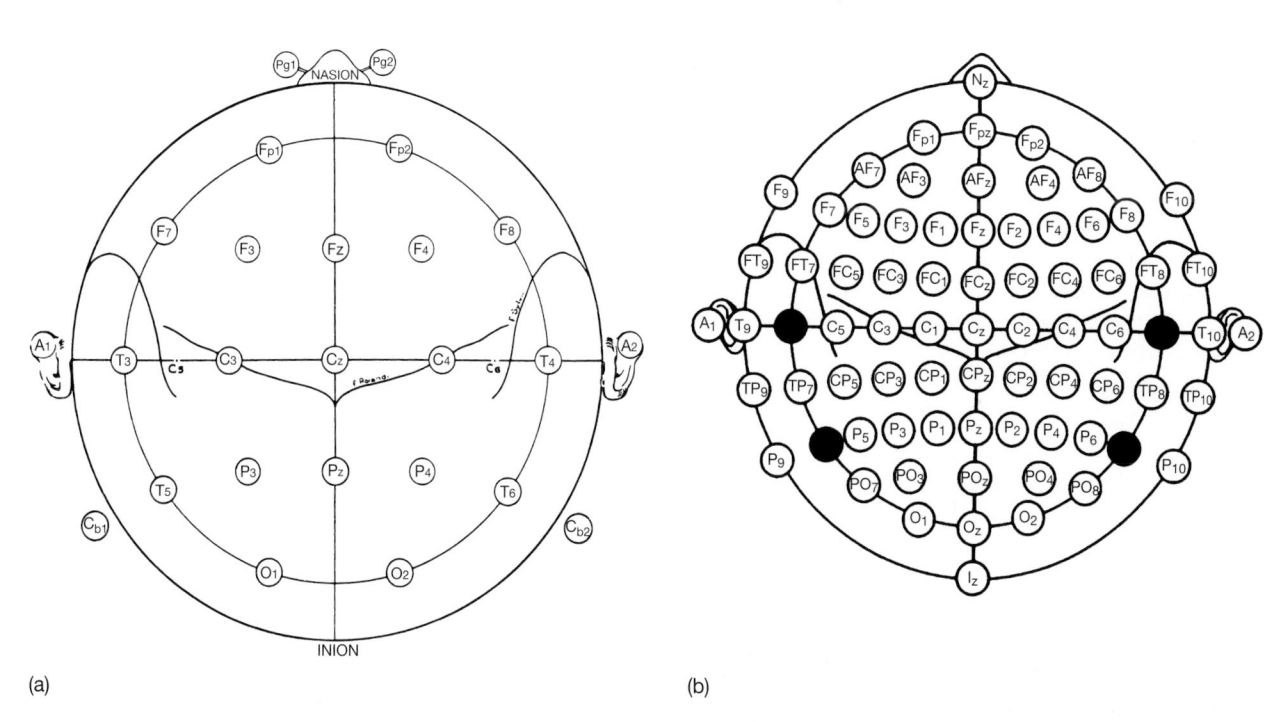

(a)   (b)

Fig. 2.11 Comparison between (a) formal schema of the international 10–20 electrode placement system (Jasper 1958) and (b) the 75-electrode system of the American Electroencephalographic Society (1994c) as compared by Klem et al (1999).

at the vertex and the occiput. Again, if the same (in-phase) activity occurs at the two electrodes it will not be recorded. Nor is it possible to tell from which of the electrodes activity is arising. Another important point to note is that such recordings give no information about the electrical activity at any other point on the scalp – even in the parietal region which lies between the two electrodes – or in the tissue between the central core of the concentric EMG needle and the surrounding sheath, tiny though it may be. The only signal collected is the difference in potential at the two electrodes; any other interpretation is inference based on some knowledge of field spread or volume conduction or brain function.

At this point it would be appropriate to look at the polarity of the displayed data. There is considerable ignorance and confusion about the way that voltages applied to the input terminals cause the output of the recording system (pen, oscilloscope trace, etc.) to move (up or down), especially in EP systems. This is vital in the interpretation of recordings, so it is prudent to take a little time to understand it.

As we have just seen, the differential amplifiers used in neurophysiology have two inputs and an earth. These inputs have been variously known as grid 1 and grid 2 (dating from the days of thermionic valves), 'black and white' because of the colour of the wires used, 'positive and negative' according to the convention of electrical engineering, and lead 1 and lead 2. Here we will use lead 1 and lead 2. Now imagine lead 1 connected via a switch to the negative pole of a battery and lead 2 to the positive pole. Let us also imagine that the potential difference applied to the two inputs is 100 µV. This can be obtained using a potential divider as in *Fig. 2.12*. When the switch is closed, the output will be deflected (say) up (depending on how the manufacturer has done his wiring) (*Fig. 2.12(a)*). If the polarity of the battery is now reversed (positive to input 1; negative to input 2), the output will be displaced downwards (*Fig. 2.12(b)*) – this is how a differential amplifier works. In neurophysiology, where we are trying to locate sources from the deflections of the display (the inverse problem to that in *Fig. 2.12*), we can say that, if the amplifiers are connected as in *Fig. 2.12(a)*, an upward deflection of the display means that the electrode connected to lead 1 is negative with respect to that connected to lead 2. This is the same as describing lead 2 as being positive to lead 1. It is not possible to be more specific.

From 1945 to about 1966 there was a generally accepted convention that if a potential difference was applied to the two inputs of an amplifier with lead 1 connected to the negative pole, then the output display (usually pen writers) would be deflected upwards. The same negative signal applied to lead 2 would deflect the display downwards. This convention was adopted because the early method of recording the nerve action potential ('killed end recording') produced a negative potential, which was regarded as more pleasing if displayed as an upward deflection (Smith 1997b). However, with the arrival into

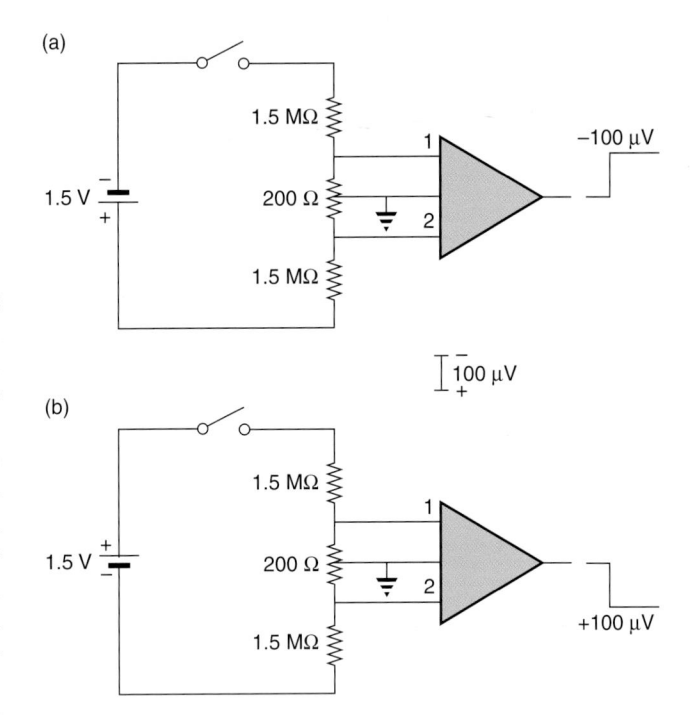

Fig. 2.12 Showing polarity convention used in EEG work (negative up). (From Binnie et al (2004), by permission.)

clinical neurophysiology of techniques other than EEG (particularly EPs), this convention was sometimes changed to match engineering practice where positivity is usually displayed upwards.

This caused confusion and a straw poll at an international meeting in 1976 showed that about 60% of workers in EPs displayed a negative signal applied to input 1 as an upward deflection on the output and 30% a downward deflection (the rest did not know!). No one was prepared to change their convention, so both were deemed acceptable provided that the polarity was clearly indicated on any diagram (Donchin et al 1977). The matter was not improved by some manufacturers labelling their input terminals positive and negative, red and green or active and passive! It is interesting to note that Grey Walter and George Dawson deliberately chose 'black' and 'white' to avoid any connotation with other conventions. The matter remains contentious, although in clinical EEG negative at lead 1 (grid 1, black lead) up is the accepted convention (BUN: black up negative); in EP work it usually depends on the type of EP being recorded. It really does not matter which way up the display is – we can all learn to read a map with south at the top – so long as we know.

To summarize:
1. both inputs of a differential amplifier are 'active';
2. identical in-phase signals are not amplified;
3. only the difference of potential (differential signal) at the two inputs is amplified;
4. a particular polarity signal applied to lead 1 (with respect to 2) will deflect the output in a certain direction; when applied to input 2 (with respect to 1) the deflection will be in the opposite direction; and

5. most workers in neurophysiology use the negative up convention, that is, when a negative signal is applied to lead 1 (with respect to 2) the output display is deflected upwards.

Now let us return to the various ways of connecting electrodes to amplifiers.

As already described, with a single channel recording it is impossible to locate the generator of a particular displayed activity. It may be close to one or the other electrode; or it may be close to neither but so positioned as to produce a potential difference between them (for instance, the ECG may be recorded from electrodes on each wrist). As one of the objectives of recording neurophysiological signals is to determine the source of the activity, it became obvious to the early workers in this field that multichannel recording systems were essential. In 1939, 3-channel pen recorders were available, 6 and 8 channels in the late 1940s and 16 or more channels from about 1956. Multichannel systems are able simultaneously to record the activity from many electrodes on the head and the strategies for connecting them to the various channels (called methods of derivation) are a matter of great importance. Within these general strategies there is also scope for considerable ingenuity in selecting the combinations and sequences of electrodes to be connected to each channel (montages) in order to highlight particular features of the activity. The fewer the number of available recording channels (and conventional EP systems have fewer channels than modern EEG machines), the more critical is the placement of electrodes and the selection of appropriate montages. In digital EEG and EP equipment, all the data are stored and different montages can be reconstructed 'off-line'. It should be noted that in multichannel recordings there is only one 'earth' electrode.

There are conventions for labelling electrode connections on a diagram of the head. The electrode connection to lead 1 is shown as a solid line, that to lead 2 as a broken line (*Fig. 2.13(a)*). Alternatively, the connections may be shown as an arrow joining the two electrodes on a particular channel, the tip pointing towards the electrode on lead 2 (*Fig. 2.13(d)*).

The simplest multichannel montage to understand is that in which each electrode is connected to one input of the amplifiers (usually lead 1) and all the lead 2 inputs are joined together and connected to another electrode (*Fig. 2.13(a)*). This, understandably, is known as common reference derivation. The common reference electrode is usually placed so as to minimize the possibility of picking up signal and/or artefact, but the lack of activity can never be assumed. The terms 'passive' or 'inactive' are wrongly used for this electrode. A good deal of effort has been expended, to no avail, in searching for an inactive reference. If there is little electrical activity at the site of the reference electrode the multichannel recordings are very easy to interpret as each channel displays the activity at the electrode to which it is connected. If there is activity at the reference electrode this will appear (inverted) in all channels intermixed with the activity arising at the individual electrode sites, making interpretation more difficult.

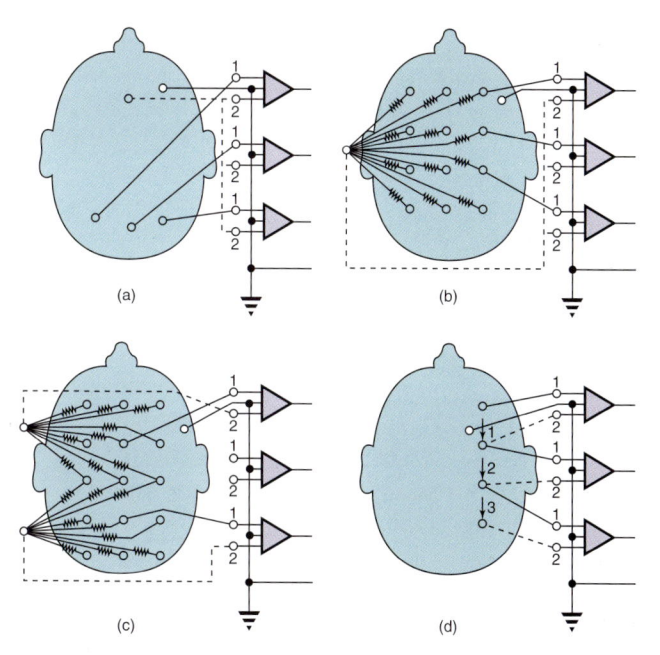

Fig. 2.13 Various types of montages used in electroencephalography. (a) Common reference: three occipital channels referred to a frontal electrode. (b) Average reference: the average reference is formed by connecting electrodes to a common point through equal high resistors. (c) Source reference: separate local average references are formed from neighbouring electrodes (only two channels shown). (d) Bipolar recording. (From Binnie et al (2004), by permission.)

Another type of reference recording uses the average of the electrical activity at many or all electrodes as the reference against which individual activities are measured. This, as may be expected, is termed common average (or just average) reference derivation. The average activity is obtained by connecting all, or nearly all, electrodes through equal individual high resistances to a common point which is then connected to lead 2 of all channels (*Fig. 2.13(b)*). One of the consequences of this method is that if a large potential change occurs under just one electrode, this significantly affects the average so that equal deflections of one Nth its magnitude and of opposite polarity occur in the channels connected to each of the other electrodes, where N is the total number of electrodes in the system.

Yet another type of reference derivation is known as source reference, in which each electrode is referred to the weighted average of the activities of the immediately surrounding electrodes. This is a kind of local average reference for each electrode (*Fig. 2.13(c)*).

Quite different in concept from referential derivations is bipolar recording. In this method, which is widely used in EEG, each channel is connected to two electrodes (usually close together) both of which may be affected by the electrical activity of interest. Although in principle all recording methods are the same, in that there are two (differential) inputs to each and every channel, the deliberate mixing of signals from adjacent electrodes significantly alters the appearance of the final display. In particular, the linking of serial pairs of electrodes in a bipolar chain, as shown in *Fig. 2.13(d)*, can be used to locate localized peaks of activity by what is called a phase reversal. The various derivations each have their own advantages and disadvantages, and are used for particular purposes.

In summary, the amplifiers used in clinical neurophysiology record the differences between the electrical activities applied to the two input terminals. To facilitate the location and interpretation of neurophysiological data, multichannel recording systems are used, and the way these channels are connected to the electrodes depends upon the particular problem being investigated and, it should be admitted, on the personal preference of the neurophysiologist who has to analyse the record. The primary data, the electrical activities arising at the electrodes, are the same in all the reference or bipolar techniques used – the output displays are different in that the various derivations emphasize this or that aspect, and this is where personal experience becomes an important factor.

Amplifiers

Neurophysiological signals are small (microvolts (μV) or millivolts (mV)) and have to be amplified before they can be displayed. They are often recorded in places where there are electrical fields from lighting or other mains-operated equipment. These fields are usually at mains frequency (50 or 60 Hz) but can be at 100 and 120 Hz, all of which are within the neurophysiological signal range. They can be large compared with the signals from nerves or brain. Physiological amplifiers are designed to reduce interference, but before considering how that is achieved it is necessary to understand how such contamination arises.

The electrical fields that interfere with electrophysiological recordings are electrostatic and/or electromagnetic. Electrostatic fields are differences of potential in the environment like the familiar static electricity (but in neurophysiology they fluctuate). For example, if two metal plates, separated by, say, 1 m, are connected to an oscillator, the field between them will be electrostatic (little or no current will be conducted by the air). Electrodes on a patient in an electrostatic field, in the vicinity of a fluorescent light for instance, will pick up the local field potential. This may be amplified and appear as interference in the output.

On the other hand, electromagnetic interference is created by current flowing in a conductor. For example, a coil connected to an oscillator will create a fluctuating magnetic field (at the oscillator frequency) which will induce an AC (at the same frequency) in any nearby conductor. If the head or input leads of the amplifier are in a fluctuating magnetic field, currents will be induced that may be amplified with the electrophysiological signal. Interference can also be due to a nearby radiofrequency (RF) source such as a staff paging system. In this case the interference is not due to the RF signal (the frequency is too high) but to the amplitude modulation of the RF carrier by speech or music which is within the frequency range of some electrophysiological investigations.

Electrostatic fields are much reduced within a metallic container and such interference may be reduced by screening the recording room by an earthed conducting sheet or mesh. Electromagnetic interference is more difficult to contain and screening may even make it worse. Being confined to recording in screened rooms is very inconvenient and an important step was made in the 1930s by Jan F. Tönnies (1902–1970) which enabled neurophysiological data to be recorded in the presence of electrostatic and electromagnetic fields. This was the development of the *balanced* or *differential amplifier* (Tönnies 1933).

An amplifier increases the amplitude of the signal applied across its two input leads. In simple amplifiers, used for domestic audio equipment, only one lead is regarded as a signal input; the other serves as a reference lead connected to earth (assumed to be potentially immovable), usually via the chassis of the machine. If such amplifiers were used in an EEG machine the situation would be as shown in *Fig. 2.14(a)*.

Each channel would display the potential difference between one recording electrode and the (same) reference point. This has two drawbacks: first, it would not be possible to display the potential difference between a particular pair of electrodes; secondly, and much more important, each amplifier would amplify not only the EEG but any difference in potential between the patient's head and the chassis of the EEG machine. In a building containing electrical equipment and AC mains, the patient and machine would most probably be in different electrostatic and electromagnetic fields, and large interference signals would be recorded, which would swamp the EEG.

In the differential amplifier (*Fig. 2.14(b)*), there are two 'active' inputs to each amplifier and the *difference* of potential between them is amplified. In the diagram this is the difference between the brain activities at the two electrodes. Such activities occurring differently at the two inputs comprise the *differential input* to the amplifier. In contrast, any interference signal between patient and earth is identical in amplitude and phase in both active leads and is not amplified because the difference is zero. Similar in-phase potentials occurring on the two inputs are called *common-mode* potentials. Differential amplifiers reject common mode and amplify differential potentials.

This explanation is probably sufficient for most neurophysiologists, but for those interested in knowing more, the differential amplifier is in effect two separate amplifiers each recording with respect to the third (usually earth) electrode. When combined, the output of this pair of amplifiers is the amplified difference of the potentials between the two input leads with respect to earth. If V_A and V_B are the biological signals at electrodes A and B with respect to earth, and V_C is the in-phase component, the output of the amplifier is an enlarged version of

$$(V_A + V_C) - (V_B + V_C) = V_A - V_B$$

Recording the electrical activities of the nervous system

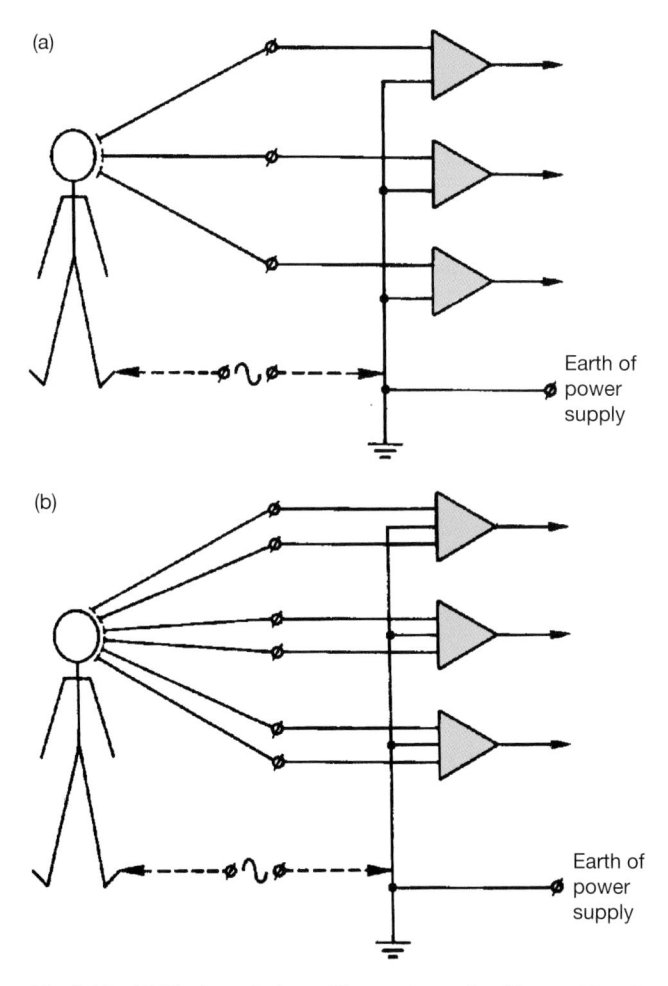

Fig. 2.14 (a) Single-ended amplifiers: only one lead is considered active and the other is connected to earth; any signal between patient and earth will appear in all channels. (b) Differential amplifiers: each channel registers the potential difference between the two 'active' leads; signal between patient and earth will be the same in both halves of the amplifier (in-phase) and will not be amplified. (From Binnie CD, Rowan AJ, Gutter Th 1982b *A Manual of Electroencephalographic Technology*, Cambridge University Press, by permission.)

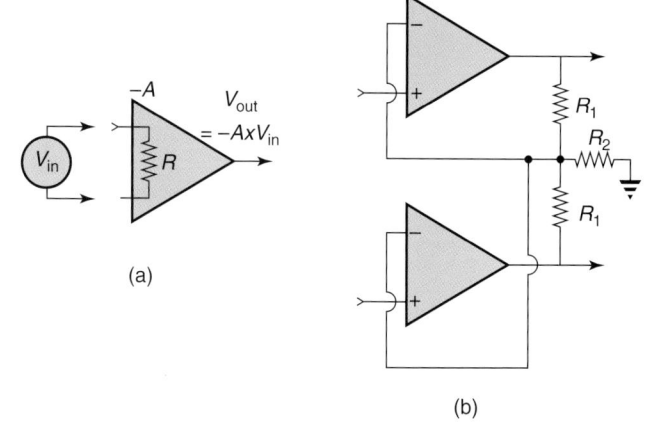

Fig. 2.15 Schematic diagram of (a) an amplifier without feedback and (b) two amplifiers with feedback network to form a balanced amplifier with in-phase discrimination. (From Binnie et al (2004), by permission.)

Thus the difference signal is amplified and the common-mode signal is cancelled out.

Another way of considering the rejection process is that the two halves are connected such that the in-phase signal produces negative feedback which reduces the gain of the amplifier, whereas there is no feedback from the differential signal which is amplified in full. This may be seen from *Fig. 2.15* which shows (a) a non-differential amplifier with no in-phase discrimination (sometimes called single-ended) and (b) a differential amplifier. If a signal is applied between each input of the differential amplifier and earth, the following conditions apply:

1. If the input signals are in phase, that is, both inputs go equally (say) positive with respect to earth, an opposing (negative) voltage is developed across the resistor R_2 and, when fed back to the inputs, reduces the output of each amplifier (negative feedback).
2. If the signals on the two inputs are not equal, the (difference) voltages developed across R_2 by each amplifier are of opposite polarity and cancel out. There is no negative feedback and the amplifier operates at full gain. Thus signal *differences* at the input terminals are amplified.
3. In practice, the input signal usually has both in-phase components from the interference and differential components from the biological preparation – they are amplified as if each component were applied separately. The differential signal is amplified more than the in-phase component – the amplifier has in-phase discrimination. It should also be noted that if similar *biological* signals occur at the two inputs they will also cancel out – only the difference of the signals applied to the inputs is amplified *whatever their origin*.

The ratio of the amplification factors for differential and common-mode signals is called the *discrimination* or *common-mode rejection ratio* (CMRR). It is determined by the design and components of the amplifier and represents the ability to record in the presence of in-phase (interference) signals. It should be large: 100 000:1 (100 dB) is not unreasonable. With such an amplifier, 100 mV in-phase interference will be attenuated 100 000 fold and appear in the display the same size as a 1 μV differential signal.

The gain of one differential amplifier is insufficient to amplify neurophysiological signals to a level at which they can be displayed, and most recorders have several stages of amplification for each channel. These stages are usually linked together with coupling capacitors, which also behave as high-pass (low-cut) filters (*Fig. 2.16(a)*) and set a limit to the low-frequency response of the system. The capacitors also determine the blocking time of the system (see below).

Interference can be picked up by any part of the amplifying chain, but the loop formed by the patient and electrodes, the input circuit of the amplifier, and the leads connecting them, are all particularly susceptible to electromagnetic fields. Currents induced here produce potential differences across the electrodes, which are

33

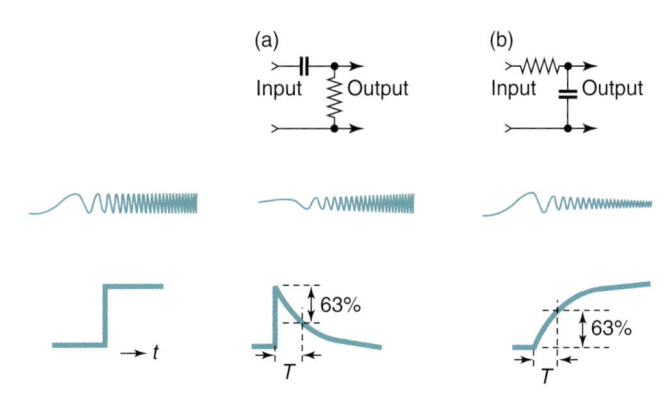

Fig. 2.16 Single pole filter circuits and their effect on a sine wave of increasing frequency and a step function. (a) The high-pass capacitor–resistor (CR) circuit attenuates the low frequencies and (b) the low-pass resistor–capacitor (RC) circuit attenuates the high frequencies. T is the time constant of the circuit. (From Binnie et al (2004), by permission.)

amplified as differential signals. The solution lies in (a) using amplifiers of high input impedance to reduce the induced current, (b) reducing the electrode impedance to minimize the resulting potential, and (c) keeping the input leads as short as possible (and in difficult conditions grouping them together to make the induced current equal in each lead). It is common for the first stage of amplification (*preamplifiers*) to be located on, or as close as possible to, the patient. These preamplifiers provide little amplification, but convert the impedance of the signal source from several thousands of ohms at the electrodes to a few ohms, so that interference is greatly reduced during transmission down the input cable to the main stages of amplification in the recorder.

A last resort for reducing 50 or 60 Hz interference is by the use of a sharply tuned filter at mains frequency. These *notch filters* do not significantly distort the biological signals at other frequencies if they are well designed. However, such 'sharp' filters can resonate if activated by a large transient such as can occur with electrical or auditory (click) stimulation. These filters should be used only after all other methods of eliminating mains interference, such as reduction of electrode impedances and/or appropriate earthing and positioning of mains equipment, have been fully explored.

An important characteristic of a biological amplifier is its *input impedance*, which should be high (at least 5 MΩ). As noted above, a high impedance reduces the susceptibility to interference. Furthermore, it reduces the possibility of signal loss from high impedance electrodes. For example, if the input impedance is only 1 MΩ and the combined impedance of the electrodes is 10 kΩ (not unusual in EEG recording), then 1% of the signal will be lost (the ratio of 10 000 to 1000 000). However, if the electrode impedance is 100 kΩ, as can happen with small needle electrodes, then about 10% of the signal will be lost. Modern EEG and EMG machines have input impedances of 50 MΩ or more and little loss occurs even with very small electrodes.

In some amplifiers, coupling capacitors are not used and the high-pass filter (time constant) can be switched out and records taken that have no low-frequency attenuation. These are called DC (direct current; cf. AC, alternating current) amplifiers and are necessary if steady potentials have to be measured, for example to monitor eye position from the electro-oculogram. Modern DC amplifiers are very stable, but there can be large (millivolts) differences of potential between electrodes (arising at the electrode–skin interface) which change over relatively short periods of time (minutes). This causes a drift of the baseline, which has to be corrected before the channels reach the limits of their amplitude excursions. These *steady* potentials are 'backed off' by injecting an equal and opposite potential in the amplifier, either manually or automatically. DC amplification is rarely used in clinical neurophysiology, but is preferred in polygraphy when measuring slowly changing variables, such as respiration.

AC amplifiers suffer from *blocking* if a large steady potential difference (e.g. from differing electrodes) exists between the two input terminals. The amplifiers return only slowly to their normal operating range and no data can be recorded during this time. This can occur from movement artefact or when switching from one configuration of electrodes to another to change montage. Blocking is caused by the coupling capacitors, which separate the stages of an AC amplifier, becoming charged and is worse when using long time constants. Many neurophysiological recorders (especially those for recording the EEG) are fitted with amplifiers which have a manually controlled *antiblock* device, which equalizes the potentials on the plates of each of the coupling capacitors. Blocking does not occur in DC amplifiers and a large steady potential applied to the input permanently deflects the channel to one extreme of its range.

In any sensitive electrical measuring system, a small fluctuating output is observed even when there is no input signal. This is called *noise* (*Fig. 2.17*). The noise adds to the signal which, like speech on a telephone line, is distorted by it. There are many sources of electrical noise – switch contacts (input selector switches in particular), wiper arms of potentiometers inside the amplifying system – as well as the fundamental thermal noise generated in the input circuit of the amplifier by random movement of electrons.

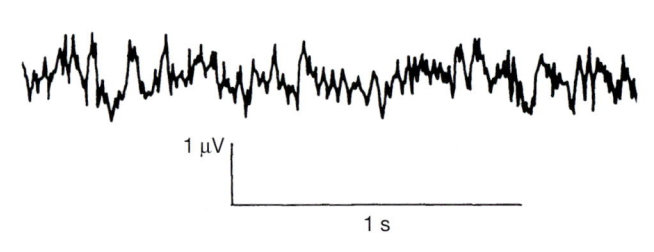

Fig. 2.17 Recording of electrical noise in an EEG amplifier (bandwidth 1–700 Hz). (From Binnie et al (2004), by permission.)

Thermal noise is proportional to the square root of the (absolute) temperature, the bandwidth of the amplifier and the source (electrode) impedance. In practice, there is not a lot that can be done about the temperature, but the bandwidth should be kept to a minimum consistent with preservation of the signal. For example, there would be little point in starting the attenuation of the high frequencies at say 1 kHz for clinical EEG, as the noise might be mistaken for, or worse, obscure brain activity. Electrode impedances should be kept low, although mains interference will be more troublesome than thermal noise if the electrode impedances are high. The plugs on the leads from the electrodes are also a source of noise; they should be kept clean and dry.

The noise in a clinical EEG machine with shorted input and low-pass filters set to 70 Hz should be not more than 1 µV peak to peak. The noise in amplifiers used for recording the EMG depends upon the bandwidth used. It should not exceed 15 µV peak to peak when the filter controls are set to give a bandwidth from 10 to 10 000 Hz.

Signal bandwidth and filters

Like the audio signals from a tape head or stylus of a hi-fi system, the electrical activity of the nervous system needs amplification before it can drive the output device – loudspeaker, pen recorder or oscilloscope. As for the reproduction of music, it is important that this process does not distort the signal. Distortion in amplifiers can arise in a number of ways, but the one that has most influence on the fidelity of the output is the frequency response of the system. This can be changed by the 'tone' controls, and in a hi-fi system they are set so that the full frequency range of the music recorded on the tape or disc is amplified equally. Similarly in neurophysiological amplifiers the bandwidth controls should be set to reproduce that range of frequencies in the data that are of interest.

The shape and time-course of a waveform determine its frequency content. A continuous sine wave consists of one frequency only, whereas a repetitive square wave needs many (high) frequencies to generate its shape – the sharper the 'corners', the greater the bandwidth required. Jean Baptiste Joseph Fourier (1768–1830), the French mathematician, showed that any periodic signal can be formed by summating a series of harmonically related sine waves. *Figure 2.18* shows how eight harmonically related sine waves when added together with particular time relationships can make a pulse wave (really a series of pulse waves since the sine waves before and after the epoch shown will coincide elsewhere to make similar pulses). The 'narrower' the pulse, the higher is the fundamental frequency and its harmonics.

It should be noted that the starting point (phase) of each component is not necessarily zero and that a change of phase of any component will change the shape of the composite waveform. It is well recognized that a complex waveform can be described in terms of component frequencies; here we are concerned with the frequency content and the

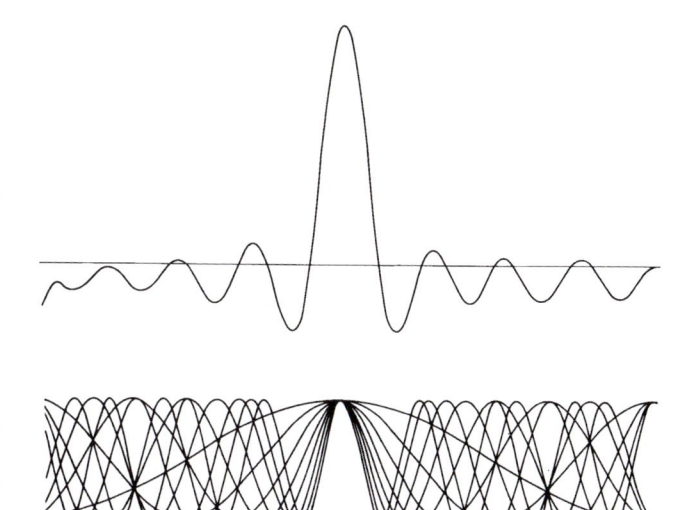

Fig. 2.18 Pulse wave formed by the summation of eight harmonically related sine waves. (From Binnie et al (2004), by permission.)

bandwidth required for accurate reproduction of neurophysiological signals. Wide bandwidths are needed to reproduce 'sharp' waveforms. A 1 ms square pulse would need a pass band extending up to more than 10 kHz for its shape to be well preserved. *Table 2.4* shows the approximate bandwidth of some neurophysiological signals.

The frequency characteristic of an amplifier can be expressed by its frequency response curve. This is a graph of the percentage of the signal passed through the filter at different frequencies. *Figure 2.19* shows three response curves of neurophysiological amplifiers that have high-pass filter (filters that attenuate the low frequencies) and low-pass filters (high frequency attenuators), which correspond to three rows of *Table 2.4*. Note that in order to have a wide range of frequencies on the *x*-axis, the scale is logarithmic (the scale is compressed at the high frequencies). It might be thought that the amplitude of the signal between the filter frequencies 0.5 and 70 Hz would be constant at the full value (100%), but this is not so because the filter circuits used in neurophysiological amplifiers do not cut off abruptly at the specified frequency. Particularly towards the limits of the range from 0.5 to 70 Hz there is considerable reduction of amplitude, which means that

Table 2.4 Approximate bandwidth of some neurophysiological signals

Signal	Lowest frequency	Highest frequency
Clinical EEG	0.5 Hz	70 Hz
Visual evoked potentials (VEPs)	1 Hz	100 Hz
Auditory brainstem EPs	100 Hz	3 kHz
EMG	2 Hz	500 Hz
Single fibre EMG	500 Hz	10 kHz

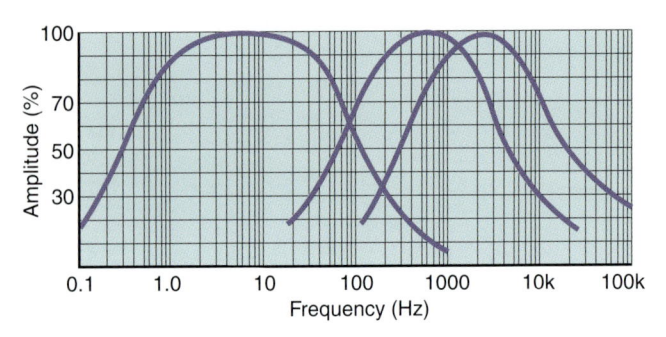

Fig. 2.19 Response curves of amplifiers for EEG (left), brainstem auditory evoked potentials (middle), and single-fibre electromyography (right). These have the corner frequency characteristics shown in *Table 2.4*. (From Cooper et al (2005), by permission.)

some filtering occurs within the bandwidth selected. Indeed, in the response curve shown, the full amplitude is achieved only over a narrow range between 4 and 10 Hz.

The amplitude sensitivity of a filter is often expressed in decibels. This, in many ways, is an inconvenient measure and it is generally sufficient to remember that –3 dB is about 70% and –6 dB is 50% of full signal. For the mathematically minded, the relationship of amplitude and decibels is given by

$$\text{decibels} = 20 \log_{10} \frac{\text{amplitude out of filter}}{\text{amplitude into filter}}$$

Thus a 50% reduction of amplitude through a filter can be expressed as

$$20 \log_{10}(50/100) = 20(-0.301) = -6.0 \text{ dB}$$

For a 3 dB loss, the ratio of amplitudes is given by

$$-3 = 20 \log_{10}(\text{ratio of amplitudes})$$

Therefore, the amplitude ratio = $10^{(-3/20)} = 10^{-0.15} = 0.707$, or a reduction to 70.7%.

The pass band of a filter is defined by its *turnover points* – the lower and upper frequencies at which 3 dB attenuation (70% amplitude) occurs. The rate of attenuation (slope) of a filter as a function of the signal frequency is expressed in dB/octave. In a simple resistor–capacitor filter (called a single pole filter) the slope of the response curve at the –6 dB (50%) point is 6 dB/octave. The larger the decibel value, the steeper the slope and the more rapid is the attenuation outside the pass band. It might be thought that filters having a slope of say 48 dB/octave would be better than those having a slope of 6 dB/octave, but steeper slopes can have undesirable features such as phase distortion or they may 'ring' and produce spurious signals when excited by a sharp transient. For most neurophysiological recordings, filters having slopes of 6 dB/octave are sufficient.

In recording systems with pen writers, the mechanical response of the pen sets the upper limit to the frequency response of the system (120 Hz or so), although there are usually amplifier output terminals with a wider pass band for connection to other equipment. The low inertia of the ink-jet writer permits a wider pass band (up to 1500 Hz) and the absence of inertia of oscilloscope displays allows upper limits of 10 kHz or more.

In digital systems (see *Digital systems*, p. 37), the upper limit of the bandwidth is determined first by the sampling frequency and secondly by the resolution of the display device. If 8 seconds of signal are displayed across the screen of a visual display unit (VDU) 1024 pixels wide, only 128 data points can be displayed per second. This imposes a maximum frequency resolution of 64 Hz, but in practice waveforms at about 40 Hz will be subject to some distortion. In digital systems analogue filters are used to condition the signals before sampling, but the eventual bandwidth of the display is determined by mathematical operations on the data, the procedures being known as digital filters.

All electrophysiological recorders have several high- and low-pass filter settings. *Figure 2.20* shows the frequency response curves for some filter settings of the amplifiers of a jet-ink writer system.

In electroencephalography the low-frequency characteristic of the amplifier is often designated by a different measure – the *time constant* (TC) – a term which has a precise mathematical value and is convenient to measure with a simple test signal. If a voltage is applied to the input of

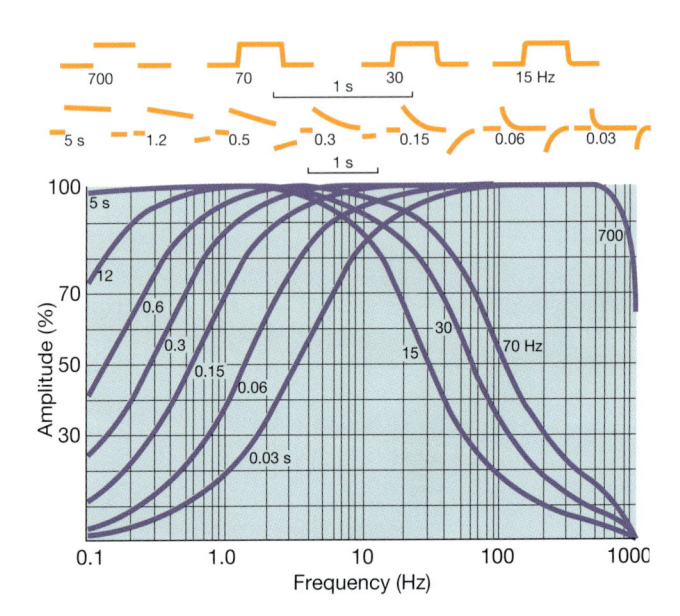

Fig. 2.20 Overall characteristics of an ink-jet EEG recorder. Response curves on the right are for low-pass filters of 700, 70, 30 and 15 Hz. Those on the left are for high-pass filters designated in terms of time constant values. The intersections with the 70% amplitude level gives the turnover frequencies. The effect of the various filters on a square-wave calibration signal is shown at the top of the figure. (From Cooper, Osselton and Shaw (1980), by permission.)

an amplifier with low-frequency attenuation, the write-out will show the step change and then return to the baseline in an exponential manner, even though the voltage is still applied to the input terminals (Fig. 2.21). The time (T) taken to return to 37.7% of the initial deflection is the TC. The TC and turnover frequency (F) at −3 dB are different ways of expressing the low-frequency characteristic of an amplifier with a single pole filter. Their relationship is given by the equation TC = $1/(2\pi F)$. Equivalents of common values in EEG are given in Table 2.5. Any equivalent value of time constant and turnover frequency, for 3 dB or 6 dB loss of sensitivity, can be obtained from Fig. 2.22.

Designation of the low-frequency response by time constant is common in electroencephalography, although digital systems increasingly use the actual values of the turnover points in hertz (or give the user the choice). In electromyography and clinical EP studies (where low frequencies are less common), turnover frequency is more often used.

The simplest analogue filters consist of a resistor and capacitor (see Fig. 2.16). When connected as in Fig. 2.16(a), the capacitor 'blocks' the steady potentials, attenuates the low frequencies and passes the high frequencies. This is a high-pass filter (sometimes called low-cut) or time constant circuit; the value of the time constant in seconds is the product of the resistance in ohms and the capacitance in farads. For example, if the resistance is 1 MΩ (10^6 Ω) and the capacitance 1 microfarad (μF; 10^{-6} F) the time constant is 1 s and the filter will have a turnover frequency at 0.16 Hz (see Table 2.5). The effect of this type of filter on a sine wave and a square wave is shown in Fig. 2.16.

If the resistor and capacitance are connected as in Fig. 2.16(b), the high frequencies are attenuated because the impedance of the capacitor decreases as the frequency increases and 'shunts' the high frequencies. This is a low-pass filter (sometimes called a top-cut, high-frequency (HF) or HF cut filter); the effect on a sine wave and a square wave is shown. The effect of filters with the response curves given in Fig. 2.20 on a square wave calibration signal is shown at the top of that figure.

Filters are used to reduce unwanted signals or emphasize particular features in the recordings that have different frequency components. However, care must be taken in their use, as filters can cause phase shifts which introduce an error in time measurements of electrical activity, for example, in relating high-frequency components (spikes) and slow waves or in the measurement of the latencies of EPs.

Digital systems

In digital systems, each socket of the input connection box has its own amplifier; one lead is connected to the 'active' electrode, the other to a common reference point, usually an additional electrode. The output of each amplifier is measured (digitized) by an analogue to digital converter to give a numeric value. Binary notation is used and the number of voltage levels that can be distinguished, the amplitude resolution, is determined by the number of binary digits available. Modern equipment usually samples at least 4096 levels. This is known as 12-bit digitization (2^{12} = 4096) and, as Picton et al (1984) have shown, is adequate for most applications, giving, for example, a resolution of 1 μV over a range of ±2 mV. The numerical values

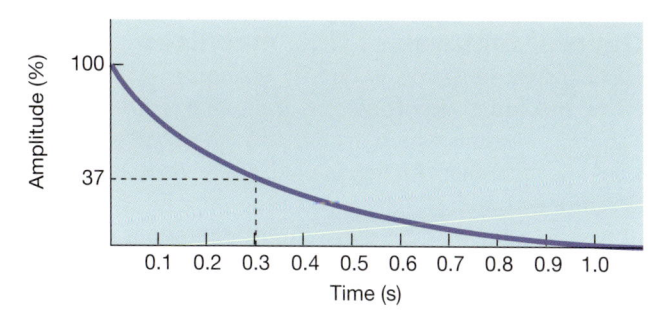

Fig. 2.21 The time constant is the time (T) for the amplitude to fall to 37% of the value attained by the application of a step function (calibrator), in this case 0.3 s. (From Binnie et al (2004), by permission.)

Table 2.5 Relationship between turnover frequency and time constant

Turnover frequency (Hz) for a loss of 3 dB (down to 70%)	Time constant (s)
0.016	10.0
0.027	6.0
0.053	3.0
0.16	0
0.53	0.3
1.6	0.1
5.3	0.03

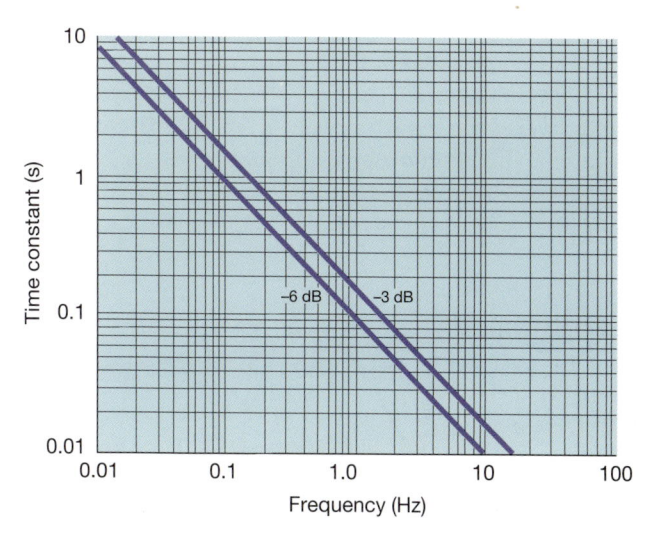

Fig. 2.22 Graph showing the relation between frequency and time constant for 70% (−3 dB) and 50% (−6 dB) loss of sensitivity. (From Binnie et al (2004), by permission.)

are multiplexed and transmitted sequentially by a single cable to an interface card in the standard computer that serves as the recorder.

The signal voltages are sampled repeatedly at a constant rate, the sampling frequency. The presence of any signals at more than half the sampling frequency of the analogue to digital conversion produces spurious slower activity (aliasing) (*Fig. 2.23*). To overcome this, anti-aliasing filters are used to eliminate high frequencies before analogue to digital conversion and often the EEG is 'oversampled' at a rate several times more samples per second than will eventually be stored. An everyday example of aliasing is the slow, sometimes backward, rotation of the spokes of wagon wheels shown in films or on TV. This occurs when the number of frames per second is too low to resolve the rapid rotational movement of the spokes. In neurophysiology it can be troublesome when sampling rhythmic waveforms such as occur in steady-state responses to flicker, for example. It does not occur often in recording responses to transitory stimuli. Note that there is risk of aliasing whenever activity is present at more than half the sampling frequency, irrespective of whether or not that activity is of interest. A 60 Hz mains interference will cause aliasing if sampled at 100 Hz, even if the wanted signal is an event-related slow potential, containing little information above 40 Hz.

The time resolution, that is, the maximum frequency that can be measured in the signal, depends upon the sampling rate. The *theoretical* maximum resolution of a sampling system is half the sampling rate; that is, if the sampling rate of 2 kHz is used for say, a particular EP study, then a component of 1 kHz is theoretically resolvable. This sampling rate is known as the *Nyquist frequency* and is based on a *sine wave* being specifiable by 2 points (its peak and trough). In practice this is not sufficient and a minimum of 5, and preferably 10, points per cycle are necessary to define a component, an EP wave or an EEG spike, for instance. If this criterion is adopted, a sampling rate of 2 kHz would resolve wave components up to 200 Hz – components of 5 ms duration or longer. For waveforms other than sine waves – square waves, sharp waves, waves that start rapidly from the baseline, for example – high-frequency components are necessary to preserve the waveshape and allow accurate measurements of latency.

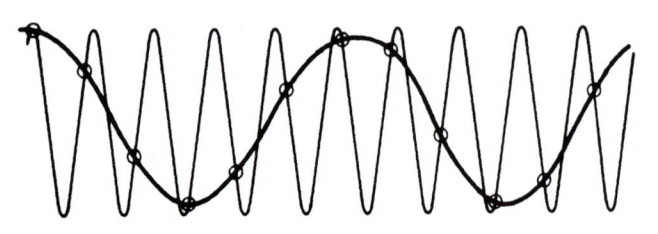

Fig. 2.23 Generation of a low-frequency wave by inadequate sampling of a high-frequency (regular) wave. (From Binnie et al (2004), by permission.)

The highest frequency contained in the data displayed is also influenced by the settings of the frequency response controls of the amplifiers. The use of a top-cut filter of 70 Hz, for example to reduce the muscle artefact of an anxious patient, will determine the limit of the system and a sampling rate of 2 kHz with a theoretical limit of 1 kHz and a practical limit of 200 Hz will then be more than is necessary.

As described above, the signals are recorded in referential format, as the potential difference between each electrode and a common reference, converted to digital form and input to a computer. Thereafter, all operations on the data are mathematical functions. The various display montages are set up by simple subtraction. For instance the potential difference between electrodes A and B, V_{A-B} is given by:

$$V_{A-B} = V_{A-ref} - V_{B-ref}$$

where V_{A-ref} and V_{B-ref} are the potential differences recorded between A and B and the common reference. The re-montaged signal values calculated in this way are plotted on a suitable display device, usually a VDU. The sensitivity of the display, the scale of trace deflection to signal amplitude (in µV/cm), the time scale (mm/s) and any frequency filtration, are all determined by digital operations.

EEG RECORDING EQUIPMENT

General features of EEG machines

EEG instrumentation is rapidly changing. Traditional EEG machines recorded on a moving paper chart. Paperless systems which display the signals on workstations and store them on digital media were an exciting new development in the 1990s, but now account for most sales of new equipment. Currently, three generations of EEG machines are in use. The oldest designs use analogue technology and their functions are selected by mechanical switches. The best machines of this type were robust and continue to do good service in many departments; similar products are still manufactured to high specification and at low cost in some developing countries. The second generation used digital technology for signal processing and switching functions were controlled by electronics or microcomputers, but signals were still displayed on paper charts. Modern machines use entirely digital storage and display and, apart from the amplifiers, may contain only a standard PC. First- and second-generation machines are likely to remain in use in many departments for some years, and not only on economic grounds: for some applications (research, ICU and intraoperative recording) there are considerable advantages in having a paper chart available for annotation of events, experimental conditions, etc.

Conventional EEG machines and digital workstations share many basic features. The present text describes these first in the context of traditional machines, which illustrate many of the basic principles, offers some incidental commentary on the different features of digital systems, and concludes with a section on digital technology.

The EEG machine must be capable of recording electrical activity from a level of one or two microvolts up to one or two millivolts within a bandwidth of about 0.1–100 Hz. Exceptionally, DC amplification (infinite time constant) should be available at the microvolt level. Current machines amplify and display 8, 16, 21 or more EEG channels with additional channels for polygraphic recording (respiration, ECG, etc.). For routine use 21 EEG channels may be considered the minimum standard as this allows all the electrodes of the 10–20 system to be included simultaneously in common reference montages. In special circumstances, for example, intracerebral recording (when many electrodes are usually available), more channels are required. Machines having 16 or more channels with pen writers are transportable in that they can be wheeled but not carried; for recordings away from the base hospital 8-channel portable machines are available, but digital systems of up to 32 channels based on laptop computers are much to be preferred.

In most laboratories there are considerable electrical fields from mains (50/60 Hz) operated equipment including the EEG machine itself. These can cause interference in various ways, as any amplifier working at microvolt level is liable to be affected by mains-borne interference as well as artefact arising from movement of connecting wires and cables. Some machines incorporate a suppressor circuit which amplifies the potential difference between a sensing electrode on the head and the chassis of the machine, inverts the phase, and applies the resulting signal to another electrode elsewhere on the head; this amounts to a negative feedback system which reduces common mode interference by keeping the patient equipotential with the machine. There are many mechanisms and possible sources of mains interference; see Screening Commission of the New University Hospital, Copenhagen, Denmark (1967).

Input circuits for EEG recording

Most mains interference arises in the input circuits and the system is less susceptible if the leads between patient and amplifiers are kept as short as is practicable. Modern machines have preamplifiers in a headbox that can be placed close to the subject. The purpose of these preamplifiers is not so much for amplification of the signal but to convert the impedance of the signal source from several thousands of ohms at the electrodes to a few ohms, so that any subsequent transmission along cables to the main amplifiers in the EEG machine is done at low impedance, which reduces the possibility of pick-up and cable movement artefact. In digital systems, analogue to digital conversion and possible multiplexing take place in the headbox.

The sockets on the headbox should provide for not only the standard electrode placements but also for additional EEG and possibly polygraphy electrodes. A reasonable provision is for 21 standard placements, sphenoidal or nasopharyngeal electrodes, at least three earths (useful when connecting ancillary equipment), 5–12 non-standard microvolt inputs and two ECG sockets, possibly connected directly to an ECG amplifier. Sockets should be located on a head diagram, designated by the standard abbreviations of the 10–20 system and preferably by numbers also. For applications using entirely non-standard placements, such as electrocorticography or depth recording, a rectangular matrix of numbered sockets is more convenient than irrelevant letter codes on a head outline, and more inputs may be required than for routine EEGs.

One of the problems when preamplifiers are mounted in a headbox is that calibration is more complex, and when purchasing a new machine the buyer should ascertain whether the preamplifiers are tested by the internal calibration circuit.

In first-generation machines, electrode montages were selected by mechanical, multipole switches. These are prone to make poor contact after years of use and generate electrical noise. On second generation machines all electronic switching is controlled by an internal microcomputer, and montages and other control settings are displayed on a VDU. A connector may be provided to allow remote control by a computer, or conversely to output the values of the settings to an analysis system. Information about control settings may also be written out on the chart as pulse codes on a marker channel or otherwise. On digital machines, there are no switched functions: the digitized values from each electrode socket are reformatted, filtered and scaled entirely by mathematical operations.

Amplifiers for EEG recording

After the preamplifiers, there are several further stages of amplification in a traditional EEG machine in order to provide sufficient power to drive the output device. If the writer sensitivity is 10 mm deflection per volt, the amplifier must have a total gain of 40 000 for a sensitivity of 2.5 µV at the input per mm deflection of the writer. The various stages of amplification are normally coupled together by capacitors that act as high pass circuits and care must be taken in design to ensure that the overall time constant is as specified. When several stages are coupled in this way, it is usual to make all time constants except one much larger than the required TC of the channel; the TC is then mainly determined by one adjustable, shorter time constant circuit. If this is not done and two or more coupling time constants have similar values, there will be phase distortion of complex waves (Saunders and Jell 1959). This can be seen in the response to a step function input, such as a calibration signal. Instead of a smooth exponential decay the trace will cross the base-line (undershoot) and then return to it only slowly. The later stages of amplification are often directly coupled and access is provided so that transducers for monitoring other slowly changing physiological variables (e.g. respiration), which have relatively high (mV) output signals can be displayed without distortion alongside the EEG.

Each amplifier has a continuously variable control for fine adjustment during calibration. There are also stepped gain controls, each step changing the sensitivity, typically by a factor of √2 – two steps therefore double or halve the sensitivity. Each channel can be altered separately or all channels can be changed together with a master gain control. There may also be switches to toggle between an EEG sensitivity range, say 5–200 μV/cm, and another for polygraphy and intracranial recording, perhaps 50–2000 μV/cm. Thus, the sensitivity of any channel may be determined by one continuous and up to three stepped controls, a situation that invites confusion and error. This problem no longer arises with digital machines that simply show the actual display settings on the VDU.

Filters for EEG recording

EEG machines have a number of high pass filter settings encompassing a range of time constants from a few seconds to 0.01 s (3 dB points, or turnover frequencies from about 0.05 to 10 Hz). The relation between time constant (TC) and turnover frequency (F) is TC = $1/(2\pi F)$. These filters have an effect on the phase of signals as shown in *Fig. 2.24*.

EEG machines also have a number of selectable low pass filters ranging from about 15–120 Hz (3 dB points) for conventional pen recorders and up to 700 or 1000 Hz for machines with ink-jet galvanometers. The upper limit of these filter settings should be determined by the frequency characteristic of the writing system. If the machine has sockets so that the outputs can be sent to other equipment (e.g. computers) an additional, much higher setting is provided (2 kHz or more). Most, but unfortunately not all, manufacturers specify filter settings as the frequencies at which 3 dB attenuation occurs. The filters of each channel can be altered separately or all channels can be changed together with master filter controls. Filters on most machines have a slope of 6 dB/octave. For recording in an environment where mains interference cannot be eliminated, a notch filter may be provided, giving marked attenuation over a narrow band around the mains frequency. Some machines also have a low pass filter with a very steep slope starting just below mains frequency.

Analogue EEG systems
EEG write-out systems

Practically all traditional EEG machines had ink writers but modern systems display the EEG on a VDU (digital or 'paperless' recording) and may provide some option for producing a hard copy of a selected section of the record, for instance on a laser printer. The following section concerns traditional machines only; digital recorders are considered on page 42.

Each channel of an ink (or carbon paper) recorder has its own pen motor (also called a galvanometer). These are electromagnetic systems of moving coil or moving iron type. The moving coil type has the pen attached to a coil carrying current and so behaves like a magnet whose strength is proportional to the current. The coil rotates in the field of a permanent magnet and turns through an angle determined by the magnetic interaction of this field and the field of the coil, and a restoring force due to the method of mounting the coil. The angle of rotation is proportional to the current in the coil, which is supplied by the driver amplifier and modulated by the input signal.

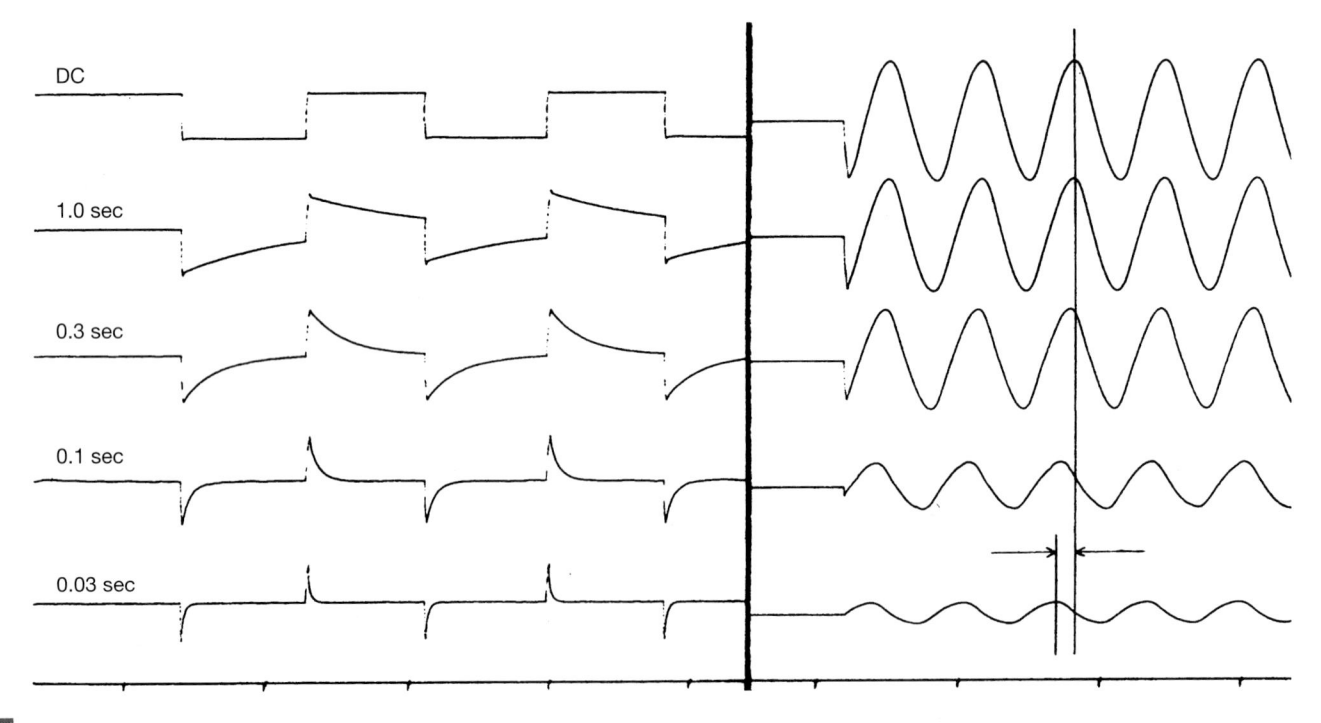

Fig. 2.24 Effect of different time constant values: (a) on a square wave input; (b) on a 1.5 Hz sine wave input. (From Cooper et al (1980), by permission.)

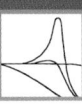

The moving iron type of writer has a stationary coil and rotating magnet.

The sensitivity of the pen writer is dependent on frequency and falls off at high values because of inertia, mechanical resonance and friction. Four possible characteristics are shown in *Fig. 2.25*. Curve (a) shows sensitivity that is constant up to a particular frequency after which it falls; (b) shows a slight increase in sensitivity and then a fall; (c) shows a large increase in sensitivity before the fall; and (d) shows the extended frequency response of an ink-jet galvanometer (which has little inertia because of its very small mass). The system having the characteristics of (a) is described as overdamped and would respond slowly to a square wave input. That of curve (c) is an underdamped system and would respond to a step input with an oscillation (because of the excessive high frequency sensitivity). In between (a) and (c) there is a condition known as 'critical damping' in which the system returns to rest in the shortest possible time without overshoot after a step function input. If the damping factor can be set by an electrical or mechanical control, the manufacturer's literature should give instructions for setting the degree of damping.

The friction due to the pressure of the pen arm on the recording paper also influences the damping. The friction to be overcome in order to start a stationary system moving (static friction or 'stiction') is usually greater than when moving. This has the effect of selectively reducing the response to small amplitude signals. Pens are sometimes made with jewel tips to reduce friction and wear.

Pen arms move in arcs and there are both amplitude and timing errors that increase in a non-linear manner with the deflection and cause some distortion of waveforms. If the pen has a length of L cm and rotates through an angle of θ radians (*Fig. 2.26*), it can be shown (Cooper et al 1980) that the amplitude error introduced by measuring along a straight line PN perpendicular to the baseline instead of along the arc BP is $1/6\ (D^3/L^2)$ cm, where D cm is the deflection measured along PN. The displacement BN is equivalent to a timing error of $D^2/2Ls$ seconds, where s is the paper speed in cm/s.

The ink-jet writer and other systems such as a stylus pressure on a knife edge at right angles to the paper imprinting the trace through carbon paper do not suffer from arc distortion but are non-linear and require correction circuits in the amplifiers to avoid exaggeration of high-amplitude signals.

The convention in recording the EEG is that when a negative signal is applied to the black input lead (grid 1, lead 1) and the white input lead (grid 2, lead 2) is connected to the positive side of the source of signal, the display will be deflected upwards. The convention is followed (the BUN convention) in digital systems also: the value displayed on each channel is the value recorded from the input socket designated lead 1, minus that from the socket designated lead 2.

Paper transport and time marking

The recording paper travels in a direction at right angles to that of the pen deflection and so provides the time scale for the record. An accurate and stable paper drive mechanism is necessary and it is normal practice for several paper speeds to be available. Speeds of 15, 30 and 60 mm/s are essential and values outside this range are useful.

A timescale is usually registered on the record by one or two time-marker pens, which make a mark once per second. These channels can also be used to record the occurrence of events and digital codes for control settings.

Calibration of analogue EEG systems

The traditional method of calibrating an EEG machine is by applying a square wave to the input of all channels. This is normally an integral part of the machine and the calibration voltage can be selected from a number of microvolt values. This is an excellent test, especially if the paper is run at 60 mm/s, as at one press of the calibrator switch the operator can check amplitude calibration (by the deflection of the pens), the high pass filters or time constants (by the exponential fall), the low pass filters and pen damping (by the amount of over- or undershoot) and pen alignment. Any 'stiction' can be assessed by using a very small amplitude signal. Linearity and dynamic range can

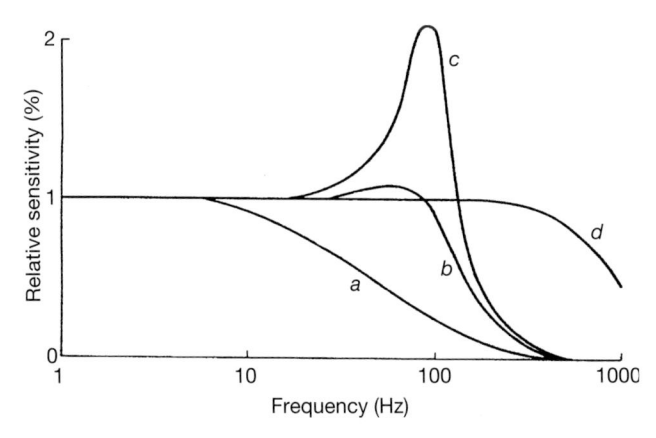

Fig. 2.25 Frequency response characteristics of a pen writer which is (a) overdamped, (b) slightly underdamped, and (c) very underdamped. (d) Frequency response characteristic of a jet galvanometer. (From Cooper et al (1980), by permission.)

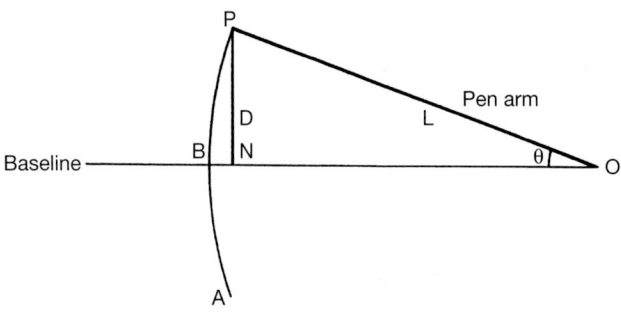

Fig. 2.26 Arc distortion: diagram for calculating amplitude and timing errors. (From Cooper et al (1980), by permission.)

be determined by measuring the deflections when the calibrator voltage is increased in steps to a maximum deflection. A sinusoidal calibration signal is also available on many machines. This provides less information overall about machine function but may be easier to measure than a square wave for purposes of sensitivity calibration and for setting up data processing systems.

Testing other characteristics such as frequency response curves, discrimination and input impedance needs special equipment such as a low-frequency oscillator.

Digital EEG systems
General features of EEG recording workstations

Modern EEG workstations convert the EEG to digital form and display it on a VDU (or computer screen) instead of paper. This development offered a great advantage over chart recorders of allowing digital processing of the signals after they had been recorded, so that the same data could be reviewed with any montage or settings of filters, sensitivity or time scale. The technology is rapidly changing, as are the products commercially available; major developments may well take place between the writing and publication of this text. For technical details of digital recording of biological signals, see *Digital systems* (p. 37).

Digital EEG first appeared as an almost incidental feature of some special purpose brain mappers, evoked response or telemetry systems. It rapidly became apparent that purpose-built machines were not commercially viable; the new generation of digital systems are based on standard personal or laptop computers, with no special purpose hardware apart from the input circuits. This has brought large savings in cost, as the price of a powerful microcomputer together with a headbox containing preamplifiers and an analogue to digital converter is considerably less than that of traditional EEG machines. Attempts by manufacturers to sell relabelled, standard PCs as 'EEG machines' at an inflated price is unjustified and have largely proved unsustainable. Some insist on setting the start-up parameters to initiate the EEG acquisition software and stipulate that the use of any other standard applications software on the same system invalidates the maintenance contract. The purchaser would be well-advised to choose other products as competition for market position is keen and now, rightly, based on software. Despite an understandable concern of both manufacturers and many users to maintain the distinction between EEG, EMG, telemetry, brain mapping, intensive monitoring systems, polysomnographs and EP machines, these are progressively being replaced by general purpose digital bioelectric data acquisition stations. These run different software packages for the various modalities of investigation, and have optional hardware modules for particular applications: a telemetric input and video capture or synchronization capability for long-term recording, and a multimodality stimulator unit for EPs.

Signals are generally reviewed on separate workstations and increasingly the acquisition and review stations are linked by a network. Compatibility between the various units is therefore essential: it is no longer reasonable to purchase individual recorders on an *ad hoc* basis; the modern clinical neurophysiology laboratory is planned as a single, integrated system for recording, displaying, reporting and archiving neurobiological signals of all kinds. In a department where several physicians and technologists may wish to inspect EEGs at the same time, the cost saving on the acquisitions systems may be offset by the need for multiple review stations and software licenses.

At the time of writing, most EEG machines commercially available are already entirely digital and paperless. However, some recent products represented a transitional phase: essentially conventional EEG machines with thermal printers in place of pen recorders, but with hard discs for additional, digital storage. These features offer no great advantage, but are preferred in some countries where only hard copy is trusted for archival purposes. Thermal printer output has a limited storage life (currently about 5 years under controlled environmental conditions), which is insufficient for permanent archiving. It is difficult to imagine many busy clinicians provided with a paper record taking advantage of the digital capabilities by spending extra time reformatting and reviewing EEGs from disc, except for research purposes. Moreover, some manufacturers provided digital output only from the final amplifier stage (after montaging and filtration); reformatting of the signals off-line was then possible only if the montage displayed during the recording used common reference derivation.

Inputs for digital EEG recording

The signals are recorded from each headbox socket referred to a common reference, usually an additional, designated scalp electrode but on some machines the average of two or more standard electrodes (this last feature presents problems if one unwittingly tries to record without one of the reference electrodes in place during a special procedure such as corticography) were used. The signals are then digitized in the headbox (analogue to digital conversion), usually multiplexed and transmitted through a cable to an interface card in the standard computer that serves as the recorder. The presence of any signals at more than half the sampling frequency of the analogue to digital conversion produces spurious activity (aliasing). To overcome this, anti-aliasing filters are used to eliminate high frequencies and often the EEG is 'oversampled' at a rate several times more samples per second than will eventually be stored.

Data storage for digital EEG

The data are stored as they are recorded, either on a local hard disc or reusable optical disc, or over a network to a file server. Ideally, even if the server option is generally used, local storage facilities should be available, so that recording is possible when the network is shut down or at sites without access to a network socket. As an hour of 30-channel digital EEG recorded with acceptable spatial and temporal resolution typically occupies about 40 megabyte, a modest hard drive built into the recorder will have a storage capacity upwards of 200 h, and the server may store many months' output of a busy department. For long-term archiving, records are backed-up onto optical discs, CD ROMs or cassette tapes. A small filing cabinet of discs replaces the storeroom with its metres of shelving used to archive paper traces. It is prudent to take advantage of the reduced cost and volume of storage to keep duplicate archives on different sites in case of fire or other mishaps.

Displaying digital EEG

The EEG can be displayed during recording, scrolling across the screen in much the same way as the conventional chart rolls across the table of the recorder. It can also be recalled from disc for subsequent off-line inspection and interpretation, or for computer analysis, brain mapping, etc. Sections of particular interest can be plotted on a high resolution printer/plotter, or transcribed onto a chart recorder, if available.

The quality of display which can now be achieved with modern VDUs is much superior to that provided by conventional video technology, as used, for instance, in the older telemetry systems. To show the smallest (say 5 μV) components on a tracing of ±100 μV requires at least 40 lines per channel. Thus, the 768-line screens now available on laptop computers will display a 16-channel EEG with as much detail as some pen recorders, and a 1024-line VDU can present up to 64 channels with fair legibility. Indeed, provided one is prepared to take time to examine the same section of recording repeatedly with different values of sensitivity and bandwidth, a far more detailed assessment is possible than with a conventional recording, the quality of which is forever fixed by the control settings in use when the event of interest occurred. Digital EEG displays are free of the non-linearities, arc distortion and possible interchannel differences of bandwidth imposed by pen recorders. When transients of unexpectedly high amplitude occur, one channel can cross another and software can also limit deflections to pre-set maximum values. Colour coding, for instance of left- and right-sided channels, can be used to make the montage easier to follow. Vertical time markers can be inserted at selected intervals, or suppressed, and cursors can be used to compare temporal relationships of features on different channels, or to measure EEG waveforms. Users accustomed to conventional EEGs may prefer the signals to be displayed with the usual aspect ratio of amplitude to time, and with a time scale on the screen similar to that on a paper chart (e.g. 50–100 μV/cm and 15–30 mm/s). However, as the height/width ratio of a standard VDU is considerably less than that of the pages used on a 21-channel chart recorder, the user will often need to use lower sensitivities. The time scale can be expanded or compressed at will, to display slow trends or improve resolution and it is possible to 'zoom' in and magnify interesting details. Some form of time index indicates the location of the current display within the entire recording and also elapsed time and time of day. Events are marked logged, displayed on the trace as text or icons, and can be listed and indicated by markers along a time axis. Random access to any particular section of the record is possible, so that a couple of keystrokes call up; for instance, when the patient is asked to open his or her eyes or the time at which a seizure occurred. Additional event markers may be generated by software that recognizes, for instance, seizures, spikes or apnoeic episodes.

The facility for modifying the original record before display creates the possibility of viewing the data with a montage different from that used during recording; indeed, the same section of EEG can be repeatedly reviewed with various montages. The primary record is made and stored with common reference derivation, even if a different, possibly bipolar, montage is displayed during recording. The data can be displayed 'as recorded' using the montage employed during acquisition. It can also be reformatted to provide any possible bipolar, common reference or common average reference montage involving the electrodes in use. The software may or may not permit the creation of montages using a mixture of bipolar and common reference derivations. There may be a facility for the sophisticated user to select the electrodes to be included in the average reference or to fine tune the weighting factors used in a source reference to suit non-standard electrode placements. A conservative approach is recommended: it is not recommended that the entire record should routinely be displayed repeatedly using a large number of different montages. The possibility of recording only one montage and subsequently reformatting does not justify reducing the total registration time (as is, regrettably, suggested by some sales literature). Many significant EEG features occur only intermittently or in a specific state of vigilance or drowsiness, and the probability of detecting them depends on the length of recording. Nevertheless, when a feature of interest is found, it is valuable to be able to inspect it with different methods of derivation, particularly using references carefully selected to highlight specific aspects. The montages available should include that used during recording (so that the reader knows what the technologist saw at the time) and the referential montage used for the primary data acquisition and storage (for calibration and quality assurance purposes).

Power spectra of selected sections of the tracing, brain mapping and equivalent dipole modelling may be available for detailed examination of features of interest during review (*Fig. 2.27*). The authors have enjoyed experimenting with these features, but are not yet able to comment on their value for assessment of routine EEGs. The discovery that one commercial system had for several years

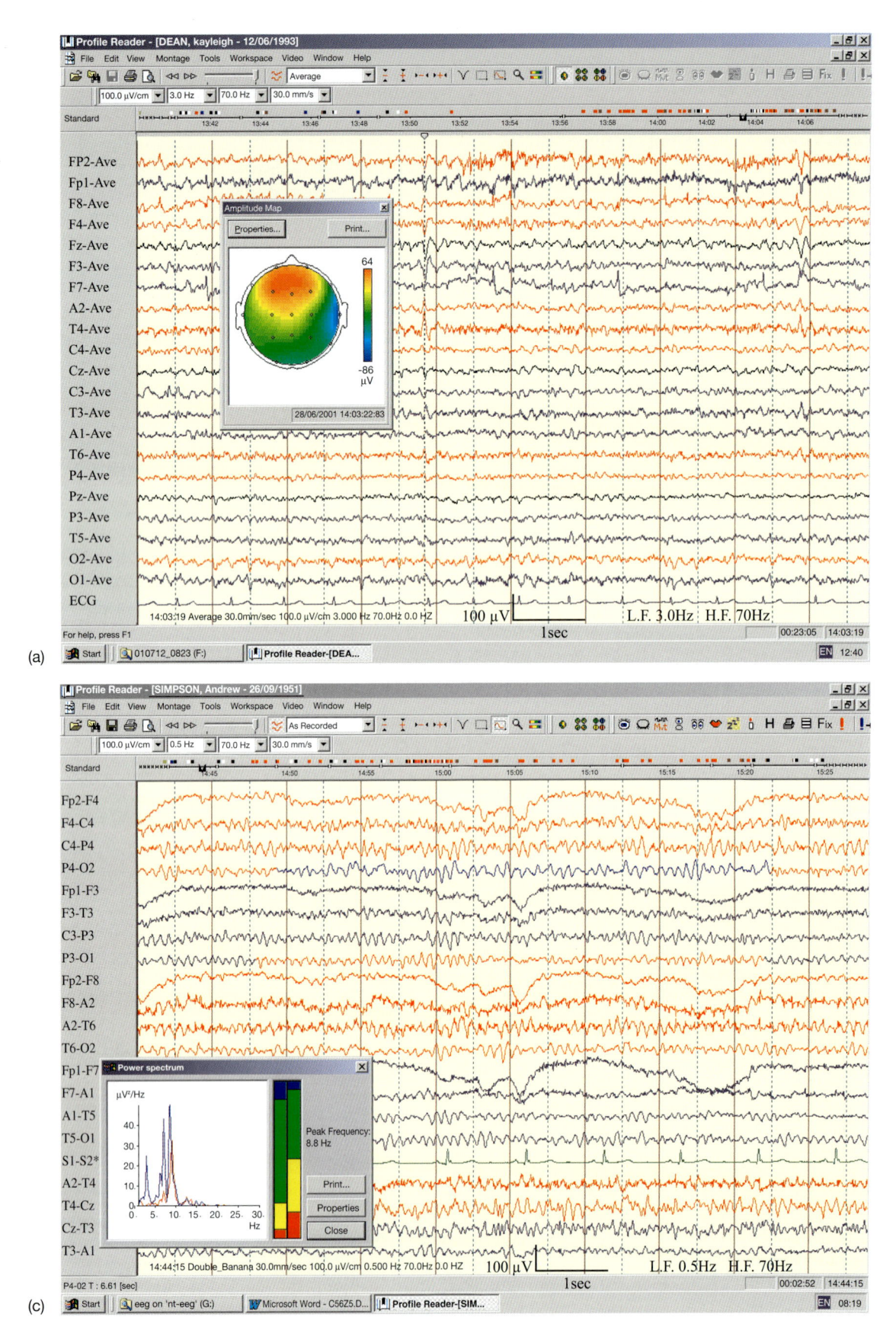

Chapter 2 Neurophysiological Instrumentation, Connection with Patients and Recording Methods

Fig. 2.27 Use of computational aids to interpretation. Isopotential mapping to distinguish typical Rolandic spike of benign childhood epilepsy (a) from frontal spike due to cortical dysplasia (b). Spectra confirm (c) asymmetry of ongoing activity after a right-sided stroke (reduced alpha frequency and excess theta activity right sided spectrum blue and left sided red), and (d) that prominent theta activity is at a subharmonic of the alpha frequency and most probably a slow alpha variant. (From Binnie et al (2003), by permission.)

EEG recording equipment

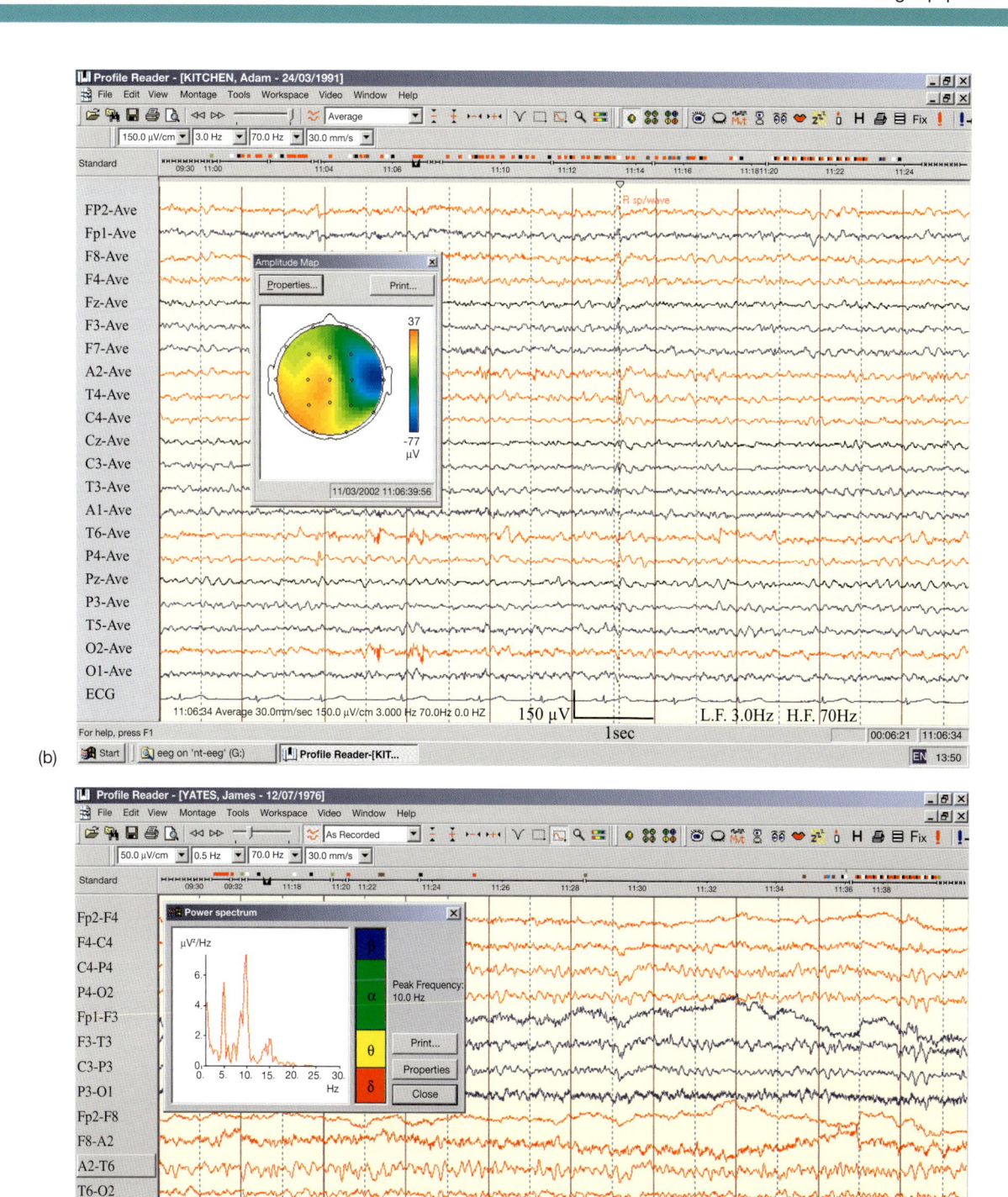

45

suffered from a cursoring error, so that the isopotential maps did not correspond to the features selected, apparently unnoticed by the purchasers, leads one to doubt how seriously these facilities are used. Certainly they highlight clinically significant features (such as complex topographies of epileptiform discharges or subtle asymmetries of alpha frequency) not obvious to the novice and are a valuable educational aid, but the display produced can usually be predicted from the primary display by an experienced reader.

Any large department with a teaching commitment should have at least one multimedia projection system, to allow EEGs to be viewed by a large audience. This facility greatly enhances the educational value of EEG case conferences.

Calibration of digital EEG systems

Some form of automatic calibration is usual. When the computer is switched on, or when the recording program is started, calibration signals are injected into all the preamplifiers. The resulting digital signals are compared with the expected values and corrections calculated for DC offset and sensitivity. If these are within tolerated limits, the necessary corrections are applied to signals subsequently recorded, either by adjusting the amplifier characteristics or by simple digital manipulation of the input values (subtracting the DC offset and multiplying by a sensitivity correction factor). If the errors are outside tolerated limits a message is issued indicating an amplifier fault. This approach is reliable and efficient but totally invisible to the user, who may regret the inability personally to check amplifier performance, as on a traditional machine. 'Calibration' now tells us only that the manufacturer's DC and sensitivity correction procedures are working adequately and that no amplifier has totally failed. It is now recommended (Electrophysiological Technologists' Association 2004) that a new system should be checked on arrival with an external signal generator and specially designed attenuator, and that similar checks should be repeated every 3 months and after each software update. The detailed procedure can be found in Cooper et al (2005, pp. 142–144).

Limitations, problems and opportunities of digital EEG

Various ergonomic problems are peculiar to digital machines; potential solutions are available but not necessarily provided as yet by particular commercial products. Technologists will need to learn to identify transients (and particularly artefacts) very quickly as they pass across the VDU screen. Considering evaluation of these and other brief events during the course of the investigation may require the recording to be temporarily suspended whilst the feature of interest is recalled to the VDU. This is soluble by use of a split-screen format, so that whilst recording and online display continue, an earlier section of the record is displayed for inspection in a second window. It may be desirable, during either recording or review, to display previous EEGs. This too is achievable by the use of multiple windows. There arises from time to time a need to show the findings to a neurophysiologist or senior technologist to discuss further action whilst the recording continues. Again, if disc access is fast enough and a large RAM memory is provided, it should be possible to review the EEG interactively in one window, whilst displaying the current recording in another. If the data are stored online over a network, review of an ongoing recording should be possible from a remote workstation, to permit consultation on emergency EEGs.

Annotation of the record during registration demands a new approach on the part of the technologist. Present systems already allow at the least for common events, such as eye closure or the start of hyperventilation, to be signalled by dedicated keys or software buttons, possibly for additional keys to have definitions assigned by the user, and for the typing in of free text which can be displayed with the EEG. Some allow freehand annotation of the record with a light pen or mouse, and touch-sensitive screens are beginning to appear. Voice recognition will probably become available and prove more user friendly. The displays should include all control settings and, optionally, a graphic display of the montage. Real or elapsed time should be indicated, and during review any point in the record should be directly accessible by specifying the required time or event mark. A synoptic display should be available during review so that the locations in the tracing of all events can be identified, selected with a cursor and immediately accessed. As the physical dimensions of the display depend on screen size, suitable provision must be made for indicating, and specifying, amplitude and time scales. The software should be customized to the VDU, so that conventional physical units (e.g. mm/s, µV/cm) can be used. Some current machines take the easy option of specifying time in seconds per screen, and amplitude as microvolts per interchannel interval.

There are considerable ergonomic differences between commercial systems, some of which are much less convenient to use than others. Many, for instance, require the user to position the cursor over a 'button' on the screen to move the display forward or back. Consequently, when the cursor has been used for any other purpose one has to divert one's attention from the EEG to locate the button; moreover, repeated clicking on the mouse (perhaps 180 times to view a 30-minute EEG) carries a real risk of repetitive strain injury. A more sensible design allows the page to be moved forwards or back by clicking anywhere on the right, or left side of the main display area, with the alternative of using the Page-Up and Page-Down keys for the same functions. On some systems, changes in sensitivity, time scale, etc. can be selected only from 'pick-lists' or by explicitly typing the required values; others include these options but also allow changes of one pre-set interval to be made more conveniently by clicking on + and – screen buttons and keys. Flexibility, customization options, and a choice of alternative ways of achieving the same effect is the hallmark of well-designed display software. This in turn depends to a considerable degree on the operating system and the skill of the manufacturer in exploiting its capabilities. It should be possible to use a word processor to write a report whilst viewing the EEG. These requirements demand an operating system with a graphic interface that allows multitasking and simultaneous displays in different windows.

Other features that are desirable but often not available include: ease of changing montages and control settings of individual channels; control of maximum trace deflection to prevent over-large signals (usually artefacts) from masking other channels; ease of adding, repositioning, annotating and moving to event marks; split-screen displays to call up previous EEGs of the same patient; cursors which automatically display the amplitude, duration and frequency of any selected EEG feature; adjustable paging speed; and display of patient details to ensure that the correct file is being viewed. Doubtless many more are yet to be devised.

Storage of the EEG records on computer-readable media can readily be combined with storage of administrative details and text reports to provide a single archive. In a fully integrated system, data acquisition and review stations, and the workstations used by secretaries and receptionists are linked via a local area network (LAN). Management software allows transfer of patient ID information, etc., between the appointments system, the EEG record files and report writing modules, billing system, and administrative databases, which must of course be stored securely to ensure patient confidentiality

Hard copy on paper is still required for examination at sites remote from laboratory workstations. The provision of illustrative examples with, or indeed as integral parts of printed EEG reports will increase their value and interest for the recipient. Some current hard copy facilities are mere screen dumps; properly annotated printouts in an appropriate format should be provided. Hard copy should be available, not only as single pages, but on continuous stationery, preferably generated by a laser printer rather than a pen or thermal recorder. Few manufacturers provide an output in a standard digital graphic format with text and line thickness optimized for display on a multimedia projector or publication in a scientific journal.

New methods are required for training and assessing technologists. They no longer have to be able to adjust pen damping, but need to learn new tasks, such as checking free disc capacity, backup, housekeeping, and disc recovery procedures. Conventional chart recorders will probably be retained for some years for teaching purposes.

There is an urgent need for agreed industry standards on storage formats, to allow portability of data and analysis software between different systems and to ensure that records do not become irretrievable when a department changes its equipment. Paperless EEG machines are not susceptible to many of the hardware faults that demand continual vigilance and repeated calibration when conventional equipment is used. They may, however, suffer from other kinds of defect, notably due to software errors. Thus, control of display sensitivity, montages or bandwidth settings become mathematical operations, immune to switch faults or inaccuracies of components, but liable to gross errors (which may not easily be recognized) if incorrectly programmed. It is the quality of software which determines the ease of operation of paperless systems and consumer demand should lead to program development less costly and more rapid than the past evolution of hardware-based EEG machines.

It is at the time of writing too early in the evolution of digital EEG machines to allow any conclusion about the outcome of these developments. It is almost impossible to propose technical standards, as reasonable demands can become more exacting with improved computer technology, but recommendations were published by the American Electroencephalographic Society appropriate to the situation in 1991 (American Electroencephalographic Society 1991) and by the International Federation for Clinical Neurophysiology (IFCN) in 1998 (Nuwer et al 1998). In summary, minimum requirements include:

1. Storage of basic demographic and administrative information about the patient and the investigation within the digital record.
2. Recording of a 100 µV square wave calibration signal on the reference montage used for data acquisition.
3. Provision for full annotation during recording of artefacts, state of patient and clinical events etc., including free text.
4. At least 24 recording channels, preferably 32.
5. A digitization rate selectable of at least 200 Hz, at frequencies that are multiples of 50 or 64, for example 500 or 512 samples/s. Appropriate anti-aliasing filters, for example 70 Hz cut-off for sampling at 200/s with a roll-off of at least 12 dB/octave. Digitization with 12-bit resolution to provide a reasonable dynamic range from 0.5 to ±1023 µV. (Fourteen bits may be required for intracranial recording.)
6. Preamplifier input impedances more than 100 MΩ. Interchannel cross-talk less than 1%. Common mode rejection ratio at least 110 dB. Noise level less than 1.5 µV peak to peak and 0.5 µV root mean square (r.m.s.) at any frequency from 0.5 to 100 Hz.
7. Various commercial recording media are acceptable but the user must take account of their limited durability and make arrangements for archives to be copied onto new storage media as necessary.
8. Both video and paper displays should preferably be available. Display on a screen or printout should approximate the temporal and spatial resolution of traditional paper recordings.
9. Additional digital filtering must be available during review with at least high pass filters at 0.5, 1.0, 2.0 and 5.0 Hz, and low pass at 15, 30 and 70 Hz. Playback systems must show montages, sensitivity and filter

settings, comments by technologists and event markers. Every screen or page should display a time stamp. The display should indicate when hyperventilation is in progress throughout the procedure.

10. A standard display setting should be provided in which 1 s occupies approximately 30 mm along the horizontal axis, with a minimum resolution of 120 data points, per second, per channel. Other more compressed or expanded horizontal scales should be available. A standard vertical scaling with a minimum interchannel spacing of 10 mm should be available on a screen with a minimum vertical resolution of 4 pixels/mm. The horizontal and vertical scales should be displayed.
11. The display system must allow presentation of separate segments of EEG, from within the same recording or from different records.
12. Paper printout should have a resolution of at least 300 dots per inch (dpi).
13. Each manufacturer must make available a means of sending the digital record to any other users who have a need to review it. The vendor should make available a means to put the record into a standard generally accepted format, and must publish and permit unrestricted use of their own data format to read and translate the EEG record into a form readable by another vendor's equipment.

EVOKED POTENTIAL RECORDING SYSTEMS

Electrodes for EP recording

If slow potentials such as the contingent negative variation (CNV) or Bereitschaftspotential are being recorded, only high-quality, non-polarizable electrodes should be used. Electrodes for recording other EPs (composed of frequencies above a few hertz) are not so critical, but electrodes made from dissimilar metals should be avoided. Electrodes are usually attached to the skin by means of a special paste which serves as an adhesive but also establishes good contact with the skin. Common reference montages with the reference electrode on the ear lobe or mastoid process are commonly used. The choice of reference electrode site depends upon the type of EP to be recorded and the scalp distribution of the component under investigation. Specific montages adapted to each kind of EP study are detailed in the appropriate sections. As with the EEG, a reference electrode, even if it is off the head, cannot be assumed to be inactive.

Amplifiers for EP recording

Equipment for recording EPs must be suitable for amplifying electrical signals with amplitudes down to 0.1 μV and bandwidth from below 1 Hz to 10 kHz. Most investigations require multichannel recordings.

Recordings are usually made in laboratories in which there is a high level of electrical field (electrostatic and electromagnetic) which gives rise to 50 or 60 Hz interference. In modem equipment the first stage of amplification uses preamplifiers in a 'head-box' which can be placed close to the patient so that the leads to the electrodes can be kept short and the cable between the pre- and main amplifiers is carrying signals from a low-impedance source – both features help to reduce pick-up of mains and other interference. The impedance of skin electrodes is typically less than 10 kΩ and the input impedance of the amplifiers should be 10 MΩ or more. The CMRR should be high (100 dB or more) so that in-phase (mains) interference is minimized. The amplifiers should have a wide range of stepped gain controls; machines with many channels should have master gain controls. Calibration of the whole system from preamplifier to display should be possible. A system for measuring the impedance of each electrode should also be available.

Filters for EP recording

Both low- and high-pass filters should have a number of set positions so that the upper frequency limit can be reduced from 10 or 20 kHz to 30 Hz and the low-frequency limit raised from 0.01 to 300 Hz. Filters with roll-off slopes of 6 dB/octave are adequate. Excessive filtering can give rise to significant distortion of the waveform and changes of latency.

Averaging

Practically all EPs are small in amplitude compared with the ongoing EEG and artefacts such as muscle potentials; some form of signal enhancement is therefore required before they can be displayed and measured. This usually is done by averaging: successive trials are summed and, after sufficient trials have been collected, the sum is divided by the number of trials to give the average EP. The sampling rate must be sufficiently high to resolve the fastest frequency component in the EP. This is known as the Nyquist frequency and is twice the frequency of the fastest component; sampling rates below the Nyquist frequency can generate spurious low frequencies by aliasing. The duration of the epoch is selected according to the investigation. The epoch length and number of trials range from about 10 ms and 1000 or more trials, for brainstem auditory potentials, to several seconds and 20 trials, for the CNV. Trials contaminated by artefact such as eye movement, blink and muscle potentials, must be rejected during data acquisition.

There can be considerable variability of the EPs from set to set and within populations of normal people or apparently similar patients. Grand averages across sets of trials and groups of people are often used to provide data to which individual patients can be compared. The facility to average averages is available on most instruments. The number of trials making up each individual average should be the same. Normalization of latencies of the individual trials may be required to avoid smearing of the components and reduction of amplitude in the grand average, due to variability of individual peak latencies. Amplitude normalization may also be used to avoid the grand average being excessively influenced by a few traces of exceptionally high amplitude.

EP display

The EPs are displayed on an oscilloscope after averaging. In most modern systems the data are manipulated by digital techniques before being displayed in colour on a VDU. Further manipulation can be done in the display mode before the data are stored. They can be written on a plotter, with details of the recording conditions in numerical form (e.g. gain and filter settings). Several traces can be displayed on the VDU at the same time, so that comparisons can be made with previously stored data or sequential acquisitions in the same investigation. Split-screen displays enable the data to be viewed simultaneously on different time scales. Latencies and amplitudes can be obtained from the displayed data either automatically or by cursor measurement – reliable measures depend upon a low level of residual noise. The spatial distribution of the EP at particular times in the epoch can be displayed, usually in colour.

For purposes of quality control, it is useful to have a facility for displaying the ongoing activity, and not only the averaged EPs, both while setting up and during data acquisition. This is particularly important when recording in circumstances where interference is likely or the electrodes may become detached (ICU, operating theatre, actively moving subjects). For some applications, it may be necessary to store the raw data and then select epochs for averaging after completion of the experiment. This arises, for instance, when a very large amount of oculographic or movement artefact is anticipated, or when selection of epochs to be averaged is contingent on the behaviour or state of the subject (e.g. maintaining fixation, remaining alert, responding correctly to a cognitive task). To monitor artefacts, behaviour, vigilance, task performance, etc., several polygraphic channels may be required in addition to those used for the EPs. Due to such considerations as the capacity (speed and disc space) of the averaging computer, it may not be convenient to store these extra signals digitally and a chart recorder may then be used. Obviously, if events recorded on a paper tracing are to be used for epoch selection, the chart must in some way be annotated, for instance by marker pulses, to indicate the timing of EP epochs.

Calibration of EP recorders

While it is customary, almost mandatory, for a calibration signal to be recorded on all channels of each and every EEG record, the amplitude calibration on averaging systems is rarely checked; indeed, calibration facilities are not provided on most modern averagers. The user is entirely dependent on the manufacturer setting up the machine accurately and on the sensitivity not changing for the next few months or years! The only satisfactory way to check the amplitude calibration is to average a known amplitude square pulse from an external calibrator that is triggered from (or triggers) the averager. A suitable delay will cause the pulse to be displayed in the centre of the epoch, and will also provide a check on the accuracy of the time base.

EP stimulators

Practically all EP investigations require sensory stimulation (the only exception is recording the Bereitschafts-potential). Stimulators in modern machines are an integral part of the equipment. The stimuli are synchronized to the sampling epoch and the trigger point advanced or delayed from the start of the epoch to facilitate viewing the data on the screen. Various types of stimulation likely to be used in ICUs and during operations are described in the following sections.

Visual stimulation

Flash The most used type of visual stimulation is the stroboscope flash, introduced in clinical neurophysiology in the 1940s, when flash discharge tubes became available. The stroboscope produces an intense, brief pulse of light and is used in clinical neurophysiology mainly to evoke the electroretinogram (ERG) and visual evoked potentials (VEPs) or to provoke seizure activity in the EEG. It can be used as a single flash or repetitively; some stroboscopes can be triggered from the EEG. The intensity of the flash may decrease as the repetition frequency increases because the capacitors that are discharged through the lamp to produce the light do not fully recharge in the time between flashes. Considerable current flows in the lamp circuit during discharge and this can cause flash artefact in the EEG or EP recording. In stroboscopes designed for clinical use, the intensity of the flash can be changed; several levels are usually available. Few stroboscopes are calibrated in terms of intensity and the level used in clinical practice is set empirically. The reason for this lack of calibration is that the intensity of the flash is difficult to measure. In photography, where flash intensities and pulse widths are greater than in neurophysiology, the intensity is, when specified, usually in joules. This is a measure of energy and is usually calculated from the energy stored in the capacitor ($\frac{1}{2}CV^2$, where C is the capacitance and V is the voltage to which it is charged). This calculation does not take into account the efficiency of conversion of electrical to visual energy in the lamp. The calibration of stroboscopes and other visual stimulators requires special equipment and is discussed in Binnie et al (2004, Section 3.2.2.1.6).

LED goggles Flash stimulation can also be presented via goggles fitted with a mosaic of light-emitting diodes (LEDs). These are usually red in colour and less intense than a stroboscope, even though they are closer to the eye.

Each eye can be stimulated separately. In some circumstances, such as in the operating theatre, they may be more convenient to use than a stroboscope (Keenan et al 1987).

Auditory stimulation

Click A narrow pulse of electrical current passing through the coil of a loudspeaker or earphone such that the diaphragm is displaced towards a listener causes a pulse of compressed air ('condensation') to be propagated away from the diaphragm with the velocity of sound (about 340 m/s). It will be heard as a click. If the diaphragm is 'sucked' away from the listener by a current pulse of opposite polarity, the reduced pressure is also propagated, this time as a 'rarefaction'. It will be heard as a (similar) click.

The loudness of the click is determined by the sound pressure (force per unit area) – in the SI system of units newton/square metre are called pascal (Pa). The quietest sound at 1000 Hz that can be heard by the average person has a pressure of 20×10^{-6} Pa or 20 µPa, and this pressure level has been adopted as the normal threshold of hearing against which other pressure levels are measured. The threshold of pain is about 100 Pa, giving a dynamic range of the normal ear of several million. The ear responds logarithmically to sound and for this reason (and to keep the numbers smaller) the sound pressure level (SPL) is converted to a logarithmic scale. SPL is defined as the log of the ratio of the measured *intensity* to the threshold *intensity*. As the intensity (power) is proportional to the square of the pressure:

$$\text{sound pressure level} = \log_{10} (P/P_0)^2$$

where P is the measured pressure and P_0 is 20 kPa. The unit of sound level is bel (after Alexander Graham Bell), but this is rather 'heavy' and a unit of one-tenth of a bel – the decibel (dB) – is in general use.

$$\text{sound pressure level in dB} = 10 \log_{10}(P/P_0)^2$$
$$= 20 \log_{10}(P/P_0)$$

Therefore:

$$\text{dynamic range} = 20 \log_{10}[100/(20 \times 10^{-6})]$$
$$= 134 \text{ dB}$$

It should be noted that the threshold of hearing varies with frequency; it is about 20 dB at 100 Hz, i.e. the intensity must be much higher for the lower frequency sound to be heard. It will be seen later that the threshold for clicks is also much higher than for a 1000 Hz tone. Subjectively, an increase of 3 dB can just be distinguished and an increase of 10 dB sounds twice as loud.

The clicks used to evoke brainstem auditory potentials are generated by passing short-duration (10–250 µs) square pulses of current through earphones. The threshold of hearing for clicks is not the same as for a 1000 Hz tone (20 µPa). To compare results in different laboratories, the intensity of the clicks presented to the subject has to be measured on a scale that can be easily reproduced. Unfortunately there is, as yet, no standard technique for calibrating the intensity of click stimuli. Both behavioural and acoustical calibrations are used.

There are two methods of standardizing recordings using behavioural testing. In the first the threshold of each individual subject is measured by presenting a number of clicks (singly or repetitively at one or two per second) around threshold. This gives the 'sensation level' (SL) in decibels to which the required intensity is added. This is a very useful and often adequate measure for producing standard brainstem responses when they are used to test neurological (as distinct from audiological) function, but varies to some extent with the ambient noise level and, more importantly, depends on the ability of the subject to respond to the test procedure. To avoid this, some laboratories use the average threshold of 10 normally hearing young subjects in the 'standard' environment. This gives a 'normal hearing level' (nHL) against which the hearing of patients can be measured using brainstem auditory evoked potentials (BAEPs). This standardization has the disadvantage of being dependent on the ambient noise remaining the same. If the stimulus is changed (for whatever reason), recalibration means that another 10 normal subjects have to be tested – a time-consuming procedure.

An acoustic method of measuring click intensity would be very useful, but there is no simple technique for calibrating a very brief stimulus. The easiest way is to measure the peak sound pressure level – the 'peak SPL' – using a 'peak-hold' sound level meter. Alternatively, the 'peak equivalent SPL' can be measured. This is the r.m.s. SPL of a continuous pure tone having the same amplitude as the pulse.

Stapells et al (1982) measured the normal hearing threshold for clicks in these terms for 40 young adults, varying the duration of the listening period, the presentation rate, polarity and symmetry of the clicks. They showed that the average threshold using 100 µs square wave clicks presented at 10/s through a TDH-49 earphone was 36.4 dB peak SPL and 29.9 dB peak equivalent SPL; i.e. the threshold for clicks is about 30 dB higher than for a continuous tone at 1000 Hz. Thus, using a peak-hold sound level meter, clicks of any intensity as referred to the normal threshold of 36.4 dB can be presented to the patient. These measurements were done in a single-walled sound-attenuated chamber; ambient noise in another laboratory could alter the threshold. Accurate calibration of clicks is difficult.

There are small differences between the averaged brainstem EPs to condensation and rarefaction clicks (Maurer 1985) – rarefaction seems to be preferred, although Stapells et al (1982) showed that there is no difference in threshold for the two types. Some instruments alternate rarefaction and condensation clicks. Checking that the polarity is as marked on the auditory stimulator (condensation or rarefaction) can be done by displaying on an oscilloscope the output from a small loudspeaker (acting as a microphone) placed near each earphone in turn. When the loudspeaker microphone is 'hearing' condensation clicks coming from the energized earphone, the oscilloscope deflection should be in the same direction as when lightly tapping the cone of the speaker with the finger. (Most earphones 'ring' when energized with a narrow pulse and this may be seen on the oscilloscope – the initial direction is the one to note.) The EPs depend upon a sharp onset of the click; the earphones should have a good high frequency response.

The ear not being tested can be masked with 30–40 dB white (wideband) noise to prevent cross-hearing; though the effect is small it is important if the tested ear is deaf.

Tone Tone bursts or pips of 30–100 ms duration at 1000 Hz are also used as auditory stimuli. To avoid a sharp onset, the start of the tone is sometimes 'tapered', i.e. the intensity increases from zero to the required level over a few waves or milliseconds. The end of the tone is also tapered to zero. Relatively cheap sound level meters can be used to check the sound level of continuous tones.

Somatosensory stimulation

Electrical Somatosensory evoked potentials (SEPs) are usually evoked by electrically stimulating the nerve. The objective of the stimulation is to depolarize a small area of the membrane from its resting potential of 80 mV (inside negative) to its threshold of say 50 mV (inside negative), thus initiating an action potential which will be propagated along the nerve. Classical physiologists did this by applying a positive potential to an electrode inserted within the nerve (with respect to a cathode outside). The same effect can be obtained by generating a potential difference in the medium adjacent to the nerve using electrodes on the skin. The high resistance of the nerve membranes and myelin sheath prevents significant current flow within the nerve, and a potential difference is established across the membrane, thus causing a depolarization close to the site of the electrode connected to the negative pole of the stimulator (cathode). A nerve close to the positive pole (anode) will be hyperpolarized. Because of their higher resistance, myelinated fibres are easier to stimulate than unmyelinated fibres. Single monophasic pulses of about 100 µs duration are usually used to evoke the SEP.

When single pulses, or more importantly repetitive stimuli, are applied to needle electrodes, there can be a net flow of current in one direction giving rise to electrolytic action and disturbance of the chemical balance close to the electrodes. This can be avoided by using the so-called Lilly waveform which is composed of two consecutive square wave pulses of equal amplitude but of opposite polarity.

Stimulators can be designed to provide a constant voltage or a constant current output. A constant voltage stimulator will apply the selected voltage to the electrodes irrespective of their contact impedances – they could even be disconnected. A constant current stimulator passes the selected current through the electrodes whether they are in good contact (low impedance) or poor contact (high impedance) with the tissue. These different features are obtained by designing the stimulator to have a high output impedance for constant current and a low output impedance for constant voltage. The manufacturer sets limits to the maximum current that a constant voltage stimulator can generate and to the maximum voltage that can be generated by the constant current device. The constant current device is preferred, as the biological effect is more consistent. In both types the impedance of the stimulating electrodes should be kept low (as with recording electrodes).

Transformer coupled devices have output impedances somewhere between the two stimulators described above. Their main advantages are that there is no net DC through the electrodes and that the patient is isolated by the output transformer from the mains-operated part of the stimulator. The latter is beneficial both from the safety point of view and because it reduces the mains interference that can get into the amplifiers via the stimulating and recording electrodes. Isolation in the constant current and constant voltage stimulators is achieved by optically coupling the mains-driven side into the output stage which has its own battery power source.

Leads from the stimulator to the electrodes can carry considerable and rapidly changing currents and are a source of 'stimulus artefact'. Although the pulses are very narrow, any leakage can cause the amplifiers to go outside their working range and early SEP components can be lost while the amplifiers recover from the overload. The artefact can be reduced by twisting the stimulating leads together; they should not be shielded with a screened cable that is connected to the earth of the recording system. An earthed plate or ring electrode around the limb between the stimulating and recording electrodes can reduce the artefact.

Electrical stimuli bypass the peripheral encoding of natural stimuli (pressure, vibrations, joint movements) by the receptors. The different categories of fibres, which subserve the different types of sensation, can be individualized on the basis of their diameter and thickness of their myelin sheets. There is an inverse relation between the fibre diameter on the one hand and its threshold to electrical stimulation and conduction velocity on the other. In most clinical applications of SEPs electrical stimuli are delivered at intensities equivalent to 3 to 4 times the sensory threshold, which produce a twitch in the muscles innervated by the stimulated nerve when it contains a contingent of motor fibres. At this stimulus intensity the rapidly conducting large myelinated fibres, including fibres conducting skin and joint and also muscle inputs, are activated because of their higher resistance.

Transcutaneous stimulation of the motor cortex

Measurement of conduction in central motor pathways became feasible as a clinical tool, with the development by Merton and Morton (1980) of a high-voltage electrical stimulator. Prior to this, investigation of corticospinal tract conduction in the conscious intact subject had either relied on indirect techniques such as F-wave or long-loop studies (Eisen 1986), or had proved unacceptably painful because of the necessity for rapid trains of electrical stimuli (Gualtierotti and Paterson 1954). Direct stimulation of the exposed motor cortex had of course been in use for many years during neurosurgical procedures. Merton and Morton's device enabled the latency and amplitude of compound muscle action potentials to be measured, permitting physiological studies of a type which previously could only be performed in animals. Furthermore, a technique thus became available for clinical studies of the integrity of central motor pathways. However, despite modification of the original technique in various ways, local pain has limited its clinical applicability. The introduction by Barker and coworkers of a magnetic stimulator capable of exciting the motor cortex transcutaneously without causing pain, has revolutionized clinical studies (Barker et al 1985).

Electrical stimulation A high-voltage, low-output impedance electrical device is used, delivering stimuli of 300–700 V, with a brief rise time (50 μs or less) and a time constant of decay of 50–100 μs. As was originally shown by Fritsch and Hitzig and also by Ferrier in the 1870s, anodal stimulation of the motor cortex is more effective than cathodal stimulation (Day et al 1989). The anode is a surface electrode placed over the motor area of the scalp and the cathode may be a similar electrode placed a few centimetres anterolateral to the anode or at the vertex; alternatively, the cathode may take the form of a band or chain of plates around the scalp. The use of a large band cathode ('unifocal' method (Rossini et al 1985)) reduces local discomfort to some extent.

Percutaneous electrical stimulation of the corticospinal tracts at the level of the pyramidal decussation may be performed by placing the anode and cathode on the posterior aspects of the mastoid processes (Ugawa et al 1991). Placement of the electrodes over the spinal column enables the corticospinal tracts to be stimulated at cervical and thoracic level (Marsden et al 1982, Snooks and Swash 1985) using the same device.

Magnetic stimulation The magnetic stimulator relies on the fact that a stimulating electrical current is induced within conductive tissue by a brief and time-varying magnetic field. A current of several thousand amps is passed through a copper coil when a capacitor charged up to 4 kV is discharged, producing a magnetic field of the order of 2 tesla (T) with rise time up to 200 μs. The induced current in the underlying brain tissue when the coil is held close to the scalp excites the motor cortex and enables muscle action potentials to be recorded in limb or trunk muscles. The magnetic field passes unattenuated through the scalp and skull and current densities in these tissues are low, causing no pain.

Magnetic stimulation can also be used to stimulate other parts of the brain. Currently available devices cannot stimulate the spinal cord but, when positioned over the spinous processes, can produce muscle responses by stimulating the ventral roots a few centimetres from the cord itself. Phosphenes are evoked by stimuli over the occipital region, and magnetic stimuli presented 80–100 ms after a visual stimulus can suppress visual perception (Amassian et al 1989, Maccabee et al 1991). Effects on the oculomotor areas of the cortex and on the cerebellum are also demonstrable (Britton et al 1990, Priori et al 1991).

Safety considerations with magnetic stimulation Current clinical experience indicates that adverse effects of magnetic stimulation of the motor cortex are extremely rare. However, there have been isolated reports of focal seizures occurring during or immediately after magnetic stimulation in patients with ischaemic lesions of the cortex (Homberg and Netz 1989), and epilepsy should be regarded at present as a relative contraindication. Paradoxically, a recent study (Hufnagel et al 1990) suggests that magnetic stimulation may possibly be a useful tool for localization of epileptic foci in patients undergoing presurgical evaluation, since repeated stimuli may activate foci. However, seizure activation which would be of greater value for preoperative assessment is very rarely achieved (Hufnagel and Elger 1991). Dhuna et al (1991) found that trains of stimuli at 8–25 Hz failed either to activate the epileptogenic focus or to induce the habitual seizures in any of the eight patients, but a partial motor seizure, arising contralateral to the known temporal focus, was induced in the only patient subjected to maximal stimuli. Implanted metal structures within the brain, such as aneurysm clips, are subject to mechanical forces, albeit small, from the magnetic field, and patients who have previously undergone neurosurgery should not be stimulated without consideration of this, and nor should patients with cardiac stimulators (Cadwell 1990). The low frequency of repetitive stimulation (around 0.3 Hz) means that the theoretical risk of kindling an epileptic focus is remote with either magnetic or electrical devices: the total energy delivered is small in comparison with the electrical stimuli administered during electroconvulsive therapy (Barker et al 1988). A comprehensive review of safety recommendations for repetitive transcranial magnetic stimulation has been published by the International Federation of Clinical Neurophysiology (Hallett et al 1999).

The passage of current through certain stimulating coils causes a loud 'crack', giving rise to concern that this noise could cause acoustic damage. Experiments by Counter et al (1990) on rabbits showed that the acoustic artefact reached peak levels of 140 dB or more at the tympanic membrane and that morphological damage could occur. They recommend the wearing of ear protectors by patient and clinician during magnetic brain stimulation. There is some disagreement about the need for this in adult subjects (Barker and Stevens 1991, Boyd et al 1991), but ear plugs may be advisable when babies are tested.

EMG AND NERVE CONDUCTION STUDIES

Electrodes for EMG and nerve conduction studies

Surface electrodes

Choice of metals for use as surface electrodes for recording EMG or nerve conduction studies (NCS) are not critical because of the short time constants used, but pairs of electrodes made of dissimilar metals should be avoided. Traditionally the recording surfaces of the metal recording electrodes were coated with thin film of conducting gel and attached to the patient's skin with medical adhesive tape. Recently disposable 'peel off' adhesive conducting electrodes have been introduced by many manufacturers which can be connected to the preamplifiers by a crocodile-type clip. These electrodes have many advantages in avoiding any smearing of the electrode gel and are disposable after each patient, thus reducing any possible patient to patient cross-infection risk.

Electrodes used for recording compound muscle action potentials should not be placed too closely, as this will lead to activity common to both that will be attenuated by the differential amplifier. Surface electrodes used to record nerve action potentials should be placed over the nerve being investigated (*Tables 2.6 to 2.8*), with a separation of between 30 and 90 mm. Gilliatt et al (1965) showed that peak-to-peak amplitude altered and duration increased as the inter-electrode recording distances increased.

Concentric needle electrodes

The electrode most commonly used in the EMG is the concentric needle electrode (CNE). This was described in a paper by Adrian and Bronk, in 1929, as being made from 'a central enamel covered wire passed through a hypodermic needle and connected to the amplifier input'. This essential design has endured well and the modern CNE is is constructed on very similar lines (*Fig. 2.28*).

The recording surface is made by grinding the tip at 15° to give an elliptic area of 580 × 150 µm. This picks up activity from fibres that lie within a hemisphere of about 0.5 mm radius. Since muscle fibres have diameters of between 25 and 100 µm, the needle records activity from about 20 fibres; fibres further away make a minimal contribution to the recorded potentials. The number of motor units recorded depends both on the local arrangement of muscle fibres within the motor unit and on the level of contraction of the muscle.

Single fibre needle electrodes

The single fibre needle was developed in Uppsala, Sweden, by Ekstedt and Stålberg in the 1960s and knowledge of its structure and recording characteristics are crucial to an understanding of single fibre EMG. A single fibre needle electrode has similar external proportions to a concentric needle electrode, being made of a steel cannula 0.5–0.6 mm in diameter with a bevelled tip. However, instead of having the recording surface at the tip, a fine, insulated platinum or silver wire embedded in epoxy resin is exposed through an aperture on the side of the needle, 1–5 mm behind the tip (*Fig. 2.29(a)*). This recording surface is on the side opposite to the bevel to avoid recording from fibres that could have been damaged by insertion. The steel cannula acts as the reference electrode.

Table 2.6 Stimulation and recording sites of compound nerve action potentials

Nerve	Stimulation site	Recording site
Median	Wrist	Flexor surface at elbow
Ulnar	Wrist	Ulnar groove behind elbow
Common peroneal	Ankle	Fibular neck

Table 2.7 Sensory action potentials recorded with surface electrodes

Upper limb	Lower limb
Median	Sural (proximal and distal)
Ulnar	Plantar (medial and lateral)
Superficial radial	Saphenous
Dorsal ulnar	Superficial peroneal
Musculocutaneous	Lateral cutaneous nerve of thigh

Table 2.8 Small distal muscles suitable for recording CMAPs with surface electrodes

Muscle	Innervated by
Abductor digiti minimi	Ulnar
First dorsal interosseous	Ulnar
Abductor pollicis brevis	Median
Extensor digitorum brevis	Deep peroneal
Flexor hallucis brevis	Tibial
Abductor digiti minimi	Tibial
Extensor indicis	Radial

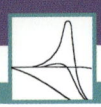

Needle recording electrodes	Needle tip and recording surface	Pick-up	Needle diameter	Filter settings	Activity recorded
Concentric needle electrode Central insulated platinum wire inside a steel cannula		Hemisphere radius 0.5 mm	0.3–0.65 mm	10 Hz to 10 kHZ	Motor units
Single fibre needle electrode Fine platinum wire (25 μm diameter) inside a steel cannula, which records from a steel aperture		Hemisphere radius 250–300 μm	0.5–0.6 mm	500 Hz to 10 kHZ	Individual muscle fibres of motor units. In health the potentials are either single or pairs; after reinnervation the potentials have multiple components
Monopolar needle electrode Sharpened stainless steel needle insulated down to 25–50 μm from tip Subcutaneous or surface reference		Sphere radius 200 μm	0.3–0.5 mm	2 Hz to 10 kHZ	Motor units of higher amplitude and more complex waveforms than those recorded with a concentric needle, but similar duration

Fig. 2.28 Three different types of recording electrode, their physical characteristics, the filter settings required for use and the nature of the activity that each records. (From Binnie et al (2004), by permission.)

The platinum wire which forms the recording surface has a diameter of 25 μm and will pick up activity from within a hemispherical volume, 300 μm in diameter (Stålberg and Trontelj 1994). This is very much smaller than the volume of muscle tissue from which a concentric needle electrode (with an uptake area of 1 mm diameter) records. Because of the arrangement of muscle fibres in a normal motor unit, the geometry of the single fibre needle will record from only 1–3 single muscle fibres which belong to the same motor unit (*Fig. 2.29(b)*). Because of the recording characteristics of the needle, the potentials are of comparable or even higher amplitude than individual motor units, it being not unusual to record potentials of up to 10 mV from single muscle fibres.

Amplifiers for EMG and nerve conduction

Recording equipment for EMG studies and NCS must be capable of amplifying signals with amplitudes of between a few microvolts to around 100 mV and a bandwidth of 1 or 2 Hz to 20 kHz.

Recordings are made in laboratories in which both electrostatic and electromagnetic fields are often sufficient to introduce interference in the recordings. In EMG equipment, the first stage of amplification is done with a preamplifier which is placed close to the patient so that electrode leads can be kept short. The cable between the preamplifier and the main amplifiers carries signals from low-impedance sources. Both these features help to reduce pick-up of mains and other interference.

The impedance of small needle electrodes can be large (mega-ohms) and the input impedance of the amplifiers should be very high, 100 MΩ or more, to avoid loss of the signal. The CMRR should also be high, so that the in-phase mains interference is minimized, especially when using high-impedance electrodes. The amplifiers should have a wide range of stepped gain controls. Calibration of the whole system should be possible, although drift of gain

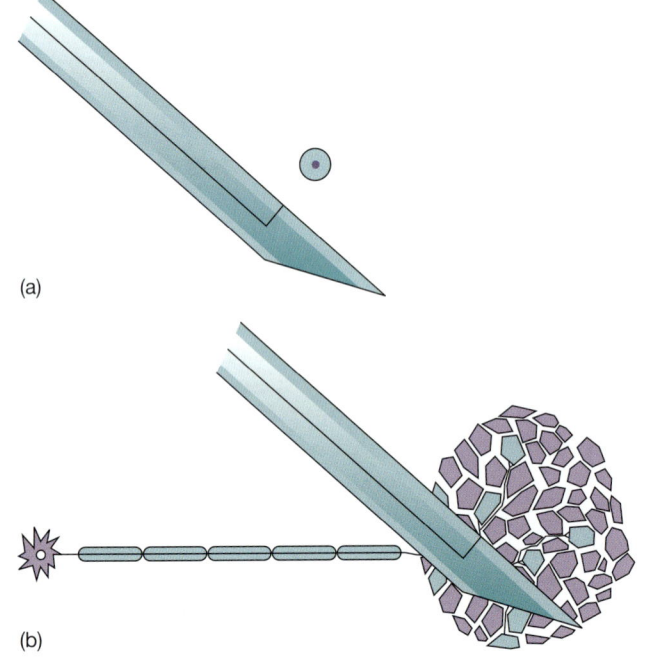

Fig. 2.29 (a) Single fibre recording electrode. The small circle with the central 'dot' shows the recording surface of this type of needle electrode. (b) Because in health the muscle fibres of a motor unit are usually not adjacent to one another, a single fibre needle electrode with a highly restricted recording surface mostly records only single or double potentials. (From Binnie et al (2004), by permission.)

of amplifiers is unusual in modern EMG equipment. Most investigations can be done using one or two recording channels, although multiple channels may be useful for detecting evanescent phenomena such as fasciculations or neuromyotonic discharges.

Both low-pass and high-pass filters should have a number of set positions, so that the upper frequency limit can be reduced from 10 or 20 kHz to 100 Hz and the low frequency limit raised from 0.1 Hz to 1 kHz. Filters with roll-off slopes of 6 dB/octave are adequate.

Displaying EMG and nerve conduction data

The electrical activity is displayed on an oscilloscope either as the 'raw' signal or after some processing such as averaging. There should be a wide range of time bases from milliseconds to seconds.

In modern systems the data are digitized and manipulated by digital techniques before being displayed on a screen. Modern computer based machines allow digital storage of the data and also the acquisition and display parameters and can produce high quality printouts using modern laser or ink jet print technology. Multiple traces can be displayed on the screen at the same time so that sequential acquisitions in the same investigation can be examined. Multiple windows allow the data to be viewed simultaneously on different time scales and display modes, for example a stacked raster or superimposed display is now frequently used for F wave recordings. Latencies and amplitudes are obtained by measurement from the displayed data often using algorithms automatically to set the cursors. It is important that those using this facility assure themselves of the accuracy of the automatic cursoring and also that the expected parameter is being measured, e.g. the baseline to negative peak amplitude and not the peak to peak amplitude, or vice versa.

The new range of EMG machines are 'menu driven' and on selecting a test to run, all filters, gains and display modes are automatically set. Some flexibility must be preserved, and although it remains possible to alter most of the recording and stimulation parameters, there is often less choice as to the display format. In addition to automatic settings, many of the new machines have integral processing units so that on-line analysis is also available. Resident programs can perform individual motor unit analysis, turns/amplitude analysis or power spectrum analysis of EMG, and jitter and fibre density measurements from single fibre needle recordings.

Stimulation for nerve conduction studies

Nerve conduction studies require a nerve to be electrically stimulated using a narrow pulse (about 100–200 μs wide); some investigations require repetitive stimulation and the machine should have a range of stimulus repetition rates available (0.5–50 pulses/s). The stimulation pulse can be from a constant current or constant voltage device and the relative merits of each are discussed by Smith (1997a).

When using single pulse stimulation, the trigger point can be advanced or delayed from the start of the time base to facilitate viewing data on the screen. Techniques for recording F waves and the H reflex are given in *Tables 2.9* and *2.10*.

PRACTICAL ASPECTS OF INTRAOPERATIVE AND ICU NEUROPHYSIOLOGICAL RECORDING

Safety during neurophysiological recording in the ICU and operating theatres

The whole ethos of development of modern monitoring equipment for work in the ICU, and indeed also in the operating theatre, is towards the goal of improved safety of care and quality of outcome. This aim may be focused on the improved management of specific medical conditions and procedures or directed towards the overall acquisition of information for decision making. Work should be clinically relevant and of evidence-based efficacy (i.e. with high specificity and sensitivity for the purpose in hand) and provided at reasonable cost in terms of equipment and staffing requirements. At a more specific level, safety implies careful attention to the principles of elec-

Table 2.9 Method of recording F waves

1. Place surface electrodes over a chosen intrinsic hand or foot muscle, as for motor conduction studies
2. Orientate the stimulator so that the cathode is closest to the spinal cord, i.e. the opposite direction to that used for motor conduction studies
3. Set the time base of the EMG machine to 5 or 10 ms/division, with an amplitude gain of 200 or 100 μV/division
4. Adjust the stimulator to give supramaximal stimuli at a rate of 1 Hz or less
5. Record 20 responses using either a cascade or superimposition display
6. Measure the subject's height or limb length

Table 2.10 Method of recording the H reflex

1. The subject lies prone with ankle supported
2. Surface electrodes are placed over soleus
3. Set the time base of the recording unit to 10 ms/division
4. Stimulate the tibial nerve in the popliteal fossa with a pulse of 0.5–1.0 ms duration and at a rate of 0.5 Hz or less, while gradually increasing the intensity
5. With low-intensity stimulation (usually less than that which is required to elicit the M response), the H reflex appears. With increasing stimulus intensity, the amplitude of the M response increases while that of the H reflex diminishes, but the latency of the H reflex will remain

trical and physical safety and infection control. No patient, relative or member of staff should be put at immediate or long-term risk from any form of hazard posed by investigations or care in the ICU. Safety considerations pose particular problems when electronic equipment and even minimally invasive techniques are used in the ICU. It is always essential to plan each and every neurophysiological investigation with ICU staff and to check carefully for unexpected hazards.

General care of the patient

Training in appropriate safety routines and general conduct in the ICU should be arranged with the head of the unit; this usually leads to a reciprocal request from ICU staff for regular updates on neurophysiological matters relevant to their work! Such a dialogue certainly results in better mutual understanding of what can be achieved; it also makes a strong basis for introductory discussion and planning when neurophysiological procedures are requested in an individual patient. Intensive care is undertaken because patients are critically ill and dependent upon external support for one or more vital functions. The aim of ICU staff is to assist patients through this period until they are well enough to be discharged from the unit with the prospect of recovery with a good quality of life. It is essential that any neurophysiological procedure, whether diagnostic or to help hour-by-hour management, should not jeopardize these aims. Great attention should be paid to avoidance of physical interference with the medical and nursing care of the patient. It is neither safe nor professionally responsible to remove a patient's head dressings without permission, or accidentally to disturb vascular lines or airway connectors, or indeed to demand that other equipment be turned off 'to reduce interference'.

It is essential to avoid undue stress to the patient by overenthusiastic neurophysiological procedures which might cause distress or unwarranted risk to precarious function of the heart or brain. Likewise, it is inappropriate to alter medication specifically to 'clean up' the EEG, for example by use of muscle relaxant drugs to reduce excessive myogenic artefact. One report details the catastrophic consequence of this practice when a patient developed a fatal cardiac arrest attributable to such a procedure (Verma et al 1999). Even therapeutic trials under EEG cover, for example to establish a diagnosis of possible non-convulsive status epilepticus in an apparently comatose patient by testing the effect of intravenous diazepam, should be limited to occasions when the clinicians in charge of the patient specifically agree to and share management of the proceedings.

Great sensitivity and good communication skills are required for neurophysiological work in the ICU. Problems may arise when the classical neurophysiological approach of demonstrating reproducibility of findings in a 'steady state' is adopted, whether it is for EEG, EP or EMG work. The ICU provides constant change – in the clinical condition of the patient, in the environment and in the level of tolerance of individuals (patients, relatives or ICU staff) to prolonged examinations on critically ill patients. Rightly, there is a high level of protectiveness towards the best interests of the patient; this is accompanied by an unavoidable level of emotional tension when frivolous demands for 'just one more go' conflict with the more basic needs or dignity of a critically ill or dying person.

Electrical safety

It cannot be emphasized too strongly that it is essential to obtain specific expert advice on electrical safety from the appropriate hospital safety officer before conducting any form of monitoring work in the ICU or operating theatres. Practical, clinically orientated, accounts are given by Bruner and Leonard (1989) and Billings (1994).

The widespread use of electromedical equipment has increased the possibility of electrical faults and injury to patients. Modern equipment is extremely reliable and the risk of injury in a modern neurophysiological laboratory is extremely small. The danger is greater during intensive care monitoring and recording in the operating theatre when several electrical instruments may be connected to the patient for long periods, but it is still very small compared with the hazards of surgery and anaesthesia. However, although the risk is small, in view of the expensive legal proceedings that can be initiated for even trivial accidents, it is essential to take 'reasonable precautions'. Safety specifications for electromedical equipment are detailed in the International Electrotechnical Commission Standard IEC 60601-1 (1988) (in the UK the British Standard BS EN 60601-1 (1990) applies). Particular requirements for EEG machines were published as IEC 60601-2-26 (2002) and adopted as European Standard 60601-2-26 (2003). Note: the latest IEC publications can be found at: http://www.iec.ch. American standards are described in American Electroencephalographic Society Guideline Nine (1994a).

The main hazard, as in any piece of household electrical equipment, is that arising from the inadvertent passage of electrical current through the body. Normally the high resistance of the skin offers sufficient protection, provided that the body does not touch those parts of the apparatus which are at high voltage. Low-voltage sources, such as torch batteries, can be handled without harm. However, during electrophysiological recording the situation is very different, as the equipment is deliberately connected to the body with low contact resistance.

Electrical currents passing through the body may cause pain, burns, respiratory failure or ventricular fibrillation. The current depends upon the magnitude of the applied voltage and the impedance of the tissue through which it flows. Considerable current has to flow (tens of mA) through skin to produce a burn, but less than 100 µA of 50 Hz current applied directly to the right ventricle is sufficient to disturb the cardiac rhythm. This is substantially less than the current that would flow if only 1 V was connected across a pair of ECG electrodes carefully applied to each arm (not all passing through the right ventricle of course).

IEC 60601-1 (1988) states: 'Equipment shall be so designed that the electric shock in normal use and in single fault condition is obviated as far as practicable.' There are two types of faults that may give rise to injury. In the first an electrode may become connected to a relatively high voltage source (with respect to earth) so that there is current flow through the electrode on the body to the earth electrode. This could be caused by a fault in the equipment. The second kind of fault is due to the earth connection becoming disconnected within the apparatus, at the mains outlet or in the electrical wiring system. This defect can easily escape detection, unless routine safety checks are performed. If a second fault then develops it is possible that the equipment casing or chassis or patient 'earth' connection could be at mains voltage without the fuse blowing. If the patient (or operator) then touches items of equipment or water pipes that are earthed, lethal current will flow.

Note that an additional hazard exists if a possible path to earth of especially low resistance is created when the skin is breached by the use of intravenous catheters, needles, etc. Under such circumstances extra precautions are desirable, such as the use of optically isolated equipment.

In most modern electrophysiological apparatus such accidents are prevented by isolating the electrodes and preamplifiers from the main amplifiers, displays and power supplies (where high voltages exist) by optical or high-frequency transformer coupling. IEC 60601-1 (1988) recommends that the maximum current which leaks to earth should not exceed 100 μA at 50 Hz in normal circumstances and 10 μA where the current could pass directly through the heart. A single fault, such as an interruption of one of the supply conductors or a protective earth conductor, may increase the leakage current. The leakage current should be checked *and recorded* at regular intervals by an authorized person. Note that leakage current may increase when mains leads are lengthened by the use of extension cables, or when the recorder is connected to other apparatus (computers, tape recorders, etc.) not designed for biomedical use.

When two or more items of mains-operated equipment are connected to a patient they should be plugged into the same mains supply. This is because different earth points within a building may not be at the same potential and because of the risk of an accidental shorting of one of the mains supplies to earth. Apart from the safety aspect, this single-point earthing helps to minimize the interference that is developed from currents in the 'earth loops'. Only one earth electrode should be on the patient and the bed should be positioned so that the patient cannot reach out and touch water pipes or the metal cases of equipment.

IEC 60601-1 (1988) also specifies standard symbols that should be affixed to modern electromedical equipment to indicate the type of protection afforded and, for example, whether it is sufficient for use in patients with indwelling cardiac pacemakers, etc.

Pulsed electrical stimulation as used in nerve conduction studies and in the evocation of SEPs could affect the operation of cardiac pacemakers, especially those that detect atrial and ventricular excitation separately and those that detect and correct tachyarrhythmias (Smith 2000). It is recommended that when the above procedures are to be implemented in patients with programmable pacemakers, further advice is sought, preferably from a cardiologist.

For further information see Bruner and Leonard (1989).

Particular physical hazards and safety considerations

In addition to the general safety measures routinely observed when recording the EEG, there are also some particular hazards that may be encountered during EEG monitoring in the operating theatre (Prior and Maynard 1986, Bruner and Leonard 1989). These include surgical diathermy (electrocautery), defibrillators and fibrillators, possible hazards from widely spaced electrodes, flammable vapours, static electricity, sparks, high temperatures and mechanical hazards. The general principles underlying electrical safety during EP monitoring are expounded with great clarity by Burke et al (1999) are highly pertinent to EEG monitoring. Protection of the patient from the power supply by input isolation is important and is mandatory if non-medical equipment, such as a computer, is connected to a non-isolated medical device.

Special knowledge and technical advice from medical physicists about safety in the operating theatre is needed and recordings should be made by experienced neurophysiology technologists with expertise in operative monitoring. The potential problems facing the inexperienced are such that some workers, even in centres with a strong tradition of monitoring, have suggested that it is virtually impossible to obtain 'clean' EEG recordings in the operating theatre or ICU because of external electrical interference from other equipment. One group has cited surprisingly high levels of up to 40% of recordings being contaminated (Bashein et al 1992). Naturally surgery involving extensive use of electrocautery (surgical diathermy) will lead to considerable contamination from that source, for example when opening the chest for 'open heart surgery', but the times of risk to the brain are mainly when the need for diathermy is minimal. In neurosurgery those with experience from EP monitoring have identified some rarer sources of interference, such as head rests and self-retaining retractors, both of which, in spite of proper grounding procedures, seemed to pick up electromagnetic radiation originating from electrocautery and other sources (Moritake et al 1986). These authors reported that such artefacts could be minimized by grounding the head pins at each point of insertion into the patient's scalp; however, they still had to use additional artefact rejection by voltage-overload sensing to eliminate the effect of extraneous mechanical and electrical interference which 'occur inevitably' in surgical procedures. The general opinion of those working regularly in an intraoperative neurophysiological monitoring team is that, apart from difficulties during periods when surgical diathermy or cardiac fibrillation or defibrillation are in use, experienced technologists have an armamentarium of checks and routines which enable them to ensure an acceptable level of stable and reliable EEG and EP recordings.

Since some flammable vapours, for example from volatile anaesthetic agents such as ether or cyclopropane, may still be encountered in the operating theatre, monitoring equipment must be checked for markings as to whether or not it is safe to be used in the presence of flammable vapours (the AP or APG marks of the IEC (IEC-60601 1990). Specific professional advice about explosion risks during anaesthesia is available (e.g. Association of Anaesthetists of Great Britain and Ireland 1971). Precautions must also be taken to avoid static electricity, sparks and unduly high temperatures which might ignite flammable vapours. Reduction of static charges on patients or staff are usually avoided by antistatic drapes and gowns and the use of a fairly humid atmosphere. Sparks and high temperatures may arise from VDUs, thermal chart recorders and computer printers, quite apart from problems due to accidental ingress of metal objects or fluids into electronic equipment. National standards should be followed on these issues.

Operating theatres are often crowded with personnel and equipment with associated cables and tubes, usually with multiple connections to the patient. These can pose considerable mechanical problems for the neurophysiology team who, in addition to adding to the difficulties by their own presence and equipment, may be unfamiliar with local theatre routines. A preliminary induction to appropriate behaviour should be sought from the chief nurse in charge of the operating suite so that specific dangers to the patient from disconnection of intravenous or other lines, ventilator tubes and monitors can be avoided along with the parallel danger of distracting or irritating surgical and anaesthetic teams. Neurophysiology cables should be clearly labelled and placed so as to avoid risk of impeding access to the patient or of detachment.

Infection control

The acquired immune deficiency syndrome (AIDS) epidemic has served only to highlight the long-standing problem of the risk of transmitting infection to patients or personnel through the use, and particularly the re-use, of electrodes. Apart from human immunodeficiency virus (HIV), the infections most likely to be transmitted are Creutzfeldt–Jakob disease (CJD) and viral hepatitis. More recently, methicillin-resistant *Staphylococcus aureus* (MRSA) has become a major concern among hospital infections, as it is difficult to treat and can be transmitted directly from patient to patient, or via staff or infected electrodes. Detailed recommendations for sterilization of electrodes and other measures have been published by the American Electroencephalographic Society (1994d) and by the Association of British Clinical Neurophysiologists and the British Society for Clinical Neurophysiology (Evans et al 1993). Approved practices are likely to change and to differ between countries and institutions and (not least for medico-legal reasons) the reader is advised to determine and follow current local regulations. Only some very general principles and typical practices will be summarized here.

Although some special precautions may be taken in dealing with known carriers of infection, the only safe practice is to assume that a risk is always present. Electrodes which are intended to penetrate the skin expose the technologist to the risk of infection by needle stick or by blood coming into contact with broken skin. There is also obviously a risk of patient-to-patient transmission by the electrodes themselves. Skin preparation before use of surface electrodes may result in bleeding or oozing of serum which may contaminate both the electrodes and the hands of the operator. Needles used for applying jelly under scalp electrodes are particularly likely to become contaminated.

Any lesions (cuts, abrasions, burns, eczema) on the operator's hands should be covered with a waterproof dressing. Disposable gloves should be worn during electrode application, removal and cleaning. Particular care should be taken to avoid touching other equipment when wearing contaminated gloves. If possible, sinks used to clean electrodes should not be used for hand washing. Electrode application to patients known, or suspected to be infected with MRSA should be carried out with full physical protection: mask, gown, plastic gloves and plastic apron. EMG or scalp needle electrodes should be cleaned in an ultrasonic cleaning bath and autoclaved (e.g. at 121°C for 15 min) after every use (see Al-Seffar (1990) for an evaluation of the economics of EMG needle maintenance). Needles for introducing jelly require similar precautions and should be flushed with clean water before sterilization, or should be disposable. The use of needle electrodes in high-risk patients should be avoided if possible. Scalp disc electrodes should be cleaned in hot detergent after every use, including brushing to remove fragments of jelly or adhesive, and the contacts, but not the leads, immersed in 10 000 ppm sodium hypochlorite for 10 min after each use. Pad electrodes should be immersed in boiling water for 3 min after each use. After use in patients known or suspected to be infected with CJD or MRSA, disc electrodes should be autoclaved at 134°C at 30 psi for 18 min or for 6 cycles of 3 min each. This is practicable only if the insulation of the leads has a high melting point (e.g. Teflon); it may be simpler to discard the electrode into a suitable container for incineration. All electrodes should be washed thoroughly after immersion in disinfectant. Intracranial electrodes carry a high risk of transmitting CJD, particularly if used for chronic recordings, and should be disposed of after use unless they can be autoclaved as indicated above.

Equipment should be swabbed with 10 000 ppm sodium hypochlorite if contaminated by 'high-risk' biological fluids (blood, semen, female genital tract secretions, CSF, amniotic fluid, pericardial fluid, synovial fluid, etc.; saliva, urine and faeces are not normally considered risky for blood-borne viruses unless visibly blood-stained, although other infections such as *Salmonella* and tuberculosis could, of course, be present). As sodium hypochlorite corrodes metal it should be washed off after 10 min. Glutaraldehyde (2%) is a potent and useful disinfectant for all except CJD, but is not suitable for routine swabbing as it is likely to sensitize staff. Contaminated bedding, etc. should be handled according to locally recommended procedures.

Finally, it should hardly be necessary to mention the importance of following infection precautions in a manner which respects the dignity of the patient. This is more likely to be achieved where the measures are routine than when they are adopted only exceptionally when the perceived risk is high.

Basic checkpoints for avoidance of transmission of infection (to patients or staff) by electrodes include:
1. Swabbing all apparatus with the appropriate disinfectant before and after it is taken to the ICU or operating theatre.
2. Ensuring routine use of proper infection control procedures (such as hand-washing, wearing sterile gloves, no touch techniques) by staff applying or attending to electrodes or other devices attached to the patient.
3. Using only disposable electrodes and associated materials.
4. Avoiding skin penetration, e.g. by needle electrodes or by scalp abrasion to improve contact impedance of disc electrodes used for EEG or EP monitoring.

Intraoperative and ICU EEG recording
Electrodes for intraoperative EEG recording

Electrode application requires both expertise and time both for ICU monitoring work that often continues for many hours or days at a time and for intraoperative recording when access may be limited during surgery and electrodes may be at risk of displacement when the patient is moved. Electrodes should be attached by experienced clinical neurophysiology technologists; when this is impossible, they should provide regular training and support for designated ICU or anaesthetic staff. Suitably low electrode contact impedance is required for avoidance of artefacts, e.g. those due to pulsation during cardiopulmonary bypass (Levy et al 1980). In the ICU, attention to electrodes is needed at least once a day to ensure good quality, trouble-free, recordings and their contact impedance should be continuously recorded on the monitoring apparatus.

In all cases, preliminary skin preparation and careful measurement and marking of electrode placement sites are required. For intraoperative work, this is usually undertaken in the anaesthetic room or, in some circumstances, at the patient's bedside in the ward in the course of preparation for surgery or pre-anaesthetic baseline recording. In spite of the possible difficulty involved in placing electrodes in seriously ill ICU or surgical patients, conventional methods of proven efficacy should be used. Most commonly this means securely self-retaining EEG electrodes, such as the traditional cup or disc electrodes held in place with collodion or one of the modern combined contact and adhesive pastes. Modified electrocardiography (ECG) electrodes are generally inadequate for ICU and surgical work, although comparative studies by Seitsonen et al (2000) suggest that with careful skin preparation with both abrasion and electrode paste, inexpensive pre-jelled ECG electrodes can achieve low skin–electrode impedances comparable to those with EEG electrodes for monitoring depth of anaesthesia with a Bispectral (BIS) Index Monitor (Aspect Medical Systems Inc., Newton, MA. USA. http://www.aspectmedical.com). This is certainly an area where new developments of simpler disposable but secure electrodes would be welcome, but as yet no ideal solution has been found. Care must be taken to ensure that performance at least matches that of traditional types and that there are no unexpected adverse reactions (Litscher et al 1996). Consideration of the requirements for electrode removal after intraoperative monitoring include the avoidance of flammable vapours in the operating suite and the need to leave the patient's scalp and hair clean and tidy after surgery.

Because of the dangers of bleeding and transmission of infection in ICU and surgical patients, neither needle electrodes nor scalp abrasion to reduce electrode contact impedance should be used unless there are exceptional circumstances, for example, major injury of the scalp when disposable needles may be considered as being the least damaging. Reusable needles should never be used because of dangers of transmission of infective agents; even single-use needle electrodes hold the potential danger of needle-stick transmission of infection to staff. Infection control measures are discussed in *Particular physical hazards and safety considerations* (p. 57).

The limitation in the number of channels for EEG monitoring, most commonly to two or four, means that electrode placement for ICU work also has to be planned with care. Whilst standardization of electrode positions at carefully measured sites reduces difficulties in interpretation, important considerations are the likely distributions of both the changes to be monitored and of the biological artefacts likely to be encountered. Eye movements, frontalis muscle potentials and nursing or medical attention to the upper part of the face may increase artefact and make the choice of frontal electrode placements unsuitable, even though convenient to access (*Fig. 2.30*) (De Deyne et al 1998). Likewise, for prolonged monitoring, occipital placements may be subject to pressure, causing discomfort and artefacts. The amount of artefactual contamination at the standard electrode sites of the 10–20 system has been studied in 150 ICU recordings; C3–P3 and C4–P4 were the least affected (Schultz et al 1992).

Comparative studies between the full 10–20 system placements and limited montages suggest that bilateral centroparietal placements are appropriate for most monitoring during anaesthesia in patients without intracranial

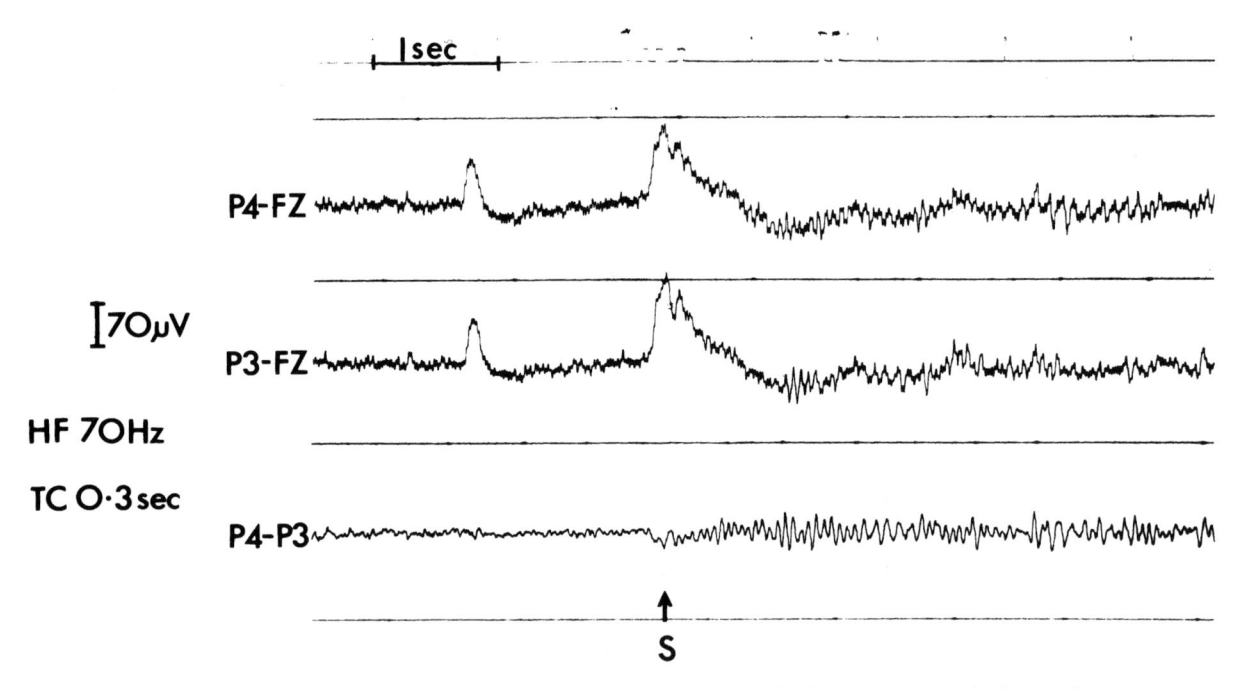

Fig. 2.30 Scalp distribution of corneoretinal potentials associated with blinks and other eye movements. S, eye closure. (From Prior and Maynard (1986), by permission.)

or carotid artery pathology (Schultz et al 1995). However, a larger number of electrodes and appropriate montages for detection of lateralized or regional changes are required during carotid artery surgery; preoperative screening is also recommended to provide a presurgical baseline. Limiting electrode placement to the forehead for convenience of access, as some have suggested, cannot be recommended. This is because (a) alterations with induction of and arousal during anaesthesia are less well seen there, (b) sensitivity to early cerebral ischaemic changes is greatest in the parietal regions overlying the triple arterial boundary zone (*Fig. 2.31*) and (c) there is a danger of erroneous interpretation when 'rhythms at a distance' such as frontal intermittent rhythmic delta activity (FIRDA) or paradoxical delta arousals responses (see *Rhythms at a distance*, p. 117, and *Altered reactivity*, p. 118) occur. These projected rhythms are seen both with cerebral dysfunction or as a form of arousal response; in these circumstances a sudden increase in delta activity may be misinterpreted as an increase in depth of anaesthesia when it is actually due to surgical stimulation during light anaesthesia or during emergence from it (Kochs et al 1994, Scherer et al 1994).

When the EEG is processed for monitoring, difficulties can arise when intermittent frontal delta bursts affect stationarity and frequency domain measures unpredictably, depending on their duration in relation to epoch length. Indeed, because they cause a lack of stationarity, averaged spectra and their derivatives will be invalid (Bischoff et al 1993). The possibility of such intermittent phenomena must be recognized in the design of any form of EEG signal processing for monitoring, as well as in the selection of suitable electrode placement sites. Likewise, it should be noted that quantitative comparisons such as interhemispheric correlation and coherence usually depend upon assumptions about suitable electrode placement and symmetry and unexpected difficulties may arise when these conditions are not fulfilled (Andrew and Pfurtscheller 1996), for example due to surgical access, wound sites, etc.

Interference due to surgical diathermy

Surgical diathermy presents one of the most difficult problems in intraoperative neurophysiological monitoring, particularly in cardiovascular surgery when electrocautery may be used extensively, especially during opening of the chest. Modern diathermy equipment contains safety circuits which switch off the system in the event of earthing faults. Before monitoring during surgery when diathermy is likely to be used, it is necessary to check what level of protection is provided in neurophysiological monitoring equipment and, indeed, whether disconnection during diathermy is advised. EEG monitoring equipment should have appropriately designed amplifiers to minimize the risk of damage during diathermy. Even if the diathermy current does not damage the amplifiers, it can cause problems since RF waves can be rectified by the amplifier input transistors and hence add a DC offset to the amplifier output, thereby causing blocking and signal distortions. One can try to miminize this by: (a) using an RF filter to short the RF waves to the independent reference (ground), or (b) clamping the RF waves before they reach the level where the input amplifier will rectify them (Dr DE Maynard, personal communication, 2005). The latter will still result in some blocking but may recover more quickly. A deblocking switch is highly desirable. In the absence of a deblocking circuit, technologists need to be taught to short the input leads when the amplifier is

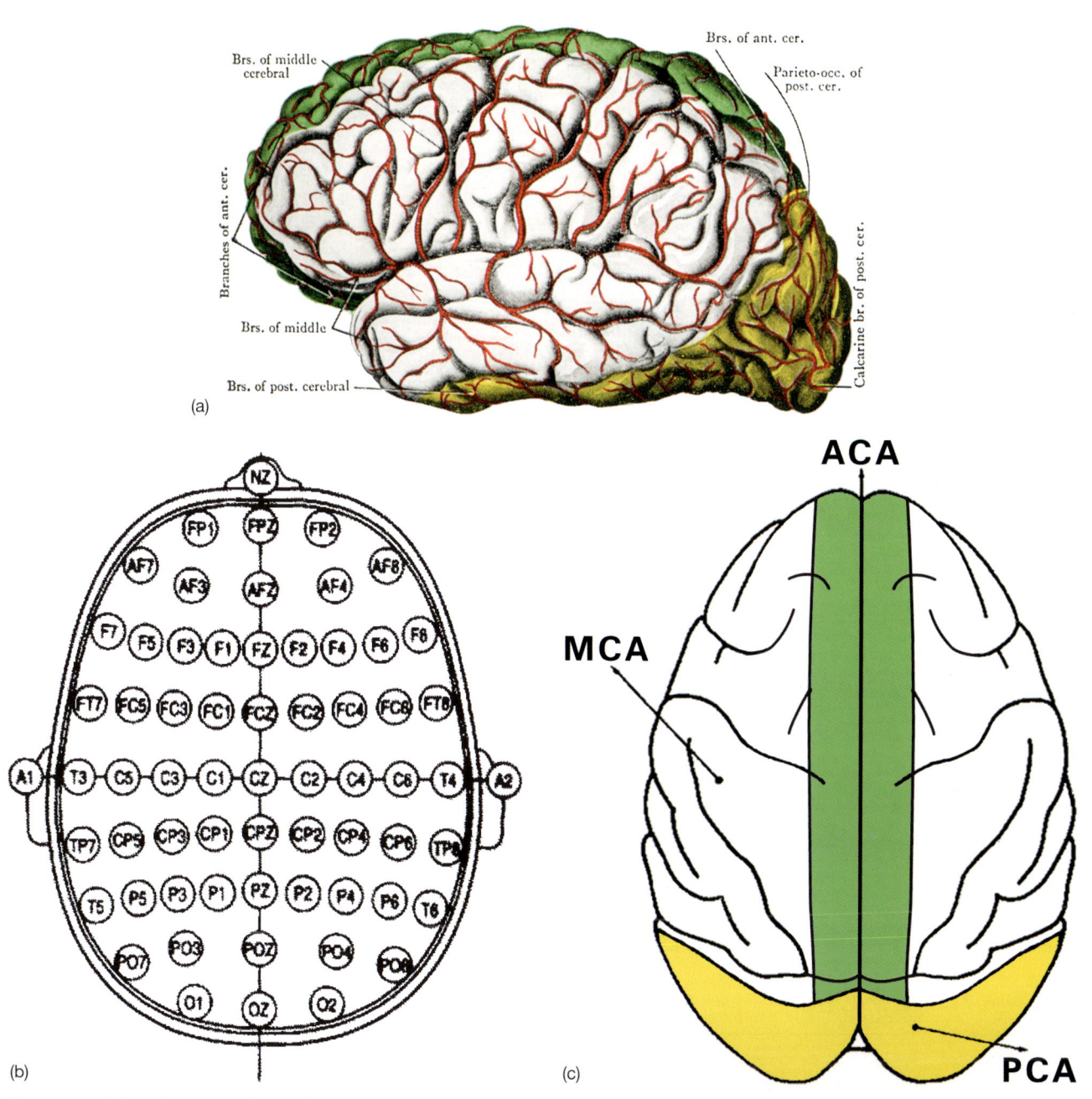

Fig. 2.31 (a) Superior–lateral view of the brain to show the territories of the major cerebral arteries (that of anterior cerebral artery in green, middle cerebral artery in grey, posterior cerebral artery in yellow). (From Jamieson (1946), by permission.) (b) Extended 10–20 system (Nuwer et al (1998), by permission.) (c) Cerebral arterial territories seen from above. ACA, anterior cerebral artery, MCA, middle cerebral artery, PCA, posterior cerebral artery.

blocked. Some amplifiers develop a standing potential between the input terminals after several bursts of diathermy; connecting a 1 MΩ resistor across the input terminals has been recommended to avoid this problem.

Approaches to the problem of EEG contamination by diathermy are exemplified by the special preamplifiers with input floating from earth and RF shielding and filtering (Van der Weide and Pronk 1979) (*Fig. 2.32*). Another method using a switching system which disconnects all channels during surgical diathermy was described by Barlow (1985). Those using neurophysiological monitoring which includes EPs have resorted to automated cut-outs during periods of unavoidable diathermy interference and subsequent blocking of EEG preamplifiers (Kalkman et al 1991). More recently an isolated, battery-powered, preamplifier unit for EP monitoring 'largely immune to diathermy interference due to excellent isolation via a digital fibre optical link, small size and RF screening and filtering' has been described by Jordan et al (1995).

Artefactual contamination of the EEG can be reduced by good technique, including placing the diathermy return plate as close as possible to the operating site to reduce the extent of current flow in the rest of the body. Similarly the extent of any stray current flowing in the electrode leads is reduced by keeping electrode leads as far as possible both from any possible return current path and from the diathermy cables. Because of the hazard of a faulty diathermy ground and of theatre staff simultaneously touching both the patient and grounded metalwork, monitoring equipment should be kept out of range and, ideally, should have no exposed metal or other such conducting surfaces.

To prevent the potentially serious problem of the diathermy return plate becoming detached from the patient, the diathermy equipment manufacturers include suitable safety circuits in their products (Dr PR Weller, personal communication, 1999). The European standards for diathermy (International Electrotechnical Commission, IEC 60601-2-2, 1998) specify that a plate continuity circuit, ensuring that plate and cable are intact, is the only compulsory monitoring circuit required. However, many manufacturers include additional safety circuits that check conditions such as the connection between plate and patient or the accidental connection between patient and earth. The use of these forms of circuitry are usually identified by a split patient return plate on the diathermy equipment. The method by which these circuits function varies with manufacturers, but it usually involves the periodic delivery of a check pulse to test the required parameter. This signal can cause problems in EEG and EP recordings (Carson et al 2000). It may be evident even if the diathermy equipment is not being activated but is merely connected to the patient and in standby mode. The problem typically manifests itself as a regular high amplitude artefact in the EEG recording. Some diathermy machines produce a continuous monitoring signal at a frequency not displayed by neurophysiological recorders, but having an amplitude sufficient to affect amplifier input stages, significantly reducing the amplitude of EEG and EPs (Taskey and Seaba 1997). In the worst case it can offset the amplifiers in the EEG headbox and data can be lost until normal operation returns (*Fig. 2.33*).

Defibrillators, fibrillators, pacemakers and balloon pumps
Defibrillators and fibrillators can pose problems similar to those of diathermy, and preliminary checks must be made that any neurophysiological monitoring equipment to be used has isolation amplifiers that are protected against potentials from these sources. Risks are greater in the presence of widely spaced electrodes and this may require separate isolation for each recording channel. During cardiac surgery, in addition to the use of diathermy, fibrillators, defibrillators, pacemakers and balloon pumps, even external cardiac massage may lead to unfamiliar forms of artefact (*Fig. 2.34*). Given that patients with widespread vascular disease may have endocardial pacemakers, interference may also pose problems during vascular surgery such as carotid endarterectomy (Clark et al 1989). Clark's group considered that potential misinterpretation of a processed EEG display at the time of carotid clamping could occur if changes due to artefacts had not been monitored by simultaneous display of the unprocessed signals. Similar problems have been encountered when using automated systems for identification of ischaemic changes in the EEG during testing of implantable cardioversion defibrillators (Adams et al 1995).

Dangers of misinterpretation due to unrecognized artefacts in EEG monitors displaying only processed signals
This provides a salutary reminder that considerable care is needed in the use of monitors which seek to automate interpretation during surgical procedures in the course of which both well-recognized and unusual artefacts may be encountered. This topic forms part of the detailed analytical review of misleading factors during surgical EEG

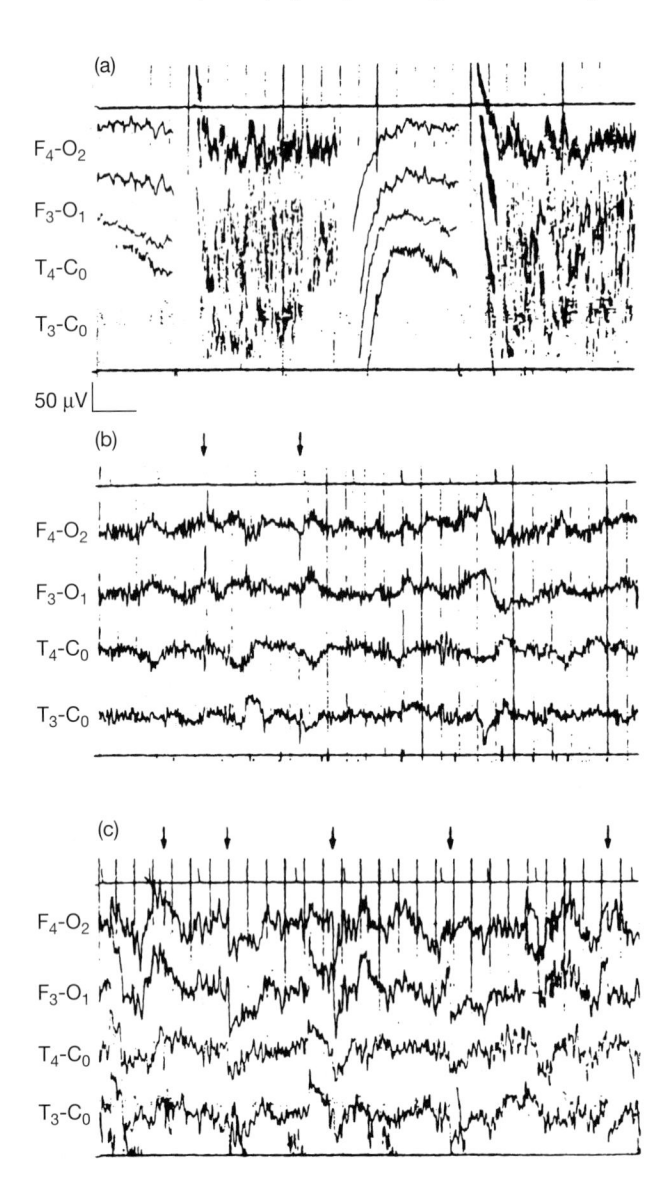

Fig. 2.32 Minimization of surgical diathermy artefacts by preamplifier with floating input and RF shielding and filtering in EEG monitor during open heart surgery, (a) Amplifier overload during electrocautery; (b) artefact reduction when a preamplifier is used (arrows indicate when diathermy is turned on and off); and (c) as in (b) under unfavourable circumstances. (From Van der Weide and Pronk (1979), by permission.)

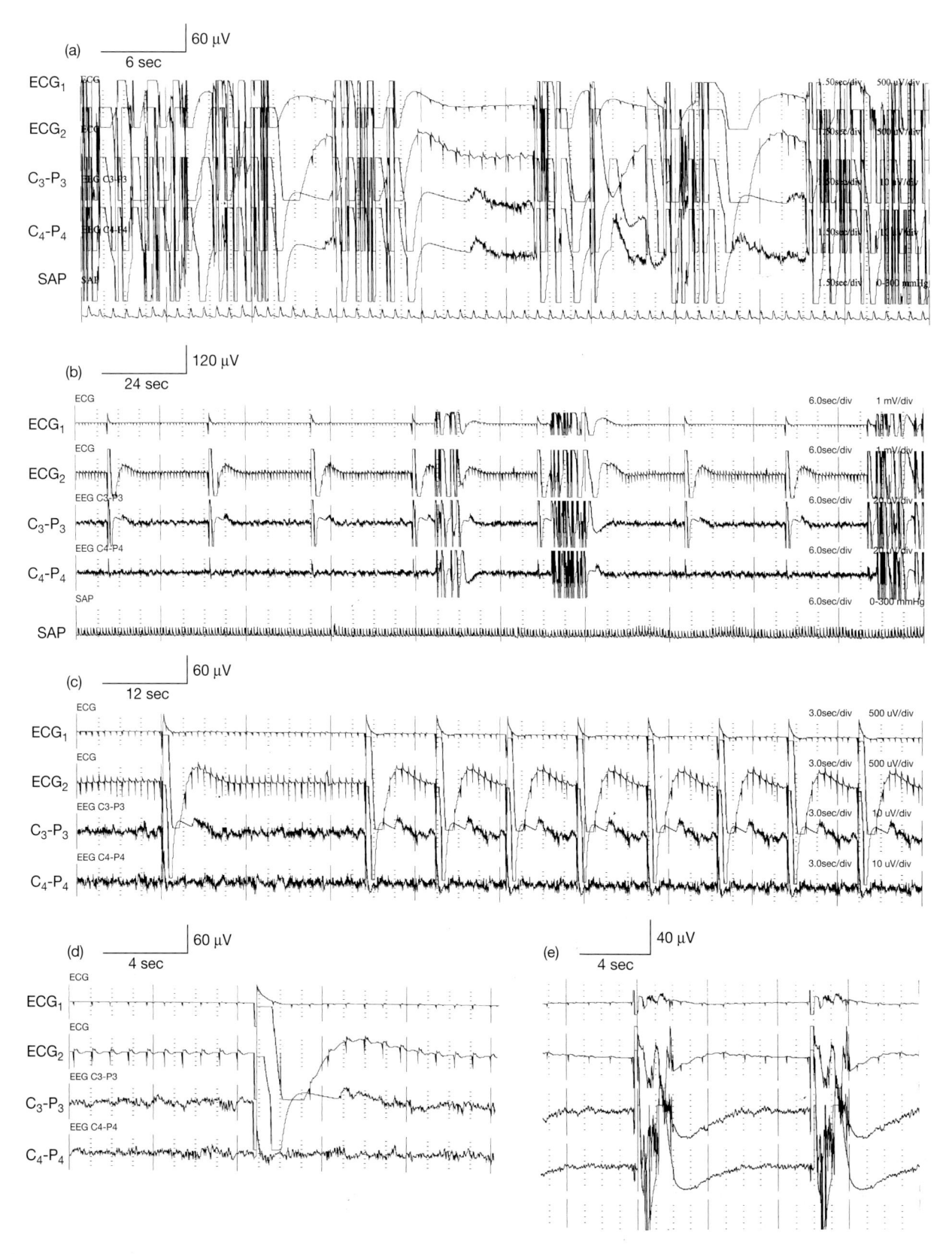

Fig. 2.33 Examples of interference from diathermy apparatus which includes patient safety circuitry to avoid electrical burns (see text). Recording during cardiac surgery in two patients undergoing coronary revascularization. Polygraphic recording with two ECG and two EEG channels and, in samples (a) and (b), systemic arterial pressure. Note different sensitivity and time scales. Sample (e) from second patient in whom a different diathermy machine was used. Artefacts comprise: (a) classical diathermy artefact; (b–e) deflections produced by intermittent signals from safety circuits shown between periods of electrocautery. Note that in sample (b) the latter artefact is of greater amplitude from the left-sided EEG lead (channel 3) than the right, probably due to electric field spread from the diathermy plate on the patient's left leg. (Data from the EU-Biomed-2 project IBIS (Carson et al 2000), by permission.) (From Binnie et al (2003), by permission.)

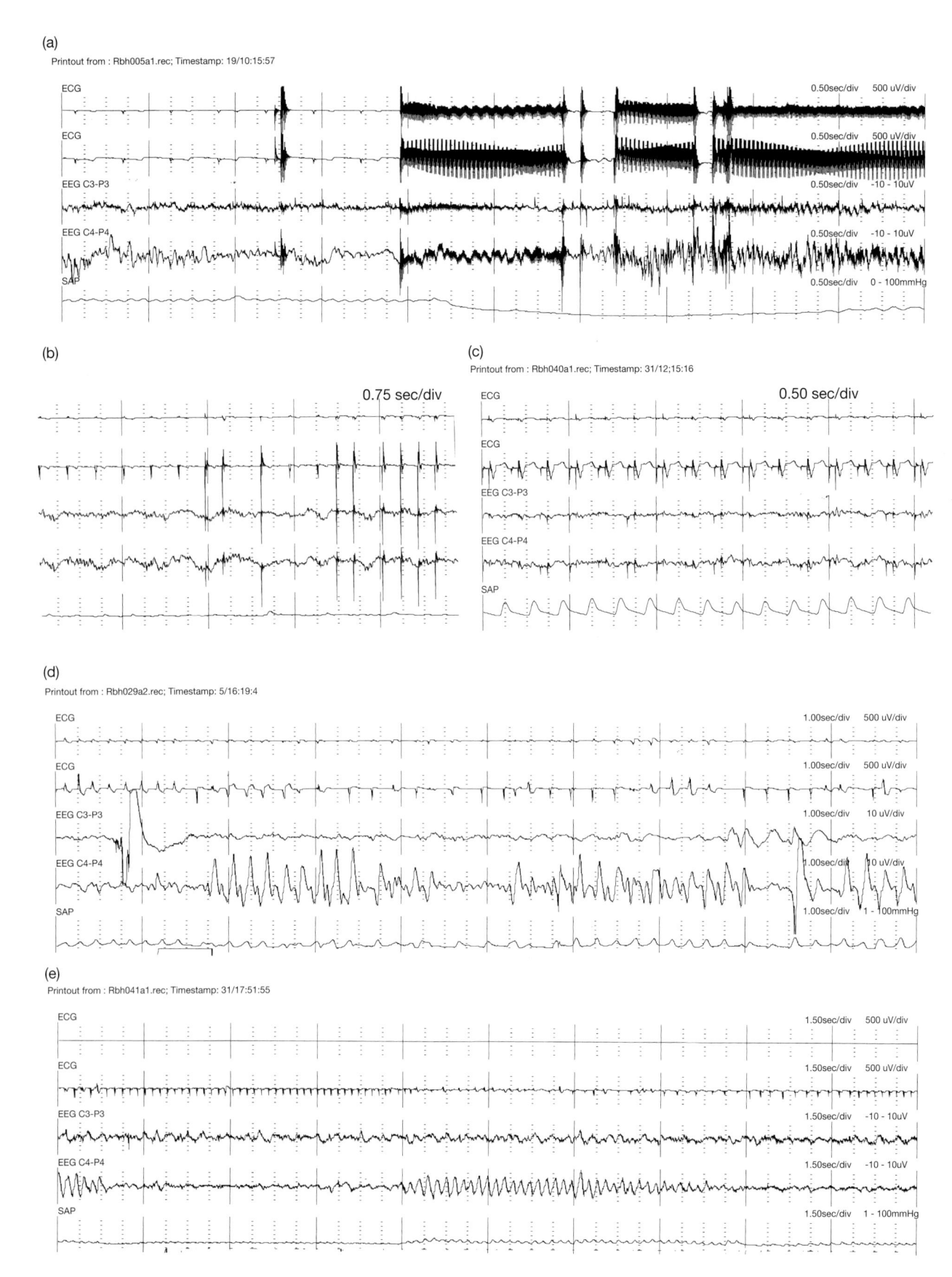

Fig. 2.34 Examples of artefacts during open cardiac surgery (note different sensitivity and time scales): (a) fibrillator, (b, c) pacemaker, (d) balloon pump and (e) external cardiac massage. Note that these artefacts, in contrast to those during surgical diathermy, often preferentially affect right-sided EEG leads. This may reflect the right lateral tilt position of the patient. (Data from the EU-Biomed-2 project IBIS (Carson et al 2000), by permission.) (From Binnie et al (2003), by permission.)

monitoring by Dahaba (2005). It is one of the strongest arguments for the universal inclusion of facilities to view the unprocessed EEG signal throughout monitoring and to have parallel quality checks by means of continuous display of electrode impedance, line frequency and other external potentials. In addition, special attention is required in the design of algorithms for computerized EEG monitoring systems to deal with the specific problem of artefacts. Some groups have used automatic artefact detection by simple voltage threshold detection and/or smoothing to reduce contamination, particularly during trend detection (Hinrichs et al 1992, 1996, Eckert 1998), whilst others working with intensive care monitoring have used artefact classification and labelling systems (Van de Velde et al 1999). More work is required in this area, especially in relation to dealing with artefact from surgical diathermy and related signals.

Intraoperative and ICU EP recording

The different monitoring techniques are learned most effectively during a period of apprenticeship with an acknowledged expert, but useful introduction can be obtained from many books and reviews (Nuwer 1986, Grundy and Villani 1988, Møller 1988a, Desmedt 1989, Jones 1993, Beck 1994, Loftus and Traynelis 1994, Sebel and Fitch 1994, Grundy 1995, Legatt 1995, Møller 1995, Russell and Rodichok 1995, Lopez 1996, Sloan 1996, Linden at al 1997, Guérit 1998, Kumar et al 2000, Zouridakis and Papanicolaou 2001, Deletis and Shils 2002). Adequate training is only one aspect of providing a reliable monitoring service; a standardized protocol for each type of operation is equally important. This should be based on accepted standards of practice, which are summarized in the recent reports of a committee of the IFCN (Nuwer et al 1993, Burke et al 1999) and guidelines produced by the American Electroencephalographic Society (1994b), the International Federation of Clinical Neurophysiology (Burke et al 1999) and the International Organisation of Societies for Electrophysiological Technology (OSET) (1999), details being modified in accordance with local requirements and experience.

Preoperative recordings are essential to identify any pre-existing abnormality (Haisley 1992). The monitoring team should draw up a list of items to be taken into the operating theatre, and equipment checks to be completed before the operation. Mistakes will be minimized if the procedure for setting up the recordings and the criteria for recognizing a significant deterioration are written on cards, preferably with a plasticized washable surface for use in the operating theatre. Similar written reminders are desirable for the action to be taken following a deterioration in the recording; this includes confirming that an adequate stimulus has been given, electrode impedances are low and electrodes properly connected to the amplifiers, and that no extraneous factors, such as drugs or hypotension, are responsible for the alteration. Useful examples of operative monitoring protocols are given by Harper and Daube (1989). The monitoring system (equipment, personnel and procedures) should be designed to minimize the risk of serious error (Leape and Berwick 2000, Nolan 2000). The safest arrangement is for monitoring to be undertaken by a team consisting of at least two competent persons, one of whom checks each action as it is performed, in the same way that airline pilots check each other's actions (Hunt 1988). This will ensure that disastrous mistakes are avoided, and that false positive and false negative results are minimized.

The principal purpose of monitoring is fulfilled by information given verbally to the surgeon during the operation. However, in accordance with standards maintained in other branches of clinical neurophysiology, a written report should be produced after every operation; this is particularly important in view of the medico-legal considerations associated with operations in which monitoring is required. Representative traces with full annotation should be printed out and stored safely, even if they do not form part of the report. After uneventful operations only a brief report is necessary, but sufficient traces should be retained to provide a defence against any allegation that a postoperative deficit resulted from failure to recognize changes in EPs during the operation. In complex cases the neurophysiologist responsible for monitoring should write a comprehensive report, documenting problems encountered and action taken, supplemented by appropriate traces to justify all decisions. The report must be written as soon as possible after the operation while memories are fresh, since it may feature prominently in any subsequent legal action. Many operating departments now have facilities for video recording of operations, particularly where a video camera is attached to the operating microscope, and the audio channel of the video recorder can be used to indicate any change in monitoring. This provides further valuable documentation in the event of a deficit being detected postoperatively, and may permit identification of particularly hazardous manoeuvres which can be avoided in subsequent operations (Levine et al 1978).

An important requirement of any monitoring system is that it should do the patient no harm (Saunders 1997). Current standards for electrical safety in susceptible patients should always be observed, and intensities of stimulation should not be excessive, particularly in view of the need to stimulate continuously at high rates for long periods (American Electroencephalographic Society 1994b). Patients with pacemakers may be in extra danger, since the current from a nerve stimulator can interfere with the pacemaker, producing either a dangerous tachycardia or cardiac arrest (Regan et al 1986, Merritt et al 1988, O'Flaherty et al 1993). The artefact from a nerve stimulator often appears in the ECG trace, and may be misinterpreted as a cardiac arrhythmia or pacemaker malfunction when the pacemaker is working perfectly (Diachun et al 1998). Even if the patient is protected from electrical hazards of recording equipment, there is still a danger that watching the monitor diverts attention from the patient (Saunders 1997). Monitoring should therefore be undertaken by a dedicated team, and not the surgeon or anaesthetist (anaesthesiologist) responsible for the operation.

Any EP averaging equipment can be used for monitoring, provided it is electrically isolated, and robust enough to withstand use in operating theatres and ICUs. However, recording techniques must be modified to provide reliable monitoring, and certain design features are desirable in any equipment intended for operative monitoring rather than routine EP recording. No single commercial machine incorporates all the facilities that would be ideal for every monitoring application, and some can only be provided by the addition of a computer running specially written programs. If a computer, or any other additional item, is attached to recording equipment, then this equipment should ideally be powered from an isolation transformer, and the electrical safety of the combination should be checked by a competent engineer before being connected to a patient.

The special requirements for monitoring are listed below:

1. During a single operation several electrodes may be used to record different EPs. The small patient connection boxes supplied with many EP machines are therefore not satisfactory, since individual electrode leads have to be inserted and removed whenever a different response is to be recorded, which may be necessary many times during a complex procedure. The connection box should have a large number of sockets, the appropriate electrodes for each recording being selected either by the computer program, or manually by a switch similar to the montage selector of an EEG machine; the appropriate stimulator output should be selected in a similar way. Ideally the choice of stimulating and recording electrodes should be clearly visible, either from the switch settings, or from the screen display, so as to minimize the possibility of recording from electrodes not appropriate to the stimulus being administered.

2. Monitoring is carried out in less than perfect surroundings for electrophysiological recording, so (electrical) noise elimination must be superior to that found in equipment for routine use. This should include improved common mode rejection, screened leads with sufficient capacitance between screen and earth to ensure filtering of RF interference (MacGillivray 1974), and the addition of high frequency filters rolling off at 18 dB/octave, rather than the usual 6 dB (Møller 1988a, 1995). Although the sharper attenuation of high frequencies distorts the response somewhat, any distortion is constant throughout the operation, so recognition of changes will not be impaired and may be improved by greater elimination of unwanted high frequencies.

In addition, further noise elimination by digital filtering is often desirable after averaging. Digital filtering can take several forms, from simple smoothing (low pass filtering) to complex weighted time domain filters (Møller 1988b), and adaptive filters to remove spectral components due to noise (John et al 1985, Bertrand et al 1987, Sgro et al 1989). Although not widely used, the simpler forms of digital filtering should be easily implemented in modern equipment, particularly where a computer is available independent of the averager.

3. An important feature of any recording performed under adverse circumstances should be some indication of the residual noise in the average. The simplest approach is to include a prestimulus interval in each average. However, this interval may not be entirely free of stimulus-related activity, particularly if the stimulation rate is high, and assumes that noise remains constant before and after the stimulus. A preferable alternative is to record the '(±) average' (Schimmel 1967). In this average, responses are alternately added and subtracted. The (±) average of N responses therefore consists of the difference between the means of the $N/2$ even numbered and the $N/2$ odd numbered responses. Components time-locked to the stimulus, including the EP, will be eliminated, and any deflection from the baseline represents noise that has not been removed by averaging. Using appropriate equipment a (±) average can be recorded at the same time as the conventional average. The variance (r.m.s. value) of the (±) average is an estimate of the residual noise, and the ratio of the variance of the conventional average to the variance of the (±) average is an estimate of the signal-to-noise ratio (Wong and Bickford 1980, Chaudhri and Smith 1999, Van de Velde 2000). Some commercial EP equipment provides a display of the separate averages of odd numbered and even numbered responses, but the (±) average is not displayed, and the variances cannot be calculated. Several alternative methods of estimating the signal-to-noise ratio are reviewed by Özdamar and Delgado (1996).

4. In the presence of a high noise level, the raw trace (analogue signal after amplification and filtering) should be visible at all times to permit recognition of artefacts which may not be rejected, but which could contaminate the average. In addition, the raw trace permits identification of amplifier blocking, which is very common after use of the diathermy. Ideally the switch by which the surgeon operates the diathermy apparatus should disconnect the amplifiers from the patient, and halt averaging until the amplifiers are no longer blocked (Simard and Friedman 1984). In the absence of such a sophisticated arrangement, a manual deblocking circuit is highly desirable in monitoring equipment.

As well as eliminating a prominent artefact, disconnection of the patient during use of the diathermy is desirable to prevent any risk of burns at electrode sites, and is essential if electrodes are placed directly on any part of the nervous system.

5. In many operations the time taken to acquire an average seriously limits the usefulness of EP monitoring, since only a delayed report can be given as to whether a particular manoeuvre has had any adverse effect. In many circumstances, use of an adaptive digital filter, as described above, will considerably reduce the number of responses required for each average. Alternatively, the acquisition and display of averages must be modified. For example, 'moving block' (Boston 1985) or 'pseudorecursive' averaging (Nuwer 1986) can be used, in which separate averages are acquired, each consisting of a small number of responses, and the display is the mean of a number of recently acquired averages. Thus, if it is desired to display the average of 1000 responses, separate averages each of 250 responses are acquired sequentially, and the display consists of the mean of the last four such averages (*Fig. 2.35(a)*). A new average can therefore be displayed four times as often as would be the case for a conventional average of the same number of responses. The same effect can be obtained by setting a number of averagers running at increasing intervals; with four averages of 1000 responses, the second averager begins when the first has acquired 250 responses, the third when the first has acquired 500, and so on (*Fig. 2.35(b)*). A logical extension of this idea is a running average, where the display is the average of the last N responses, updated as each new response is acquired (*Fig. 2.35(c)*). This has the disadvantage that all N responses must be stored individually, so considerably more memory is required than for conventional averaging. Nevertheless, use of a modern computer with more than 100 Mbyte of memory should permit implementation of a running average. A compromise requiring considerably less memory is the exponential average, in which the display consists of a weighted mean of the existing contents of the averager memory and each newly acquired response (Svensson 1993). The contribution of earlier responses to the display declines exponentially, hence the name. This form of averaging was included in some early electronic averagers, and has been incorporated in at least one commercial monitoring device.

6. Effective monitoring depends upon prompt recognition of changes in averages acquired over a period of time, and this is facilitated by displaying successive averages one above the other on the same screen, preferably with an early response retained for comparison throughout the operation. This is a feature of the monitoring programs supplied with several commercial machines. Alternatively, a conventional averager can be connected to an independent computer, which is programmed to produce the desired display.

7. Repeated recording of EPs during a long operation generates a large amount of data, and it is undesirable for individual traces to be printed separately in the operating theatre. Some machines produce a rolling hard copy, with the stacked traces plotted one above the other, just as they appear on the screen; this results in a considerable saving of paper, and permits recognition of long-term trends. In addition, modern apparatus should include data storage on magnetic tape or discs (magnetic or optical), which must of course be stored securely to ensure patient confidentiality. This allows the recordings to be analysed and annotated at leisure after the operation, and provides a back-up in the unfortunate event of the original hard copy being lost or damaged. Ideally, the processes of storing and printing should occur automatically on completion of each average, and should not interrupt data acquisition.

8. In most monitoring applications the amplitude and latency of EP components can be regarded as simple physiological variables similar to blood pressure or intracranial pressure. Some form of automatic peak recognition is therefore desirable, with measurement of the appropriate latencies and amplitudes, which can be plotted on a chart. Systems of this type have been designed for BAEP monitoring, but are not yet commercially available (Boston et al 1985, Bertrand et al 1987).

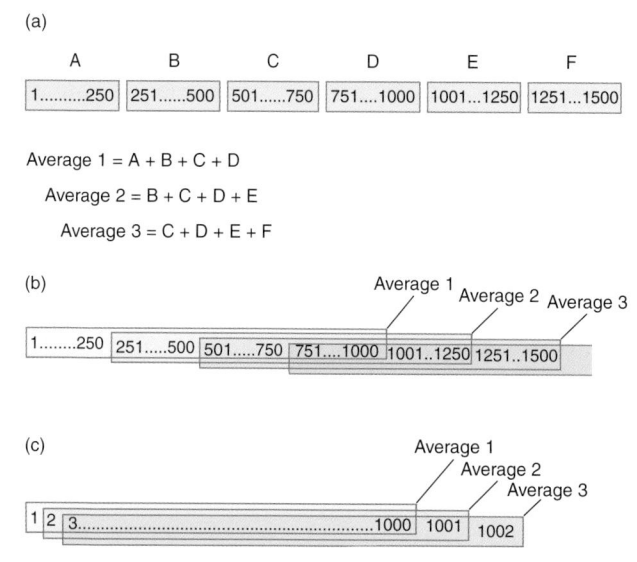

Fig. 2.35 Moving block averaging. (a) The mean of four successive small averages is updated every 250 responses. (b) Successive averages of 1000 responses are initiated every 250 responses, each overlapping its predecessor by 750. This method gives exactly the same results as (a). (c) A running average of 1000 responses is updated each time a new response occurs. (From Binnie et al (2003), by permission.)

9. In many conventional evoked response machines there is only a limited choice of rates of stimulation. This is a serious disadvantage for operative recording, since most of the possible rates (1, 2, 5, 10 Hz) are fractions of 50 Hz (and 60 Hz); thus prominent mains interference will inevitably be time-locked to the stimulus. Rates of 3 Hz or 7 Hz can be selected on some machines; either will reduce 50 Hz interference, but only a rate of 7 Hz will reduce 60 Hz interference. A greater choice of rates is desirable so that the interstimulus interval can be set at an odd multiple of half the period of the mains frequency, or any other rhythmical interference, so that the noise will appear in opposite phases in successive responses (Sgro and Emerson 1985).

10. Monitoring during many operations requires simultaneous recording of two or more EPs, often in different modalities. A machine designed for monitoring should be able to deliver stimuli through at least two different stimulators, and add the appropriate responses to separate averages. Completed averages are acquired much more rapidly if stimuli in the different modalities can be either interwoven or overlapped (Plourde et al 1988).

EMG and nerve conduction studies

Peripheral nerve studies in the ICU are generally considered to be complex and labour intensive. The reader is referred to Binnie et al (2004) for a detailed account of all basic nerve conduction and EMG techniques. Here we shall consider some practical aspects which are helpful in carrying out these studies successfully in the ICU; subsequent sections will address the investigation of the commonly encountered clinical problems.

Electrical studies in the ICU often involve physical transportation of the equipment, which was not necessarily designed to be portable, and all accessories such as connecting leads, electrodes, electrode jelly, etc. It is therefore expedient, and indeed saves time, to be properly organized and to maintain a tried and tested list of the items that are necessary. Where possible, it is best to keep a separate set of electrodes and leads for ICU use in a small multidrawer carrying case. Long leads are convenient, as there is often very limited access to the patient and very limited sites available for NCS.

Equally important is to be absolutely sure about the identity of the patient, who may be unconscious, and to check what restrictions are necessary because of infection control, life support systems, specific lesions or recent surgical operation sites, important infusion lines, drug therapy and other connected electrical equipment.

Coping with electrical interference is perhaps the part of the procedure most fraught with problems. There is no easy solution but it can be minimized by paying special attention to site of earth placement, avoiding contact with other electrically active equipment, avoiding sweaty skin surfaces and having impeccably clean, well applied electrodes.

Sensory action potentials (SAPs) may be difficult to obtain purely as a result of increased stimulus artefacts and the automatic artefact rejection feature that is incorporated in most modern systems. In this situation it may be worth trying with a lower amplification setting or simply acquiring the responses without averaging. A reduced SAP amplitude may also be the result of tissue oedema.

An absent motor response is not uncommon and can result from a number of reasons, including technical failures and neuromuscular blockade. It is always worth checking the stimulus on oneself, if this happens, before making any clinical conclusions about the relevance of such findings. A comparable situation exists in the case of the isoelectric EEG (electrocerebral silence), it being much more difficult to demonstrate a high degree of proof of the absence of a biological potential than to document its presence.

The ambient temperature in the ICU is generally higher than in other areas of the hospital and recorded motor conduction velocities generally reflect this difference. It is important to make appropriate corrections otherwise a misleading interpretation may be made, particularly when serial data are assessed as part of long-term follow-up studies.

Recording the EMG to assess acute denervation changes such as fibrillations and positive sharp waves is not a problem, but further information such as motor unit abnormalities, firing frequencies and interference patterns are all generally unobtainable or difficult due to the patient's inability to cooperate. Patients may be unconscious, sedated, paralysed, confused or have co-existing upper motor neuron lesions.

It is worth remembering that multiple pathology is common in the ICU setting. This may be the result of the nature or severity of the illness itself or the result of prolonged immobilization and the consequences of treatment.

REFERENCES

Adams DC, Heyer EJ, Emerson RG, et al 1995 The reliability of quantitative electroencephalography as an indicator of cerebral ischemia. Anesth Analg 81:80–83.

Adrian ED, Bronk DW 1929 The discharge of impulses in motor fibres. J Physiol 67:119–151.

Al-Seffar J 1990 Never mind the quality, what's the cost? An evaluation of EMG needle maintenance. J Electrophysiol Technol 16:179–191.

Amassian VE, Cracco RQ, Maccabee PJ, et al 1989 Suppression of visual perception by magnetic coil stimulation of human occipital cortex. Electroencephalogr Clin Neurophysiol 74:456–462.

American Electroencephalographic Society 1991 *Guidelines: Recording Clinical EEG on Digital Media.* American Electroencephalographic Society, Bloomfield.

American Electroencephalographic Society 1994a Guideline nine: Guidelines on evoked potentials. J Clin Neurophysiol 11:40–73.

References

American Electroencephalographic Society 1994b Guideline eleven: guidelines for intraoperative monitoring of sensory evoked potentials. J Clin Neurophysiol 11:77–87.

American Electroencephalographic Society 1994c Guideline thirteen: Guidelines for standard electrode position nomenclature. J Clin Neurophysiol 11:111–115.

American Electroencephalographic Society 1994d Report of the Committee on Infectious Diseases. J Clin Neurophysiol 11:128–132.

Andrew C, Pfurtscheller G 1996 Dependence of coherence measurements on EEG derivation type. Med Biol Eng Comp 34:232–238.

Association of Anaesthetists of Great Britain and Ireland 1971 Explosion hazards. Recommendations of the Association of Anaesthetists of Great Britain and Ireland. Anaesthesia 26:155–157.

Barker AT, Stevens JC 1991 Measurement of the acoustic output from two magnetic nerve stimulator coils. J Physiol 438:301.

Barker AT, Jalinous R, Freeston IL 1985 Non-invasive magnetic stimulation of human motor cortex [letter]. Lancet i:1106–1107.

Barker AT, Freeston IL, Jalinous R, et al 1988 Magnetic and electrical stimulation of the brain: safety aspects. In: *Non-invasive Stimulation of Brain and Spinal Cord: Fundamentals and Clinical Applications* (eds PM Rossini, CD Marsden). AR Liss, New York, pp. 131–144.

Barlow JS 1985 A general-purpose automatic multichannel electronic switch for EEG artifact elimination. Electroencephalogr Clin Neurophysiol 60:174–176.

Bashein G, Nessly ML, Bledsoe SW, et al 1992 Electroencephalography during surgery with cardiopulmonary bypass and hypothermia. Anesthesiology 76:878–891.

Bauer H, Korunka C, Leodolter M 1989 Technical requirements for high-quality scalp DC recordings. Electroencephalogr Clin Neurophysiol 72:545–547.

Baumann SB, Noll DC 1999 A modified electrode cap for EEG recording in MRI scanners. Clin Neurophysiol 110:2189–2193.

Beck DL (ed) 1994 *Handbook of Intraoperative Monitoring*. Singular Publishing Group, San Diego, CA.

Bertrand O, Garcia-Larrea L, Artru F, et al 1987 Brainstem monitoring. I: A system for high rate BAEP sequential monitoring and feature extraction. Electroencephalogr Clin Neurophysiol 68:433–145.

Billings RJ 1994 Some thoughts on electrical safety. J Electrophysiol Technol 20:156–200.

Binnie CD, Dekker E, Smit A, et al 1982a Practical considerations in the positioning of EEG electrodes. Electroencephalogr Clin Neurophysiol 53:453–458.

Binnie CD, Rowan AJ, Gutter Th 1982b *A Manual of Electroencephalographic Technology*. Cambridge University Press, Cambridge.

Binnie CD, Cooper R, Mauguière F, et al (eds) 2003 *Clinical Neurophysiology*. Volume 2: *EEG, Paediatric Neurophysiology, Special Techniques and Applications*. Elsevier, Amsterdam.

Binnie CD, Cooper R, Mauguière F, et al (eds) 2004 *Clinical Neurophysiology*. Volume 1: *EMG, Nerve Conduction and Evoked Potentials*, revised and enlarged edition. Elsevier, Amsterdam.

Bischoff P, Kochs E, Droese D, et al 1993 Topographic-quantitative EEG-analysis of the paradoxical arousal reaction. EEG changes during urologic surgery using isoflurane/N$_2$O anesthesia [in German]. Anaesthesist 42:142–148.

Boston JR 1985 Noise cancellation for brainstem auditory evoked potentials. IEEE Trans Biomed Eng 32:106–1070.

Boyd SG, Kandler RH, Stevens WR 1991 A comparison of noise levels produced by different magnetic stimulators. J Physiol 438:368.

Britton TC, Brown P, Day PL, et al 1990 Can the cerebellum be stimulated through the intact scalp in man? J Physiol 420:19.

Bruner JMR, Leonard PF 1989 *Electricity, Safety and the Patient*. Year Book Medical Publishers, Chicago.

Burke D, Nuwer MR, Daube J, et al 1999 Intraoperative monitoring. In: *Recommendations for the Practice of Clinical Neurophysiology. Guidelines of the International Federation of Clinical Neurophysiology* (EEG Suppl 52) (eds G. Deuschl, A. Eisen). Elsevier Science, Amsterdam, pp. 133–148.

Cadwell J 1990 Principles of magneto-electric stimulation. In: *Magnetic Stimulation in Clinical Neurophysiology* (ed S Chokroverty). Butterworth, London, pp. 13–32.

Carson ER, van Gils M, Saranummi N (eds) 2000 IBIS: Improved monitoring for brain dysfunction in intensive care and surgery. Comput Method Program Biomed 63:157–235.

Chaudhri Q, Smith NJ 1999 Effect of residual noise on peak-to-peak amplitude of averaged responses: a computer simulation study. J Physiol 518:77P–78P.

Clark S, Goldberg M, Gorman R, et al 1989 Interference of automated electroencephalographic processing by an endocardial pacemaker. J Clin Monit 5:22–25.

Cooper R 1962 Electrodes. Proc Electrophysiol Technol Assoc 9:22–32.

Cooper R, Osselton JW, Shaw JC 1980 *EEG Technology*, 3rd edition. Butterworth, London.

Cooper R, Binnie CD, Billings RJ 2005 *Techniques in Clinical Neurophysiology: A Practical Manual*. Churchill Livingstone, Edinburgh.

Counter SA, Borg E, Lofqvist L 1990 Acoustic trauma in extracranial magnetic brain stimulation. Electroencephalogr Clin Neurophysiol 78:173–184.

Dahaba AA 2005 Different conditions that could result in the Bispectral Index indicating an incorrect hypnotic state. Anesthesiology 101:765–773.

Day BL, Dressler D, Maertens de Noordhout A, et al 1989 Electric and magnetic stimulation of human motor cortex: surface EMG and single motor unit responses. J Physiol 412:449–173.

De Deyne C, Struys M, Decruyenaere J, et al 1998 Use of continuous bispectral EEG monitoring to assess depth of sedation in ICU patients. Intensive Care Med 24:1294–1298.

Deletis V, Shils J 2002 *Neurophysiology in Neurosurgery: A Modern Intraoperative Approach*. Academic Press, New York.

Desmedt JE (ed) 1989 *Neuromonitoring in Surgery*. Elsevier, Amsterdam.

Deuschl G, Eisen A (eds) 1999 *Recommendations for the Practice of Clinical Neurophysiology: Guidelines of the International Federation of Clinical Neurophysiology*, 2nd revised and enlarged edition (EEG Suppl 52). Elsevier, Amsterdam, pp. 3–6.

Dhuna A, Gates J, Pascual-Leone A 1991 Transcranial magnetic stimulation in patients with epilepsy. Neurology 41:1067–1071.

Diachun CA, Brock-Utne JG, Lopez JR, et al 1998 Evoked potential monitoring and EKG: a case of a serious cardiac arrhythmia. Anesthesiology 89:1270–1272.

Donchin E, Callaway E, Cooper R, et al 1977 Publication criteria for studies of evoked potentials in man. In: *Attention, Voluntary Contraction and Event-related Cerebral Potentials* (ed JE Desmedt). Karger, Basel, pp. 1–11.

Eckert O 1998 Automatic artefact detection in intraoperative EEG monitoring [in German]. Biomed Technol 43:236–242.

Eisen A 1986 Non-invasive measurement of spinal cord conduction: review of presently available methods. Muscle Nerve 9:95–103.

Electrophysiological Technologists Association 2004 EPTA guidelines for checking digital EEG machines. http://www.epta.50megs.com/resources/cg/digi_cal_eeg.pdf

Evans B, Kriss A, Jeffries D, et al 1993 British Society for Clinical Neurophysiology guidelines for preventing transmission of infective agents and toxic substances by clinical neurophysiology procedures: an update. J Electrophysiol Technol 19:129–135.

Gilliatt RW, Melville ID, Velate AS, et al 1965 A study of normal nerve action potentials using an averaging technique (barrier grid storage tube). J Neurol Neurosurg Psychiatry 28:191–200.

Goldman RI, Stern JM, Engel J, et al 2000 Acquiring simultaneous EEG and functional MRI. Clinic Neurophysiol 111:1974–1980.

Grahame DC 1952 Mathematic theory of faradaic admittance. J Electrochem Soc 99:370C.

Grundy BL 1995 The electroencephalogram and evoked potential monitoring. In: *Monitoring in Anesthesia and Critical Care Medicine* (eds CD Blitt, RL Hines), 3rd edition. Churchill Livingstone, New York, pp. 423–489.

Grundy BL, Villani RM 1988 *Evoked Potentials. Intraoperative and ICU Monitoring*. Springer-Verlag, New York.

Gualtierotti T, Paterson AS 1954 Electrical stimulation of the unexposed cerebral cortex. J Physiol 125:278–291.

Guérit JM 1998 Neuromonitoring in the operating room: why, when and how to monitor. Electroencephalogr Clin Neurophysiol 106:1–21.

Haisley L 1992 Evoked potential intraoperative monitoring: troubleshooting. Am J EEG Technol 32:308–317.

Hallett M, Wassermann EJ, Pascual-Leone A, et al 1999 Repetitive transcranial magnetic stimulation. In: *Recommendations for the Practice of Clinical Neurophysiology* (eds G Deuschl, A Eisen). Electroencephalogr Clin Neurophysiol Suppl 52:105–113.

Harper CM, Daube JR 1989 Surgical monitoring with evoked potentials: the Mayo Clinic experience. In: *Neuromonitoring in Surgery* (ed JE Desmedt). Elsevier, Amsterdam, pp. 275–302.

Hinrichs H, Heinze HJ, Gaab MR 1992 Neurophysiologic monitoring in neurosurgical vascular operations: specific technical requirements and their conversion [in German]. Z Elektroenzephalograph Elektromyograph Gebiete 23:195–202.

Hinrichs H, Feistner H, Heinze HJ 1996 A trend-detection algorithm for intraoperative EEG monitoring. Med Eng Phys 18:626–631.

Homan RW, Herman J, Purdy P 1987 Cerebral location of international 10–20 system electrode placement. Electroencephalogr Clin Neurophysiol 66:376–382.

Homberg V, Netz J 1989 Generalised seizures induced by transcranial magnetic stimulation of the motor cortex. Lancet ii:1223.

Hufnagel A, Elger CE 1991 Induction of seizures by transcranial magnetic stimulation in epileptic patients. J Neurol (Berlin) 238:109–110.

Hufnagel A, Elger CE, Durwen HF, et al 1990 Activation of the epileptic focus by transcranial magnetic stimulation of the human brain. Ann Neurol 27:49–60.

Hunt P 1988 Safety in aviation. Perfusion 3:83–96.

International Electrotechnical Commission (IEC) 1988 *Medical Electrical Equipment. Part 1: General Requirements for Safety*, 2nd edition. Publication 60601-1. IEC Secretariat, Geneva.

International Electrotechnical Commission (IEC) 1998 *Medical Electrical Equipment. Part 2-2: Particular requirements for the safety of high frequency surgical equipment*, 3rd edition. Publication 60601-2-2. IEC Secretariat, Geneva.

International Electrotechnical Commission (IEC) 2002 *Medical Electrical Equipment. Part 2-13: Particular Requirements for the Safety and Essential Performance of Anaesthetic Systems*. Publication 60601-2-26. IEC Secretariat, Geneva. [Adopted as European Standard 60601-2-26 (2003).]

International Organisation of Societies for Electrophysiological Technology (OSET) 1999 Guidelines for performing EEG and evoked potential monitoring during surgery. Am J Electroneuro-diagnostic Technol 39:257–277.

Jamieson EB 1946 *Illustrations of Regional Anatomy*, 6th edition. E & S Livingstone, Edinburgh, p. 8.

Jasper HH 1958 Report of the Committee on Methods of Clinical Examination in Electroencephalography. Electroencephalogr Clin Neurophysiol 10:370–375.

John R, Prichep LS, Ransohoff J, et al 1985 Real-time intraoperative monitoring of evoked potentials using optimized digital filtering. Electroencephalogr Clin Neurophysiol 61:20P.

Jones SJ 1993 Evoked potentials in intra-operative monitoring. In: *Evoked Potentials in Clinical Testing*, 2nd edition (ed AM Halliday). Churchill Livingstone, Edinburgh, pp. 565–588.

Jordan C, Weller C, Thornton C, et al 1995 Monitoring evoked potentials during surgery to assess the level of anaesthesia. J Med Eng Technol 19:77–79.

Kalkman CJ, Romijn K, Denslagen W 1991 Eliminating diathermy-induced artifacts during intraoperative monitoring of somatosensory-evoked potentials: a hardware solution. J Clin Monit 7:320–324.

Keenan NK, Taylor MJ, Coles JG, et al 1987 The use of VEPs for CNS monitoring during continuous cardiopulmonary bypass and circulatory arrest. Electroencephalogr Clin Neurophysiol 68:241–246.

Klem GH, Lüders HO, Jasper HH, et al 1999 The twenty-electrode system of the International Federation. In: *Recommendations for the Practice of Clinical Neurophysiology: Guidelines of the International Federation of Clinical Neurophysiology* (eds G Deuschl, A Eisen), 2nd revised and enlarged edition. Electroencephalogr Clin Neurophysiol Suppl 52:3–6.

Kochs E, Bischoff P, Pichlmeier U, et al 1994 Surgical stimulation induces changes in brain electrical activity during isoflurane/nitrous oxide anesthesia. A topographic electroencephalographic analysis. Anesthesiology 80:1026–1034.

Kumar A, Bhattacharya A, Makhija N 2000 Evoked potential monitoring in anaesthesia and analgesia. Anaesthesia 55:225–241.

Lagerlund TD, Sharbrough FW, Jack CR, et al 1993 Determination of 10–20 system of electrode locations using magnetic resonance image scanning with markers. Electroencephalogr Clin Neurophysiol 86:7–14.

Le J, Lu M, Pellouchoud E, et al 1998 A rapid method for determining 10/10 electrode positions for high resolution EEG studies. Electroenceph Clin Neurophysiol 106:554–558.

Leape LL, Berwick DM 2000 Safe health care: are we up to it? BMJ 320:725–726.

Legatt AD 1995 Intraoperative neurophysiologic monitoring: some technical considerations. Am J EEG Technol 35:167–200.

Levine RA, Montgomery WW, Ojemann RG, et al 1978 Evoked potential detection of hearing loss during acoustic neuroma surgery. Neurology 28:339.

Levy WJ, Shapiro HM, Meathe E 1980 The identification of rhythmic EEG artefacts by power-spectrum analysis. Anesthesiology 53:505–507.

Linden RD, Zappulla R, Shields CB 1997 Intraoperative evoked potential monitoring. In: *Evoked Potentials in Clinical Medicine* (ed KH Chiappa), 3rd edition. Lippincott-Raven, Philadelphia, pp. 601–638.

Litscher G, Kehl G, Schwarz G, et al 1996 Neurophysiologic signal recording. New technical and general practice aspects of EEG recording electrodes [in German]. Biomed Technol (Berlin) 41:106–110.

References

Loftus CM, Traynelis VC (eds) 1994 *Intraoperative Monitoring Techniques in Neurosurgery*. McGraw-Hill, New York.

Lopez JR 1996 Intraoperative neurophysiological monitoring. Int Anesthesiol Clin 34:33–54.

Maccabee PJ, Amassian VE, Cracco RQ, et al 1991 Stimulation of the human nervous system using the magnetic coil. J Clin Neurophysiol 8:38–55.

MacGillivray BB 1974 Notes on planning an EEG department. In: *Handbook of Electroencephalography Clinical Neurophysiology* (ed A Rémond), volume 3, Part C, pp. 109–123.

Marsden CD, Merton PA, Morton HB 1982 Percutaneous stimulation of spinal cord and brain: pyramidal tract conduction velocities in man. J Physiol 328:6.

Maurer K 1985 Uncertainties of topodiagnosis of auditory nerve and brain-stem auditory evoked potentials due to rarefaction and condensation stimuli. Electroencephalogr Clin Neurophysiol 62:135–140.

Merritt WT, Brinker JA, Beattie C 1988 Pacemaker-mediated tachycardia induced by operative somatosensory evoked potential stimuli. Anesthesiology 69:766–768.

Merton PA, Morton HB 1980 Stimulation of the cerebral cortex in the intact human subject. Nature 285:227.

Mirsattari SM, Lee DH, Jones D, et al 2004 MRI compatible EEG electrode system for routine use in the epilepsy monitoring unit and the intensive care unit. Clin Neurophysiol 115:2175–2180.

Møller AR 1995 *Intraoperative Neurophysiologic Monitoring*. Harwood Academic, Luxembourg.

Møller AR 1988a *Evoked Potentials in Intraoperative Monitoring*. Williams and Wilkins, Baltimore.

Møller AR 1988b Use of zero-phase digital filters to enhance brain-stem auditory evoked potentials (BAEPs). Electroencephalogr Clin Neurophysiol 71:226–232.

Moritake K, Takebe Y, Konishi T, et al 1986 Technical problems in intra-operative monitoring of sensory evoked potentials [in Japanese]. No Shinkei Geka 14:135–141.

Morris HH, Lüders H 1985 Electrodes. In: *Long-term Monitoring in Epilepsy* (eds J Gotman, JR Ives, P Gloor). Electroencephalogr Clin Neurophysiol Suppl 37:3–25.

Nolan TW 2000 System changes to improve patient safety. BMJ 320:771–773.

Nuwer MR 1986 *Evoked Potential Monitoring in the Operating Room*. Raven, New York.

Nuwer MR, Comi G, Emerson R, et al 1998 IFCN standards for digital recording of clinical EEG. Electroencephalogr Clin Neurophysiol 106:259–261.

Nuwer MR, Daube J, Fischer C, et al 1993 Neuromonitoring during surgery. Report of an IFCN committee. Electroencephalogr Clin Neurophysiol 87:263–276.

O'Flaherty D, Wardill M, Adams AP 1993 Inadvertent suppression of a fixed rate ventricular pacemaker using a peripheral nerve stimulator. Anaesthesia 48:687–389.

Özdamar Ö, Delgado RE 1996 Measurement of signal and noise characteristics in ongoing auditory brainstem response averaging. Ann Biomed Eng 24:702–715.

Picton TW, Hink RF, Perez-Abalo M, et al 1984 Evoked potentials: how now? J Electrophysiol Technol 10:177–221.

Plourde G, Picton T, Kellett A 1988 Interweaving and overlapping of evoked potentials. Electroencephalogr Clin Neurophysiol 71:405–414.

Prior PF, Maynard DE 1986 *Monitoring Cerebral Function. Long-term Monitoring of EEG and Evoked Potentials*. Elsevier, Amsterdam.

Priori A, Bertolasi L, Rothwell JC, et al 1991 Evidence that transcranial magnetic stimulation delays saccadic eye movements by interfering with activity in oculomotor areas of cortex in man. J Physiol 438:302.

Regan JJ, McAfee PC, Achuff SC 1986 The induction of cardiac arrhythmia and hypotension from spinal cord monitoring. A case report. Spine 11:1031–1032.

Rémond A, Torres F 1964 A method of electrode placement with a view to topographical research. I: Basic concepts. II: Practical applications. Electroencephalogr Clin Neurophysiol 17:577–578; 19:187–189.

Rossini PM, Marciani MG, Caramia M, et al 1985 Nervous propagation along 'central' motor pathways in intact man: characteristics of motor response to 'bifocal' and 'unifocal' spine and scalp non-invasive stimulation. Electroencephalogr Clin Neurophysiol 61:272–286.

Russell GB, Rodichok LD 1995 *Primer of Intraoperative Neurophysiolgic Monitoring*. Butterworth-Heinemann, Oxford.

Saunders DA 1997 On the dangers of monitoring. Or, *primum non nocere* revisited. Anaesthesia 52:399–400.

Saunders MG, Jell RM 1959 Time distortion in electroencephalograph amplifiers. Electroencephalogr Clin Neurophysiol 11:814–816.

Scherer GB, Mihaljevic VG, Heinrichs MA, et al 1994 Differences in the topographical distribution of EEG activity during surgical anaesthesia and on emergence from volatile anesthetics. Int J Clin Monit Comput 11:179–183.

Schimmel H 1967 The (±) reference: accuracy of estimated mean components in average response studies. Science 157:92–94.

Schultz A, Schultz B, Grouven U, et al 1995 Channel selection for EEG-monitoring in anesthesia [in German]. Anaesthesist 44:473–477.

Schultz B, Bender R, Schultz A, et al 1992 Reduktion der Anzahl von EEG-Ableitungen fur ein routinemassiges monitoring auf der Intensivstation. Biomed Technol (Berlin) 37:194–199.

Screening Commission of the New University Hospital, Copenhagen, Denmark 1967 Reduction of electrical interference in measurements of bioelectrical potentials in a hospital. Acta Polytech Scand, Electrical Eng Ser 15:1–37.

Sebel PS, Fitch W 1994 *Monitoring the Central Nervous System*. Blackwell Science, Oxford.

Seitsonen E, Yli-Hankala A, Kortilla K 2000 Are electrocardiogram electrodes acceptable for electroencephalogram Bispectral Index monitoring? Acta Anaesthesiol Scand 44:1266–1270.

Sgro JA, Emerson RG 1985 Phase synchronized triggering: a method for coherent noise elimination in evoked potential recording. Electroencephalogr Clin Neurophysiol 60:464–468.

Sgro JA, Emerson RG, Pedley TA 1989 Methods for steadily updating the averaged responses during neuromonitoring. In: *Neuromonitoring in Surgery* (ed JE Desmedt). Elsevier, Amsterdam, pp. 49–60.

Simard M, Friedman WA 1984 Automatic artefact rejection during intra-operative recording of somatosensory evoked potentials. J Neurosurg 61:609–611.

Sloan TB 1996 Evoked potential monitoring. Int Anesthesiol Clin 34:109–136.

Smith NJ 1997a Electric nerve stimulators: assessment and performance. J Electrophysiol Technol 23:4–17.

Smith NJ 1997b The recording convention and the nerve action potential. J Electrophysiol Technol 23:80–83.

Smith NJ 2000 Patient safety in NCS clinics and theatres. Electrical nerve stimulation in patients with cardiac pacemakers. J Electrophysiol Technol 26:177–180.

Snooks SJ, Swash M 1985 Motor conduction velocity in the human spinal cord: slowed conduction in multiple sclerosis and radiation myelopathy. J Neurol Neurosurg Psychiatry 48:1135–1139.

Stålberg E, Trontelj JV 1994 *Single Fiber EMG: Studies in Healthy and Disease Muscles*. Raven, New York.

Stapells DR, Picton TW, Smith AD 1982 Normal hearing thresholds for clicks. J Acoust Soc Am 72:74–79.

Svensson O 1993 Tracking of changes in latency and amplitude of the evoked potential by using adaptive LMS filters and exponential averagers. IEEE Trans Biomed Eng 40:1074–1079.

Taskey B, Seaba PJ 1997 Technical tips: what is REM? Am J Electroneurodiag Technol 37:276–280.

Tönnies JF 1933 Der Neurograph, ein Apparat zur unmittelbar sichtbaren registrierung bioelektrischer Erscheinungen. Deutsche Z Nervenheilkund 130:60–67.

Towle VL, Bolaños J, Suarez D, et al 1993 The spatial location of EEG electrodes: locating the best-fitting sphere relative to cortical anatomy. Electroencephalogr Clin Neurophysiol 86:1–6.

Ugawa Y, Rothwell JC, Day BL, et al 1991 Percutaneous electrical stimulation of corticospinal pathways at the level of the pyramidal decussation in humans. Ann Neurol 29:418–427.

Van de Velde M 2000 Signal validation in electroencephalography research. Thesis, Technische Universiteit, Eindhoven.

Van de Velde M, Ghosh IR, Cluitmans PJM 1999 Context related artefact detection in prolonged EEG recordings. Comput Method Progam Biomed 60:183–196.

Van der Weide H, Pronk RAF 1979 Interference suppression for EEG recording during open heart surgery. Electroencephalogr Clin Neurophysiol 46:609–612.

Verma A, Bedlack RS, Radtke RA, et al 1999 Succinylcholine induced hyperkalemia and cardiac arrest: death related to an EEG study. J Clin Neurophysiol 16:46–50.

Wong PKH, Bickford RG 1980 Brain stem auditory evoked potentials: the use of noise estimate. Electroencephalogr Clin Neurophysiol 50:25–34.

Zouridakis G, Papanicolaou AC 2001 *A Concise Guide to Intraoperative Monitoring*. CRC Press, Boca Raton, FL.

Chapter 3

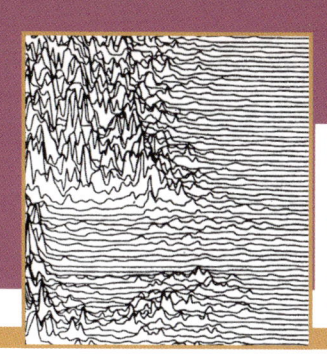

Introduction to Methods for Continuous EEG and Evoked Potential Monitoring

INTRODUCTION

This chapter provides a brief introduction to the development of equipment for continuous monitoring of the electroencephalogram (EEG) and of evoked potentials (EPs) in the intensive care unit (ICU) and during surgical operations. It sets out how methods for quantitative analysis of the neurophysiological signals were refined and gradually adapted in an attempt to characterize features (described in subsequent chapters) which indicate adverse events and/or responses to medical or surgical procedures in patients who are sedated or anaesthetized. A more detailed account of the scientific basis of data acquisition and processing methods for EEG and EP monitoring is provided in Chapter 8.

In the early days of clinical neurophysiology, 'continuous monitoring' meant recording non-stop in the operating theatre or the ICU for hours on end with a traditional multichannel ink-writing EEG machine and an enormous pile of paper. It provided valuable experience of the phenomenology (and bizarre artefacts) in these unfamiliar clinical settings as well as the possibility of relating EEG changes to clinical events or to changes in recordings (e.g. of arterial or intracranial pressures or oxygenation) on other clinical monitoring equipment. Some groups consider that traditional multichannel recording will remain the 'gold standard', for example for the sometimes complex and unpredictable changes during carotid artery surgery (Ahn et al 1988, Grady et al 1998).

Before the advent of convenient digital recorders with facilities for quantitative EEG analysis (qEEG), on-line interpretation of EEGs on multifold paper charts could be enhanced by use of a slower paper speed; however, slow trends over the several hours, for example during cardiac surgery with cardiopulmonary bypass and profound hypothermia, were difficult to assess. Indeed, trends often had to be judged by inspection of the patterns of variability (width of trace) of the ink on the edge of the large stack of folded paper! By the mid-1960s, biomedical engineers were working with several clinical teams to provide some degree of automated signal processing, including visualization of long-term trends and quality control. With the advent of more accessible computing power in the 1970s and 1980s, several approaches to neurophysiological monitoring were developed to help in various clinical and pharmaceutical applications. In practice, clinical acceptance of purpose-built EEG monitoring tools lagged far behind research demonstrations of their utility. Although some had a considerable impact, many overcomplex or idiosyncratic systems remained under dust covers because the doctors and nurses whose patients should have benefited from their help either found them mysterious or an unwelcome intrusion (Van Hemel and Pronk 1977, Pronk 1978). Since that time, the emphasis has been on making the tools more appropriate for the desired clinical application and easier for the non-expert to use. There is a greater recognition of the importance of training standards and teamwork, with experienced clinical neurophysiology monitoring staff working alongside the clinical teams in the ICU or surgical operating suites.

EP monitoring began to develop for use in intensive care and surgery when equipment for EEG monitoring was already well established. Typical early applications were spinal and posterior fossa surgery. It was soon followed by the use of combined displays of EEG and EPs together with other biological parameters to highlight timing of trends and detect causal events in the ICU or during carotid artery surgery.

Early neurophysiological monitoring systems often suffered from oversimplification and ignored factors such as the inherent variability of the signals and the difficulties this made for prompt and reliable detection of significant change. This proved a particular problem with EP monitoring with its requirement for averaging the responses to large numbers of stimuli. Several ingenious methodological approaches have allowed us to reduce many of the early difficulties. There is now a large body of high quality research demonstrating the unique capacity of EEG and EP signals to provide otherwise unobtainable, clinically relevant, warning and control information to improve care of patients during intensive care and surgery. However, further work on the development and understanding of tools is needed. To this end, it is valuable to take stock of the previous history of joint work by clinicians, neurophysiologists and biomedical engineers to produce appropriate and effective tools which are fit for the chosen purpose (*Table 3.1*).

EEG MONITORING

Historical development of EEG monitors
Hans Berger and early workers: 1920s to 1960

The EEG has always been used for documenting changing functional states of the brain and various clinical approaches to quantification of the signals for monitoring changes were already being explored soon after Hans Berger's first recordings in man. These were carried out in 1924, but not reported until 1929 (see Gloor (1969) for a translation from the original German of the complete series of Berger's reports on the human EEG between 1929 and 1938).

Berger's reports include detailed examination of the nature of the EEG signal and early use of *Fourier analysis* carried out by his colleague Dietsch (1932) from the raw data magnified ×30 via an epidiascope and projected onto paper (*Figs 3.1* and *3.2*). This work included examination of changes due to anaesthesia with chloroform and other agents. It is salutary to note how the analysis of the precise nature of the signals progressed over the 6 years of the reports illustrated in these figures.

Other groups soon began to study the effects of anaesthesia and also the effects of changes in oxygen, carbon dioxide and glucose levels with quantitative assessment of the EEG (see *Table 3.1*). One early example of an attempt to produce a clinical EEG monitor was the *quantitative electrospectrogram* used by Drohocki and Drohocka (1939a,b) to study the effects of narcosis (*Fig. 3.3*). The evolution and other clinical applications of this system, developed with the help of the engineer M Gillard, were reported in a series of papers up to 1973.

Within 10 years of Drohocki's quantitative electrospectrogram, Bickford (1949, 1950, 1951a) reported a *servo-control system* described as 'an automatic anaesthetic regulator controlled by brain potentials' using an EEG integrator (*Fig. 3.4*) (see p. 169). The system was based on

Table 3.1 Development of clinical neurophysiological monitoring – some key dates

Decade	Developments	Key references
1920s	First reports of human EEG – Hans Berger attempted recordings in man from 1920, succeeding in 1924. In 1927 he noted that the alpha rhythm disappeared during chloroform anaesthesia	Berger 1929 (see Gloor 1969)
1930s	Hand-calculated Fourier analysis of EEG and electrode resistance measurements	Dietsch 1932
	Detailed EEG changes with chloroform anaesthesia	Berger 1934 (see Gloor 1969)
	Effects of ischaemia on the EEG	Adrian and Matthews 1934
	Anoxia produces a decrease in frequency and an increase in amplitude in the EEG	Gibbs et al 1935
	Isoelectric EEG in deep anaesthesia can be reversible	Beecher et al 1938
	The electrospectrogram (integrated amplitude in different spectral bands) during narcosis	Drohocki and Drohocka 1939a,b
1940s	Sequential changes in EEG with coma	Romano and Engel 1944
	Servo-control of anaesthesia using integrated EEG amplitude	Bickford 1949
1950s	EEG stages of nitrous oxide/ether anaesthesia defined	Courtin et al 1950
1960s	Cerebral function monitor (CFM) – amplitude and variability analysis of filtered EEG + electrode impedance	Maynard et al 1969
1970s	Compressed spectral array (CSA)	Bickford et al 1972, 1973
	Density spectral array (DSA; later CDSA)	Fleming and Smith 1979a,b
1980s	Derived spectral variables – median frequency (MF), spectral edge frequency (SEF), etc.	Many authors
	Zero-crossing based spectral measures	Pronk and Symons 1982, 1984
	Cerebral Function Analysing Monitor (CFAM)	Maynard and Jenkinson 1984
	BAEP monitoring	Bertrand 1985 Bertrand et al 1987
	Combined displays of EEG and EPs with other variables, e.g. heart rate, arterial and intracranial pressures	Pfurtscheller et al 1987
	Knowledge-based system for anaesthetic monitoring (Advanced Depth of Anaesthesia Monitor (ADAM))	Thomsen et al 1989
1990s	Bispectral analysis (BIS)	Sigl and Chamoun, 1994
2000s	Narcotrend	Schultz et al 2000, 2002
	Patient State Index	Drover et al 2002
	Entropy	Viertiö-Oja et al 2004

EEG monitoring

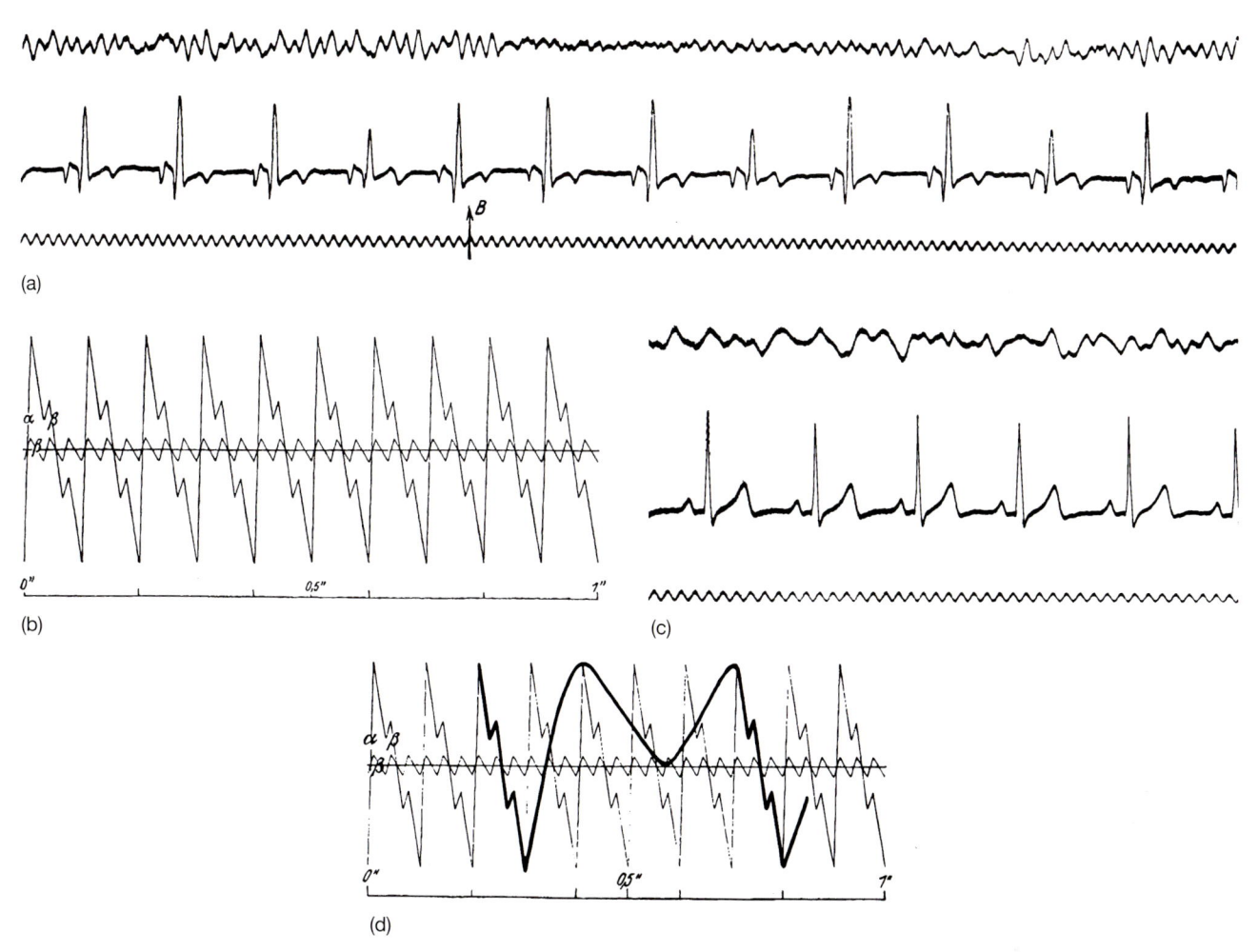

Fig. 3.1 (Berger's original descriptions from translation by Gloor (1969).) (a) (*Fig. 5 from his second report, 1930*) Dr V, 30-year-old physician. Double coil galvanometer. Condenser inserted. Recording from forehead and occiput with chlorided silver needle electrodes. Electrocardiogram with silver foil electrodes from the left arm and the left leg. At the top: the electroencephalogram; in the middle: the electrocardiogram; at the bottom: time in 1/10ths sec. B: time at which the dorsum of the right hand was touched and stroking with a glass rod along the latter began. (b): (*Fig. 6 from his second report, 1930*) Alpha waves do not have a smooth course – they display on their descending lower limb notches, usually two, which are easily recognizable as beta waves superimposed upon the alpha waves. Thus there are usually three beta waves for each alpha wave. Because the length of one second has been indicated on the lowermost line, the time relationships can be read from the curve. (c) (*Fig. 24 from his third report, 1931*): 38-year-old woman. Skull fracture with cerebral contusion three weeks earlier. EEG with needle electrodes from the left frontal and the right occipital quadrants of the head; ECG recorded from both arms. Time in 1/10ths sec. (d) (*Fig. 29 from his third report, 1931*): Diagram of the time course of the alpha waves and beta waves of the normal EEG. Heavy line: pathological alpha waves from Figure 24. (From Gloor (1969), by permission.)

an *EEG scoring system*, initially consisting of five stages for nitrous oxide/ether anaesthesia defined by inspection of the EEG and related clinical observations. It characterized (1) the resting state, (2) light, (3) moderate, (4) surgical and (5) deep anaesthesia. The scoring was later refined by Courtin et al (1950) to increase the ratings scale for burst suppression activity and then later extended to cover intravenous anaesthesia with thiopental (Kiersey et al 1951, 1954). By 1959, Bickford's group had not only shown that such sequential changes were common to all general anaesthetics, but that a typical asymmetrical parabolic curve could be constructed of the changes in integrated EEG amplitude which formed a steep rise with induction up to the peak of high amplitude slow wave activity and then declined with onset of burst suppression activity (Martin et al 1959). Of interest is the design of the filter used for the servo-control system, which was used to reduce artefacts from unwanted biological signals or from external interference (*Fig. 3.5*). Several similar servo-control devices appeared whilst single agent anaesthesia was the norm (e.g. Verzeano 1951, Bellville et al 1954), but they all fell into disuse with the advent of 'balanced anaesthesia' using combinations of agents. The great clinical interest in more informed control of anaesthetic depth led to many studies describing characteristic EEG stages of anaesthesia associated with a range of agents (see Faulconer and Bickford 1960) and led to several remarkable early endeavours with quantitative methods which anticipated later developments by several decades (see Burch 1959).

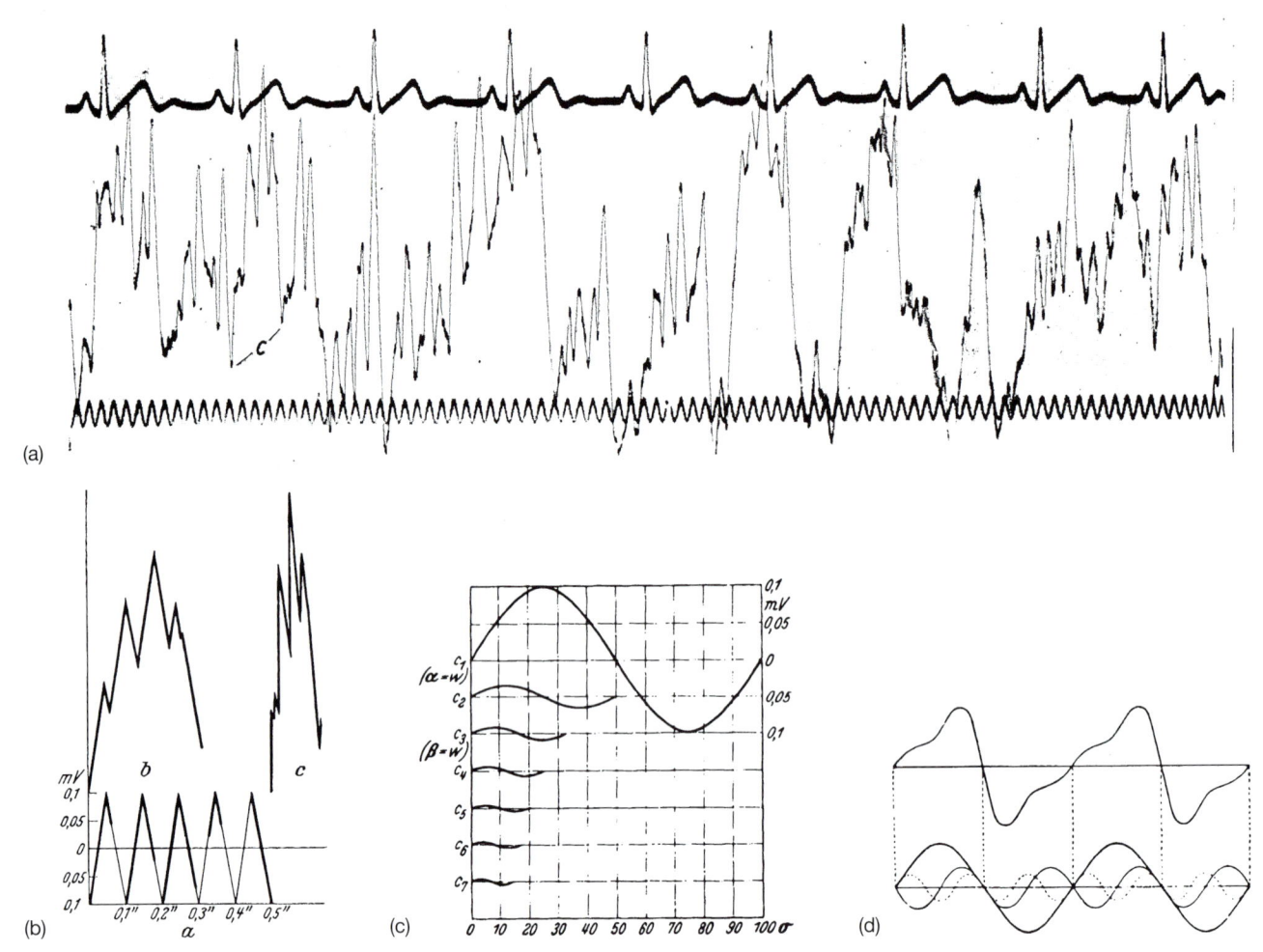

Chapter 3 Introduction to Methods for Continuous EEG and Evoked Potential Monitoring

`Fig. 3.2` (Berger's original descriptions from translation by Gloor (1969).) (a) (*Fig. 2 from his eighth report, 1934*) Miss EG, 30 years old; dementia praecox of many years' duration. Fifteen minutes after intravenous administration of 5.0 ml (of 10% aqueous solution) of Pernocton (*butyl-β-bromoallylbarbituric acid – a short-acting barbiturate much used in psychiatry in the 1930s*). At the top: ECG; in the middle: EEG; at the bottom: time in 1/10th of a second. (b) (*Fig. 5 from his eighth report, 1934*) Sketch *a* normal sequence of the alpha wave with a wavelength of 100σ and with equal amplitudes of the individual waves; *b*, group formation of the alpha wave caused by a loss of those parts of individual alpha waves drawn in with a thin line in a; *c*, group c of Figure 2 (Pernocton narcosis), which has been transferred to the sketch without alteration of its amplitude, but with some shortening of the time scale. (From Gloor (1969), by permission). (c) (*Fig. 9.1 from his ninth report, 1934*) Single oscillations of the EEG presented in the form of a graph according to the results of Dr G Dietsch's frequency analysis. (d) (*Fig.1 from his eleventh report, 1936*) Breakdown of a composite oscillation (top) into three single sinusoidal oscillations (Fourier analysis). (From Gloor (1969), by permission.)

Both medical and surgical applications of continuous EEG surveillance began to appear during the 1940s and 1950s. Development of *EEG scales for stages of coma*, typified by the work of Romano and Engel (1944) and Fischgold and Mathis (1959), led to prolonged EEG surveillance in patients after drug overdose and with metabolic encephalopathies. Quantitative relationships, generally between measures of EEG frequency and clinical estimates of depth of coma, were found to provide a basis for simple monitoring systems to aid clinical management (see Maynard et al 1969, MacGillivray 1969, 1976). Continuous EEG recording was also used during 'open heart surgery' in an effort to give an early warning of deterioration in brain function and impending damage in these high risk clinical situations where there was some possibility of prompt and effective remedial action (Bellville et al 1955, Pampiglione and Waterston 1958, 1961, Davenport et al 1959, Fischer-Williams and Cooper 1964).

Early time- and frequency-domain monitors

Early EEG monitors reflected the basic division of signal processing methods into *time and frequency domains*. This led to two main 'families' of early EEG monitoring equipment becoming available commercially from the early 1970s. One example from each type will be detailed to give a general indication of the key features that have led to their successful use in clinical monitoring and their subsequent evolution into increasingly advanced systems.

The Cerebral Function Monitor (CFM) (RDM Consultants, Uckfield, Sussex, UK; http://www.cfams.com) (Maynard et al 1969) represents a typical time domain system using mea-

EEG monitoring

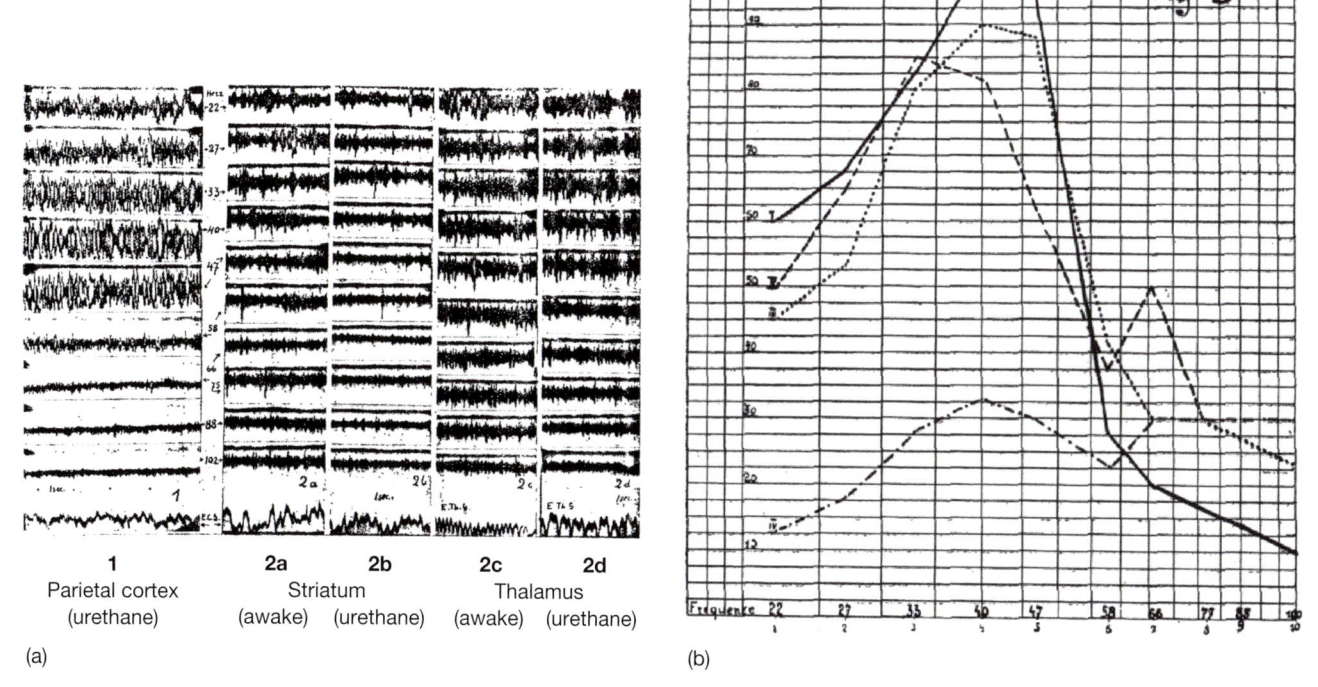

Fig. 3.3 Drohocki's electrospectrogram from rabbit brain to show changes with urethane anaesthesia. (a) Displays of analysis from circuits resonating at 10 frequencies from 21 to 102 Hz (values shown between first two blocks of spectra; 1 sec marks on lower part of spectral displays). In each case the unprocessed recording is displayed below the amplitude spectra (see original numbering at the bottom of each recording strip). (1) Spectra from the parietal cortex during anaesthesia. (2a) Spectra from the striatum in the waking state and (2b) during anaesthesia. (2c) Spectra from the thalamus in the waking state and (2d) during anaesthesia. (b) Graphical representation of the electrospectrogram from the parietal cortex during urethane anaesthesia as shown in part (1). The abscissa gives the frequencies 22–102 Hz of outputs in blocks 1–10 of the display on the left. The ordinate gives: for curve 1 (continuous line) the amplitude values as a percentage of the total; for curve 2 (dashed line) the number of waves traversing a filter expressed as a percentage of the frequency in curve 3; curve 3 (dotted line) the surface determined by the planimetric method corresponding to each frequency band (expressed in mm × μV × sec; curve 4 (dashed/dotted line) the absolute number of waves corresponding to each spectral band. (From Drohocki and Drohocka (1939b), by permission.)

sures of amplitude distribution of the filtered EEG. The basic principles underlying the design of the CFM are given in *Fig. 3.6*. The strengths of this early EEG monitor lay, first, in its robust simplicity, the clinical relevance and ease of measurement of EEG amplitude range and moment-to-moment variability (especially the clear depiction of burst suppression activity) as a reflection of the functional state of the brain in a wide range of critical

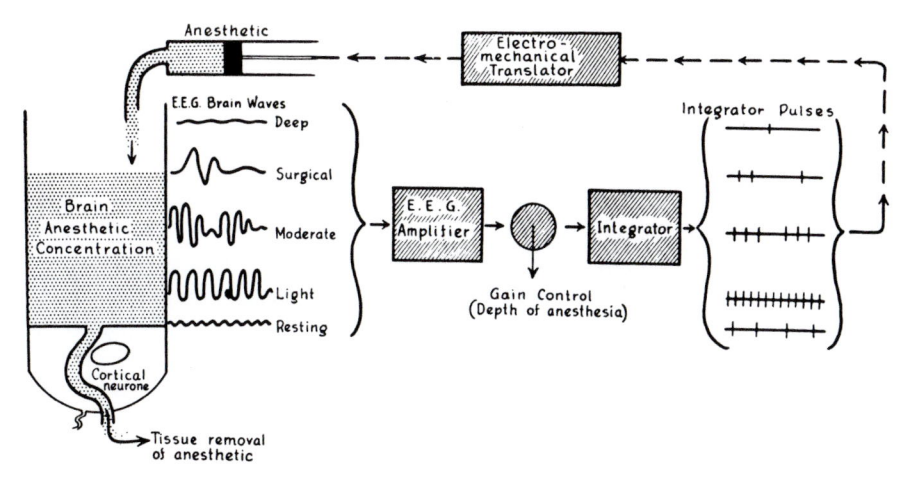

Fig. 3.4 Servo-anaesthesia. Schematic representation of the servo-loop for control of ether anaesthesia based on integrated amplitude of the EEG (From Bickford (1951b), by permission.)

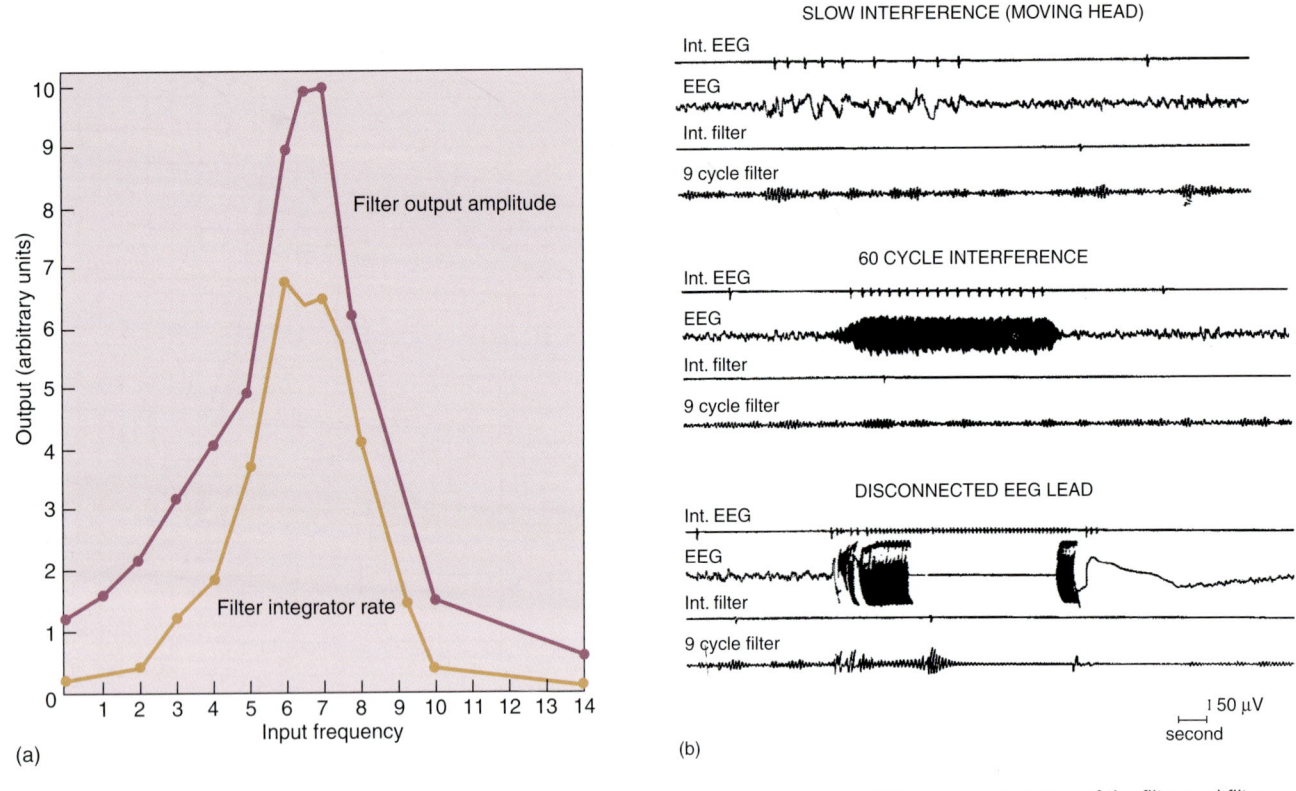

Fig. 3.5 Filter designed to reduce artefactual interference when monitoring during surgery. (a) Frequency response of the filter and filter integrator. (b) Comparison of wide and narrow filter with respect to discrimination against interference from slow components due to head movement; 60 cycle interference; a disconnected EEG lead. Int. EEG, direct integration of the EEG. Int. filter, integration of filter output. (From Bickford (1951b), by permission.)

conditions (Prior and Maynard 1986), and, second, the value of facilities such as continuous electrode contact impedance monitoring (*Fig. 3.7*). Equipment comparable to the original CFM remains popular for monitoring in the neonatal ICU, with the proviso that it should be 'calibrated' by means of conventional EEGs to interpret individual patterns and localization of neonatal seizure discharges (Toet et al 2002, Burdjalov et al 2003, Hellström-Westas et al 2003, Rennie et al 2004). Criticisms of the CFM have centred on the limitation to a single recording channel and the lack of frequency information. EEG frequency content is highly important to anyone trained in traditional clinical EEG evaluation, although it was not the top priority (burst suppression detection being considered more important) for the high risk ICU and operative monitoring applications in the 1960s.

The original CFM had a wide dynamic range, being able to represent EEG signals up to 1mV peak-to-peak. This is necessary in order to display the high amplitude signals seen during seizures (see e.g. *Fig. 3.12(c)*). Later versions of the CFM, produced by different manufacturers, had a lower dynamic range, in some cases not extending beyond 150 μV peak-to-peak, and would therefore not be able to display high amplitude seizure discharges (DE Maynard, personal communication, 2005). The output of the CFM was linear up to 10 μV peak-to-peak, and logarithmic above 10 μV, but in a later device, the Cerebral Function Analysing Monitor (see p. 83), logarithmic compression was applied to all amplitudes.

Fourier analysis with compressed spectral arrays (CSA) (Bickford et al 1972, 1973)   This method represents the classical frequency domain approach. The CSA displays sequential amplitude or power spectra using 'hidden line suppression' to avoid the ambiguities of overlapping lines – a presentation practised in other disciplines such as engineering and cartography. The basic methodology is shown in *Fig. 3.8(a)*. The strengths of the CSA lie in the ease of evaluating the complex patterns of frequency distribution over time and the effects on these CSA patterns of clinical events (*Fig. 3.8(b)*), features of value for those experienced in conventional EEG. The typical two-channel displays with time running vertically can produce striking insights into asymmetries and trends as well as revealing subtle patterns of prognostic value, and have proved valuable clinical tools in a wide range of ICU and surgical applications (e.g. Bricolo et al 1978, 1987) (*Fig. 3.9(a,b)*). The appearances necessarily depend on the length of the epochs of time which are averaged to produce each spectrum (varying between seconds to a minute or two). Events which are brief in relation to the duration of the epoch will either be missed altogether or, if detected, not amenable to precise identification of onset time compared with that available with time domain measures (*Fig. 3.9(c)*).

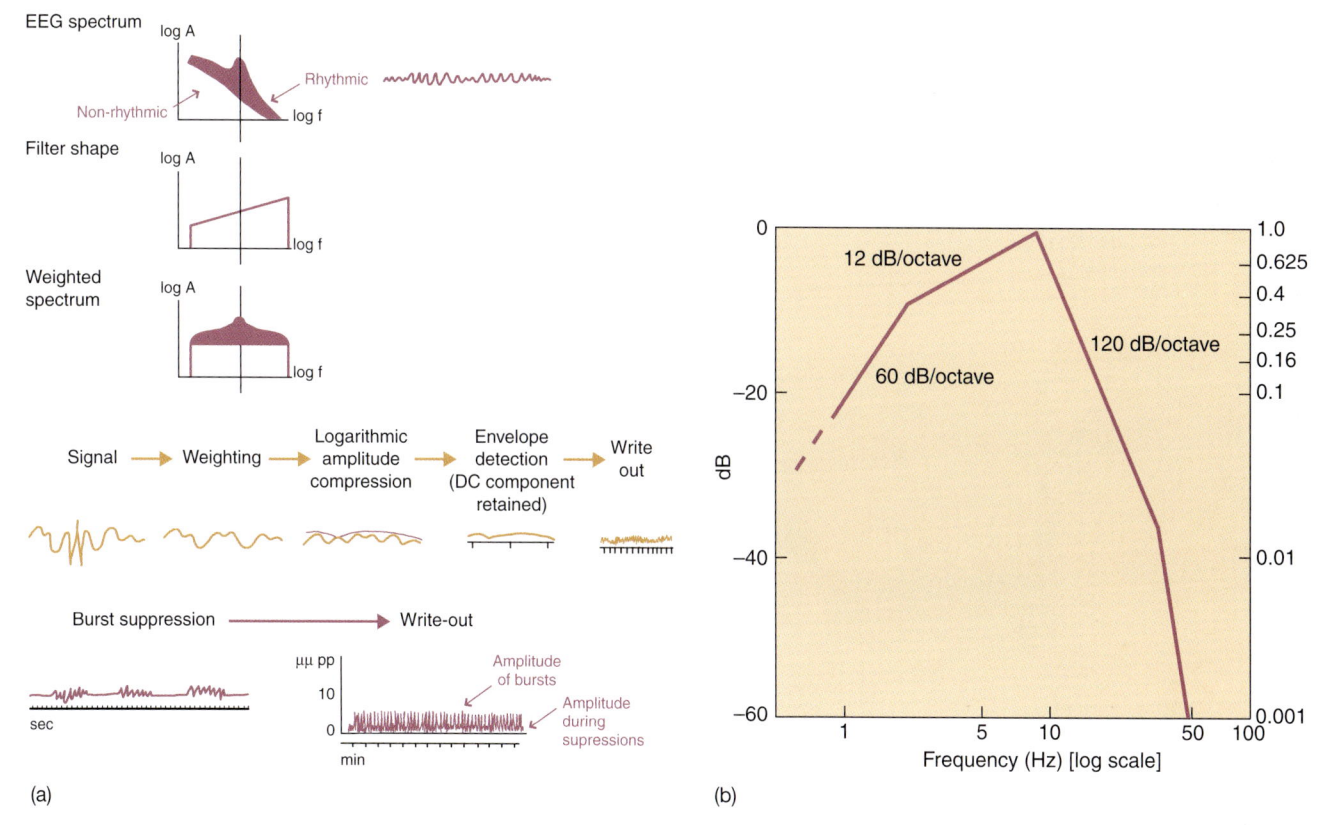

Fig. 3.6 (a) General principles for the Cerebral Function Monitor (CFM) (Maynard et al 1969). From above downwards: the 'pre-whitening' filter, which weights the EEG spectrum to counteract the normal tendency of slow components to be of larger amplitude than faster ones, is typical of most EEG monitors and is given in more detail in (b). The subsequent sequence of processing (logarithmic amplitude compression and envelope detection) governs the appearance of the write-out which continually moves between minimum and maximum peak-to-peak amplitudes. In burst suppression the processed amplitude display resembles a comb lying on the 0 μV baseline with 'teeth' formed by peaks related to the amplitude of the bursts. (b) Diagram of CFM filter: the design is based on the results of analysis of the EEG as normal subjects drowsed and fell asleep (Maynard 1969, 1972) which showed that the sinusoidal components, alpha and beta, appeared to be superimposed on a background which attenuated with frequency. This suggested a background corresponding to the spectrum produced by randomly occurring excitatory or inhibitory synaptic potentials which, as sleep deepened, was associated with a smaller proportion of excitatory potentials, i.e. it appeared that the EEG had a background from randomly occurring synaptic potentials, on top of which were the sinusoidal waveforms. To properly measure the latter it would be necessary to pre-whiten the spectrum so that the random background was the same at all frequencies. This led to the choice of the pre-whitening being about +12 dB per decade of frequency. (Redrawn from Prior and Maynard (1986), by permission.)

Spectral measures

Criticisms of spectral analysis have been three-fold. First, the premise of stationarity, on which the averaging technique for spectral analysis is based, cannot be justified in many acute monitoring situations. Secondly, any processing based on averaging will mask events that are of short duration in relation to that of the epoch averaged. Thirdly, the traces are difficult to interpret for the less experienced and impossible to measure by hand. Improvements in display, such as the *density modulated array* (DSA) (Fleming and Smith 1979a,b) and the *colour density spectral array* (CDSA) (Salinsky et al 1987), have made for easier understanding and communication but not measurement of the display. Unfortunately, the basic problems of non-stationarity and loss of brief events bedevil applications in the ICU and surgery where extremes of moment-to-moment variability may occur, for example in the burst suppression patterns of major sedation and anaesthesia, or in status epilepticus in the ICU. These matters were explored by Levy (1984a,b, 1987) in relation to surgical applications of EEG monitoring, particularly in cardiac surgery with hypothermia. The essential conflict is between the need for short epochs to enhance immediate detection of change, but longer ones to reduce the effect of EEG variability. Some of the problems may be overcome by the use of zero-crossing computations to estimate frequency (Pronk and Simons 1982) (see *Fig. 3.20*) although this too has limitations in some circumstances such as complex waveforms (see Rampil (1998) and *Zero crossing computations*, p. 333). This was overcome in the CFAM1, a development of the CFM mentioned above, by using two baseline crossing

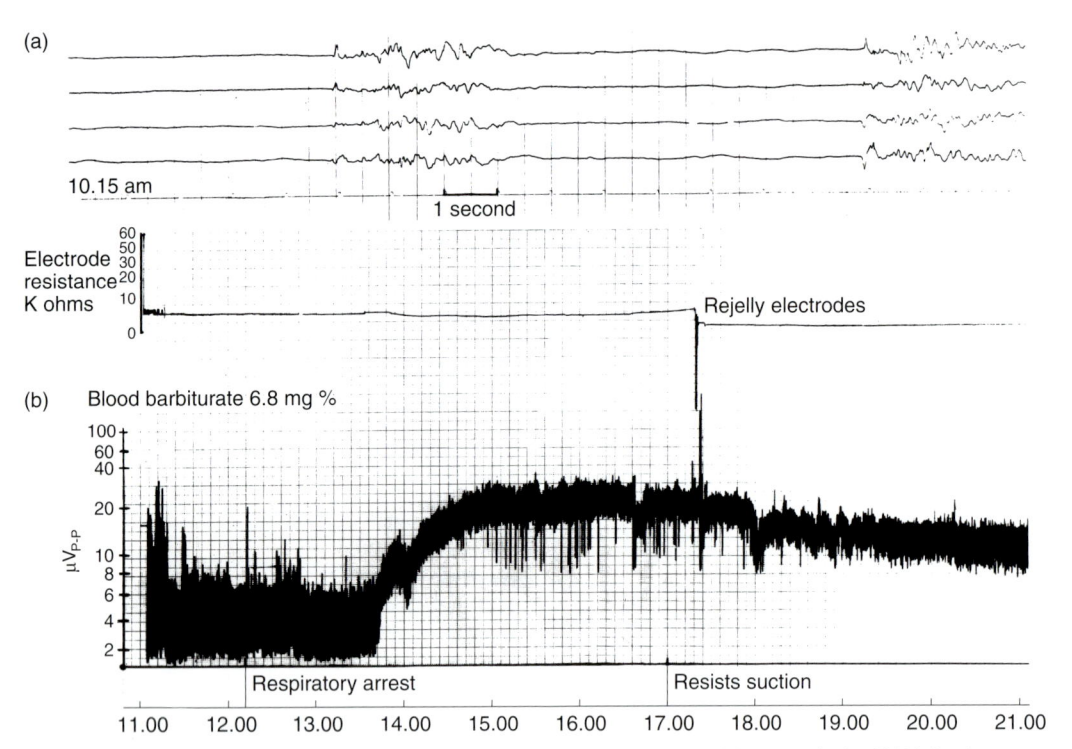

Fig. 3.7 Initial EEG sample and subsequent CFM monitoring in a 48-year-old woman in the ICU following a barbiturate overdose. (a) 18-sec EEG sample (anteroposterior parasagittal derivations) showing marked burst suppression from preliminary recording 4 h after an isoelectric EEG showed her to be deeply unconscious. (b) Continuous CFM recording began 50 min later; the first 10 h are shown here. The two simultaneous outputs from the CFM are electrode impedance and amplitude analysis of the EEG. CFM showed a low-level broadband trace (note the (semi-)logarithmic scale) during EEG burst suppression which was unchanged on respiratory arrest at 12.12 h and subsequent ventilatory support. Peak-to-peak EEG amplitudes increased from 13.43 h and the trace gradually became more variable as the patient recovered consciousness. Note reduction of electrode resistance after re-jellying of electrodes at 17.20 h. (Modified from Prior (1973), by permission.)

detectors, one applied to all frequencies and the other only to higher frequencies, to give percentage amplitudes in EEG frequency bands very close to the percentages produced by Fourier analysis (DE Maynard, personal communication, 2005). Understanding of spontaneous short-term variability is essential for interpretation of spectral EEGs (Oken and Chiappa 1988).

Univariate spectral measures It is possible to quantify and display various spectral indices such as median frequency or spectral edge frequency from the qEEG in order to simplify presentation into univariate measures for comparison with other variables. These proved highly effective in pharmacodynamic studies of anaesthetic agents during the middle ranges of anaesthetic depth, providing the burst suppression pattern was not present in the EEG (Hudson et al 1983, Schwilden et al 1985) (*Fig. 3.10*). However, it is doubtful if derived univariate spectral measures are ever legitimate in the presence of rapid alterations in the nature of the EEG which lead to non-stationarities, for example pronounced forms of its inherent *moment-to-moment variability* (see below), abrupt *arousal phenomena* or *burst suppression pattern*. Similarly, *alternating patterns*, including *sleep cycles*, in which amplitude alters more than frequency may not be well characterized by univariate spectral measures alone (*Figs 3.11* and *6.5*). These EEG features are sufficient to explain why the early optimism regarding close relationships between anaesthetic concentration and univariate spectral parameters were not followed by a universally applicable monitor for depth of anaesthesia or sedation in the ICU. Tracking the full range changes through the whole range of anaesthesia or sedation from induction to deep surgical anaesthesia or the heavy sedation to the level of burst suppression used in brain protection regimens for trauma patients with raised intracranial pressure involves more sophisticated signal processing. Nonetheless, in many circumstances, the ability to select groups of effective univariate spectral measures and frequency ratios for a particular clinical purpose may provide an invaluable qEEG facility, particularly when used as an adjunct to conventional EEG monitoring. The study by Claassen et al (2004) provides a good example of methodology where the aim was to identify the qEEG parameters most sensitive and specific for detection of changes due to delayed cerebral ischaemia in stuporous or comatose patients after subarachnoid haemorrhage (SAH). Parameters calculated from the EEG after a standardized alerting stimulus from four recording sites (overlying areas perfused by different cerebral arteries) included absolute power, relative power ratios, coherence and aver-

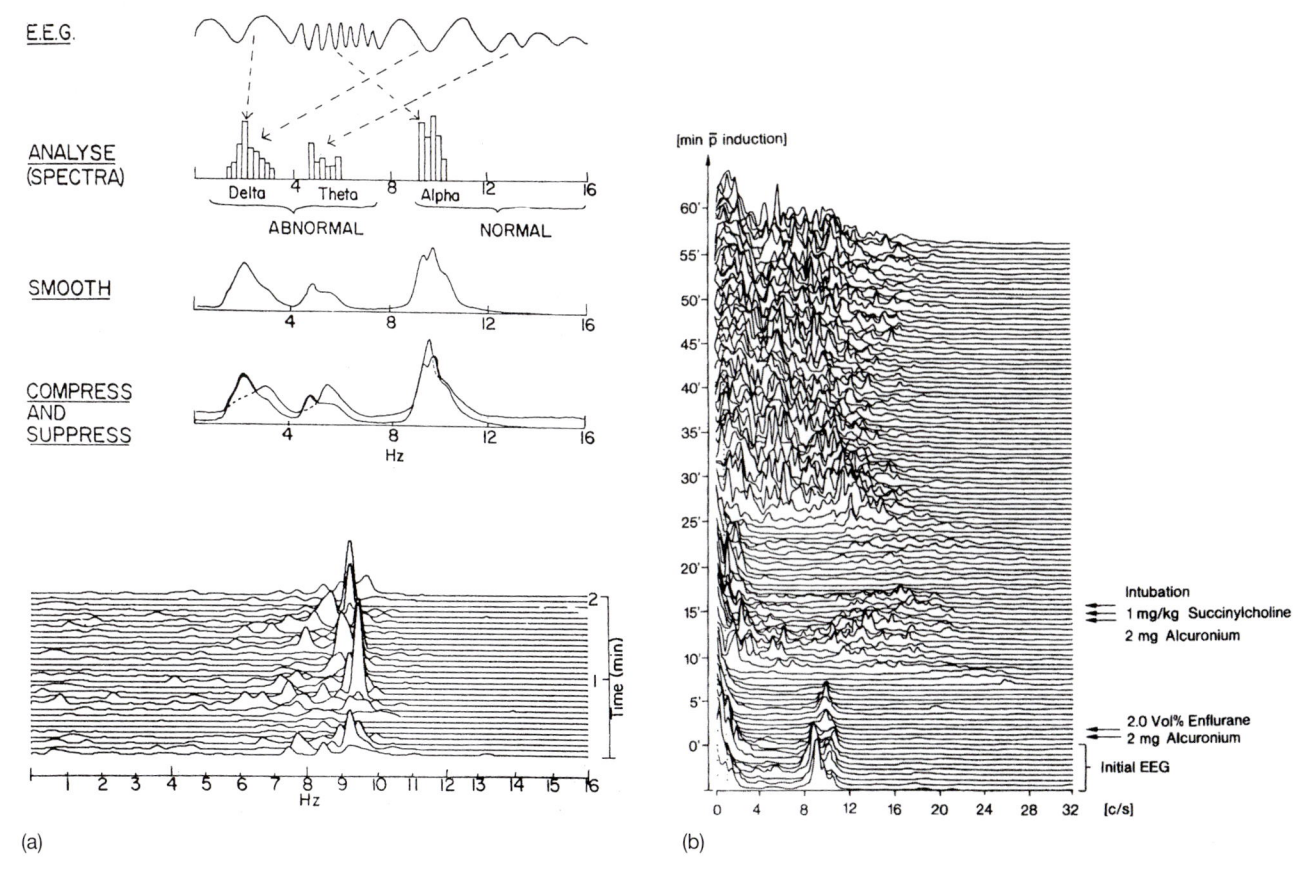

Fig. 3.8 (a) General principles for the CSA. Frequency spectra averaged from each analysis period ('epoch') are plotted sequentially as histograms. Duration of epochs may vary between a few seconds to a minute or more; here a 4 s epoch has been used. After smoothing and suppression of 'hidden lines', the write-out resembles a series of mountain ranges. The 'height' of the mountains usually represents power or amplitude. (From Bickford et al (1973), by permission). (b) Effect of intubation in a 32-year-old woman following enflurane induction. CSA display of EEG from C3-P3; TC 0.3 s, HF 70 Hz; FFT in 30 sec periods with time running upwards from the bottom of the figure. (From Pichlmayr et al 1984 *The Electroencephalogram in Anesthesia. Fundamentals, Practical Applications, Examples* (English edition), © Springer-Verlag, with kind permission of Springer Science and Business Media.)

age frequency values. In the event, the alpha/delta power ratio proved the most effective tool for monitoring to detect local cerebral ischaemia after SAH (see *EEG recording and interpretation in coma and related states*, p. 189); the paper is cited here as a general model for setting up a study to choose appropriate parameters for continuous monitoring for particular clinical needs.

More recent approaches to the development of monitoring equipment

Since 1980s, approaches to the development of EEG monitoring equipment have been led by three factors: (i) changes in clinical practice of anaesthesia, (ii) increased understanding of the physiological importance of clusters of cyclical fluctuations in biological variables and (iii) the increasing sophistication of data processing tools. Advances in EEG monitoring of sleep and epilepsy have led to imaginative developments in pattern recognition for transient features (including artefacts; see *Artefact identification and handling*, p. 90), cyclical changes (see *Inherent variability of the EEG and other biological rhythms*, pp. 113–114) (Rosa et al 1999) and clusters of patterns which could characterize a particular physiological or clinical 'state'. The general aims for monitoring have focused on clinical quality control and cost-effectiveness, being supported by a range of *pattern recognition* and *knowledge-based* approaches (see *Integration of features into the clinical context – pattern classification*, p. 346). Such systems seek to replicate the significant clusters of (clinico-)neurophysiological patterns found in best clinical practice, and to guide the practitioner accordingly. *Expert systems* can take the process of pattern recognition a step further by reference to analysis of databases of cases where other variables or outcome are known. Although demanding in terms of establishing appropriate learning databases collected under strictly controlled clinical conditions for methodology, they are having an increasingly important role in clinical neurophysiology. Applications include systems to guide clinicians in the optimal control of anaesthesia and sedative regimens using various algorithms (Hudson et al 1983, Frank et al 1984, Schwilden et al 1985, Stoeckel and Schwilden 1986, Thomsen et al 1989, Veselis et al 1991, Webber et al 1994, Liu et al 1997,

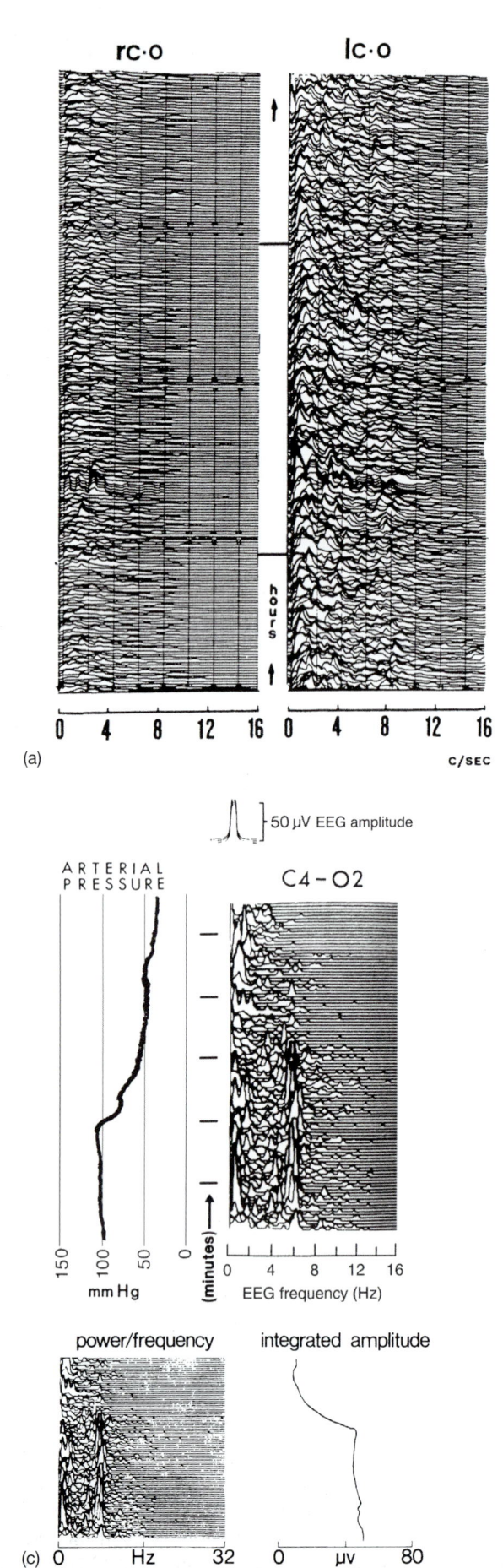

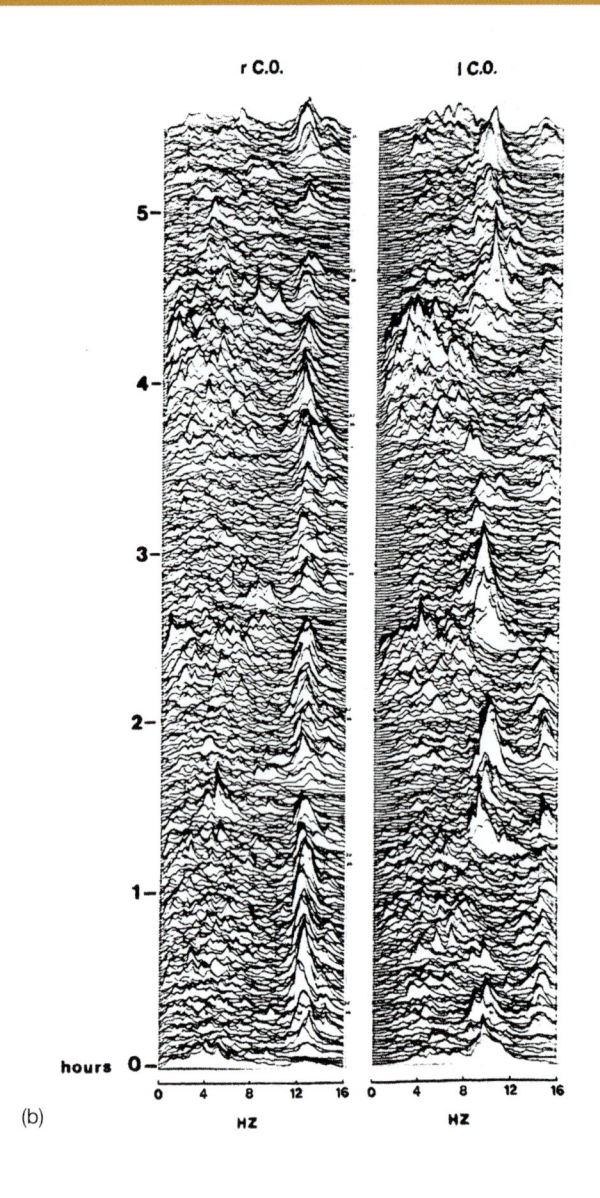

Chapter 3 **Introduction to Methods for Continuous EEG and Evoked Potential Monitoring**

Fig. 3.9 Compressed spectral arrays. (a) CSA from right and left centroparietal derivations (time in hours shown between spectra) in a 40-year-old patient after removal of a right-sided subdural haematoma; it shows an interhemispheric asymmetry consisting of diminished power and slowed frequency on the right side. (From Bricolo et al (1978), by permission.) (b) A more subtle interhemispheric asymmetry in a comatose patient with a sleep-like pattern recorded from the right hemisphere and a changeable pattern of high amplitude slow waves alternating with faster lower amplitude periods. The changeable pattern is usually revealed by prolonged monitoring (from Bricolo et al (1987), by permission). (c) Features of CSA amplitude trace in relation to (top) arterial pressure and (bottom) integrated EEG amplitude in a 66-year-old woman undergoing mitral valve replacement. Recording samples include a period of hypotension in which mean arterial pressure falls from 100 to 35 mmHg following onset of cardiopulmonary bypass. Loss of higher frequency components and overall decrease in amplitude occurs, the latter more clearly displayed on the integrated trace (below). (From Myers RR, Stockard JJ, Saidman LJ 1977 Monitoring of cerebral perfusion during anesthesia by time-compressed Fourier analysis of the EEG. *Stroke* 8:331–337, by permission.)

and frequency domain analysis, together with continuous display of unprocessed EEG signals and electrode impedance for quality control. The more recent CFAM monitors have facilities to display two or four channels with analysis of amplitude, relative frequency content (assessed in the earliest version (CFAM1) with a modified baseline crossing technique, but Fourier analysis in subsequent versions) and burst suppression (*Fig. 3.12(a–c)*). They incorporate statistical facilities for compiling 'normative' databases (e.g. of data from particular anaesthetics, neonatal recordings) to enable on-line statistical comparisons between chosen parameters so that the clinician can, for example, view the position and statistical significance of new data in relation to the limits of the relevant training dataset (*Fig. 3.12(d)*). This has the additional advantage that when non-typical patterns are detected in new patients the user can be warned that they are unsuitable for such comparative classification.

The Advanced Depth of Anaesthesia Monitor (ADAM) (Thomsen et al 1989, 1991) uses pattern recognition techniques based on unsupervised learning and hierarchical analysis to classify patterns characterizing various depths of anaesthesia. Basic amplitude and frequency information is extracted by a multiparametric method from an initial training set of patients under controlled conditions. Patients were anaesthetized with different agents given in standardized concentrations. Autoregressive modelling of 2 s segments enabled classification and learning. The unsupervised learning process involved hierarchical analysis to derive sets of patterns for different anaesthetic drugs studied at standardized concentrations. Each of the corresponding clinical states was assigned a colour code which could be used to guide control of anaesthetic depth with agents on which the algorithm had been trained, at a depth chosen for the type and stage of a particular surgical operation. Results with the ADAM system improved when a burst suppression marker was included as a variable (*Fig. 3.13*); it was then possible to demonstrate a more rational sequence of changes following the course of anaesthesia than with simpler univariate methods (Thomsen and Prior 1996) (see *Fig. 3.24*). *Figure 3.13* also demonstrates the value of both the suppression and asymmetry indices in distinguishing, in this example, lateralized changes during surgical procedures such as carotid endarterectomy from those due to deepening anaesthesia. Use of the ADAM system for anaesthetic studies has also emphasized the fact that individual patients exhibited different EEG patterns for effective depth of anaesthesia whilst at the same end-tidal concentration of a particular volatile agent (Lloyd-Thomas et al 1990) (see *Fig. 5.13*), a fact well recognized in relation to the age of patients receiving intravenous agents (Homer and Stanski 1985). This has implications for design of systems for monitoring depth of anaesthesia based on group data, and perhaps means that there was some merit in the early attempts at individual servo-control of anaesthesia based upon EEG feedback.

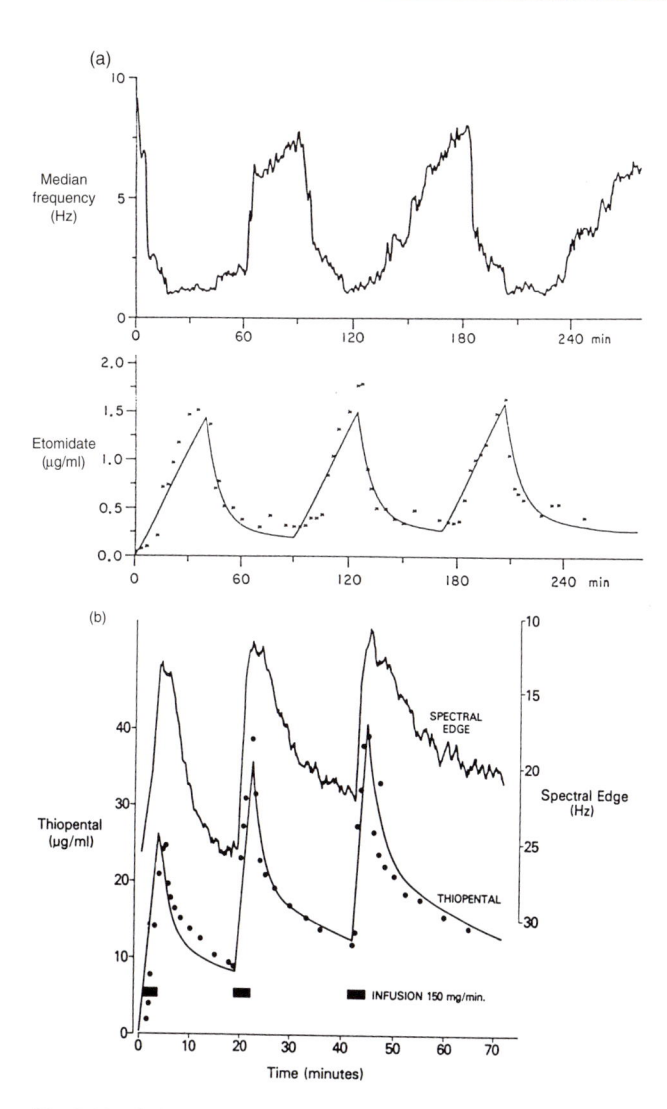

Fig. 3.10 Relationship between anaesthetic concentration and derived EEG measures. (a) Median frequency versus plasma concentrations of etomidate with repeated cycles of infusion. (From Schwilden et al (1985), by permission.) (b) Spectral edge frequency versus serum thiopental levels with intermittent infusions. (From Hudson RJ, Stanski DR, Saidman LJ, Meathe E 1983 A model for studying depth of anesthesia and acute tolerance to thiopental. *Anesthesiology* 59:301–308, by permission.) Such close relationships can be demonstrated during light to moderate anaesthesia when a single anaesthetic agent is used, but are less evident when several anaesthetic and analgesic drugs are used. (Reproduced with permission of the authors and publishers, on behalf of the European Academy of Anaesthesiology from Schwilden H, Schüttler J, Stoeckel H 1985 Quantitation of the EEG and pharmacodynamic modelling of hypnotic drugs: etomidate as an example. *European Journal of Anaesthesiology* 2:121–131, Cambridge University Press, by permission.)

Eckert et al 1997, Sebel et al 1997, Schmidt et al 2003, Prichep et al 2004, Vakkuri et al 2004).

The Cerebral Function Analysing Monitor (CFAM) (RDM Consultants, Uckfield, Sussex, UK; http://www.cfams.com) (Maynard and Jenkinson 1984) was developed from the original CFM of the late 1960s (see *The Cerebral Function Monitor*, p. 76). It provides combined displays from time

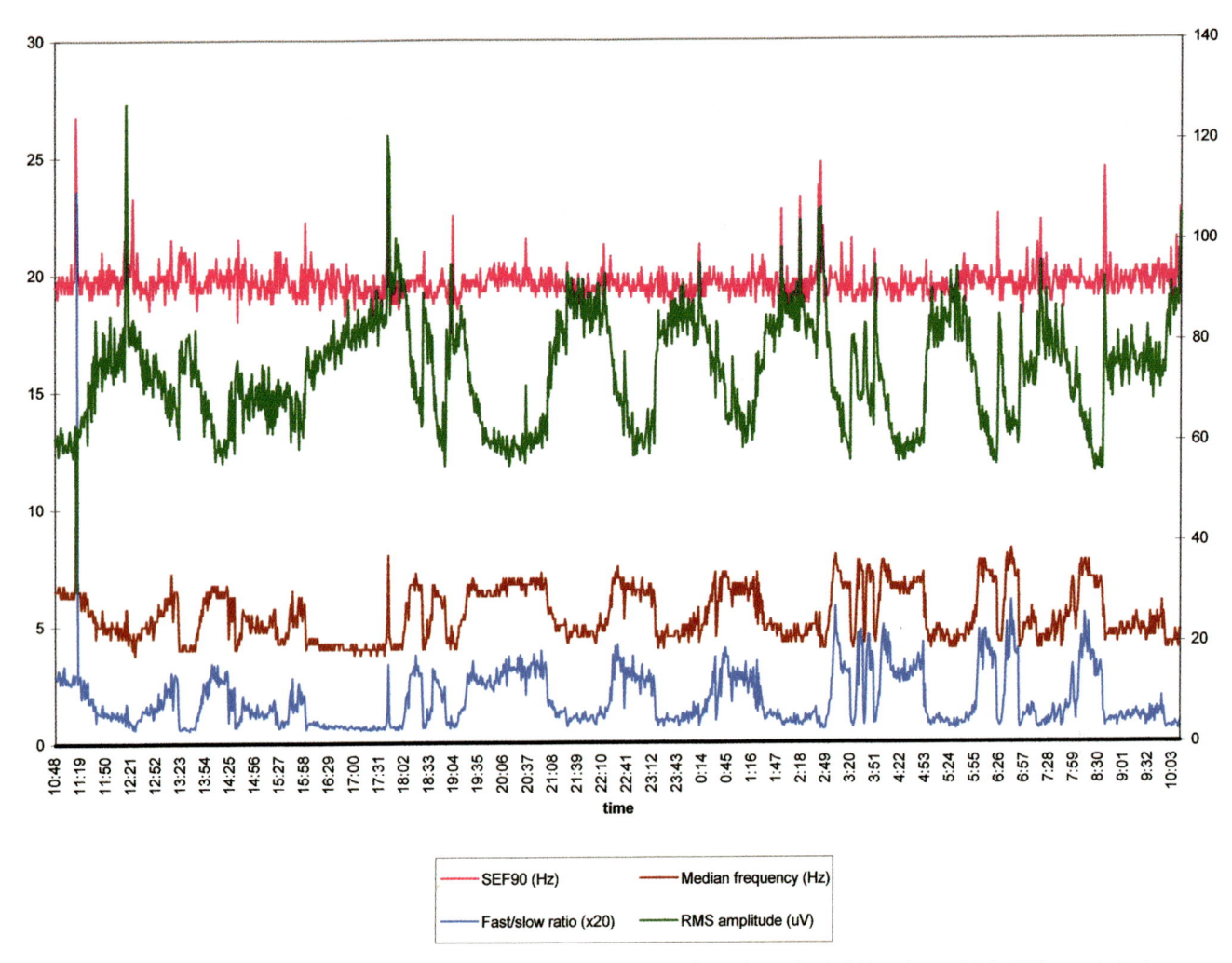

Chapter 3 Introduction to Methods for Continuous EEG and Evoked Potential Monitoring

Fig. 3.11 Comparison of three univariate EEG measures, root-mean-square (r.m.s.) amplitude (μV scale on right), 90% spectral edge frequency (SEF90) and median frequency (Hz scale on left) with a fast/slow ratio, in a 24-hour recording from an 18-year-old man who exhibited unusual cyclical oscillations in multisystem parameters including EEG following high dose thiopental therapy for status epilepticus secondary to herpes zoster encephalitis (see also *Fig. 6.5* which shows the relationship between r.m.s. amplitude, spectral edge frequency, burst suppression, mean arterial pressure, external stimuli and arousal patterns during the same period of recording. (Unpublished data from the patient reported by Ghosh et al (2000), from the EU Biomed-1 Project IMPROVE, Cerutti and Saranummi (1997), courtesy of Professor N Saranummi. From Binnie et al (2003), by permission.))

Bispectrum analysis uses a calculation of bicoherence to measure the extent to which two frequency components in a signal are related to a third, their sum (Dumermuth et al 1971). Thus it might be used to indicate how much of a second harmonic is related to a fundamental, and to examine their phase relation (biphase). The methodology has been applied to clinical monitoring in an attempt to overcome assumptions about linear sources of signals and to include phase information. The Bispectral Index (BIS) was developed by Sigl and Chamoun (1994) to simplify EEG interpretation during monitoring; this resulted in a clinical monitor, the BIS Monitor (Aspect Medical Systems Inc., Newton, MA, USA; http://www.aspectmedical.com). A surge of interest developed, particularly for applications such as anaesthesia for short day-case surgical procedures and sedation in the ICU (Rampil 1998). Reports suggest that monitoring with BIS can have considerable benefits, for example the amount of anaesthetic or sedative drug and subsequent recovery time can be reduced along with hospital costs but with the risk of unwanted awareness under anaesthesia retained (Gan et al 1997, De Deyne et al 1998, Song et al 1998, Dexter et al 1999, Yli-Hankala et al 1999, Recart et al 2003, Myles et al 2004) although not all studies confirm its discriminatory power (e.g. in predicting adequacy of sedation (Gill et al 2004)). Many comparative studies have suggested that BIS, like various other modern monitors, has considerable advantages over monitoring with conventional EEG recording and visual assessment of the signal. Any new method with extensive impact leads to strong and healthy discussion or even controversy; scientific journals and their editors have an important role in publishing both sides of a case and using editorial comment to assist clinicians in taking an informed view. A few examples follow. Anxieties were expressed about limitations in the published information about methodology for BIS, for example some reports merely described the use of a 'black box' (see Todd 1998) but a commissioned review of methodology (Rampil (1998), with the assistance of the manufacturer) has helped to clarify the details (*Fig. 3.14(a,b)*). Some reports have concerned a possible confounding influence on BIS from facial electromyographic (EMG) potentials in some circumstances, e.g. related to high frequency (gamma) activity in the EEG, to neuromuscular blockade and to fentanyl-induced rigidity (Sleigh et al 2001, Renna et al 2002, Messner et al 2003, Vivien et al 2003), a point of interest in view of the recommended electrode placement on the forehead (see *Electrodes for intraoperative recording*, p. 59). Others question whether BIS adds complexity but not extra performance compared to power spectral analysis in defining depth of anaesthesia (Miller et al 2004, Tempe and Satyanarayana 2004) or note its failure, in common with spectral edge and median frequencies, to perform as well as auditory evoked potential (AEP) indices in tasks such as prediction of movement in response to laryngeal mask airway insertion (Doi et al 1999). Similar discussions have related to the ability of the signal processing and simple numerical display in the BIS system to provide the anaesthetist with a consistent means of reducing awareness during routine general anaesthesia (e.g. Bevacqua and Kazdan 2003, Sneyd 2003, Sebel 2004). An important, detailed, analysis of different conditions that could result in the BIS indicating an incorrect hypnotic state has been provided by Dahaba (2005). It includes examples, categorized by the BIS Monitor model number, of paradoxical BIS changes with anaesthetics, electrical devices interfering with BIS monitoring, effect of different clinical conditions (including hypoglycaemia, cardiac arrest, hypovolaemia, cerebral ischaemia, hypo- and hyperthermia), modification by abnormal EEG patterns (including post-ictal, Alzheimer's dementia, cerebral palsy, severe brain injury, brain death and low voltage EEGs) and the effects of EMG and neuromuscular blocking agents. Two recent large prospective studies have now confirmed its superiority compared to routine anaesthetic management or historical controls in this regard (Ekman et al 2004, Myles et al 2004) – but beg the question as to whether this facility is specific to BIS or whether other methods for EEG monitoring would be equally effective.

The Patient State Index (PSI) (Physiometrix, North Billerica, MA, USA; now Hospira Inc., Lake Forest, Illinois, USA; http://www.hospira.com) uses the derived qEEG (relative power in delta, theta, alpha and beta frequency bands after a z-score transformation for age) in a multivariate algorithm that varies as a function of hypnotic state. The development was based on a study of 306 patients in a multicentre study (Drover et al 2002). Data are recorded from two anterior, one midline central, and one midline posterior scalp locations because of regional variations in the effects of anaesthetics and responses to peroperative stimuli (Rundshagen et al 2004). The performance of the algorithm shows how the variables chosen for this index are not agent specific but behave reasonably similarly across different classes of anaesthetic agents (*Fig. 3.15*) (Prichep et al 2004).

The Narcotrend monitor (Monitor Technik, Bad Bramstedt, Germany; http://www.narcotrend.de) (Schultz et al 2002) uses an algorithm based on the classification of EEG changes during anaesthesia by Kugler (1981) which is comparable to that of Bickford (1951b) (see *Fig. 3.5*) except for the addition of a burst suppression stage (*Fig. 3.16(a)*). The processing algorithm (*Fig. 3.16(b)*) originally provided a six-letter classification, A (awake) to F (increasing burst suppression), which included 14 substages (Schmidt et al 2003, Kreuer et al 2003) but has now, with the latest version, Narcotrend 4.0, moved to a 100–0 scale (Schultz et al 2004). So far only limited studies are available (Bauerle et al 2004, Kreuer et al 2004a,b, Schneider et al 2004, Kreuer et al 2005).

Entropy is a measure that describes the irregularity, complexity or unpredictability characteristics of a signal (Shannon 1948) and can be applied to the power spectrum (Johnson and Shore 1984). It is used in the algorithm for a recently developed monitoring system (M-ENTROPY Module, GE Healthcare, Helsinki, Finland; http://www.gehealthcare.com) (Viertiö-Oja et al 2004, Vakkuri et al 2004) to assess EEG and EMG signals from the frontal region (forehead) in the steady state (state entropy (SE)) and (response entropy (RE)) as might occur during anaesthesia and surgery. SE and RE originate from the same signal picked up at the same sensor at the same time – but RE is calculated using a different (higher) frequency range from the signal than SE. Thus SE captures (mostly) EEG activity, whereas the higher frequency range for RE includes both EEG and frontalis muscle EMG components. As the proportion of EMG rises, for example near emergence from anaesthesia or in reaction to noxious stimuli, RE will change rapidly. So, sudden increase of RE then indicates emergence or maybe inadequate analgesia (for details of the method see *Measures of complexity – entropy*, p. 345). In a study of 70 patients (30 anaesthetized with propofol, and 20 each with sevoflurane or thiopental) there was a comparable degree of distinction between conscious and unconscious states with RE, SE and BIS; RE, however, gave the clearest indication of emergence from anaesthesia. During burst suppression (only present in the patients anaesthetized with propofol or thiopental), RE and SE decreased monoton-ously whilst BIS showed biphasic behaviour with a clear linear relation with BSR when suppressions were increasingly prominent, i.e. as a burst suppression ratio (BSR) of > 50% (*Fig. 3.17*) (Vakkuri et al 2004).

Technical requirements for EEG monitoring in the ICU and during surgical operations

With the present surge of interest in clinical monitoring and in partnerships between biosignal analysis experts and

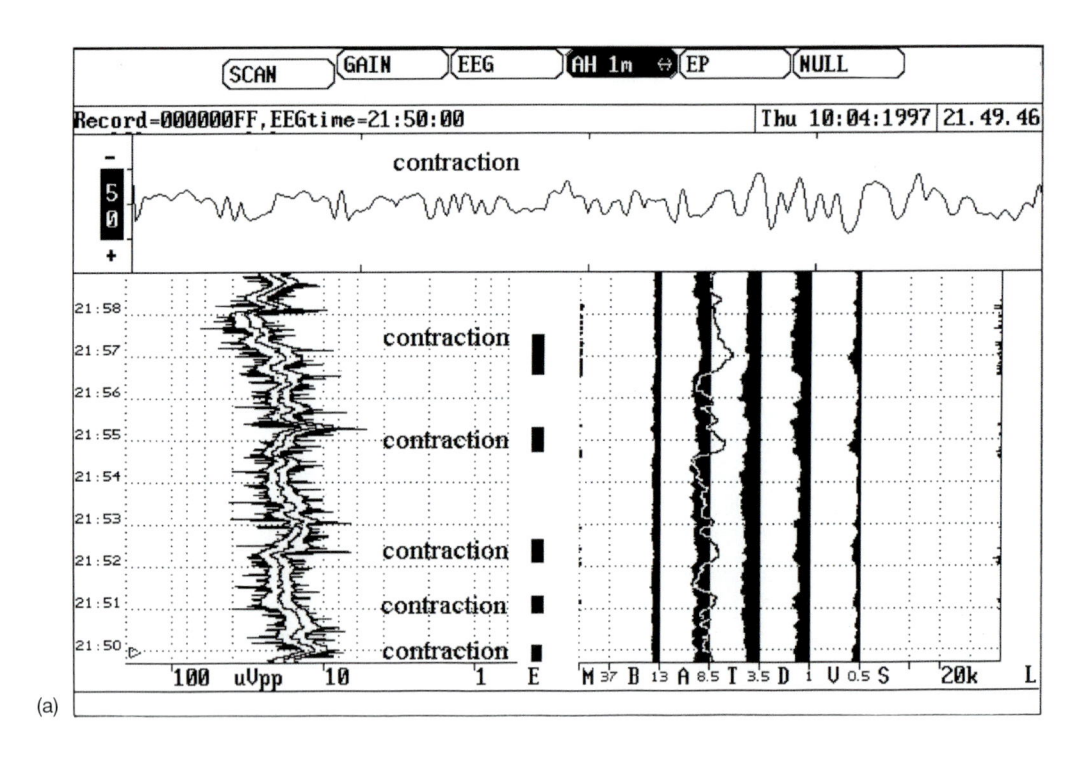

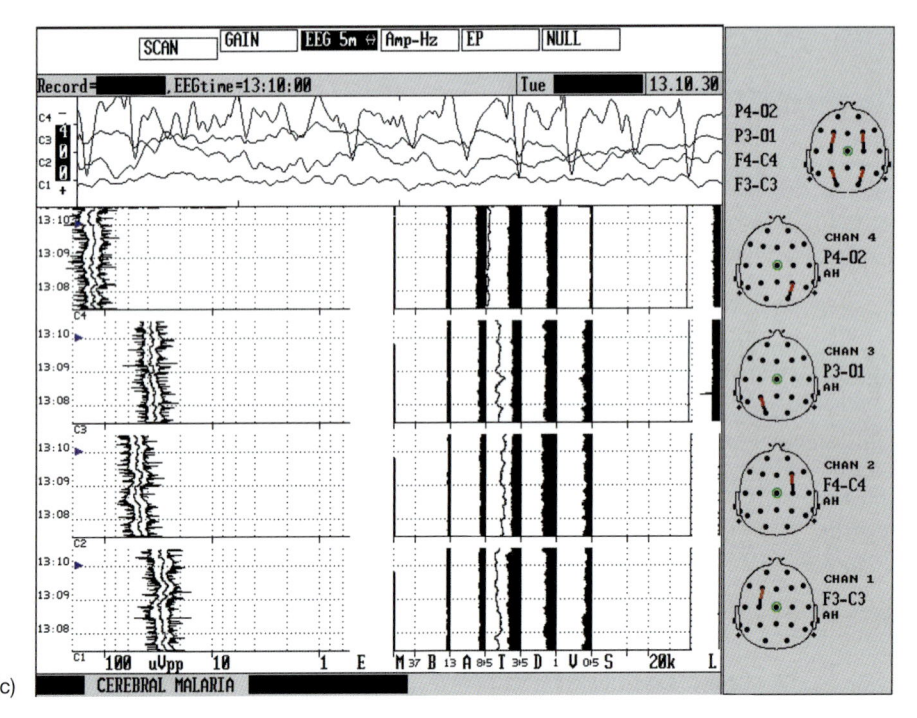

Chapter 3 Introduction to Methods for Continuous EEG and Evoked Potential Monitoring

Fig. 3.12 (a) The Cerebral Function Analysing Monitor (CFAM2) showing intrapartum fetal EEG (FEEG) during uterine contractions. Note the standard CFAM3 display with, from the top, function buttons, patient and recording identification details, and a quality control sample of unprocessed EEG taken at 21.50 h and marked with a triangle at 21.50 h. on the time scale of the vertical processed trace below. This displays amplitudes as maximum (90th centile), mean (l0th centile) and minimum values, calibrated as μVpp at l0 Hz. To the right are shown: percentage values (relative EEG frequency content) for scalp muscle potentials (M), beta (B), alpha (A), theta (T), delta (D), subdelta (V) and suppression (S). A plot of mean frequency is superimposed. The duration of each contraction is marked; note that these are associated with an increase in %delta leading to a reduction in mean frequency and, in some instances, a reduction in FEEG amplitude. Electrode impedance and line interference (L) are indicated on the far right. (b) Asymmetrical CFAM3c recording from a 3-week-old, 28 week gestation, neonate with an extensive unilateral subpial haemorrhage (confirmed at autopsy). Periods of higher amplitude correspond to times during which the comb-like EEG waves seen above are occurring. There is 50 Hz interference particularly on the right which has been filtered from the EEG and is outside the range of frequencies used for amplitude–frequency analysis. (From Hellström-Westas L, De Vries L, Rosén I 2003 *An Atlas of Amplitude-Integrated EEGs in the Newborn*, Parthenon Press, by permission. Courtesy of Dr M Thoresen.) (c) Status epilepticus in a 3-year-old boy with cerebral malaria. He had been admitted unconscious following a 5-day history of fever, with two generalized convulsions within 24 h of admission. He was unconscious (unable to localize a painful stimulus), but without localizing or lateralizing signs and within 2 h of admission (12.08 h), he had another generalized convulsion lasting 5 min. This 3 min extract from a four-channel CFAM3 monitor starts an hour later and shows, from above downward, simultaneous 4 s EEG samples from alternate right and left sided parieto-occipital and frontocentral regions, followed by simultaneous CFAM amplitude and frequency analysis outputs from the same derivations (mean frequency superimposed on frequency analysis), the timing of the EEG samples being marked on each analysis trace by an arrow. Note the continuous repetitive sharp wave discharge arising from the right parieto-occipital region and the markedly increased amplitude analysis output from the same area (note the amplitude scale; the dynamic range of CFAM3 permits capture of signals in excess of 1 mV peak-to-peak) (Dr DE Maynard, personal communication, 2005). The child was able to localize pain after 8 h and had one further short fit at 18 h, his recovery thereafter was uneventful. Although this four-channel monitor recording highlights localization and the nature of the localized activity, a full diagnostic EEG is also essential. (Courtesy of Dr CRJC Newton.) (d) Graphical display of a record from a 30-weekgestational age infant retrospectively compared with a CFAM database, the dashed lines show the boundaries of the available database. In this example values depict neonatal EEG states; the same thing can be done in real time. Similar data sets can be derived from carefully controlled databases, e.g. for depth of anaesthesia or sedation with different agents, and used to indicate whether new data are typical or show unexpected characteristics (with outlying or atypical values for a particular dataset from a CFAM3 database). (Courtesy of Dr DE Maynard and Dr C Sutton.)

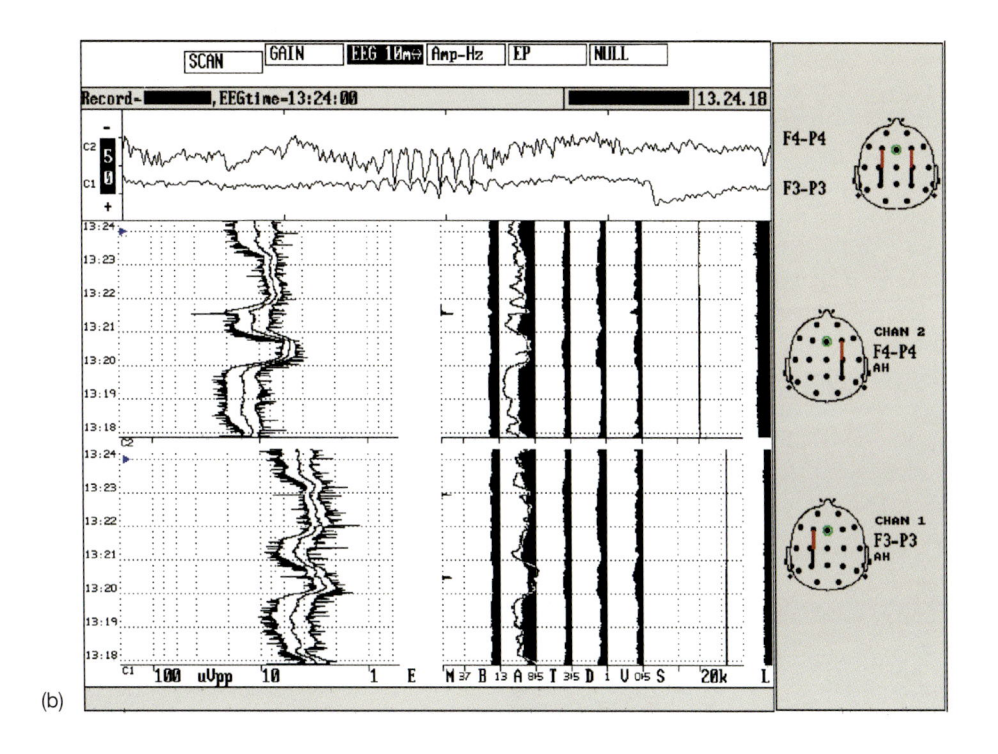

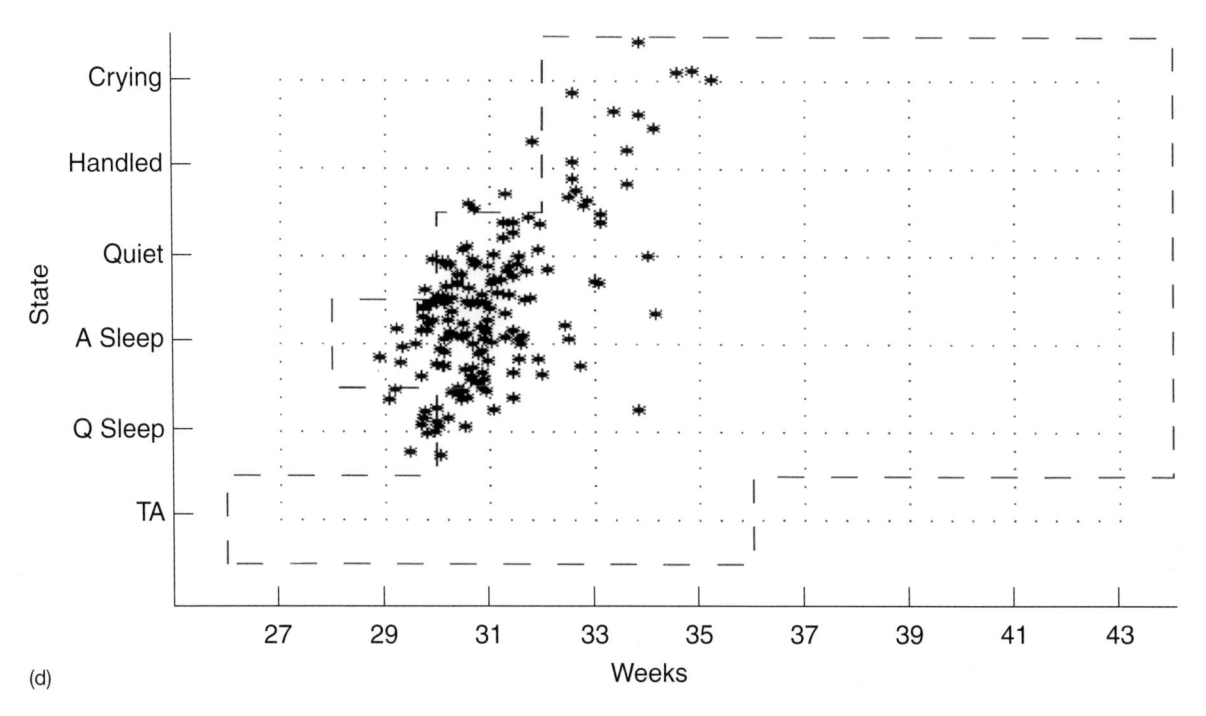

Chapter 3 Introduction to Methods for Continuous EEG and Evoked Potential Monitoring

clinicians, it is probable that effective technologies will continue to evolve rapidly. Cheaper computing power has led to a relative fall in costs for resultant commercial developments and hence wider availability. In such a rapidly changing scene, we have sought to set out some general principles about what to look for in an EEG monitoring device, whether it is a stand-alone monitor or a module in a general anaesthetic or ICU monitoring system, to be sure that it will be able to provide valid information for successful clinical use. Before choosing a monitor, an experienced biomedical/electronics engineer should check the environment where the monitoring will take place. Although hospitals are constructed in such a manner that electrical interference is minimized (the general principles outlined by the Screening Commission of the New University Hospital, Copenhagen (1967) still hold good) not all are adequately designed to meet the demanding task of recording the extremely low amplitude signals of the EEG and EPs. Special arrangements may be required to optimize the quality of monitoring.

The general criteria and the design objectives for neurophysiological monitoring are given in *Table 3.2*. International standards for digital EEG and its analysis (Nuwer et al 1999a,b) and for EEG monitoring in the ICU (Guérit et al 1999) or during anaesthesia and surgery (Burke et al 1999) should be followed. All the relevant international, national and local regulations and specific practical guidelines on safety (see *Safety during neurophysiological recording in the ICU and operating theatres*, p. 55) must be fully understood and adhered to by

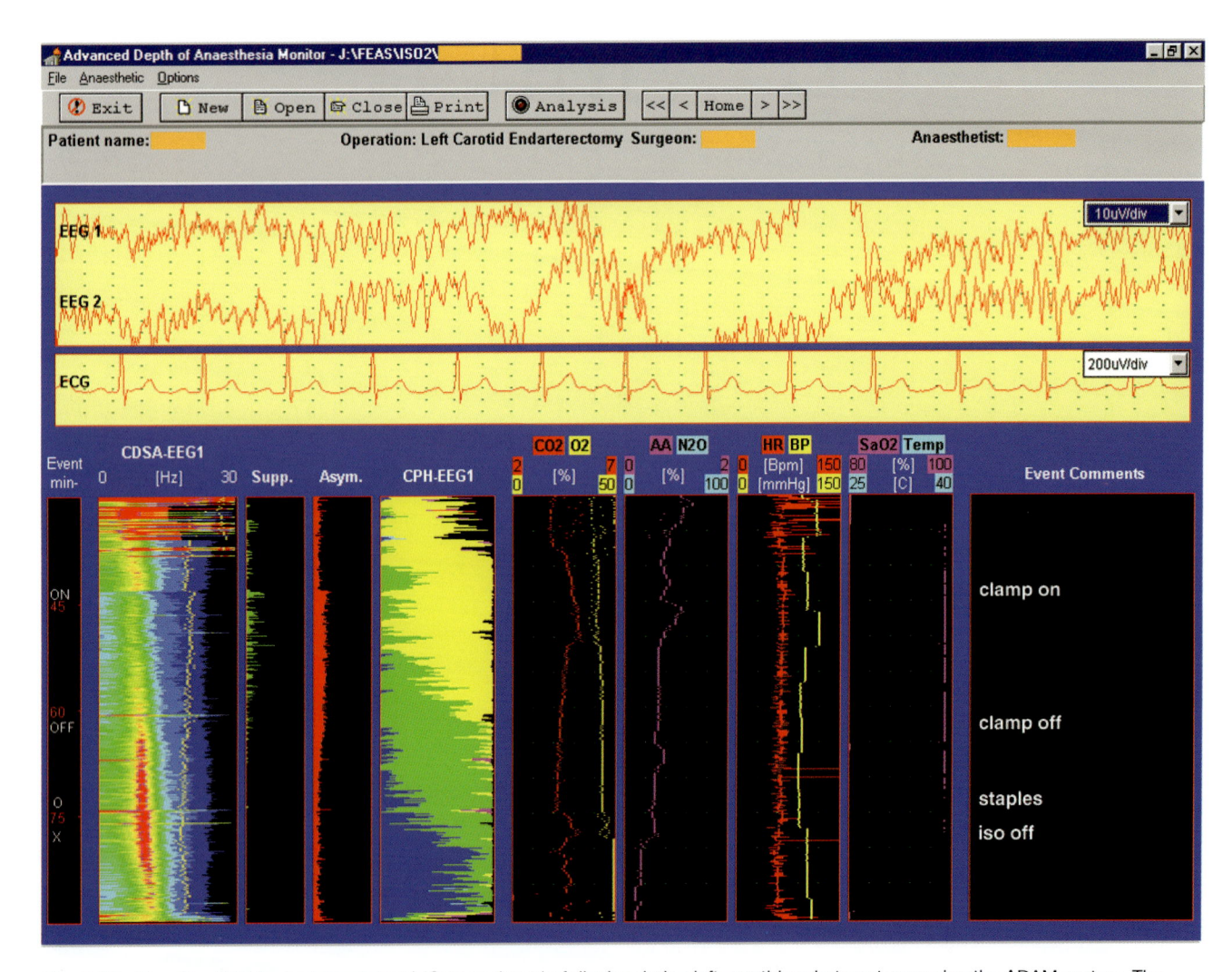

Fig. 3.13 (See *Fig. 7.30* for typical printout.) 'Screen dump' of display during left carotid endarterectomy using the ADAM system. The upper part shows a compressed display of EEG from left and right centroparietal (C3–P3 and C4–P4) derivations and ECG is provided for quality control. The lower part shows, from left to right, EEG analyses from the left (clamped) side with a colour density spectral array (CDSA) (the superimposed yellow dots show spectral edge frequency), a suppression index, an asymmetry index, a class histogram (CPH) for depth of anaesthesia (scaled so that dark blue equates to drowsiness, light blue to very light, green to light, yellow to 'normal', purple to deep, and red to very deep anaesthesia), then CO_2, O_2 and anaesthetic concentrations, heart rate and mean arterial blood pressure, arterial O_2 saturation and temperature. Time runs from the top downwards and surgical events are noted on the extreme right. Note appearance of EEG suppressions and increase in pre-existing asymmetry at the time of decrease in alpha–beta activity during carotid clamping. (Courtesy Dr RM Langford, Dr CE Thomsen and Professor J Lumley.)

Table 3.2 General criteria and design objectives for EEG monitors for the ICU and surgery

Features	Important criteria
1. Patient safety	Must comply with international, national and local safety standards
2. Sensors	Easy application; standard locations; training materials available
3. Data acquisition and basic feature extraction	Should follow international recommendations for bandwidth, sampling and basic processing
4. Detection of asymmetries and lateralized abnormalities	(a) Minimum of two channels of recording (b) Significant asymmetries indicated
5. Polygraphic approach	Useful to synchronize and display EEG output together with other signals (e.g. EMG, EPs, SAP, ICP, etc.)
6. Quality control	(a) Electrode contact impedance monitored continuously (b) Unprocessed ('raw') signals accessible on- and off-line to confirm validity (c) Minimize external interference, detect and mark biological artefacts (d) Derived measures validated by access to raw signal or frequency spectra
7. Annotation facilities	Easy on-line annotation and event marker tools
8. Advanced analysis and display of clinically important features reflecting disease and/or medical or surgical interventions	(a) Steady state (amplitude, frequency, morphology, symmetry, continuity, etc.) and changing features (spontaneous or iatrogenic) (b) Brief events: arousals, seizure discharges, short depressions (c) Trends: over minutes, hours, days (d) EEG variability: e.g. 2–4/min cyclical changes, sleep features, BRAC cycles, burst suppression, etc. (e) Isoelectric EEG: indicated immediately
9. Output	(a) Measurable display/tracing (b) Digital output available for further analysis or signal processing (c) Facility to switch between primary signals, digital data and trend plots (d) Statistical comparison with appropriate, validated, database available on line (e) Permanent record with text report meeting medico-legal requirements
10. Feedback or alarms?	Follow local policy or establish needs with clinicians
11. Fitness for purpose	Monitors with high specificity for quantifying CNS effects of anaesthetic or sedative agents may be ineffective in detecting brain or spinal cord ischaemia or epileptiform discharges. Alternative processing may be needed for different clinical tasks

EMG, electromyogram; EPs, evoked potentials; SAP, systemic arterial pressure; ICP, intracranial pressure.

all personnel developing and using neurophysiological monitoring techniques.

Methodology and recent trends in EEG analysis are described in Chapter 8 and cross-references will be made here to the appropriate sections. Current equipment for EEG monitoring includes a wide range of feature extraction and basic signal processing methods with increasing use of advanced techniques for pattern recognition and expert systems (see reviews by Gade et al 1996, van Gils et al 1997, Rampil 1998). Techniques are evolving rapidly as clinical aims are defined more precisely as the results of evidence-based studies about efficacy of different methods become available and those undertaking monitoring should check the current advice as to optimal methods for the clinical problem concerned. Here we give a few general notes.

Preprocessing (see *Filtering methods*, p. 318)

The design principles for filters in EEG monitoring apparatus concern the balance between a bandwidth which is adequate for faithful recording of the signals of interest and yet sufficiently restricted to minimize unwanted signals. When evaluating a new monitor it has to be borne in mind that filtering to reduce artefacts will lead to the loss of some genuine EEG components – a necessary price to pay when recording in difficult environments such as the ICU or operating theatre where potentials due to both active and passive movements, as well as external electrical interference, may contaminate the primary signals. Examples of filters are shown in *Figs 3.5(a)* and *3.6(b)* and methods are described in detail in *Filtering methods* (p. 318). Frequency selective filters improve the quality of the signal by means of a sharp reduction of signals of less than 1–2 Hz (to avoid contamination by the common low frequency artefacts encountered during ICU and surgical monitoring) and those above 20 Hz to (to reduce high frequency artefacts such as muscle potentials, but also some higher frequency EEG activity such as gamma waves). In addition, most monitors now include a 'pre-whitening' filter to weight the EEG spectrum to counteract the normal

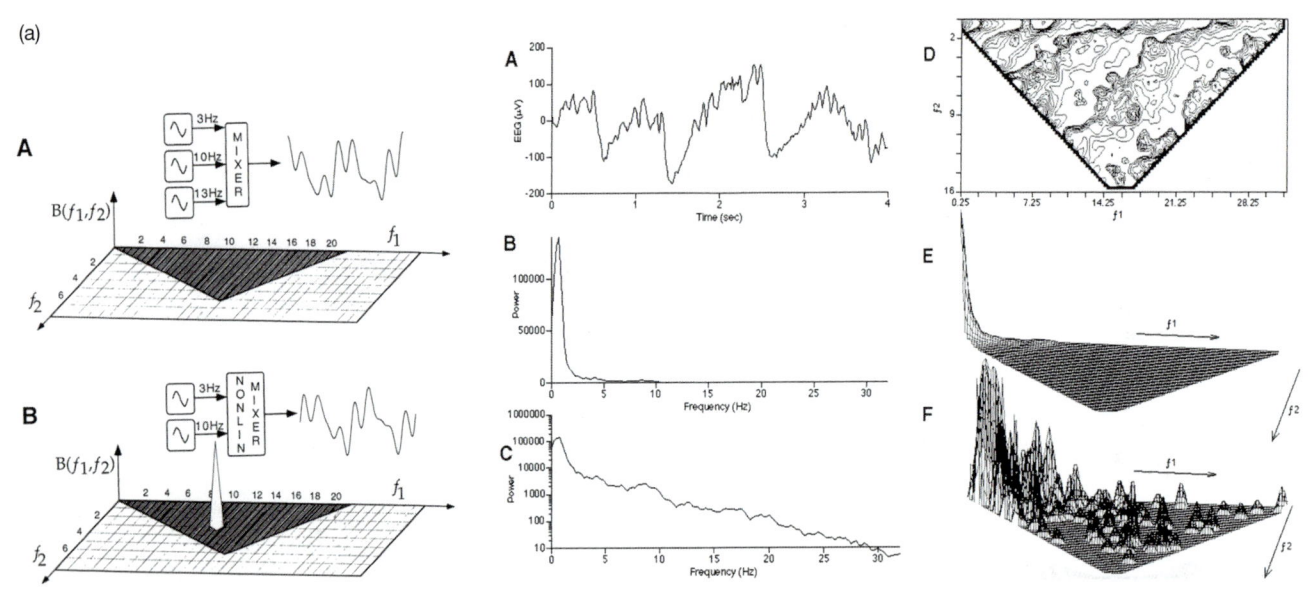

Chapter 3 Introduction to Methods for Continuous EEG and Evoked Potential Monitoring

Fig. 3.14(a) Left: The bispectrum is calculated in a two-dimensional space of frequency$_1$ versus frequency$_2$ as represented by the coarsely cross-hatched area. Because of the symmetric redundancy and the limit imposed by the sampling rate, the bispectrum need only be calculated for the limited subset of frequency combinations illustrated in the darkly shaded triangular wedge. A strong phase relationship between f_1, f_2, and f_{1+2} creates a large bispectral value $B(f_{1+2})$ represented as a vertical spike rising out of the frequency versus frequency plane. Left, A: Three waves having no phase relationship are mixed together producing the waveform shown in the upper right. The bispectrum of this signal is everywhere equal to zero. Left, B: Two independent waves at 3 and 10 Hz are combined in a non-linear fashion, creating a new waveform that contains the sum of the originals plus a wave at 13 Hz, which is phase-locked to the 3- and 10-Hz components. In this case the bispectrum reveals a pool of high bispectral energy at f_1 3 and f_2 10 Hz. Centre and right: Examples of bispectral analysis. (A) An epoch of raw EEG that has been anti-alias filtered and sampled at 125 Hz. (B) The power spectrum resulting from Fourier analysis of the epoch in part A. In this epoch, the large contribution from frequencies around 1 Hz overwhelm the contributions from higher frequencies. (C) The same data as in (B) are plotted with a logarithmic power scale, allowing the contribution from the faster waves to be identified despite the presence of larger slow waves. (D) After computing the complex spectrum X(f) for this epoch of data, the Bispectrum may be calculated. The bispectrum is computed as a function of two frequencies f_1 and f_2. (D) Illustrates the resulting two-dimensional contour plot of the bispectrum of the same epoch. The largest bispectral features here are in the upper left corner, relating pairs of delta-range frequency pairs. The amplitude scale is again logarithmic with the heavy lines indicating powers of 10. (E) The same bispectral data as in (C) plotted in three-dimensional perspective. The frequency scales are the same as part (C). A plot of the real triple product function does not appear to be significantly different than this plot. The amplitude scale here and in (E) is linear. (F) However, when the bicoherence (BIC) is plotted here, phase features are visible in a wide distribution of frequency. (From Rampil IJ 1998 A primer for EEG signal processing in anesthesia. *Anesthesiology* 89:980–1002, by permission.)

tendency of slow components to be of larger amplitude than faster ones (see *Fig. 3.6*) and hence ensure equal representation across the frequency range of interest (see *Digital filters*, p. 318). Pre-whitening prevents high amplitude low frequency activity from dominating the output of the monitor at the expense of lower amplitude activity of higher frequency, and minimizes errors in frequency analysis caused by high amplitude low frequency activity appearing in the side lobes of higher frequency filters (DE Maynard, personal communication, 2005).

To reduce line interference, there is usually an additional notch filter at mains frequency (50 or 60 Hz); specific measures for reduction of interference from surgical diathermy (electrocautery) and related equipment are discussed in *Interference due to surgical diathermy* (p. 60). In reality, external interference from special equipment used in the ICU and during surgery can still cause unexpected intermittent artefacts or disrupt recording, and a routine has to be agreed that due warning about checking new equipment will be given to avoid unexpected problems.

Artefact identification and handling (see *Artefact detection and rejection methods*, p. 327)

A major point of discussion is whether monitor recordings should be 'cleaned up' by removal of all possible artefacts or whether these should merely be identified and marked on the relevant portions of the output display. The present trend is three-fold: first, avoidance of external artefact by good electrode technique; secondly, appropriate filtering to remove unavoidable external interference; and, thirdly, the use of pattern recognition algorithms to identify, classify and mark biological artefacts. Clearly, electrode placement near to the eyes bears the risk of recording large deflections from corneo-retinal potentials during eye movements (see *Fig. 2.30*); none-the-less, biological artefacts in EEG monitoring can sometimes provide invaluable behavioural information about changes in patient state. For example, such artefacts (e.g. *Figs 3.18* and *3.19*), especially when followed by alterations in EEG pattern, can provide evidence of arousal responses to ICU events or draw attention to subtle epileptic seizures. Patients often become restless as they emerge from coma,

EEG monitoring

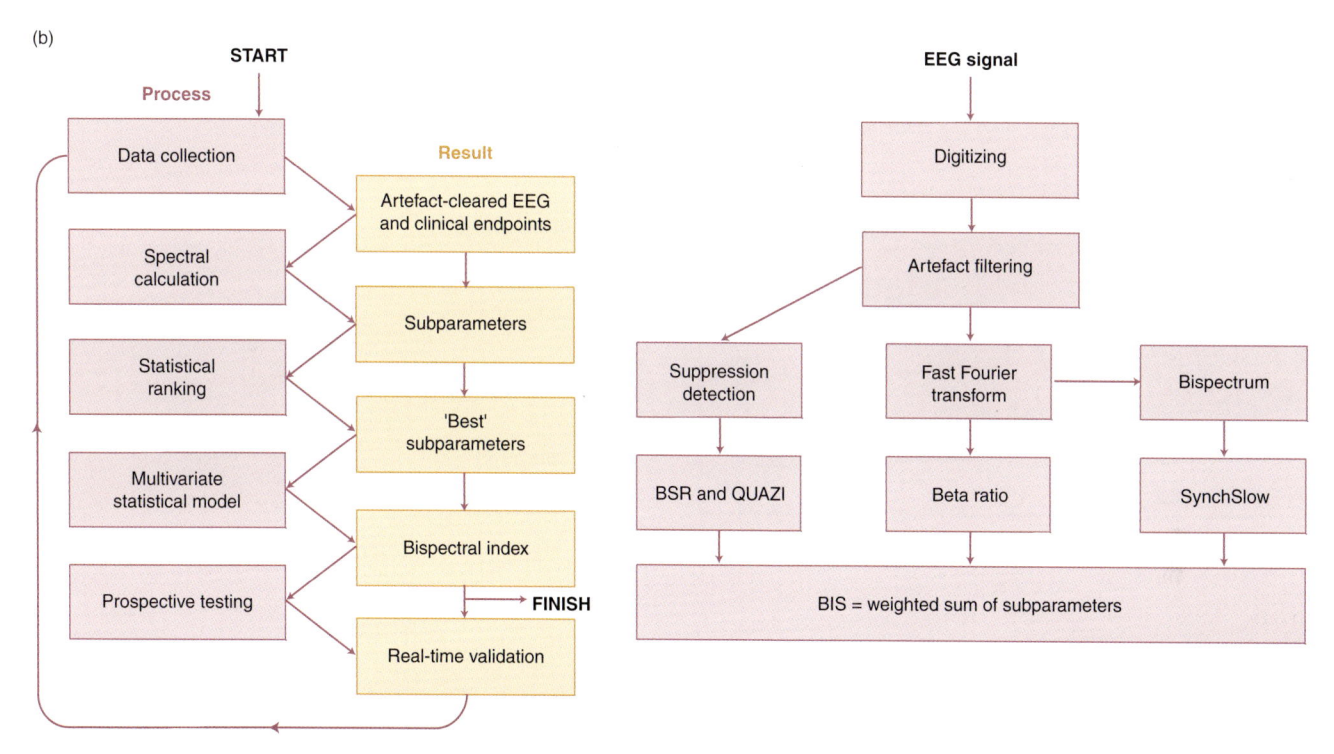

Fig. 3.14(b) Left: The development of BIS by Aspect Medical Systems, Inc., proceeded in a stepwise fashion. First a library of artefact-free EEG (with concurrent behavioural correlates) was accumulated. A range of prospective subparameters were calculated, and their correlation with behaviour was tested. The parameters with the best performance were entered into a multivariate analysis for the creation of a final composite parameter, the BIS. The performance of the BIS has been enhanced by an iterative process that involved at least three major phases of new data collection, modelling, and parameter refinement. Right: Flow chart for the calculation of the Bispectral Index (BIS). (From Rampil IJ 1998 A primer for EEG signal processing in anesthesia. *Anesthesiology* 89:980–1002, by permission.)

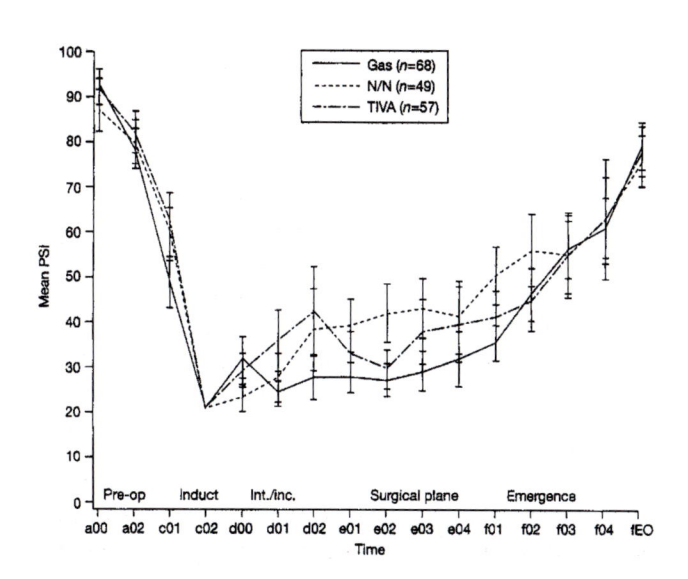

Fig. 3.15 Patient State Index (PSI). Group average curves of PSI values (mean ± 95% confidence interval) as a function of state throughout the surgical procedure in 176 patients. Data shown separately for inhalational anaesthetic (GAS), nitrous oxide/narcotic anaesthesia (N/N) and total intravenous anaesthesia (TIVA). The states plotted include: a00, day before surgery, with no medication; a02, day of surgery, with preoperative sedation given (taken as baseline in the study since a00 was only obtained in a subset of patients); c01, beginning of induction, patient starts counting; c02, at point where patient stops counting, loss of consciousness; d00, just before intubation; d01, just after intubation; d02, incision; e01–e04, anaesthetic maintenance in surgical plane; f01–f04, emergence, decreasing anaesthetic; fEO, eyes open, return of consciousness. (From Prichep LS, Gugino LD, John ER, Chabot RJ, Howard B, Merkin H, Tom ML, Wolter S, Rausch L, Kox WJ 2004 The Patient State Index as an indicator of the level of hypnosis under general anaesthesia. *British Journal of Anaesthesia* 92:393–399. © The Board of Management and Trustees of the *British Journal of Anaesthesia*. Reproduced by permission of Oxford University Press/*British Journal of Anaesthesia*.)

Chapter 3 Introduction to Methods for Continuous EEG and Evoked Potential Monitoring

Fig. 3.16 The Narcotrend monitor. (a) Electroencephalographic patterns observed at different levels of anaesthesia and the respective Narcotrend stages. (b) Narcotrend algorithm processing. EEG during anaesthesia. (From Kreuer S, Biedler A, Larsen R, Altmann S, Wilhelm W 2003 Narcotrend monitoring allows faster emergence and a reduction of drug consumption in propofol–remifentanil anesthesia. *Anesthesiology* 99:34-41, by permission.)

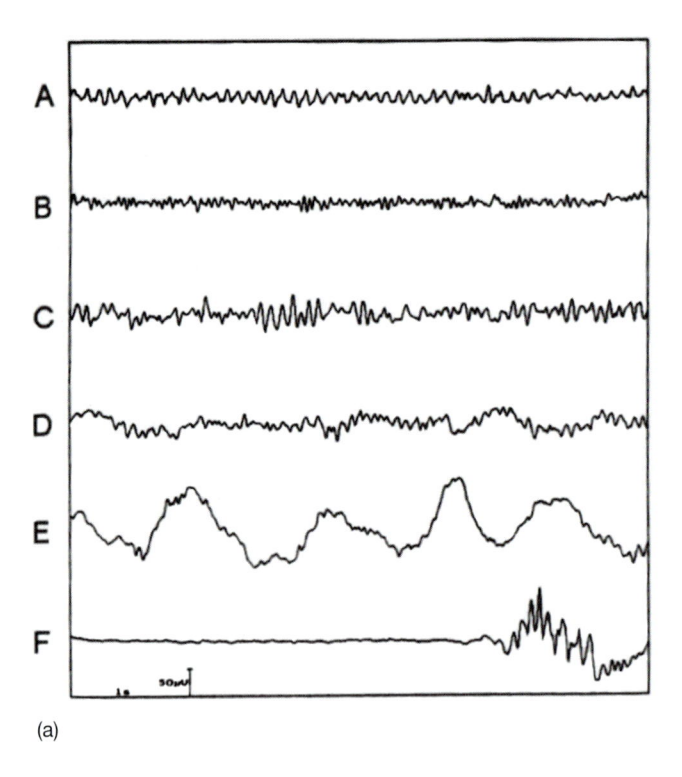

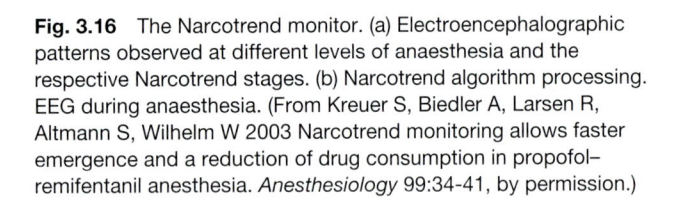

sedation or anaesthesia in an unfamiliar environment and pain or discomfort from routine postsurgical or ICU procedures may lead to complex, temporally related, biological artefacts. Careful annotation of monitor recordings and study of nursing care records, may confirm clustering of artefacts at such times. Cyclical increases of artefacts may relate to periods of 'arousal' and 'sleep' even in comatose patients. Similar event-related artefacts during surgery, although uncommon, may provide invaluable early warnings of inadequate analgesia or anaesthetic depth (and the potential for 'awareness') for a particular patient during a painful part of a procedure.

Monitoring work during major surgery has the additional problem of interference from other equipment near or attached to the patient. The particular problems of surgical diathermy, fibrillators, defibrillators, pacemakers and other cardiac support systems are discussed in Chapter 2 (see *Figs 2.32, 2.33* and *2.34*). Most of these produce artefacts which have little clinical relevance and strenuous efforts are required to identify and eliminate or mark them, not least as they may confuse qEEG assessment of possible cerebral ischaemic changes (Adams et al 1995).

Feature extraction and basic signal processing and interpretation (see *Signal processing and interpretation*, p. 327)

The fundamentals of modern biomedical signal processing and interpretation related to EEG are set out in Chapter 8. From the preceding historical account the reader will have learned that expectations about the performance of equipment depend on (a) ensuring that high quality, valid signals are recorded and (b) that the method of handling them in the monitor's algorithms is relevant to the clinical problem. Factors affecting the performance of processing algorithms for EEG monitors (*Fig. 3.20*) include their sensitivity to abnormalities, to changing abnormalities, to artefacts and to alterations in level of consciousness including drowsiness (Matousek et al 1978, Makeig and Jung 1995), sedation, anaesthesia or coma. From a practical point of view, when undertaking a new monitoring service it is important to define the type of abnormalities expected to provide early warning of unwanted events and then to establish whether the algorithms can detect them in the presence of possible confusing factors such as sedation or anaesthesia. Another approach has been the detection of a significant change from the previous state, as provided by Barlow (1984) using adaptive segmentation

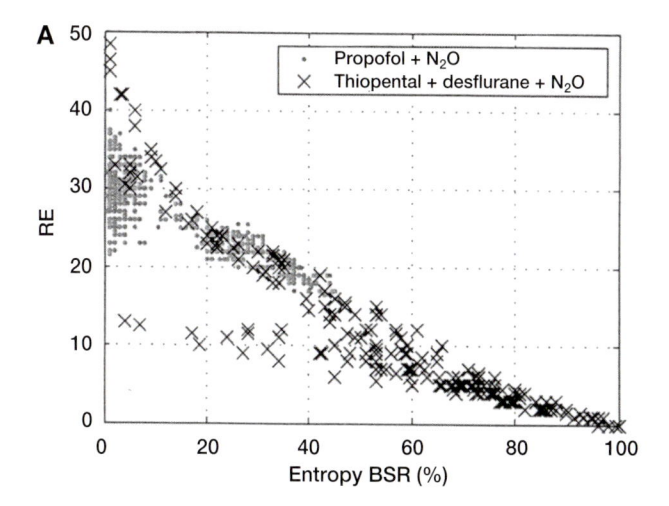

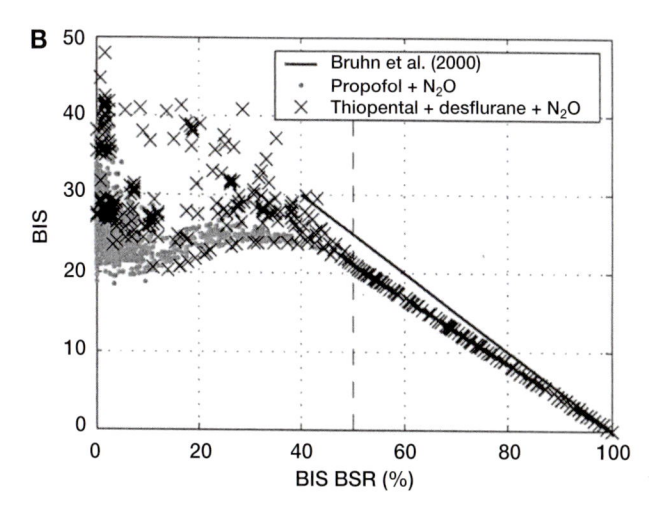

Fig. 3.17 Processing during EEG suppressions. (A) Response entropy (RE) as a function of burst suppression ratio (BSR) measured by M-ENTROPY (Entropy BSR). (✗) Thiopental–desflurane–N$_2$O, (●) propofol–N$_2$O. (B) BIS as a function of BSR measured by BIS monitor (BIS BSR). (✗) Thiopental–desflurane–N$_2$O, (●) propofol–N$_2$O. In the range of BSR > 50, the BIS value follows the linear equation BIS = (42.010 ± 0.005, 1 − BSR/100). Solid line indicates the relation BIS = 50 (1 − BSR/100) obtained by Bruhn et al (2000) with an earlier BIS version, 3.22 (A-1000). (From Vakkuri A, Yli-Hankala A, Talja P, Mustola S, Tolvanen-Laakso H, Sampson T, Viertiö-Oja H 2004 Time–frequency balanced spectral entrop as a measure of anesthetic drug effect in central nervous system during sevoflurane, propofol, and thiopental anesthesia. *Acta Anesthesiologica Scandinavica* 48:145–153. Blackwell Publishing Ltd, by permission.)

in a comparative study of monitoring during carotid artery surgery, and Lipping et al (1995) for detection of burst suppression pattern. Evidence-based studies to define sensitivity and specificity in detection of types of abnormality relevant to the clinical problem are a prerequisite to successful monitoring and such studies are increasingly available in the clinical research literature. Once appropriate processing methods have been found, decisions are required as to how to relate findings to the clinical context and the level and type of decision support that is required (see *Integration of features into the clinical con-text – pattern classification*, p. 346, and *Use of the tools*, p. 358).

Requirements for monitor displays (see *Display of results*, p. 360)

Most clinical, nursing and technological staff working in the ICU or the operating room are experienced in reading complex on-line displays of human physiological variables and related items. The addition of yet another screen or cluster of signals requires thoughtful implementation. The clinical neurophysiology team can and should work together with the front line team to provide expertise in interpretation and help in providing evidence to assist decision-making, as happens for example during spinal or carotid artery surgery. This may be more convenient when separate equipment is used for the neurophysiological monitoring. In contrast, in applications such as routine monitoring of depth of anaesthesia, EEG monitoring modules may be added to the regular anaesthetic monitoring apparatus; generally these are used by the anaesthetist after appropriate training and support in conjunction with the neurophysiology team.

Displays should be as informative and as all-embracing as is consistent with the user's ability to view changing signals of several types simultaneously. The neurophysiologist often needs to show the surgeon a 'before and after' display after an intervention such as clamping a blood vessel as well as showing overall trends after interventions by the anaesthetist, for example to improve level of arterial blood pressure. Fusion of data from multiple signals may be provided, when appropriate, for multivariate trend analysis to define the sequence of a cascade of events or to generate alarms (see *Summary and examples of usage*, p. 353). Facilities for annotation should be easy to use and should be accompanied by a means of providing a printed summary report at the end of surgery to return to the ward with the patient at the end of surgery (*Fig. 3.21*). Such documents are not only of immediate clinical value to guide continuing care of the patient, but also provide essential documentation of the quality of management of high risk situations for research, training and medico-legal purposes.

Quality control and validation

Quality control has many aspects! *Table 3.2* indicates the basic requirements. In addition monitoring equipment and procedures must be safe and regular checks logged by the authorized safety officer. The same applies to training for neurophysiology technological staff who should have accredited training and continuing educational updates for the ICU and intraoperative monitoring work that they undertake. Interpretation of the monitoring information is normally led by an experienced medical clinical neurophysiologist who should be trained at an approved level. When automated decision-making support algorithms are used, it is essential to check their performance against the observations and measurements used in decision-making

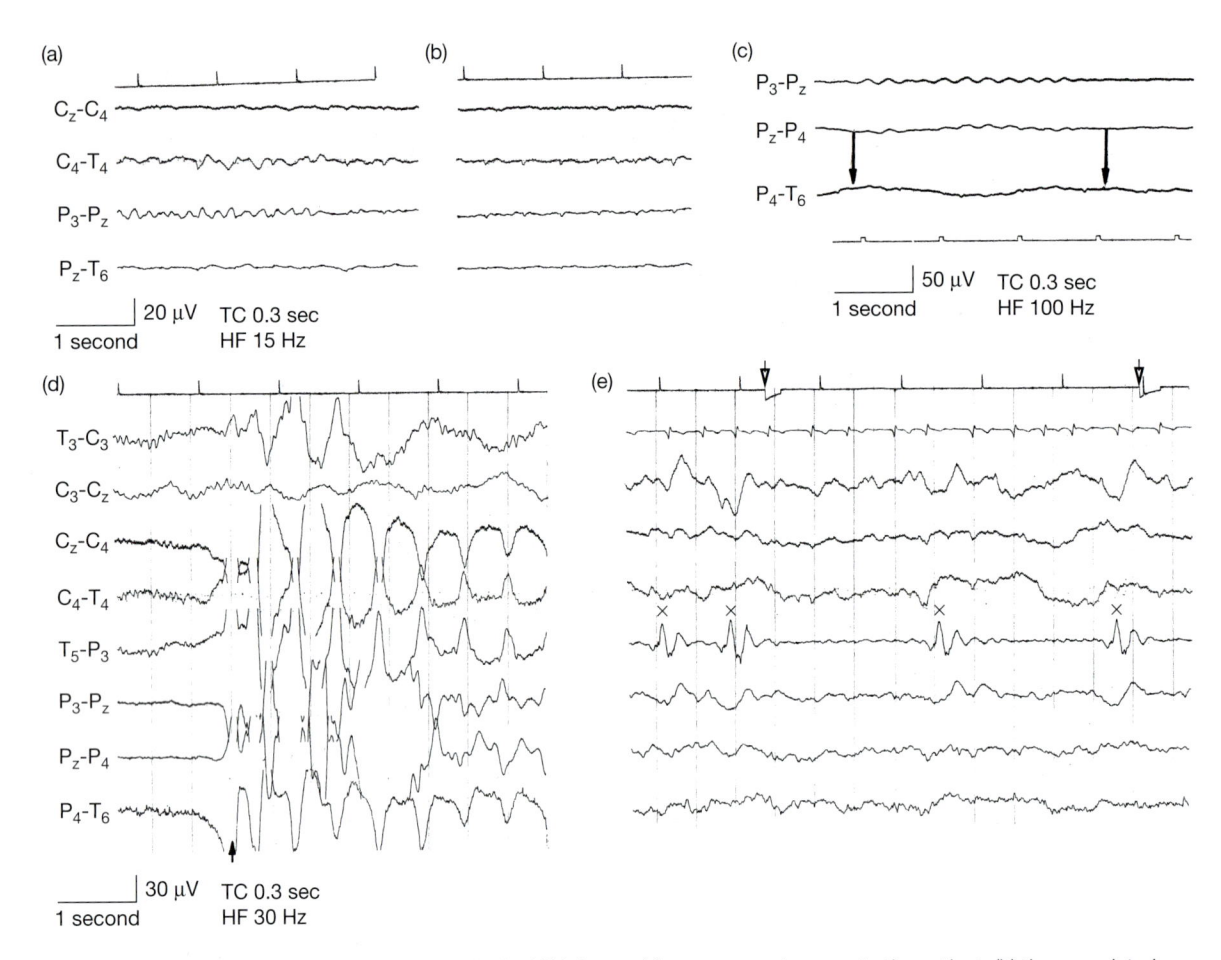

Fig. 3.18 Examples of artefacts encountered in the ICU due to: (a) a nurse moving near to the patient; (b) the associated deflections disappear when the nurse keeps still but ECG components are then evident (channels 2 and 3); (c) a technologist attempting to elicit a pain response by pinching the patient's ear lobe (at arrows); (d) inflation of a sphygmomanometer cuff (at arrow) to check the patient's blood pressure; (e) patient hiccups (×); chest movements monitored in channel 5; arrows in upper trace indicate respirations. (Modified from Prior (1973), by permission.)

by an experienced clinician. Agreement should be equal or better than that found in other human inter-rater studies, whether in EEG interpretation, sleep staging, detection of epileptiform spikes or other areas (Volavka et al 1973, Kemp et al 1987, Hostetler et al 1992, Wilson et al 1996). If monitoring systems are too complex or have ambiguous displays, or the operator is naive or inadequately trained, misinterpretation may negate the whole purpose of monitoring to improve quality of outcome.

EVOKED POTENTIAL MONITORING

Most of the pioneering developments for EP monitoring have been in the ICU and surgical fields (Bertrand et al 1987, García-Larrea et al 1987, Pfurtscheller et al 1987, Fischer 1989). Further information on signal processing aspects is given found in Chapter 8.

Technical requirements for EP monitoring

The enumeration and discussion of all methodological requirements for EP monitoring devices would be long and tedious. Nevertheless it is important to draw attention to several technical points that need special consideration before attempting EP monitoring in comatose patients (Chatrian et al 1996).

Automation

Automation must include iteration of recordings, storage and sequential display of the EPs. Different acquisition rates should be available, since optimal EP recording rates may vary from one patient to another, but all monitoring systems should allow the collection of a new short-latency EP every few minutes if needed. Monitoring parameters such as filter settings, number of responses averaged, intensity of stimulation, etc. must be automatically saved with every new EP, and easily accessible when required. Ideally, systems with the possibility of recording several types of EPs should allow the user to predetermine a sequence of acquisitions (e.g. 1 brainstem auditory evoked potential (BAEP), 1 middle latency auditory evoked potential (MLAEP), 1 somatosensory evoked potential (SEP)) to be repeated automatically throughout a monitoring session.

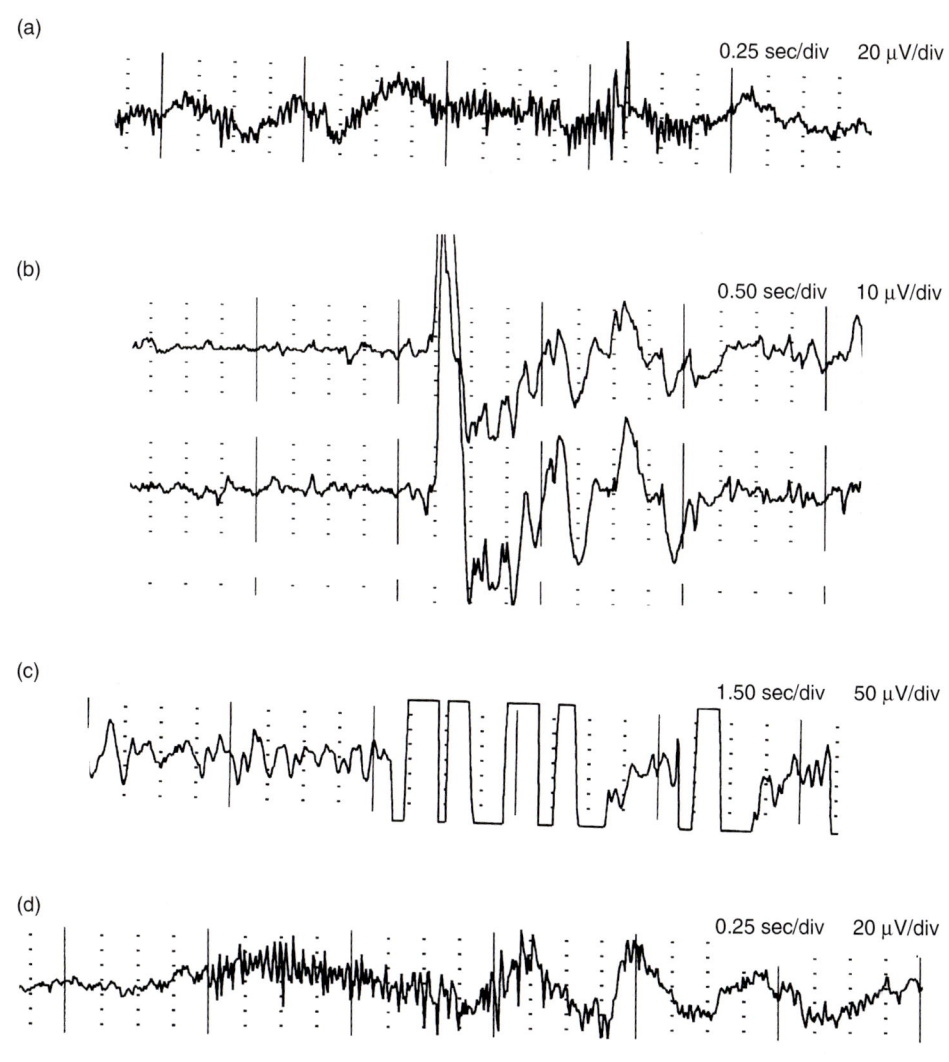

Fig. 3.19 Artefacts in ICU monitoring: (a) scalp muscle potentials (EMG); (b) patient coughing; (c) coughing during airway suction; (d) mixed scalp muscle potentials and movement artefacts during physiotherapy. Recordings from the EU Biomed-1 IMPROVE project (Carson and Saranummi 1996, Cerutti and Saranummi 1997). (Courtesy of Professor N Saranummi and Dr IR Ghosh.) (From Binnie et al (2003), by permission.)

Filtering

If automatic iteration of recordings is relatively easy to implement with microcomputer-based technology, feature extraction and detection of aberrant EP patterns have proved so far relatively resistant to automation. One main reason for this is difficulty in obtaining high-quality recordings in the ICU, a problem particularly enhanced during long periods of EP monitoring since there might well not be a technologist present throughout to ensure immediate correction of system failures or recording artefacts.

Good quality recordings with an excellent signal-to-noise ratio (SNR) are mandatory before attempting automatic feature extraction. Of course, the SNR can be improved by simply increasing the number of averaged responses (and this is the most frequently used method in standard EP recording); however, this solution is far from optimal in long-term monitoring, when the number of stimuli delivered can be extremely high, and reducing the time required for collection of each EP run becomes important. Accurate filtering is then of paramount importance in improving the SNR and may have the advantage of substantially decreasing the number of stimuli needed to obtain reproducible EPs. This is important since the total number of stimuli delivered over several days can be substantial, and also because reduction of actual EP recording time lessens the probability of contamination by intermittent artefacts.

Bandpass filtering aims at attenuating frequencies outside a predefined band, and is the most common technique for standard EP studies. However, conventional bandpass filters often prove insufficient for monitoring purposes and several other methods have been proposed by different groups. Some workers have tried to 'optimize' bandpass filters to make them suitable for monitoring, while others have preferred to develop non-conventional, adaptive filtering strategies. Among the

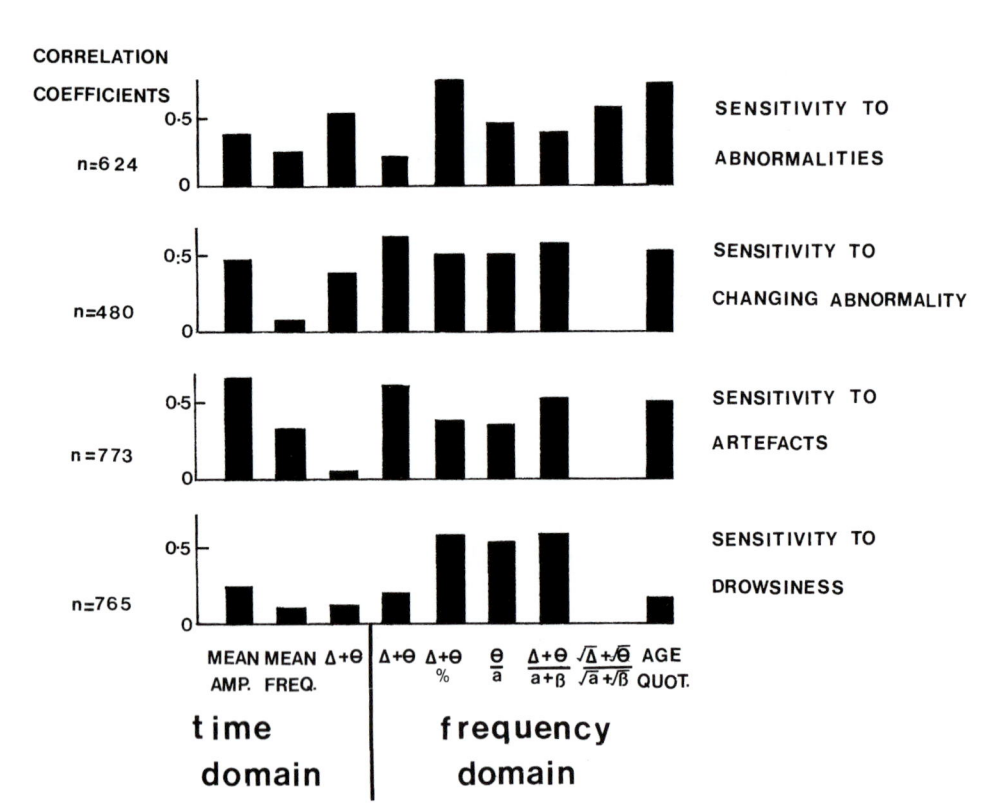

Chapter 3 Introduction to Methods for Continuous EEG and Evoked Potential Monitoring

Fig. 3.20 Factors affecting performance of various signal processing algorithms for qEEG monitoring include sensitivity to static and changing abnormalities, to artefacts and to changing level of arousal. (From Matousek et al (1978), by permission.)

former, Fridman et al (1982), working with BAEPs, tried to estimate which frequencies were preferentially present in the EP signal and which in the background noise. They compared the frequency spectrum of BAEPs obtained after only a few stimuli (containing both signal and noise frequencies) with that of the grand average of at least 2000 individual responses (in which frequencies carrying noise were substantially attenuated); on the basis of their estimates these authors proposed a relatively narrow digital bandpass (450–1300 Hz) for BAEP monitoring. This necessarily entails the loss of a substantial part of the amplitude and morphology information contained in the low frequencies, but allows satisfactory automatic latency calculation (John et al 1982); use of this method should therefore be limited to instances where amplitude information is considered non-relevant.

A different technique for improving SNR was developed by Boston (1985), who used a second channel to estimate the noise content for each response in the average. A noise-cancellation function was derived for each individual response, and successive 'cancelled' responses were averaged. The major difficulty with this technique is that the 'noisy' channel should record only noise, for if a part of the signal is present it will also tend to be cancelled. This implies spatially separated signal and noise derivations, from which only widespread noise affecting both channels simultaneously will be effectively cancelled. This is probably the reason why this technique results in good SNR improvements when noise is due to diffuse electrical interference, but not if it is mainly muscular or electroencephalographic in origin. Consequently, it is not of much value operationally.

Among the non-bandpass filters, those of Wiener type belong to the family of 'optimal' filters, so called because they minimize the squared error between the filtered response and the actual evoked signal, provided that the spectra of signal and noise are previously known. Wiener filters are designed to emphasize those frequency regions in which the signal power spectrum is larger than the background noise spectrum (Walter 1969, Doyle 1975, De Weerd and Martens 1978). The performance of standard Wiener filters when applied to averaged EPs has been shown to be rather unsatisfactory in conditions with low SNRs (De Weerd 1981, Wastell 1981) such as those usually encountered in ICU recordings. A second, important drawback of Wiener filtering results from the variation of the frequency content of EPs with latency, from high frequencies in the earlier components to lower frequencies in the later ones. When short- and long-latency EPs are simultaneously monitored as, for example, in the case of SEPs, different filters are optimal for different latency ranges. This leads to the solution proposed by De Weerd and Kap (1981) who devised a method, derived from Wiener's, that considers simultaneously the frequency and time domains, and thus entails changes of the filtering characteristics that are a function of latency ranges. This 'time-varying filter' has already been implemented in commercial systems, but not yet used for monitoring purposes.

Bertrand (1985) and Bertrand et al (1987) proposed a modification of Wiener filtering to be applied to BAEP

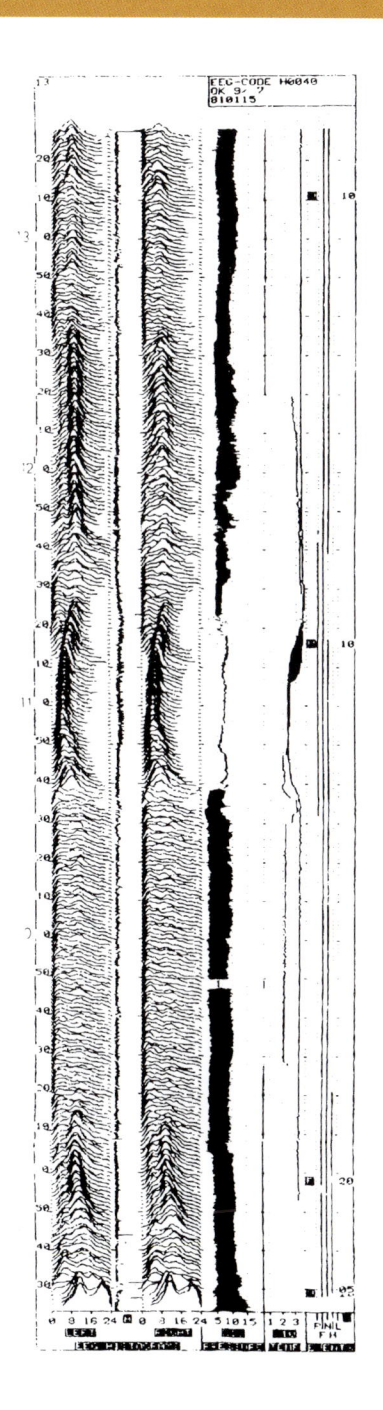

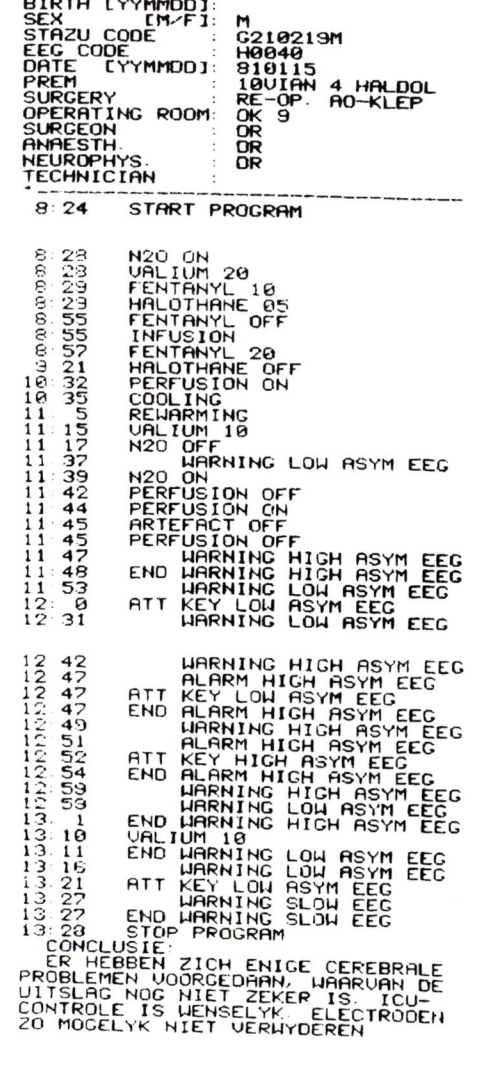

Fig. 3.21 Display and reporting. Example of complete report of monitoring during cardiac surgery, including compressed display of all recorded data, as sent back to the ward with the patient. Real time from 08.24 to 13.28 h runs from the bottom upwards on the left. The recordings comprise left and right-sided zero-crossing period histograms (computed every minute) with mean EEG amplitude between them, arterial pressure, nasopharyngeal temperature, and anaesthetic data. The clinical comment (in Dutch) at the end of surgery includes a recommendation that, because of significant EEG asymmetry (see alarm onset at 11.47 h in the operation notes), the electrodes should remain attached and monitoring continue in the ICU. (From Pronk and Simons (1982), by permission.)

monitoring in the ICU. Since conventional Wiener filters are not reliable when spectra are estimated under low SNR conditions, he proposed (a) calculation of the spectra from small groups of averages of 200 stimuli instead of individual responses, thus ameliorating SNR of data from which the filter is derived, and (b) application of this modified Wiener filter to the small averages which stand for 'individual' responses rather than to the grand-averaged EP. This greatly enhances SNR in small averages and allows reproducible BAEPs to be obtained after only 200–400 stimuli (Bertrand et al 1987). In its application to monitoring this can be considered an adaptive filter, in the sense that the frequencies eliminated are not necessarily the same from one EP to the next, and thus 'adapt' to changes in the noise and signal content of successive small averages. Since the frequency content of BAEPs does not vary significantly from the beginning to the end of the response (within a 10 ms window), a time variation of the filter is not necessary in this context. Most of the results on BAEP monitoring that will be presented here have been obtained with this adaptive filtering technique which, in the author's experience, has proved highly satisfactory for BAEP monitoring in the ICU and operating theatre. Although further improvements have been made to this technique, notably with the use of 'wavelet' transformation (Bertrand et al 1990, 1994), these have not yet been applied to ICU monitoring in a large series of patients.

Feature extraction and display

Accurate online interpretation of EP changes depends on the quality of feature extraction. Given the enormous number of EPs collected after several hours of recording, there is a need for some sort of data compression; this is usually obtained by the extraction of selected parameters from raw tracings. Algorithms for automatic detection of peaks are relatively easy to implement, provided the SNR is acceptable (Billings 1981, Fridman et al 1982, Maresch and Pfurtscheller 1983), but reduction of EP data to latency and amplitude values is not enough for monitoring purposes. Derived parameters such as the stability (reciprocal of variance) of latency and amplitude over time, or the slope (first derivative) of their changes, which measures the 'aggressiveness' of the causative process, are more specifically relevant to monitoring. Such techniques permit the chronobiological correlation of changes in monitored EPs with those in other clinical data. This requires the construction and display of 'trend curves' (*Fig. 3.22*) where information is presented sequentially over a time axis, making its evolution over hours or days readily accessible; such a process can be extended to include EEG parameters with the EPs and polygraphic data (*Fig. 3.23*).

Since the raison d'être of EP monitoring is to provide rapid warning in the event of central nervous system (CNS) homeostatic changes, systems should be equipped with alarms to call for the clinician's attention if any one of the monitored parameters is no longer detected, or if a recorded value goes beyond accepted normal limits.

More recently, a combined AEP and EEG monitor has been developed commercially based on experience of monitoring depth of anaesthesia with the *Alaris AEP monitor* system. The *AEP Monitor/2* (Danmeter A/S, Odense, Denmark, www.danmeter.dk) changes from EP monitoring to EEG monitoring at deeper levels of anaesthesia when EPs disappear. Clinical studies are starting to appear (Vereecke et al 2005, Weber et al 2005). The original Alaris system utilized a fast system for extraction of the MLAEPs by an autoregressive model with exogenous

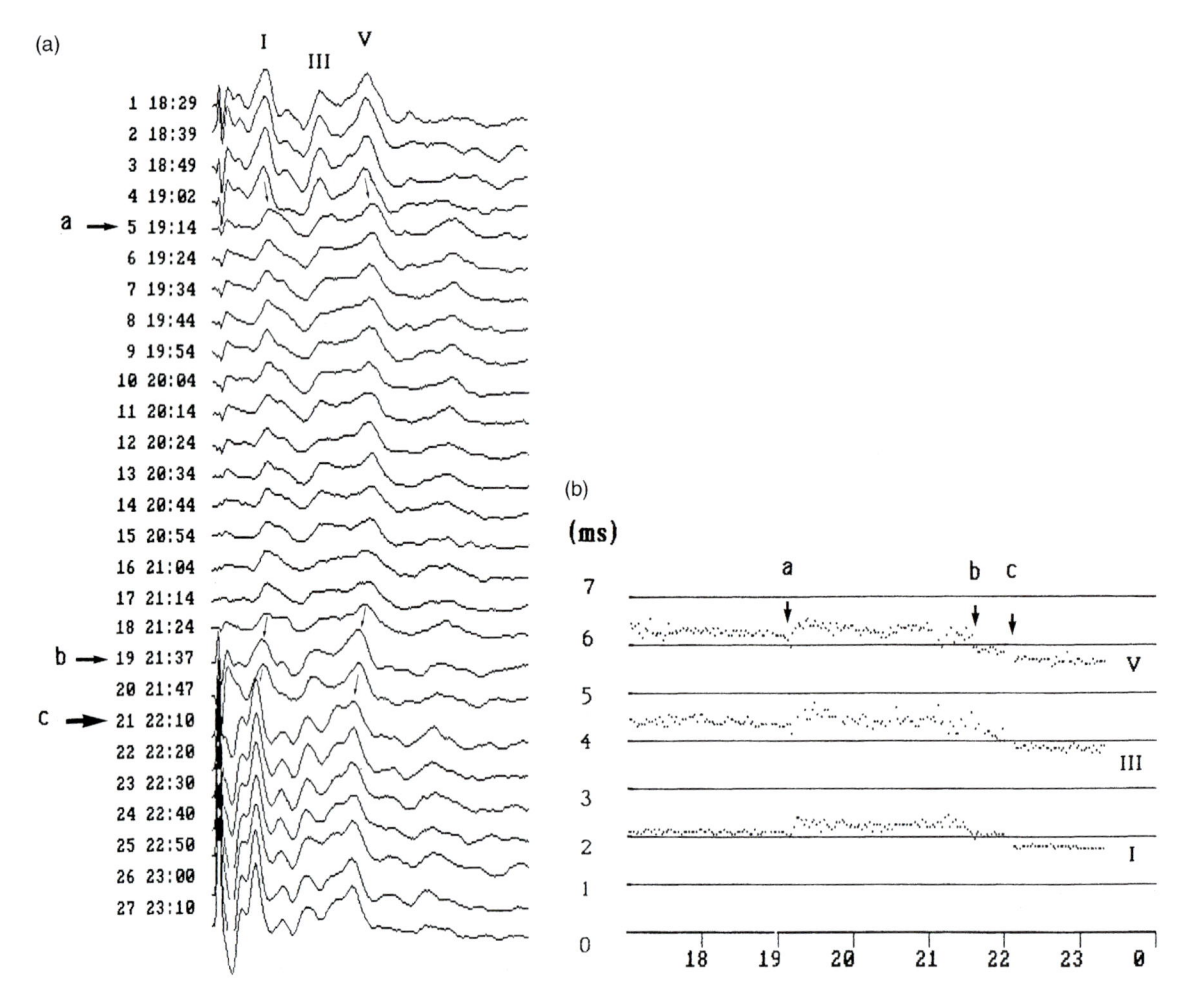

Fig. 3.22 BAEP monitoring: sequential displays of BAEPs at 10-min intervals (time runs from the top downwards). On latency trend plots (right) each dot corresponds to a measurement performed every 2 min. Sudden BAEP latency shifts occurred during monitoring due to earphone movements; the inserted earphone was pulled accidentally at a, and pushed in again at b and c. Note that the I–V interpeak interval remains constant. (From Bertrand et al (1987), by permission.)

Evoked potential monitoring

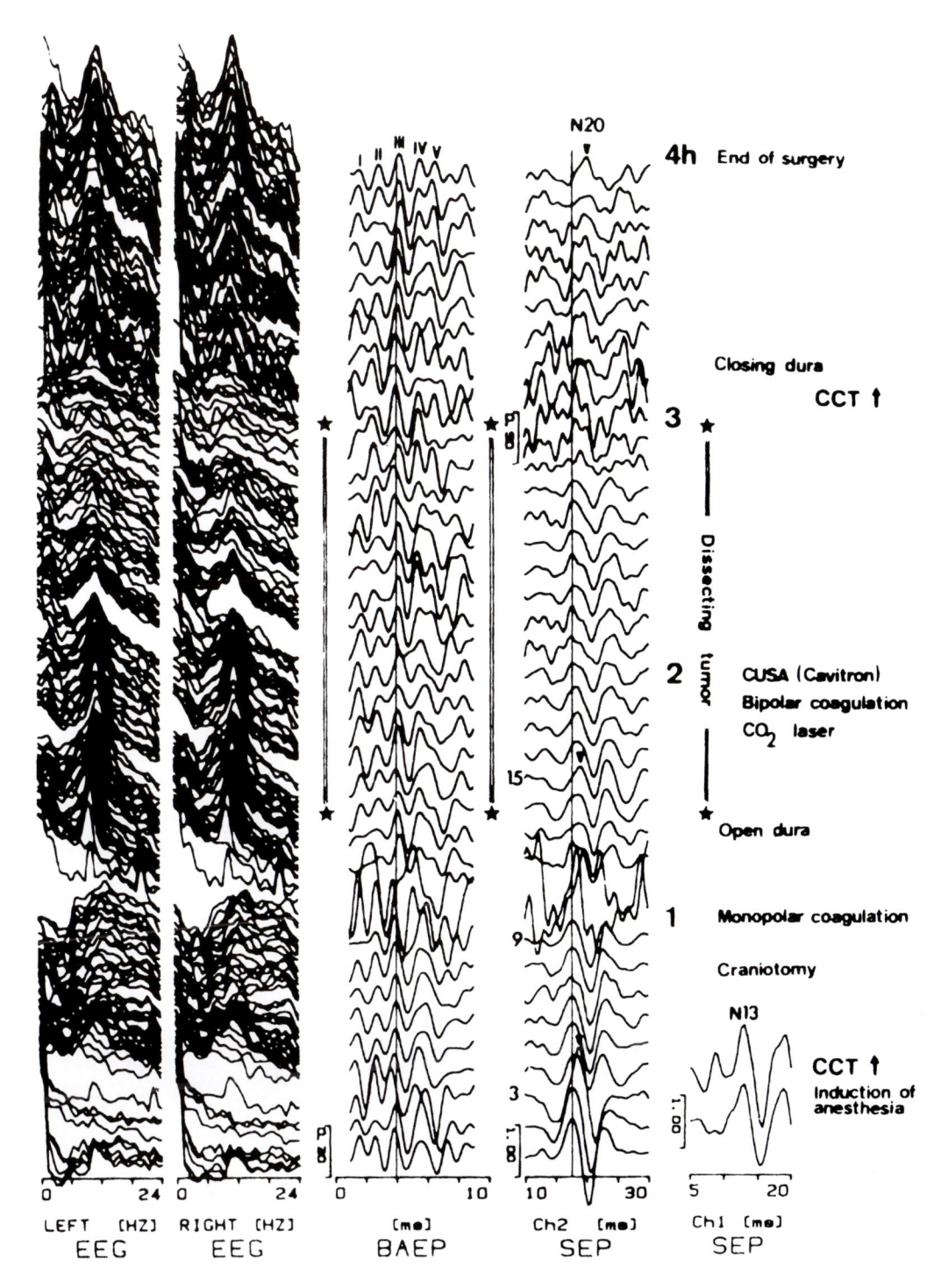

Fig. 3.23 Combined EEG and EP monitoring during posterior fossa surgery for removal of a meningioma from a 52-year-old patient showing both anaesthetic and surgically induced alterations. The display, in which time runs from the bottom upwards, shows left and right-sided EEG-CSAs (logarithmic scales), BAEPs, cortical SEPs, and two trials of cervical SEPs permitting calculation of central somatosensory conduction time. Operation notes are given on the right. The combined display aids distinction between surgically- and anaesthetic-induced changes. (From Pfurtscheller G, Schwarz G, Schroettner O, Litscher G, Maresch H, Auer L, List W 1987 Continuous and simultaneous monitoring of EEG spectra and brainstem auditory and SEPs in the ICU and the operating room. *Journal of Clinical Neurophysiology* 4:389–396. Lippincott, Williams and Wilkins Inc. Philadelphia © 1987 American Clinical Neurophysiology Society, by permission.)

input enabling extraction of the AEP within 1.7 s, thus allowing the depth of anaesthesia to be monitored at almost real-time (Alpiger et al 2002, Litvan et al 2002a,b). This proved useful for monitoring effects of tracheal intubation (Urhonen et al 2000). Comparisons between this AEP system and BIS monitors have been reported (see *Fig. 5.20*) (Nishiyama et al 2004, Struys et al 2002, 2003).

COMPARISONS BETWEEN CURRENTLY USED METHODS

Some comparisons between different approaches to early forms of signal processing for EEG monitoring have been described above. With increasing understanding of the characteristics of the signals involved, more recent monitoring systems have been more effective in tracking the complex multivariate changes in temporal EEG patterns seen during ICU treatment and with sedation or anaesthesia. Evidence-based studies have begun to define the sensitivity and specificity of such methods in a range of clinical circumstances – although, mainly for sedation and anaesthesia. Some groups have addressed problems of monitoring for early detection of signs of regional or global cerebral ischaemia but rather few broad-based comparative studies are yet available. A few examples related to anaesthetic applications will be given to indicate some of the approaches used.

For monitoring anaesthesia or sedation in the ICU, it is now generally accepted that none of the univariate qEEG measures such as median frequency, spectral edge frequency (SEF90%), total power or frequency-band power ratios, used in isolation, are sufficiently robust to serve as a completely reliable, sole, monitoring measure through the whole range of anaesthetic depth, mainly because of disruption of the signals by periods of EEG suppression (Levy 1984a,b, 1987, Drummond et al 1991, Dwyer et al 1994). Separate identification of burst suppression activity is essential and, when implemented, has improved performance (e.g. Thomsen and Prior 1996) (*Fig. 3.24*). This study also illustrates the agent specific nature of the sequence of changes during anaesthesia; individual anaesthetics have different qEEG profiles according to their rate of action and potency, both of which influence also the appearance and degree of periods of EEG suppressions.

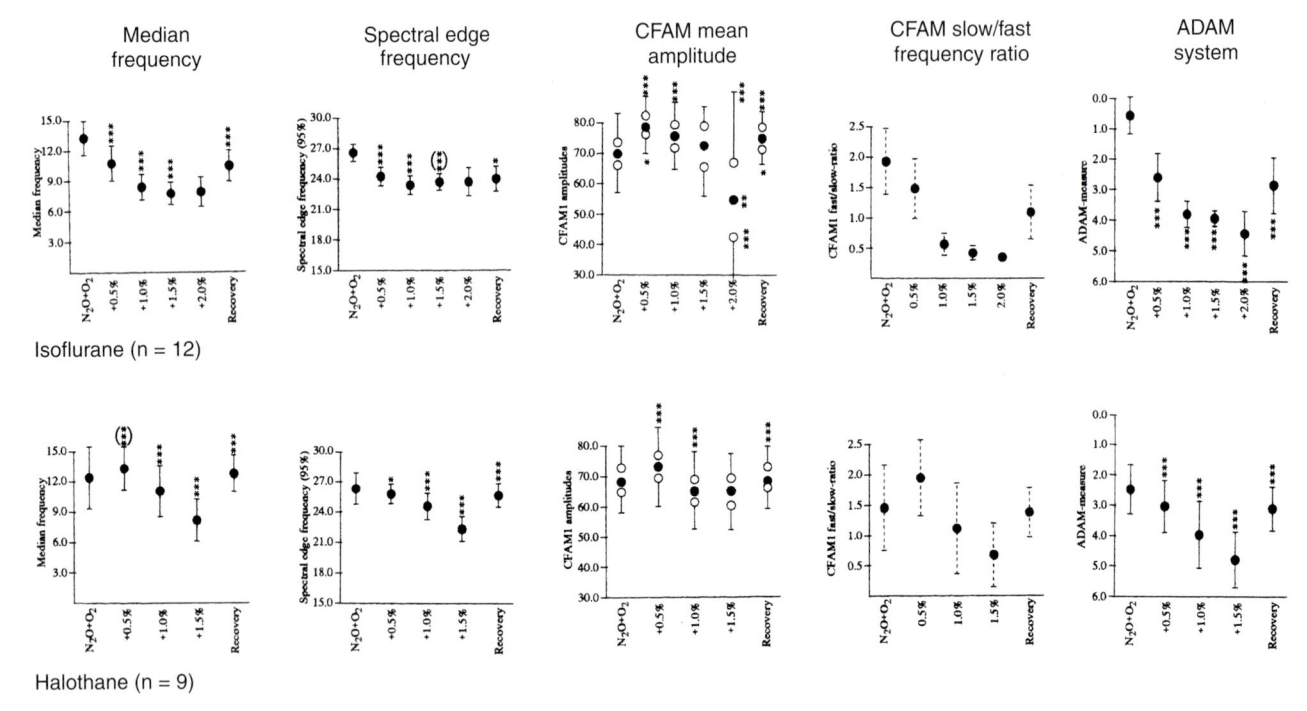

Fig. 3.24 Comparative performance of various EEG signal analysis methods during anaesthesia with stepwise changes in end-tidal concentrations of isoflurane and halothane. Group data calculated from study of Lloyd-Thomas et al (1990). The significance of changes in mean values compared with the previous state is given above each reading using the following notation: *$p < 0.01$; **$p < 0.001$; ***$p < 0.0001$; symbols in parentheses indicate significant changes in the opposite direction to the overall trend. The CFAM1 fast/slow ratio is calculated as (beta2 + beta1 + alpha2 + alpha1)/(theta2 + theta1 + delta2 + delta1). Note that standard deviations for fast/slow-ratio on CFAM1, although indicated with dotted lines, are not legitimate, because the frequency measures are based on pooled individual data and are not independent observations. For anaesthetic depth as indicated by the EEG pattern recognition system ADAM, the scale on the vertical axis is based on a conversion of the class probability histograms to a single numerical value such that 0 is equivalent to the awake state, 1 to drowsiness, 2 to 'very light' anaesthesia, 3 to 'light' anaesthesia, 4 to 'normal' surgical anaesthesia, 5 to 'deep' and 6 to 'very deep' anaesthesia. The ADAM system comes nearest to a linear scale for both anaesthetic agents. (From Thomsen CE, Prior PP 1996 Quantitative EEG in assessment of anaesthetic depth: comparative study of methodology. *British Journal of Anaesthesia* 77:172–178. © The Board of Management and Trustees of the *British Journal of Anaesthesia*. Reproduced by permission of Oxford University Press/ *British Journal of Anaesthesia*.)

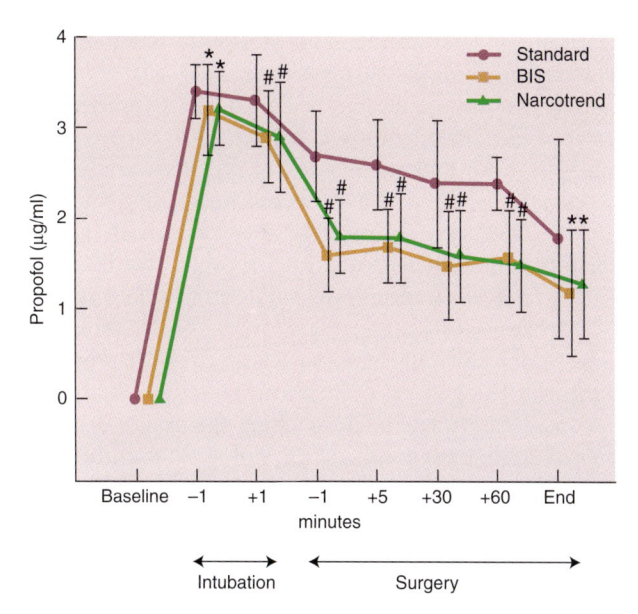

Fig. 3.25 Propofol target-controlled infusion concentrations (μg/ml) at various time points of anaesthesia in 50 orthopaedic patients. Data are mean (SD. #$p < 0.001$ for BIS or Narcotrend groups versus standard practice group. *$p < 0.05$ for BIS or Narcotrend groups versus standard practice group. No significant differences between Narcotrend and BIS groups (analysis of variance with Student–Newman–Kreuls test for multiple comparisons). (From Kreuer S, Biedler A, Larsen R, Altmann S, Wilhelm W 2003 Narcotrend monitoring allows faster emergence and a reduction of drug consumption in propofol–remifentanil anesthesia. *Anesthesiology* 99:34–41, by permission.)

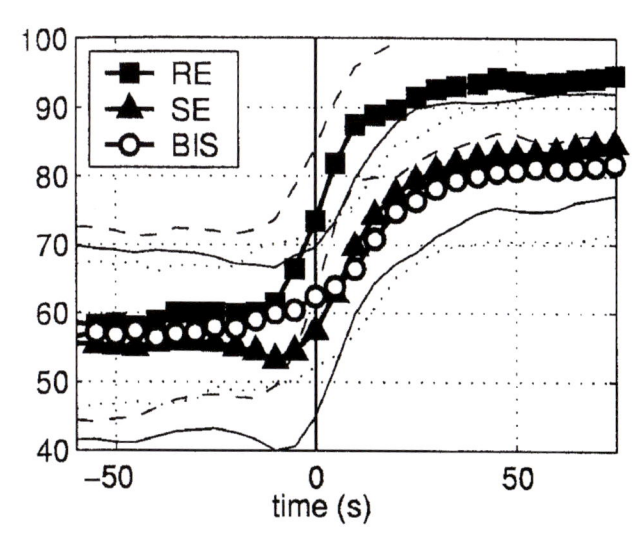

Fig. 3.26 Emergence from anaesthesia in 70 patients (30 anaesthetized with propofol and 20 each with thiopental or sevoflurane) monitored with different EEG monitoring measures. Time behaviour (mean ± SD) of response entropy (RE; dashed curves), state entropy (SE; continuous curves) and Bispectral Index (BIS; dotted curves) during emergence from anaesthesia. (From Vakkuri A, Yli-Hankala A, Talja P, Mustola S, Tolvanen-Laakso H, Sampson T., Viertiö-Oja H 2004 Time–frequency balanced spectral entropy as a measure of anesthetic drug effect in central nervous system during sevoflurane, propofol, and thiopental anesthesia. *Acta Anesthesiologica Scandinavica* 48:145–153, Blackwell Publishing Ltd, by permission.)

The performance of two different modern EEG monitors (BIS and Narcotrend) was compared to see how they facilitated a significant reduction of recovery times and propofol consumption when used for guidance of propofol titration during propofol–remifentanil anaesthesia. Both proved equally effective and both outperformed standard clinical assessment of anaesthesia (Kreuer et al 2003) (*Fig. 3.25*).

Studies of the period of emergence from anaesthesia have been of interest because of the increasing use of day-case surgery and the need to ensure a patient's fitness to return home. A comparison between two modern monitors showed that the general trends are similar between BIS and entropy algorithms, but that 'response entropy' (which assesses EEG and EMG recorded from the forehead; see *Entropy*, p. 85) indicated emergence earlier than 'state entropy' or BIS (Vakkuri et al 2004) (*Fig. 3.26*).

Evoked potentials have a considerable relevance in relation to anaesthetized or sedated patients, not least because short latency components are resistant to change (and so can be used to test integrity of somatosensory and auditory functions during surgery or in the ICU). At the same time, middle latency auditory EP components have proved effective in evaluating depth of anaesthesia and, in common other paradigms such as mismatch negativity, transitions between unconsciousness and consciousness, and vice versa (see Chapter 5). There are now some studies comparing the performance of auditory EP measures and modern EEG monitoring systems (e.g. Sharpe et al 1997, Doi et al 1999, Gajraj et al 1999, Kochs et al 1999, Kreuer et al 2003, Nishiyama et al 2004, Vanluchene et al 2004). It must be noted that simultaneous EEG and EP studies make considerable technical demands for high quality recording and care needs to be taken in methodology to avoid influence by the effects of auditory or somatosensory stimulation itself in arousing a lightly anaesthetized patient (see *Fig. 5.23*). To avoid such problems, Nishiyama et al (2004) studied three separate groups of patients undergoing comparable types of anaesthesia and surgery to compare BIS, processed EEG (spectral edge frequency, SEF90) and MLAEPs (Alaris system), following the manufacturers stated index threshold values. Both middle latency auditory EPs and modern EEG measures have useful roles in the evaluation of some specific aspects of depth of anaesthesia, such as transitions between consciousness and unconsciousness (e.g. Gajraj et al 1999).

It would be invidious to draw any firm conclusions on the basis of the few examples given in this chapter but two general comments can be made:

- *For work in monitoring sedation or anaesthesia*, the evidence encourages optimism that the raw signals are suitable for processing and that appropriate and effective groups of advanced algorithms are available. Comparative studies suggest that their performance in most modern commercially available EEG monitors is satisfactory and, used sensibly, provide a convenient tool to assist the anaesthetist with routine monitoring of depth of anaesthesia or sedation.

- *For monitoring for global or regional brain ischaemia*, more evidence from comparative studies is required to validate the performance of modern EEG monitors. The case is not yet proven as to whether these monitors are adequate for all the more complex tasks of general monitoring during cardiac surgery, or operations on the carotid artery or brainstem, or in unconscious patients in the neuro-ICU, where a sequence or cascade of unusual signals may provide the only early warning of impending disaster from cerebral ischaemia. In the interim, proven neurophysiological methods displaying sufficient raw *and suitably processed* data from several recording channels, give the neurophysiologist a more flexible means of interpreting significant changes.

CONCLUSIONS

The use of neurophysiological monitoring is recognized as a requirement in many high risk situations. However, as Professor Yli-Hankala has pointed out in his Foreword to this volume, there are some unresolved creative tensions between protagonists of two quite different views of monitoring. There is the understandable desire of many clinicians for an 'intelligent' monitor which displays a simple numerical scale which will provide a valid basis for decisions to inform and improve their management, for example of anaesthesia. This viewpoint is set against the neurophysiologists' insistence that the derived parameter(s) must be clearly explained and demonstrably related to the, often complex, behaviour of the original signal – both for making best use of the physiological or pathophysiological information and for quality control, to ensure the safety of surgery in a child with spinal deformity or an older patient undergoing carotid artery surgery to prevent transient ischaemic attacks becoming a major stroke.

The brief historically based review given above provides insight into the general objectives that should be taken into consideration when approaching the design or suitability of equipment for neurophysiological monitoring. Happily there are also robust international guidelines to help us set standards for training and practice of neurophysiological monitoring. The recommendations of the International Federation for Clinical Neurophysiology (Deuschl and Eisen 1999) include those for digital recording and topographic and frequency analysis of the EEG (Nuwer et al 1999a,b) and for intraoperative and ICU monitoring with EEG and EPs (Burke et al 1999, Guérit et al 1999). These indicate the neurophysiological parameters to be monitored (*Table 3.3*).

Setting up or developing a neurophysiological monitoring service requires a strategy for decisions about choice and tools, personnel and procedures for the task. It usually needs an exercise in multidisciplinary consultation comparable to that for a military assault, aimed at defining objectives and deciding what equipment and procedures are most fit for the purpose. Key guidelines are given in *Table 3.4*.

Table 3.3 Neurophysiological parameters to be monitored and procedures (based on recommendations of Guérit et al 1999)

Steps	Procedure for EEG	Procedure for EPs
1. Quality screening, followed by automatic feature extraction	(a) Frequency analysis of steady EEG segments (b) Time analysis of transients	(a) Automatic identification of peaks
2. Selecting and combining parameters to define patterns	(a) Recombination of Fourier parameters to classical spectral measures (absolute or relative power, median frequencies, spectral edges, etc.) and user-defined indices (asymmetry index, power ratios, etc.) (b) Time analyses to identify specific patterns such as burst suppression, complex arousal patterns, epileptiform discharge, EP and EEG encephalopathy grades, ECS, etc. (c) Combination of neurophysiological recordings to reflect overall CNS function (BAEPs and SEPs to differentiate pontine from midbrain involvement – for early detection of brainstem involvement in transtentorial herniation)	
3. Integration of neurophysiological features into the clinical context	Reference to evidence-based data libraries showing clusters or patterns of high specificity and sensitivity	

Table 3.4 Strategy for setting up a neurophysiological monitoring service

Process	Main factors	Also consider
Defining purpose of monitoring	What difference it will make	Cost-effectiveness of implementing service for clinical purpose
Choice of equipment	Group decision depending on purpose, available personnel and biomedical engineering support services	Evidence of specificity and sensitivity of tools from comparative studies in current literature and international recommendations
Set-up of agreed routine	– Roles for staff – Positioning of equipment – Specific safety requirements – Criteria for alerting designated clinician	Important to have clear written record of chain of responsibilities
Preparation	– Training – Electrode placement, etc. – Safety checks – Quality control routines	– Who is responsible for what – Where and when – Who records, checks, and identifies need for changes
Liaison with anaesthetic and surgical teams about specific procedure or requirements for the particular patient	– Pre-existing abnormalities – Special safety risks – Unusual questions to be addressed by monitoring	Routine preoperative check sheet
Data collection and annotation	– Which signals to be monitored – Which other variables to be recorded or noted in annotations	Designate codes for frequently used annotations
Display	What does the clinician need to see or know 'on-line'	Psychology of a good display and labelling to reduce observer errors
Reporting at time	Attributable written notes to support verbal comments	Value of dialogue between clinical team and expert neurophysiologist
Permanent record in patient's notes and departmental archive	Medico-legal responsibilities must be fulfilled	Prompt reporting helped by annotation and reporting facilities provided in monitoring equipment

REFERENCES

Adams DC, Heyer EJ, Emerson RG, et al 1995 The reliability of quantitative electroencephalography as an indicator of cerebral ischemia. Anesth Analg 81:80–83.

Adrian ED, Matthews BHC 1934 The Berger rhythm: potential changes from the occipital lobes in man. Brain 57:354–385.

Ahn SS, Jordan SE, Nuwer MR, et al 1988 Computed electroencephalographic topographic brain mapping. A new and accurate monitor of cerebral circulation and function for patients having carotid endarterectomy. J Vasc Surg 8:247–254.

Alpiger S Helbo-Hansen HS, Jensen EW 2002 Effect of sevoflurane on the mid-latency auditory evoked potentials measured by a new fast extracting monitor. Acta Anaesthesiol Scand 46:252–256.

Barlow JS 1984 Analysis of EEG changes with carotid clamping by selective analog filtering, matched inverse filtering and automatic adaptive segmentation: a comparative study. Electroencephalogr Clin Neurophysiol 58:193–204.

Bauerle K, Greim CA, Schroth M, et al 2004 Prediction of depth of sedation and anaesthesia by the Narcotrend EEG monitor. Br J Anaesth 92:841–845.

Beecher HK, McDonough FK, Forbes A 1938 Effects of blood pressure changes on cortical potentials during anesthesia. J Neurophysiol 1:324–331.

Bellville JW, Artusio JF, Bulmer MW 1954 Continuous servo-motor integrator of the electrical activity of the brain and its application to the control of cyclopropane anesthesia. Electroencephalogr Clin Neurophysiol 6:317–320.

Bellville JW, Artusio Jr JF, Glenn F 1955 The electroencephalogram in cardiac manipulation. Surgery 38:259–271.

Bertrand O 1985 Système informatisé d' enregistrement sequentiel des potentiels évoqués auditifs précoces adapté à la surveillance des malades comateux PhD thesis, University of Lyon, France.

Bertrand O, García-Larrea L, Artru F, et al 1987 Brainstem monitoring I: A system for high rate BAEP sequential monitoring and feature extraction. Electroencephalogr Clin Neurophysiol 68:433–145.

Bertrand O, Bohórquez J, Pernier J 1990 Technical requirements for evoked potentials monitoring in the intensive care unit. In: New Trends and Advanced Techniques in Clinical Neurophysiology. Electroencephalogr Clin Neurophysiol (Suppl 41):51–69.

Bertrand O, Bohórquez J, Pernier J 1994 Time-frequency digital filtering based on an invertible wavelet transform: an application to evoked potentials. IEEE Trans Biomed Eng 41:77–88.

Bevacqua BK, Kazdan D 2003 Is more information better? Intraoperatavie recall with a Bispectral Index monitor in place. Anesthesiology 99:507–508.

Bickford RG 1949 Neurophysiological applications of an automatic anesthetic regulator controlled by brain potentials. Am J Physiol 159:562–563.

Bickford RG 1950 Automatic electroencephalographic control of general anaesthesia. Electroencephalogr Clin Neurophysiol 1:93–96.

Bickford RG 1951a Use of frequency discrimination in the automatic electro-encephalographic control of anaesthesia (servo-anaesthesia). Electroencephalogr Clin Neurophysiol 3:83–86.

Bickford RG 1951b The use of feedback systems for the control of anesthesia. Electr Eng 70:852–855.

Bickford RG, Billinger TW, Fleming NI, et al 1972 The compressed spectral array (CSA) – a pictorial EEG. Proc San Diego Biomed Symp 11:365–370.

Bickford RG, Brimm J, Berger L, et al 1973 Compressed spectral array in clinical EEG. In: *Automation of Clinical Electroencephalography* (eds P Kellaway, I Petersén). Raven Press, New York, pp. 55–64.

Billings RJ 1981 Automatic detection, measurement and documentation of the visual evoked potential using a commercial microprocessor-equipped averager. Electroencephalogr Clin Neurophysiol 52:214–217.

Binnie CD, Cooper R, Mauguière F, et al (eds) 2003 *Clinical Neurophysiology*, Volume 2: *EEG, Paediatric Neurophysiology, Special Techniques and Applications*. Elsevier, Amsterdam.

Binnie CD, Cooper R, Mauguière F, et al (eds) 2004 *Clinical Neurophysiology*, Volume 1: *EMG, Nerve Conduction and Evoked Potentials* (revised and enlarged edition). Elsevier, Amsterdam.

Boston JR 1985 Noise cancellation for brainstem auditory evoked potentials. IEEE Trans Biomed Eng 32:106–1070.

Bricolo A, Turazzi S, Faccioli F, et al 1978 Clinical applications of compressed spectral array in long-term EEG monitoring of comatose patients. Electroencephalogr Clin Neurophysiol 45:211–225.

Bricolo A, Faccioli F, Grosslercher JC, et al 1987 Electrophysiological monitoring in the intensive care unit. In: *The London Symposia* (EEG Suppl 39) (eds RJ Ellingson, NMF Murray, AM Halliday). Elsevier Science, Amsterdam, pp. 255–263.

Bruhn J, Bouillon TW, Shafer SL 2000 Bispectral index (BIS) and burst suppression: revealing a part of the BIS algorithm. J Clin Monit Comput 16:593–596.

Burch NR 1959 Automatic analysis of the EEG. Electroencephalogr Clin Neurophysiol 11:827–834.

Burdjalov VF, Baumgart S, Spitzer AR 2003 Cerebral function monitoring: a new scoring system for the evaluation of brain maturation in neonates. Pediatrics 112:855–861.

Burke D, Nuwer MR, Daube J, et al 1999 Intraoperative monitoring. In: *Recommendations for the Practice of Clinical Neurophysiology Guidelines of the International Federation of Clinical Neurophysiology* (EEG Suppl 52) (eds G Deuschl, A Eisen). Elsevier Science, Amsterdam, pp. 133–148.

Carson E, Saranummi N (eds) 1996 Improving control of patient status in critical care: the IMPROVE project. Comput Methods Programs Biomed 51(1,2):1–130.

Cerutti S, Saranummi N (eds) 1997 Improving control of patient status in critical care. IEEE Eng Med Biol Mag 16:19–79.

Chatrian GE, Bergamasco B, Bricolo A, et al 1996 IFCN recommended standards for electrophysiologic monitoring in comatose and other unresponsive states. Report of an IFCN committee. Electroencephalogr Clin Neurophysiol 99:103–122.

Claassen J, Hirsch LJ, Kreiter KT, et al 2004 Quantitative continuous EEG for detecting delayed cerebral ischemia in patients with poor-grade subarachnoid hemorrhage. Clin Neurophysiol 115:2699–2710.

Courtin RF, Bickford RG, Faulconer Jr A 1950 Classification and significance of electro-encephalographic patterns produced by nitrous oxide–ether anesthesia during surgical operations. Proc Staff Meetings Mayo Clin 25:197–206.

Dahaba AA 2005 Different conditions that could result in the Bispectral Index indicating an incorrect hypnotic state. Anesthesiology 101:765–773.

Davenport HT, Arfel G, Sanchez FR 1959 The electroencephalogram in patients undergoing open-heart surgery with heart–lung bypass. Anesthesiology 20:674–684.

De Deyne C, Struys M, Decruyenaere J, et al 1998 Use of continuous bispectral EEG monitoring to assess depth of sedation in ICU patients. Intensive Care Med 24:1294–1298.

Deuschl G, Eisen A (eds) 1999 *Recommendations for the Practice of Clinical Neurophysiology: Guidelines of the International Federation of Clinical Neurophysiology* (EEG Suppl 52). Elsevier, Amsterdam.

De Weerd JPC 1981 Facts and fancies about '*a posteriori* Wiener' filtering. IEEE Trans Biomed Eng BME-28:252–257.

De Weerd JPC, Kap JI 1981 A posteriori time-varying filtering of averaged evoked potentials (Parts I and II). Biol Cybern 41:211–134.

De Weerd JPC, Martens WLJ 1978 Theory and practice of '*a posteriori* Wiener' filtering of average evoked potentials. Biol Cybern 30:81–94.

Dexter F, Macario A, Manberg PJ, et al 1999 Computer simulation to determine how rapid anesthetic recovery protocols to decrease the time for emergence or increase the phase I postanesthesia care unit bypass rate affect staffing of an ambulatory surgery center. Anesth Analg 88:1053–1063.

Dietsch G 1932 Fourier-analyse von Elektroencephalogrammen des Menschen. Pflúger's Arch Ges Physiol 230:106–112.

Doi M, Gajraj RJ, Mantzaridis H, et al 1999 Prediction of movement at laryngeal mask airway insertion: comparison of auditory evoked potential index, Bispectral Index, spectral edge frequency and median frequency. Br J Anaesth 82:203–207.

Doyle DJ 1975 Some comments on the use of Wiener filtering for the estimation of evoked potentials. Electroencephalogr Clin Neurophysiol 38:533–534.

Drohocki Z, Drohocka J 1939a L'electro-spectrogramme du cerveau. C R Soc Biol (Paris) 130:95–98.

Drohocki Z, Drohocka J 1939b L'electro-spectrographie quantitative du cerveau a 1'etat de veille et pendant la narcose. C R Soc Biol (Paris) 132:494–498.

Drover DR, Lemmens HJ, Pierce ET, et al 2002 Patient State Index: titration of delivery and recovery from propofol, alfentanil, and nitrous oxide anesthesia. Anesthesiology 97:82–89.

Drummond JC, Brann CA, Perkins DE, et al 1991 A comparison of median frequency, spectral edge frequency, a frequency band power ratio, total power, and dominance shift in the determination of depth of anesthesia. Acta Anaesthesiol Scand 35:693–699.

Dumermuth G, Huber PJ, Kleiner B, et al 1971 Analysis of the interrelationships between frequency bands of the EEG by means of the bispectrum. A preliminary study. Electroencephalogr Clin Neurophysiol 31:137–148.

Dwyer RC, Rampil IJ, Eger EI, et al 1994 The electroencephalogram does not predict depth of isoflurane anesthesia. Anesthesiology 81:403–409.

Eckert O, Werry C, Neulinger A, et al 1997 Intraoperative EEG monitoring using a neural network [in German]. Biomed Tech (Berlin) 42:78–84.

Ekman A, Lindholm ML, Lennmarken C, et al 2004 Reduction in the incidence of awareness using BIS monitoring. Acta Anaesthesiol Scand 48:20–26.

Electrophysiological Technologists Association 2004 EPTA guidelines for checking digital EEG machines. http://www.epta.50megs.com/resources/cg/digi_cal_eeg.pdf

Faulconer Jr A, Bickford RG 1960 *Electroencephalography in Anesthesiology*. Thomas, Springfield, IL.

Fischer C 1989 Brainstem auditory evoked potential (BAEP) monitoring in posterior fossa surgery. In: *Neuromonitoring in Surgery* (ed JE Desmedt). Elsevier, Amsterdam, pp. 191–207.

Fischgold H, Mathis P (eds) 1959 Onubliations Comas et Stupeurs. Electroencephalogr Clin Neurophysiol (Suppl 11):13–26.

Fischer-Williams M, Cooper RA 1964 Some aspects of electroencephalographic changes during open-heart surgery. Neurol (Minneapolis) 14:472–482.

Fleming RA, Smith NT 1979a An inexpensive device for analyzing and monitoring the electroencephalogram. Anesthesiology 50:456–460.

Fleming RA, Smith NT 1979b Density modulation – a technique for the display of three-variable data in patient monitoring. Anesthesiology 50:543–546.

Frank M, Maynard DE, Tsanaclis LM, et al 1984 Changes in cerebral electrical activity measured by the Cerebral Function Analysing Monitor following bolus injections of thiopentone. Br J Anaesth 56:1075–1081.

Fridman J, John ER, Bergelson M, et al 1982 Application of digital filtering and automatic peak detection to brain-stem auditory evoked potential. Electroencephalogr Clin Neurophysiol 53:405–416.

Gade J, Rosenfalck A, van Gils M, et al 1996 Modelling techniques and their application for monitoring in high dependency environments – learning models. Comput Method Programs Biomed 51:75–84.

Gajraj RJ, Doi M, Mantzaridis H, et al 1999 Comparison of bispectral EEG analysis and auditory evoked potentials for monitoring depth of anaesthesia during propofol anaesthesia. Br J Anaesthesia 82:672–678

Gan TJ, Glass PS, Windsor A, et al 1997 Bispectral index monitoring allows faster emergence and improved recovery from propofol, alfentanil, and nitrous oxide anesthesia. BIS Utility Study Group. Anesthesiology 87:808–815.

García-Larrea L, Bertrand O, Artru F, et al 1987 Brainstem monitoring in coma II: dynamic interpretation of preterminal BAEP changes observed until brain death in deeply comatose patients. Electroencephalogr Clin Neurophysiol 68:446–457.

Ghosh IR, Langford RM, Nieminen K, et al 2000 Repetitive cyclical oscillations of multisystem parameters subsequent to high-dose thiopental therapy for status epilepticus secondary to herpes encephalitis. Br J Anaesth 85:471–473.

Gibbs FA, Davis H, Lennox WG 1935 The electro-encephalogram in epilepsy and conditions of impaired consciousness. Arch Neurol Psychiatry (Chicago) 34:1133–1148.

Gill M, Haycock K, Green SM, et al 2004 Can the Bispectral Index monitor the sedation adequacy of intubated ED adults? Am J Emerg Med 22:76–82.

Gloor P 1969 *Hans Berger on the Electroencephalogram of Man*. (EEG Suppl 28). Elsevier, Amsterdam.

Grady RE, Weglinski MR, Sharbrough FW, et al 1998 Correlation of regional cerebral blood flow with ischemic electroencephalographic changes during sevoflurane–nitrous oxide anesthesia for carotid endarterectomy. Anesthesiology 88:892–897.

Guérit J-M, Fischer C, Facco E, et al 1999 Standards of clinical practice of EEG and EPs in comatose and other unresponsive states. In: *Recommendations for the Practice of Clinical Neurophysiology: Guidelines of the International Federation of Clinical Neurophysiology* (EEG Suppl 52) (eds G Deuschl, A Eisen). Elsevier, Amsterdam, pp. 117–131.

Hellström-Westas L, De Vries LS, Rosén I 2003 *An Atlas of Amplitude-Integrated EEGs in the Newborn*. Parthenon Press, Boca Raton, FL.

Homer TD, Stanski DR 1985 The effect of increasing age on thiopental disposition and anesthetic requirement. Anesthesiology 62:714–724.

Hostetler WE, Doller HJ, Homan RW 1992 Assessment of a computer program to detect epileptiform spikes. Electroencephalogr Clin Neurophysiol 83:1–11.

Hudson RJ, Stanski DR, Saidman LJ, et al 1983 A model for studying depth of anesthesia and acute tolerance to thiopental. Anesthesiology 59:301–308.

John ER, Baird H, Fridman J, et al 1982 Normative values for brain stem auditory evoked potential obtained by digital filtering and automatic peak detection. Electroencephalogr Clin Neurophysiol 54:153–160.

Johnson RW, Shore JE 1984 Which is the better entropy expression for speech processing: –S log S or log S? IEEE Trans Acoust ASSP-32:129–137.

Kemp B, Groeneveld EW, Janssen AJ, et al 1987 A model-based monitor of human sleep stages. Biol Cybernet 57:365–378.

Kiersey DK, Bickford RG, Faulconer Jr A 1951 Electro-encephalographic patterns produced by thiopental sodium during surgical operations; description and classification. Br J Anaesth 23:141–152.

Kiersey DK, Faulconer Jr A, Bickford RG 1954 Automatic electroencephalographic control of thiopental anesthesia. Anesthesiology 15:356–364.

Kochs E, Kalkman CJ, Thornton C, et al 1999 Middle latency auditory evoked responses and electroencephalographic derived variables do not predict movement to noxious stimulation during 1 minimum alveolar anesthetic concentration isoflurane/nitrous oxide anesthesia. Anesth Analg 88:1412–1417.

Kreuer S, Biedler A, Larsen R, et al 2003 Narcotrend monitoring allows faster emergence and a reduction of drug consumption in propofol–remifentanil anesthesia. Anesthesiology 99:34–41.

Kreuer S, Wilhelm W, Grundmann U, et al 2004a Narcotrend index versus Bispectral Index as electroencephalogram measures of anesthetic drug effect during propofol anesthesia. Anesth Analg 98:692–697

Kreuer S, Bruhn J, Larsen R, et al 2004b Application of Bispectral Index and Narcotrend index to the measurement of the electroencephalographic effects of isoflurane with and without burst suppression. Anesthesiology 101:847–854.

Kreuer S, Bruhn J, Stracke C, et al 2005 Narcotrend or Bispectral Index monitoring during desflurane-remifentanil anesthesia: a comparison with a standard practice protocol. Anesth Analg 101:427–434.

Kugler J 1981 *Elektoenzephalogram in Klinik und Praxis*. Georg Thieme, Stuttgart.

Levy WJ 1984a Quantitative analysis of EEG changes during hypothermia. Anesthesiology 60:291–297.

Levy WJ 1984b Intraoperative EEG patterns: implications for EEG monitoring. Anesthesiology 60:430–434.

Levy WJ 1987 Effect of epoch length on power spectrum analysis of the EEG. Anesthesiology 66:489–495.

Lipping T, Jäntti V, Yli-Hankala A, et al 1995 Adaptive segmentation of burst suppression pattern in isoflurane and enflurane anesthesia. Int J Clin Monit Comput 12:161–167.

Litvan H, Jensen EW, Revuelta M, et al 2002a Comparison of auditory evoked potentials and the A-line ARX Index for monitoring the hypnotic level during sevoflurane and propofol induction. Acta Anaesthesiol Scand 46:245–251.

Litvan H, Jensen EW, Galan J, et al 2002b Comparison of conventional averaged and rapid averaged, autoregressive-based extracted auditory evoked potentials for monitoring the hypnotic level during propofol induction. Anesthesiology 97:351–358.

Liu J, Singh H, White PF 1997 Electroencephalographic Bispectral Index correlates with intraoperative recall and depth of propofol-induced sedation. Anesth Analg 84:185–189.

Chapter 3 Introduction to Methods for Continuous EEG and Evoked Potential Monitoring

Lloyd-Thomas AR, Cole PV, et al 1990 Quantitative EEG and brainstem auditory evoked potentials: comparison of isoflurane with halothane using the cerebral function analysing monitor. Br J Anaesth 65:306–312.

MacGillivray BB 1969 An EEG monitor incorporating simple pattern recognition. J Physiol 201:65–67P.

MacGillivray BB 1976 EEG monitoring in metabolic disease. In: *Handbook of Electroencephalography and Clinical Neurophysiology*, Volume 15C: *Metabolic Endocrine and Toxic Diseases* (ed A Rémond). Elsevier, Amsterdam, pp. 1–15.

Makeig S, Jung TP 1995 Changes in alertness are a principal component of variance in the EEG spectrum. Neuroreport 7:213–216.

Maresch H, Pfurtscheller G 1983 Simultaneous measurements of auditory brainstem potentials and EEG spectra. Electroencephalogr Clin Neurophysiol 56:531–533.

Martin JT, Faulconer Jr A, Bickford RG 1959 Electroencephalography in anesthesiology. Anesthesiology 20:359–376.

Matousek M, Arvidsson A, Friberg S 1978. Implementation of analytical methods in daily clinical EEG. In: *Contemporary Clinical Neurophysiology* (EEG Suppl 34) (eds WA Cobb, H van Duijn). Elsevier, Amsterdam, pp. 199–204.

Maynard DE 1969 A note on the nature of the non-rhythmic components of the electroencephalogram. Activnervsup (Praha) 11:238–241.

Maynard DE 1972. Separation of the sinusoidal components of the human electroencephalogram. Nature 236:228–230.

Maynard DE, Jenkinson JL 1984 The cerebral function analysing monitor. Initial clinical experience, application and further development. Anaesthesia 39:678–690.

Maynard D, Prior PF, Scott DF 1969 Device for continuous monitoring of cerebral activity in resuscitated patients. BMJ 4:545–546.

Messner M, Beese U, Romstock J, et al 2003 The Bispectral Index declines during neuromuscular block in fully awake persons. Anesth Analg 97:488–491.

Miller A, Sleigh JW, Barnard J, et al 2004 Does bispectral analysis of the electroencephalogram add anything but complexity? Br J Anaesth 92:8–13.

Myers RR, Stockard JJ, Saidman LJ 1977 Monitoring of cerebral perfusion during anesthesia by time-compressed fourier analysis of the electro-encephalogram. Stroke 8:331–337.

Myles PS, Leslie K, McNeil J, et al 2004 Bispectral index monitoring to prevent awareness during anaesthesia: the B-Aware randomized controlled trial. Lancet 363(9423):1757–1763.

Nishiyama T, Matsukawa T, Hanaoka K 2004 A comparison of the clinical usefulness of three different electroencephalogram monitors: Bispectral Index, processed electroencephalogram, and Alaris auditory evoked potentials. Anesth Analg 98:1341–1345.

Nuwer MR, Comi G, Emerson R, et al 1999a IFCN standards for digital recording of clinical EEG. In: *Recommendations for the Practice of Clinical Neurophysiology Guidelines of the International Federation of Clinical Neurophysiology* (EEG Suppl 52) (eds G Deuschl, A Eisen). Elsevier Science, Amsterdam, pp. 11–14.

Nuwer MR, Lehmann D, Lopes da Silva F, et al 1999b In: *Recommendations for the Practice of Clinical Neurophysiology Guidelines of the International Federation of Clinical Neurophysiology* (EEG Suppl 52) (eds G Deuschl, A Eisen). Elsevier Science, Amsterdam, pp. 15–20.

Oken BS, Chiappa KH 1988 Short-term variability in EEG frequency analysis. Electroenceph Clin Neurophysiol 69:191–198.

Pampiglione G, Waterston DJ 1958 Preliminary EEG observations during partial and complete occlusion of cerebral blood flow. Electroencephalogr Clin Neurophysiol 10:354.

Pampiglione G, Waterston, DJ 1961 Observations during changes in venous and arterial pressure. In: *Cerebral Anoxia and the Electroencephalogram* (eds H Gastaut, JS Meyer). Thomas, Springfield, IL, pp. 250–255.

Pfurtscheller G, Schwarz G, Schroettner O, et al 1987 Continuous and simultaneous monitoring of EEG spectra and brainstem auditory and somatosensory evoked potentials in the intensive care unit and the operating room. J Clin Neurophysiol 4:389–396.

Pichlmayr I, Lips U, Künkel H 1984 *The Electroencephalogram in Anesthesia Fundamentals Practical Applications Examples*, English edition. Springer-Verlag, Heidelberg.

Prichep LS, Gugino LD, John ER, et al 2004 The Patient State Index as an indicator of the level of hypnosis under general anaesthesia. Br J Anaesth 92:393–399.

Prior PF 1973 *The EEG in Acute Cerebral Anoxia*. Excerpta Medica, Amsterdam.

Prior PF, Maynard DE 1986 *Monitoring Cerebral Function Long-term Monitoring of EEG and Evoked Potentials*. Elsevier, Amsterdam.

Pronk RAF 1978 *Peri- and Postoperative Computer Assisted Patient Monitoring*. Report M-1978-1. Institute of Medical Physics, TNO, Utrecht.

Pronk RAF, Simons AJR 1982 Automatic recognition of abnormal EEG activity during open heart and carotid surgery. *Kyoto Symposia* (EEG Suppl 36) (eds PA Buser, WA Cobb, T Okuma). Elsevier Biomedical Press, Amsterdam, pp. 590–602.

Pronk RAF, Simons AJR 1984 Processing of the electroencephalogram in cardiac surgery. Comput Prog Biomed 18:181–190.

Rampil IJ 1998 A primer for EEG signal processing in anesthesia. Anesthesiology 89:980–1002.

Recart A, Gasanova I, White PF, et al 2003 The effect of cerebral monitoring on recovery after general anesthesia: a comparison of the auditory evoked potential and Bispectral Index devices with standard clinical practice. Anesth Analg 97:1667–1674.

Renna M, Wigmore T, Mofeez A, et al 2002 Biasing effect of the electro-myogram on BIS: a controlled study during high-dose fentanyl induction. J Clin Monit Comput 17:377–381.

Rennie JM, Chorley G, Boylan GB, et al 2004 Non-expert use of the cerebral function monitor for neonatal seizure detection. Arch Dis Child Fetal Neonatal Ed 89:F37–40.

Romano J, Engel GL 1944 Delirium: I. EEG patterns. Arch Neural Psychiatry (Chicago) 51:356–377.

Rosa AC, Parrino L, Terzano MG 1999 Automatic detection of cyclic alternating pattern (CAP) sequences in sleep: preliminary results. Electroencephalogr Clin Neurophysiol 110:585–592.

Rundshagen I, Schröder T, Prichep LS, et al 2004 Changes in cortical electrical activity during induction of anaesthesia with thiopental/fentanyl and tracheal intubation: a quantitative electroencephalographic analysis. Br J Anaesth 92:33–38.

Salinsky M, Sutula T, Roscoe D 1987 Representation of sleep stages by color density spectral array. Electroenceph Clin Neurophysiol 66:579–582.

Schmidt GN, Bischoff P, Standl T, et al 2003 Narcotrend and Bispectral Index monitor are superior to classic electroencephalographic parameters for the assessment of anesthetic states during propofol–remifentanil anesthesia. Anesthesiology 99:1072–1077.

References

Schneider G, Heglmeier S, Schneider J, et al 2004 Patient State Index (PSI) measures depth of sedation in intensive care patients. Intensive Care Med 30:213–216.

Schultz B, Schultz A, Grouven U 2000 Sleeping stage based systems (Narcotrend). In: *New Aspects of High Technology in Medicine 2000* (eds HP Bruck, P Koeckerling, F Bouchard, C Schug-Pass). Monduzzi Editore, Bologna, pp. 285–291.

Schultz B, Grouven U, Schultz A 2002 Automatic classification algorithms of the EEG monitor Narcotrend for routinely recorded EEG data from general anaesthesia: a validation study. Biomed Tech 47:9–13.

Schultz A, Grouven U, Beger FA, et al 2004 The Narcotrend Index: classification algorithm, correlation with propofol effect-site concentrations, and comparison with spectral parameters. Biomed Tech (Berlin) 49:38–42.

Schwilden H, Schüttler J, Stoeckel H 1985 Quantitation of the EEG and pharmacodynamic modelling of hypnotic drugs: etomidate as an example. Eur J Anaesthesiol 2:121–131.

Screening Commission of the New University Hospital, Copenhagen, Denmark 1967 Reduction of electrical interference in measurements of bioelectrical potentials in a hospital. Acta Polytech Scand, Electr Eng Ser 15:1–37.

Sebel PS 2004 Comfortably numb? [editorial]. Acta Anaesthsiol Scand 48:1–3.

Sebel PS, Lang E, Rampil IJ, et al 1997 A multicenter study of bispectral electroencephalogram analysis for monitoring anesthetic effect. Anesth Analg 84:891–899.

Shannon CE 1948 A mathematical theory of communication. Bell System Techn J 27:379–423, 623–656.

Sharpe RM, Nathwani D, Pal SK, et al 1997 Auditory evoked response, median frequency and 95% spectral edge during anaesthesia with desflurane and nitrous oxide. Br J Anaesth 78:282–285.

Sigl JC, Chamoun NG 1994 An introduction to bispectral analysis for the electroencephalogram. J Clin Monit 10:392–404.

Sleigh JW, Steyn-Ross DA, Steyn-Ross ML, et al 2001 Comparison of changes in electroencephalographic measures during induction of general anaesthesia: influence of the gamma frequency band and electromyogram signal. Br J Anaesth 86:50–58.

Sneyd JR 2003 How low can we go? [editorial]. Br J Anaesth 91:771–772.

Song D, van Vlymen J, White PF 1998 Is the Bispectral Index useful in predicting fast-track eligibility after ambulatory anesthesia with propofol and desflurane. Anesth Analg 87:1245–1248.

Stoeckel H, Schwilden H 1986 Methoden der automatischen Feedback-Regelung fur die Narkose. Konzepte und klinische Anwendung. Anasthesiologie, Intensivmedizin, Notfallmedizin, Schmerztherapie 21:60–67.

Struys MM, Jensen EW, Smith W, et al 2002 Performance of the ARX-derived auditory evoked potential index as an indicator of anesthetic depth: a comparison with Bispectral Index and hemodynamic measures during propofol administration. Anesthesiology 96:803–816.

Struys MM, Vereecke H, Moerman A, et al 2003 Ability of the Bispectral Index, autoregressive modelling with exogenous input-derived auditory evoked potentials, and predicted propofol concentrations to measure patient responsiveness during anesthesia with propofol and remifentanil. Anesthesiology 99:802–812.

Tempe DK, Satyanarayana, L 2004 Is there any alternative to the Bispectral Index Monitor? [editorial]. Br J Anaesth 92:1–3.

Thomsen CE, Prior PF 1996 Quantitative EEG in assessment of anaesthetic depth: comparative study of methodology. Br J Anaesth 77:172–178.

Thomsen CE, Christensen KN, Rosenfalck A 1989 Computerized monitoring of depth of anaesthesia with isoflurane. Br J Anaesth 63:36–43.

Thomsen CE, Rosenfalck A, Nørregaard Christensen K 1991 Assessment of anaesthetic depth by clustering analysis and autoregressive modelling of electroencephalograms. Comput Methods Progr Biomed 34:125–138.

Todd MM (ed) 1998 EEGs, EEG processing, and the Bispectral Index. Anesthesiology 89:815–816.

Toet MC, van der Meij W, de Vries LS, et al 2002 Comparison between simultaneously recorded amplitude integrated electroencephalogram (cerebral function monitor) and standard electroencephalogram in neonates. Pediatrics, 109:772–779.

Urhonen E, Jensen EW, Lund J 2000 Changes in rapidly extracted auditory evoked potentials during tracheal intubation. Acta Anaesthesiol. Scand 44:743–748.

Vakkuri A, Yli-Hankala A, Talja P, et al 2004 Time-frequency balanced spectral entropy as a measure of anesthetic drug effect in central nervous system during sevoflurane, propofol, and thiopental anesthesia. Acta Anesthesiol Scand 48:145–153.

Van Gils M, Rosenfalck A, White S, et al 1997 Signal processing in prolonged EEG recordings during intensive care. Methods for analyzing and displaying EEG signals. IEEE Eng Med Biol Mag 16:56–63.

Van Hemel NM, Pronk RAF 1977 *Monitoring of the Seriously Ill.* Report M-1977-3. Institute of Medical Physics, TNO, Utrecht.

Vanluchene AL, Vereecke H, Thas O, et al 2004 Spectral entropy as an electroencephalographic measure of anesthetic drug effect: a comparison with Bispectral Index and processed midlatency auditory evoked response. Anesthesiology 101:34–42.

Vereecke HE, Vasquez PM, Jensen EW, et al 2005 New composite index based on midlatency auditory evoked potential and electroencephalographic parameters to optimize correlation with propofol effect site concentration: comparison with Bispectral Index and solitary used fast extracting auditory evoked potential index. Anesthesiology 103:500–507.

Verzeano M 1951 Servo-motor integration of the electrical activity of the brain and its applications to the automatic control of narcosis. Electroencephalogr Clin Neurophysiol 3:25–30.

Veselis RA, Reinsel R, Sommer S, et al 1991 Use of neural network analysis to classify electroencephalographic patterns against depth of midazolam sedation in intensive care unit patients. J Clin Monitoring 7:259–267.

Viertiö-Oja H, Maja V, Särkelä M, et al 2004 Description of the entropy algorithm as applied in the Datex-Ohmeda S/5 entropy module. Acta Anesthesiol Scand 48:154–161.

Vivien B, Di Maria S, Ouattara A, et al 2003 Overestimation of Bispectral Index in sedated intensive care unit patients revealed by administration of muscle relaxant. Anesthesiology 99:9–17.

Volavka J, Matousek M, Feldstein St Roubicek J, et al 1973 Die Zuverlassigkeit der EEG-Beurteilung. EEG EMG Z Elektroenzephalogr Elektromyogr Verwandte Geb 4:123–130.

Walter DO 1969 A posteriori Wiener filtering of averaged evoked responses. Electroencephalogr Clin Neurophysiol Suppl. 27:61–70.

Wastell DG 1981 When Wiener filtering is less than optimal. An illustrative application to the brain stem evoked potential. Electroencephalogr Clin Neurophysiol 51:678–682.

Weber F, Seidl M, Bein T 2005 Impact of the AEP-Monitor/2-derived composite auditory-evoked potential index on propofol consumption and emergence times during total intravenous anaesthesia with propofol and remfentanil in children. Acta Anaesthesiol Scand 49:277–283.

Webber WR, Litt B, Wilson K, et al 1994 Practical detection of epileptiform discharges (EDs) in the EEG using an artificial neural network: a comparison of raw and parameterized EEG data. Electroencephalogr Clin Neurophysiol 91:194–204.

Wilson SB, Harner RN, Duffy FH, et al 1996 Spike detection. I. Correlation and reliability of human experts. Electroencephalogr Clin Neurophysiol 98:186–198.

Yli-Hankala A, Vakkuri A, Annila P, et al 1999 EEG Bispectral Index monitoring in sevoflurane or propofol anaesthesia: analysis of direct costs and immediate recovery. Acta Anaesthesiol Scand 43:545–549.

Chapter 4

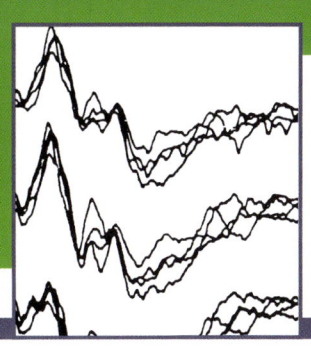

Normal and Pathological Phenomena in EEG, Evoked Potential, EMG and Nerve Conduction Studies

GENERAL INTRODUCTION

In this chapter we provide a brief account of the main features of the electroencephalogram (EEG), evoked potentials (EPs), electromyogram (EMG) and nerve conduction studies and the general patterns of abnormality likely to be encountered in the context of patients during intensive care and surgery. Specific changes due to sedation and anaesthesia are described in Chapter 5. The reader can find more detailed information on phenomenology in Binnie et al (2003, 2004) and, for EEG, in Rowan and Tolunsky (2003). The features described are also relevant to the design of monitoring equipment.

The concepts of 'normality' and 'abnormality' are increasingly used in a statistical sense as quantitative studies based on digital measurements replace descriptive data in clinical neurophysiology. Two caveats are necessary:

(i) normative data, particularly for EPs, are related to recording conditions and hence apply to the laboratory where they were obtained; and
(ii) many features of clinical importance are not readily amenable to simple quantification, for example some specific abnormal patterns such as periodic complexes, short duration events such as arousal responses or interictal seizure discharges in the EEG, and also fluctuations in the EEG and EPs relating to innate biological periodicities such as sleep cycles and other circadian rhythms.

primarily involving the diencephalon, or a metabolic or pharmacologically induced encephalopthy. Interpretation, as with most medical investigations, must be made within the clinical context

The EEG of any individual undergoes dramatic changes as a function of age and state of awareness, and there are also marked interindividual differences between the EEGs of people of the same age and in the same state of vigilance; such differences tend to decrease during sleep, sedation and anaesthesia. Some of these interindividual differences are quantitative, for instance with respect to the mean amplitude, frequency, symmetry and responsiveness of the waking rhythms; others are qualitative, reflecting the occurrence of certain phenomena in some persons but not in others. Likewise, some pathological EEG findings are unequivocally abnormal and are not encountered in health, whilst others involve features within the range of variation between normal subjects but different from the individual's premorbid EEG. Disease (and medication) may produce cerebral dysfunction leading to subtle slowing or reduced reactivity of a degree only evident in serial recordings or acute monitoring or, alternatively, may be associated with dramatic acute changes including asymmetries that are typical (but rarely specific) for a particular pathophysiological process. An important consideration is the fact that the EEG reflects cerebral function, and structural abnormality will be manifest only in so far as it results in functional changes.

EEG WAVEFORMS AND INTERPRETATION

INTRODUCTION

The scalp EEG is in the main recorded from generators in the superficial parts of the cerebral cortex. The activities generated by these superficial cortical neurones depend in part on their own functional state, but are also influenced by afferents from deep structures. More than one mechanism can produce essentially similar phenomena. Thus abnormal slow waves recorded at the front of the head may reflect dysfunction of superficial neurones in the frontal lobes, a structural or functional abnormality

DESCRIBING EEG PHENOMENA

The recommendations of specialist committees of the International Federation for Clinical Neurophysiology (IFCN) are published in its journal, *Clinical Neurophysiology* (formerly *Electroencephalography and Clinical Neurophysiology*). The most recent comprehensive update of all IFCN recommendations is available in a single volume edited by Deuschl and Eisen (1999) which includes a glossary of terms defining specific EEG phenomena (Noachtar et al 1999). In this section describing EEG phenomena, all illustrations use the international 10:20 system electrode position designations described in Chapter 2 (see *Figs 2.9–2.11*) to label the recordings. First we consider the basic categories of descriptors and then indicate how these are modified in disease. A brief excerpt from a typical multichannel EEG recording obtained from a healthy awake adult is shown in *Fig. 4.1* to demonstrate the main features used to describe EEG phenomena. These features are described below.

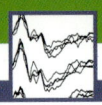

Wave shape (morphology)

When the eyes are closed, there is a rhythmical, though not sinusoidal, waveform, the amplitude of which fluctuates with time in a regular way. In *Fig. 4.1* this is the alpha rhythm seen in the occipital regions in the awake state with the eyes closed. Activity that persists for much of the time during a particular part of a recording (e.g. the alpha rhythm in this example) is described as 'ongoing' or 'dominant' activity.

Rhythmicity

The regular fluctuation about the baseline gives a rhythmical nature to the alpha rhythm illustrated in *Fig. 4.1* and this is a typical characteristic of the EEG. Often the waveform is more complex, particularly in pathological conditions, but the wave pattern may still be rhythmical, often with two rhythmical components intermixed. In other conditions, the wave pattern has no obvious rhythmicity, it is then described as irregular.

Frequency of repetition

An ongoing EEG signal can usually be described in terms of the number of times the waveform repeats itself in one second. In this way we refer to the frequency of repetition, or just frequency, of the signal. When the EEG is recorded using a digital system and displayed on a VDU, it is possible to align cursors with the waves in such a way that their frequency is automatically measured and displayed. The repetition rate is usually measured in hertz (Hz), or cycles per second (c/s). There is a subtle difference of meaning between these two terms, and there is a long-standing dispute over the appropriate units to describe repetition rates of EEG phenomena. Hertz are normally used by physicists to specify the frequencies of sinusoidal signals. Some neurophysiologists use hertz to describe EEG rhythms, others object on the grounds that the EEG phenomena in question are not sinusoidal, and prefer cycles per second. For transients which are discontinuous and not necessarily regular, it may be appropriate just to state the rate, for example as 'spikes per second'; isolated non-rhythmic events may also be described in terms of equivalent frequency, for example a wave of 300 ms duration can be regarded as having a frequency of 3 Hz.

The frequencies most commonly seen in the ongoing EEG are classified into four arbitrary frequency ranges or bands (*Fig. 4.2*). Frequencies below 4 Hz are *delta* waves, from 4 to less than 8 Hz they are called *theta* waves, the 8–13 Hz range is the *alpha* band, and frequencies from more than 13 to 35 Hz are called *beta*. For the purposes of quantitative EEG (qEEG) studies, but not in routine clinical practice, alpha and beta are sometimes subdivided into two or more ranges. A fifth range, 35–70 Hz, called *gamma*, is of interest to neurophysiologists, but may be difficult to detect because of its low amplitude and the limited high frequency response used in conventional recordings; at present it has little clinical significance except at the onset of some epileptic seizures.

Amplitude

Implicit in the features described so far is the concept of amplitude. EEG signal amplitude may vary over a wide range from very small, a few microvolts on the scalp, up to several hundred microvolts. The clinical interpretation of EEG records must take amplitude into account, but manual assessment of amplitude can be very arbitrary and few EEG texts give any guidance. The amplitude with

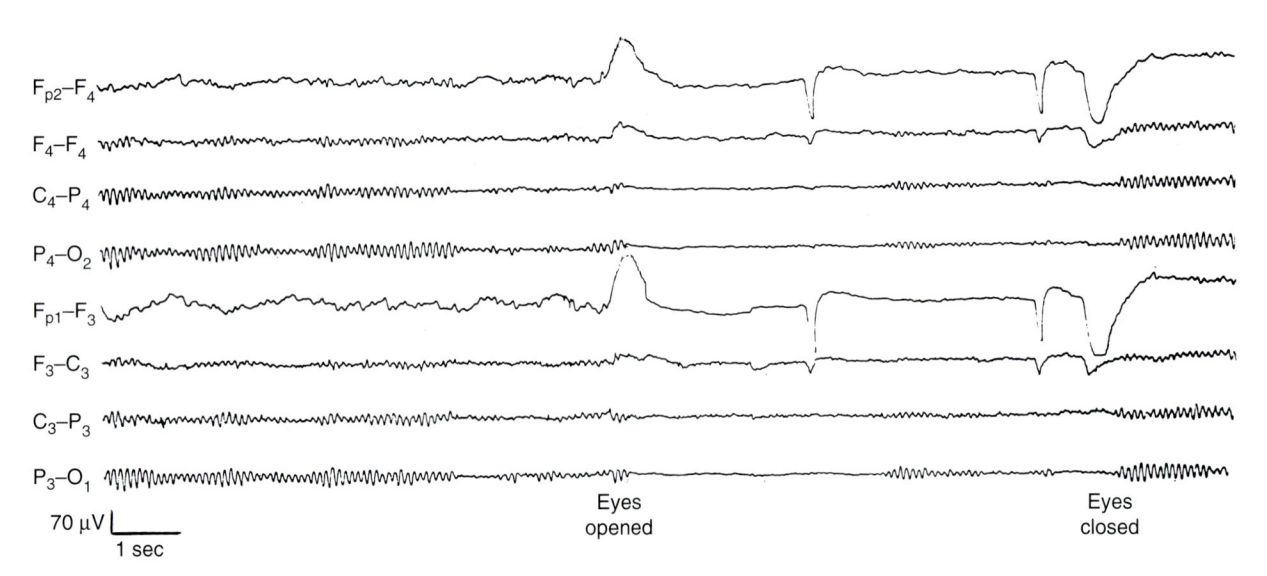

Fig. 4.1 Normal EEG. Posterior alpha rhythm waxing and waning posteriorly attenuates initially on eye opening and reappears on eye closure. Notice large frontal deflections from eye movements and blinks. (From Binnie et al (2003), by permission.)

EEG waveforms and interpretation

(a) [waveform]

(b) [waveform]

(c) [waveform]

(d) [waveform]

sec

[50 μV

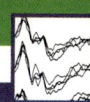

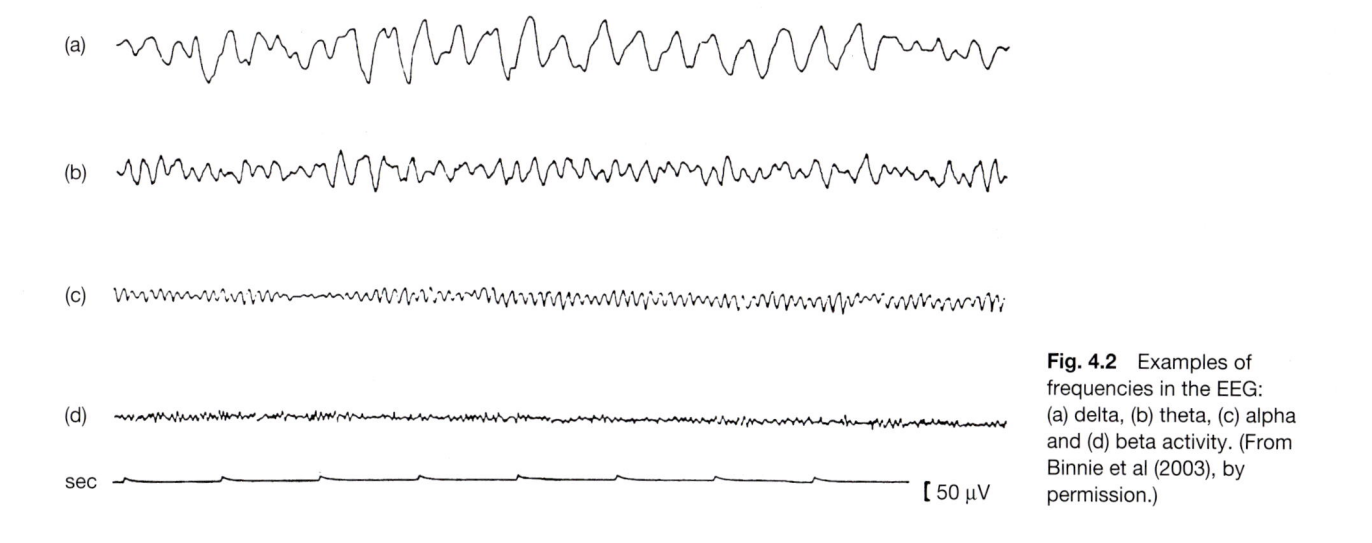

Fig. 4.2 Examples of frequencies in the EEG: (a) delta, (b) theta, (c) alpha and (d) beta activity. (From Binnie et al (2003), by permission.)

which a given phenomenon is displayed depends on its spatial distribution, the location of the electrodes and the method of derivation used.

When visually interpreting an EEG from a chart recording, or screen, it is usual to estimate the average peak-to-peak amplitude of the waves and this will result in a somewhat larger value than is mathematically correct.

This arises because it is easier to judge the overall amplitude of the envelope formed by an ongoing activity than to estimate the mean peak-to-peak values of individual component waves. For manual measurement, the peak-to-peak amplitude should be assessed using a pair of dividers to measure the waves over a length of record in which the phenomenon is clearly present. *Figure 4.3* compares sev-

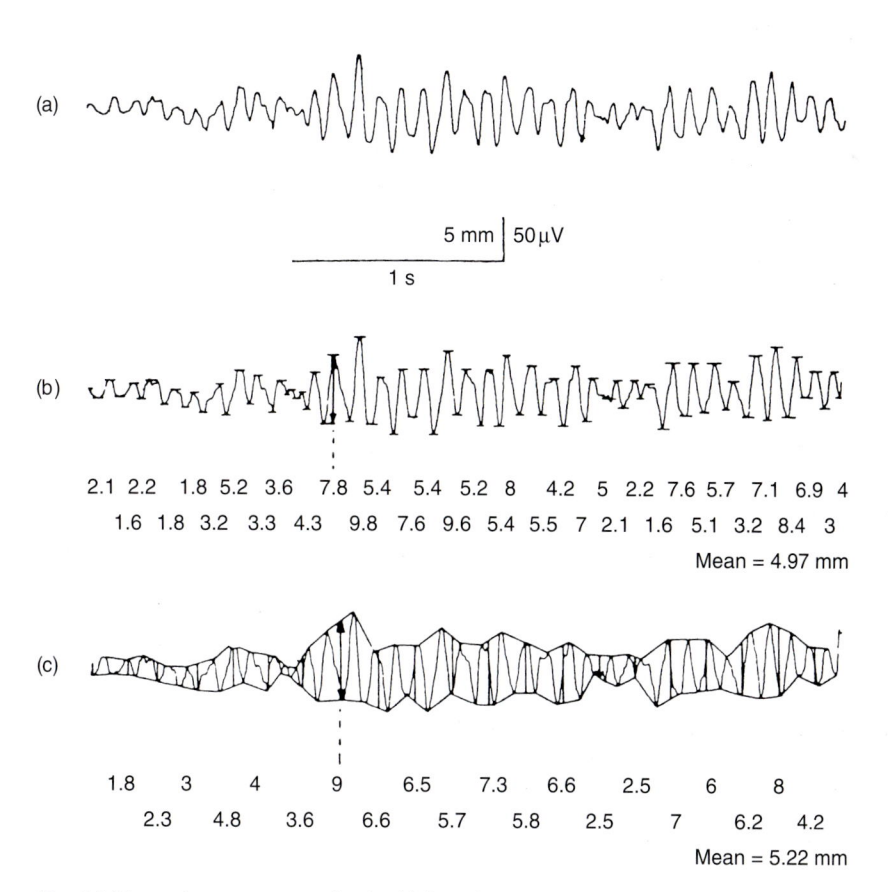

(a) [waveform]

5 mm | 50 μV
1 s

(b) [waveform with markers]

2.1 2.2 1.8 5.2 3.6   7.8 5.4   5.4 5.2 8   4.2 5   2.2 7.6 5.7   7.1 6.9 4
1.6 1.8 3.2 3.3 4.3   9.8 7.6   9.6 5.4 5.5   7   2.1 1.6 5.1 3.2   8.4 3

Mean = 4.97 mm

(c) [waveform envelope]

1.8   3   4   9   6.5   7.3   6.6   2.5   6   8
   2.3   4.8   3.6   6.6   5.7   5.8   2.5   7   6.2   4.2

Mean = 5.22 mm

Fig. 4.3 Measuring average amplitude. (a) By eye: 'probably about 60 μV'. (b) By measuring and then averaging individual waves: about 50 μV. (c) By measuring and then averaging the widths of the envelope at regular intervals: 52 μV. (From Binnie et al (2003), by permission.)

111

eral ways of measuring the amplitude of a sample of EEG. Because of the alternating current (AC) coupling in the amplifiers of EEG machines, the fluctuating EEG output signal from the amplifiers will have a zero mean amplitude over a duration of at least three times the time constant value. Therefore, instrumental analysis of EEG amplitude usually describes the amplitude deviations from zero (regardless of sign). Suitable measures are the root mean square amplitude (equivalent to the standard deviation (SD) of the amplitude fluctuations about the base line), or the mean square amplitude (equivalent to the variance of the fluctuations about the baseline). As with frequency, when the EEG is recorded using a digital system and displayed on a VDU, it is possible to align cursors with the peak and trough of a wave and read off the amplitude value, or a chosen amplitude function of a series of waves may be displayed automatically.

The waxing and waning in amplitude of ongoing activity is often described as 'spindling' (Fig. 4.4(g)). Spindles also occur in the central regions during light sleep. A group of waves which appear and disappear abruptly and are distinguished from background activity by differences in frequency, form and/or amplitude, is called a 'burst' (Fig. 4.4(i)).

Transients

Against a background of ongoing, more or less sustained activities, discrete transient waves, or brief stereotyped sequences of two or more waves (complexes) may occur. Some are apparently spontaneous, others are evoked by extrinsic or intrinsic stimuli. Those elicited by known stimuli or cognitive events are designated evoked, or event-related, potentials. Some transients are normal, others are pathological phenomena (or even artefactual); it is therefore important when interpreting records to identify different types of transients and to recognize them for what they are. This is helped by the annotation of the record provided by the technologist who should note

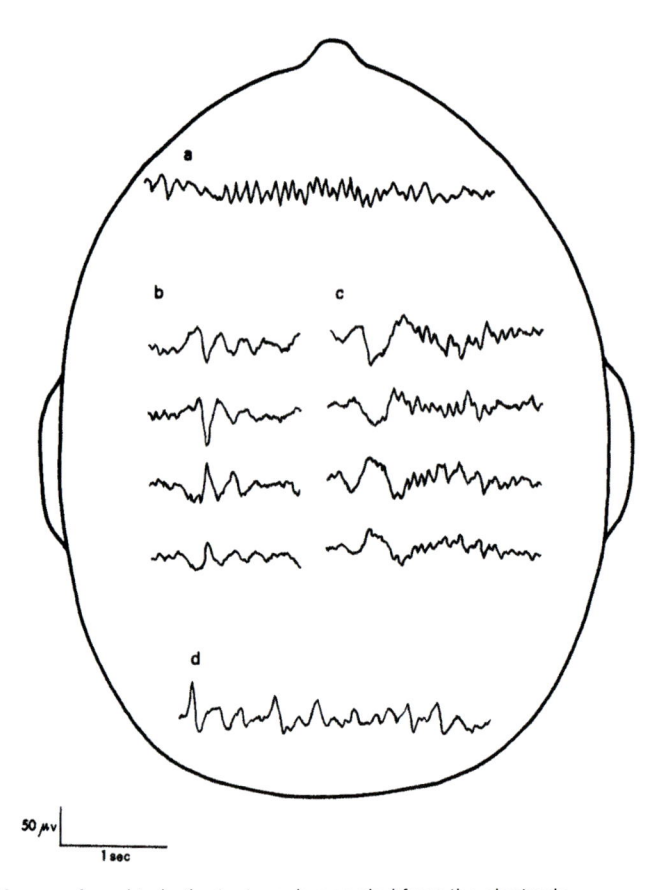

Fig. 4.4 Various named activities in the EEG. Left: Examples of waveforms referred to in the text, each recorded from the electrode indicated with respect to an electrode on the chin: (a) K-complex; (b) lambda wave; (c) mu rhythm; (d) spike; (e) sharp waves; (f) spike and wave; (g) sleep spindle; (h) vertex sharp waves; (i) polyspikes. In all examples, an upward deflection corresponds to negativity of the specified electrode with respect to the chin. Calibration marks = 100 μV. (From Binnie et al (2003), by permission.) Right: Transient events during sleep: (a) sleep spindles, which are prominent over the frontal and central regions; (b) vertex sharp wave; (c) K-complex; (d) lambdoid waves which appear only over the occipital regions in sleep; they should be distinguished from occipital lambda waves seen in the waking state with the eyes open as a reponse evoked by saccadic eye movement. (From Scott (1976), by permission.)

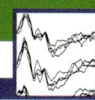

intrinsic (e.g. patient moving – the eye movement deflections in channels 1 and 5 in *Fig. 4.1*) or extrinsic (e.g. extraneous noise) events. Transient potentials may particularly be evoked by internal or external events, especially if the subject is drowsy or asleep (e.g. the lambda waves in response to visual stimuli and the K-complex in response to sounds, shown in *Fig. 4.4*).

Spatial distribution

The amplitude of the alpha rhythm in *Fig. 4.1* is larger towards the occipital area, decreasing anteriorly, but is approximately symmetrical over the two hemispheres. In many disorders affecting the EEG, whether intracranial or systemic in origin, activity is also symmetrical but with abnormal distribution, for example with the highest amplitude is towards the front of the head. Changes localized to a particular part of the head are usually suggestive of localized brain dysfunction or pathology.

Spatiotemporal patterns

A particular activity may be characterized by a combination of frequency and spatial distribution. For example, as well as the posterior alpha rhythm, there is another activity with a frequency in the alpha band, but with a more central site of origin known as the mu rhythm (*Fig. 4.4(c)*) which is reactive to proprioceptive stimuli such as clenching the contralateral fist. Localized activity of a similar pattern may also occur at the site of a previous burr-hole or craniotomy rendering the EEG markedly asymetrical; it is described as a breach rhythm. It is important to be able to recognize particular localized spatiotemporal patterns of EEG activity because they may indicate either localized pathology or a non-cerebral artefact, for example from a poorly applied electrode. The recognition of specific spatial patterns, and their localization, requires the use of a wide coverage of electrodes, and both bipolar and reference electrode derivations. This is one reason for performing an initial multichannel EEG as a preliminary to continuous monitoring of trends and detections of transients such as epileptiform discharges (which usually involve a substantial degree of data reduction) in the neuro-ICU or for intraoperative monitoring during carotid artery or intracranial surgery. This also has implications for continuous montoring systems where the number of electrodes and recording channels must be sufficient to provide valid information for the required decision support. An example is monitoring during carotid artery surgery where it is desirable to be able to detect not only significant asymmetries but also regional changes.

Symmetry and synchrony

Two other aspects of EEG spatiotemporal patterns are synchrony and symmetry and it is necessary to appreciate the different meanings that these terms can have, depending on the interval of time over which the EEG is being considered. For example, if an EEG feature such as a spike occurs apparently simultaneously in two locations, for purposes of routine clinical reporting it will be described as synchronous. There may, however, still be a small time difference between the two deflections, detected only by expanding the time scale or the use of instrumental analysis, which indicates that they are not strictly synchronous. Such subtle asynchronies may be of clinical importance, for instance when an abnormal activity is rapidly propagated from one hemisphere to the other or when observing the immediate effect of a surgical intervention such as clamping an internal carotid artery. Similarly, a rhythmic component that occurs bilaterally, may have a small time displacement between comparable waves on either side. Special measurement techniques usually show that the degree of synchrony between the alpha peaks in the two hemispheres varies continually, with one side leading for a few seconds, then the other. More extremely, if the rhythm on one side is of different frequency from that on the other (measured over an interval of about a second), then the two are unequivocally asynchronous; such differences may occur after stroke or head injury.

Similar considerations apply to amplitude asymmetry. The peak-to-peak amplitudes of a transient recorded in two or more channels can be measured and compared, but this method is difficult to apply visually to intermittent or ongoing activity, such as the alpha rhythm. The *average* amplitudes on either side are then estimated, from which it can be decided whether or not the record is significantly asymmetrical. However, sleep spindles, which have a tendency to occur in discrete, non-simultaneous runs on either side, may still be regarded as 'symmetrical' if the overall *amounts* of activity are substantially the same. The 'amount' here refers to a subjective visual assessment based on a combination of mean amplitude and the proportion of recording time during which the activity is present. The term *abundance* is sometimes used in much the same way, but once had a more specific meaning as a measure used for quantifying EEGs in the pre-computer era.

An apparent asymmetry of the EEG may, of course, be due to asymmetrical electrode placement or spacing, or other technical errors, and this possibility should be checked by the technologist during the recording.

Inherent variability of the EEG and other biological rhythms

Second-by-second or minute-by-minute variability is a key feature of the normal EEG. Classic examples include the 2–4/min fluctuations in EEG akin to those in heart rate, arterial blood pressure or intracranial pressure (ICP). There are two other main cyclical patterns of interest that occur in normal individuals and may persist in patients in the ICU; these are the basic rest and activity cycles (BRACs) (Kleitman 1963) with a period of about 90 min that occur throughout the 24 h and the much faster 2–60 s cyclical alternating patterns (CAP) in sleep. In CAP, the two phases are designated A and B, the former consisting of relatively high amplitude phasic activity corresponding to arousal, and the latter of low amplitude and unstructured activity. Intervening non-CAP phases of similar duration follow phases A and B. The fluctuations in arousal associated with these cycles modulate various physiological and pathological CNS events (Terzano et al 1987) and form a continuum with some of the phasic variations in arousal in comatose patients (Evans 1992, 2002, Evans and Bartlett 1995). The alternation of the EEG in burst suppression may form an extreme case where the variance of inter-burst intervals differs between drug-induced and ischaemic forms (Beydoun et al 1991).

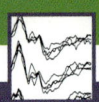

Reactivity

The pattern of the EEG alters, or reacts, to many factors, both intrinsic (drowsiness) and extrinsic (noises). A common reaction in the awake subject is illustrated in *Fig. 4.1* where the alpha rhythm occurs when the eyes are closed and is not seen in the initial eyes-open period. Sometimes the reactivity is paradoxical, for example alpha activity usually disappears during extreme drowsiness, but returns when the subject is alerted and opens the eyes. Reactivity is a characteristic of various EEG activities, both normal and abnormal, and may be elicited by external events such as sounds or tactile stimuli or by internal events such as pain.

GENERAL CATEGORIES OF ABNORMALITY IN THE EEG OF IMPORTANCE FOR ICU AND INTRAOPERATIVE MONITORING

Change in frequency content
Slowing

An increase in lower frequency components of the EEG is seen in a variety of cerebral disorders. Slowing is a non-specific abnormality which can reflect a variety of pathological processes including cerebral hypoxia, oedema, raised ICP, cerebral inflammatory or degenerative processes, the postictal state and various intoxications. Sometimes the slow activity arises *de novo*, sometimes it appears to represent an increase in the amount of slow rhythms which were previously present, on occasion it may appear to reflect a change in frequency of a pre-existing activity such as the alpha rhythm. These phenomenologically distinct patterns probably reflect different pathophysiological mechanisms; slowing of the alpha generators, dysfunction of deep midline structures, and partial cortical deafferentation similar to the one which produces polymorphic delta activity (Steriade et al 1990). Sometimes it may appear that more than one type of change occurs sequentially with increasing dysfunction. Thus serial EEGs in a patient with progressive metabolic disturbance may at first show an alpha rhythm slower than that previously exhibited by the same person in health, then activity with similar characteristics but in the upper theta range, then an abrupt transition to generalized delta waves of totally different topography (*Fig. 4.5*) and, in hepatic encephalopathy, eventually a triphasic waveform (see *Triphasic complexes (or waves)*, p. 120).

Minor changes in ongoing frequencies may be recognized as pathological only if they are also asymmetrical. The normal EEG is much more symmetrical in terms of frequency than of amplitude. Thus asymmetries in the dominant alpha frequency of only 0.5 Hz may be clinically significant, and can be found not only within the territory of the alpha rhythm but also in the central areas, for example after a cerebrovascular accident. The slower component which is likely to reflect the side of greater cerebral dysfunction may be of either greater or lesser amplitude than the corresponding normal activity over the other hemisphere. The lateralizing significance of an asymmetry of alpha amplitude may therefore be misinterpreted unless frequency is also taken into account.

There is indeed a general problem in interpreting asymmetrical pathological slow activities. These may be expected to be of greater amplitude over the more disturbed hemisphere, but the reverse will apply if the underlying pathological process also leads to amplitude

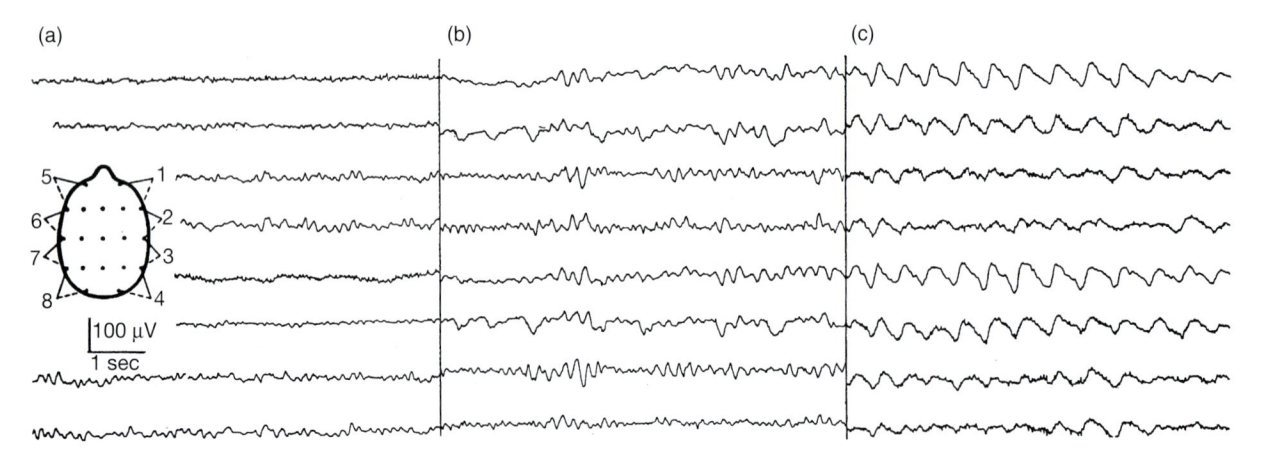

Fig. 4.5 Progressive EEG changes in 33-year-old man with deteriorating hepatic encephalopathy. (a) Blood ammonia 200 μg/100 ml. Mental state normal. Occasional slowing of dominant posterior rhythm below 8 Hz. (b) Blood ammonia 280 μg/100 ml. Some clouding of consciousness. Widespread theta activity at 5–6 Hz. (c) Blood ammonia 500 μg/100 ml. Patient very confused. High voltage frontally predominant delta activity at 2 Hz. (From Kiloh et al (1981), by permission.)

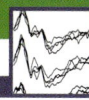

reduction. Thus, if a subdural haematoma produces bilateral slowing due to raised ICP and an asymmetry of amplitude, it may be impossible to lateralize from the EEG. The hypothesis that unilateral amplitude reduction is present may, however, be supported by the finding of a parallel asymmetry of beta activity. Because the alpha frequency of most alert adult subjects is somewhat higher than 8 Hz, essentially similar considerations apply to the interpretation of slowed alpha and of excess theta activity. It may be helpful in describing such pathological EEGs to employ the concept of the 'postcentral dominant rhythm' regardless of whether this is within or below the alpha frequency range.

Excessive beta activity

The range of normal variation in the amount of beta activity is very great, particularly in childhood. Beta activity is moreover increased in spontaneous drowsiness and by many sedative drugs, notably barbiturates and benzodiazepines. Because notions of normality are generally based on clinical material in which prominent beta activity is often associated with medication, there may be a tendency to overlook the fact that prominent beta activity can be normal, and wrongly to assume that patients showing this feature are taking drugs not mentioned by the referring physician. Drug-induced beta activity commonly has the same frontocentral emphasis as that occurring normally (*Fig. 4.6*).

Amplitude reduction

The most unequivocal sign of cerebral dysfunction is the reduction in amplitude of normal activities. Amongst other causes, this may result from neuronal loss, or widespread suppression of neuronal activity by toxic agents including drug overdose, acute hypoxia and hypothermia. Extreme amplitude reduction (i.e. absence of recordable activity, so-called 'electrocerebral silence' (ECS)) is observed in neocortical death, after massive barbiturate overdose, during profound hypothermia, and often for some tens of seconds after a major convulsive seizure.

Transient generalized amplitude reduction may occur as a brief reproducible response for a second or so following an external stimulus in a stuporose or lightly comatose patient or a generalized epileptic seizure. In contrast, the rarer, or underreported, generalized 'episodic low-amplitude events' (ELAEs) lasting 0.5–4 s (Rae-Grant et al 1991) occur in patients with coma of various aetiologies and usually with fatal outcome. They should be distinguished from the low amplitude periods of the burst suppression pattern (which are usually of longer duration, interrupted

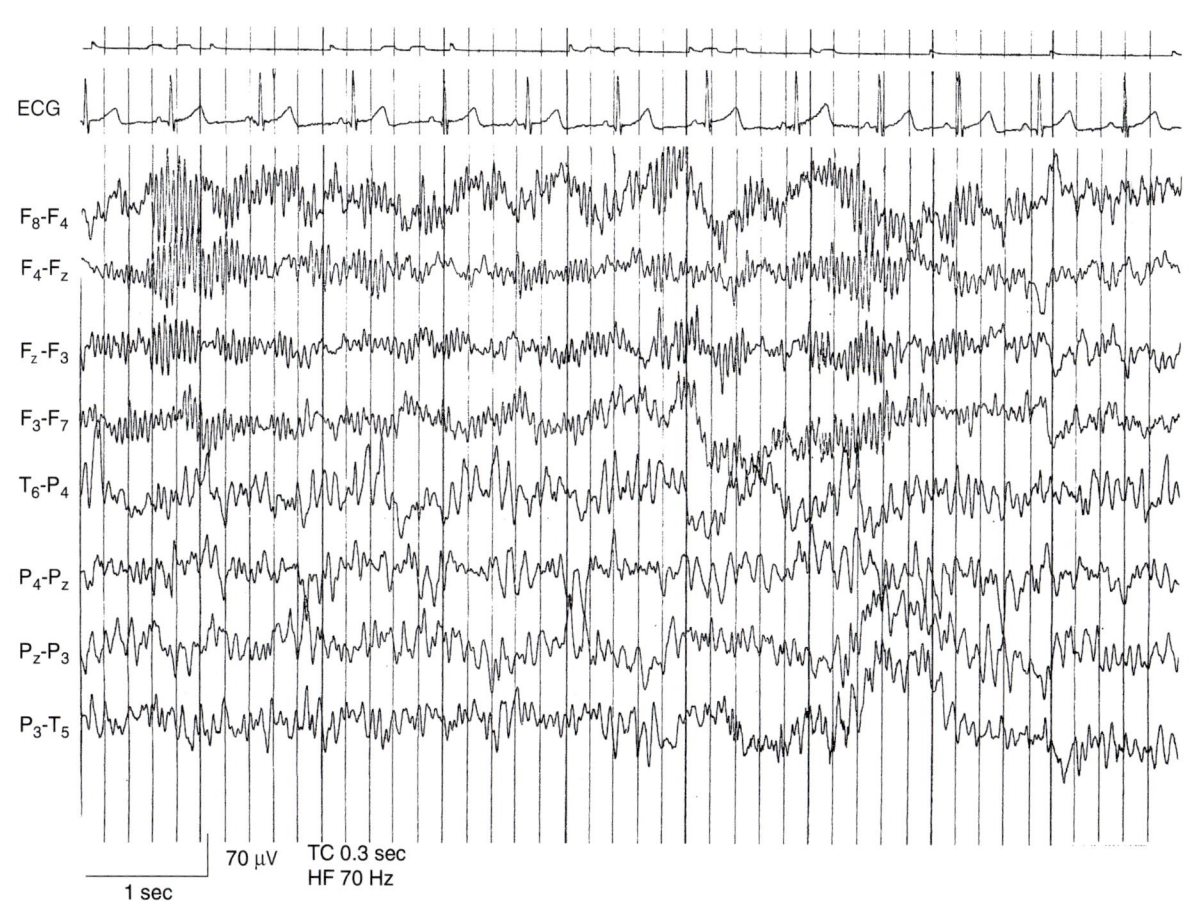

Fig. 4.6 Prominent beta activity in a 6-year-old girl who was receiving clonazepam and sodium valproate to control absence seizures. Note the anterior distribution and spindled pattern of the beta activity. (Modified from Binnie et al (2003), by permission.)

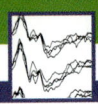

by separate clear bursts of high amplitude activity) and from the effects of arousal by stimulation, to which they are not apparently related.

An asymmetry greater than 50% in amplitude of ongoing activities will generally reflect disease on the side where the amplitude is lower. An exception may occur in patients with a skull defect (e.g. a burr hole or craniectomy) when, in addition to the breach rhythm mentioned above, all on-going EEG activity may be of greater amplitude over a wide area around the defect, particularly with respect to fast components which are ordinarily reduced by conduction through the skull. Unilateral localized amplitude reduction, leading to signicant asymmetries, may be seen after stroke and in association with some superficial cerebral tumours or haematomas. Technical causes of asymeties in amount or amplitude of activity, such as poor electrode contact in an unconscious patient lying on one side, must always be excluded. Localized amplitude reduction in the surface EEG may also result from impaired conduction of signals from the cortex to the scalp, for instance due to the presence of an intervening extra- or subdural haematoma which may 'mask' the activity or, by local pressure, produce localized cortical ischaemia.

An important, and underrecognized, abnormality is bilateral localized amplitude reduction affecting both parieto-occipital regions, sometimes asymmetrically; this is an important warning sign of impending or actual cerebral ischaemic damage due to cerebral hypoperfusion (*Fig. 4.7*) (Prior 1973). Classically, during a period of profound arterial hypotension or raised ICP, hypoperfusion first affects the boundary zone areas between the territories of the anterior, middle and posterior cerebral arteries (see *Fig. 2.31*). The localized reduction in amplitude of the EEG (and cortical somatosenosry evoked potentials) may be reversible if adequate cerebral perfusion is restored before the threshold for ischaemic neuronal damage is reached (Brierley and Graham 1984).

Localized and lateralized abnormalities

Localized changes characterized by localized irregular slow waves (polymorphic delta activity) and a reduction of normal ongoing rhythms generally reflect structural abnormality in the cortex underlying the electrodes from which they are recorded (*Fig. 4.8*). Space-occupying lesions of acute or subacute onset (such as rapidly growing tumours, abscesses and intracerebral haematomata) associated with localized destruction and oedema of white and grey matter may produce prominent triangular delta waves or biphasic 'zeta' waves (Magnus and van der Holst 1987). In general, acute destructive processes causing gross local cerebral dysfunction produce frequent, steep, high voltage slow delta waves. The occurrence of polymorphic focal delta activity at the site of a lesion probably requires both cortical damage and partial deafferentation due to white matter involvement (Gloor et al 1977). Pathology confined to cortical grey matter produces only reduction of amplitude, without delta activity either overlying or at the margins of the lesion.

Theta activity and sporadic delta waves of lower amplitude and duration may be seen during recovery or in association with lesions less disruptive of local cerebral function. Amplitude reduction may be the eventual outcome, when those damaged cells that have not died have recovered. Lesions which are static or only slowly progressive (porencephalic cysts, mengingiomata) may pro-

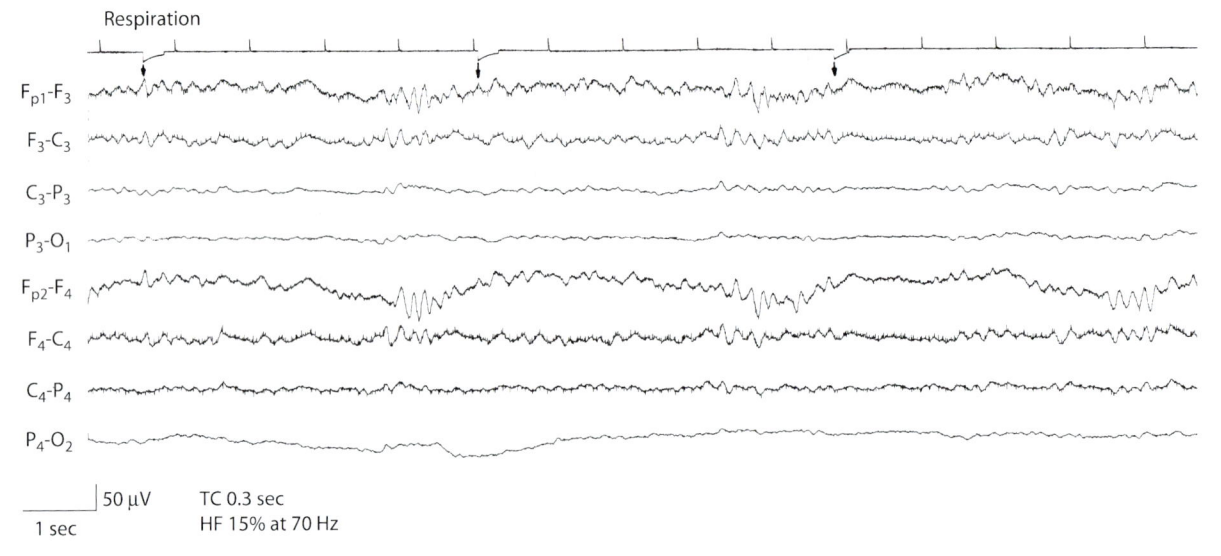

Fig. 4.7 Relative flattening of activities over both parieto-occipital regions in a man comatose 2 days after a cardiac arrest following crush injury to the chest. This pattern is often associated with cerebral arterial boundary zone ('watershed') infarcts following acute profound arterial hypotension. Note the periodic changes in the amount of activity anteriorly, which showed some relationship to respirations. (From Prior (1973), by permission.)

EEG waveforms and interpretation

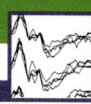

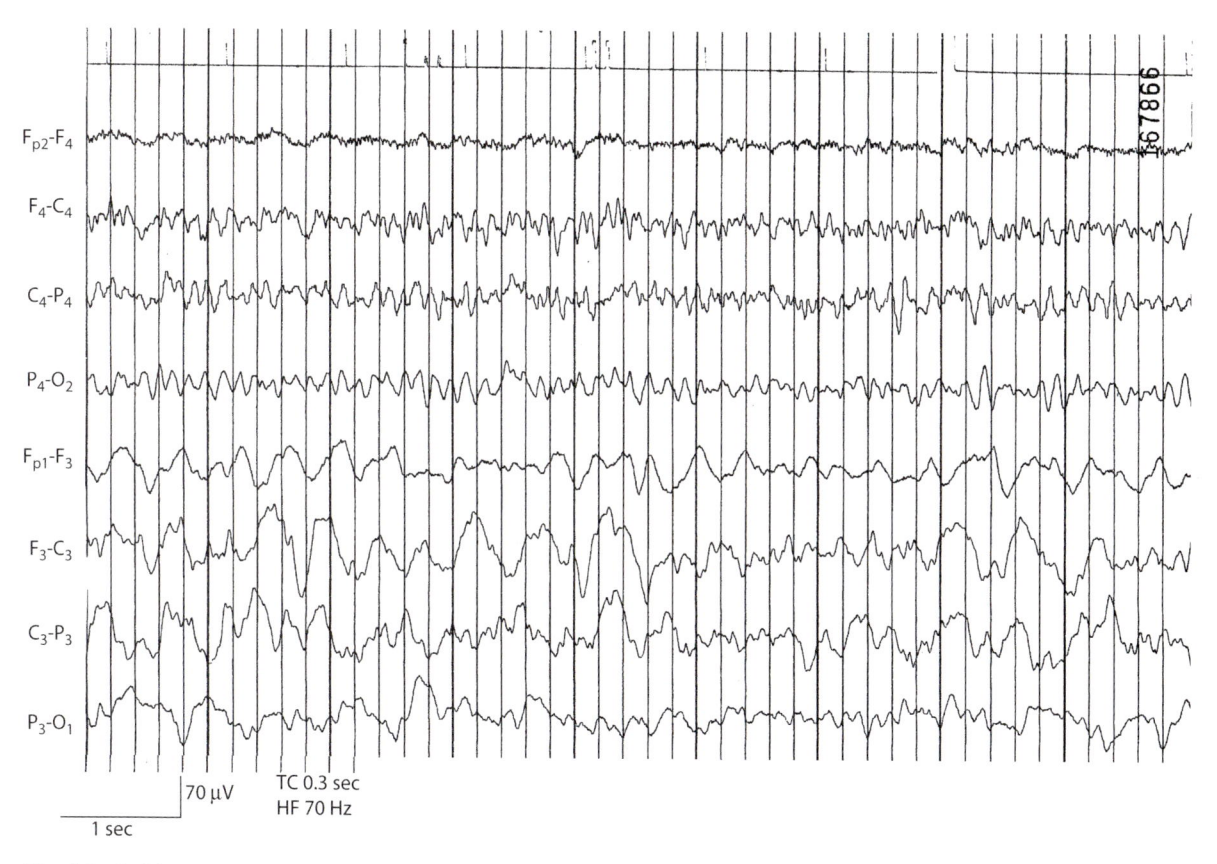

Fig. 4.8 Striking asymmetry after left internal capsule and parietal infarcts in a 63-year-old man with a 1 week history of dysphasia. High voltage irregular delta activity over the left hemisphere with marked diminution of faster components on this side. Marginally slow 7–8 Hz rhythmic activity on the right side posteriorly. (Modified from Binnie et al (2003), by permission.)

duce little slowing, even if large. However, small chronic lesions giving rise to epilepsy may produce surprisingly large amounts of focal slow activity even when no seizure has occurred for several days. In partial epilepsies, slowing may be related to the 'hypofunctional zone', an area more extensive than that from which the seizures arise, characterized by various forms of dysfunction, including hypometabolism (Koutroumanidis et al 1998).

Note that it is not cerebral lesions but dysfunctional neurones which generate abnormal EEG activity; thus slowing may be seen around the periphery of a space-occupying lesion but not over its centre where, if it is sufficiently large, amplitude reduction may be detected. Another important category of abnormal phenomena which may be localized, epileptiform activity, is considered in *Epileptiform activity* (p. 119).

Rhythms at a distance (projected rhythms)

As indicated in the introduction to this section, structurally normal cortex may generate abnormal activities in response to altered afferent inputs from deep structures; these are termed 'rhythms at a distance'. The most common is frontal intermittent rhythmic delta activity (FIRDA), seen over the frontal regions, and usually bilateral and synchronous (*Fig. 4.9*). The frequency is in the range 1.5–2.5 Hz, the slower type tending towards an anterior temporal distribution. The waveform is typically sinusoidal (hence the term formerly used: 'monorhythmic sinusoidal frontal delta activity'), or saw-toothed, and occurs in rhythmic bursts ranging from single waves to an almost continuous rhythm, but typically lasting 2–5 s. It is markedly dependent on arousal level, being seen during drowsiness or stupor but not in the fully alert subject nor in sleep or coma. Its demonstration may therefore depend upon allowing an alert subject to become slightly drowsy or, more often, arousing a stuporous patient to a level at which eye opening occurs. Note the difference in relationship to state of awareness from that shown by paradoxical arousal.

Frontal intermittent rhythmic delta activity occurs most often in association with cerebral pathology but can be seen in metabolic and toxic disorders, status epilepticus, and occasionally postictally. A similar though less clearly intermittent quasi-sinusoidal response sometimes occurs during hyperventilation in normal subjects. Formerly it was claimed that when FIRDA was found in a patient with structural brain disease, there was likely to be an abnormality of the diencephalon. It is thus found, for instance, with thalamic tumours, metastases presumed to include deep lesions, with obstruction of the aqueduct, and in ischaemia of the anterior thalamus due to vascular spasm following rupture of an anterior communicating artery aneurysm. However, the most common pathological correlate of FIRDA is diffuse disease involving grey and white matter (Gloor et al 1968); CT scans may be negative and discrete lesions can occur in any lobe of the brain (Fariello et al 1982). There is no known experimental model of this phenomenon and its pathophysiology remains unclear. Its occurrence appears to depend on a dysfunctional diencephalon, and functional, if damaged, cerebral cortex and corticothalamic projection pathways. Thus in deep frontal tumours FIRDA may be of greater amplitude contralateral to the lesion, possibly encouraging incorrect lateralization unless the lesion also gives rise to a distinct focus of slow activity, or an area in which all activity is reduced.

Similar 'rhythms at a distance' may occur in the occipital regions (occipital intermittent rhythmic delta activity (OIRDA)) with brainstem dysfunction. Much slower occipital delta waves (>1 s duration) are sometimes seen in association with mass lesions in the posterior fossa and also with haemorrhage from the vertebrobasilar system. They occur most often in children where they may attain an amplitude of several hundred microvolts; in adults they are generally inconspicuous and easily mistaken for electrode artefacts from the head moving on the pillow. This slow activity too tends to show a right-sided preponderance and is therefore non-lateralizing, and like FIRDA it occurs most readily in a state of slight drowsiness.

Altered reactivity

Arousal in sleep, from Stage II or deeper, produces various slow wave transients, usually maximal at the vertex, often followed by changes in ongoing activity reflecting a change of state (see *Fig. 4.4*). However, some drowsy or lightly comatose patients show persistent or prolonged runs of bifrontal slow activity accompanied by tachycardia and hyperpnoea when stimulated, for example by calling their name or tactile stimulation (*Fig. 4.10*). This phenomenon of 'paradoxical slow wave arousal' occurs with presumed brainstem or midbrain lesions after head injury (Schwartz and Scott 1978) and occasionally with metabolic encephalopathies. Surprisingly perhaps, the prognosis is generally favourable; the opportunity has not therefore arisen to establish the neuropathological basis.

An EEG which is unresponsive to external stimulation is a feature of moderate to deep coma whether due to cerebral pathology or to a toxic or metabolic encephalopathy. An unusual variant is 'alpha coma', in which a strikingly unresponsive activity of alpha frequency, often of widespread distribution, occurs in states of altered consciousness, usually due to midbrain lesions, following cerebral hypoxia, or in some intoxications.

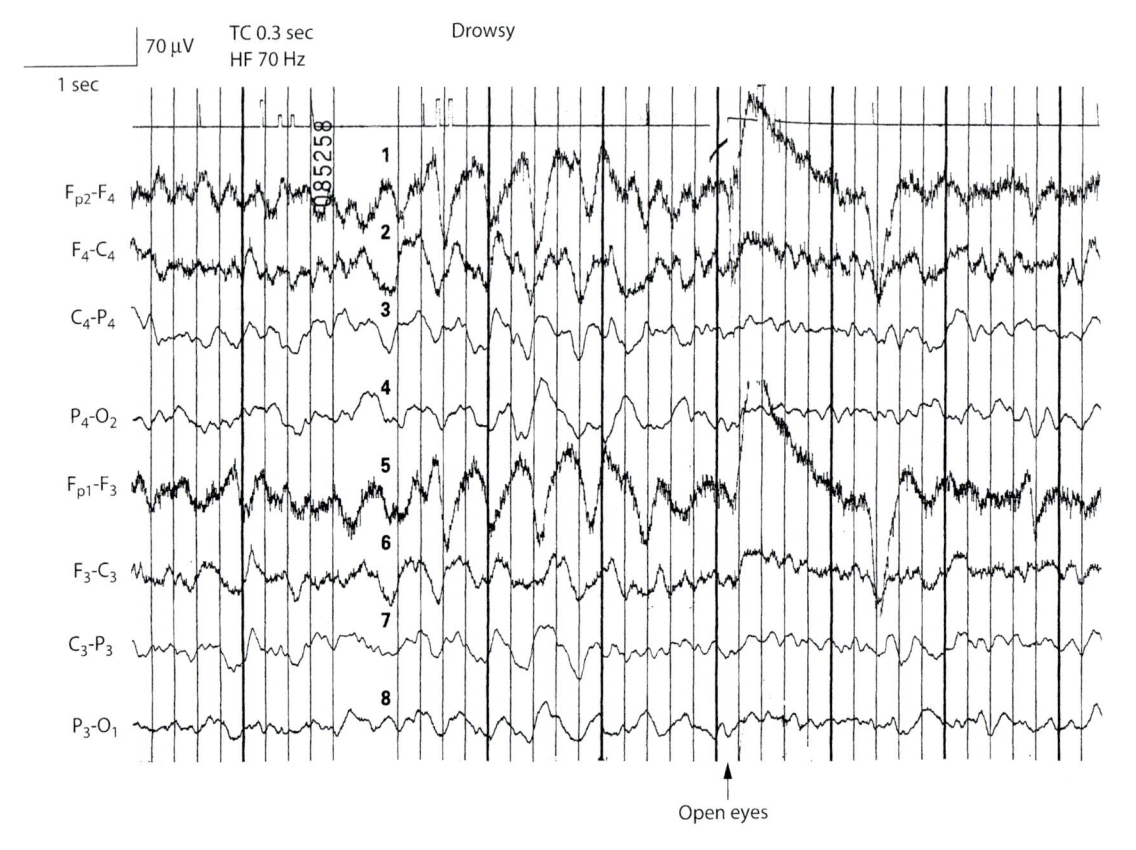

Fig. 4.9 Frontal intermittent rhythmical delta activity (FIRDA) in a 50-year-old man with an 8-year history of chronic renal failure who was found to be hypocalcaemic; the EEG shows intermittent frontally predominant delta bursts which are attenuated on eye opening. (From Binnie et al (2003), by permission.)

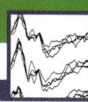

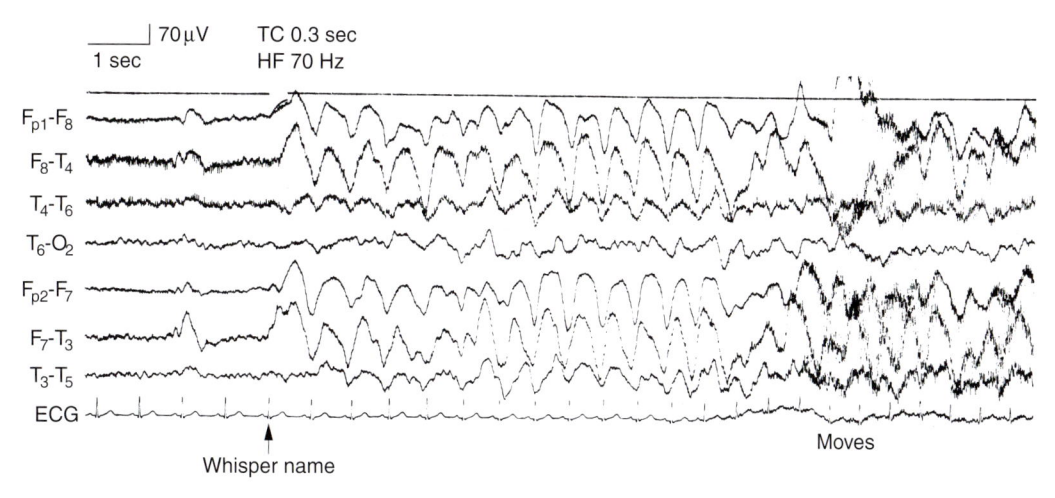

Fig. 4.10 Paradoxical delta arousal response to whispering the patient's name in a lightly comatose 19-year-old man after blunt head injury. Note the very low frequency of the delta waves (1.5–2 Hz); also the increase in heart rate and onset of muscle potentials on the right side of the sample. (From Binnie et al (2003), by permission.)

Epileptiform activity

Cerebral electrical activity changes dramatically during epileptic seizures and characteristically spiky waveforms may be seen, some of which are illustrated in *Fig. 4.11*. Waves of sharp outline standing out from the ongoing background rhythms and lasting < 70 ms are conventionally called 'spikes'. Those lasting 70–120 ms are termed 'sharp waves'. It will be noted that this definition begs the question of how obviously the waveform should stand out from the background and also implies that a wave which would be indistinguishable from ongoing beta or alpha activity may be designated as spike or sharp wave when it arises from a background in which these rhythms are not prominent.

Official terminology recognizes that spikes are often followed by delta waves either singly ('spike-and-slow-wave complex') or in runs ('spike-and-slow-wave activity'; see *Fig. 4.4(f)*). In fact most spikes are polyphasic but, where this is a conspicuous feature, the terms 'multiple spike complex', 'multiple spike-and-slow-wave complex', etc. may be used.

The presumption that (with some exceptions) spikes are negative at the cortical surface may be an important consideration in deciding whether a sharp transient is artefactual, and in locating dipolar sources. For instance, a spike which is positive below the sylvian fissure and negative above probably arises from the superior aspect of the first temporal convolution.

Sharp or spiky waveforms occur in the interictal state, between overt seizures in most people with epilepsy. Similar phenomena are also found in some patients with other cerebral disorders who are not thought to suffer from seizures. Normal subjects also exhibit a variety of sharp transients, the majority of which are generally distinguishable with little difficulty from those associated with epilepsy.

There is no agreement as to a suitable collective term to describe this category of phenomena. 'Epileptic activity'

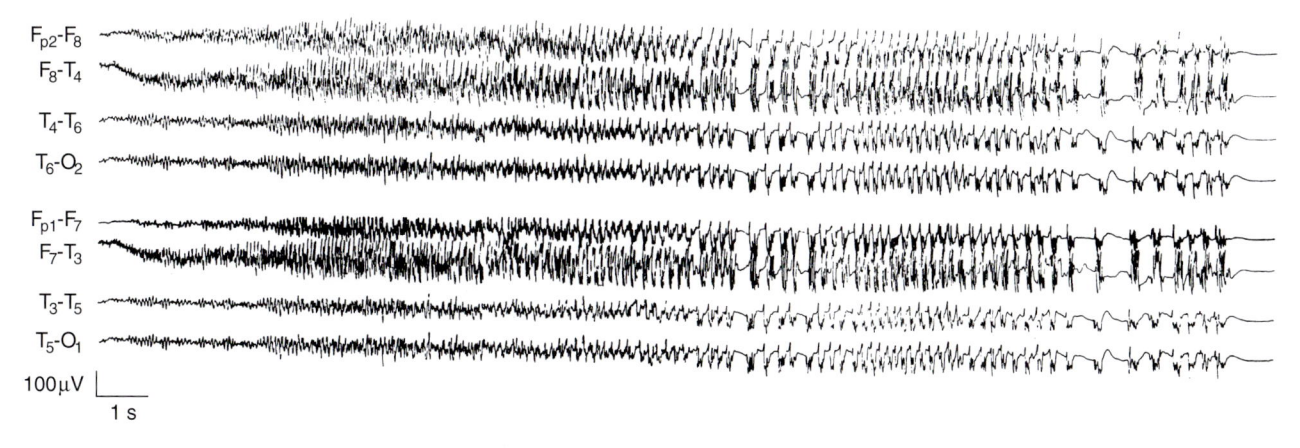

Fig. 4.11 Spontaneous tonic–clonic seizure in a 28-year-old man. Initial crescendo of generalized 10 Hz activity obscured by increasing muscle action potentials of the tonic phase. The clonic phase is associated with bilateral spike-and-wave discharges and synchronized bursts of muscle potentials. (From Binnie et al (2003), by permission.)

is unacceptable as it wrongly attaches to these waveforms diagnostic significance which they do not possess. 'Seizure discharges', widely used in North America, appears even more inappropriate, as even in people with epilepsy the discharges may be interictal. 'EEG seizure pattern' is countenanced by the IFCN (Noachtar et al 1999), but only to describe repetitive discharges with abrupt onset and termination, lasting several seconds – as seen during seizures. If not accompanied by a clinically overt seizure, they should be designated 'subclinical'. 'Paroxysmal activity', favoured in the United Kingdom, is unsuitable, as many normal EEG phenomena have a paroxysmal quality, for instance K-complexes or lambda waves. In continental Europe, 'irritative activity' is a popular euphemism which nevertheless implies epileptogenesis. 'Epileptiform pattern, discharge, or activity' is a compromise supported by the IFCN and favoured by many authorities on epilepsy in Europe and North America, which at least acknowledges the statistical association with epilepsy underlying the concept, whilst stressing that the term refers to a category of waveforms and not a diagnosis.

During seizures, EEG discharges occur with relatively abrupt onset and termination and characteristic pattern of evolution, lasting at least several seconds (Fig. 4.11). The component waves or complexes vary in form, frequency and topography. They are generally rhythmic and frequently display increasing amplitude and decreasing frequency during the same episode. When focal in onset, they tend to spread subsequently to other areas. EEG seizure patterns unaccompanied by overt clinical epileptic manifestations should be referred to as 'subclinical'. In ICU, patients who are unresponsive may be suffiereing from subclinical status epilepticus, the electrical accompanimernts of which are shown in *Fig. 4.12(a)*; both the prolonged atypical geenrlaized discharges and the obtunded state, may respond to small, fractionated i.v. administration of diazepam as shown in the figure.

Periodicity

Various pathological EEG phenomena occur repetitively at more or less constant intervals. They can be localized, lateralized or generalized and have a distinctive, often dramatic, appearance.

Triphasic complexes (or waves)

Diffuse, large amplitude repetitive complexes dominate the EEG and replace on-going rhythmic activities. Each complex consists of three waves of alternating polarity, one at least of which lies in the delta range. Typically the first and third components are electronegative and of greatest amplitude anteriorly, but the whole complex may spread across the head from front to back with a small time lag. The complexes occur in bilateral runs at 1.5–2.5 Hz, generally appearing as a continuous sequence rather than as isolated complexes or bursts (*Fig. 4.13*). This phenomenon, which is a sign of severe diffuse cerebral dysfunction, is almost always associated with a metabolic disorder and typically

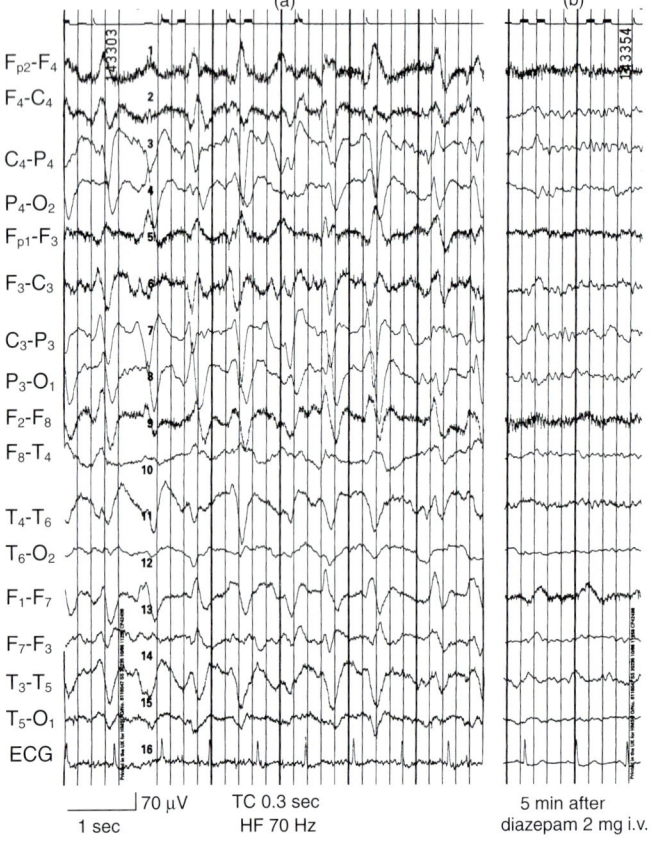

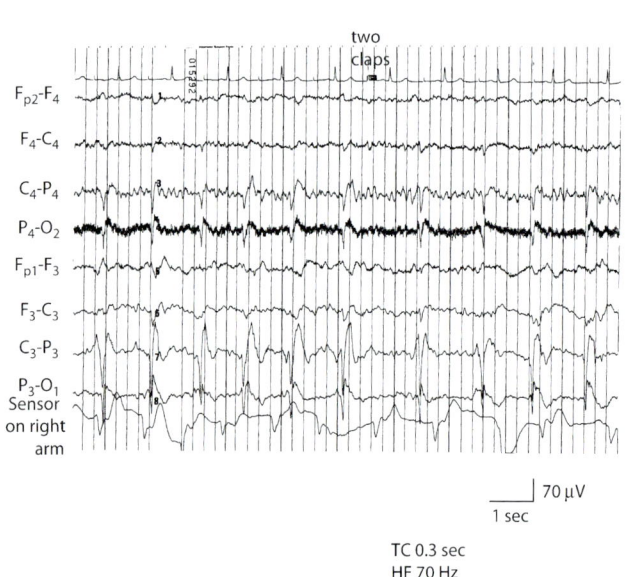

Fig. 4.12 Periodic complexes. Left side: (a) Periodic complexes in non-convulsive status epilepticus in a 73-year-old woman with a 2 week history of confusion, restlessness and increased tone, (b) EEG normalizes following cautious intravenous injection of 2 mg diazepam. Right side: PLEDs in a 78-year-old man with an a episode of jargon dysphasia; flailing and episodes of rhythmic jerking of right arm and abdomen 2 months after a cerebrovascular accident producing left-sided paresis. (From Binnie et al (2003), by permission.)

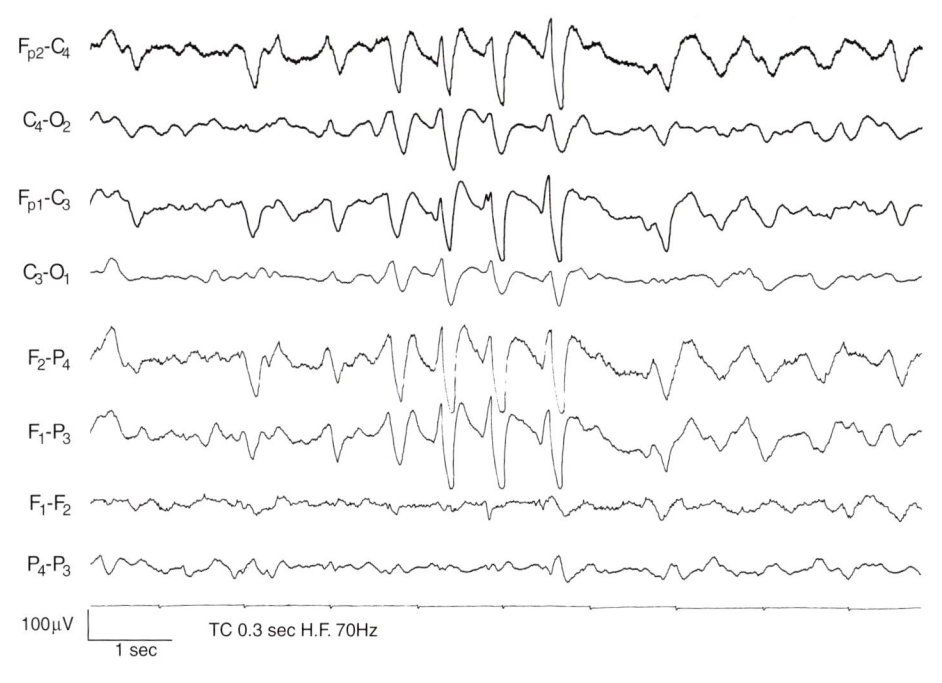

Fig. 4.13 Typical triphasic waves in hepatic encephalopathy, here appearing in paroxysmal runs. Typical of the pre-comatose stage. (From Binnie et al (2003), by permission.)

with hepatic encephalopathy. Triphasic waves appear only after the EEG has considerably slowed (see *Fig. 4.5*) and the patient is in manifest coma or precoma.

Periodic lateralized epileptiform discharges (PLEDs)
These are lateralized stereotyped slow waves or complexes including sharp waves repeating at fairly regular intervals, typically of about 1 s (see *Fig. 4.12(b)*). They generally show a localized maximum but appear widely over one hemisphere. Occasionally the phenomenon is bilateral (bi-PLEDs), or multifocal, although usually asynchronous. PLEDs are seen in a variety of conditions, both acute and chronic, but always associated with localized structural disease. Examples include rapidly growing tumours, cerebrovascular accidents, cerebral abscess and following head injury. A particularly common correlate is herpes simplex encephalitis. As the interval between two consecutive PLEDs may be seen as a slow wave bounded by two sharp components of opposite polarity, it is understandable that they are sometimes mistaken for runs of triphasic waves. Their periodicity varies according to underlying pathology (most periodic with viral encephalitides (Gross et al 1999)), as well as with their course over time. PLEDs tend to be a self-limiting acute phenomenon, even with progressive pathology, typically becoming less frequent and then fading away in the course of a matter of a week or so. They may also fluctuate within the course of a single recording, increasing during drowsiness and disappearing when the patient is alerted (Chatrian et al 1964, Markand and Daly 1971, Schwartz et al 1973). They may be associated with partial seizures, including subtle localized twitching (see *Fig. 4.12(b)*), and can often be suppressed by antiepileptic medication, such as intravenous diazepam.

Other repetitive patterns
Generalized repetitive phenomena are encountered in the subacute encephalopathies, subacute sclerosing panencephalitis (SSPE) and Creutzfeldt–Jakob disease (CJD). The complexity of the discharges is greater, their morphology more variable and the repetition rate slower in SSPE than in CJD. Early in the illness the SSPE complexes, often recurring at about 1 per 10–20 s, may be difficult to appreciate against the background of high amplitude ongoing activity in a child or teenager. Complexes may consist merely of a run of fast waves, later with an associated slow wave. They tend to be bilaterally synchronous and similar in morphology, being stereotyped in any one area but often of different waveform and amplitude in different regions. In CJD the complexes tend to be nearer 1/s and hence may even be mistaken for ECG pick-up initially. In both disorders, there is a gradual and relentless loss of ongoing activity, making the complexes increasingly obvious as the disease progresses (*Fig. 4.14(a)*). This picture of complexes on a denuded background contrasts with those in generalized non-convulsive status epilepticus where plentiful on-going activity including fast components is preserved (*Fig. 4.14(b)*).

Less elaborate transients with simple monophasic or diphasic morphology (*Fig. 4.14(c)*) are encountered in comatose patients, for example with hypoxic encephalopathy after cardiac arrest (Binnie et al 1970, Prior 1973). They should be carefully distinguished from regular artefacts (*Fig. 4.14(d)*). Occasionally, in patients in a vegeta-

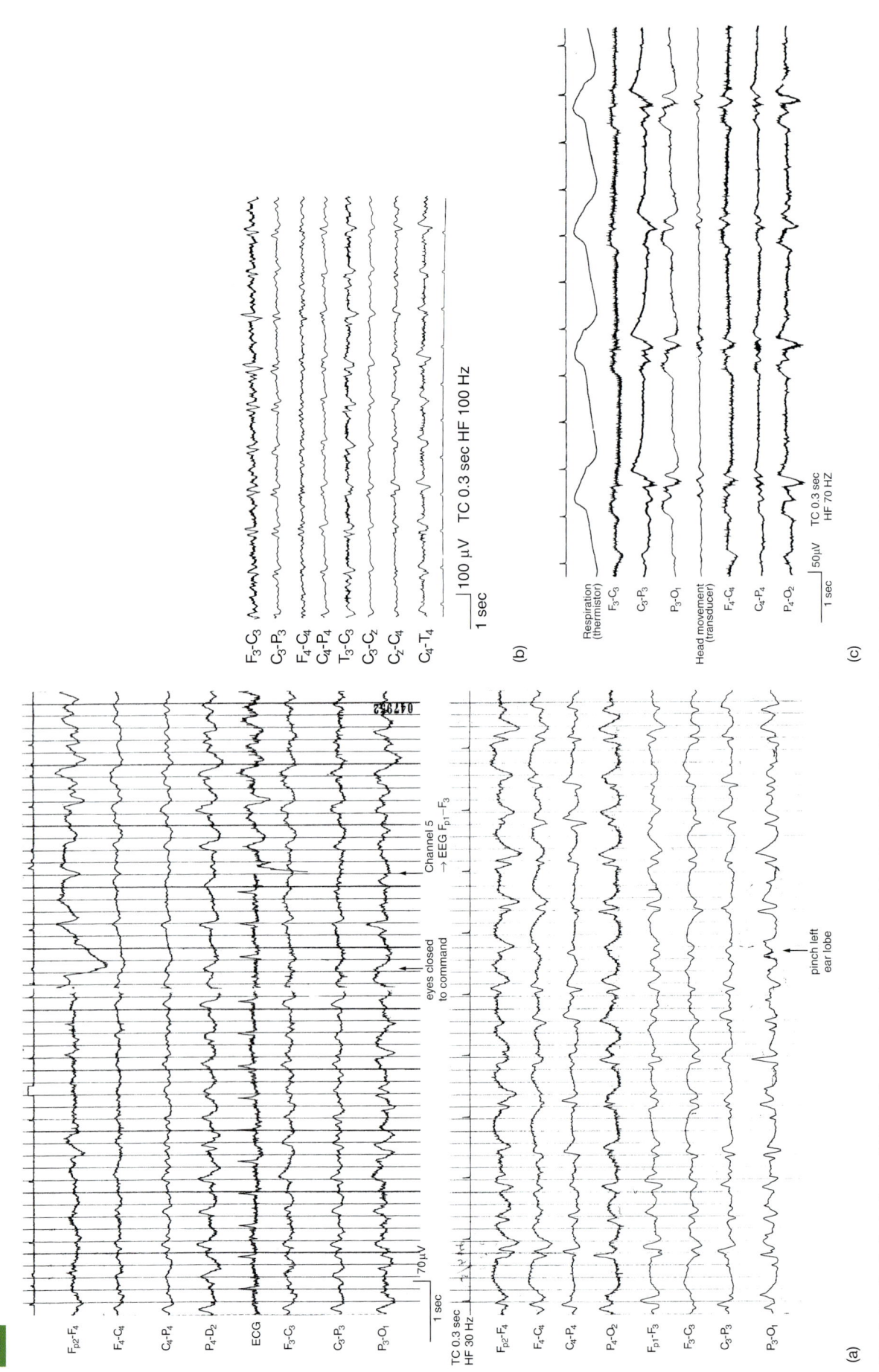

Chapter 4 Normal and Pathological Phenomena in EEG, Evoked Potential, EMG and Nerve Conduction Studies

Fig. 4.14 Periodic features. (a) Periodic complexes in a 54-year-old patient with CJD. Deterioration in background activity occurs between samples at 6 and 10 weeks after onset of illness; death occurred 1 week later. Brief reduction in complexes on eye opening at 6 weeks; none occurs with pain at 10 weeks. Note the coincidental similarity in timing of EEG and ECG complexes in the 6 week sample and their lower repetition rate at 10 weeks. (From Binnie et al (2003), by permission.) (b) Periodic transients with simple, generally diphasic, morphology in an unconscious patient 2 days after cardiac arrest. Note depletion of background activity. (From Prior (1973), by permission.) (c) Apparent periodic features in a patient after cardiac arrest. Monitoring of respiration and head movement indicates their artefactual origin. (From Prior (1973), by permission.)

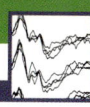

tive state after cardiac arrest, periodic transients may persist for many months with a gradual increase in interdischarge intervals – a course not dissimilar to that of CJD (Takahashi et al 1993).

Burst suppression pattern

The appearance of periods of amplitude suppression or total loss of activity (*Fig. 4.15*) in the EEG has long aroused interest. Early descriptions were in the context of circumstances as diverse as deep barbiturate anaesthesia (Brazier and Finesinger 1945) and physical isolation of the cerebral cortex from underlying white and deep grey matter influences by surgical undercutting (Henry and Scoville 1952) or by pathological processes such as tumours. In general terms, the phenomenon of burst suppression in the EEG may be considered as a potentially reversible expression of reduced cortical neuronal metabolic function. Burst suppression pattern occurs with the progressive failure of oxygen delivery of hypoxic–ischaemic encephalopathy (Brierley et al 1980), and as a dose-related event with deepening anaesthesia (Bickford 1950, Michenfelder 1974, Shapiro 1985) and profound hypothermia (below 20°C) (Mizrahi et al 1989). These circumstances are all associated with progressive reduction of neuronal metabolic activity, albeit potentially reversible. During moderately deep surgical anaesthesia, the onset and offset of the bursts may be associated with accelerations and decelerations of heart rate (Yli-Hankala et al 1990) and there is evidence that they may be induced by vibratory or visual stimuli (Yli-Hankala et al 1993, Hartikainen et al 1995). In patients with hypoxic–ischaemic coma, the variance of interburst intervals has been shown to differ from that during thiopental treatment of status epilepticus, suggesting different mechanisms between drug-induced and brain damage forms of the phenomenon (Beydoun et al 1991). Studies on the nature and mechanisms of the burst suppression pattern have aroused renewed interest of late, modelling suggesting a non-linear dynamic system at the transition to chaos (Rae-Grant and Kim 1994), probably from near complete disconnection of thalamocortical brain circuits implicated in the genesis of the EEG (Steriade et al 1994).

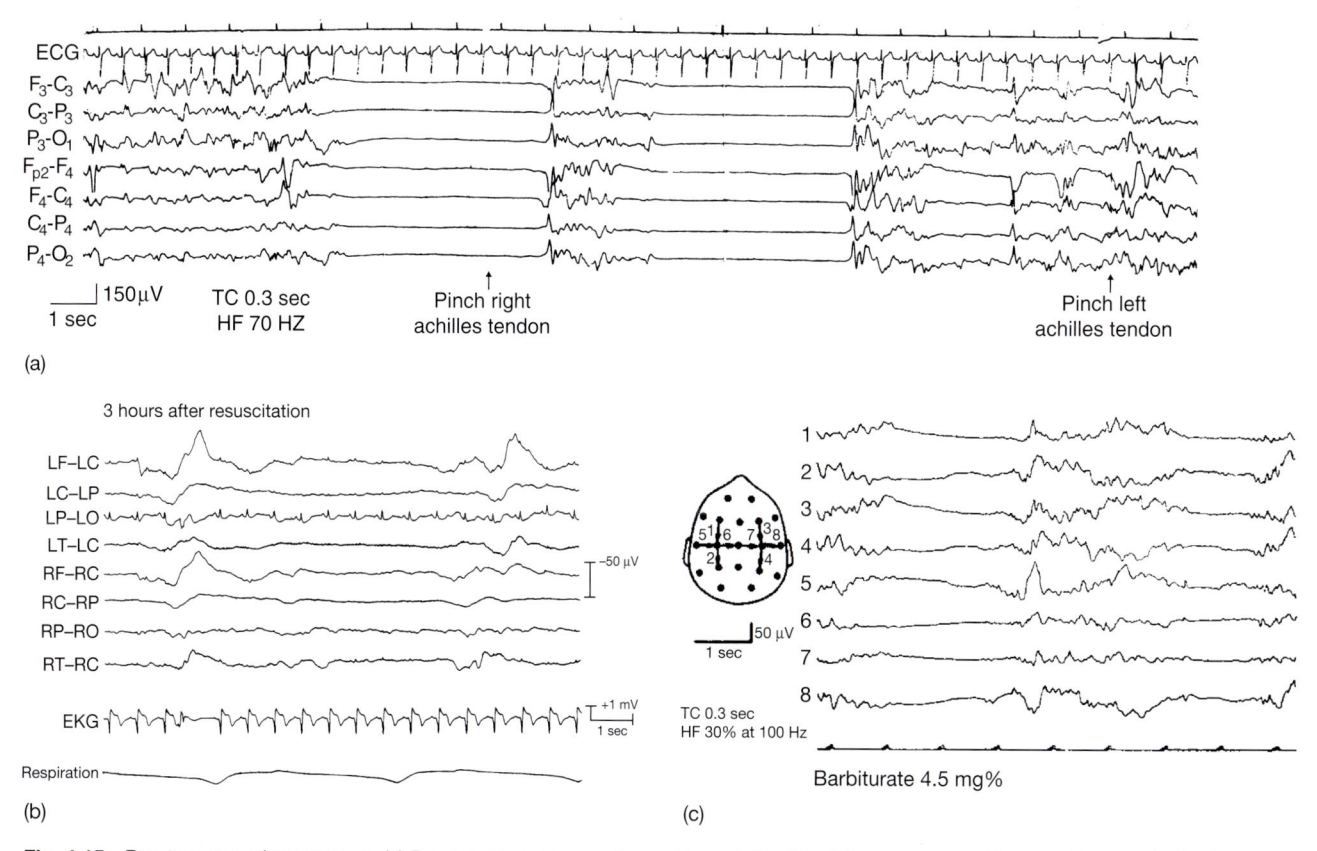

Fig. 4.15 Burst suppression patterns. (a) Burst suppression unaffected by painful stimuli in a comatose 40-year-old man who had sustained an intraoperative cardiac arrest 16 h earlier. He was ventilated but died 8 h later. (From Binnie et al (2003), by permission.) (b) Recordings in a 59-year-old man after resuscitation from circulatory arrest following myocardial infarction; initially there was no detectable cortical activity in the EEG but all cranial nerve reflexes and decerebrate posturing could be elicited. At 3 h, intermittent delta activity appeared with large amplitude in the frontal leads and cutaneous stimulation elicited decorticate posturing. The patient recovered consciousness after 29 days and gradually began to utter syllables, but remained unable to sit, stand or walk and required assistance with all personal needs until final asystole 79 days after resuscitation. (From Jørgensen and Holm (1999), by permission.) (c) A 46-year-old man with repeated suicide attempts, 14 hours after resuscitation after cardiac arrest following overdose with 30 tablets of amytal; blood barbiturate 4.5 mg% at time of recording. The patient recovered fully. (From Prior (1973), by permission.)

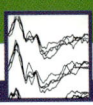

Electrocerebral inactivity – electrocerebral silence (ECS) – the isoelectric EEG

The recommendations on terminology of the IFCN discourage use of the term 'isoelectric EEG', preferring that of 'record of electrocerebral inactivity' (Noachtar et al 1999). An alternative term, 'electrocerebral silence' (ECS), is also commonly used, as here.

It is essential to pay great attention to recording technique. Considerable work from highly experienced experts has led to the American Electroencephalographic Society Guidelines (1994) and the IFCN Recommendations (Chatrian et al 1996, Guérit et al 1999); they form essential reading. It is much more demanding to prove the absence of activity (*Fig. 4.16(a,b)*) than to characterize active EEG phenomena. The essentials of the American Electroencephalographic Society Guidelines (1994) give considerable detail as to the general standard expected:

1. All 21 of the 10–20 system scalp electrodes should be utilized.
2. Interelectrode impedances should be less than 10 kΩ but over 100 Ω.
3. A minimum of 12 and preferably 16 or more channels should be available.
4. The integrity of the entire recording system should be tested.

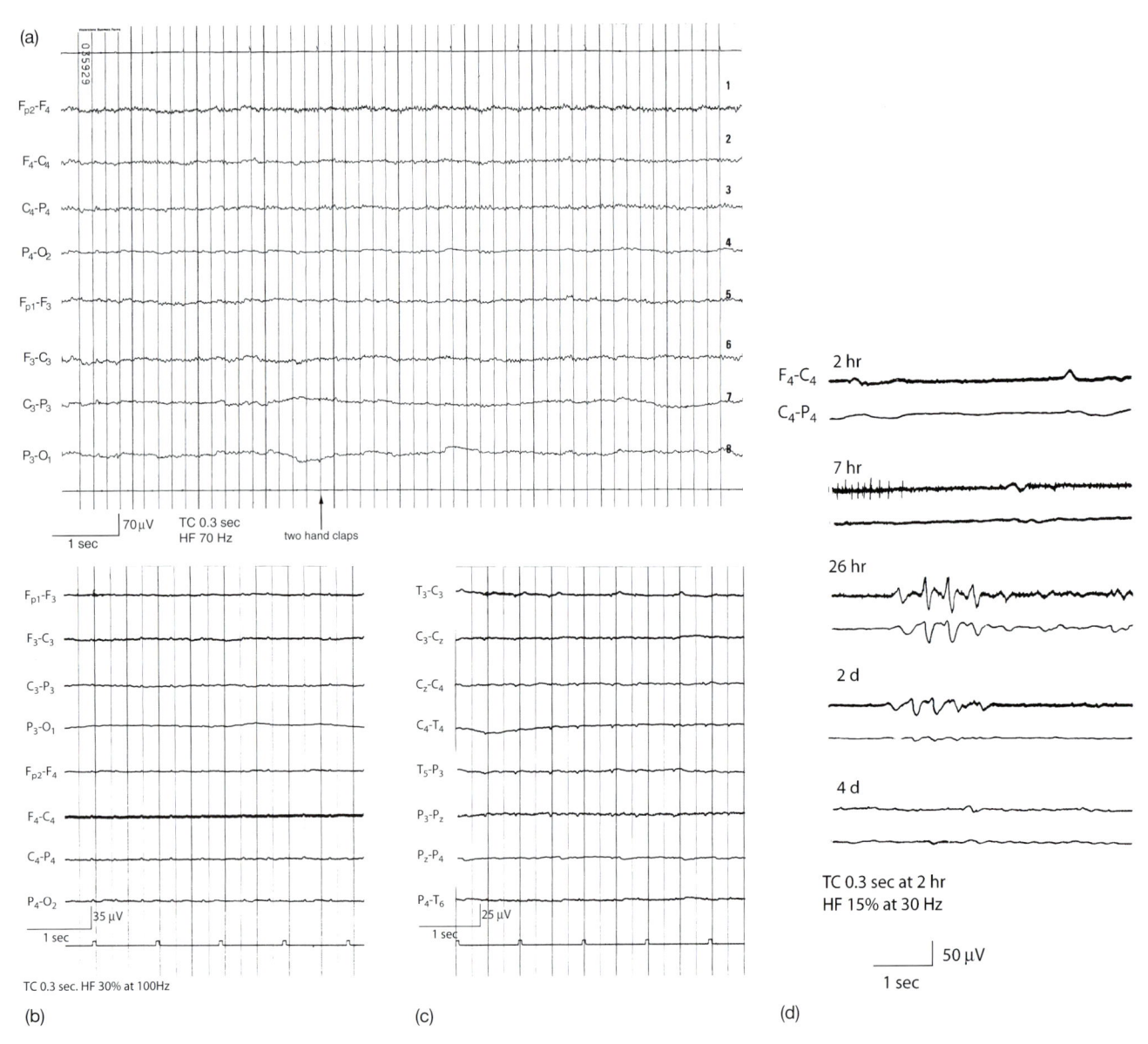

Fig. 4.16 Strategy for investigating electrocerebral silence (ECS or 'isoelectric EEG'). (a) Temporarily unresponsive and very low voltage fast EEG has to be distinguished from ECS in a 59-year-old man during a short episode of depressive stupor; the patient eventually blinked to an auditory stimulus but neither EEG nor ECG (not shown) changed. (b, c) ECS in a 43-year-old man 12 h after cardiac arrest in whom irreversible hypoxic brain damage was demonstrated neuropathologically. Note that increasing sensitivity enhances head tremor and ECG artefacts. (d) Serial recordings after cardiac arrest showing transient return of activity in a patient after cardiac arrest with initial ECS who had sustained severe hypoxic damage and died a few hours after the sample on the 4th day. (From Prior (1973), by permission.)

5. Some derivations should be from pairs of electrodes at least 10 cm apart.
6. Sensitivity must be increased from 7 µV/mm to at least 2 µV/mm for at least 30 min of the recording, with inclusion of appropriate calibrations.
7. Filter settings should be appropriate for the assessment of ECS.
8. Additional monitoring techniques should be employed when necessary.
9. There should be no EEG reactivity to intense somatosensory, auditory or visual stimuli.
10. Recordings should be made only by a qualified technologist.
11. A repeat EEG should be performed if there is doubt about ECS.

The current IFCN Guidelines (Guérit et al 1999, embracing the earlier report of Chatrian et al 1996) give excellent guidance and specify appropriate technical requirements specific to neurophysiological recording in the ICU; their advice on EEG recording covers the need for polygraphic recordings, the number of channels and type of electrodes, and the adaptation of technique to the clinical circumstances. The emphasis on matching technique to the clinical questions being asked carries the important rider that the clinical neurophysiologist should always be present to guide the investigation and discuss its interpretation with the ICU clinicians.

In the comatose patient, ECS is always a cause for concern, even though various reversible causes, such as high levels of central nervous system (CNS) depressant drugs (due to overdose or therapeutic regimens in the ICU), or hypothermia, are not uncommon. Rigorous checks must be made to discover any reversible cause; reports should draw attention to this and to the need for further neurophysiological investigation, including repeat EEG examination and testing for auditory and somatosensory evoked potentials (SEPs). The demonstration of short latency EP components, which are fairly resistant to major CNS depressant drugs sufficient to render the EEG isoelectric, provides crucial evidence for functional integrity of some central pathways, thereby refuting any suggestion of irreversible damage (Ganes and Lundar 1983).

EVOKED POTENTIAL WAVEFORMS AND INTERPRETATION

GENERAL DEFINITION – LIMITATIONS – CLINICAL UTILITY OF EVOKED POTENTIALS

Any neuronal response triggered by stimulating sensory receptors or peripheral nerves, and also any neuronal activity time-related to cognitive processes or motor programming, can be viewed as an EP. This comprehensive definition covers more than the scope of those EP techniques detailed in this chapter, which have been selected because they have proved to be clinically useful.

Some of the many limitations of EP recording in human subjects deserve emphasis because of their relevance to clinical applications. First, only skin electrodes can be routinely used which, among other inconveniences, precludes the study of unitary responses and impairs discrimination between postsynaptic and action potentials on the basis of their respective time courses and voltage profiles. Secondly, the averaging of bioelectrical responses to repeated stimulation, as pioneered by Dawson (1947) for extracting EP signals from biological noise, presupposes that the stimulus and the response are tightly time-locked one with another. In more practical terms this means that the reliability of EPs, and consequently their potential clinical applications, will be greatest for responses to brief, reproducible and mon-otonous stimuli which reflect early stages of input processing via anatomically identified sensory pathways with few synapses. Thirdly, the degree of interindividual variability among normals is a crucial factor which must be considered when clinical application of EPs is envisaged, since it determines the size of the control population needed for establishing normative data and the possibility of discriminating between normal and abnormal responses in any particular individual. EPs reflecting neuronal activities potentially modulated by factors such as vigilance, volition or attention are obviously less consistent than those generated peripherally or in the so-called primary sensory areas which are believed to be independent of cognitive processes or action programming. The above statements suffice to explain the prevailing opinion that most, if not all, of the clinically validated EP techniques concern early sensory responses.

However, in spite of their limitations, EPs represent an inexpensive and excellent means for assessing, as it occurs, the processing of sensory information in the human CNS. Indeed, time resolution is far better for EPs than for other available functional investigations, such as measures of cerebral blood flow or metabolic rates for oxygen or deoxyglucose, which require at least a few seconds to detect changes in brain activity.

The clinical applications of EPs began to emerge shortly after Halliday demonstrated for the first time, in the early 1970s, that visual evoked potentials (VEPs) could disclose clinically silent lesions of the optic tracts in patients with multiple sclerosis or a previous history of optic neuritis. EPs have proved to be helpful: (a) to test sensory functions when clinical examination is not reliable; (b) to investigate purely subjective symptoms and detect whether they are related to any dysfunction of organic origin; (c) to appraise the causative mechanisms of neurological deficits or the degree of functional recovery; and (d) to monitor various cerebral functions when the patient's condition is critical or at risk in the operating theatre or during intensive care.

RESPONSES AND EP COMPONENTS

The term 'response' refers to any change in brain electrical activity related to the processing of information by the nervous system, while for each of the sequential electrical waves recorded at the exploring electrodes the term EP 'component' is used. Each EP component is characterized by features such as latency, amplitude and topography, but also by extrinsic factors such as the particular stimulus by which it is elicited or the particular electrode montage used for the recording. This distinction between the response, viewed as a physiological concept, and the EP component considered as an electrical field phenomenon is more than a linguistic nicety. It is justified by four main arguments:

1. Surface recordings may be 'blind' to some elements of the neuronal response.
2. Some EP components may not correspond to any discrete synaptic or axonal response and are due to physical changes of the volume conductor in which an axonal volley is propagating, or to changes in the direction of the propagation with respect to the site of the recording electrode (*Fig. 4.17*).
3. A single surface EP component may result from the combination of responses of separate cell populations activated through parallel pathways and overlapping in time.
4. An EP component is a composite signal resulting from the difference between activities picked up separately by the so-called 'active' and 'reference' electrodes.

The averaging technique for EP recording is based on the assumption that the sequence of neuronal responses generating EPs remains invariable in time over long runs of repeated stimuli delivered under stable conditions. This does not mean, however, that EP sources are necessarily activated sequentially during the processing of sensory information. One often assumes that, when all EP components are missing after a given latency, conduction must be interrupted beyond the anatomical level at which the latest of the preserved components is supposed to be generated. This assumption is valid in most cases; however, in some instances late events can persist even though earlier responses have disappeared. This would be in apparent contradiction with the model if one did not consider the possibility that cortical generators can be activated sequentially through parallel pathways with independent processing times, synaptic relays and cortical targets. Moreover, the absence of a given component means that the response cannot be recorded but not necessarily that it is abolished; a well-known example is the absence of recordable nerve action potentials with persisting cortical SEPs in severe neuropathies (see Desmedt (1984) for a review).

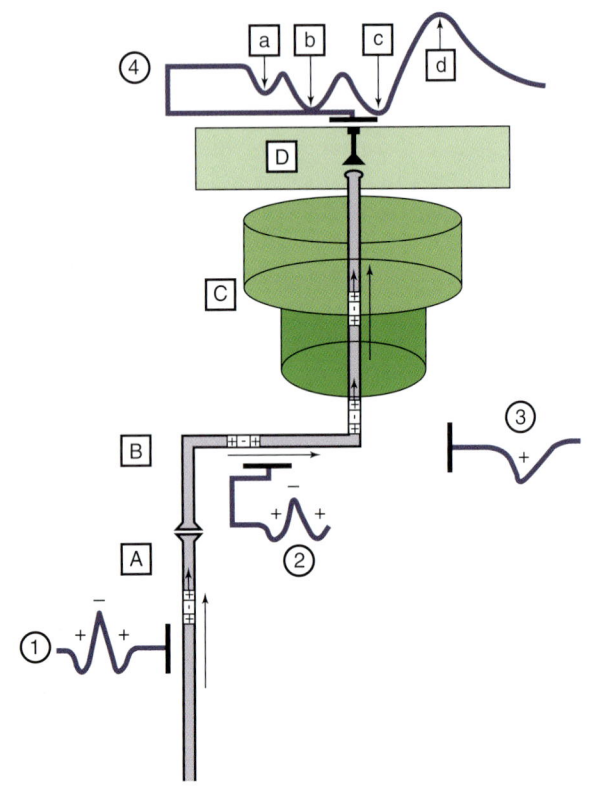

Fig. 4.17 Schematic representation of EP generators, illustrating a model of (A) a sensory pathway with one subcortical synaptic relay, (B) a midline crossing of the fibres, (C) a change in the physical properties of the volume conductor and (D) an intracortical synapse. Any electrode situated close to the trajectory of the action potential volley (1 and 2) picks up a triphasic wave. Electrodes situated at a distance but in the axis of the propagation (3 and 4) 'see' the moving positive front of the action potential volley and pick up far-field positivities, which may be interrupted, with a return to baseline, by (A) a synaptic gap, (B) a change in the direction of the axon bundle or (C) a change in the volume conductor. Thus the scalp electrode records a sequence of three far-field positivities followed by (D) a negative potential reflecting the postsynaptic response of the cortical synapse. This figure illustrates the notion that the term 'generator' does not refer only to the anatomy of the active pathways but also to their orientation in space and their electrophysical environment. (From Binnie et al (2004), by permission.)

THE ELECTRORETINOGRAM AND ELECTROCOCHLEOGRAM

The electroretinogram (ERG) may be regarded as a special type of EP which does not arise in neuronal tissue as such. A single bright white flash elicits a highly consistent waveform comprising two main components which are generated at receptor and postreceptor retinal levels. Manipulation of stimulus and recording parameters permits the assessment of different retinal levels and cell types. Additionally, use of both flashes and pattern stimuli (with no overall luminance change) allows separate assessment of both peripheral and central (macular) retinal function.

The electrocochleogram (ECochG) is a complex of signals arising from the cochlear and auditory nerve, most readily elicited by impulsive auditory stimuli (clicks) (*Fig. 4.18*). It comprises the cochlear microphonic which

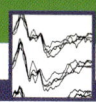

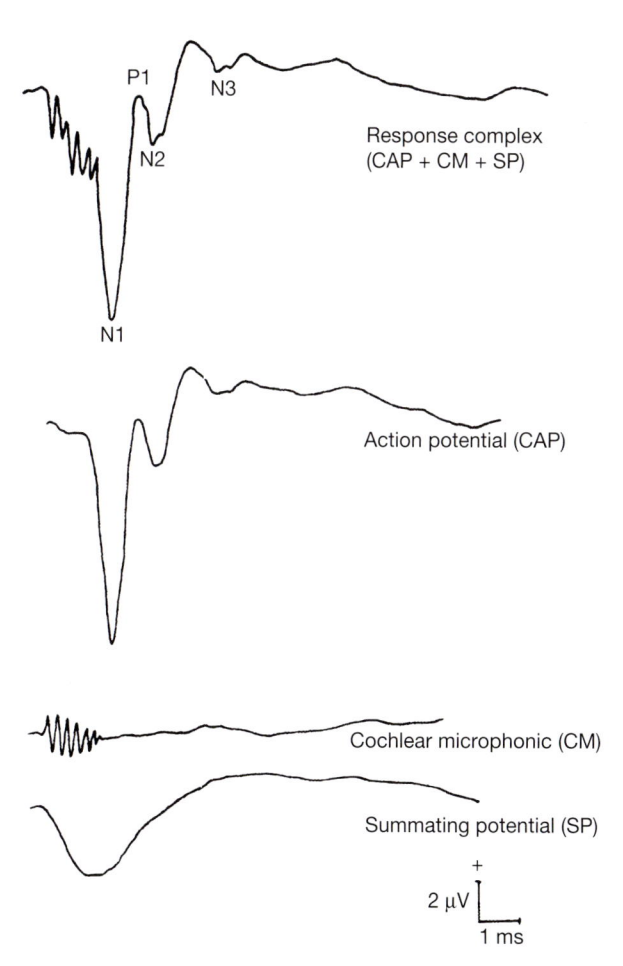

Fig. 4.18 Electrocochleographic response complex, recorded with a promontory electrode, and comprising the compound action potential (CAP), cochlear microphonic (CM) and summating potential (SP). (From Prasher D, Luxon LM 1988 Methods of examination – audiological and vestibular. In: *Mawson's Diseases of the Ear*, 5th edition (ed. H Ludman). Edward Arnold, London, pp. 116–189, reproduced by permission of Edward Arnold.)

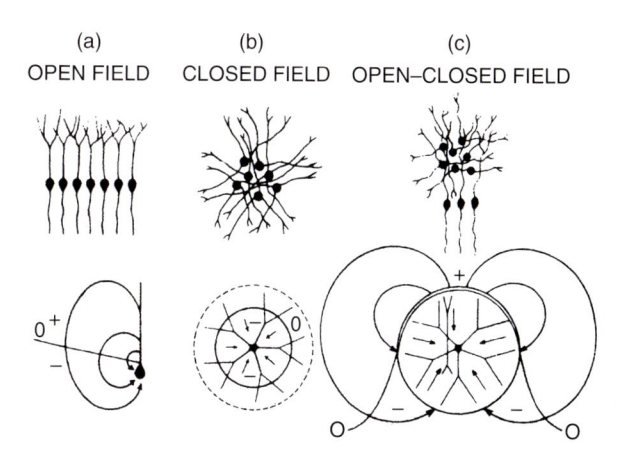

Fig. 4.19 Lorente de Nò's models of potential field generators. Predicted current flows and potential fields generated by the synchronous depolarization of the cell bodies. (A) In the open-field model, which is a schematic representation of the arrangement of pyramidal cortical neurones, the cell bodies, axons and dendritic tree are supposed to be aligned in a row. Depolarization of the soma produces an inward current flow of the soma (sink) and an outward current flow from the dendrites (source), whereas depolarization of the dendrites would produce the reverse. The field generated by the activation of such a structure can be picked up at a distance from the source and sink, provided that the two electrodes are not located on the same isopotential line. Note that the current flow from the axon to the cell bodies is not represented; this flow is considered to be negligible compared with that arising from apical dendrites which have a broader surface. (B) In the closed-field model, which corresponds to a nuclear structure, the somas are gathered in the centre of the structure and dendrites extend radially outwards. Somatic depolarization produces an inward current and maximal negative potential at the centre of the nucleus. This radial inward current causes the potential gradient to be zero anywhere outside the nucleus. Consequently, the electrical activity cannot be recorded using electrodes situated at the periphery of the nucleus. (From Lorente de Nò (1947), by permission.)

arises from the organ of Corti, the summating potential produced by displacement of the basilar membrane, and the compound action potential (CAP) of the cochlear nerve.

RELATION BETWEEN NEURONAL RESPONSES AND SURFACE EVOKED POTENTIALS

Two types of neuronal activity are generated after a stimulus is applied to any part of the nervous system: propagated action potentials and synaptic potentials. Volume conduction of these potentials accounts for the observed changes in potential difference between surface electrodes, according to models first proposed by Lorente de Nò (1947), shown schematically in *Fig. 4.19* (see also Schlag (1973) for a review).

Action potentials

The extracellular potentials recorded when a CAP travels along an axonal bundle are illustrated in *Fig. 4.20*. The action potential is an all-or-none transitory change in membrane potential, which travels with a constant voltage (without space decrement) and at a constant velocity. The voltage of the action potential is determined by the physical properties of the membrane and consequently is the same for all the spikes emitted by a given cell; thus any change in axonal activity can manifest itself only by a variation in the rate at which action potentials are emitted.

An action potential can be represented schematically as an axonal segment of depolarization where the intracellular space is positive and the extracellular space negative and which moves along an axon where the relative distribution of electrical charges across the membrane is the reverse (intracellular space negative, extracellular space positive; *Fig.4.20(B)*). As the action potential approaches the recording electrode a positive potential is recorded reflecting the flow of negative charges towards the approaching positive segment (*Fig. 4.20(C)*); as it reaches the electrode a negativity is recorded, reflecting the reversal of the membrane potential (*Fig. 4.20(D)*); and as it passes the electrode the recorded potential becomes transiently positive (flow of negative charges towards the departing positive segment; Fig. 4.20(E) before returning to zero. The resulting waveform is a triphasic potential with successively positive–negative–positive components, which can be picked up under routine conditions by any surface electrode placed along the route of a stimulated nerve. An important concept is that recordings from the end of a nerve or beyond the termination of a fibre tract will yield only the first positivity of this triphasic complex, since the segment of depolarization approaches but never reaches the recording electrode (Fig. 4.20(A)). Under appropriate conditions (see below) this positivity, reflecting the moving positive front of an afferent volley of action potentials, can be volume conducted, and consequently recorded, at very large distances from the nerve. Since it is not possible with surface electrodes to record directly in the close vicinity of intracranial tracts of fibres, such 'far-field' recordings of the axonal activity, first described by Jewett (1970), have received and will continue to receive considerable attention in clinical neurophysiology.

Postsynaptic potentials

Postsynaptic potentials (PSPs) may be excitatory or inhibitory. During excitatory postsynaptic potentials (EPSPs) positive ions enter the cell, reducing the negative intracellular potential and causing a negative shift of the extracellular potential. The reverse occurs during inhibitory postsynaptic potentials (IPSPs): negative ions enter the cell, increasing the intracellular negativity and producing a positive shift of the extracellular potential. During PSPs the regions of the cell outside the postsynaptic area exhibit similar potential changes, but with a lesser magnitude due to the rapid decrease of current flow with distance. An extracellular electrode close to the cell membrane, but distant from the synaptic cleft, will record a positive-going potential shift during EPSPs (as during action potentials) and a negative one during IPSPs. When the relative spatial coordinates of the synapse and of the recording electrode are unknown, as is the case with surface recording of EPs, any shift of extracellular potential can reflect either type of PSPs. Knowing only the polarity of the extracellular potential at a given location does not allow any inference about the underlying synaptic event that produced it. For instance, a negative potential picked up on the surface of the cortex can reflect either the superficial dendritic EPSPs or the deep somatic IPSPs generated by the same pyramidal cell.

NEAR-FIELD VERSUS FAR-FIELD EVOKED POTENTIALS

Synaptic potentials in superficial sources were originally thought to account for the changes in potential difference between electrodes applied to the skin, but it is now recognized that activity in nuclei some distance from the surface can also be detected, the so-called 'far-field' potentials (Jewett 1970, Jewett and Williston 1971). This only happens if all or most of the neurones in a nucleus are orientated in the same direction, so that synaptic potentials from a large number of cells summate to produce an appreciable potential field outside the nucleus, the openfield system of Lorente de Nò (1947). Other cellular orientations result in little or no net external potential, and are known as closed-field systems. More recently it has been appreciated that ascending volleys in sensory pathways produce widespread potentials. Why moving action potential volleys should produce potentials at fixed latencies has not been fully explained, but physical models (Nakanishi 1982), computer simulations (Cunningham et al 1986) and human recordings (Kimura et al 1983, 1984) suggest that a stationary potential is generated when propagated impulses cross the boundary between two media with different conductivities.

Potentials recorded close to their sources (near-field) may be of axonal or of postsynaptic origin. Most of them have a dipolar field distribution on the scalp and their sources, at any given instant, can be modelled as a single resulting dipole. If this dipole is tangentially orientated and close to the surface, both its positive and negative poles can be observed by using an appropriate electrode array. Consequently, the source of the EP should not be located at either of the maxima of the electric field, but between the two (at a depth that can be indirectly evaluated from the distribution of isopotential lines, or by measurements of the magnetic field). It will be equidistant from each maximum in the case of a theoretical tangential dipole (Wood et al 1985, Deiber et al 1986, Desmedt et al 1987) (Figs 4.21 and 4.22). The 'paradoxical' lateralization of the P100 VEP to half-field checkerboard stimulation results from a tangentially orientated dipole (Halliday 1982).

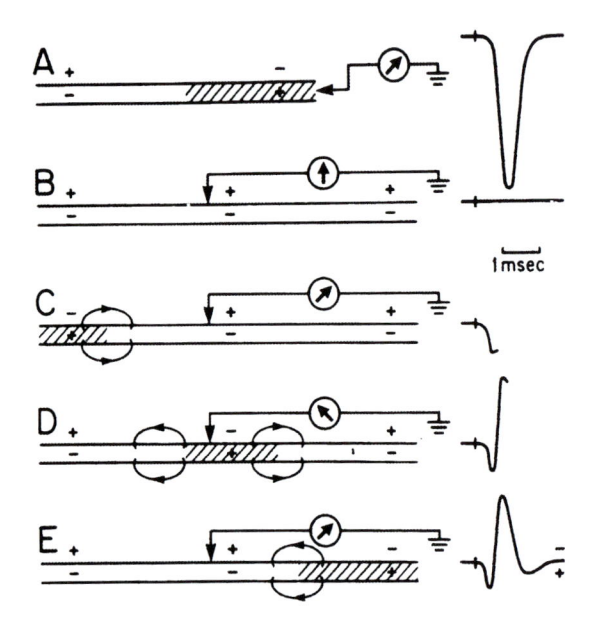

Fig. 4.20 Schematic illustration of potentials generated by an action potential travelling along an axon, or a CAP travelling along a nerve. (From Schlag (1973), by permission.)

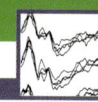

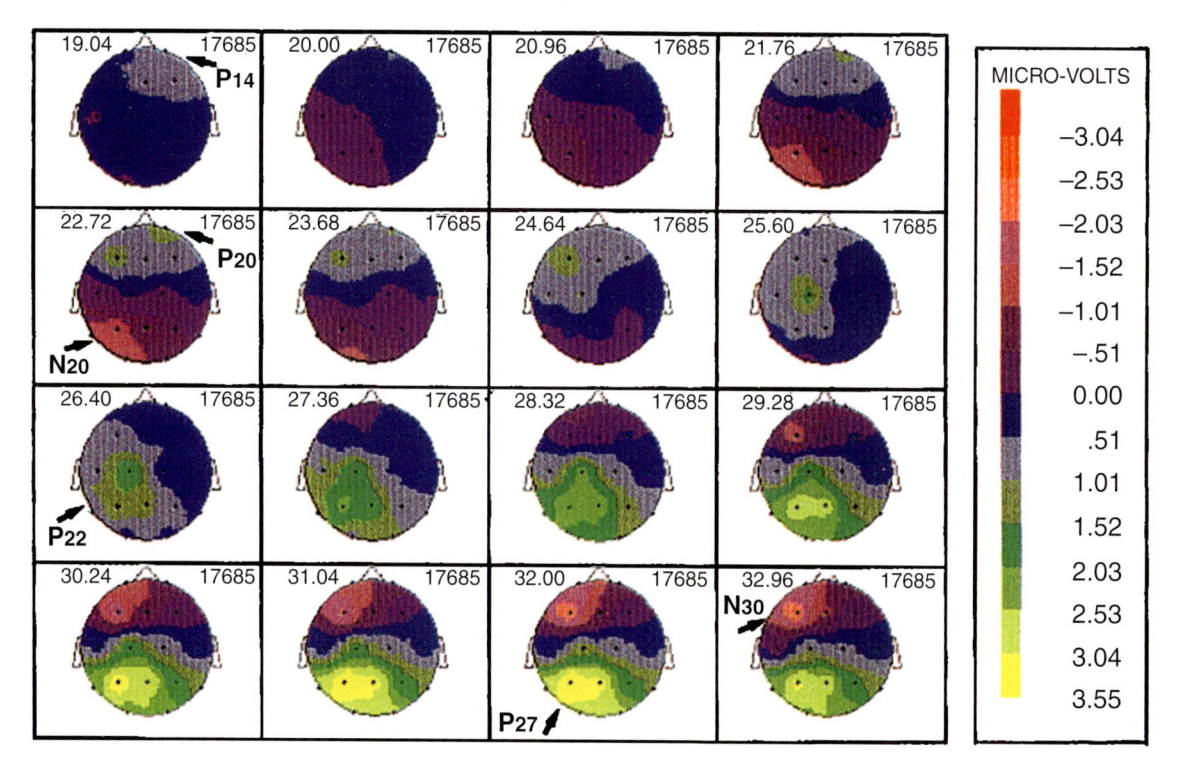

Fig. 4.21 Sequential maps of normal SEPs. Frozen maps of SEPs to stimulation of the right thumb are displayed for the latencies (in ms) indicated in the corresponding upper left corner. The reference electrode is on the ear lobe on the non-stimulated side (left). Note that, because of conduction time from finger to wrist, SEPs are peaking 2–3 ms later than after median nerve stimulation. The widespread P14 far-field positivity ends at 19 ms and is maximal frontally, while the contralateral N20–P20 dipolar profile begins at 21 ms, peaks at 23 ms and has a distribution compatible with a dipolar generator orientated tangentially to the scalp surface. The isopotential lines corresponding to P22 (maximal at 26 ms) are concentric around the contralateral central electrode; this central positivity should not be confused with the frontal P20; its spatial distribution on the scalp suggests a dipolar generator perpendicular to the scalp surface. At 32 ms the contralateral parietal P27 reaches its maximum, while the frontal N30 is still increasing to reach its maximum after the time range illustrated in this figure. (From Deiber et al (1986), by permission.)

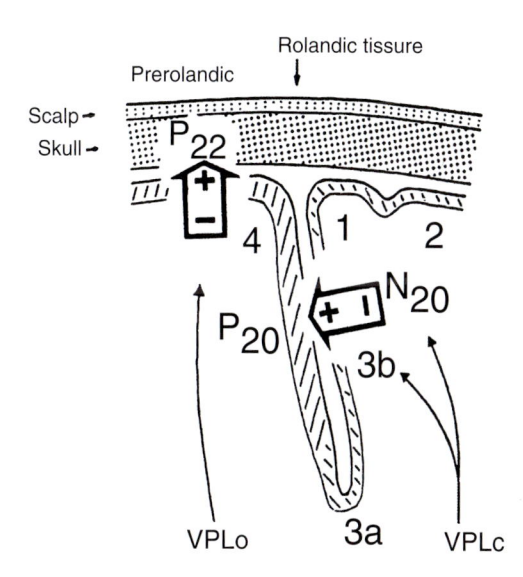

Fig. 4.22 Schematic models of early cortical SEP generators. This model was proposed to explain the field distributions of early SEPs illustrated in *Fig. 4.21*. The N20–P20 dipolar configuration is supposed to reflect the activation of the cells in Brodmann's area 3b, which receive sensory afferents from the caudal part of the ventroposterolateral nucleus of the thalamus VPLc. Due to the orientation of the functional columns perpendicular to the surface of the cortex, the depolarization of pyramidal cells can be modelled as an equivalent dipole perpendicular to the plane of the Rolandic fissure and thus parallel to the scalp surface. The P22 is supposed to reflect the response of motor area 4 receiving sensory inputs from the oral part of the VPL (VPLo also known as the ventralis intermedius nucleus (VIM) in man). The radial orientation of the equivalent dipole is compatible with the anatomy and the field distribution illustrated in *Fig. 4.21*. The question whether this radial generator is situated in front of the Rolandic fissure, as illustrated in this figure, or immediately behind it in Brodmann's area 1 is still debated. Clinical data favour the former hypothesis while dipole modelling and magnetic field studies are more equivocal. (From Desmedt et al (1987), by permission.)

If the dipole is radially orientated, the pole which is closer to the surface will produce a field of concentric isopotential lines over a restricted area, and no second maximum of opposite polarity will be found on the scalp (*Figs 4.21* and *4.22*). The EP source is then assumed to be situated under the maximum positivity or negativity, at a depth that can be determined neither by electrical recording from surface electrodes, nor by magnetic recording, which is blind to radial electrical sources. In many cases the orientation is intermediate between a purely radial and a purely tangential dipole.

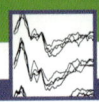

HOW TO LOCALIZE EVOKED POTENTIAL SOURCES FROM SURFACE RECORDINGS

The relationship between the surface distribution of electrical potential and the underlying EP generators can be considered from two points of view: (a) the 'forward' problem consists of calculating the distribution on the surface of potential fields generated by electrical sources of known location in the volume conductor; (b) the 'inverse' problem is to infer the putative location of neural generators of EPs from the distribution of field potentials on the surface (see Wood (1982) for a review). The inverse problem has no unique solution; it is impossible to calculate the position, orientation and magnitude of a dipole source from the surface potential distribution alone (Helmholz 1853), but methods have been developed for source localization if the approximate location can be assumed (Michel et al 2004). These are applicable particularly to sources in primary sensory cortex, where the approximate location is well established. However, it must not be forgotten that it is the equivalent dipole which is identified; the equivalent dipole of a convoluted area of cortex may not be in the cortex at all, and two separate sources active simultaneously may sum to produce an equivalent dipole distant from both sources. A related problem is that a dipolar scalp distribution is usually assumed to be due to a tangentially orientated source, but an identical scalp distribution could be produced by simultaneous activation of two radially orientated sources producing potentials at the surface of opposite polarity. *Figure 4.23* shows how a midline predominance of the early cortical SEPs can arise from simultaneous stimulation of both median nerves, leading to activation of primary sensory areas in both hemispheres.

The basic principle of potential field mapping rests on the assumption that the baseline from which voltages are measured remains stable over the analysis time. In other words, the reference baseline should not be affected by the potentials recorded at active electrode sites, which are used for computing maps of the electrical fields. Theoretically this zero level can be obtained by using the 'average' reference computed as the average over time of the potentials picked up by all electrodes (Lehmann and Skrandies 1984). At a given instant, the spatial integral of the potentials recorded on the surface of a sphere, in which the source is enclosed, is zero. However, this is only true if the recording electrodes are infinitely numerous, regularly

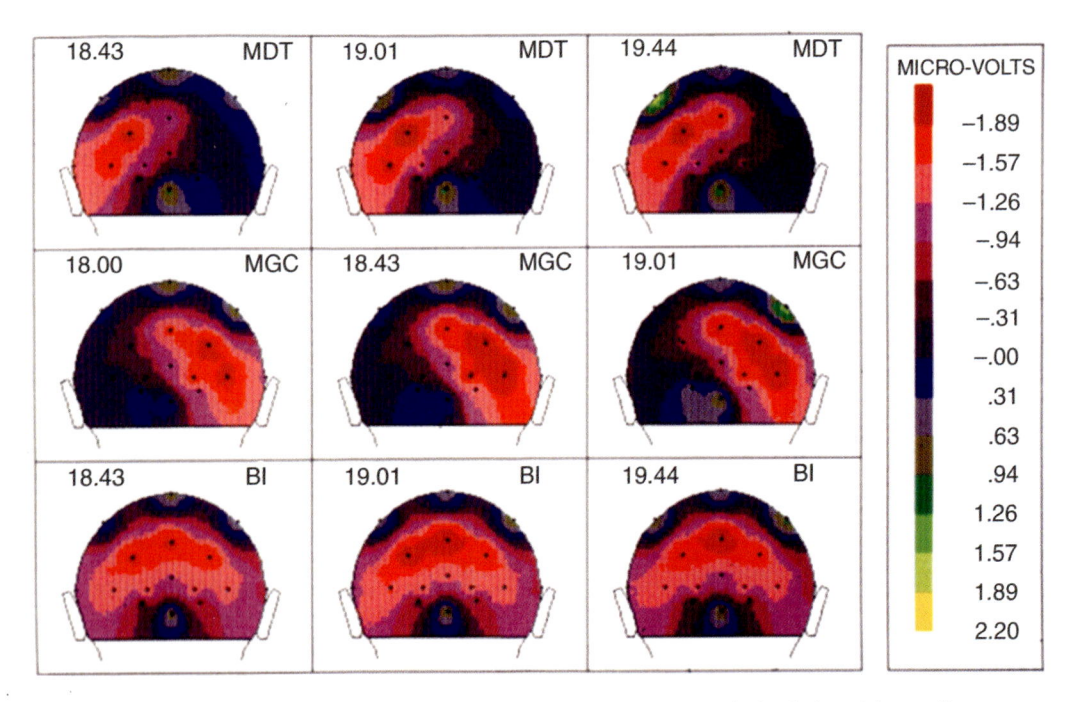

Fig. 4.23 Field distributions of the N20 potential after unilateral and bilateral stimulation of the median nerve at the wrist. In this figure the head is seen from behind to show the scalp distribution of the contralateral N20. After stimulation of the right (upper row) or left (middle row) median nerves, only the hemisphere contralateral to the stimulation is activated and the N20 potential is distributed in the corresponding parietal region. When the somatosensory inputs arrive simultaneously in both parietal areas after bilateral stimulation (lower row), the potential fields generated in each hemisphere merge and the resulting field distribution shows a maximum at the midline in the parietal region. Such a combination of potentials generated synchronously in symmetrical and functionally homologous cortical areas explains the vertex predominance of cortical auditory EPs, since both auditory areas are simultaneously activated, even by monaural stimulation. (From Binnie et al (2004), by permission.)

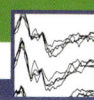

spaced and cover the whole surface of the sphere. These impossible conditions are not even approximated in scalp EP recording, where electrodes explore only the upper half of the head (see Desmedt et al (1990) for a discussion of this point). With the development of commercial mapping systems, visual analysis of electrical field distributions on the scalp, possibly assisted by dipole modelling software, may well become routine, but the exercise will remain liable to errors and misinterpretations.

NORMAL FINDINGS BY MODALITY

Flash VEPs
Normal waveforms in adults

The so-called 'transient response' to diffuse light flashes is a complex waveform which contains up to seven distinct components peaking within the 250 ms following the stimulus. Roman and Arabic numerals have been proposed, respectively, by Cigánek (1961) and Gastaut and Régis (1965), to label each of these peaks as illustrated in *Fig. 4.24*. Using the polarity latency nomenclature, Halliday et al (1979) have labelled peaks II to VI of Cigánek, respectively, P60, N70, P100, N125 and P160. Of these different peaks, only N70, P100 and P160 are identifiable in most normal subjects (*Fig. 4.25*) and the W-shaped P100–P160 waveform is very variable between individuals; either one of these two positivities may predominate, and in many subjects there is only one positive peak in this latency range. For diagnostic purposes each laboratory usually selects one, or a few, of the least variable components. The latency of the flash VEP can be measured at the peak of the N70 or P100 waves, but SDs are 15–22% of the mean values in normals (Halliday 1982). In the experience of the author (Mauguière et al 1979), the peak latency of the N70 potential (wave III of Cigánek or wave 4 of Gastaut and Régis) shows the smallest degree of interindividual variability with a SD/mean ratio of 13%. In the 250–1500 ms latency range, a train of rhythmical waves, at or near the frequency of the alpha rhythm, may be observed; this is of doubtful clinical utility, but may be reduced on the side of a hemisphere lesion.

Latency and amplitude values of the normal flash VEP

Because of the great variability of flash VEPs, the complete absence of response to monocular stimulation by the conventional strobe flashes used for intermittent photic stimulation is often considered to be the only reliable criterion of abnormality in many laboratories of clinical neurophysiology (Spehlmann 1985). However, latencies of waves N70 and P100, which correspond to Cigánek's waves III and IV and are consistently identifiable in normals, can be useful for assessing flash VEPs in patients (*Table 4.1*). As an indication, the values obtained in 17 normal subjects by Halliday et al (1972) and by the authors in 25 normal subjects (Mauguière et al 1979) are respectively of 73 ± 15 ms and 63.5 ± 8 ms for N70 and of 103 ± 15 ms and 104 ± 19 ms for P100. In our experience the N70 latencies in normals have a Gaussian distribution and we accept an upper normal limit of 82.5 ms (mean + 2.5 SD) for this variable. Absolute amplitudes are practically unusable, with SDs exceeding 50% of the mean values; for this reason flash VEPs are often interpreted for clinical purposes only in terms of their presence or absence. However, an interocular difference of more than

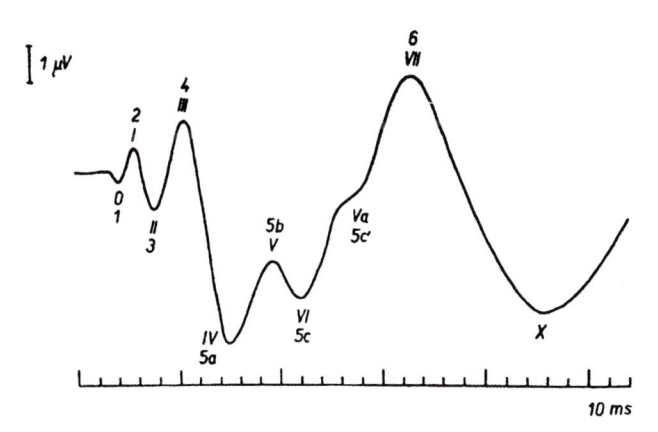

Fig. 4.24 Schematic drawing and nomenclature of flash VEP components. The nomenclature proposed by Cigánek (1961) uses Roman numerals I–VII, whereas in the classification of Gastaut and Régis (1965) peaks are identified by Arabic figures. The three earliest peaks are inconstantly obtained as is wave VII (or 6). The subcomponents of the main positivity labelled 5 a, b, c and c' show large interindividual variations. Negativities are represented upwards. (From Binnie et al (2004), by permission.)

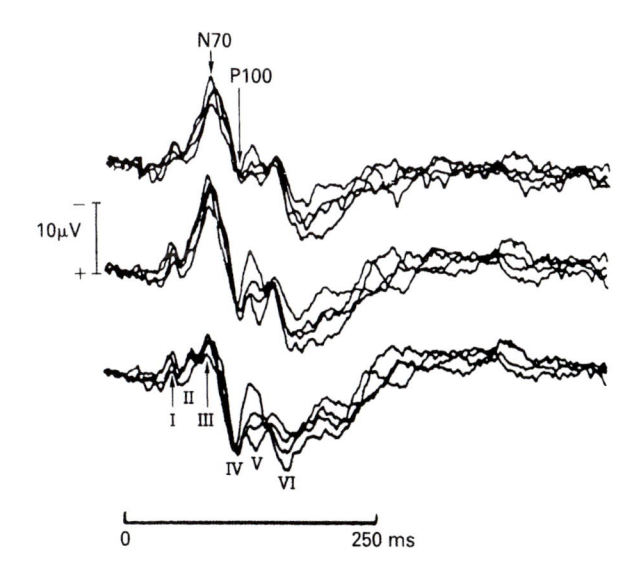

Fig. 4.25 Test–retest variability of the flash-evoked response. Traces recorded on four successive runs of 100 flash stimulations in a normal individual are superimposed (filter bandpass 1–160 Hz). Waveforms are very similar in the midoccipital (middle traces), left (upper traces) and right (lower traces) occipital regions. In this subject, the early negative wave I (2 in the Gastaut and Régis nomenclature – see upward arrow) are fairly reproducible. The most consistent components are represented by Cigánek's waves III and IV (N70 and P100). (From Binnie et al (2004), by permission.)

Table 4.1 VEP normative data in adults

	Latency (mean ± SD) (ms)	Interocular difference latencies	Amplitude (μV)
FLASH VEPs			
Wave III (N70)			
Halliday et al (1972)	73 ± 16 (LE)	–	5 ± 4 (LE)
	73 ± 14 (RE)	–	7 ± 5 (RE)
Mauguière et al (1979)	63 ± 8 (WL)	> 13 ms	9 ± 4 (LE)
	71 ± 12 (RL)	> 20 ms	10 ± 6 (RE)
Wave IV (P100)			
Halliday et al (1972)	103 ± 15 (LE)	–	9 ± 4 (LE)
	102 ± 14 (RE)	–	10 ± 6 (RE)
Mauguière et al (1979)	104 ± 19 (WL)	–	3–12 μV*
	122 ± 30 (RL)	–	3–13 μV*

LE, Left eye; RE, right eye; WL, white light; RL, red light.
*Range of observed values.

50% for the N20–P100 deflection amplitude is not usual in normal subjects and may be considered abnormal.

Auditory evoked potentials: BAEPs and MLAEPs
Waveform and origins of normal adult BAEPs

Brainstem (auditory evoked potentials (BAEPs) are composed of six successive peaks, labelled I–VI (or occasionally VII) in Roman figures. All of these peaks can be obtained using a simple montage in which the electrode connected to lead 1 of the amplifier is placed close to the stimulated ear, on the mastoid (Mi) or the ear lobe (Ai), while the other, connected to lead 2 of the amplifier, is positioned at the vertex, usually at the Cz site. Many factors (electrode position, filters, click intensity, frequency spectrum, polarity and click rate, etc.) may affect BAEP waveforms and will be considered later. However, given the robust nature of this signal and considering the fact that the above-mentioned sources of variability cause only minor changes without major distortion of the response, a schematic 'ideal' BAEP waveform can be outlined which illustrates what is now obtained in all clinical neurophysiology laboratories.

The traces in *Fig. 4.26* weren obtained with a high-pass filter at 150 Hz which gives an oscillatory appearance by eliminating the contribution of slower components to the overall waveform. According to Jewett's scheme (Jewett et al 1970) waves I–VI refer to upward (vertex positive) deflections in the Ai-Cz montage. Wave I is a localized negative-going change close to Ai (lead 1), whereas waves II to VI are far-field positivities picked up at Cz (lead 2); all therefore produce upward deflections. Due to the complexity of the network of auditory fibres and to the small area in which fibres and nuclei are packed and intermingled in the brainstem, any attempt to match each of the BAEP waves with a single source is naive and is likely to be unsuccessful. However, specific 'generators' can be

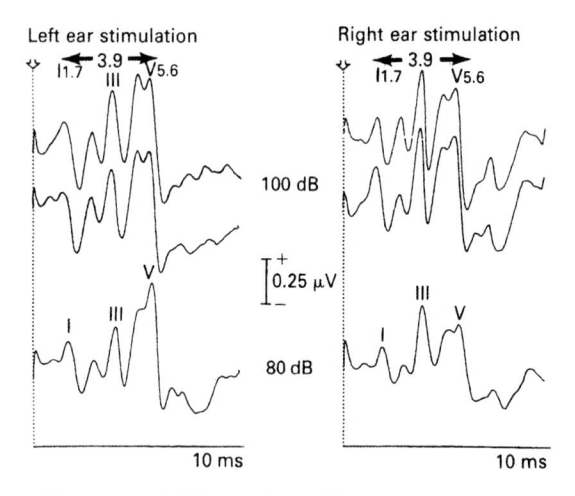

Fig. 4.26 Normal BAEP waveforms. These responses were evoked by rarefaction clicks of 100 dB HL (two upper traces) and 80 dB HL (lower traces). Note the test–retest stability of the waveform and peak latencies (two upper traces) and symmetrical peak latencies. With 80 dB HL, all peaks are clearly identifiable and increasing the stimulus intensity up to l00 dB HL adds very little to waveform definition. Note also that, despite negligible inter-ear differences in amplitude, the I/V amplitude ratio is less than 1 in both ears and at both stimulus intensities (arrows). Derivations: Cz–M1 and Cz–M2, respectively, for left and right ear stimulation. Filter settings: 150–3000 Hz. (From Binnie et al (2004), by permission.)

attributed to each BAEP wave on the basis of intraoperative recordings and of correlations between abnormal waveforms and lesion sites in patients with discrete cochlear nerve or brainstem lesions (see Hashimoto 1989, Pratt et al 1999 for reviews). In fact most of these generators are compound activities originating in several fibre bundles and relay nuclei and are a mixture of action and PSPs.

Wave I originates from the peripheral portion of the cochlear nerve. It peaks with a latency similar to that of the CAP recorded directly from the cochlear nerve; it is virtually absent or drastically reduced in recordings from the ear lobe contralateral to stimulation (Ac–Cz). It is the only component to be preserved when conduction is blocked or slowed in the proximal portion of the cochlear nerve, and also in patients with irreversible deterioration of brainstem function and clinical evidence of brain death.

Wave II, although constantly obtained at click intensities of more than 60 dB HL, is not commonly used in clinical interpretation of BAEPs. It most probably consists of two separate components overlapping in time and reflecting, respectively, action potentials in the proximal portion of the cochlear nerve and postsynaptic responses of cochlear nucleus cells. The change in physical characteristics of the volume conductor between the inside and outside of the internal auditory meatus (see *Fig. 4.17*) probably explains why the same volley of action potentials propagating along the cochlear nerve may produce two stationary far-field positivities (i.e. wave I and part of wave II).

Wave III is, with waves I and V, one of the most constant and robust components of the BAEP complex. It is positive in the Ai–Cz traces, but reverses its polarity or is substantially reduced when the activity is recorded between the contralateral ear lobe and Cz (Ac–Cz). This strongly suggests that wave III is generated by a horizontal dipolar source which, on the basis of intraoperative recordings, has been localized in the pontine portion of the brainstem auditory pathways.

Waves IV and V can be clearly separated or combined to form a single IV/V complex. Wave V reflects the activity of a vertically orientated dipolar source in the midbrain which is rostral positive and caudal negative. It is most probably generated both by propagation of action potentials in the lateral lemniscus (LL) fibres and by postsynaptic responses of midbrain auditory nuclei including the inferior colliculus (IC), in particular that contralateral to the stimulated ear. The origin of wave IV is not firmly established; both a fixed generator in the pons and action potentials ascending in LL could contribute to this component. Wave V can be detected even at near-threshold levels and has proved to be the most reliable and useful of BAEP components in clinical practice.

A *wave VI* and even a *wave VII* can be identified in some individual traces. They are present in up to 70% of normal BAEPs in our laboratory, but are poorly reproducible. Their origin is unclear and their clinical utility has not yet been demonstrated.

Latency and amplitude values of normal adult BAEPs

Intra- and interindividual variability of waves I–V recorded in the same conditions is recognized to be very small in all published series of normative values. Measurements and calculations routinely performed in clinical testing include: peak latencies of waves I, III and V; inter-peak latencies I–III, III–V and I–V; I/V amplitude ratio; and inter-ear latency differences, using either absolute peak latencies or inter-peak latencies. The normal values of these different variables used in our own laboratory are given in *Table 4.2*. Due to the influence of recording conditions, these values should not be viewed as gold standards; however, they indicate that SDs of peak and inter-peak latencies are very similar from one laboratory to another. For click intensities between 60 and 90 dB HL, the mean and SD values for the I–V interval vary from only 3.9 to 4.1 ms and from 0.20 to 0.26 ms, respectively, in spite of appreciable differences in recording procedures (see Allison (1987) for a review).

Peak-to-peak amplitudes can be measured from the peak of each wave to the trough which immediately follows. Their mean values vary for the different components from 0.2 to 0.5 µV, the highest values being obtained for the peak of wave V or for the IV/V complex. Absolute amplitudes are highly variable between individuals, with SD values of 30–50% of the mean. They therefore have no clinical utility even in inter-ear comparisons. To overcome this difficulty the I/V amplitude ratio, expressed in per cent, can be used. In control subjects this ratio is generally less than 100% and a value over 200% is above the upper limit of normality (mean + 3 SD) for click intensities over 80 dB HL. By comparing the peripheral input in the cochlear nerve (wave I) to the main brainstem component (wave V), this ratio depends both on the synchrony of ascending inputs in the proximal part of the cochlear nerve and brainstem auditory pathways and on the number of cells activated by the stimulus in the brainstem. In principle, it is very similar to the P9/ PI4 ratio of median nerve SEPs (see later). The I/V amplitude ratio changes with the click intensity and is less reliable in patients suffering from peripheral hearing loss.

Middle latency auditory evoked potentials

MLAEPs are elicited within the 10–50 ms post-stimulus latency range and are made up of a sequence of five waves usually labelled No, Po, Na, Pa and Nb (Picton et al 1974). Of these five peaks, only the Na and Pa waves, peaking respectively in the 16–20 ms and 27–33 ms latency ranges, are consistent enough to be clinically useful. The earlier No and Po waves peaking, respectively, at about 10 and

Table 4.2 BAEP normative data in adults

BAEPs 80 dB HL	Mean (ms)	SD (ms)	Mean + 2.5 SD (ms)
Absolute latencies			
Peak I	1.5	0.17	1.93
Peak III	3.8	0.15	4.18
PeakV	5.6	0.20	6.10
Intervals			
I–III	1.6	0.20	2.10
III–V	1.7	0.20	2.20
I–V	3.9	0.20	4.40
Right–left difference I–V	0.17	0.10	0.42

13 ms are of very low amplitude and are not identifiable in all normal subjects. The Nb wave shows substantial interindividual variation. Some authors include in MLAEPs a Pb wave which peaks after 50 ms, but others describe it as the P1 component of the long latency AEP. Because MLAEPs overlap in time with a reflex response of postauricular muscle, their neurogenic origin was questioned in the 1960s (Bickford et al 1964). However, their persistence under curarization demonstrated that this myogenic response, although causing recording difficulties in the awake and alert patient, does not explain all of the auditory evoked activities recorded on the scalp within the MLAEP latency range.

When recorded with a non-cephalic reference, the Na and Pa components are largest in the frontocentral regions and of lower amplitude in the parietal and temporal regions, but there is wide interindividual variation. Rarefaction clicks of 100 µs, similar to those used for BAEP recording, are the most convenient stimuli for eliciting MLAEPs. They have the advantage of being produced by the commercially available EP systems used in most laboratories of clinical neurophysiology. Moreover, clicks produce Na and Pa potentials of greater amplitude than do tone bursts of longer duration. Na and Pa attain their maximal amplitudes and shortest latencies at 70–80 dB HL. Above 80 dB HL these variables are stable; reducing the click intensity to below 70 dB HL causes a latency increase of Na and an amplitude reduction of all components. Na and Pa are recordable for click intensities down to 30 dB HL, but are not widely used for audiometric testing. A click rate of 3–4/s yields consistent responses, the amplitude of which drops by a factor of 2 when the rate increases to 10/s.

The major Na and Pa components of the MLAEP are not detectable in all normal-hearing children until the age of 10 years (Kraus et al 1985). In subjects over the age of 60 years the Pa component is enlarged and delayed compared with adults aged 20–40 years (Woods and Clayworth 1986).

The precise origin of the MLAEP components is not known. Scherg and Von Cramon (1985) suggested that both Na and Pa were related to the activation of the primary auditory cortex, while others have concluded from the scalp distribution, maturation and effects of lesions that Pa is generated in the auditory cortex, while Na has a subcortical origin (Ibañez et al 1989a).

Somatosensory evoked potentials
Fibre tracts involved in the genesis of SEPs
SEPs to electrical stimulation of peripheral nerves differ from other EP modalities because the non-physiological stimulus depolarizes nerve fibres directly and bypasses the peripheral encoding of the physical information (pressure, vibration, joint movement, etc.). The different types of sensation are conveyed by fibres of different diameters, having myelin sheaths of different thicknesses. Fibre conduction velocity and threshold to electrical stimulation are inversely related to fibre diameter, and non-noxious electrical stimuli only exceed the thresholds of the fastest conducting fibres, which are also the most heavily myelinated.

Electrical stimulation of pure sensory nerves, such as digital nerves for the upper limb and the sural nerve for the lower limb, activates exclusively skin and joint peripheral and dorsal column fibres. At the upper limb, stimulation of the finger tip (Desmedt and Osaki 1991) or distal phalanx of the fingers (Restuccia et al 1999) activates selectively skin fibres, while stimulation of digital nerves at the level of the first or second phalanx activates joint afferents as well as those from the skin. SEPs evoked in man by non-painful stimulation of peripheral nerves are produced by impulses ascending in the dorsal columns, and are not affected by spinal lesions which interrupt the spinothalamic tracts but spare the dorsal columns (Halliday and Wakefield 1963, Cusick et al 1978). When high-intensity stimuli are used, other ascending tracts in the spinal cord may also be stimulated, especially the anterolateral spinothalamic tracts (Powers et al 1982). Normal values are given in *Table 4.3*.

In most clinical applications the electric stimulus is applied to mixed nerve, containing motor and sensory fibres, at an intensity 3 to 4 times the sensory threshold, which is sufficient to produce a twitch of the muscles innervated by the nerve. Stimuli of this intensity activate all large myelinated fibres, which convey touch, joint sensation and afferent impulses from muscles. Cortical responses to upper limb stimulation probably include only a small contribution from muscle afferents (Kunesch et al 1995, Restuccia et al 2002), but muscle afferents make a greater contribution to cortical responses from stimulation of lower limb nerves (Macefield et al 1989).

Short latency SEPs to electrical stimulation of the upper limb
The main early SEP components evoked by the stimulation of the median nerve at the wrist are illustrated in *Fig. 4.27*. Most of the SEP components of proven clinical utility peak before 50 ms.

In the following sections each of the SEP components will be described as they appear in recordings using a non-cephalic reference electrode placed at the shoulder on the non-stimulated side. The modifications of the waveform and latency related to the use of other sites for the reference electrode will be described as necessary.

Peripheral components Stimulation of a mixed nerve, such as the median or ulnar nerve, elicits a CAP corresponding to the peripheral ascending volley that can be recorded at different levels of the forearm and arm. The amplitude of peripheral CAP is unaffected by interfering stimuli applied on the stimulated hand or when the stimulus rate is increased up to 50 Hz (Ibañez et al 1989b). In most SEP studies, and particularly in those carried out on patients with central lesions, the ascending volley is recorded from the supraclavicular fossa, at Erb's point, where the activa-

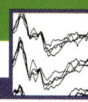

Evoked potential waveforms and interpretation

Table 4.3 SEPs – normative data in adults

	Component or interval	Topography	Amplitude Mean ± SD (μV)	Side-to-side asymmetry	Latency Upper normal value (mean + 2.5 SD) (ms)	Maximal side-to-side difference (ms)
Median nerve SEPs						
Absolute latencies	N9	Erb's point (1)	6 ± 3	< 50%	12.5	–
	P9	Cerv. and scalp	1.6 ± 0.5	–	11.5	–
	N11	Cervical	–	–	13.9	–
	P11	Scalp	1.9 ± 0.6	–	13.3	–
	N13	Cv4 to Cv6	2.0 ± 0.6	–	14.8	–
	P14	Scalp	3.0 ± 0.6	–	17.0	0.7
	N20	Contra. parietal	2.2 ± 0.9	< 41%	23.4	1.3
	P22	Contra. central	2.3 ± 1.3	< 58%	25.1	3.7
	P27	Contra. parietal	2.4 ± 1.5	< 60%	33.0	5.2
	N30	Contra. frontal	3.7 ± 2.0	< 58%		
Intervals	N9–N13	Cervical	–	–	4.5	1.3
	P9–P14	Contra. parietal	–	–	5.9	0.9
	P9–N20	Contra. parietal	–	–	11.3	0.9
	P14–N20	Contra. parietal	–	–	6.8	1.0
	P14–P22	Contra. frontal	–	–	10.1	1.3
	P14–P27	Contra. parietal	–	–	17.9	1.4
Amplitude ratios	N13/P9	Cervical	> 1	–	–	–
	P9/P14	Contra. parietal	< 1.3	–	–	–
Tibial nerve SEPs						
Absolute latencies	N22	L2	1.4 ± 0.7	–	24.8	1.2
	P30	Frontal	0.6 ± 0.2 mV	–	33.3	–
	P39	Pz	2.9 ± 1.3	–	44.3	1.5
Interval (CCT)	N22–P39	–	–	–	21.3	2.0

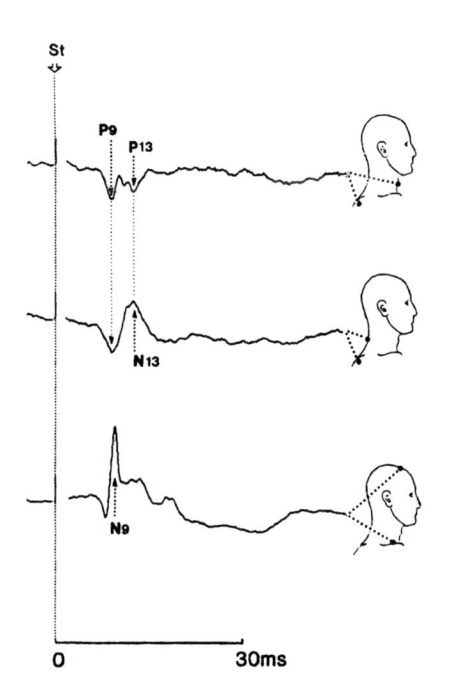

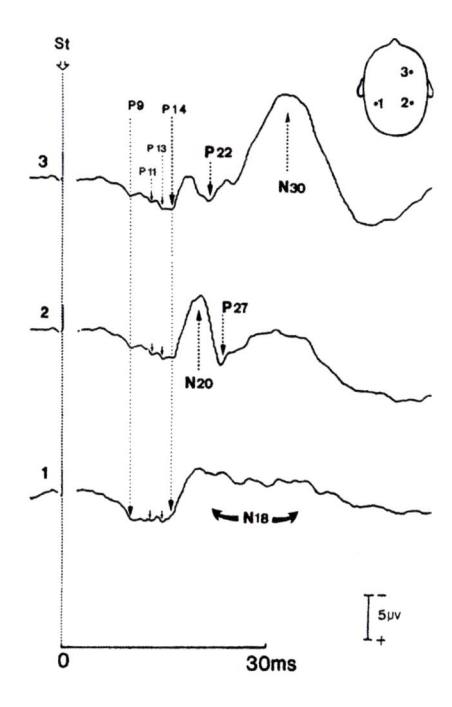

Fig. 4.27 Normal SEPs to median nerve stimulation. Early SEPs recorded with a shoulder reference following stimulation of the left median nerve at the wrist. The nuchal N13 component is recorded as a P13 positivity by an anterior cervical electrode. Contralateral parietal N20–P27 and frontal P22–N30 are consistently recorded in normal subjects. Note that in the parietal region ipsilateral to the stimulation (trace 1), there is a long-lasting negative shift which immediately follows the widespread P9–P11–P13–P14 scalp far-field positivities and corresponds to the N18 potential (see text for details). (From Mauguière F 1989 Evoked potentials in nondemyelinating diseases. In: *Advanced Evoked Potentials* (ed H Lüders). Kluwer, Boston, MA, pp. 181–221. © Kluwer Academic Publishers, with kind permission of Springer Science and Business Media.)

tion of the brachial plexus trunks appears as a triphasic positive–negative–positive waveform, with a negative peak culminating at about 9 ms in normal subjects (N9). The peak latency of N9 varies slightly with arm length. The peripheral ascending volley can also be recorded more distally over the trajectory of the nerve at the elbow (N6).

After stimulation of a mixed nerve, both potentials are a mixture of motor antidromic and sensory orthodromic responses, and thus are qualitatively different from the sensory SEP components generated in the CNS. Both can be obtained using runs of 500 stimuli delivered at 2/s.

Spinal components

The N11 potential The nuchal N11 potential following stimulation at the wrist is recorded all along the posterior aspect of the neck, where it usually appears to encroach upon the ascending slope of N13, from which it is often not clearly differentiated) (Cracco 1973). In noncephalic reference recordings its onset latency increases from Cv6 to Cv1 spinal processes, consistent with the suggestion that N11 is generated by the ascending volley of action potentials in the cervical cord (Desmedt and Cheron 1980a, Mauguière 1983). Because of the difficulty of distinguishing N11 from the following N13 its clinical usefulness is limited, except in the few cases of cervical cord lesions which selectively obliterate the N13 potential.

The spinal segmental N13 potential The N13 potential is picked up on the posterior aspect of the neck, with a maximum amplitude at the Cv5–Cv7 levels. When recorded directly from the surface of the cord, it decreases in amplitude at more rostral or caudal electrode positions. No latency shift is observed for N13 between Cv6 and Cv2 recordings in normals (Desmedt and Cheron 1980a, Mauguière 1983, Desmedt and Nguyen 1984). N13 reverses into a P13 when recorded anterior to the cord (Desmedt and Cheron 1980a, Mauguière and Ibañez 1985, Cioni and Meglio 1986, Jeanmonod et al 1989, 1991) (*Fig. 4.28*). The distribution of N13/P13, its behaviour with high stimulus rates (Ibañez et al 1989b) and the fact that it persists in lesions at the cervicomedullary junction (Mauguière et al 1983a, Mauguière and Ibañez 1985, Buchner et al 1986) lead to the conclusion that it arises from a fixed dipolar generator in the dorsal horn of the cord, perpendicular to the cord axis. The

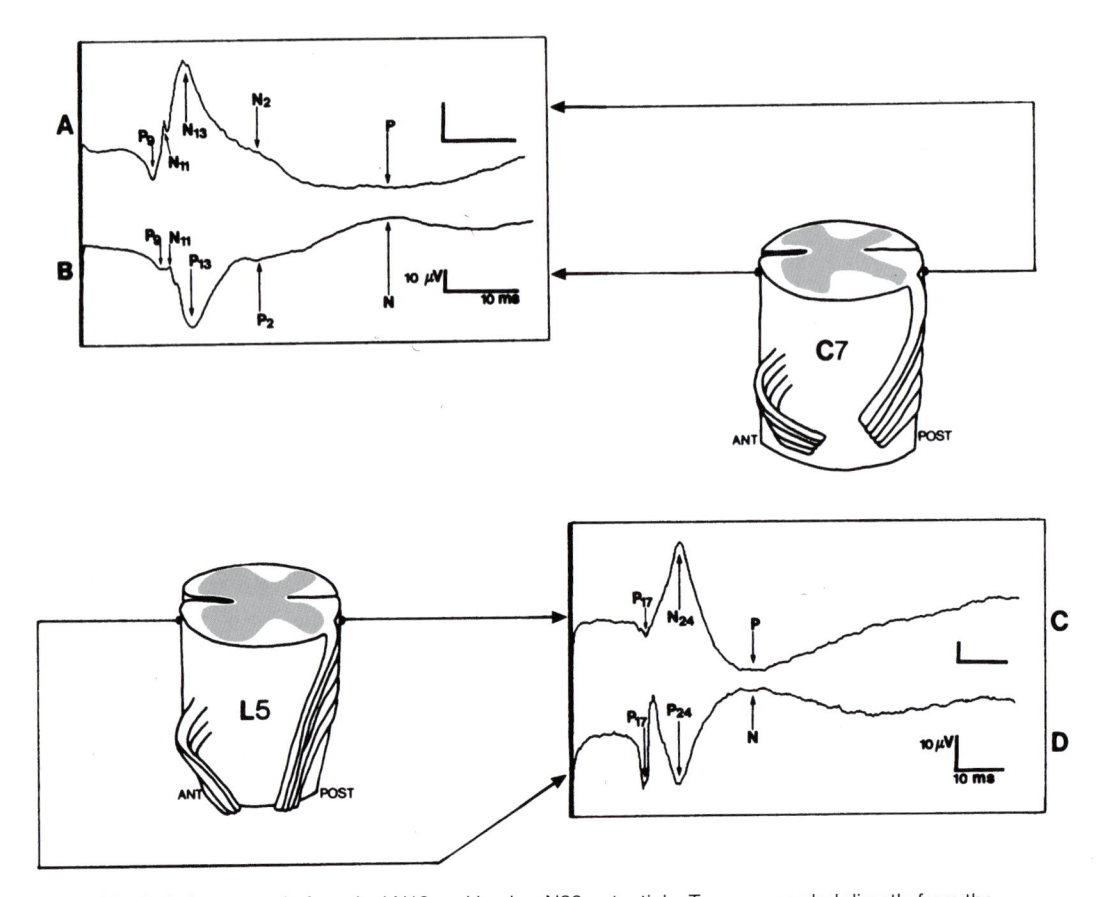

Fig. 4.28 Polarity reversal of cervical N13 and lumbar N22 potentials. Traces recorded directly from the cord surface during surgery show the dorsoventral organization of the cervical N13–P13 and lumbar N22–P22 potentials. The far-field potentials P9 and P17 originating, respectively, in the proximal roots of the cervical and lumbosacral plexuses do not show any polarity reversal between anterior (Ant) and posterior (Post) recording sites. On cervical recordings at Cv7 level there is an N11 wave; no equivalent is recorded at L5 level in this patient because of time dispersion of the dorsal column volley at this level due to the pathological process (multiple sclerosis). (From Jeanmonod et al (1989), by permission.)

generator is most likely to be the compound PSP triggered in the dorsal horn grey matter by the afferent impulses arriving from fast-conducting myelinated fibres. The human N13 therefore corresponds to the N1 spinal potential described in animal studies, which reflects the postsynaptic neuronal response to inputs conveyed by groups I and II peripheral afferent fibres in laminae IV and V of the dorsal horn (Gasser and Graham 1933, Austin and McCouch 1955, Beall et al 1977). Because of the spatial distribution of the potential field corresponding to this dorsal horn response, a transverse montage connecting two electrodes located, respectively, over the Cv6 spinal process and above the laryngeal cartilage is the most suitable for recording the spinal N13/P13 activity. N13 is enhanced in the waveform obtained with this montage due to injection of anterior P13 positivity at lead 2 of the amplifier, and the resulting waveform is not contaminated by SEP components generated above the foramen magnum, unlike the more conventional Cv6-Fz montage (Mauguière and Restuccia 1991).

Runs of 1000 stimuli delivered at 2/s are appropriate in most cases for obtaining well-defined cervical potentials using either a non-cephalic or anterior cervical reference site. The spinal N13/P13 must not be confused with the negative potential peaking at a latency of about 13 ms recorded intraoperatively at the dorsal aspect of the cervicomedullary junction (Allison and Hume 1981, Lesser et al 1981, Møller et al 1986, Morioka et al 1991), which is supposed to be generated by the presynaptic volley in the dorsal column fibres close to the cuneate nucleus.

Far-field scalp positivities (P9, P11, P14) With an appropriately wide filter bandpass (e.g. 1 Hz to 3 kHz), a fast sampling rate, a large number of averaged runs (at least 1000) and a non-cephalic reference montage, three stationary positivities are recorded on the scalp after median or ulnar nerve stimulation; these are widely distributed with a mid-frontal predominance. They peak with mean latencies of 9, 11 and 14 ms, respectively, in normal subjects and are labelled P9, P11 and P14 according to the polarity–latency nomenclature. There are some interindividual variations in the waveform of the last of these potentials, which can contain two subcomponents labelled P13 and P14, respectively.

The P9 potential can be recorded both on the neck and on the scalp. It reflects the afferent volley in the trunks of the brachial plexus traversing the supraclavicular fossa (Cracco and Cracco 1976, Nakanishi et al 1981, Desmedt et al 1983). The peak latency of P9 varies with arm length.

The P11 potential begins in synchrony with the cervical N11 potential at the Cv6 level (Desmedt and Cheron 1980a) and is considered to reflect the ascending volley in the dorsal columns. P11 cannot be identified in about 20% of normal controls, but in patients with lesions of the medulla oblongata or at the cervicomedullary junction its persistence is a reliable indicator of intact dorsal column function.

The P14 far-field potential is consistently recorded in normal subjects (see Restuccia 2000, for a recent review), but with some degree of interindividual variability in its shape, due to the presence in most subjects of a superimposed positivity around 13 ms, usually labelled as P13 (*Fig. 4.29*). P14, but not P13, always peaks later than the cervical near-field N13 potential (Mauguière 1987). The P13–P14 complex is the only scalp far-field SEP to be reduced when interfering stimulation, such as vibration, is applied to the hand on the stimulated side (Ibañez et al 1989b). This suggests that it is generated after the synaptic relay in the nucleus cuneatus, whereas P9 and P11 are produced by the first-order neurones of the dorsal column pathways). Studies in patients indicate that P14 originates above the level of the foramen magnum (Mauguière et al 1983b, Mauguière and Ibañez 1985, Buchner et al 1986, 1988) but below the level of the thalamus (Nakanishi et al 1978, Anziska and Cracco 1980, Mauguière et al 1982).

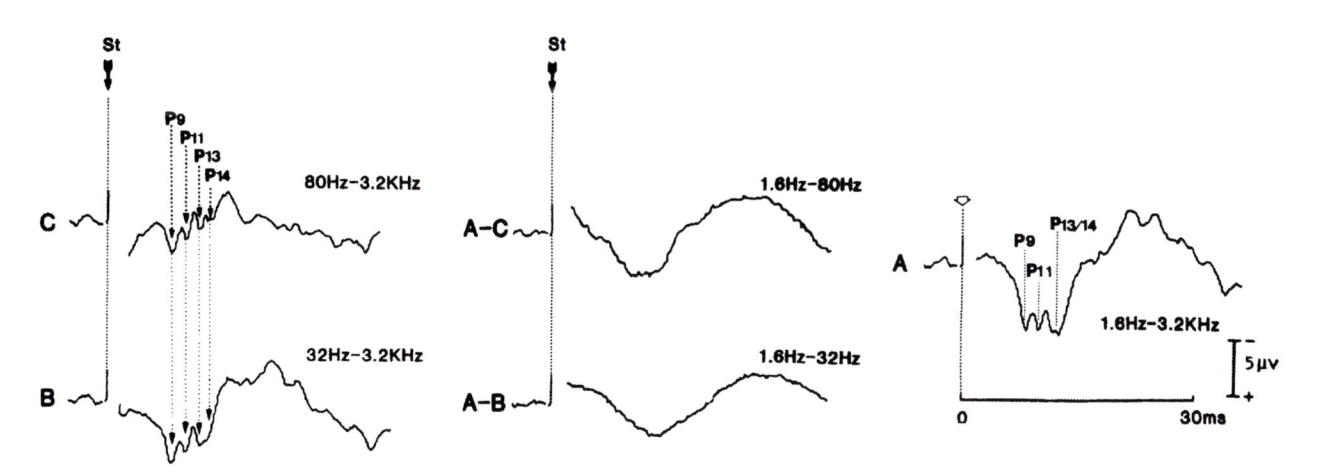

Fig. 4.29 Changes in the amplitude of scalp far-field SEPs related to the filter bandpass. Like BAEPs, the SEP far-field potentials contain a set of fast-frequency peaks superimposed upon a 'slow wave'. Cutting the lower frequencies causes a reduction in scalp far-fields, which is more pronounced for P14 than for the other components of the response. (By courtesy of L García-Larrea, from Binnie et al (2004), by permission.)

The most probable generators of the P13–P14 potentials are:
1. the ascending volley in upper cervical dorsal column fibres at the cervicomedullary junction or in medial lemniscus fibres in the brainstem;
2. the postsynaptic response of nucleus cuneatus neurones; and
3. junctional stationary potentials related with the moving action potential volley across the border between neck and posterior fossa (Kimura et al 1983, 1984).

In patients with lesions in the pons or mesencephalon P13 is sometimes preserved when P14 is lost, suggesting that P13 and P14 may be produced by separate generators at different levels (Nakanishi et al 1983, Delestre et al 1986, Kaji and Sumner 1987, Mavroudakis et al 1993, Restuccia et al 1995, Wagner 1991, 1996). However, the precise level at which P14 is generated remains uncertain because it is usually preserved in patients with lesions of the pons, in whom cortical responses are absent or reduced (Convers et al 1989).

The early cortical components Two sets of early cortical potentials, each made up of two components, are superimposed on the widespread N18 (see *Figs 4.27* and *4.30*). The first is an N20–P27 complex located in the parietal region contralateral to the stimulation; the second is composed of the P22 and N30 potentials which are recorded in the contralateral, central and frontal regions, although the N30 potential often spreads to the frontal region ipsilateral to stimulation. These components are also present after finger stimulation and their spatial distribution on the scalp is not distorted in ear lobe reference recordings (see *Fig. 4.21*).

The N20–P20 and P22 potentials One of the most debated issues concerning identification of these cortical components was to determine whether a single tangential dipolar source could account for both parietal N20 and central P22, or whether each of these two components has its own generator. Early investigators favoured the former hypothesis and proposed that a dipolar generator situated in the posterior bank of the Rolandic fissure (area 3b) could be responsible for this negative/positive configuration (Broughton 1969, Goff et al 1977, Allison et al 1980, Allison 1982). However, the mean onset and peak latencies of P22 recorded in the central and frontal region (i.e. anterior to the projection of the central sulcus on the scalp) exceed by 1–2 ms those of the parietal N20 potential (Desmedt and Cheron 1980a, 1981, Papakostopoulos and Crow 1980, Mauguière et al 1987) (see *Fig. 4.3*). This latency shift has been taken as an argument for labelling the centrofrontal positivity as 'P22', and for considering it to be a genuine component with a separate generator distinct from that of N20 (Desmedt and Cheron 1981).

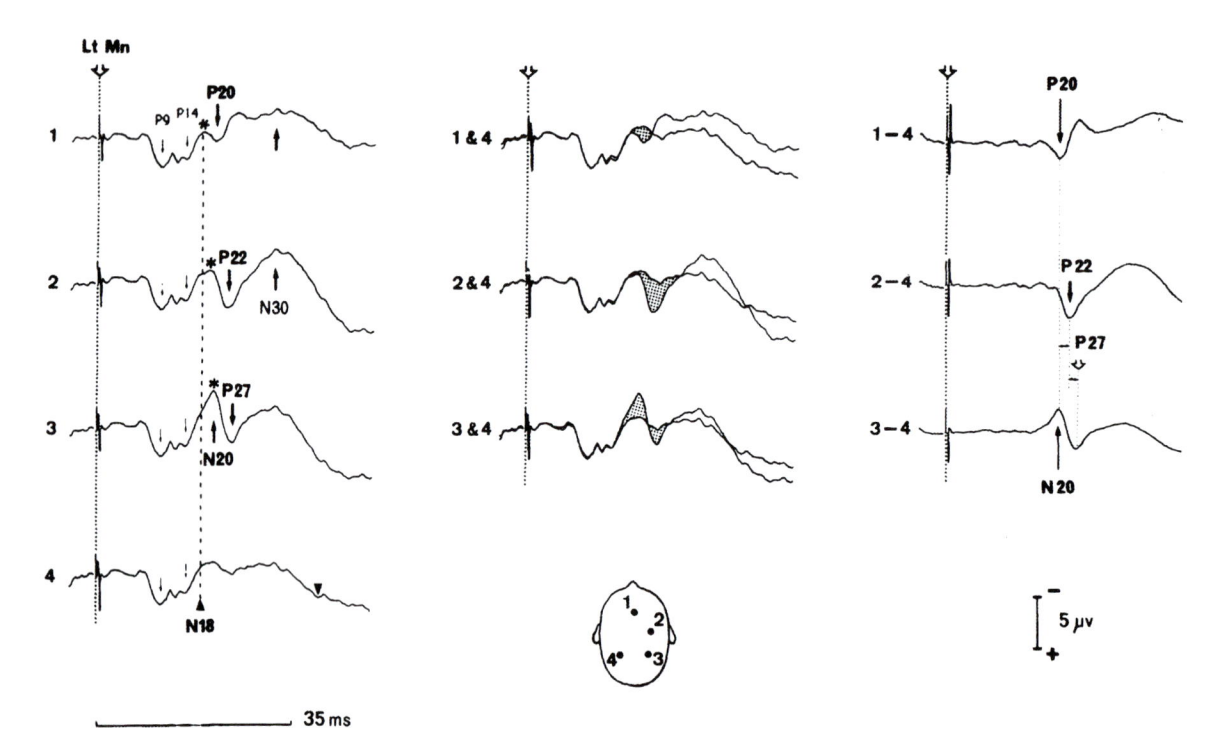

Fig. 4.30 Identification of early cortical SEP components (left median nerve stimulation; non-cephalic reference recordings). All the SEPs recorded in the parietal region ipsilateral to median nerve stimulation have a subcortical origin since they persist long after hemispherectomy. The cortical SEPs can be identified either by superimposing the ipsilateral parietal traces on contralateral responses (middle traces) or by subtraction of the former from the latter (right traces). The negativities (asterisks) recorded in the contralateral central (2) and midfrontal (1) regions completely disappear in the difference waveforms. Due to the filter setting of 32–3200 Hz used here, the amplitude of the P14 potential is less than that of the P9 (see trace B in *Fig. 4.29*). (From Mauguière et al (1987), by permission.)

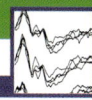

Two-dimensional maps with high time resolution show that there is a frontal positivity (P20) at the same latency as the parietal N20, and in the central region contralateral to the stimulus, a positive field peaking 1.0–2.5 ms later than the N20–P20, with little or no negative counterpart on the scalp (see *Fig. 4.21*) (Desmedt and Cheron 1980b, 1981, Papakostopoulos and Crow 1980, Deiber et al 1986). N20–P20 is produced by a dipolar generator in Brodmann's area 3b of the primary somatosensory cortex (area SI), tangential to the scalp surface and situated in the posterior bank of the Rolandic fissure (Broughton 1969, Goff et al 1977, Allison et al 1980). The spatial distribution of the central P22 positivity suggests that the source is close to the surface of the scalp and radial rather than tangential.

The parietal P24 and P27 potentials These potentials are recorded in the parietal region contralateral to stimulation, their peaking latencies show large interindividual variations between 24 and 27 ms. In some subjects two distinct P24 and P27 can be identified while only one of the two peaks is observed in others. This explains why, according to the polarity–latency nomenclature, the first parietal positive potential following N20 has received various labels in literature (P24, P25 or P27). These variations reflect the fact that the activities of several parietal sources overlap in time in this latency range. The P27 potential often culminates more centrally than N20 and was found to be abnormal in patients with focal lesions of the parietal cortex, presenting with astereognosis and normal N30 frontal responses (Mauguière et al 1982, 1983a). A P27 source in the primary somatosensory area (Brodmann's area 1), radial to scalp surface, would accord with these findings. The P24 potential on its own is associated with a frontal N24 potential. The dipolar source of the P24–N24 field is tangential to the scalp surface, and thus presumed to be orientated perpendicular to the Rolandic fissure. There are some arguments favouring a common source for N20–P20 and P24–N24 fields. Moreover, there is no report of a lesion affecting the P24 potential with preserved N20 and P27 potentials.

The frontal N24–N30 complex The frontal potential, labelled as 'N30' in most clinical studies, is picked up in the frontal region contralateral to stimulation; however, it often spreads to the midfrontal region and to the frontal region ipsilateral to stimulus. Its waveform shows two distinct components of which the earlier one, peaking at about 24 ms (N24), appears as notch on the ascending slope of the later one, which peaks at 30 ms (N30) and has the larger voltage of the two in normal young adults.

There is presently some consensus on the conclusions that the frontal negativity with its two N24 and N30 peaks cannot be generated by a single source (Delberghe et al 1990, García-Larrea et al 1992, Ozaki et al 1996a), and that the early part (N24) of the frontal negativity corresponds to the polarity reversal of the parietal P24 potential across the central sulcus. The latter statement is supported by several observations arising from intracranial (Allison et al 1989) and scalp (García-Larrea et al 1992, Ozaki et al 1996a) recordings, and the different effects on P24 and P30 of changes in stimulus rate (Delberghe et al 1990, García-Larrea et al 1992, Valeriani et al 1998). Based on these findings it has been proposed that the N30 and P27 potentials originate in the precentral cortex and in the primary sensory (SI) area (Brodmann's area 1), respectively; each of these two dipolar sources being orientated perpendicular to the scalp surface.

Some observations (for a review see Cheron et al (2000)) are compatible with the view that the N30 potential could be related with motor programming and generated in the premotor cortex of area 6, and more precisely in the supplementary motor area (SMA).

Short latency SEPs to lower limb stimulation
Electrical stimulation of the tibial nerve at the ankle is adopted by most authors for the testing of the sensory pathways of the lower limb. However, electrical stimulation can also be applied to the sural nerve at the ankle, or to the peroneal nerve at the knee, without major changes in the general waveform of the spinal or scalp responses. Only the peak latencies will be modified, being lengthened by about 3 ms, or shortened by about 5–6 ms, respectively, for sural and peroneal nerve stimulation, compared with those after tibial nerve stimulation. In the following account, tibial nerve SEPs will be taken as the example for describing the normal waveforms.

Peripheral components A CAP corresponding to the activation of tibial nerve fibres running through the popliteal fossa may be recorded at the posterior aspect of the knee using a bipolar montage. The negative peak of this near-field CAP, peaks at a latency of about 7 ms. This N7 potential reflects a mixed response of motor and sensory fibres, which may be clinically useful to assess the integrity of the peripheral segment of the pathway. The afferent volley in cauda equina roots can be recorded using skin electrodes placed at the L5–S1 level with a distant reference site, for instance at the knee opposite to the stimulation.

Spinal potentials An electrode situated on the spinal process of T12 or L1 vertebrae and referred to a more distal electrode records a negative potential peaking at between 21 and 24 ms in normal subjects (*Fig. 4.31*). There is no consensus as to the name that should be given to the negative potential recorded in the lumbar region. Following the polarity–latency nomenclature, N20 (Tsuji et al 1984), N21 (Small and Matthews 1984), N22 (Lastimosa et al 1982, Riffel et al 1984), N23 (Yamada et al 1982) and N24 (Desmedt and Cheron 1983) have been proposed. Differences in mean latencies between authors are due mainly to differences in filter settings and in mean body height of the subjects sampled for normative studies.

The lumbar negativity originates in the spinal segment receiving fibres from the S1 root (Delbeke et al 1978,

Jones and Small 1978, Dimitrijevic et al 1978, Lastimosa et al 1982, Desmedt and Cheron 1983, Small and Matthews 1984, Tsuji et al 1984, Delwaide et al 1985). As with the cervical N13 potential elicited by median nerve stimulation, Desmedt and Cheron (1983) demonstrated a polarity reversal of this negativity by means of prevertebral recordings. This finding is fully confirmed by data from direct recording from the cord during surgery, as illustrated in *Fig. 4.28*. The field distribution of this lumbar negativity is consistent with a horizontally orientated dipolar source in the cord, which probably corresponds to the postsynaptic responses of dorsal horn neurones to incoming inputs. This potential has a segmental distribution; its amplitude is maximal close to the entry zone of the S1 root in intraoperative recordings and decreases steeply without any latency shift at more rostral or caudal electrode sites. Both posterior negativity and anterior positivity show maximal amplitude at T12–L1 vertebral level, decreasing without any latency shift at more rostral or caudal electrode positions.

In some subjects a small negative peak is seen on the ascending limb of the lumbar negativity, which reflects activity in the dorsal columns, analogous to N11 recorded over the cervical spine following median nerve stimulation. However, this potential is not consistently present in recordings from surface electrodes, unlike N22–N24.

When recorded through electrodes applied directly to the cord, N22–N24 is preceded by a positive potential with peak latency at about 17 ms. This P17 positivity is a far-field potential originating in lumbosacral plexus trunks which can also be recorded occasionally by scalp electrodes. Its voltage may be more than 10 μV when it is recorded directly from the anterior aspect of the lumbosacral spinal cord (see *Fig. 4.28*), but is only about 1 μV on the skin surface. It is best seen if lead 2 is connected to the knee (or iliac crest) on the non-stimulated side (Yamada et al 1982, Desmedt and Cheron 1983). Invasive electrodes placed directly on the cord dorsum record a slow positivity following N22–N24, known as the 'P wave', which reverses into a negativity on the anterior aspect of the cord (Shimoji et al 1977, Jeanmonod et al 1989), and may reflect presynaptic inhibition affecting the primary afferent fibres in the dorsal horn.

Far-field scalp positivities With appropriate non-cephalic reference montages (see below), a widespread positivity is consistently recorded in normal subjects. The peak latency depends upon body height, and different authors have labelled it P27, P28, P30 or P31 (Yamada et al 1982, Desmedt and Cheron 1983, Kakigi et al 1982, Kakigi and Shibasaki 1983, Seyal et al 1983) (*Fig. 4.31*). The P30 potential is widely distributed on the scalp but predominates in the frontal region (Desmedt and Bourguet 1985, Guérit and Opsomer 1991), therefore, it is drastically reduced in scalp-reference recordings (Seyal et al 1983). P30 is easier to record if the scalp electrode is placed in the midfrontal region at Fz or Fpz, where it has its maximal

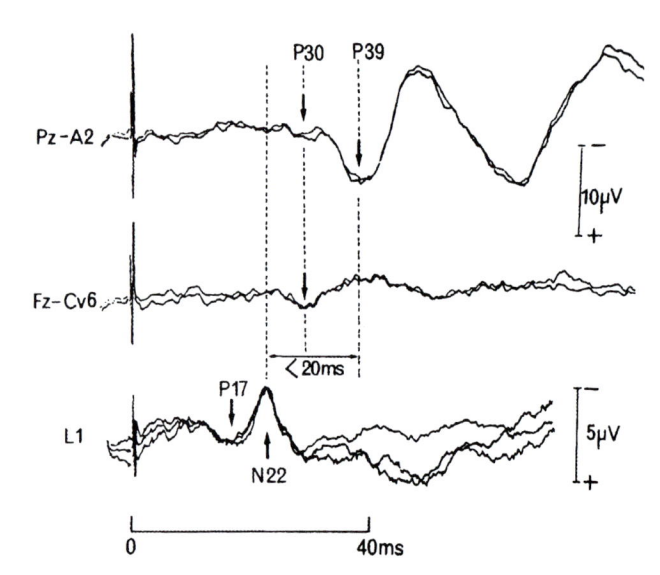

Fig. 4.31 Tibial nerve SEPs. The lumbar responses (L1) were recorded between the spinous process of L1 and the knee contralateral to stimulation. Three traces obtained by averaging successive runs of 1000 stimuli are superimposed. Note that the P30 component is recorded at the parietal (Pz) or frontal midline (Fz) electrodes; this widespread scalp distribution favours a deep subcortical generator for this component. Moreover, P30 is usually larger when lead 2 of the amplifier is connected to a cervical (Cv6) than to an ear lobe electrode (A2); it may be missing in traces recorded using the Pa–(A1–2) montage in normals. This suggests that part of the P30 positivity recorded with the Fz–Cv6 montage reflects a small negative potential picked up at the cervical level. Thus P30 probably reflects the action potential volley ascending in cervical dorsal columns. (From Binnie et al (2004), by permission.)

amplitude with minimal contamination by subsequent cortical potentials. If the electrodes are too posteriorly situated, P30 may be lost in the descending slope of the cortical P39 potential. The ideal reference site for recording P30 is the knee on the non-stimulated side. This site is caudal to the P30 generator and distant from the axis of propagation of the ascending volley. However, due to the technical recording difficulties with such a montage, the most practical reference site for recording of P30 is the spinous process of the 6th cervical vertebra (Seyal et al 1983, Tinazzi and Mauguière 1995). P30 is thought to be generated in the brainstem (Yamada et al 1982, Desmedt and Cheron 1983) and can be viewed as the homologue, for the lower limb, of the far-field P14 recorded on the scalp after median nerve stimulation. The utility of the P30 potential has been validated for clinical applications of tibial SEPs in patients with spinal cord and brainstem lesions (Tinazzi and Mauguière 1995, Tinazzi et al 1996b), and is not appreciably affected by anaesthesia, so it has been used for monitoring spinal cord function intraoperatively (see *Spinal cord function monitoring*, p. 268).

Cortical potentials

The P39 potential The first cortical potential elicited by stimulation of the tibial nerve at the ankle is a positive component, with peak latency 37–40 ms, variously labelled

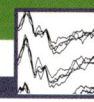

P37, P39 or P40 (see Yamada (2000) for a recent review) (see *Fig. 4.31*). The latency depends on the subject's height and is 6–7 ms shorter after stimulation of the tibial nerve at the popliteal fossa instead of the ankle, and about 3 ms longer after stimulation of the sural nerve at the ankle. A similar response can be obtained by stimulating the peroneal nerve at the knee, but the amplitude is lower and shows greater interindividual variability than responses to either tibial or sural nerve stimulation. In all normals, P39 is present at the vertex and can most often be reliably obtained midway between Pz and Cz using a scalp to ear lobe montage. Frequently, however, the maximum of P39 to tibial nerve stimulation, although close to the midline, is slightly shifted to the side of the scalp *ipsilateral* to the stimulus (Cruse et al 1982); this paradoxical lateralization is only seen when nerves are stimulated distally, and is not apparent when stimulation is proximal, for example the femoral or lateral femoral cutaneous nerves (Wang et al 1989, Yamada et al 1996). The reason for the paradoxical lateralization of P39 is that the somatotopic representation of the distal lower limb in the somatosensory SI, and in particular the foot area, is situated on the medial aspect of the hemisphere. SEP mapping in normals shows a dipolar field distribution for P39, with a maximum of positivity close to the vertex, but often shifted on the scalp towards the side of stimulation, and a maximum of negativity (N39) in the contralateral parietal region (Cruse et al 1982, Seyal et al 1983, Desmedt and Bourguet 1985, Kakigi and Jones 1986, Tinazzi et al 1996a). Dipole fields reflecting the activity of cortical cell layers are expected to be perpendicular to the cortical surface, but the orientation with respect to the scalp itself varies with the folding of the brain surface. The negative N39 maximum of this dipole is not constantly obtained, due probably to intersubject differences in the orientation of the grey matter of the leg area in SI with respect to the scalp surface. Moreover, there are some interindividual variations of the anatomy of the foot and leg representations in SI, so that one can envisage a range of conditions intermediate between a vertical radial dipole with a positive maximum at the vertex, and a nearly tangential one with P39 and N39 culminating, respectively, in the ipsilateral and contralateral parietal regions (MacDonald 2001, Miura et al 2003, MacDonald et al 2004).

Despite these topographical variations, one never misses P39 in normals when recording between Pz and the ear lobe ipsilateral to the stimulus (Tinazzi et al 1997a). In a few cases the P30 far-field potential (see above) is obtained with this derivation because its amplitude is higher on the scalp than at the ear. An easy way of differentiating P30 and P39 is to use a Pz–Fz derivation which cancels the widespread P30 far field but not the cortical P39. An abnormal P39 may be poorly delineated at Pz, particularly when it is dispersed in time because of demyelination, and the second vertex positivity, which culminates at a latency of about 60 ms (see *Fig. 4.31*), can be confused with a delayed P39. In such cases, topographical studies may be useful, not merely to confirm that SEPs are abnormal, since this conclusion can be reached with a simple montage, but better to analyse the abnormal pattern.

The N50 and P60 potentials The 'W' profile of the cortical response recorded on the vertex and in the centroparietal region is made of three sequential responses P39, N50 and P60. This waveform has been reported in all studies of tibial nerve SEPs (see Tinazzi et al 1997a, for a recent review). Electrical stimulation of the L5 and SI dermatomes also evokes a consistent, W-shaped response on the scalp (Katifi and Sedgwick 1986). The N50 and P60 potentials culminate at the vertex (Cz) and do not show the same clear dipolar distribution as the P39–N39 response on the scalp, thus suggesting that the W shaped P39–N50–P60 waveform could be generated by several sources with distinct orientations (Desmedt and Bourguet 1985, Guérit and Opsomer 1991). However, there is no clinical study demonstrating that N50 or P60 can be selectively affected by a single focal lesion, and these two potentials show the same attenuation as P39 during active movement (Tinazzi et al 1997b).

Central conduction time in the dorsal column system

One of the advantages of SEP recording in clinical routine is to permit an evaluation of the transit time of the ascending volley in the central segments of the somatosensory pathways. In most studies this transit time is referred to as the 'central conduction time' (CCT).

Upper limb CCT

Critical review of the different methods A distinction should be made between the different techniques for measuring the CCT according to whether they aim at detecting slowing of conduction in patients or at locating accurately the site where conduction velocity is reduced. The techniques which provide the investigator with an index of abnormality are considered to provide enough information in most clinical situations; among these, the measurement of the interpeak interval between the cervical N13 and the parietal N20 components is the most widely used. Nevertheless, this technique is open to criticism in that it evaluates the CCT from the peak of the postsynaptic dorsal horn potential (N13), which is produced in parallel to the ascending volley in the dorsal columns. Thus, a normal conduction in the dorsal columns may coexist with a clearly abnormal, or even absent, N13 potential. Despite this major drawback, the N13–N20 interval became synonymous with CCT, probably because these two peaks are easily obtainable, and their latencies considered reliable – a view which cannot be endorsed without some reservations.

The different methods which have been proposed for measuring the CCT can be classified according to the recording procedure, and in particular according to the site of the reference electrode.

Frontal reference recordings In midfrontal (Fz) reference montages the CCT is usually measured from the peak of the cervical negativity recorded at Cv6 to the peak of the contralateral parietal N20 (Hume and Cant 1978, Eisen and Odusote 1980). However, the use of a frontal reference introduces significant uncertainty as to the identification of the maximal negative component which can be either the spinal N13 or an 'N14' negativity that is the P14 picked up by the frontal reference (*Figs 4.32* and *4.33*). Moreover, in these recording conditions the frontocentral P22 is combined with the parietal N20 and can modify its peak latency. Thus the so-called 'CCT' can reflect either the delay between the peaks of the spinal N13 or the brainstem P14 on the one hand and the peaks of the parietal N20 or the centrofrontal P22 on the other. Such uncertainties can be accepted when the CCT is calculated for monitoring the somatosensory pathways in a given individual since, in these circumstances, SEP waveforms are usually stable enough to ensure that one is measuring the latencies of the same components, whatever they may be, at the neck and on the scalp, on each occasion. Interindividual variations of waveform resulting from interference between potentials picked up at the neck and on the scalp may be problematic when sampling normative data, but these difficulties increase when analysing distorted waveforms in patients.

The delay between the onset of the spinal N11 and that of the scalp N20 reflects more reliably the total transit time from spinal entry to contralateral parietal cortex. The onset of N11 corresponds to the spinal entry time of the afferent volley and the onset of N20 to the arrival time of the fastest action potentials at the parietal cortex. Since the frontocentral P22 begins later than the onset latency of the N20–P20, the onset of the N20 recorded with a parietal-to-frontal derivation corresponds to the true onset of the parietal response. The only major drawback of this method is that the N9 potential, which is a combination of the P9 potentials picked up, respectively, at the neck and at Fz, may prevent reliable measurement of the onset latency of N11.

Non-cephalic and ear lobe reference recordings Using non-cephalic and ear lobe reference sites, three transit times can be calculated by measuring interpeak intervals. The interval between the supraclavicular N9 (or far-field P9) and the spinal N13 may be helpful in assessing conduction in the proximal segment of brachial plexus roots. Conduction from the brachial plexus roots to the lower brainstem at the cervicomedullary junction, and from the brainstem to the parietal cortex, can be explored, respectively, by calculating the P9–P14 and the P14–N20 intervals.

In theory, the interval between the onsets of N11 and of N20 would be the only means of assessing the transit time from spinal entry to the onset of the cortical response (Desmedt and Cheron 1980a). In practical terms, this method is limited, even when using a non-cephalic reference site, by the difficulty in identifying an onset latency for N11. This component immediately follows the P9 positivity and its onset can be evaluated only by reference to a baseline. This limitation does not hold for N20 onset which corresponds to the time when the superimposed ipsilateral and contralateral parietal traces are diverging, the N18 potential being used as the baseline to measure onset latencies as well as amplitudes of all cortical components peaking before 35 ms. Ozaki et al (1996b) have recently shown that the onset CCT is the only procedure which is able to show that the ulnar nerve CCT is longer than the median nerve CCT, and that the CCT correlates with body height in normal subjects.

Practical recommendations For measuring the 'conventional' N13–N20 CCT, the Cv6-AC (anterior cervical) and contralateral parietal to ear lobe derivations are recommended since they permit the peak latencies of the dorsal horn N13 and contralateral parietal N20 to be determined with more accuracy than by midfrontal Fz reference recordings. By adding to this two-channel montage an Erb's point derivation for recording the N9 potential, it becomes possible to measure the three interpeak intervals, namely N9–N13, N9–P14 and P14–N20, which explore conduction, respectively, in proximal dorsal roots up to the dorsal horn of the spinal cord (N9–N13), in proximal dorsal roots and dorsal columns up to the cervicomedullary junction (N9–P14), and in the intracranial segment (P14–N20). These conduction times should be measured routinely in patients.

Lower limb CCT By analogy with median nerve SEPs, it was originally proposed that the transit time from cervical cord to cortex be evaluated by measuring the interval between the negative peak recorded at the neck with a cervicofrontal derivation and the peak of the P39 potential at the vertex (Lüders et al 1981, Small and Matthews, 1984). The cervical negativity evoked by tibial nerve stimulation in these recording conditions is represented by the P30 potential picked up by the frontal reference electrode. It must be noted here that this scalp far-field P30 potential has a brainstem origin and does not reflect the ascending volley in the dorsal columns of the cervical cord (see above).

Measurement of conduction velocity of the propagated volley along the cord by recording tibial SEPs at different levels of the spine proves to be relatively unsatisfactory. It is indeed possible to record a dorsal column CAP, the latency of which increases from caudal to rostral levels of the spine as reported by Yamada et al (1982), Kakigi et al (1982), Seyal et al (1983) and Desmedt and Cheron (1983). However, these reports mostly concern young normal volunteers, in whom very small potentials can be recorded up to the cervical level by averaging responses to 2000–6000 stimulations. The amplitude of the spinal CAP drops dramatically (by more than 60%) from T12 to T6 (Delbeke et al 1978, Seyal et al 1983); moreover, cervical responses recorded with skin electrodes are small (Sherwood 1981, Desmedt and Cheron 1983, Tsuji et al 1984) due to the

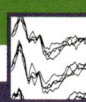

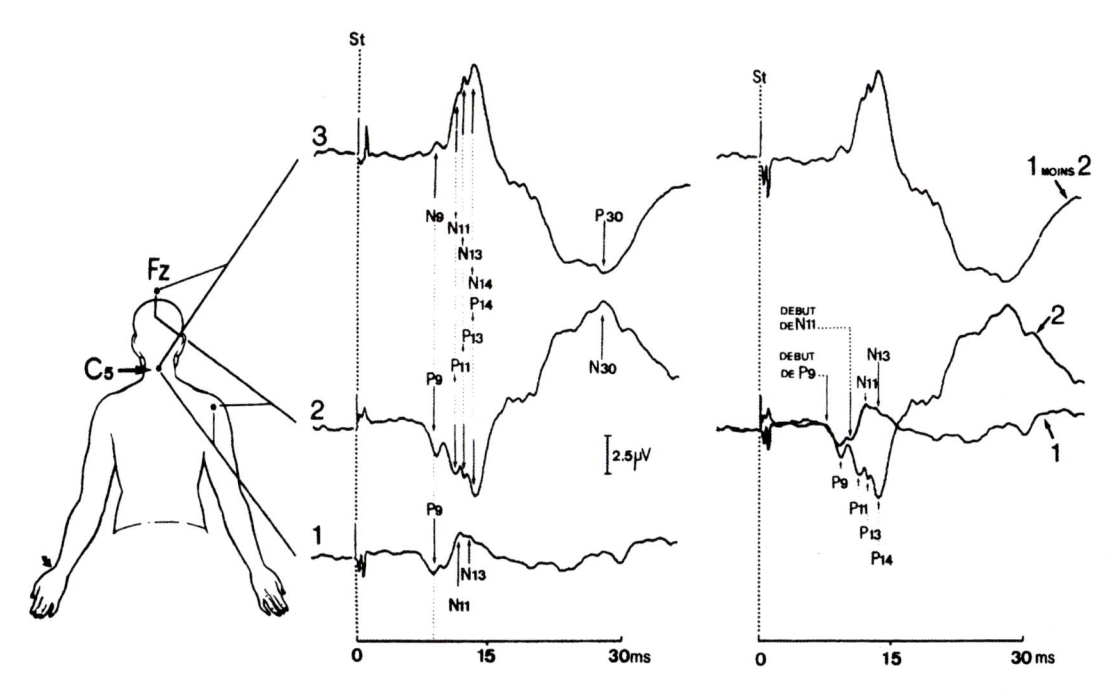

Evoked potential waveforms and interpretation

Fig. 4.32 The cervical to frontal response to median nerve stimulation. Most of the activity recorded with the cervical to frontal derivation (trace 3) is picked up at the scalp and consists of the succession of P9 to P14 far-field positive waves followed by the frontal cortical N30 component (trace 2). Note that the contribution of the N13 spinal potential (trace 1) to the resulting cervical Fz waveform is so small that the disappearance of N13 in patients with intramedullary cervical lesions will not cause any significant change in the cervical Fz response if the dorsal columns are unaffected, with normal scalp far-field potentials. (From Mauguière F 1989 Evoked potentials in nondemyelinating diseases. In: *Advanced Evoked Potentials* (ed H Lüders), Kluwer, Boston, MA, pp. 181–221, © Kluwer Academic Publishers, with kind permission of Springer Science and Business Media.)

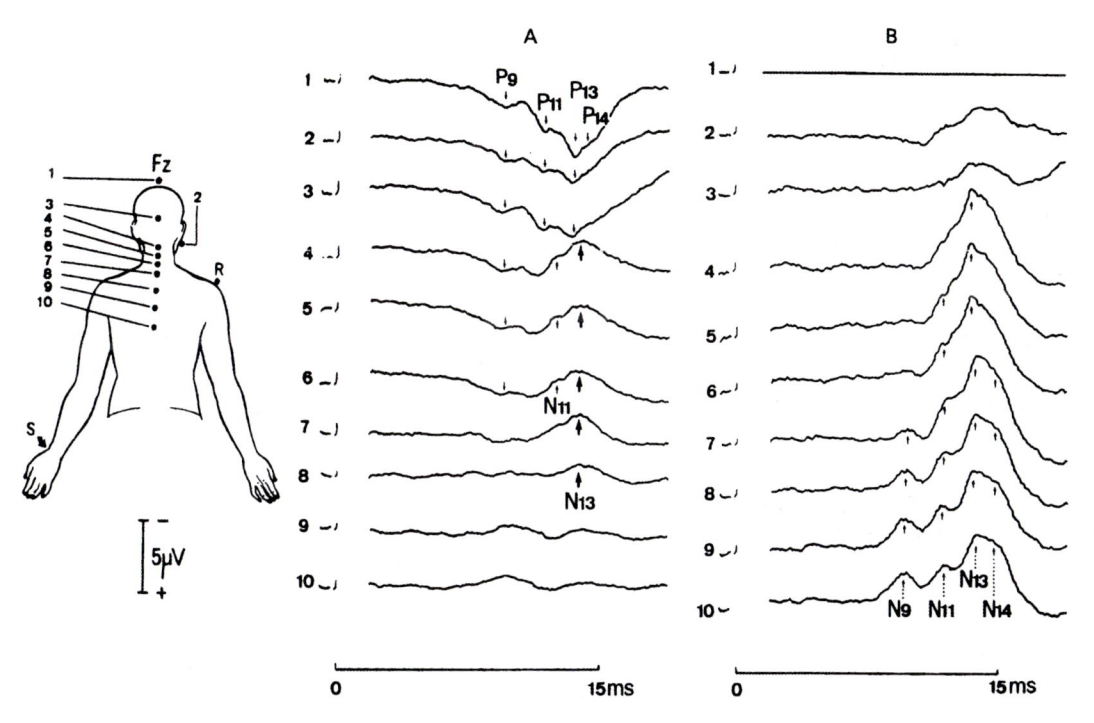

Fig. 4.33 Topography of cervical response in cervical to frontal derivation. (A) In non-cephalic reference recordings, N13 is recorded only at the lower neck; the frontal electrode picks up the far-field positive potentials P9–P14. (B) In frontal reference recordings, the scalp positive potentials recorded at Fz are injected in all traces, including those recorded at midthoracic level (electrodes 9 and 10). (From Binnie et al (2004), by permission.)

time-dispersion of the ascending volley at the cervical level, as evidenced by epidural (Macon and Poletti 1982, Macon et al 1982, Whittle et al 1986) or direct recordings from the cord (Turano et al 1995).

In most clinical applications of tibial SEPs, the CCT from lumbosacral spinal cord to cortex is evaluated by the interval between the peak of the lumbar N22 negativity on the one hand and the peak of the cortical P39 potential on the other (Small and Matthews 1984). It can be argued, as for median nerve SEPs, that the N22 potential reflects a postsynaptic dorsal horn response, which can be abnormal in patients with lesions of the lumbosacral spinal cord and normal conduction in the dorsal columns. The interval between the onsets of the lumbar negativity and of the scalp P39 would be more accurate than the interval between peaks, since the onset of the ascending slope of the lumbar negativity reflects the spinal entry time. However, because of the presence of the far-field PI7 in the lumbar region, this measurement is often problematic and only the N22–P39 interpeak interval has been validated in clinical studies to evaluate the transit time from lumbosacral cord up to the cortex, with an upper limit for normal values of about 20 ms in adults. Several attempts have been made to evaluate separately the intraspinal and intracranial CCT using lower limb SEPs. It has been shown that measuring the N22–P30 and P30–P39 interpeak intervals provides a reliable evaluation of the intraspinal and intracranial CCTs in normal subjects and patients with focal cervical cord, brainstem and hemispheric lesions (Tinazzi and Mauguière 1995, Tinazzi et al 1996b).

Maturation

The process of somatosensory development from birth to adult life is dominated by two coexisting phenomena with opposite effects on SEPs. Myelinogenesis, increasing fibre diameter and synaptic modifications cause a progressive increase in conduction velocities and synchronization of potentials and thus a latency decrease with age, whereas body growth has the opposite effect, namely latency increase and desynchronization. Interaction of these two variables explains the complex patterns of SEP development in children.

Differences in sites of stimulation, placement of the electrodes and filter settings make it difficult to compare the latency and amplitude values reported in the available studies of SEP maturation. During the first 4–5 years of life, SEP maturation is marked by a progressive synchronization and a latency reduction of all potentials (Desmedt et al 1976, Hashimoto et al 1983, Cadilhac et al 1985, Tomita et al 1986, Bartel et al 1987, Zhu et al 1987, Laureau et al 1988). Conduction velocities reach adult values before the age of 3 years in the peripheral nervous system (Thomas and Lambert 1960, Desmedt et al 1973), but this acceleration is slower in the central somatosensory pathways (Desmedt et al 1973, 1976, Cracco et al 1979). Spinal cord conduction in lower limb fibres reaches adult values at the age of 5–6 years (Cracco et al 1979). Later, changes in conduction velocity related to fibre maturation begin to interact with those of body growth, and the peak latencies progressively increase to adult values which are reached at the age of 15–17 years (Allison et al 1983, Tomita et al 1986). To sample normative data in children, it is useful to correct absolute latencies by an index according to the length of the relevant neural pathways, namely arm length or body height. After this correction, the latencies of all central SEPs decrease from birth to the age of 10 years and stabilize thereafter (Tomita et al 1986); they also showed that absolute values for CCTs can be corrected for head circumference, which correlates with other indices of growth (Bartel et al 1987). After this correction, the P14–N20 interval decreases progressively from 1 month to reach adult values at the age of 15 years. This decrease of CCT with age is not steady, but rapid between birth and 6 years and slower thereafter.

The rise time between onsets and peaks of the cortical N20 and P39 potentials, evoked respectively by median and tibial nerve stimulation, decreases steadily from birth to the age of 16 years (Zhu et al 1987). There are also some morphological changes in the waveforms which are age dependent. For instance, in children aged 7–14 years, the voltage of the cervical N13 median nerve SEP recorded with an anterior cervical electrode tends to be higher than in adults, whereas the P14 component tends to be lower with a greater P9/P14 amplitude ratio. It is essential that any laboratory involved in paediatric neurophysiology should create its own database and set up normograms. Up to now there are only two such published normograms for SEPs recorded with a non-cephalic reference montage (Tomita et al 1986, Lafrenière et al 1990), whereas several studies have been published for SEPs recorded with a frontal reference montage.

Non-pathological sources of SEP variation

Body height and arm length in adults The absolute latencies of SEPs obviously vary according to the distance between the stimulus site and the SEP generators. This effect is naturally more pronounced for lower than for upper limb SEPs. The variability related to the length of the stimulated limb or to the body height is less for interpeak than for absolute latencies. In establishing norms for conduction times based on interpeak latencies for upper limb SEPs in adults, this interindividual variability is small enough to be disregarded. This is not true for lower limb SEPs, in particular for the N22–P39 interval, the upper normal values of which range from 18 to 20 ms for body heights of 1.50 and 1.90 m, respectively (see Chiappa (1990) for a review). Moreover, absolute latencies of N22, P30 and P39 potentials evoked by lower limb stimulation correlate with body height (Chiappa 1990, Tinazzi and Mauguière 1995).

Skin and core temperature Peripheral nerve conduction velocities are affected by changes in limb temperature. Marked changes in body temperature, such as those

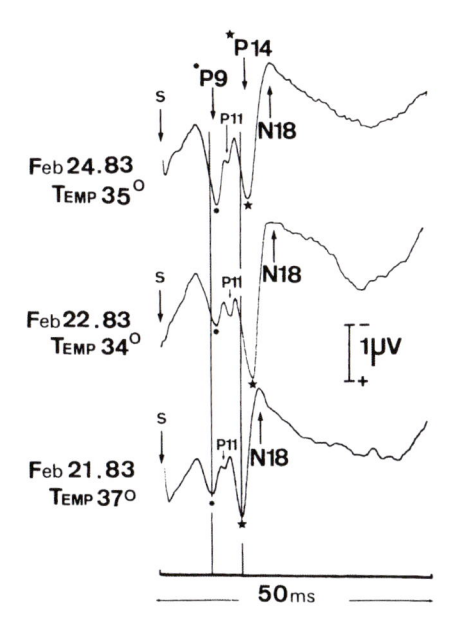

Fig. 4.34 Effect of temperature on scalp SEP far-fields. This set of traces obtained in a deeply comatose patient shows that absolute latencies of P9 and P14 as well as the P9–P14 interval fluctuate with core temperature changes. In this patient, cortical SEPs were absent, and only the P9, P14 and N18 were recorded on the scalp. (From Binnie et al (2004), by permission.)

observed during drug-induced hypothermia, can significantly affect the latencies of SEPs. Both the early cortical N20 and the far-field positive potentials have prolonged absolute and interpeak latencies during hypothermia (*Fig. 4.34*). Moreover, the curves of amplitude decrease versus body temperature during hypothermia are not the same for peripheral, spinal brainstem and cortical SEPs (Guérit et al 1990). The N20 potential disappears at temperatures ranging from 17 to 25°C, the P14 at 17 to 20°C. The peripheral N9 and spinal N13 potentials remain identifiable down to body temperature of 17°C. SEP changes related with hypothermia deserve special attention in comatose patients in whom they can combine with those induced by CNS depressant drugs.

EMG FINDINGS AND INTERPRETATION

FEATURES OF MOTOR UNITS RECORDED BY NEEDLE ELECTRODES

Electromyographic activity can be recorded distantly using surface electrodes, but for better resolution of the signal an intramuscular electrode is necessary. To facilitate introduction of recording wires into muscle tissue, electrodes have been constructed with sharpened tips. Several types of needle electrode have been developed with differing recording characteristics (see *Figs 2.28* and *2.29*). What is recorded depends upon the size of the non-insulated exposed recording surface or 'lead-off' area of the electrode.

The electrode most commonly used in electromyography is the concentric needle electrode (CNE), but for monitoring spontaneous or evoked activity either monopolar needle electrodes or a fine wire electrode are likely to be used. The recording surface of an intramuscular electrode is usually only large enough to pick up spike activity from about 20 muscle fibres lying within a hemisphere of about 0.5 mm radius. More distant fibres contribute only small blunted potentials. Surface electrodes, on the other hand, applied over the belly of the muscle and its tendon, record summated activity from a large number of muscle fibres.

For diagnostic work in conscious and cooperative patients, the needle is inserted in a weakly contracted muscle, and a few motor units are recorded. Amplitude and duration of these individual motor units can then be measured. The pattern of electrical activity during maximal voluntary contraction, known as the interference pattern, is also assessed. The method of measurement, and the results to be expected, will not be described here because they are unlikely to be relevant to surgical or ICU monitoring, but they are covered in Binnie et al (2004) and in standard reference works such as Kimura (2001). The American Association of Electrodiagnostic Medicine (AAEM) and the International Federation have prepared glossaries of terms which may be useful (AAEM 2001, Caruso et al 1999). Features of normal EMG phenomena are summarized in *Table 4.4*.

OTHER NORMAL EMG PHENOMENA

Insertion activity

At complete rest, healthy muscle should be electrically silent or almost so. Any activity that can be recorded is of such low amplitude that a high recording sensitivity of 100 μV/division should be used on needle insertion.

On first piercing the muscle with a recording needle and at subsequent adjustments of the recording tip, there may be transient activity, so-called 'insertion activity'. This consists of bursts of activity, usually less than 500 μV in amplitude, which cease within less than a second of the movement (*Fig. 4.35*). It is due to mechanical stimulation or injury of muscle fibres.

End-plate noise

Other activity that can be recorded from healthy muscle is 'end-plate' activity. Because it is generated at the motor end-plate this activity is highly localized, disappearing with minute adjustments of the recording tip (Wiederholt, 1970). Recording it will therefore be a chance occurrence, though Buchthal and Rosenfalck (1966a) reported this finding in 20% of 200 muscles without further search.

There are two types of end-plate activity. One, 'end-plate noise', sounds like the noise heard from a sea shell

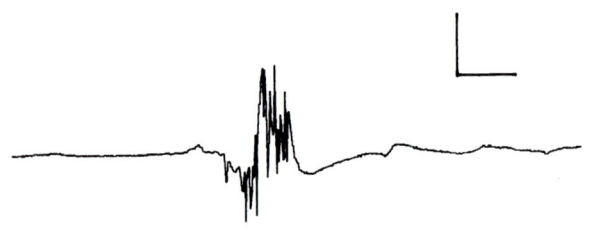

Chapter 4 Normal and Pathological Phenomena in EEG, Evoked Potential, EMG and Nerve Conduction Studies

Fig. 4.35 Insertion activity recorded as the needle enters healthy muscle. Vertical scale = 200 µV; horizontal = 10 ms. (From Binnie et al (2004), by permission.)

when held close to the ear. This type of activity, requiring a high sensitivity (100 µV/division) to be recorded, takes the form of low-amplitude, irregular disturbances of the baseline. Buchthal and Rosenfalck (1966a) suggested that this activity was miniature end-plate potential activity due to the random release of vesicles of acetylcholine at the neuromuscular junction (*Fig. 4.36*).

The other type of end-plate activity presents a significant clinical difficulty since it can easily be confused with fibrillation potentials. It is made up of biphasic potentials of moderate amplitude (100–200 µV) and short duration (2–4 ms) that fire in irregular short bursts (*Fig. 4.36*)

(Buchthal and Rosenfalck 1966a). The activity is lost by minimal repositioning of the needle tip.

Fibrillations at single sites
Although often regarded as the hallmark of denervation, fibrillations may occasionally be recorded at a single site in a normal muscle (Buchthal 1982). They are therefore included under the heading of normal phenomena to emphasize that they must be found at more than one site before they can be regarded as a significant abnormality.

Fasciculations
Fasciculations are involuntary contractions of the whole group of muscle fibres that form the motor unit. When motor units are big, this type of abnormal activity is visible through the skin as brief twitches of the muscle which can either be sporadic or quasi-rhythmic. Fasciculations can be recorded in a single muscle with a needle electrode as sporadic high amplitude potentials with a variable waveform, careful observation revealing a simultaneous movement of the needle. Alternatively, fasciculations can be recorded from several muscles using surface electrodes connected to a multichannel recorder, such as an EEG machine with good high frequency response. Fasciculations are commonly seen in anterior horn cell disease, but may

Table 4.4 EMG phenomenology in the normal muscle

Category	Definitions and key features	Figures	Key references
Insertional activity	At complete rest, healthy muscle should be electrically silent or almost so. On initial insertion of the electrode and adjustment of the recording tip there may bursts of 'insertional activity' usually less than 500 µV in amplitude, which cease within less than a second of the movement and due to mechanical stimulation or injury of muscle fibres	*Fig. 4.35*	
End-plate noise	Activity recorded from healthy muscle; generated at the motor end-plate, hence highly localized, disappearing with minute adjustments of the recording tip	*Fig. 4.36(b)*	Wiederholt (1970)
	The finding is a chance occurrence found in 20% of 200 muscles	*Fig. 4.36(c)*	Buchthal and Rosenfalck (1966a)
	Two types: 1. 'End-plate noise', sounds like the noise heard from a sea shell when held close to the ear; thought to be due to miniature end-plate potential activity due to the random release of vesicles of acetylcholine at the neuromuscular junction 2. Biphasic potentials of moderate amplitude (100–200 µV) and short duration (2–4 ms) that fire in irregular short bursts, lost on minimal repositioning of tip. Do not mistake for fibrillation potentials		
Fibrillations	Biphasic potentials, only recorded with needle electrodes, often regarded as the hallmark of denervation, but activity indistinguishable from fibrillations may be recorded at single sites from healthy muscle. Thus fibrillations only indicate significant abnormality if demonstrated at more than one sampling site in the muscle under study		Buchthal (1982)
Fasciculations	High amplitude, variable waveform, recorded with needle and surface electrodes; may occur in normal muscle, and common after administration of suxamethonium chloride (succinylcholine)		Desai and Swash (1997)

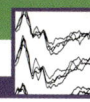

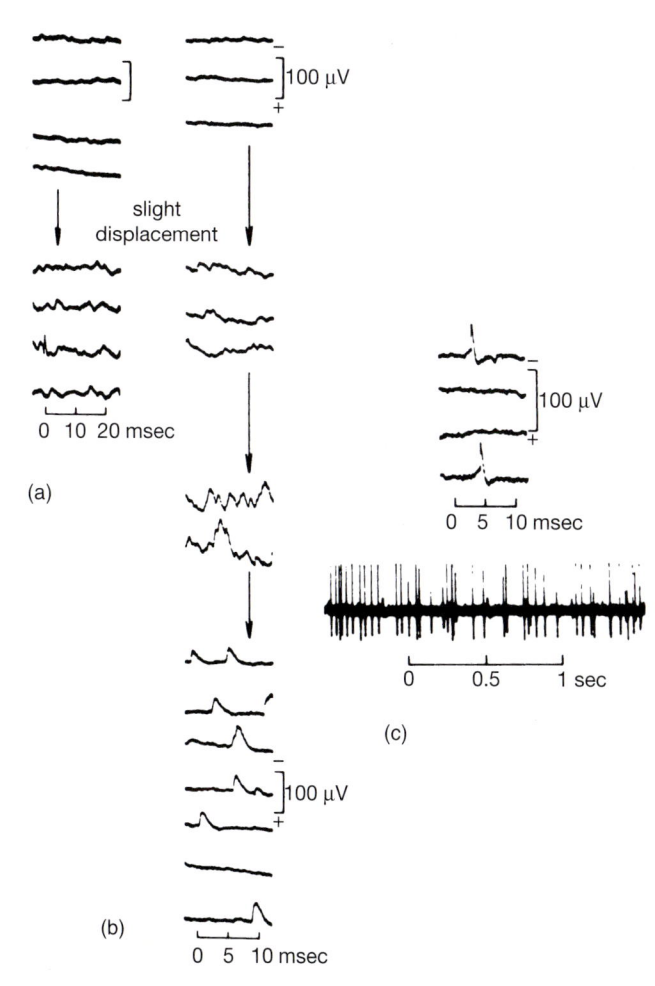

Fig. 4.36 End-plate activity. A concentric needle electrode was inserted in the end-plate zone of a normal biceps and advanced in small steps. A and B are recordings of end-plate 'noise' and in the lower part of B the discrete negative spikes indicate that the electrode was close to discharging end-plates. It has been suggested that these may be miniature end-plate potentials. C shows the other type of end-plate activity: diphasic potentials. (From Buchthal and Rosenfalck (1966a), by permission.)

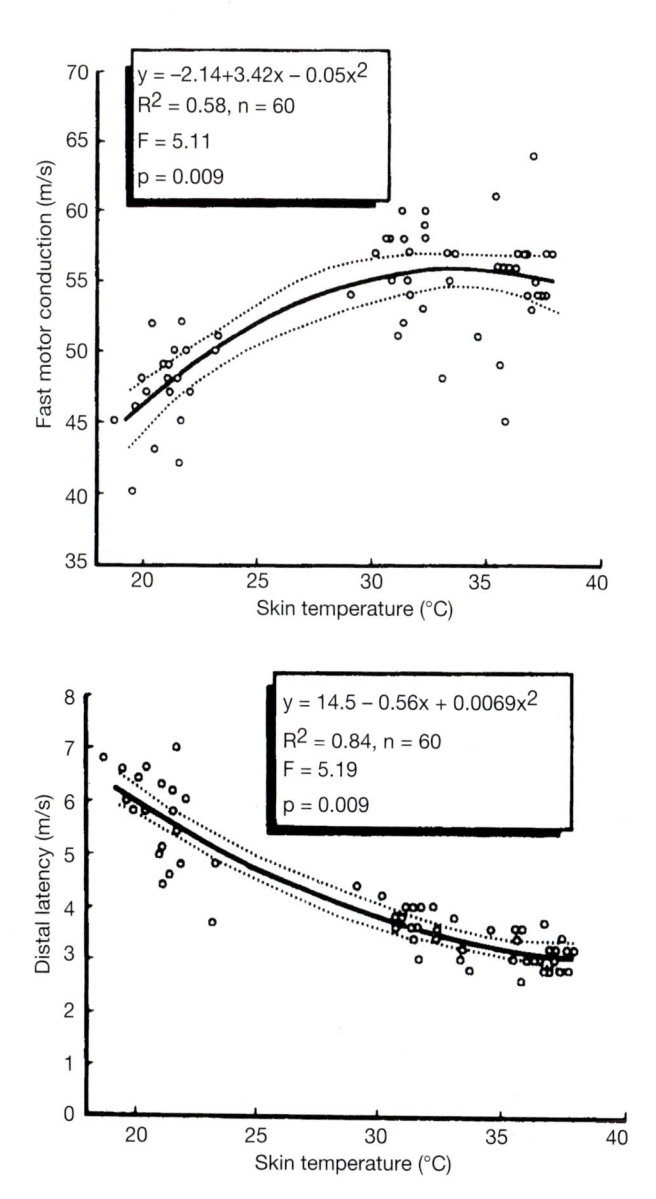

Fig. 4.37 Non-linear relationship between motor conduction velocity (upper) and distal motor latency (lower) and different skin temperatures. The least-square polynomial regression line of second power and its 95% confidence interval are given in each instance. Insets: regression equation. (From Todnem K, Kudsen G, Riise T, Nyland H, Aarli JA 1989 The non-linear relationship between conduction velocity and skin temperature. *Journal of Neurology, Neurosurgery and Psychiatry* 52:497–501, with permission from the BMJ Publishing Group.)

also occur with root lesions or peripheral nerve disease. They sometimes appear in healthy muscle, and are common after administration of suxamethonium chloride (succinylcholine). The precise mechanism of fasciculations remains uncertain; the various possibilities are reviewed by Desai and Swash (1997).

NERVE CONDUCTION

Effects of limb temperature on nerve conduction

All parameters of nerve conduction are affected by limb temperature. On cooling, conduction velocities fall, distal motor latencies become prolonged, and compound muscle and sensory action potentials increase in duration (Buchthal and Rosenfalck 1966b) and amplitude (Bolton et al 1981).

Several studies have shown a linear relationship between nerve conduction velocity and temperature (Buchthal and Rosenfalck 1966b, Ludin and Beyeler 1977), with a constant change of between 1.5 and 2.0 m/s per °C but one study has demonstrated a non-linear relationship (Todnem et al 1989), with a disproportionate reduction in motor conduction velocity at lower temperatures (*Fig. 4.37*). The implication of this finding is that the use of formulae based on a linear relationship between conduction velocity and temperature to calculate a 'corrected' value may be inaccurate.

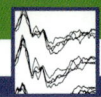

In healthy volunteers cooling produced an increase in amplitude of sensory and motor CAPs in the upper limb (*Fig. 4.38*) (Bolton et al 1981). The proposed explanation for the observed increase is that the change in duration of action potentials of single myelinated fibres which is seen on cooling results by summation in a longer duration and higher amplitude compound response. There is also probably a contributory effect from reducing the dispersion of conduction velocities between the fastest and slowest conducting fibres which becomes more prominent over longer conduction distances (*Fig. 4.38*).

Although a great deal is known about the clinical neurophysiological consequences of cooling (Denys 1991) there is no straightforward answer to what to do when patients' limbs are cold. Not only is there uncertainty as to the best theoretical correction to apply, but there also remains the question as to which site in the cool limb temperature should be measured. Consider the tibial nerve, which for some of its course lies deep beneath muscles of the posterior compartment of the leg but is superficial more distally. It is improbable that any single measurement made at the ankle or between the heads of gastrocnemius will represent the true temperature of more than a short segment of the nerve. The temperature gradient along a nerve is unlikely to be linear throughout the length of the limb. Ideally, all limbs should be warmed to 30°C and the temperature maintained with an infrared lamp, but this is often impracticable in an ICU, and there is the additional complication that patients may have poor peripheral circulation.

REFERENCES

AAEM 2001 Glossary of terms in electrodiagnostic medicine. Muscle Nerve 24(Suppl 10):S5–S28.

Allison T 1982 Scalp and cortical recordings of initial somatosensory cortex activity to median nerve stimulation in man. Ann NY Acad Sci 388:677–678.

Allison T 1987 Normal limits in the EP: age and sex differences. In: *A Textbook of Clinical Neurophysiology* (eds AM Halliday, SR Butler, R Paul). Wiley, Chichester, pp. 155–171.

Allison T, Hume AL 1981 A comparative analysis of short latency somatosensory evoked potentials in man, monkey, cat, and rat. Exp Neurol 72:592–611.

Allison T, Goff WR, Williamson PD, et al 1980 On the neural origin of early components of the human somatosensory evoked potentials. In: *Clinical Uses of Cerebral, Brainstem and Spinal Somatosensory Evoked Potentials. Progress in Clinical Neurophysiology*, vol. 7 (ed JE Desmedt). Karger, Basel, pp. 51–68.

Allison T, Wood CC, Goff WR 1983 Brainstem auditory, pattern reversal visual and short latency somatosensory evoked potentials: latencies in relation to age, sex, and brain and body size. Electroencephalogr Clin Neurophysiol 55:619–636.

Allison T, McCarthy G, Wood CC, et al 1989 Human cortical potentials evoked by stimulation of the median nerve. I: Cytoarchitectonic areas generating short-latency activity. J Neurophysiol 62:694–710.

American Electroencephalographic Society 1994 Guidelines in electroencephalography, evoked potentials and polysomnography. Guideline three: minimum technical standards for EEG recording in suspected cerebral death. J Clin Neurophysiol 11:10–13.

Anziska BJ, Cracco RQ 1980 Short-latency somatosensory evoked potentials: studies in patients with focal neurological disease. Electroencephalogr Clin Neurophysiol 49:227–239.

Austin GM, McCouch GP 1955 Presynaptic component of intermediary cord potential. J Neurophysiol 18:441–451.

Bartel P, Conradie J, Robinson E, et al 1987 The relationship between somatosensory evoked potential latencies and age and growth parameters in young children. Electroencephalogr Clin Neurophysiol 68:180–186.

Beall JE, Applebaum AE, Foreman RD, et al 1977 Spinal cord potentials evoked by cutaneous afferents in the monkey. J Neurophysiol 40:199–211.

Beydoun A, Yen CE, Drury I 1991 Variance of interburst intervals in burst suppression. Electroencephalogr Clin Neurophysiol 79:435–439.

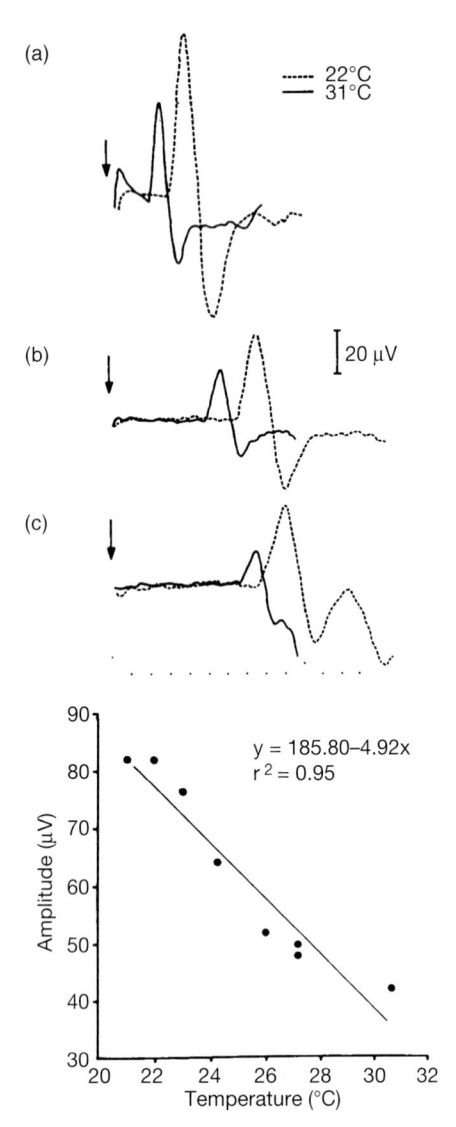

Fig. 4.38 Effect of cooling on antidromically recorded median sensory action potentials. Top: Traces following stimulation at three sites: (a) wrist, (b) elbow and (c) upper arm. Bottom: relationship between antidromically recorded median sensory action potential and rising temperature. (From Bolton CF, Sawa GM, Carter K 1981 The effects of temperature on human compound action potentials. *Journal of Neurology, Neurosurgery and Psychiatry* 44:407–413, with permission from the BMJ Publishing Group.)

Regression: $y = 185.80 - 4.92x$, $r^2 = 0.95$

References

Bickford RG 1950 Automatic electroencephalographic control of general anaesthesia. Electroencephalogr Clin Neurophysiol 2:3–96.

Bickford RG, Jacobson J, Cody D 1964 Nature of averaged evoked potentials to sound and other stimuli in man. Ann NY Acad Sci 112:204–223.

Binnie CD, Prior PF, Lloyd DSL, et al 1970 Electroencephalographic prediction of fatal anoxic brain damage after resuscitation from cardiac arrest. BMJ 4:265–268.

Binnie CD, Cooper R, Mauguière F, et al (eds) 2003 *Clinical Neurophysiology*, volume 2, *EEG, Paedirtric Neurophysiology, Special Techniques and Applications*. Elsevier, Amsterdam.

Binnie CD, Cooper R, Mauguière F, et al (eds) 2004 *Clinical Neurophysiology*, volume 1, revised and enlarged edition, *EMG, Nerve Conduction and Evoked Potentials*. Elsevier, Amsterdam.

Bolton CF, Sawa GM, Carter K 1981 The effects of temperature on human compound action potentials. J Neurol Neurosurg Psychiatry 44:407–413.

Brazier MAB, Finesinger JE 1945 Action of barbiturates on the cerebral cortex. Arch Neural Psychiatr 53:51–58.

Brierley JB, Graham DI 1984 Hypoxia and vascular disorders of the central nervous system. In: *Greenfield's Neuropathology*, 4th edition (eds J Hume Adams, JAN Corsellis, LW Duchan). Arnold, London, pp. 125–207.

Brierley JB, Prior PF, Calverley J, et al 1980 The pathogenesis of ischaemic neuronal damage along the cerebral arterial boundary zones in *Papio anubis*. Brain 103:929–965.

Broughton RJ 1969 Methods, results and evaluations. In: *Average Evoked Potentials* (eds E Donchin, DB Lindsley). NASA SP191. US Government Printing Office, Washington, DC, pp. 79–84.

Buchner H, Ferbert A, Hacke W 1988 Serial recording of median nerve stimulated subcortical somatosensory evoked potentials (SEPs) in developing brain death. Electroencephalogr Clin Neurophysiol 69:14–23.

Buchner H, Ferbert A, Sherg M, et al 1986 Evoked potential monitoring in brain death. Generators of BAEP, spinal SEP In: *Clinical Problems of Brainstem Disorders* (eds K Kunze, WH Zangemeister, A Arlt). Georg Thieme, Stuttgart, pp. 130–133.

Buchthal F 1982 Fibrillations: Clinical electrophysiology. In: *Abnormal Nerves and Muscles as Impulse Generators* (eds WJ Culp, J Ochoa). Oxford University Press, New York, pp. 632–662.

Buchthal F, Rosenfalck A 1966b Evoked action potentials and conduction velocity in human sensory nerves. Brain Res 3:1–122.

Buchthal F, Rosenfalck P 1966a Spontaneous electrical activity of human muscles. Electroencephalogr Clin Neurophysiol 20:21–336.

Cadilhac J, Zhu Y, Georgesco M, et al 1985 La maturation des potentiels évoqués somesthésiques cérébraux. Rev EEG Neurophysiol Clin 15:1–11.

Caruso G, Eisen A, Stålberg E, et al 1999 Clinical EMG, glossary of terms most commonly used by clinical electromyographers. In: *Recommendations for the Practice of Clinical Neurophysiology: Guidelines of the International Federation of Clinical Neurophysiology* (EEG Suppl 52) (eds G Deuschl, A Eisen). Elsevier, Amsterdam, pp. 189–198.

Chatrian GE, Shaw C-M, Leffman H 1964 The significance of periodic lateralized epileptiform discharges in EEG: an electrographic, clinical and pathological study. Electroencephalogr Clin Neurophysiol 17:177–193.

Chatrian GE, Bergamasco B, Bricolo A, et al 1996 IFCN recommended standards for electrophysiologic monitoring in comatose and other unresponsive states. Report of an IFCN committee. Electroencephalogr Clin Neurophysiol 99:103–122.

Cheron G, Dan B, Borenstein S 2000 Sensory and motor interfering influences on somatosensory evoked potentials. J Clin Neurophysiol 17:280–294.

Chiappa KH 1990 *Evoked Potentials in Clinical Medicine*. 2nd edition. Raven Press, New York.

Cigánek L 1961 The EEG response (evoked potential) to light stimulus in man. Electroencephalogr Clin Neurophysiol 13:163–172.

Cioni B, Meglio M 1986 Epidural recordings of electrical events produced in the spinal cord by segmental, ascending and descending volleys. Appl Neurophysiol 49:315–326.

Convers P, García-Larrea L, Fischer C, et al 1989 Les anomalies des potentiels évoqués somesthésiques precoces dans les lésions du tronc cérébral: étude de 64 observations. Rev EEG Neurophysiol Clin 19:443–468.

Cracco JB, Cracco RQ, Stolove R 1979 Spinal evoked potentials in man: a maturational study. Electroencephalogr Clin Neurophysiol 46:58–64.

Cracco RQ 1973 Spinal evoked response: peripheral nerve stimulation in man. Electroencephalogr Clin Neurophysiol 35:379–386.

Cracco RQ, Cracco JB 1976 Somatosensory evoked potentials in man: far-field potentials. Electroencephalogr Clin Neurophysiol 41:460–466.

Cruse R, Klem G, Lesser R, et al 1982 Paradoxical lateralization of cortical potentials evoked by stimulation of posterior tibial nerve. Arch Neurol 39:222–225.

Cunningham K, Halliday AM, Jones SJ 1986 Simulation of 'stationary' SAP, SEP phenomena by 2-dimensional potential field modelling. Electroencephalogr Clin Neurophysiol 65:416–428.

Cusick JF, Myklebust J, Larson SJ, et al 1978 Spinal evoked potentials in the primate: neural substrate. J Neurosurg 49:551–557.

Dawson GD 1947 Cerebral responses to electrical stimulation of peripheral nerve in man. J Neurol Neurosurg Psychiatry 10:134–140.

Deiber MP, Giard MH, Mauguière F 1986 Separate generators with distinct orientations for N20 and P22 somatosensory evoked potentials to finger stimulation. Electroencephalogr Clin Neurophysiol 65:321–334.

Delbeke J, McComas AJ, Kopec SJ 1978 Analysis of evoked lumbosacral potentials in man. J Neurol Neurosurg Psychiatry 41:293–302.

Delberghe X, Mavroudakis N, Zegers de Beffi D, et al 1990 The effect of stimulus frequency on post- and precentral short latency somatosensory evoked potentials (SEPs). Electroencephalogr Clin Neurophysiol 77:86–92.

Delestre F, Lonchampt P, Dubas F 1986 Neural generator of P14 far-field somatosensory evoked potential studied in a patient with a pontine lesion. Electroencephalogr Clin Neurophysiol 65:227–230.

Delwaide PJ, Schoenen J, De Pasqua V 1985 Lumbosacral spinal evoked potentials in patients with MS Neurology 35:174–179.

Denys EH 1991 The influence of temperature in clinical neurophysiology. Muscle Nerve 14:795–811.

Desai J, Swash M 1997 Fasciculations: what do we know of their significance? J Neurol Sci 152(Suppl 1):S43–S48.

Desmedt JE 1984 Cerebral evoked potentials. In: *Peripheral Neuropathy* (eds PJ Dyck, PK Thomas, EH Lambert, R Bunge). Saunders, Philadelphia, PA, pp. 1045–1066.

Desmedt JE, Bourguet M 1985 Color imaging of parietal and frontal somatosensory potential fields evoked by stimulation of median or posterior tibial nerve in man. Electroencephalogr Clin Neurophysiol 62:1–17.

Desmedt JE, Cheron G 1980a Central somatosensory conduction in man: neural generators and interpeak latencies of the far-field components recorded from neck and right or left scalp or earlobes. Electroencephalogr Clin Neurophysiol 50:382–403.

Desmedt JE, Cheron G 1980b Somatosensory evoked potentials to finger stimulation in healthy octogenarians and in young adults: waveforms, scalp topography and transit times of parietal and frontal components. Electroencephalogr Clin Neurophysiol 50:404–425.

Desmedt JE, Cheron G 1981 Non-cephalic reference recording of early somatosensory potentials to finger stimulation in adult or aging man: differentiation of widespread N18 and contralateral N20 from the pre-rolandic P22 and N30 components. Electroencephalogr Clin Neurophysiol 52:553–570.

Desmedt JE, Cheron G 1983 Spinal and far-field components of human somatosensory evoked potentials to posterior tibial nerve stimulation analysed with oesophageal derivations and non-cephalic reference recording. Electroencephalogr Clin Neurophysiol 56:635–651.

Desmedt JE, Nguyen TH 1984 Bit-mapped colour imaging of the potential fields of propagated and segmental sub-cortical components of somatosensory evoked potentials in man. Electroencephalogr Clin Neurophysiol 58:481–497.

Desmedt JE, Osaki I 1991 SEPs to finger joint input lack the N20–P20 response that is evoked by tactile inputs: contrast between cortical generators in areas 3b and 2 in humans. Electroencephalogr Clin Neurophysiol 80:513–521.

Desmedt JE, Noel P, Debecker J, et al 1973 Maturation of afferent conduction velocity as studied by sensory nerve potentials and by cerebral evoked potentials. In: *New Developments in Electromyography and Clinical Neurophysiology*, vol. 2 (ed. JE Desmedt). Karger, Basel, pp. 52–63.

Desmedt JE, Brunko J, Debecker J 1976 Maturation of the somatosensory evoked potential in normal infants and children, with special reference to the early N1 component. Electroencephalogr Clin Neurophysiol 40:43–58.

Desmedt JE, Tran Huy N, Carmelier J 1983 Unexpected latency shifts of the stationary P9 somatosensory evoked potential far-field with changes in shoulder position. Electroencephalogr Clin Neurophysiol 56:628–634.

Desmedt JE, Nguyen TH, Bourguet M 1987 Bit-mapped color imaging of human evoked potentials with reference to N20, P22, P27 and N30 somatosensory responses. Electroencephalogr Clin Neurophysiol 68:1–19.

Desmedt JE, Chalklin V, Tomberg C 1990 Emulation of somatosensory evoked potentials (SEP) components with the 3-shell head model and the problem of 'ghost potential fields' when using an average reference in brain mapping. Electroencephalogr Clin Neurophysiol 77:243–258.

Deuschl G, Eisen A (eds) 1999 *Recommendations for the Practice of Clinical Neurophysiology: Guidelines of the International Federation of Clinical Neurophysiology* (EEG Suppl 52). Elsevier, Amsterdam, p. 304.

Dimitrijevic MR, Larsson LE, Lehmkuhl D, et al 1978 Evoked spinal cord and nerve root potentials in humans using a noninvasive recording technique. Electroencephalogr Clin Neurophysiol 45:331–340.

Eisen A, Odusote K 1980 Central and peripheral conduction times in multiple sclerosis. Electroencephalogr Clin Neurophysiol 48:253–265.

Evans BM 1992 Periodic activity in cerebral arousal mechanisms – the relationship to sleep and brain damage. Electroencephalogr Clin Neurophysiol 83:130–137.

Evans BM 2002 What does brain damage tell us about the mechanisms of sleep? J R Soc Med 95:591–597.

Evans BM, Bartlett JR 1995 Prediction of outcome in severe head injury based on recognition of sleep related activity in the polygraphic electroencephalogram. J Neurol Neurosurg Psychiatry 59:17–25.

Fariello RG, Orrison W, Blanco G, et al 1982 Neuroradiological correlates of frontally predominant intermittent rhythmic delta activity (FIRDA). Electroencephalogr Clin Neurophysiol 54:194–202.

Ganes T, Lundar T 1983 The effect of thiopentone on somatosensory evoked responses and EEGs in comatose patients. J Neurol Neurosurg Psychiatry 46:509–514.

García-Larrea L, Bastuji H, Mauguière F 1992 Unmasking of cortical SEP components by changes in stimulus rate: a topographic study. Electroencephalogr Clin Neurophysiol 84:71–83.

Gasser HS, Graham HT 1933 Potentials produced in the spinal cord by stimulation of the dorsal roots. Am J Physiol 103:303–320.

Gastaut H, Régis H 1965 Visually evoked potentials recorded transcranially in man. In: *The Analysis of Central Nervous System and Cardiovascular Data Using Computer Methods* (eds LD Proctor, WR Adey). NASA, US Government Printing Office, Washington, DC, pp. 7–34.

Gloor P, Kalabay O, Giard N 1968 The electroencephalogram in diffuse encephalopathies: electroencephalographic correlates of grey and white matter lesions. Brain 91:779–802.

Gloor P, Ball G, Schaul N 1977 Brain lesions that produce delta activity in the EEG Neurology 27:326–333.

Goff GD, Matsumiya Y, Allison T, et al 1977 The scalp topography of human somatosensory and auditory evoked potentials. Electroencephalogr Clin Neurophysiol 42:57–76.

Gross DW, Wiebe S, Blume WT 1999 The periodicity of lateralized epileptiform discharges Clin Neurophysiol 110:1516–1520.

Guérit JM, Opsomer RJ 1991 Bit-mapped imaging of somatosensory evoked potentials after stimulation of the posterior tibial nerves and dorsal nerve of the penis/clitoris. Electroencephalogr Clin Neurophysiol 80:228–237.

Guérit JM, Soveges L, Baele P, et al 1990 Median nerve somatosensory evoked-potentials in profound hypothermia for ascending aorta repair. Electroenceph Clin Neurophysiol 77:163–173.

Guérit J-M, Fischer C, Facco E, et al 1999 Standards of clinical practice of EEG, EPs in comatose and other responsive states. In: *Recommendations for the Practice of Clinical Neurophysiology: Guidelines of the international Federation of Clinical Neurophysiology* (EEG Suppl 52) (eds G Deuschl, A Eisen). Elsevier, Amsterdam, pp. 117–131.

Halliday AM (ed) 1982 *Evoked Potentials in Clinical Testing*. Churchill Livingstone, Edinburgh.

Halliday AM, Wakefield GS 1963 Cerebral evoked potentials in patients with dissociated sensory loss. J Neurol Neurosurg Psychiatry 26:211–219.

Halliday AM, Barrett G, Halliday E, et al 1979 A comparison of the flash and pattern evoked potential in unilateral optic neuritis. Wiss Z Ernst-Moritz-Arndt-Universität, Greifswald 2:89–95.

Halliday AM, McDonald WI, Mushin J 1972 Delayed visual evoked responses in optic neuritis. Lancet i:982–985.

Hartikainen K, Rorarius M, Mäkelä K, et al 1995 Visually evoked bursts during isoflurane anaesthesia. Br J Anaesth 74:681–685.

Hashimoto I 1989 Critical analysis of short-latency auditory evoked potentials recording techniques. In: *Advanced Evoked Potentials* (ed H Lüders). Kluwer, Boston, MA, pp. 105–142.

Hashimoto T, Tayama M, Kiura K, et al 1983 Short latency somatosensory evoked potentials in children. Brain Dev 5:390–396.

Helmholz H 1853 Über einige Gesetze der Vertheilung Elektrischer Ströme in körperlichen Leitern mit Anwendung auf thierisch-elektrischer Versuche. Ann Physik und Chemie 89:211–233

Henry CE, Scoville WB 1952 Suppression-burst activity from isolated cerebral cortex in man. Electroencephalogr Clin Neurophysiol 4:1–22.

Hume AL, Cant BR 1978 Conduction time in central somatosensory pathways in man. Electroencephalogr Clin Neurophysiol 45:361–375.

Ibáñez V, Deiber MP, Fischer C 1989a Middle latency auditory evoked potentials in cortical lesions: criteria of interhemispheric asymmetry. Arch Neurol 46:1325–1332.

Ibañez V, Deiber MP, Mauguière F 1989b Interference of vibrations with input transmission in dorsal horn and cuneate nucleus in man: a study of somatosensory evoked potentials (SEPs) to electrical stimulation of median nerve and fingers. Exp Brain Res 75:599–610.

Jeanmonod D, Sindou M, Mauguière F 1989 Three transverse dipolar generators in the human cervical and lumbosacral dorsal horn: evidence from direct intraoperative recordings on the spinal cord surface. Electroencephalogr Clin Neurophysiol 74:236–240.

Jeanmonod D, Sindou M, Mauguière F 1991 The human cervical and lumbosacral evoked electrospinogram. Data from intraoperative spinal cord surface recordings. Electroencephalogr Clin Neurophysiol 80:477–489.

Jewett DL 1970 Volume-conducted potentials in response to auditory stimuli as detected by averaging in the cat. Electroencephalogr Clin Neurophysiol 28:609–618.

Jewett DL, Williston JS 1971 Auditory-evoked far-fields averaged from the scalp of humans. Brain 94:681–696.

Jewett DL, Romano MN, Williston JS 1970 Human auditory evoked potentials: possible brain stem components detected on the scalp. Science 167:1517–1518.

Jones SJ, Small M 1978 Spinal and subcortical-evoked potentials following stimulation of the posterior tibial nerve in man. Electroencephalogr Clin Neurophysiol 44:299–306.

Jørgensen EO, Holm S 1999 The course of circulatory and cerebral recovery after circulatory arrest: the influence of pre-arrest, arrest and post-arrest factors. Resuscitation 42:173–182.

Kaji R, Sumner AJ 1987 Bipolar recording of short-latency somatosensory evoked potentials after median nerve stimulation. Neurology 37:410–418.

Kakigi R, Jones SJ 1986 Influence of concurrent tactile stimulation on somatosensory evoked potentials following posterior tibial nerve stimulation in man. Electroencephalogr Clin Neurophysiol 65:118–129.

Kakigi R, Shibasaki H 1983 Scalp topography of the short-latency somatosensory evoked potentials following posterior tibial nerve stimulation in man. Electroencephalogr Clin Neurophysiol 56:430–437.

Kakigi R, Shibasaki H, Hashizume A, et al 1982 Short latency somatosensory evoked spinal and scalp-recorded potentials following posterior tibial nerve stimulation in man. Electroencephalogr Clin Neurophysiol 53:602–611.

Katifi HA, Sedgwick EM 1986 Somatosensory evoked potentials from posterior tibial nerve and lumbosacral dermatomes. Electroencephalogr Clin Neurophysiol 65:249–259.

Kiloh LG, McComas AJ, Osselton JW, et al 1981 *Clinical Electroencephalography*, 4th edition. Butterworth, London.

Kimura J 2001 *Electrodiagnosis in Diseases of Nerve and Muscle*, 3rd edition. Oxford University Press, New York.

Kimura J, Mitsudome A, Beck DO, et al 1983 Field distribution of antidromically activated digital nerve potentials. Model for far-field recording. Neurology 33:1164–1169.

Kimura J, Mitsudome A, Beck DO, et al 1984 Stationary peaks from a moving source in far-field recording. Electroencephalogr Clin Neurophysiol 58:351–361.

Kleitman N 1963 *Sleep and Wakefulness*, 2nd edition. University of Chicago Press, Chicago, IL.

Koutroumanidis M, Binnie CD, Elwes RDC, et al 1998 Interictal regional slow activity in temporal lobe epilepsy correlates with lateral temporal hypometabolism as imaged with 18FDG PET: neurophysiological and metabolic implications. J Neurol Neurosurg Psychiatry 65:170–176.

Kraus N, Smith DI, Reed NL, et al 1985 Auditory middle latency responses in children: effect of age and diagnostic category. Electroencephalogr Clin Neurophysiol 62:342–351.

Kunesch E, Knecht S, Schnitzler A, et al 1995 Somatosensory evoked potentials elicited by intraneural microstimulation of afferent nerve fibers. J Clin Neurophysiol 12:476–487.

Lafrenière L, Laureau E, Vanasse M, et al 1990 Maturation of short-latency somatosensory evoked potentials by median nerve stimulation: a cross-sectional study in a large group of children. In: *New Trends and Advanced Techniques in Clinical Neurophysiology* (EEG Suppl 41) (eds PM Rossini, F Mauguière). Elsevier, Amsterdam, pp. 236–242.

Lastimosa ACB, Bass NH, Stanback K, et al 1982 Lumbar spinal cord and early cortical evoked potentials after tibial nerve stimulation: effects of stature on normative data. Electroencephalogr Clin Neurophysiol 54:499–507.

Laureau E, Majnermer A, Rosenblatt B, et al 1988 A longitudinal study of short latency somatosensory evoked responses in healthy newborns and infants. Electroencephalogr Clin Neurophysiol 71:100–108.

Lehmann D, Skrandies W 1984 Spatial analysis of evoked potentials in man: a review. Prog Neurobiol 23:227–250.

Lesser RP, Lueders H, Hahn J, et al 1981 Early somatosensory evoked potentials by median nerve stimulation: intraoperative monitoring. Neurology 31:1519–1523.

Lorente de Nò R 1947 Analysis of the distribution of action currents of nerve in volume conductors. Stud Rockefeller Inst Med Res 132:384–477.

Lüders H, Andrish J, Gurd A, et al 1981 Origin of far-field subcortical potentials evoked by stimulation of tibial nerve. Electroencephalogr Clin Neurophysiol 52:336–344.

Ludin HP, Beyeler F 1977 Temperature dependence of normal sensory nerve action potentials. J Neurol (Berlin) 216:173–180.

MacDonald DB 2001 Individually optimizing posterior tibial somatosensory evoked potential P37 scalp derivations for intraoperative monitoring. J Clin Neurophysiol 18:364–371.

MacDonald DB, Stigsby B, Al Zayed Z 2004 A comparison between derivation optimization and Cz´-FPz for posterior tibial P37 somatosensory evoked potential intraoperative monitoring. Clin Neurophysiol 115:1925–1930.

Macefield G, Burke D, Gandevia SC 1989 The cortical distribution of muscle and cutaneous afferent projections from the human foot. Electroencephalogr Clin Neurophysiol 72:518–528.

Macon JB, Poletti CE 1982 Conducted somatosensory evoked potentials during spinal surgery: Part I: Control conduction velocity measurements. J Neurosurg 57:349–353.

Macon JB, Poletti CE, Sweet WH, et al 1982 Conducted somatosensory evoked potentials during spinal surgery: Part II: Clinical applications. J Neurosurg 57:354–359.

Magnus O, van der Holst M 1987 Zeta waves: a special type of slow delta waves. Electroencephalogr Clin Neurophysiol 67:140–146.

Markand ON, Daly DD 1971 Pseudoperiodic lateralized paroxysmal discharges in electroencephalogram. Neurology 21:975–981.

Mauguière F 1983 Les potentiels évoqués somesthésiques cervicaux chez le sujet normal. Analyse des aspects obtenus selon le siège de l'électrode de référence. Rev EEG Neurophysiol Clin 13:259–272.

Mauguière F 1987 Short-latency somatosensory evoked potentials to upper limb stimulation in lesions of brainstem, thalamus and cortex. In: *The London Symposia* (EEG Suppl 39) (eds RJ Ellingson, NMF Murray, AM Halliday). Elsevier, Amsterdam, pp. 302–309.

Mauguière F 1989 Evoked potentials in nondemyelinating diseases. In: *Advanced Evoked Potentials* (ed H Lüders). Kluwer, Boston, MA, pp. 181–221.

Mauguière F, Ibañez V 1985 The dissociation of early SEP components in lesions of the cervico-medullary junction: a cue for routine interpretation of abnormal cervical responses to median nerve stimulation. Electroencephalogr Clin Neurophysiol 62:406–420.

Mauguière F, Restuccia D 1991 Inadequacy of the forehead reference montage for detecting spinal N13 potential abnormalities in patients with cervical cord lesion and preserved dorsal column function. Electroencephalogr Clin Neurophysiol 79:448–456.

Mauguière F, Mitrou H, Chalet E, et al 1979 Intérêt des potentiels évoqués visuels dans la sclérose multioculaire (SM): étude comparative des résultats obtenus en stimulation par éclair lumineux et inversion de damiers. Rev EEG Neurophysiol Clin 9 209–220.

Mauguière F, Brunon AM, Echallier JF, et al 1982 Early somatosensory evoked potentials in thalamocortical lesions of the lemniscal pathways in humans. In: *Clinical Applications of Evoked Potentials in Neurology. Advances in Neurology*, vol. 32 (eds J Courjon, F Mauguière, M Revol). Raven Press, New York, pp. 321–338.

Mauguière F, Desmedt JE, Courjon J 1983a Astereognosis and dissociated loss of frontal or parietal components of somatosensory evoked potentials in hemispheric lesions. Brain 106:271–311.

Mauguière F, Desmedt JE, Courjon J 1983b Neural generators of N18 and P14 far-field somatosensory evoked potentials studied in patients with lesions of thalamus or thalamo-cortical radiations. Electroencephalogr Clin Neurophysiol 56:283–292.

Mauguière F, Ibañez V, Deiber MP, et al 1987 Noncephalic reference recording and spatial mapping of short-latency SEPs to upper limb stimulation: normal responses and abnormal patterns in patients with non-demyelinating lesions. In: *Evoked Potentials III* (eds C Barber, T Blum). Butterworths, Boston, MA, pp. 40–55.

Mavroudakis N, Brunko E, Delberghe X, et al 1993 Dissociation of P13–P14 far-field potentials: clinical and MRI correlation. Electroencephalogr Clin Neurophysiol 88:240–242.

Michel CM, Murray MM, Lantz G, et al 2004 EEG source imaging. Clin Neurophysiol 115:2195–2222.

Michenfelder JD 1974 The interdependency of cerebral functional and metabolic effects following massive doses of thiopental in the dog. Anesthesiology 41:231–236.

Miura T, Sonoo M, Shimizu T 2003 Establishment of standard values for latency, interval and amplitude parameters of tibial nerve somatosensory evoked potentials (SEPs). Clin Neurophysiol 114:1367–1378.

Mizrahi EM, Patel VM, Crawford ES, et al 1989 Hypothermic-induced electro-cerebral silence, prolonged circulatory arrest, and cerebral protection during cardiovascular surgery. Electroencephalogr Clin Neurophysiol 72:81–85.

Møller AR, Jannetta PJ, Burgess JE 1986 Neural generator of the somatosensory evoked potentials recording from the cuneate nucleus in man and monkeys. Electroencephalogr Clin Neurophysiol 65:241–248.

Morioka T, Shima F, Kato M, et al 1991 Direct recordings of the somato-sensory evoked potentials in the vicinity of the dorsal column nuclei in man: their mechanisms and contribution to the scalp far-field potentials. Electroencephalogr Clin Neurophysiol 80:221–227.

Nakanishi T 1982 Action potentials recorded by fluid electrodes. Electroencephalogr Clin Neurophysiol 53:343–345.

Nakanishi T, Shimada Y, Sakuta M, et al 1978 The initial positive component of the scalp recorded somato-sensory evoked potential in normal subjects and in patients with neurological disorders. Electroencephalogr Clin Neurophysiol 45:26–34.

Nakanishi T, Tamaki M, Arasaki K, et al 1981 Origins of the scalp-recorded somatosensory far field potentials in man and cat. In: *The Kyoto Symposia*, Electroencephalography and Clinical Neurophysiology Suppl 36 (eds PA Buser, WA Cobb, T Okuma). Elsevier Biomedical Press, Amsterdam, pp. 336–348.

Nakanishi T, Tamaki M, Ozaki Y, et al 1983 Origins of short latency somatosensory evoked potentials to median nerve stimulation. Electroencephalogr Clin Neurophysiol 56:74–85.

Noachtar S, Binnie CD, Ebersole J, et al 1999 A glossary of terms most commonly used by clinical electroencephalographers and proposal for the report form for the EEG findings. In: *Recommendations for the Practice of Clinical Neurophysiology* (Clin Neurophysiol Suppl 52 (eds G Deuschl, A Eisen). Elsevier, Amsterdam, pp. 21–40.

Ozaki I, Shimamura H, Baba M, et al 1996a N30 in Parkinson's disease. Neurology 47:303–305.

Ozaki I, Takada H, Shimamura H, et al 1996b Central conduction in somatosensory evoked potentials. Comparison of ulnar and median data evaluation of onset versus peak methods. Neurology 47:1299–1304.

Papakostopoulos D, Crow HJ 1980 Direct recording of the somatosensory evoked potentials from the cerebral cortex of man and the difference between precentral and postcentral potentials. In: *Clinical Uses of Cerebral, Brainstem and Spinal Somatosensory Evoked Potentials* (ed JE Desmedt). Prog Clin Neurophysiol 7:15–26.

Picton TW, Hillyard SA, Krausz HI, et al 1974 Human auditory evoked potentials: I Evaluation of components. Electroencephalogr Clin Neurophysiol 36:179–190.

Powers SK, Bolger CA, Edwards MSB 1982 Spinal cord pathways mediating somatosensory evoked potentials. J Neurosurg 57:472–482.

Prasher D, Luxon LM 1988 Methods of examination – audiological and vestibular. In: *Mawson's Diseases of the Ear*, 5th edition (ed. H Ludman). Edward Arnold, London, pp. 116–189.

Pratt H, Aminoff M, Nuwer MR, et al 1999 Short-latency auditory evoked potentials. In: *Recommendations for the Practice of Clinical Neurophysiology: Guidelines of the International Federation of Clinical Neurophysiology* (EEC Suppl 52) (eds G Deuschel, A Eisen). Elsevier Science, Amsterdam, pp. 69–77.

Prior PF 1973 *The EEG in Acute Cerebral Anoxia*. Excerpta Medica, Amsterdam.

Rae-Grant AD, Kim YW 1994 Type III intermittency: a nonlinear dynamic model of EEG burst suppression. Electroencephalogr Clin Neurophysiol 90:17–23.

Rae-Grant AD, Strapple C, Barbour PJ 1991 Episodic low-amplitude events: an under-recognized phenomenon in clinical electroencephalography. J Clin Neurophysiol 8:203–211.

Restuccia D 2000 Anatomic origin of P13 and P14 scalp far-field potentials. J Clin Neurophysiol 17:246–257.

Restuccia D, Di Lazzaro V, Valeriani M, et al 1995 Origin and distribution of P13 and P14 far-field potentials after median nerve stimulation. Scalp, nasopharyngeal and neck recordidngs in healthy subjects and in patients with cervical and cervico-medullary lesions. Electroencephalogr Clin Neurophysiol 96:371–384.

Restuccia D, Valeriani M, Barba C, et al 1999 Different contribution of joint and cutaneous inputs to early scalp somatosensory evoked potentials. Muscle Nerve 22:910–919.

Restuccia D, Valeriani M, Insola A, et al 2002 Modality-related scalp responses after selective electrical stimulation of cutaneous and muscular upper limb afferents in humans: specific and unspecific components. Muscle Nerve 26:44–54.

References

Riffel B, Stohr M, Körner S 1984 Spinal and cortical potentials following stimulation of posterior tibial nerve in the diagnosis and localization of spinal cord diseases. Electroencephalogr Clin Neurophysiol 58:400–407.

Rowan AJ, Tolunsky E 2003 *Primer of EEG. With a Mini-Atlas*. Butterworth-Heinemann, New York.

Scherg M, Von Cramon D 1985 Two bilateral sources of the late auditory evoked potentials as identified by a spatiotemporal dipole model. Electroencephalogr Clin Neurophysiol 62:32–44.

Schlag J 1973 Generation of brain evoked potentials. In: *Bioelectric Recording Techniques. Part A: Cellular Processes and Brain Potentials* (eds RF Thompson, MM Patterson). Academic Press, New York, pp. 273–316.

Schwartz MS, Scott DF 1978 Pathological stimulus-related slow wave arousal responses in the EEG. Acta Neurol Scand 57:300–304.

Schwartz MS, Prior PF, Scott DF 1973 The occurrence and evolution in the EEG of a lateralized periodic phenomenon. Brain 96:613–622.

Scott DF 1976 *Understanding EEG. An Introduction to Electroencephalography*. Duckworth, London.

Seyal M, Emerson RG, Pedley TA 1983 Spinal and early scalp-recorded components of the somatosensory evoked potential following stimulation of the posterior tibial nerve. Electro-encephalogr Clin Neurophysiol 55:320–330.

Shapiro HM 1985 Anesthesia effects upon cerebral blood flow, cerebral meta-bolism, electroencephalogram, and evoked potentials. In: *Anesthesia*, 2nd edition, vol. 2 (ed. RD Miller). Churchill Livingstone, New York, pp. 1249–1288.

Sherwood AM 1981 Characteristics of somatosensory evoked potentials recorded over the spinal cord and brain of man. IEEE Trans Biomed Engng BME-28:481–487.

Shimoji K, Matsuki M, Shimizu J 1977 Waveform characteristics and spatial distribution of evoked spinal electrogram in man. J Neurosurg 46:304–313.

Small M, Matthews WB 1984 A method of calculating spinal cord transit time from potentials evoked by tibial nerve stimulation in normal subjects and patients with spinal cord disease. Electroencephalogr Clin Neurophysiol 59:156–164.

Spehlmann R 1985 *Evoked Potential Primer. Visual, Auditory and Somatosensory Evoked Potentials in Clinical Diagnosis*. Butterworths, Boston, MA.

Steriade M, Gloor P, Llindás RR, et al 1990 Basic mechanisms of cerebral rhythmic activities. Report of the IFCN Committee on Basic Mechanisms. Electroencephalogr Clin Neurophysiol 76:481–508.

Steriade M, Amzica F, Contreras D 1994 Cortical and thalamic cellular correlates of electroencephalographic burst-suppression. Electroencephalogr Clin Neurophysiol 90:1–16.

Takahashi M, Kubota F, Nishi Y, et al 1993 Persistent synchronous periodic discharges caused by anoxic encephalopathy due to cardiopulmonary arrest. Clin Electroencephalogr 24:166–172.

Terzano MG, Parrino L, Spaggiari MC, et al 1987 Sleep-induced apneas functionally associated with cyclic alternating pattern: a polysomnographic analysis in a case of sleep apnea syndrome. Sleep Res 16:442.

Thomas JE, Lambert EH 1960 Ulnar nerve conduction velocity and H reflex in infants and children. J Appl Physiol 15:1–9.

Tinazzi M, Mauguière F 1995 Assessment of intraspinal and intracranial conduction by P30 and P39 tibial nerve somatosensory evoked potentials in cervical cord and hemispheric lesions. J Clin Neurophysiol 12:237–253.

Tinazzi M, Zanette G, Fiaschi A, et al 1996a Effects of stimulus rate on the cortical posterior tibial nerve SEPs: A topographic study. Electroencephalogr Clin Neurophysiol 100:210–219.

Tinazzi M, Zanette G, Bonato C, et al 1996b Neural generators of tibial nerve P30 somatosensory evoked potential studied in patients with a focal lesion of the cervico-medullary junction. Muscle Nerve 19:1538–1548.

Tinazzi M, Zanette G, Manganotti B, et al 1997a Amplitude changes of tibial nerve cortical SEPs when using ipsilateral or contralateral ear as reference. J Clin Neurophysiol 14:217–225.

Tinazzi M, Zanette G, La Porta F, et al 1997b Selective gating of lower limb cortical somatosensory evoked potentials (SEPs) during passive and active foot movements. Electroencephalogr Clin Neurophysiol 104:312–321.

Todnem K, Kudsen G, Riise T, et al 1989 The non-linear relationship between conduction velocity and skin temperature. J Neurol Neurosurg Psychiatry 52:497–501.

Tomita Y, Nishimura S, Tanaka T 1986 Short-latency SEPs in infants and children: developmental changes and maturational index of SEPs. Electroencephalogr Clin Neurophysiol 65:335–343.

Tsuji S, Lüders H, Lesser RP, et al 1984 Subcortical and cortical somatosensory potentials evoked by posterior tibial nerve stimulation: normative values. Electroencephalogr Clin Neurophysiol 59:214–228.

Turano G, Sindou M, Mauguière F 1995 Spinal cord evoked potentials monitoring during spinal surgery for pain and spasticity. In: *Atlas of Human Cord Evoked Potentials* (eds MR Dimitrijevic, JA Halter). Butterworth-Heinemann, Boston, MA, pp. 107–122.

Valeriani M, Restuccia D, Di Lazzaro V, et al 1998 Dipolar sources of the early scalp SEPs to upper limb stimulation. Effects of increasing stimulus rates. Exp Brain Res 120:306–315.

Wagner W 1991 SEP testing in deeply comatose and brain dead patients; the role of nasopharyngeal, scalp and earlobe derivations in recording the P14 potential. Electroencephalogr Clin Neurophysiol 80:352–363.

Wagner W 1996 Scalp, earlobe and nasopharyngeal recordings of the median nerve somatosensory evoked PI4 potential in coma and brain death. Brain 119:1507–1521.

Wang J, Cohen LJ, Hallett M 1989 Scalp topography of somatosensory evoked potentials following electrical stimulation of femoral nerve. Electroencephalogr Clin Neurophysiol 74:112–123.

Whittle IR, Johnston IH, Besser M 1986 Recording of spinal somatosensory evoked potentials for intraoperative spinal cord monitoring. J Neurosurg 64:601–612.

Wiederholt WC 1970 'End-plate noise' in electromyography. Neurology 20:214–224.

Wood CC 1982 Application of dipole localization methods to source identification of human evoked potentials. In: *Evoked Potentials*. Ann NY Acad Sci 388:139–155.

Wood CC, Cohen D, Cuffin BN, et al 1985 Electrical sources in human somatosensory cortex: identification by combined magnetic and potential recordings. Science 227:1051–1053.

Woods DL, Clayworth CC 1986 Age-related changes in human middle latency auditory evoked potentials. Electroencephalogr Clin Neurophysiol 65:297–303.

Yamada T 2000 Neuroanatomic substrates of lower extremity somatosensory evoked potentials. J Clin Neurophysiol 17:269–279.

Yamada T, Machida M, Kimura J 1982 Far-field somatosensory evoked potentials after stimulation of the tibial nerve. Neurology 32:1151–1158.

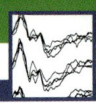

Yamada T, Matsubara M, Shiraishi M, et al 1996 Topographic analysis of somatosensory evoked potentials following stimulation tibial, sural and lateral femoral cutaneous nerve. Electroencephalogr Clin Neurophysiol 100:33–43.

Yli-Hankala A, Heikkilä H, Varri A, et al 1990 Correlation between EEG, heart rate variation in deep enflurane anaesthesia. Acta Anaesthesiol Scand 34:138–143.

Yli-Hankala A, Jäntti V, Pyykkö I, et al 1993 Vibration stimulus induced EEG bursts in isoflurane anaesthesia. Electroencephalogr Clin Neurophysiol 87:215–220.

Zhu Y, Georgesco M, Cadilhac J 1987 Normal latency values of early cortical somatosensory evoked potentials in children. Electroencephalogr Clin Neurophysiol 68:471–474.

Chapter 5

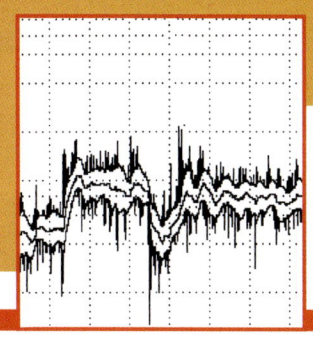

Neurophysiological Monitoring during Sedation and Anaesthesia

INTRODUCTION

There are two aspects of importance in relation to sedative and anaesthetic drugs and the use of neurophysiological monitoring in the intensive care unit (ICU) and during surgery. The first is to understand how these agents affect the monitored signals from the central and peripheral nervous system and whether they may invalidate our assessment of abnormality. The second is whether we can use the changes they produce, even in the presence of pathology, as tools to help clinical management of the adequacy and, indeed, the optimal depth of sedation or anaesthesia for the needs of the individual patient.

SEDATION: ASSESSMENT WITH EEG AND EVOKED POTENTIALS

Effects of sedative drugs on the EEG

The main agents likely to affect both diagnostic electroencephalographic (EEG) recordings and long-term monitoring in the ICU are the major sedatives, analgesics and muscle relaxants; in a European survey midazolam and propofol were the most commonly used sedatives, and morphine, fentanyl and sufentanil were the commonest analgesics (Soliman et al 2001). Their general effects on patients undergoing intensive care will be discussed here; more detail as to specific features for each group of drugs is given in *Effect of anaesthetic agents on the EEG* (p. 161) together with use in general anaesthesia.

A patient admitted to intensive care may be in pain, sleep deprived and poorly relaxed or restless because of high levels of ambient lighting or noise, or fearful about unfamiliar and potentially distressing ICU procedures. These states will affect the character of the ongoing pattern of the EEG because such patients tend to have highly variable EEGs in long-term monitor recordings (*Fig. 5.1(a)*). Abrupt, short-term, increases in amplitude often represent arousal phenomena to auditory or painful stimuli and vary morphologically from K-complexes of normal sleep (see *Fig. 4.4*) to the more prolonged, and abnormal, paradoxical slow wave arousal responses (see *Fig. 4.10*). Since arousal, for example following tracheal suction, may be accompanied by scalp or facial muscle potentials and movement artefacts, it is important that EEG monitors include facilities to distinguish them from changes in the on-going EEG. Patterns due to repeated arousals (as distinct from normal sleep–wake cyclical variations) are

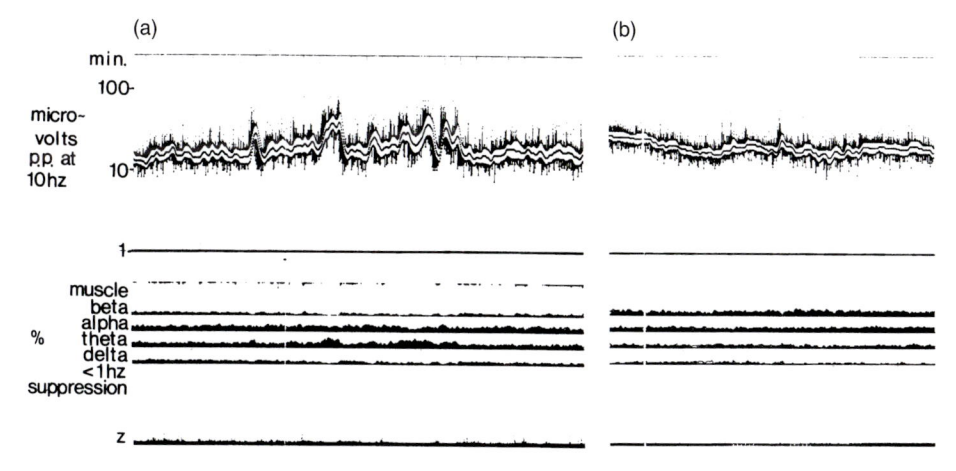

Fig. 5.1 Variability in EEG monitor recordings. (a) Variable CFAM1 (see p. 83 for a description) amplitude and frequency traces in a drowsy subject; responses to medical staff talking and moving equipment nearby. (b) Shows (in another individual) much less variability with light general anaesthesia (isoflurane 0.5% end-tidal concentration) soon after thiopentone induction. Note the move to faster frequencies and absence of scalp muscle potentials which were prominent in (a). (From Prior (1987), by permission.)

reduced by high quality nursing care, pain relief and sedation (*Figs 5.1(b)* and *5.2*). Muscle relaxants may also reduce these 'macro' variations in a non-specific way, even though they may have no direct effects on the EEG (Kovarik et al 1994). Habibi and Coursin (1996) suggested that improvements in pain control, adequacy of sedation and neuromuscular blockade could be made by a combination of clinical judgement and scoring scales, quantitative neurophysiological tools, and closed loop systems for delivery of drugs. In patients who are unresponsive to standard clinical stimulation, continuous EEG monitoring appears to provide guidance for control of sedation and avoidance of oversedation whether by simple visual observations of the monitor output (*Figs 5.1* and *5.2*) or with some forms of more advanced signal processing (De Deyne et al 1998, Brocas et al 2002), providing algorithms include appropriate capabilities. Since EEG monitoring in this context appears to be effective as much because it can indicate the presence of periods of arousal as by identifying depth of sedation (Ely et al 2004) discussions will doubtless continue about refining appropriate tools for both sedated and unsedated ICU patients (e.g. Walder et al 2001, Gilbert et al 2001, Frenzel et al 2002, Riker et al 2003, Bauerle et al 2004, Deogaonkar et al 2004).

The effects of sedative agents on the monitored EEG depend on the mode and rate of delivery (*Fig. 5.3*), and, with rapid intravenous administration, may resemble those seen during induction of anaesthesia with intravenous agents which may produce a burst suppression pattern (*Figs 5.2* and *5.4*). Classically, sedatives produce an increase in amplitude, combined with prominent fast components (especially with barbiturate or related agents), followed by an increasing amount and amplitude of slow waves as faster components decrease; with elective use of heavier sedation, periods of relative suppression occur separated by low amplitude theta and delta waves, much as during induction of anaesthesia. The changes are dose-related (see *Fig 5.3*) and in broad terms reflect the rate and degree of depression of cerebral metabolism. Some interindividual variation occurs. EEG features during withdrawal from sedative regimens are also variable, since slow metabolism and clearance, and the sedative effect of active metabolites, may lead to accumulation, and unexplained prolongation of coma in critically ill patients in the ICU (Herkes et al 1992). The EEG may show features such as consistent asymmetries, localized reduction of activities or paradoxical slow wave arousal patterns (see *General categories of abnormality in the EEG of importance for ICU and intraoperative monitoring*, p. 114) that may hint of possible neurological deficits later. Atypical triphasic waves (see *General categories of abnormality in the EEG of importance for ICU and intraoperative monitoring*, p. 114) may occur with high doses of barbiturate sedatives and should not be misinterpreted as evidence of 'electrical' status epilepticus (e.g. Lancman et al 1997) or of hepatic or other metabolic encephalopathy.

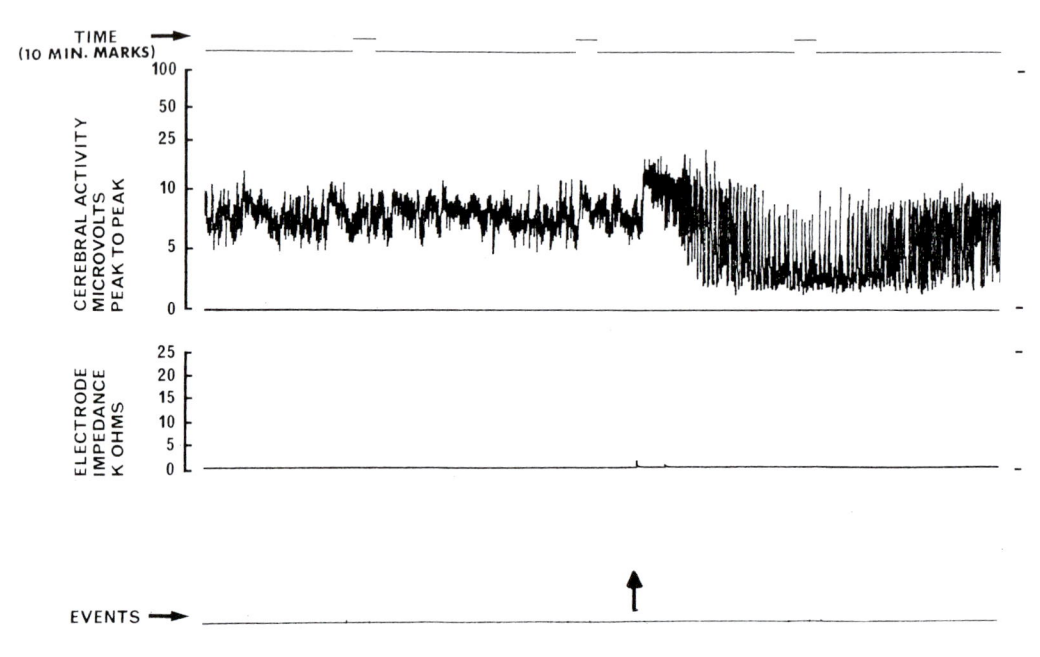

Fig. 5.2 Effect of an intravenous bolus of a short-acting sedative/anaesthetic agent althesin (alphaxalone–alphadolone, no longer in use) in a 17-year-old man lightly comatose after head injury. The Cerebral Function Monitor (for a description see p. 76) trace was variable initially but showed an abrupt increase in amplitude on at least two occasions when his mother spoke to him. The bolus (at arrow) produces an 'induction' effect; an abrupt rise in amplitude, followed by burst suppression (confirmed by multichannel EEG) which became less marked after about an hour. (From Prior and Maynard (1986), by permission.)

Sedation: assessment with EEG and evoked potentials

(a) Before

(b) 30 mg

(c) 80 mg

(d) 130 mg

100 μV
1 sec
HF 70 TC 0.30

(e) Recovery at 4 minutes

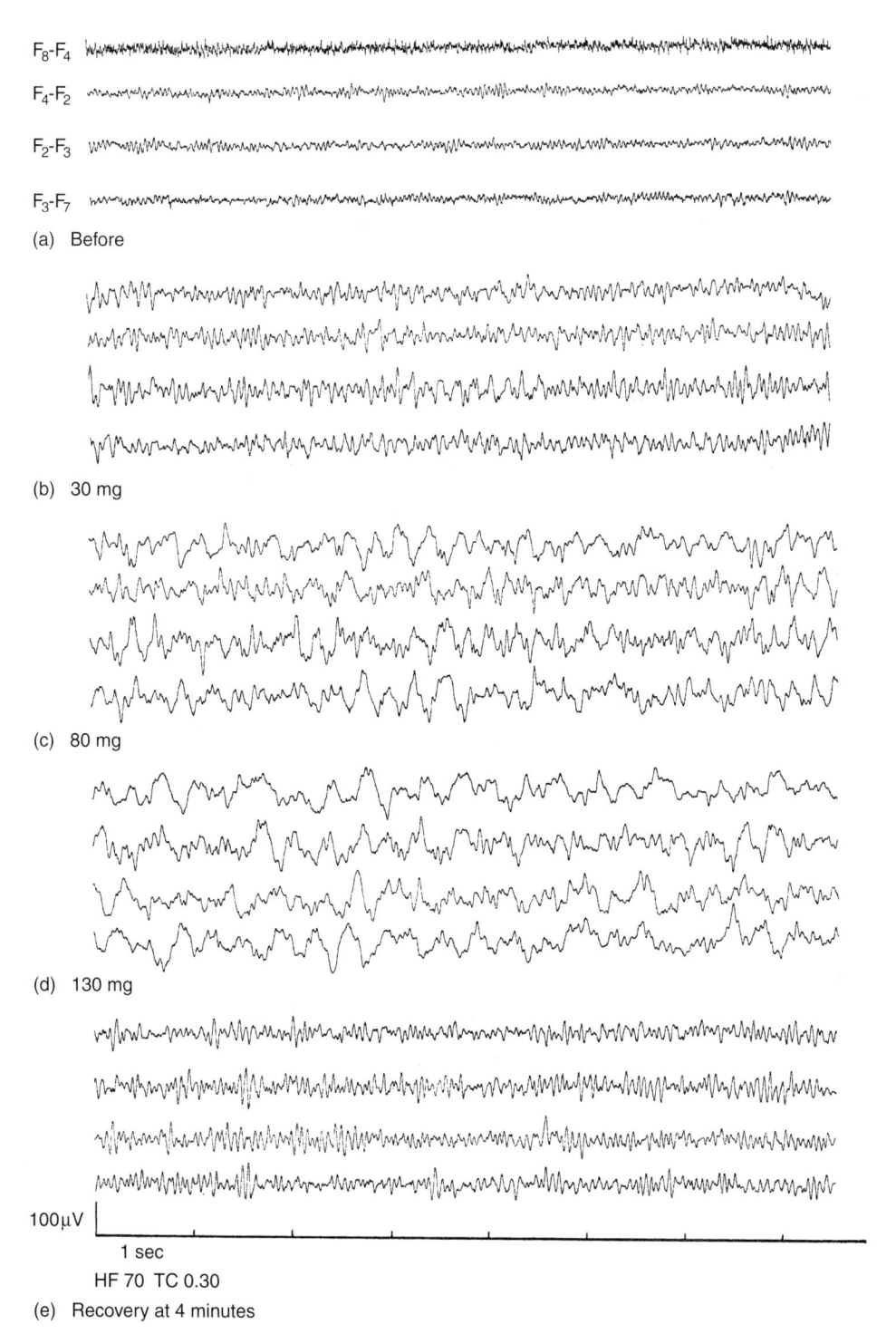

Fig. 5.3 Effects of methohexitone. Sequential EEG samples during slow induction of light anaesthesia in a 59-year-old man: (a) initial trace; (b) after 30 mg; (c) after 80 mg; (d) after 130 mg; (e) recovery 5 min later. The sequence illustrates a useful paradigm of dose–response relationships. (From Binnie et al (2003), by permission.)

Effects of sedative drugs on EPs

Evoked potentials (EPs) are more resistant to central nervous sytem (CNS) depressants than is the spontaneous EEG, this being especially true for short latency components originating in subcortical or primary cortical structures. However, even short latency EPs may be substantially distorted with increasing blood levels of sedatives and anaesthetics. Middle and long latency responses are much more sensitive to drugs, and this is a limiting factor in their use in the ICU, which will be discussed later.

As a general rule, EPs are more affected by volatile than intravenous anaesthetics, but since the former are

Chapter 5 Neurophysiological Monitoring during Sedation and Anaesthesia

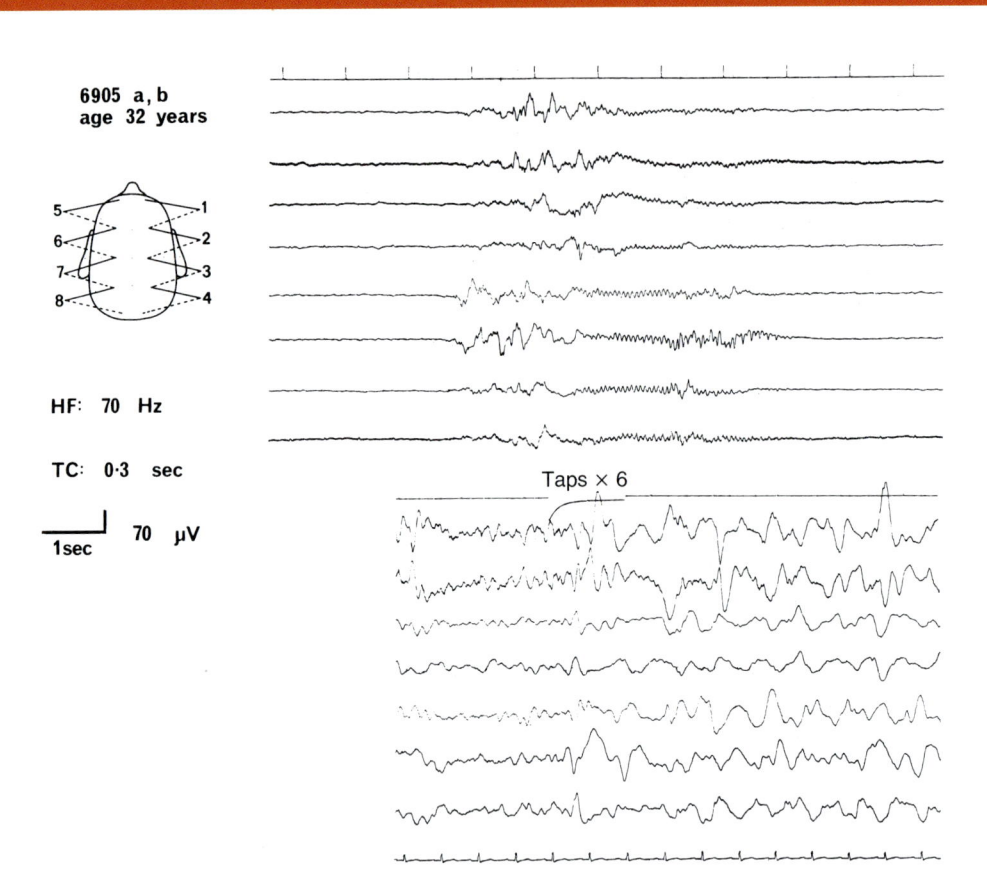

Fig. 5.4 EEG in a 32-year-old woman with herpes simplex encephalitis who was being ventilated in intensive care because of raised ICP. (a) During sedation with the short-acting intravenous sedative/anaesthetic agent althesin (alphaxalone–alphadolone, no longer in use); note burst suppression pattern with asymmetrical 15 Hz spindles. (b) After withdrawal of sedation: continuous activity with attenuation of faster components and increase in polymorphic delta activity following auditory stimulus (tapping); channel 8 monitors ECG. (From Binnie et al (2003), by permission.)

very seldom used in the management of comatose patients they will not be considered here. Intravenous CNS depressants, especially barbiturates, are used in coma to reduce both brain metabolic demands and intracranial pressure (ICP) (Marshall and Shapiro 1977, Marshall et al 1979). While their value in increasing survival rates is still controversial (Trauner 1986, Eisenberg et al 1988), their use in ICUs has progressively increased to such an extent that, in some hospitals, the majority of EPs in the neuro-ICU are recorded with the subject under barbiturate or other forms of major sedation (Hall et al 1985). Hence, a thorough knowledge of the effects of different CNS depressant agents on EPs is of paramount importance for interpretation.

Barbiturates

Barbiturates at doses commonly used for 'brain protection' in comatose patients may abolish motor and reflex functions, as well as the EEG, but have little effect on short latency EPs. Several studies in man and animals have demonstrated that brainstem auditory evoked potentials (BAEPs) are extremely resistant to high doses of barbiturates sufficient *per se* to induce coma (Sohmer et al 1978, Bobbin et al 1979, Duncan et al 1979, Sutton et al 1982, Marsh et al 1984, Hall 1985), although Shapiro et al (1984) showed that the use of extremely high doses of pentobarbital (about five times the anaesthetic level) could distort and eventually abolish BAEPs in the rat. In our experience, BAEPs are not modified, as compared to control values in the same patient, for blood levels of thiopentone up to 35 mg/l (García-Larrea et al 1987). With higher blood levels, amplitudes decline and latencies increase, but waveforms still remain within normal limits for levels that may reach 100 mg/l (Drummond et al 1985, Hall 1985).

Somatosensory central conduction time (CCT) between subcortical and primary cortical components is also resistant to barbiturates and remains within normal limits with doses producing deep anaesthesia (Hume and Cant 1981, Sutton et al 1982, Ganes and Lundar 1983, Drummond et al 1985). However, although the barbiturate levels needed to abolish short latency EPs largely exceed those currently employed in ICUs, latency increases may be observed with lower doses in both auditory and somatosensory modalities (Sutton et al 1982, Drummond et al 1985, McPherson et al 1986), and this alteration in serial recordings can be misinterpreted as a sign of neurological deterioration. Somatosensory potentials of longer latency than the parietal N20–P24 or N20–P27 complex, as well as the middle latency auditory responses Na and Pa, are much more affected by barbiturate anaesthesia. Both in animals and man, intravenous barbiturates distort middle latency auditory evoked potentials (MLAEPs) at doses that do not affect brainstem potentials (Celesia and Puletti 1971, Harada et al 1980, Buchwald et al 1981, Hall 1985). MLAEP subcomponents Na and Pa appear to be differently affected by anaesthetics, since in cats pentobarbital infusion is able to distort or even abolish Pa, while Na remains relatively stable (Harada et al 1980), although similar observations are lacking in man. The effects of anaesthesia on the MLAEPs of comatose patients have not yet been extensively studied, but the above results indicate that barbiturates may seriously hamper their interpretation. In particular, absence or abnormality of middle latency EPs should never be interpreted as an unfavourable sign without reference to the possibility of effects from administration of CNS depressant medication in the preceding hours or days.

Other intravenous CNS depressants

Phenytoin in high doses, such as those employed for status epilepticus, affects early somatosensory evoked potentials (SEPs) and auditory evoked potentials (AEPs). In rats, BAEPs became significantly depressed in amplitude and of increased latency with plasma phenytoin levels of 45 mg/l; they were abolished at a mean plasma level of 60 mg/l (Hirose et al 1986). Both the wave I latency and the I–V IPI were affected in animals and man, suggesting a mixed effect of the drug at central and peripheral levels. In normal subjects, a single loading dose of 14 mg/kg oral phenytoin did not change the I–V interval in the study of Shin et al (1988). Regarding somatosensory conduction, Green et al (1982) reported an increase of CCT, but Drake et al (1988) found no change with therapeutic levels of the drug. Mavroudakis et al (1991) have since demonstrated in a detailed study a reversible latency increase of human SEP components N13 and N20 with serum levels of phenytoin between 19 and 25 mg/l. The N9 and N11 potentials remained stable at concentrations lower than 30 mg/l, although it is known that peripheral nerve conduction may be decreased at higher serum levels (Marcus et al 1981).

Systemic infusion of lidocaine distorts the morphology and increases the latencies of BAEP components in rats (Javel et al 1982), although its oral homologue Tocainide has failed to do so in humans (Wärsterström, 1985). High dose lidocaine in association with thiopentone severely distorts BAEPs in comatose patients (García-Larrea et al 1987, 1988); lidocaine seems to be mainly responsible for this effect since it appeared at thiopentone levels known not to affect BAEP waveforms. However, the separate contributions of each drug to the BAEP depression are difficult to distinguish, for lidocaine at general anaesthetic doses is seldom administered alone because of its convulsant properties (Munson et al 1975). Intrathecal lidocaine has been shown recently to abolish short latency SEPs in man, a property not shared by the opioid agent fentanyl (Chabal et al 1988).

Etomidate and ketamine are unusual because they enhance somatosensory responses (Kalkman et al 1986, McPherson et al 1986, Schubert et al 1990). Etomidate in particular has been advocated by some to counteract the depressant effects of other drugs on these potentials during surgery (Sloan et al 1988).

It must be pointed out that most experimental studies on the effects of CNS depressants on EPs have been conducted with the use of single drugs, whereas polytherapy with sedative agents is a common clinical practice. The additive or potentiating effects of several drugs acting together may not be predictable on the bases of their actions when administered alone.

Assessment of sedation

Avramov and White (1995) made a helpful general assessment of methods for monitoring the level of sedation. They concluded that although processed EEG and AEPs were the best possible sources of objective information about brain function, no single EEG processing algorithm or AEP modality then available had been clinically proven to provide useful information to guide administration of sedative and analgesic medication in the ICU. At the time of their survey they noted the lack of commercial devices available for this purpose (this is now being remedied), but emphasized the encouraging prospects with newer algorithms, which include signal quality estimation and artefact rejection, combined with increasingly cheap and powerful computer technology. Other workers have shown similar results. Neither median power frequency nor spectral edge frequency correlated with clinical sedation scores or blood levels of sedative drugs in a study of isoflurane ($n = 15$) and midazolam ($n = 8$) in ICU patients requiring ventilation and the authors concluded that no simple univariate measure of the EEG was likely to correlate with depth of sedation in critically ill patients (Spencer et al 1994). This view is in accord with the work of Levy (1984b) which demonstrated the inadequacy of univariate spectral measures to describe EEG changes during deep anaesthesia and profound hypothermia used during cardiac surgery.

More advanced systems using various forms of multivariate pattern recognition to characterize the typical cluster of EEG features associated with each state (defined clinically or by reference to blood levels of drug) began to offer more appropriate solutions. These include the work of Veselis et al (1991) based on 63 EEG patterns in 26 ventilated patients receiving long-term sedation with midazolam. The patterns were compared with sedation levels judged by a seven-point clinical scale of responses to suction via the endotracheal tube. Both power and topographic changes were investigated and effective pattern classifiers based on artificial neural networks developed.

Chapter 5 Neurophysiological Monitoring during Sedation and Anaesthesia

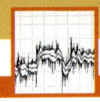

This work was advanced further in a comparative study of analysis methods in 29 volunteers receiving sedative doses of thiopental, midazolam or propofol (Veselis et al 1993). The authors reported that the three agents had different EEG profiles and that both neural network and discriminant analysis were useful in identifying these differences. They considered that EEG spectra should be analysed without using classical EEG bands (alpha, beta, etc.) and neural networks used to identify frequency bands that are 'important' in specific drug effects on the EEG. They concluded that once a classification algorithm is obtained using either a neural network or discriminant analysis, it could be used as an online monitor to recognize drug-specific EEG patterns.

Since that time, many studies have shown that data from modern EEG monitors are generally capable of showing close relationships with clinical evaluation. Comparisons between Ramsay scoring (Ramsay et al 1974) and the EEG-based Patient State Index (PSI) (Physiometrix, North Billerica, MA, USA; now Hospira Inc., Lake Forest, IL, USA, http://www.hospira.com) suggested promising possibilities for predictive information in quantifying levels of sedation in ICU patients (Schneider et al 2004) (*Fig. 5.5*). In a study of patients with severe traumatic brain injuries before and during sedation, Deogaonkar et al (2004) showed a statistically significant correlation between Bispectral Index (BIS) scores and three commonly used sedation–agitation scales: the Richmond Agitation–Sedation Scale (RASS), the Sedation–Agitation Scale (SAS) and the Glasgow Coma Scale (GCS) scores; they reported that the newer BIS XP software package was useful in this context. Another study concerning agitation in premedicated children during routine induction of anaesthesia showed different BIS levels in those who were and were not agitated at the time of induction (Constant et al 2004) (*Fig. 5.6*).

Assessment problems are somewhat different when levels of sedation sufficient to produce burst suppression are required as part of so-called 'brain protection' regimens or to help control extremely high ICP, for example after head injury (Brain Trauma Foundation 1996, Maas et al 1997). These regimens, originally involving the administration of barbiturates (Marshall et al 1979, Rockoff et al 1979, Kassell et al 1980, Michenfelder 1986), but now more commonly propofol, have been implemented with the intention of so depressing cerebral neuronal metabolism that the requirement for delivery of oxygen falls to the minimum for cell viability with the onset of electrocerebral silence (ECS). Thereafter it remains unchanged at a basal level consistent with continuing viability of neurones, even when drug levels increase (Michenfelder 1974) (*Fig. 5.7*). This has two consequences; first, the associated reduction of cerebral blood flow (CBF) leads to a decrease in ICP and, secondly, a more favourable balance between supply and demand for oxygen may be achieved which will reduce the risk of neuronal damage from ischaemia (Bingham et al 1985, Procaccio et al 1988). The practical guidelines for management have utilized a chosen degree of burst suppression which may be simply characterized in terms of a minimum number of bursts per minute (Riker et al 2003), suppression ratios (Theilen et al 2000) or an interburst interval scoring or a percentage time with suppression to below a chosen amplitude threshold. Once infusion is reduced or stopped, neuronal oxidative metabolism increases, accompanied by a return of EEG activity reflecting a more active functional state, always provided that major brain damage is not present. The sequential changes with this cycle of increasing then decreasing sedative levels are comparable to those following drug overdose; outcome likewise depends upon maintenance of an adequate supply of oxygenated blood to vital organs. The contribution of EEG has been significant in that it provides a more accurate and direct index of sedation than serum levels of sedative agents, which do not correlate well with cerebrospinal fluid (CSF) levels in patients with severe brain injuries, and helps avoid toxicity by indicating the lowest doses of drug to produce the required effect (Winer et al 1991). Since short latency EPs are less affected by CNS depressant drugs than the EEG (Ganes and Lundar 1983), they have value as a means of mapping integrity of neural pathways during brain protection regimens and recently work has begun to develop systems with a combination of EEG and EP assessment (see review by van der Kouwe and Burgess 2003). It should be noted that the concept of 'brain protection' has been controversial (see review by Drummond (1993)) and it remains an area in which there are continuing discussions

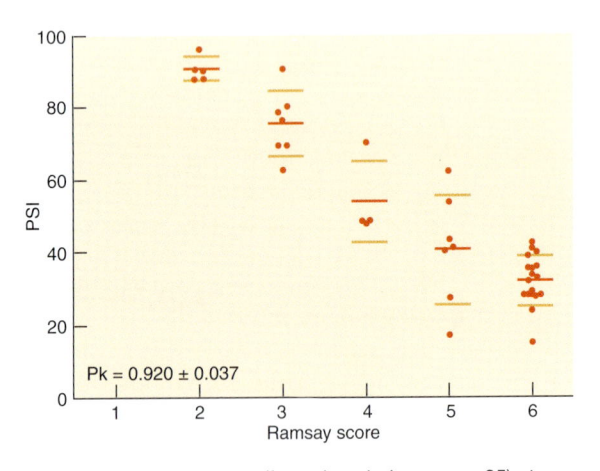

Fig. 5.5 Patient State Index (for a description see p. 85) at different levels of sedation as measured by Ramsay score in 41 intubated adult patients in a surgical ICU. The graph shows individual values (solid circles), means (red line) and standard deviations (SDs) (yellow line) at each level of sedation. Following Bonferroni-correction, paired comparisons revealed significant differences except between levels 5 and 4, and 5 and 6 (*p* < 0.077). This is also reflected by a high prediction probability (Pk) for the complete data set. (From Schneider G, Heglmeier S, Schneider J, Tempel G, Kochs EF 2004 Patient State Index (PSI) measures depth of sedation in intensive care patients. *Intensive Care Medicine* 30:213–216, by permission. © 2004 Springer-Verlag, with kind permission of Springer Science and Business Media.)

Anaesthesia: assessment with EEG and evoked potentials

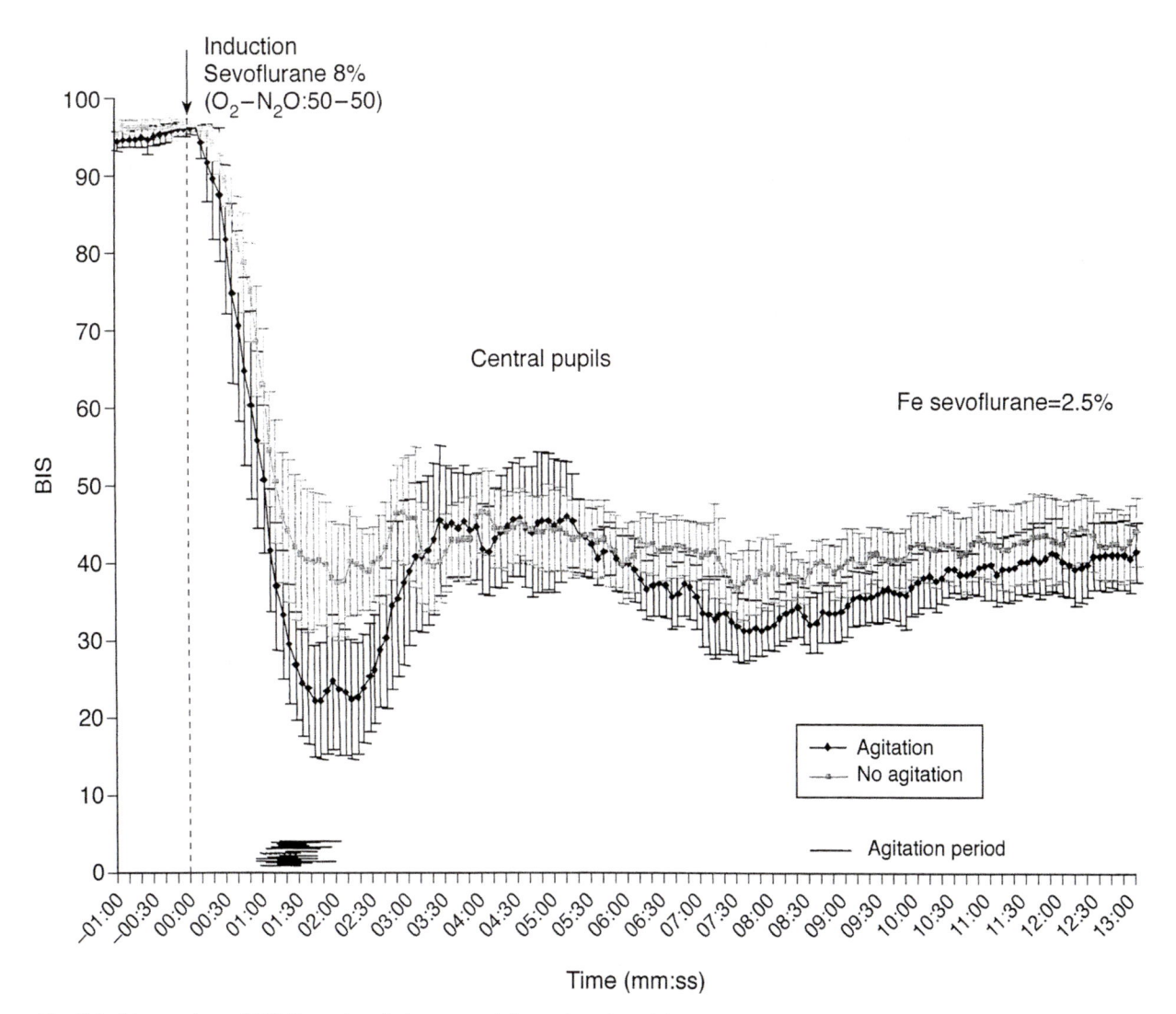

Fig. 5.6 Mean values of BIS (for a description see p. 84) as a function of time, measured during sevoflurane induction of anaesthesia for tonsillectomy in 40 children who demonstrated agitation and those who demonstrated no agitation (mean and 95% confidence interval). These traces are synchronized from the beginning of sevoflurane inhalation. Individual agitation periods are represented by black marks (lower left above time scale). (From Constant I, Leport Y, Richard P, Moutard M-L, Murat L 2004 Agitation and changes of Bispectral Index™ and electroencephalographic-derived variables during sevoflurane induction in children: clonidine premedication reduces agitation compared with midazolam. *British Journal of Anaesthesia* 92:504–511. © The Board of Management and Trustees of the *British Journal of Anaesthesia*. Reproduced by permission of Oxford University Press/*British Journal of Anaesthesia*.)

as to the overall benefits of regimens which include optimization of physiology and non-pharmacological interventions as well as sedation (Maas et al 2000, 2004, Maas 2002, Bulger et al 2002, Narayan et al 2002). Nonetheless, EEG and EP monitoring continue to have an important role when major sedatives form a part of the regimen.

ANAESTHESIA: ASSESSMENT WITH EEG AND EVOKED POTENTIALS

Effect of anaesthetic agents on the EEG

The effects of anaesthesia on the electrical activity of the brain were first described in animals (Beck 1890, Von Marxow 1890), and it was clear that major depression occurred which could be reversible. Hans Berger first reported his observations on the human EEG in 1929, he subsequently described the effects of various narcotics and anaesthetics on the EEGs of patients with intracranial lesions and with skull defects, in several of his reports from 1931 onwards (available in the excellent translations of Gloor 1969). The early clinical neurophysiologists showed considerable interest in EEG phenomenology during anaesthesia and surgery (e.g. the changes during craniotomy reported by Grey Walter in 1936 (see p. 255, *Fig. 7.1*) (Walter 1936)). Beecher and McDonough (1939) showed that deep anaesthesia with inhalational agents could produce an isoelectric EEG; when administration of anaesthetic ceased, normal EEG activity usually returned and the patient recovered without adverse cere-

Chapter 5 Neurophysiological Monitoring during Sedation and Anaesthesia

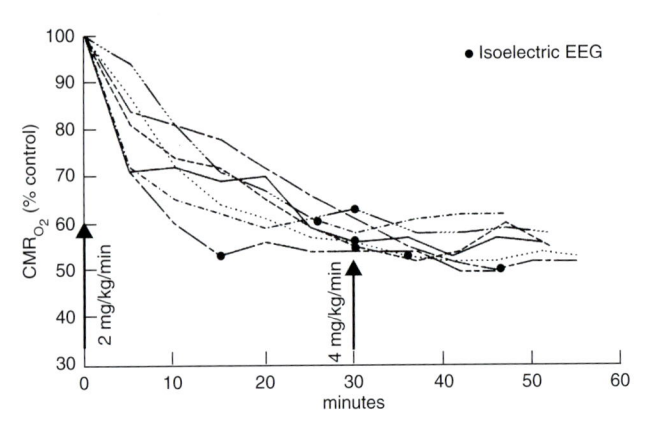

Fig. 5.7 This classic demonstration shows how the cerebral metabolic rate for oxygen ($CMRO_2$) in 7 subjects falls with increasing thiopental infusion. Once the EEG becomes isoelectric, no further depression of $CMRO_2$ occurs, even with an increased thiopentone infusion rate. The interdependency of cerebral functional and metabolic effects following massive doses of thiopentol in the dog. (From Michenfelder JD 1974 *Anesthesiology* 41:231–236, by permission.)

bral sequelae. Later, formalized stages of anaesthesia were defined in terms of stages of EEG features during deepening nitrous oxide–ether anaesthesia to parallel Guedel's (1937) 'planes' of anaesthesia for inhalational agents (Courtin et al 1950):

1. the normal alpha rhythm (of the relaxed, awake subject) is replaced by low voltage fast activity;
2. regular slow, high voltage activity;
3. irregular slow wave activity with high voltage slow waves;
4. periods of electrical silence of less than 3 s duration;
5. periods of electrical silence of more than 3 but less than 10 s duration;
6. periods of electrical silence of more than 10 s duration (levels 4–6 usually show a progressive decrease in amplitude); and
7. complete electrical silence.

In this sequence of changes, levels 2–6 were considered equivalent to Guedel's planes I–V of surgical anaesthesia, and EEG level 7 to plane IV with respiratory paralysis. By the end of the 1950s, Martin et al (1959) had shown that such sequential changes were common to all general anaesthetics and could be described by a near parabolic curve of integrated amplitude against EEG stage (*Fig. 5.8(a)*). It is of interest that with modern monitors comparable relationships with Guedel's planes of anaesthesia for ether can be demonstrated (Bhargava et al 2004) (*Fig.5.8(b)*). Subsequent work on dose–response curves for different agents was reported by Faulconer and Bickford (1960) in their monograph, and in the comprehensive reviews by Clark and Rosner (1973) and Rosner and Clark (1973) which cover effects on both the EEG and sensory EPs. These seminal publications provide an essential background to the more recent literature.

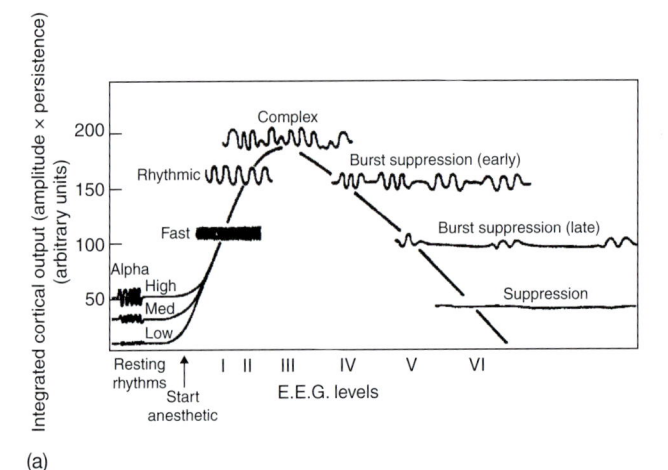

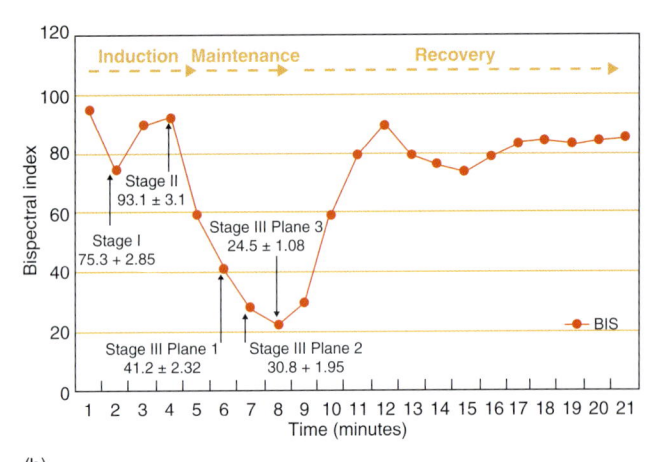

Fig. 5.8 (a) Diagram of average changes in the EEG pattern during anaesthesia. Early in the variations from normal comes an increase in frequency to 30 or 40 Hz; as consciousness is lost, this pattern of small fast waves is replaced by large (50–300 μV) slow waves (1–5 Hz) that increase in amplitude as they slow; the waves may become irregular in form and repetition time, and may have faster waves superimposed as the level of anaesthesia deepens; the amplitude next begins to decrease and periods of relative cortical inactivity (burst suppression pattern) may appear until the depression finally results in the entire loss of cortical activity and a flat or formless tracing. (From Martin JT, Faulconer A Jr, Bickford RG 1959 Electroencephalography in anesthesiology. *Anesthesiology* 20:359–376, by permission.) (b) BIS values under various stages of ether anaesthesia (mean ± SD) studied in 21 patients undergoing short surgical procedures. (From Bhargava AK, Setlur R, Sreevastava D 2004 Correlation of Bispectral Index and Guedel's stages of ether anesthesia. *Anesthesia and Analgesia* 98:132–134, by permission.)

Résumé of the main EEG effects of drugs used during anaesthesia

Nomenclature Neurophysiological staff undertaking intraoperative monitoring will need to familiarize themselves with the effects on the EEG of a wide range of drugs used during anaesthesia. To help those unfamiliar with intraoperative monitoring work, the names of some typical drugs in each category will be indicated along with their EEG effects, using the recommended International Non-proprietary Name (rINN) where available or current national formularies (e.g. British National Formulary). A preliminary briefing with the anaesthetic team is helpful regarding names and purpose of the drugs likely to be encountered, as well as the typical protocols used by each of the anaesthetists in the team for any particular type of surgical procedure. Long experience has shown that most anaesthetists have their own recognizable EEG signatures – likewise different surgeons with their individual variations of artefact-producing surgical manoeuvres!

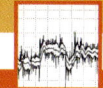

General classes of anaesthetics and EEG measures of efficacy
Anaesthetics are conventionally divided into inhalational and intravenous agents. Although the overall changes from induction to 'dangerously deep' anaesthesia have some common EEG features between the two classes of agent, marked differences arise from their rates of achieving an effective 'brain dose'. Even with administration of comparable anaesthetic 'doses' (e.g. the minimum alveolar concentration (MAC) of Eger et al (1965)), the EEG effects vary considerably with age and between individual patients in a manner which suggests that some patients are more 'lightly' or 'deeply' anaesthetized than others. This has led to suggestions that electrophysiological endpoints, such as onset of burst suppression pattern, may provide a better measure of equivalent anaesthetic depth than estimation from 'dose' administered (Hoffman and Edelmann 1995, Schwilden et al 1995), even though there may be differences in endpoint between agents (Mi et al 1999). Certainly, the onset of ECS (or isoelectric EEG) provides a clear indication of the maximum obtainable reduction in cerebral oxygen metabolism and associated changes in CBF (Michenfelder 1974, Lam et al 1995).

Combinations of drugs and 'balanced anaesthesia' In modern anaesthetic practice, a balance between the patient's need for pain relief and oblivion and the surgical need for relaxed muscles, is managed by a combination of analgesic, muscle relaxant (neuromuscular blocking), and anaesthetic drugs of inhalational and intravenous types. In addition, there may be supplementary use of a range of other drugs, including local anaesthetics for local nerve blocks and drugs for controlled arterial hypotension. This means that the aim of neurophysiological monitoring systems must be the provision of indicators of the overall effect on the brain of complex anaesthetic polytherapy in the context of variations in intensity of surgical stimulation during the procedure in the individual patient.

Potential epiletogenicity The potential epileptogenicity of some drugs used during anaesthesia is of special importance during neurosurgical monitoring and in patients with known seizure disorders, e.g. those undergoing electrocorticography. Anaesthetics have both pro- and anticonvulsant effects (discussed below with individual groups of agents or drugs) and some agents may elicit EEG discharges or clinical events even in patients without a previous history of epilepsy (Modica et al 1990, Yli-Hankala et al 1999b).

Effects of anaesthetics on CBF The effects of anaesthetics on CBF vary although the depression of metabolism produced by anaesthesia is generally reflected in a commensurate decrease in blood flow. One main objective of neurophysiological monitoring is to help reduce the risk of hypoxic–ischaemic brain or spinal cord damage at times of special risk; thus it is necessary to be aware of both advantageous and disadvantageous effects of different groups of anaesthetics on CBF and the delivery of, and metabolic demand for, oxygen in all patients in whom reduction of the supply of oxygenated blood to the brain may occur during surgery (McDowall et al 1978).

General anaesthetics

(a) Intravenous agents Intravenous agents are used for induction and for maintenance of anaesthesia. Administration is by rapid induction dose, supplementary bolus, or steady state infusion, sometimes by infusion pump which may have a feedback control or advice system based on measures including quantitative EEG (qEEG) features. These may include Target Control Infusion (TCI) systems which are, in effect, pure pharmacokinetic models aiming to maintain the drug concentration of 'the target organ' as steady as possible. Such systems have been developed to avoid prolonged/increasing emergence times due to this accumulation after prolonged infusions with opioids (remifentanil is an exception).

Most intravenous agents produce major EEG changes with an initial increase in fast components (activation), increasing amount and amplitude of theta and then delta activity replacing alpha and beta activity (inversion of slow/fast power ratio), leading to onset of intermittent periods of subtotal then total suppression (burst suppression) of increasing duration until ECS occurs (*Fig. 5.9*). The whole sequence to burst suppression pattern may occur within one arm-to-brain circulation time with rapid induction of anaesthesia. It is reversible providing supportive measures maintain ventilatory and circulatory functions. The complexity of this sequence of changes leads to problems in computer modelling of the EEG correlates of intravenous agents.

Barbiturates Intravenous injection produces prominent, frontally accentuated, fast activity, typically at 20–24Hz, which becomes mixed and sometimes phase-locked with slow components to give a so-called 'mitten pattern' or atypical triphasic waves. burst suppression pattern follows, with increasingly prolonged periods of suppression (Brazier and Finesinger 1945). Burst suppression activity can be recorded from the hippocampus as well as the neocortex, the decreasing synchronization of the bursts with increasing thiopental concentration suggesting that a depression of synaptic coupling contributes to anaesthesia (MacIver et al 1996).

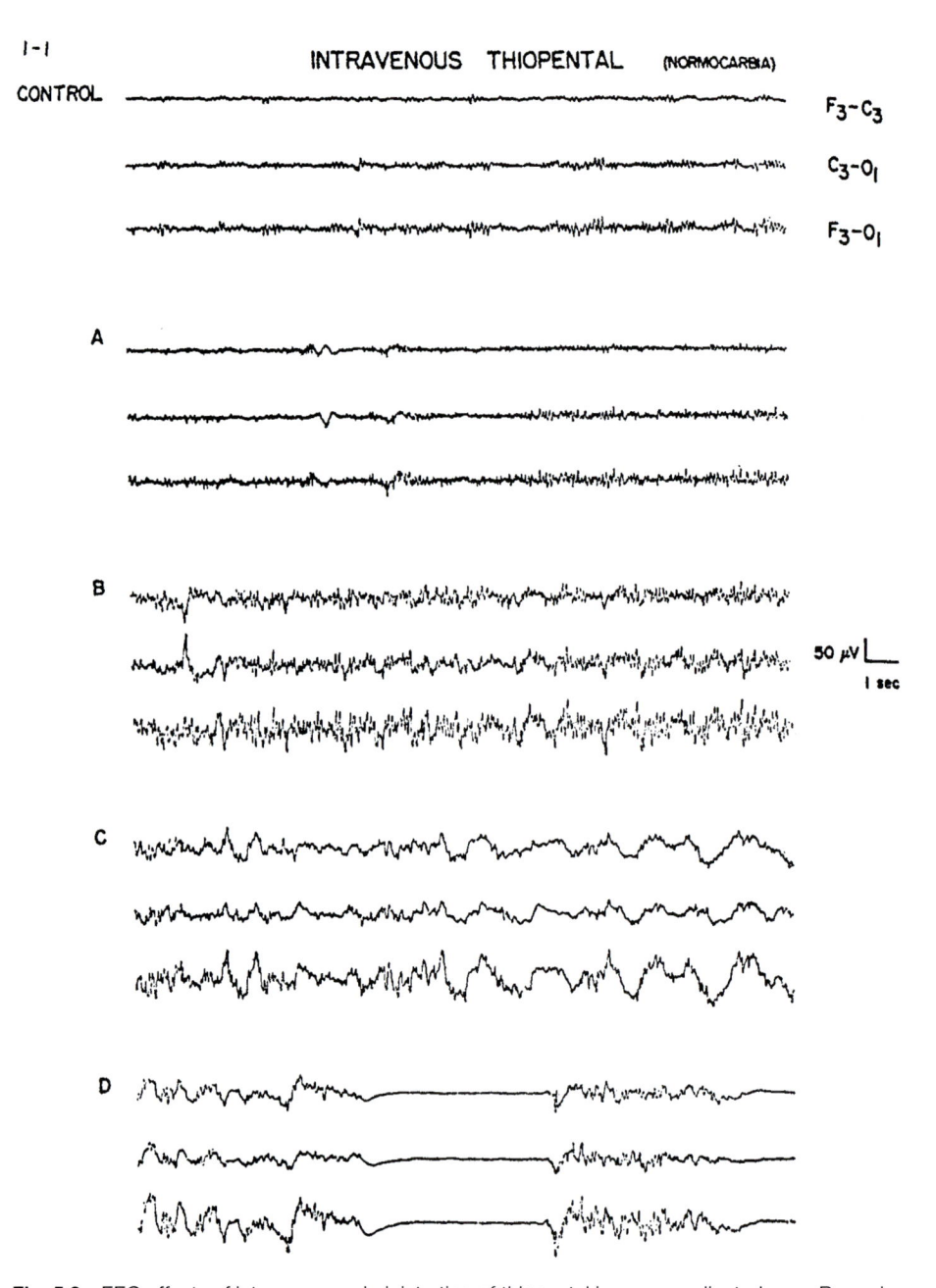

Fig. 5.9 EEG effects of intravenous administration of thiopental in unpremedicated man. Records for each condition were taken from three bipolar pairs of electrodes designated at the right of the control tracing. (A) Rapid activity at onset of administration. (B) Spindles of 7–10 Hz. (C) Slow waves. (D) Suppressions and intersuppression 'bursts'. (From From Clark DL, Rosner BS 1973 Neurophysiologic effects of general anesthetics. I. The electroencephalogram and sensory evoked responses in man. *Anesthesiology* 38:564–582, by permission.)

Thiopental sodium This is mainly used for induction of anaesthesia but has also been utilized in brain protection regimens during surgery and in the neuro-ICU in management of patients with substantially raised ICP or with intractable status epilepticus.

Other intravenous anaesthetic agents These include etomidate, ketamine and propofol, all of which produce broadly similar EEG effects to barbiturates during induction of anaesthesia, but with some individual differences of relevance to EEG monitoring.

Etomidate This is associated with a high incidence of excitatory movements, including myoclonus, tremor and dystonic posturing, during induction of anaesthesia. In one comparative study of induction in 67 unpremedicated adult patients, such movements were observed in 86.6% induced with etomidate, 16.6% given thiopental, 12.5% given methohexital and 5.5% given propofol (Reddy et al 1993). The movements usually coincided with the appearance of slow wave activity in the EEG. In those receiving etomidate, movements were myoclonic in 69%, being accompanied by multiple spikes in 22%, although none of the patients developed seizures.

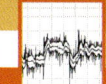

Ketamine (phencyclidine) This produces prominent rhythmic fast activity of unusually high frequency (30–40 Hz) and amplitude which becomes interspersed with periodic delta waves of low frequency (Schwartz et al 1974, Rosén and Hagerdal 1976). EEG features vary somewhat between different ketamine isomers (White et al 1985). Clinical effects in subanaesthetic doses include dream-like and hallucinatory experiences. Phencyclidine is a drug of abuse; in excessive dosage producing coma, subjects exhibit widespread sinusoidal theta activity interspersed every few seconds by periodic slow wave complexes in an appearance comparable to that in deep ketamine anaesthesia (Stockard et al 1976). Epileptiform discharges and clinical seizures were reported in many early studies; indeed, in patients undergoing surgery for epilepsy, ketamine may activate more discharges than barbiturates. EEG monitoring systems should have the capacity to identify the very high frequency components, differentiate them from muscle potentials and other forms of high frequency artefact (van de Velde et al 1998, 1999), and designers should consider developing pattern recognition algorithms similar to those used in epilepsy monitoring, to identify the periodic generalized slow waves seen with ketamine anaesthesia.

Propofol This elicits a fairly consistent sequence of EEG changes, similar to those described above for the main intravenous agents (Reddy et al 1992). The EEG features have proved amenable to simple quantification measures, such as median power frequency, for help with closed loop or manually controlled infusion schemes (Schwilden et al 1989, Forrest et al 1994). They can be tracked with various modern EEG monitors and show close dose–response relationships (e.g. Kreuer et al 2004) (*Fig. 5.10*). With dosage sufficient to produce prolonged ECS, cerebral arterial pressure autoregulation and CO_2 reactivity remain intact (Matta et al 1995). Excitatory movements during induction with propofol are infrequent in adults compared with etomidate (see above). In children, a diagnostic problem may arise because of various dystonic movements, described as cough, hiccup, hypertonus, twitching or tremors; these are associated with delta waves but not epileptiform features in the EEG and probably are of subcortical origin (Borgeat et al 1991). Propofol sedation during 'awake craniotomy' for epilepsy surgery, activated less seizures than 'neuroleptanaesthesia' (see below) using fentanyl and droperidol (Herrick et al 1997), activation at doses insufficient to produce burst suppression not being significantly different from that with thiopental (Hewitt et al 1999) in spite of an increase in spike discharges, spread to new sites or polyphasia (Smith et al 1996).

(b) Inhalational agents Inhalational anaesthetic agents are mainly used for maintenance of anaesthesia. These may be gaseous or volatile, the latter being in a liquid state at room temperature, but are vapourized in the anaesthetic apparatus where they mix with a carrier gas such as nitrous oxide (which itself has some analgesic and anaesthetic properties). The main volatile agents in present use are the halogenated agents – chlorinated and fluorinated hydrocarbons. Apart from the potential hazards of explosion inherent in open circuit inhalational anaesthetic systems, and of workplace pollution, there has been some anxiety about atmospheric contamination with associated detrimental effects on the ozone layer.

The EEG features of inhalational anaesthetics are generally less dramatic than with intravenous anaesthetics, especially with the earlier gaseous agents, mainly because of the slower build up of concentration in the brain. Nonetheless, they do produce a fairly comparable sequence of EEG changes overall (with the exception of activation of fast activity) to that with the intravenous agents (*Fig. 5.11*), although the profile of induction changes differs due to the different rate of reaching com-

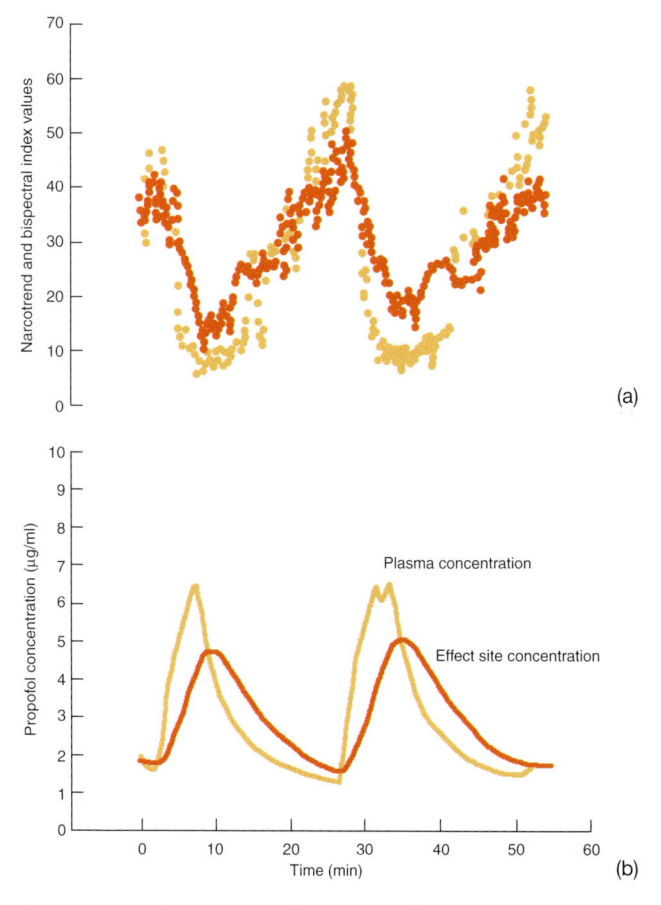

Fig. 5.10 (a) Time course of Narcotrend (Monitor Technik, Bad Bramstedt, Germany, http://www.narcotrend.de) (yellow dots) and BSI (red dots) (for a description of the monitors see p. 85) for a typical patient. Each dot represents the EEG parameter value of a 5 s epoch. (b) Propofol plasma (yellow line) and effect site (red line) concentrations as predicted by the Diprifusor data set for the same patient. (From Kreuer S, Wilhelm W, Grundmann U, Larsen R, Bruhn J 2004 Narcotrend index versus bispectral index as electroencephalogram measures of anesthetic drug effect during propofol anesthesia. *Anesthesia and Analgesia* 98:692-697, by permission.)

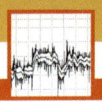

parable brain levels (*Fig. 5.12*). With establishment of anaesthesia, there is inversion of slow/fast power ratio followed by periods of subtotal then total suppression. ECS is rare except with high concentrations and there is clear interpatient variability in its appearance in different patients with comparable end-tidal concentrations (Lloyd-Thomas et al 1990a) (*Fig 5.13*).

Epileptiform activity may occur, especially with some of the halogenated agents (Yli-Hankala et al 1999b) (*Fig. 5.14*). Clark and Rosner (1973) proposed a scale of 'increasing central nervous system irritability' graded from the inert agents (xenon and nitrous oxide), the hydrocarbons (cyclopropane), the four-carbon ethers (diethyl ether, fluroxene), the three-carbon ethers (methoxyflurane, isoflurane, enflurane) to the halogenated ethers (trichloroethylene, halothane and chloroform).

Nitrous oxide is the gaseous anaesthetic agent most likely to be encountered during operative monitoring. It has both analgesic and mild anaesthetic properties and, not being adequate as a sole anaesthetic, is commonly used as part of a balanced technique with other inhalational or intravenous agents. The EEG appearances vary according to concentration and adjuvent agent; depth of anaesthesia, in the absence of burst suppression, may be judged by attainment of a chosen qEEG end-point, for example of median frequency (Schwilden and Stoeckel 1987, Röpcke and Schwilden 1996). If burst suppression has been reached, addition of nitrous oxide can change the EEG picture to that of a lighter level of anaesthesia, *pari passu* with an increase in CBF (Algotsson et al 1992, Yli-Hankala et al 1993, Matta and Lam 1995).

The main volatile liquid anaesthetics likely to be encountered during clinical monitoring nowadays are *isoflurane*, *desflurane* and *sevoflurane* (noted for rapid action and rapid emergence) all of which produce broadly similar EEG effects. Most comparative studies highlight differences in qEEG profiles between volatile agents that reflect differences in potency and speed of action (Lloyd-Thomas et al 1990a, Scholz et al 1996, Thomsen and Prior 1996, Schwender et al 1998a). Hypothermia (e.g. during open cardiac surgery) may have an additive effect and lead to the appearance of burst suppression at lower concentrations of volatile agents than would otherwise be the case. Epileptiform features, especially with sevoflurane, have been noted above (see *Fig. 5.12*). Studies during isoflurane-induced burst suppression have demonstrated the occurrence of direct current (DC) shifts during spontaneous bursts (Jäntti et al 1993) and an association between onset and offset of spontaneous bursts with accelerations and decelerations of heart rate (Yli-Hankala et al 1990). In spite of the long association of burst suppression with deep anaesthesia, there is some evidence of cortical reactivity in that visual (red light-emitting diode (LED) flashes via goggles), auditory and somatosensory stimulation during periods of suppression will elicit bursts and various evoked responses (Hartikainen et al 1995, Huotari

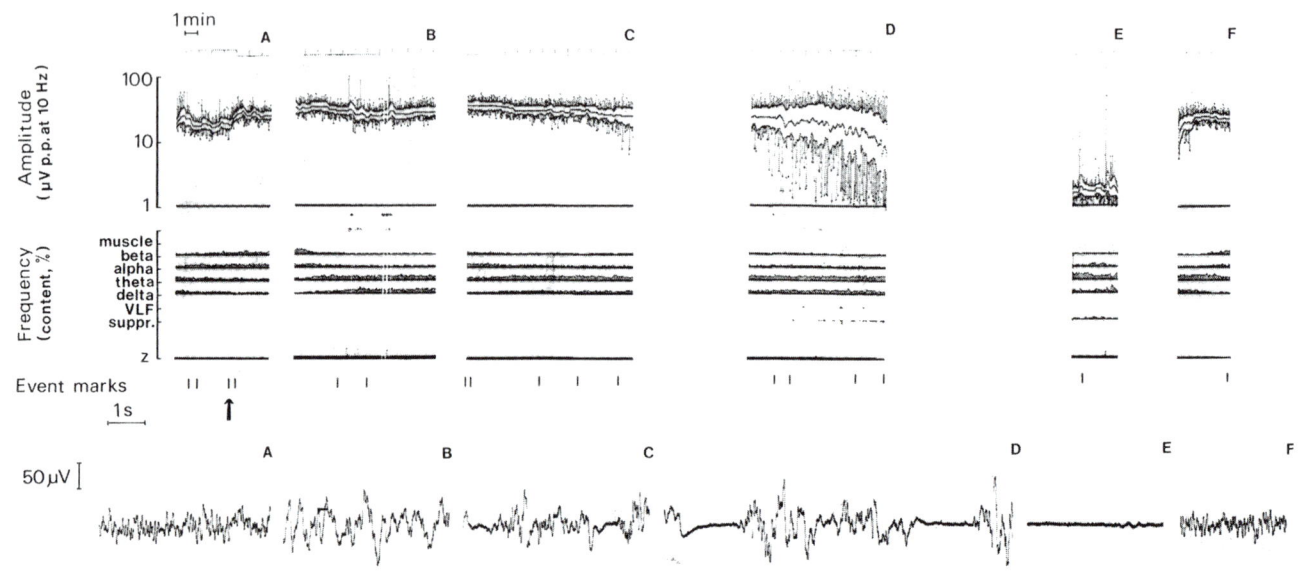

Fig. 5.11 Typical CFAM1 recordings at different end-tidal concentrations of isoflurane compared with raw EEG samples (lower strip; passband 0.8–70 Hz): (a) initially nitrous oxide and oxygen only (note the variability in amplitude and its increase on introduction of 0.5% isoflurane at arrow); (b) isoflurane 0.5% (note the decrease in variability and shift to lower frequencies; (c) isoflurane 1.0%, increased to 1.5% (note subtotal amplitude suppression up to 0.5 s); (d) stable isoflurane 1.5%, increasing to 2% (note effect on 10th centile and minimum amplitude, and the appearance of suppressions up to 1.5 min); (e) isoflurane 2.0% (note small ripple of residual EEG); (f) recovery (note rapid increase in amplitude precedes shift to faster frequencies). (From Lloyd-Thomas AR, Cole PV, Prior PF 1990a Quantitative EEG and brainstem auditory evoked potentials: comparison of isoflurane with halothane using the cerebral function analysing monitor. *British Journal of Anaesthesia* 65:306–312. © The Board of Management and Trustees of the *British Journal of Anaesthesia*. Reproduced by permission of Oxford University Press/*British Journal of Anaesthesia*.)

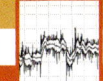

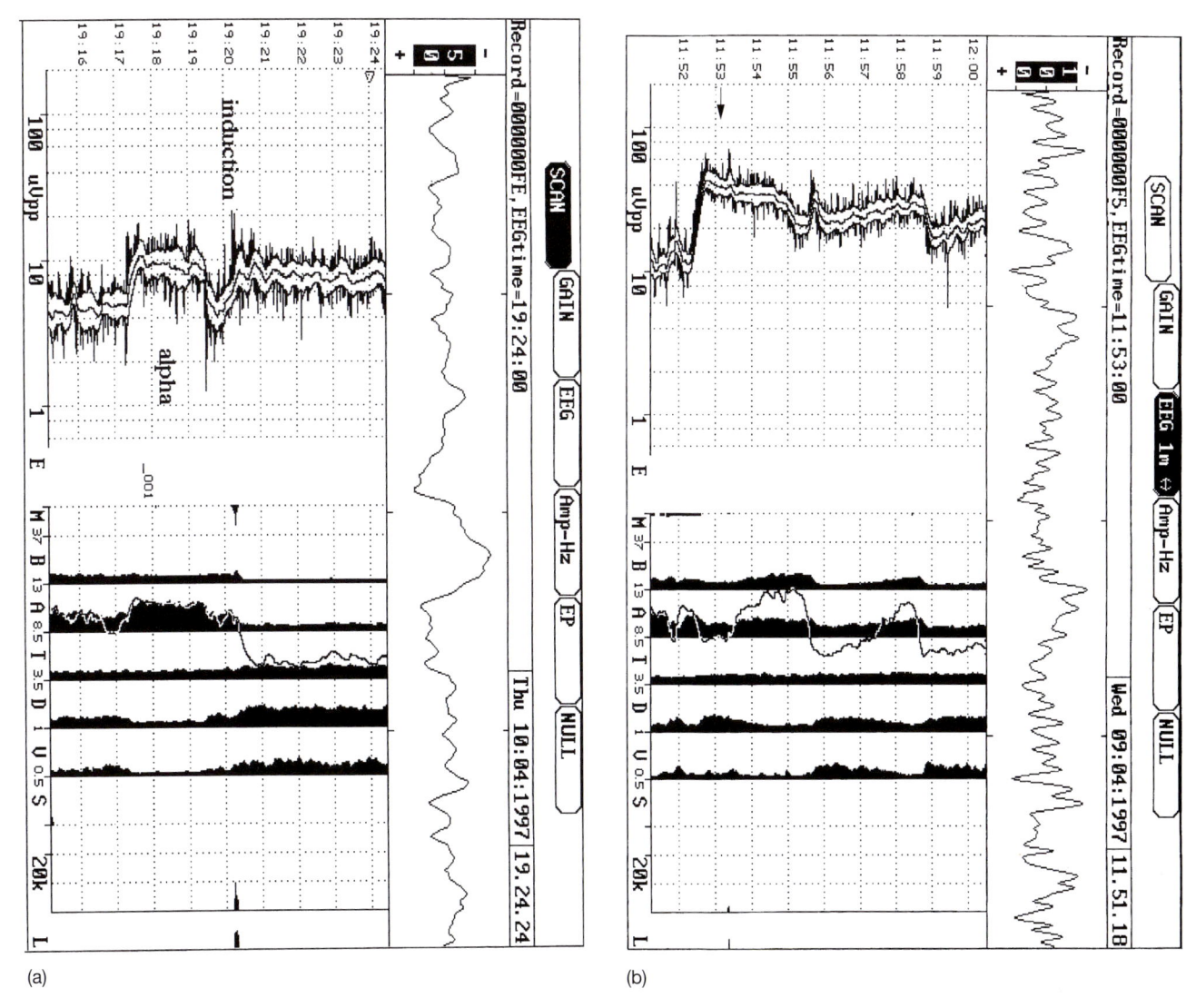

Fig. 5.12 (a) Induction of halothane anaesthesia monitored with Cerebral Function Analysing Monitor (CFAM2) (RDM Consultants, Uckfield, Sussex, UK, http://www.cfams.com). Recordings show amplitude analysis at the top, frequency analyses in the middle with superimposed mean frequency values and electrode impedance and line interference (L) at the bottom; to the right are 4 s samples of unprocessed EEG, then patient and recording identification details. Amplitudes displayed as maximum, 90th centile, mean, 10th centile and minimum values, calibrated as microvolts peak-to-peak at 10 Hz. Percentage values for scalp muscle potentials (M), beta (B), alpha (A), theta (T), delta (D), subdelta (V) and suppression (S) are shown such that full scale for each division represents 100%. As the patient drowses before induction, alpha activity appears; there is a momentary flurry of muscle potentials and artefact on application of the face mask at 19.20 h. Note the sequential changes in mean frequency. (b) Induction (at arrow on left) of anaesthesia with intravenous thiopentone (followed by two supplementary boluses) monitored with the CFAM2. Note the rapid nature and greater magnitude of induction changes typical of intravenous agents compared to the inhaled agent shown on the left. (Courtesy of Dr DE Maynard, from Binnie et al (2003), by permission.)

et al 2004). Flash visual evoked potentials (VEPs), however, can only be evoked by stimulation during bursts (Mäkelä et al 1996). When used during electrocorticography, isoflurane does not affect the number of epileptiform spike discharges, providing concentrations are kept around 0.25–1.25%, i.e. less than that producing burst suppression (Fiol et al 1993).

Sedative and analgesic perioperative drugs These include the anxiolytics (mainly the benzodiazepines *diazepam*, *lorazepam*, *midazolam* and *temazepam*) and the analgesics (mainly the opioids *alfentanil*, *fentanyl* and *remifentanil*).

Their effects are two-fold: firstly specific effects of the particular agent such as the typical fast activity produced by benzodiazepines, and secondly, the important, non-specific, effects of reducing EEG variability caused by external noise or to pain during light anaesthesia. It is important to distinguish between these two different mechanisms for EEG effects when monitoring the blockade of responses to surgical pain with analgesics as opposed to with very deep anaesthesia. In both cases specific EEG responses (including signs of arousal) to a sudden painful event will be abolished, but in the former the EEG will suggest a steady state equating to light

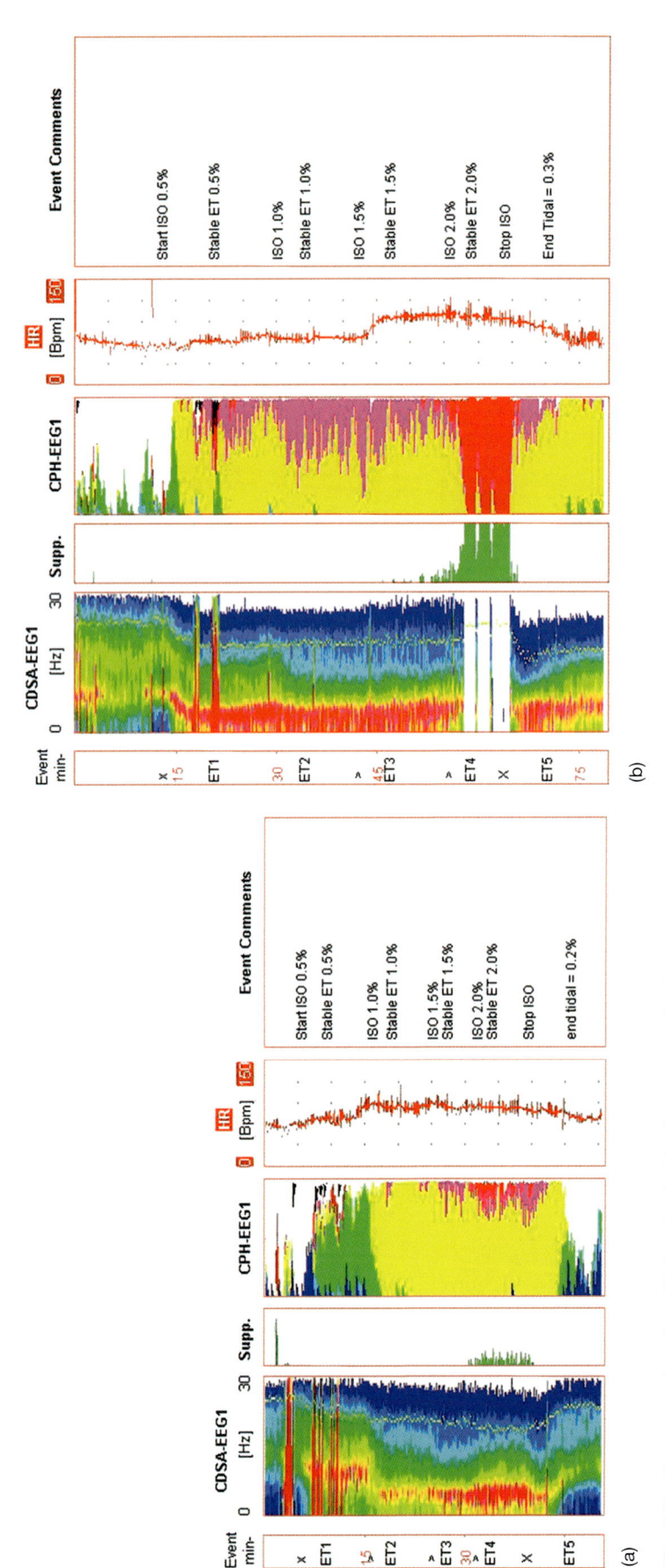

Chapter 5 Neurophysiological Monitoring during Sedation and Anaesthesia

Fig. 5.13 Two examples of monitoring depth of isoflurane anaesthesia during routine surgery using the Advanced Depth of Anaesthesia Monitor (ADAM) system (for a description see p. 83) to show variation between different patients. Note that (i) spectral edge frequency has been superimposed upon the CDSA display as a yellow dotted line, (ii) a suppression index is provided, and (iii) the quality control display of unprocessed EEG signals available on the screen is not shown. Data reprocessed from the study of Lloyd-Thomas et al (1990a) in which standardized stepwise increases in end-tidal levels from 0.5 to 2.0% (as shown in *Fig. 5.11*) were monitored prior to routine surgery. (b) In this patient, at comparable stable end-tidal concentrations of isoflurane, the appearances are of a much deeper level of anaesthesia than in the patient in (a), with more marked low frequency activity and plentiful suppressions. (Courtesy of Dr AR Lloyd-Thomas and Dr CE Thomsen, from Binnie et al (2003), by permission.)

168

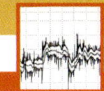

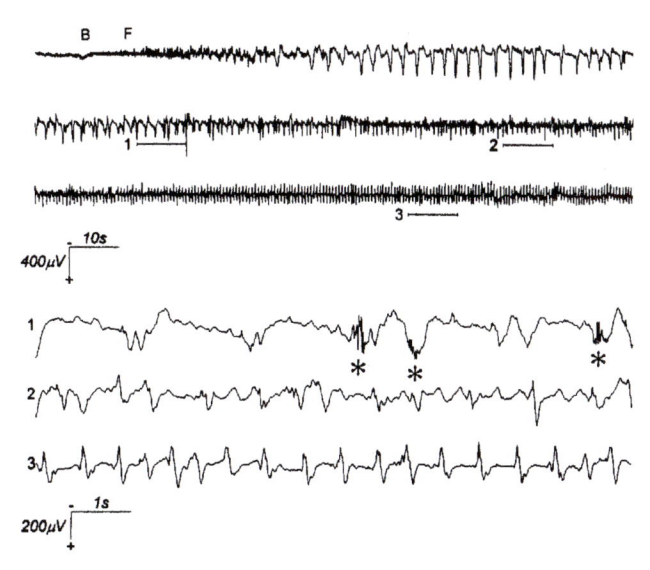

Fig. 5.14 EEG during anaesthetic induction with sevoflurane in a patient with epileptiform EEG and jerking movements of the left shoulder area. Above: six minutes of recording from Fp2-right mastoid electrodes are shown in three 2 min sections and three 10 s samples are shown below from the periods marked by numbered bars. Key points to note are the gradual build-up of a rhythmic epileptiform discharge during which jerks are indicated with a star on the close up view from period 1. B, eye blink; F, fast activity. (From Yli-Hankala A, Vakkuri A, Särkelä M, Lindgren L, Kortilla K, Jäntti V 1999b Epileptiform electroencephalogram during mask induction of anesthesia with sevoflurane. *Anesthesiology* 91:1596–1603, by permission.)

anaesthesia, whilst in the latter, marked burst suppression is more likely. Opioids, when given by infusion prior to induction of anaesthesia, produce an increase in relative delta power and a decrease in spectral edge frequency; increasing dose has little further effect on the EEG, the 'ceiling effect' coinciding with the failure to respond to verbal command (Chi et al 1991). When given in higher doses to induce anaesthesia, most opioids lead to an increase in delta activity affecting slow/fast frequency ratios to a greater extent than alterations in spectral edge frequency (Bovill et al 1982, Smith et al 1984, 1985, Hall et al 1986, Ochiai et al 1992, Schwilden and Stoeckel 1993).

'Neuroleptanalgesia' or 'neuroleptic anaesthesia', a technique described in the late 1950s, utilizes mixtures of anaesthetics, opioid analgesics and other drugs such as hypnotic or psychotropics. It is associated with bizarre combinations of EEG features, including persistent stable unreactive alpha rhythm in the 'analgesic stage', mixed theta and delta activity in the 'narcotic phase'. So-called 'dissociated states' are also described, and phenomena more commonly associated with stage 2 and 3 of sleep may be encountered. qEEG recordings may show an unusually irregular or variable pattern, or give the impression of large fluctuations in the depth of anaesthesia (Nilsson and Ingvar 1967, Barker et al 1968, Arfel and Walter 1977, Pichlmayr et al 1984).

Other drugs used during anaesthesia

Muscle relaxants (neuromuscular blocking drugs) The non-depolarizing (atracurium, cisatracurium, mivacurium, pancuronium, rocuronium, vecuronium or gallamine), the depolarizing (suxamethonium) muscle relaxants and the anticholinesterases (edrophonium and neostigmine) used to reverse the effects of non-depolarizing agents, do not have any significant direct effects on the EEG in normal clinical doses. However, they do significantly reduce or obliterate artefact from muscle potentials or patient movement. Two things follow from this: first, muscle or movement artefacts, like arousal responses to specific surgical events, may provide a warning that the patient is too lightly anaesthetized or that the effect of muscle relaxants is waning; second, muscle relaxants should not be used specifically to 'clean up' a technically poor EEG or EEG monitor recording (see p. 56 and the report of a fatality from Verma et al (1999)).

Antagonists for central and respiratory depression These agents (*nalaxone, flumazenil, doxapram*), used to counteract depressant effects of opioid analgesics, whilst not having direct effects on the EEG in normal clinical dosage, may be associated with alterations in monitor traces resulting from the reduction of the analgesic effects of the opioids. Thus pain responses may become evident, often accompanied by abrupt variations in the ongoing EEG suggesting a more responsive state.

Drugs used for local anaesthesia, including epidural and regional anaesthesia Local anaesthetics (e.g. *lidocaine, bupivacaine, levobupivacaine, prilocaine, procaine, ropivacaine, tetracaine*, etc.) in appropriate clinical concentrations have little, if any, irritant or toxic effects on the brain. Excessively high plasma concentrations can produce neurotoxicity since they are able to cross the blood–brain barrier. High dosage may lead to feelings of inebriation and light-headedness, sedation, circumoral paraesthesiae and twitching; convulsions can occur with severe reactions and following intravenous injection. When blood concentrations reach a threshold, epileptiform or ictal discharges appear and are rapidly followed by convulsions (Steen and Michenfelder, 1979).

EEG features useful in assessment of depth of anaesthesia and implications for monitoring equipment

Given the long history of concerns about the 'depth' of anaesthesia already mentioned, it is of interest that one early result of the development of qEEG analysis was 'servo-control' using frequency discrimination for automatic control of depth of general anaesthesia (Bickford 1950, 1951) (see *Figs 3.4* and *3.5*). 'Servo-anaesthesia' relied on the exceptionally close dose–response relationships demonstrated between concentration of most volatile or intravenous anaesthetics and various univariate qEEG measures (see *Fig. 3.10(a)*) (Frank et al 1982, 1984, Hudson et al 1983, Schwilden et al 1985, Stoeckel and Schwilden 1986). Although initially successful, the early servo-anaesthesia fell into disuse with the introduction of 'balanced anaesthesia', because such relationships are only robust with single agent anaesthesia in the range from 'light' to 'fairly deep' levels; they are less satisfactory with changes during induction, with the onset of burst suppression and when several drugs are being used simultaneously.

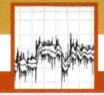

Since the essential dose–response relationships continue to be demonstrated with present day anaesthetic agents and the more complex EEG monitor algorithms, it is not surprising that a more recent revival of interest has occurred. This has led to investigations of 'informed control' of administration of various types of anaesthesia (e.g. Veselis et al 1991, Liu et al 1997, Eckert et al 1997, Sebel et al 1997, Schmidt et al 2003, Kreuer et al 2004, Prichep et al 2004, Vakkuri et al 2004, 2005). Examples are given in *Figs 5.10, 3.14* and *3.16*. Patients can show considerable differences in requirement for anaesthetics in relation to age, weight and lean body mass, as well as with different intensities of surgical stimulation (see *Fig. 5.13*); these have been demonstrated in several pharmacodynamic and pharmacokinetic studies using various EEG endpoints (e.g. Homer and Stanski 1985, Hung et al 1992, Lloyd-Thomas et al 1990a, Schultz et al 1995). Variability from the expected EEG monitor 'response' to a particular anaesthetic 'dose' can also be due to a range of medical and other conditions that may affect the performance of different algorithms (Dahaba 2005) (see p. 84). Variability occurs both within and between individual patients and between different anaesthetics (Heier and Steen 1996b); it offers a valuable and underused source of information and may be hidden by 'overly simplistic presentation of complex parameters that have no firm physiological basis' (Guérit 1998).

The clinical desire for neurophysiological monitoring tools capable of providing valid information to help guide anaesthesia, coincided with the widespread availability of increased computational power and expertise in biological signal analysis (Chapters 3 and 8). Developments applicable to EEG monitoring began to appear in parallel with the aim of producing commercially viable cost-effective, easily usable, robust clinical tools with clear applicability (Gade et al 1996, Van Gils et al 1997, 2002, Rampil 1998, Yli-Hankala et al 1999a, Carson et al 2000). Over the last decade, simple univariate spectral measures have been largely replaced by multivariate analysis incorporating pattern recognition and learning systems capable of identifying trends and generating warnings when values deviate significantly from those expected.

To summarize, when considering suitable equipment for EEG monitoring during anaesthesia, the following questions need to be considered:
1. Will it indicate if anaesthesia is 'too light'? This requires an indicator of abrupt EEG responses to specific noises or painful stimuli during surgery which will alert the anaesthetist to the possibility of impending arousal and stress responses (Blunnie et al 1983);
2. Will it indicate if anaesthetic depth is stable? This requires an indicator sensitive to moment-to-moment variability of the EEG signals as well as highlighting slow trends;
3. Will it warn if anaesthesia is 'too deep'? This requires detection of a burst suppression pattern which can also indicate the duration of interburst intervals;
4. Can it distinguish signs of emergencies such as global or regional hypoxic–ischaemic events from background anaesthetic changes? This requires detection of alterations in slow/fast ratios and the onset of periods of suppression and, at another level of sophistication, recognition of atypical patterns for the type of anaesthesia being administered; and
5. Can it provide evidence to help decisions about 'street fitness' (Doenicke et al 1967) in patients undergoing day-case surgery? This requires evidence of reversal of changes due to anaesthesia and of a return of normal reactivity.

Other factors affecting EEG assessment of depth of anaesthesia

Maturation and ageing The EEG shows substantial maturational changes through childhood and, in the later decades, with ageing. These affect (i) dominant frequency, which increases through childhood reaching the adult range around the time of puberty and slowing slightly in old age, and (ii) the amplitude range which is considerably larger in children than adults, necessitating a dynamic range in EEG monitors that can handle signals in excess of 1 mV peak-to-peak. Some abnormalities, for example epileptiform discharges in children, may exceed 800 μV peak-to-peak. Age-related effects of anaesthetics on the EEG have been reported, for example with thiopental, the dose required to achieve early burst suppression on the EEG decreased linearly and significantly with age (Homer and Stanski 1985), and with standardized propofol induction (2 mg/kg, 1.30 s^{-1}) patients older than 70 years reached significantly deeper EEG stages than younger patients (Schultz et al 2004).

Arterial hypotension Older techniques for controlled hypotension included combinations of ganglion- and beta-blockade and halothane combined with head-up tilt and positive end expiratory pressure during which the monitored EEG became abnormal, not by direct effects of these agents, but if hypotension was rapid or excessive (Patel 1981). With carefully controlled hypotension, the rate of fall in arterial pressure is relatively slow when compared with that at the onset of cardiopulmonary bypass or with massive accidental haemorrhage during surgery. This leads to EEG changes that are primarily a reduction in amplitude, eventually leading to burst suppression. There are only minor alterations in frequency content, frequency shifts being more typical of abrupt falls in arterial pressure, characterized by an initial reversal of the slow/fast frequency ratio proceeding to overall reduction in amplitude and total EEG power followed by rapidly increasing burst suppression. If hypotension is sufficient to lead to prolonged reduction of CBF in the cerebral arterial boundary zone regions, localized reduction in EEG amplitude (and cortical SEPs) will occur in the parietal regions. It follows, than when there is a risk of acute arterial hypotension, the earliest warning of potential cerebral problems will be given if electrodes are placed in the parietal regions (Brierley and Graham 1984). The EEG changes may be reversible with prompt restoration of adequate cerebral perfusion, but if persistent (see *Fig. 4.6*) they may be associated with neurological abnormality. Quantitative EEG measures become linearly related to falling mean arterial pressure (pressure dependent) when a critical degree of induced hypotension is reached at mean arterial pressures comparable to those for loss of autoregulation in CBF (*Fig. 5.15(a)*) (Thomas et al 1985). The method of induction of hypotension appears to influence the risk of EEG abnormality (Ishikawa and McDowall 1981, Ishikawa et al 1982, Thomas et al 1985), sodium nitroprusside being associated with better maintenance of the EEG than trimetaphan. Differences have also been demonstrated between patients anaesthetized with isoflurane and halothane, isoflurane preventing the EEG amplitude depression which was seen with halothane at comparable levels of trimetaphan-induced hypotension (*Fig. 5.15(b)*) (Lloyd-Thomas et al 1990b). It is now considered that EEG measures are a necessary safety procedure for techniques, such as combined controlled hypotension and haemodilution, which can decrease both the volume of blood needed for transfusion and the duration of postoperative care in the ICU (Shapira et al 1997).

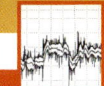

Hypothermia and hyperthermia *Hypothermia* produces EEG changes which follow a fairly consistent pattern with cooling to the levels used for profound hypothermia during cardiac surgery with circulatory arrest. Typically there is a gradual progression of changes with gradual depression

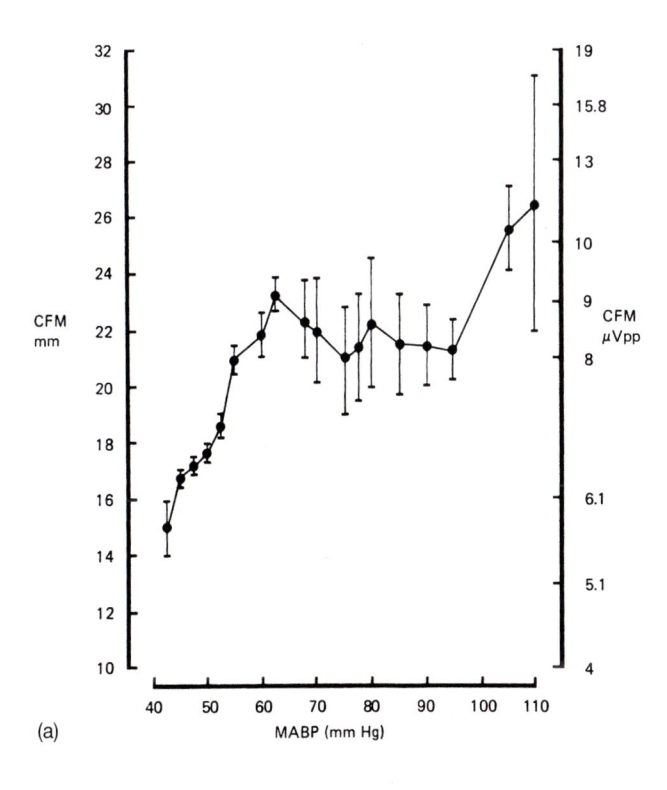

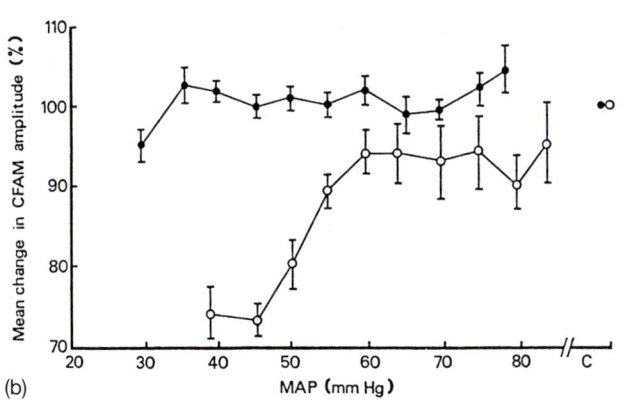

Fig. 5.15 Induced hypotension. (a) Relationship between smoothed (three-point running average) means ± 1 standard error of the mean (SEM) of mean arterial blood pressure (MABP) and average Cerebral Function Monitor (CFM) (RDM Consultants, Uckfield, Sussex, UK, http://www.cfams.com) values in a 31-year-old man during lumbar laminectomy with controlled hypotension induced by sodium nitroprusside. The dual scale for CFM amplitudes enabled measurement of the paper trace in mm and accurate conversion to μVpp before digital ouput was available. Note the point at about 65 mmHg when the EEG becomes pressure dependent. (From Thomas WA, Cole PV, Etherington NJ, Prior PP, Stefansson SB 1985 Electrical activity of the cerebral cortex during induced hypotension in man; a comparison of sodium nitroprusside and trimetaphan. *British Journal of Anaesthesia* 57:134–141.) (b) Group mean percentage change (± 1 SEM) in mean amplitude of CFAM trace at median points of 5 mmHg MAP steps during trimetaphan-induced hypotension in 10 patients (227 measurement points) receiving isoflurane (●) and 10 patients (189 measurement points) receiving halothane (o). C, control values. Note the shift to the left of the pressure-dependent point with isoflurane compared with halothane which may represent evidence for a greater degree of protection from ischaemic changes. (From Lloyd-Thomas AR, Cole PV, Prior PP 1990b Isoflurane prevents EEG depression during trimetaphan-induced hypotension in man. *British Journal of Anaesthesia* 65:313–318. Each figure © The Board of Management and Trustees of the *British Journal of Anaesthesia*. Reproduced by permission of Oxford University Press/*British Journal of Anaesthesia*.)

and slowing of ongoing rhythms; these may be followed by generalized periodic slow wave transients (sometimes almost epileptiform in morphology, Reilly et al 1974) which are gradually interrupted by periods of suppression below about 20°C and then followed by ECS at about 15°C. Total power and absolute power in each of the main frequency bands, spectral edge and median frequencies all show a reasonably tight relationship with nasopharyngeal or oesophageal temperature with changes of the order of 6% per °C quoted by some workers (Levy, 1984a, Russ et al 1987a, Grote et al 1992, Doi et al 1997b). Levy also demonstrated that, because of disruption by periods of suppression, EEG signals were too complex during hypothermia for the use of univariate measures derived from spectral analysis, (Levy 1984ab). Hypothermia may have an additive effect and lead to the appearance of burst suppression at lower concentrations of volatile agents than would otherwise be the case. Evidence during cardiac surgery suggests that there may be significant age-related changes with hypothermia; these may be more rapid or pronounced in infants than adults (Glaria and Murray 1990). On present evidence, there appear to be closer relationships between various short-latency EP measures and temperature during cooling and rewarming than have been demonstrated with qEEG measures.

Hyperthermia may occur in conditions such as heat stroke and in a severe form, with temperatures exceeding 40–41°C, as a life-threatening complication of anaesthesia. The 'malignant hyperpyrexia of anaesthesia' is linked to a genetic anomaly associated with massive discharge of calcium ions into skeletal muscle cell cytosol. This can be triggered by potent inhalational anaesthetics and depolarizing muscle relaxants, such as suxamethonium, as well as being a rare complication of idiosyncrasy to major tranquillizers (neuroleptic malignant syndrome). It can also follow poisoning with a range of drugs. If not treated promptly these conditions may proceed to a rapid onset of muscular rigidity or hyperactivity, autonomic dysfunction, multiple organ failure, coagulopathy and, ultimately, muscle and brain damage. A sequence of EEG changes culminates in ECS at about 40–42°C (Cabral et al 1977). In a study of whole body hyperthermia for a proposed treatment of metastatic carcinoma, EEG power decreased in inverse proportion to the rising temperature (Dubois et al 1980). The EEG changes of hyperthermia, like those of hypothermia, can be reversible depending on the clinical circumstances.

Effect of anaesthetic agents on EPs

Anaesthetic drugs obviously affect the nervous system since they induce a sleep-like state in which sensation and memory are impaired. They also affect the EEG, EPs and motor responses produced by CNS stimulation, to varying degrees, depending upon the nature and concentration of the drug administered. As far as EPs and motor responses are concerned, certain principles can be discerned (García-Larrea et al 1993, Banoczi 1994, Sloan 1998, Thornton and Sharpe 1998, Jäntti and Yli-Hankala 2002, Kumar et al 2000, Sloan and Heyer 2002, Banoub et al 2003):

1. Volatile halogenated agents attenuate cortical EPs (all modalities) and motor responses more than intravenous anaesthetic drugs;
2. Cortical responses are more sensitive to anaesthetic effects than subcortical potentials; and
3. Later cortical EP components are attenuated at lower anaesthetic concentrations than the primary cortical response.

Etomidate, ketamine and propofol are unusual because they do not attenuate early cortical EPs; etomidate and ketamine increase the amplitude of these responses.

Halogenated anaesthetics (desflurane, enflurane, halothane, isoflurane, sevoflurane) reduce the amplitude of cortical EPs (Wang et al 1985, Peterson et al 1986, Salzman et al 1986, Sebel et al 1987), increase interwave latencies in the BAEP and alter the waveform of the MLAEP (*Figs 5.16* and *5.17*) (Thornton and Sharpe 1998). The amplitude of the cortical SEP, particularly that from the leg, falls rapidly if the inspired concentration of any of these agents rises above 0.5% (*Fig. 5.18*), although responses are still recordable even when the EEG shows burst suppression, provided that the rate of stimulation is not greater than 1 Hz (Rytky et al 1999). These drugs increase cortical SEP latency by about 1 ms but this effect appears not to be dose related (Wang et al 1985). Spinal response amplitudes are only slightly reduced by halogenated drugs (Baines et al 1985). The increase in BAEP latencies is proportional to the end-tidal concentration (Thornton et al 1984, Kálmánchey et al 1986). For halothane the increase is about 0.2 ms for each 1% increase in concentration; the corresponding value for enflurane is 0.3 ms. With isoflurane there is a 0.7 ms increase in the latency of wave V at a concentration of 1.5%, but no further increase in latency at higher concentrations (Manninen et al 1985).

With inspired concentrations up to 1.13%, halothane increases VEP latency without appreciably affecting the amplitude (Uhl et al 1980), but the amplitude falls as the concentration rises and at concentrations over 2% the response is likely to be completely obliterated (Wang et al 1985). Enflurane, on the other hand, decreases VEP amplitude at concentrations below 2.5% (Chi and Field, 1990), but at higher concentrations the amplitude increases, and this is associated with the appearance of spike and wave discharges time-locked to the stimuli; these excitatory effects are suppressed in the presence of hypercarbia (pCO_2 65 ± 10 mmHg) (Burchiel et al 1975). During anaesthesia with isoflurane, at a concentration sufficient to produce burst suppression in the EEG, the VEP to a flash stimulus can still be recorded during the EEG bursts, but not during periods of suppression (Mäkelä et al 1996).

Low doses of barbiturates increase cortical SEP latencies slightly, and attenuate the later waves (Abrahamian et al 1963), but amplitudes of early cortical potentials are not

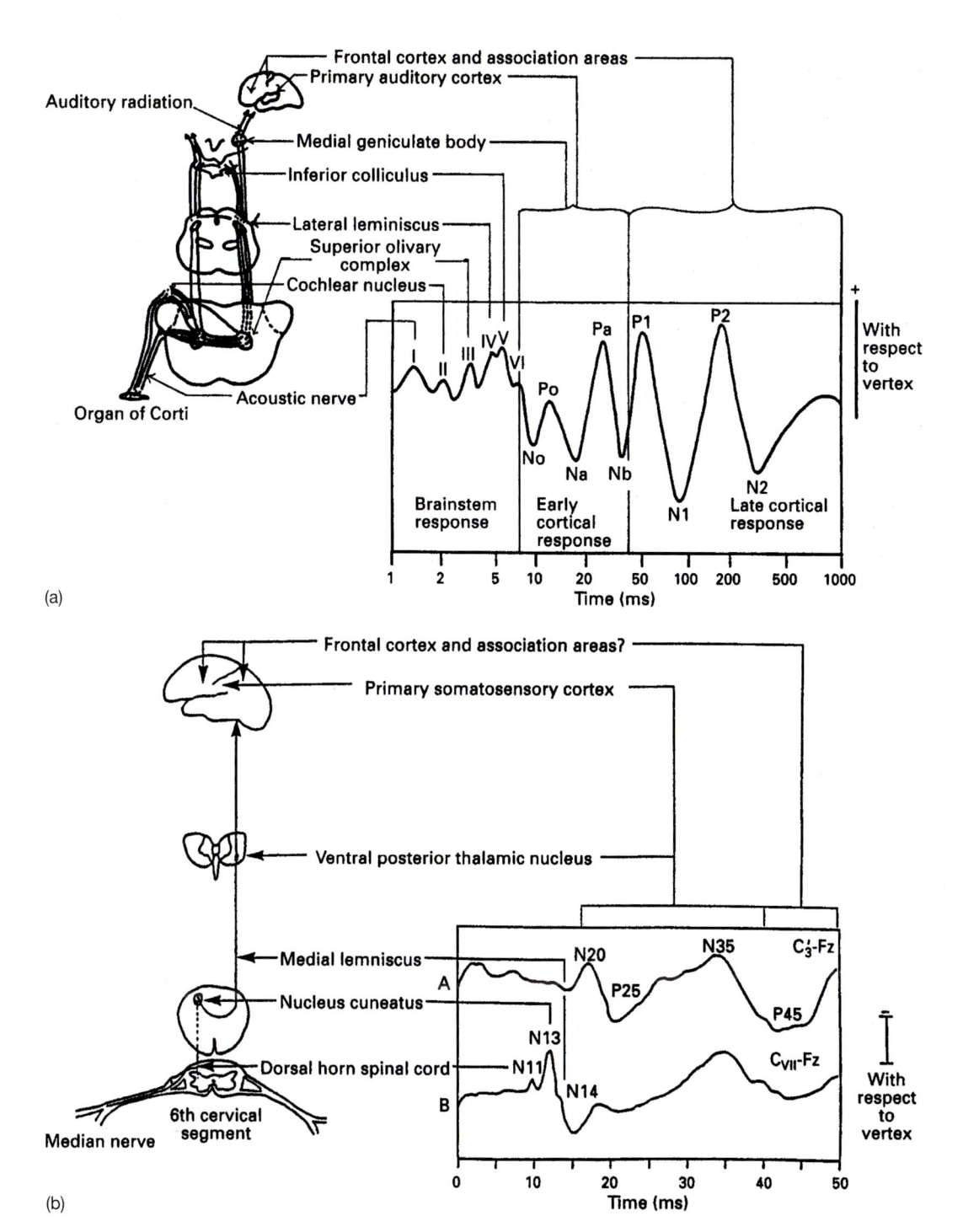

Anaesthesia: assessment with EEG and evoked potentials

Fig. 5.16 (a) Auditory evoked response. This diagram describes the nomenclature of the response and shows its anatomical relationship with the auditory neuraxis. The brainstem response has a well-documented anatomical relationship with the auditory neuraxis, whereas later responses have origins which are less easy to define. (b) Somatosensory evoked response. This diagram shows waveforms obtained when the stimulating electrode is placed at the wrist above the median nerve and the recording electrodes are above (A) the somatosensory cortex and (B) the seventh vertebra. The nomenclature of the response and its anatomical relationship with the fine/touch somatosensory neuraxis is shown. (From Thornton C, Sharpe RM 1998 Evoked responses in anaesthesia. *British Journal of Anaesthesia* 81:771–781. © The Board of Management and Trustees of the *British Journal of Anaesthesia*. Reproduced by permission of Oxford University Press/*British Journal of Anaesthesia*.)

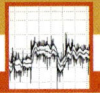

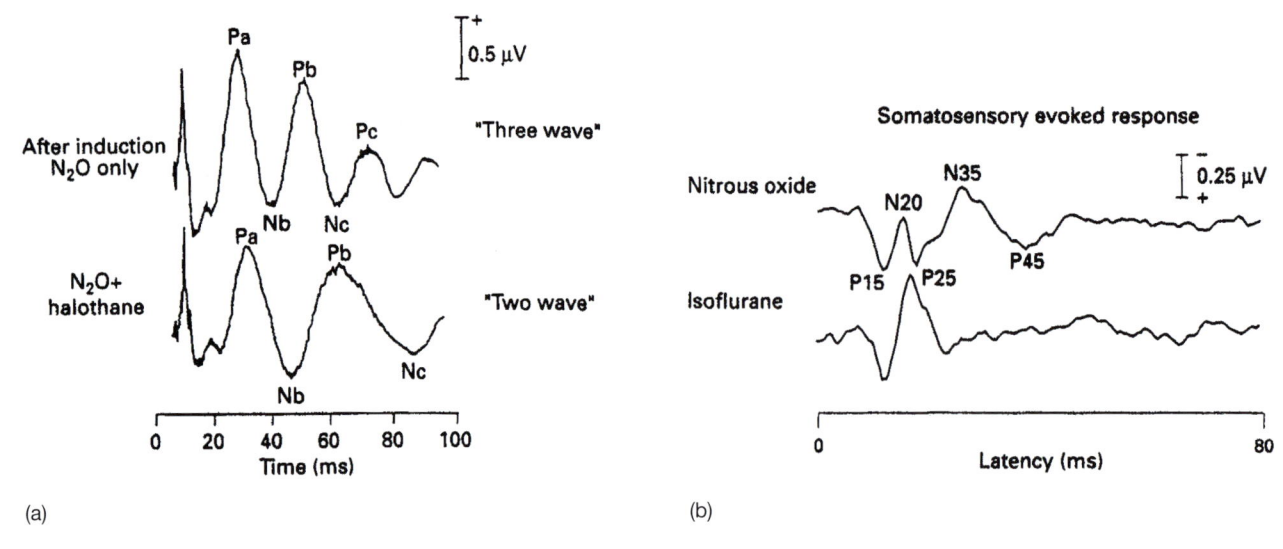

Chapter 5 Neurophysiological Monitoring during Sedation and Anaesthesia

Fig. 5.17 (a) Early cortical response. Top: The three positive waves, Pa, Pb and Pc, with latencies of 15–100 ms, suggest light anaesthesia/potential awareness. At this time anaesthesia was maintained with 70% nitrous oxide in oxygen. Bottom: After addition of a small amount of halothane, the three waves were replaced by two positive waves, Pa and Pb, suggesting no awareness. (b) Changes in SEP recorded from scalp electrodes and produced by stimulating the median nerve. The patient was given nitrous oxide or isoflurane. During the isoflurane period, P15–N20 and N20–P25 increased and P25-N35 and N35-P45 were reduced compared with the nitrous oxide period. (From Thornton C, Sharpe RM 1998 Evoked responses in anaesthesia. *British Journal of Anaesthesia* 81:771–781. © The Board of Management and Trustees of the *British Journal of Anaesthesia*. Reproduced by permission of Oxford University Press/*British Journal of Anaesthesia*.)

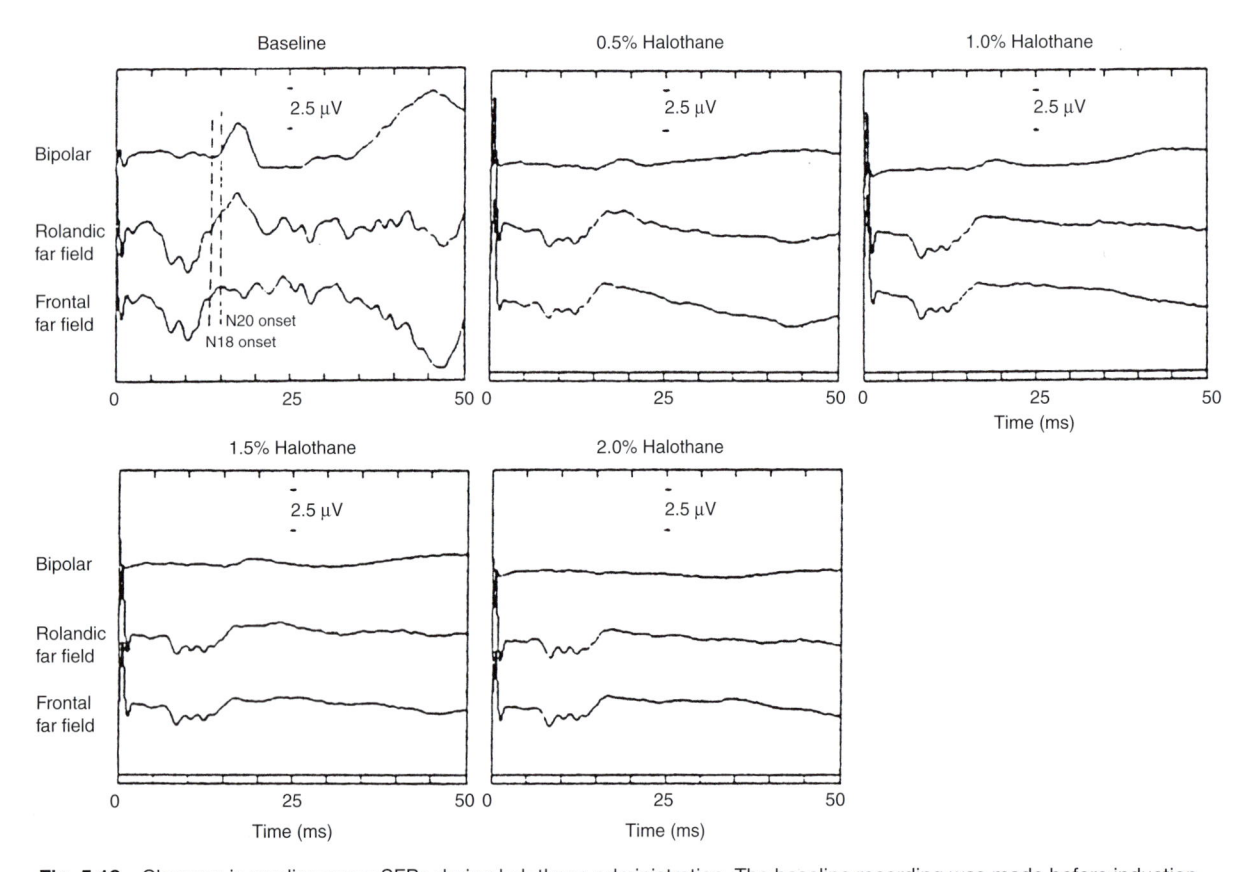

Fig. 5.18 Changes in median nerve SEPs during halothane administration. The baseline recording was made before induction of anaesthesia. Bipolar recordings are between C3′ (midway between C3 and P3) and Fpz; the Rolandic and frontal far field recordings are from C3′ and Fpz respectively, both referred to the left shoulder. With increasing concentrations of halothane, there is a progressive reduction in amplitude and increase in latency of N20 in the bipolar recording, but subcortical components in the far field recordings are not attenuated to the same degree. (From Sebel PS, Erwin CW, Neville WK 1987 Effects of halothane and enflurane on far and near field somatosensory evoked potentials. *British Journal of Anaesthesia* 59:1492–1496. © The Board of Management and Trustees of the *British Journal of Anaesthesia*. Reproduced by permission of Oxford University Press/*British Journal of Anaesthesia*.)

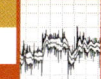

significantly reduced at a concentration sufficient to produce EEG burst suppression (Newlon et al 1983). The early cortical SEP is still present, although attenuated and delayed, at a thiopentone concentration three times that necessary to render the EEG isoelectric (Drummond et al 1985). Usual therapeutic doses of barbiturates do not affect the BAEP, and at a thiopentone concentration sufficient to suppress EEG activity completely the wave V amplitude is not significantly altered, although its latency is increased by 0.7 ms (Drummond et al 1985). Barbiturates have a biphasic effect on the VEP similar to that of the halogenated agents: low doses increase the amplitude, while higher doses attenuate and eventually abolish the response (Domino et al 1963, Brazier 1970).

Diazepam (0.1 mg/kg) decreases the amplitude of the early cortical SEP and increases its latency, and obliterates later waves (after 200 ms), but may increase the amplitude of the N35 potential to median nerve stimulation (Grundy et al 1979, Prevec 1980). However, another study failed to demonstrate any effect of 20 mg diazepam on cortical somatosensory or visual potentials (Loughnan et al 1987). Benzodiazepines have no effect on the BAEP at therapeutic doses.

Nitrous oxide (50%) causes approximately a 50% reduction in amplitude of both the cortical SEP and the VEP (Domino et al 1963, Sebel et al 1984, McPherson et al 1985, Sloan and Koht 1985). Nitrous oxide has no significant effect on either the BAEP or spinal SEPs.

Lignocaine (lidocaine) infusion increases the latencies of all BAEP waves, including wave I, and interwave intervals are also increased so that wave V may be as much as 1 ms later at blood levels over 20 mg/l (García-Larrea et al 1987). At higher blood levels of lignocaine the entire response may disappear (García-Larrea et al 1988). Lignocaine is the only drug which has been reported to produce a reversible abolition of the BAEP, but high levels of other local anaesthetic drugs might be expected to have a similar effect on the BAEP. Administration of 2% lignocaine in the lumbar extradural space has a marked effect on dorsal column function, reducing the amplitude of the spinal SEP (Loughnan et al 1990). This is attributed to interference with afferent transmission in nerve roots, but an effect on spinal cord blood flow is also possible.

Narcotic drugs such as morphine and fentanyl affect the later waves of the cortical SEP predominantly, and produce variable effects on the amplitude of the early waves (Grundy et al 1980). There may be a slight amplitude reduction in the spinal SEP, but these drugs have no effect on the BAEP.

The intravenous agents etomidate (up to 0.3 mg/kg) and ketamine (2 mg/kg) increase the amplitude of cortical somatosensory potentials to median nerve stimulation (Kochs et al 1986, McPherson et al 1986, Schubert et al 1988). A similar amplitude increase is found in the cortical response to tibial nerve stimulation during continuous infusion of propofol (Maurette et al 1988). The latency is increased with etomidate and propofol, but not affected by ketamine. The increase in amplitude is an unusual effect and in the case of etomidate may be related to the fact that this drug produces myoclonic jerks during induction of anaesthesia, at the same time as the increase in SEP amplitude. The phenomenon is perhaps analogous to the giant SEPs recorded in some forms of myoclonic epilepsy.

Muscle relaxants have no direct effect upon EPs, but inadequate neuromuscular blockade can seriously interfere with EP recording by permitting a high level of muscle activity. This is usually apparent in the raw trace, and can be suppressed by a further dose of relaxant drug. Conversely, it is essential that the patient should not be completely paralysed if motor nerves (e.g. facial) are to be stimulated. The degree of paralysis can be quantified by the compound muscle action potential amplitudes produced by a train of four stimuli applied at 2–3 Hz to a peripheral nerve. A simple device for recording these responses is usually part of the anaesthetist's armamentarium (Ali et al 1971), but in doubtful cases the same responses can be recorded on an EP machine.

During most operations several different drugs are used (anaesthetics, analgesics and relaxants) and the blood level of each will be different at different stages of the procedure. The effects of many drugs are additive, and some drugs may potentiate others, so precisely which drug is responsible for a particular effect is often difficult to determine. Close cooperation with the anaesthetist is essential so that a standardized anaesthetic technique can be established which permits reliable monitoring. Bolus doses of drugs should not be given during potentially dangerous surgical manoeuvres when monitoring is critical, and halogenated agents should only be administered at low concentrations if cortical SEPs or VEPs are to be monitored. Some groups have been able to avoid halogenated anaesthetics altogether for cortical SEP monitoring (e.g. Engler et al 1978).

EPs in assessment of depth of anaesthesia

Objective indication of the depth anaesthesia is becoming increasingly important in view of the risk of litigation following awareness during surgery (Jones 1989). Many techniques have been proposed, including oesophageal contractility, beat-to-beat variation of heart rate and several variables derived from the EEG, but the aim of providing a reliable index of depth of anaesthesia which is independent of the anaesthetic technique, while having the ability to predict responses to surgical stimulation and postoperative recall of intraoperative events, has yet to be achieved (van Gils et al 2002). Perhaps the EP candidate that comes closest to meeting these requirements is the middle latency segment (20–80 ms) of the response to auditory stimulation, commonly referred to as the MLAEP (see *Fig. 5.16*) (Thornton and Newton 1989, Thornton and Sharpe 1998). An alternative possibility is the steady-state response to high frequency auditory stimulation (Plourde and Picton 1990, Munglani et al 1993). It is of course important to remember that 'anaesthesia' consists of several components (hypnosis, analgesia and paralysis), and methods applicable to measuring hypnosis may not adequately assess analgesia or predict movement (van Gils et al 2002).

Several variables have been calculated from the MLAEP waveform to quantify the changes produced by drugs and surgical stimulation, including the average double differential (Thornton and Newton 1989), the sum of square roots of absolute difference between any two successive 0.56 ms segments of the waveform (Doi et al 1997a), and indices derived from wavelet transformation (Kochs et al 2001).

Among the many methods which have been proposed to measure the depth of anaesthesia (Jones 1989, van Gils et al 2002), neurophysiological methods based upon EPs are now commercially available, and it remains to be seen whether EP methods or those based upon the EEG, especially the BIS, or the use of combined systems, such as the new Danmeter AEP Monitor/2 (Danmeter A/S, Odense, Denmark, http://www.danmeter.dk) (see p. 98), will prove more reliable.

Other factors affecting EPs

Temperature Cooling is an important cause of EP latency increase (see p. 146 and *Fig. 4.36*). During a long operation limb temperature often falls well below core temperature; peripheral nerve conduction velocity is reduced by 2 m/s per °C, which may account for an SEP latency increase of several milliseconds. In addition, reduced core temperature slows conduction in central fibre tracts, and delays transmission across synapses. CCT therefore increases by 6.6% for each °C below a core temperature of 37°C (Kopf et al 1985, Hume and Durkin 1986, Russ et al 1987b). The latencies of all BAEP waves, and interwave intervals, are inversely related to a core temperature between 30°C and 40°C, so that there is an increase in wave V latency of 0.1–0.2 ms for each °C (Stockard et al 1978, García-Larrea et al 1987). However, below 30°C the increase in latency is related to temperature by an exponential curve (Markand et al 1987). VEP amplitudes are similarly reduced and latencies increased by reduction in core temperature (Reilly et al 1978). The VEP usually disappears at about 25°C, often before the EEG becomes isoelectric (Russ et al 1984).

Arterial hypotension Hypotension affects cortical EPs indirectly by reducing cerebral perfusion pressure and CBF. Cerebral perfusion pressure is the difference between systemic blood pressure and ICP, so a normal blood pressure may not produce adequate perfusion in the presence of high ICP. Similarly, blood flow may be reduced at an apparently adequate perfusion pressure if a major artery is partially occluded. The SEP latency and CCT are increased when CBF falls below 20 ml per 100 g/min. During cardiopulmonary bypass the CCT is prolonged by more than 50% at a mean arterial pressure of 50 mmHg, although temperature and anaesthetic effects may be partly responsible for this change (Arén et al 1985). All waves of the BAEP are delayed by a cerebral perfusion pressure below 40 mmHg (García-Larrea et al 1987), and the VEP amplitude falls at systemic blood pressures below 80 mmHg (Smith 1975). At a mean blood pressure less than 40 mmHg a reversible conduction block develops in the spinal cord (Whittle et al 1986). Mild hypotension is an important factor contributing to EP changes in some operations, particularly carotid endarterectomy and scoliosis surgery (see pp. 291 and 274, respectively). EP amplitudes and latencies must always be correlated with the blood pressure; occasionally deterioration can be reversed by a 10–20 mm Hg elevation of blood pressure, without intervention by the surgeon. The effects of different drugs combined with those of hypothermia and haemodilution, both of which may be common in some types of surgery, contribute to a great variability in the responses recorded during operations. If cortical responses are monitored, comparison of the responses from arm and leg often permits identification of systemic factors such as hypotension and drugs, which affect both responses. To eliminate any uncertainty which might arise from systemic effects if the tibial response is monitored alone, Madigan et al (1987) suggest that the ratio of median and tibial amplitudes should be calculated.

COMBINED EEG AND EVOKED POTENTIAL MEASURES

As has been indicated in *Anaesthesia: assessment with EEG and EPs* (p. 161), EPs have a considerable relevance in detection of depth of anaesthesia. Combined displays of EPs with EEG features (*Fig. 5.19*), have proved highly informative (Clark et al 1971, Clark and Rosner 1973, Pfurtscheller et al 1987). The use of simultaneous qEEG and auditory EP measures, although increasing the technical demands for high quality recording, appears useful for evaluation of some specific aspects of depth of anaesthesia, such as transitions between consciousness and unconsciousness (Sharpe et al 1997, Gajraj et al 1999, Nishiyama et al 2004, White et al 2004). Examples are shown in *Figs 5.20, 5.21* and *5.22*.

AWARENESS DURING ANAESTHESIA

Awareness during anaesthesia has been described as a state of consciousness that is revealed by explicit or implicit memory of intraoperative events (Heier and Steen 1996a). Although incidence is between 0.13% and 0.3% during anaesthesia for general surgery, patients may be permanently disabled by the experience of being awake during surgery but unable to move or otherwise communicate the fact to the anaesthetist (Bogetz and Katz 1984, Moerman et al 1993, Schwender et al 1998b, Sebel et al 2004). Medico-legal implications are also considerable (Payne 1994).

Neurophysiological features useful in prediction of possible awareness during anaesthesia

Painful anaesthetic procedures and intense surgical stimuli can induce clear EEG and other responses; typical examples include endotracheal intubation and sternotomy (Sidi et al 1990, Kearse et al 1994, Dutton et al 1996, Driessen et al 1999). Such documentation of reactivity to pain provides a warning of the possibility of inadequacy of anaesthesia (and/or analgesia) and that the patient might be or shortly become 'aware'. The state after induction of anaesthesia in which patients may be aware of their surroundings and yet unable to communicate is a relatively new problem (Desiderio and Thorne 1990) and has led to the suggestion that muscle relaxants should be kept to a minimum (Ponte 1995, Heier and Steen 1996a). Prior to the introduction of muscle relaxants in the 1940s, warning that patients were inadequately anaesthetized would be given by their movements – harking back to Snow's observations of 1847 that voluntary movement ceases with anaesthesia. The problems and some possible solutions have been reviewed in the volume edited by Lunn and Rosen (1987) and by Jones (1994) and Heier and Steen (1996a).

Although, in several studies, the AEP has appeared to provide a more sensitive indicator of inadequacy of anaesthesia than EEG measures (De Beer et al 1996, Doi et al 1999), it is salutary to note that the stimulation involved in recording EPs may of itself provide discomfort or pain sufficient to lead to EEG arousal. *Figure 5.23* shows one such instance documented at the time of adjustment of a meatal auditory stimulator prior to a period of stimulation for BAEP recording. Similar findings have been reported during high intensity somatosensory stimulation, for example during spinal monitoring. Examples during routine anaesthetic and surgical procedures are given in *Figs 5.24* and *Fig 3.8(b)*.

Not all qEEG methods are suitable for immediate recognition of abrupt, short-lived EEG responses to pain which may last less than 100 ms. In one comparison of six commonly used EEG measures (spectral-edge, 95th percentile power frequency, median power, zero crossing frequencies and total power in the alpha and delta frequency ranges), it was concluded that none predicted purposeful movements to incision, response to verbal command or the development of memory (Dwyer et al 1994). Similarly, the BSI did not perform well in detecting responses to endotracheal intubation or sternotomy during fentanyl–midazolam anaesthesia (Driessen et al 1999). Such deficiencies are hardly surprising since brief arousal responses would be lost in averaging processes during feature extraction. Moreover, detection of changes in state, for example shift from fast to slower frequencies as anaesthesia deepens, may be confused by the appearance of the prolonged 'paradoxical' slow wave arousal responses with frontal emphasis (Kochs et al 1994, Bischoff et al 1996, Litscher and Schwarz 1999). Arousal responses may vary according to anaesthetic: Wilder-

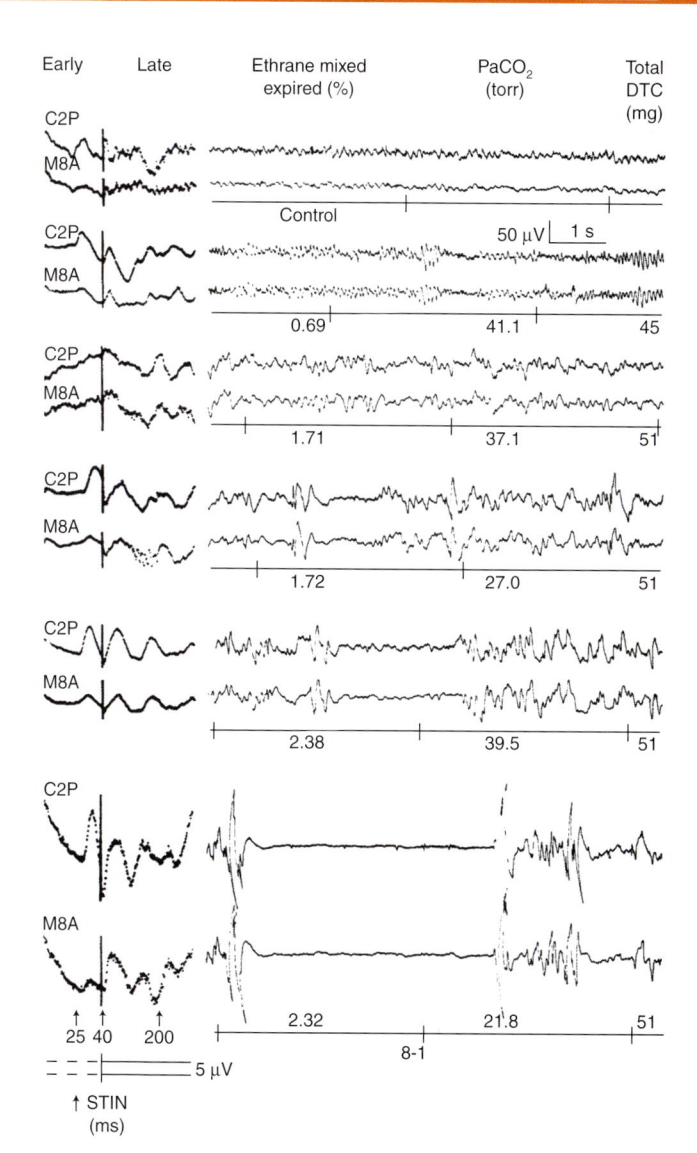

Fig. 5.19 Effects of alternating concentrations of Ethrane and carbon dioxide on SEPs and EEG. Left column: early and late evoked potentials from C2P (2 cm posterior to a point 7 cm from the vertex on a line between the vertex and the ear) and M8A (a midline point, 8 cm anterior to the vertex). 5 μV, 100 Hz calibration signal from the computer is shown at the bottom. Right column: segments of EEGs from which corresponding SEPs were derived. The small spikes on the solid lines beneath each pair of EEGs indicate stimuli. Enthrane concentrations, carbon dioxide tensions and cumulative dTc dosages are as indicated. (From Clark DL, Rosner BS 1973 Neurophysiologic effects of general anesthetics. I. The electroencephalogram and sensory evoked responses in man. *Anesthesiology* 38:564–582, by permission.)

Smith et al (1995) showed in a randomized study that responses to laryngoscopy and intubation were greater during thiopentone induction than with propofol. In another comparative study in two groups of women undergoing similar gynaecological surgery with identical anaesthetic protocols, EEG arousal responses to painful surgical stimuli were prevented in the group to whom epidural anaesthesia was also given (Kiyama and Takeda, 1997).

Chapter 5 Neurophysiological Monitoring during Sedation and Anaesthesia

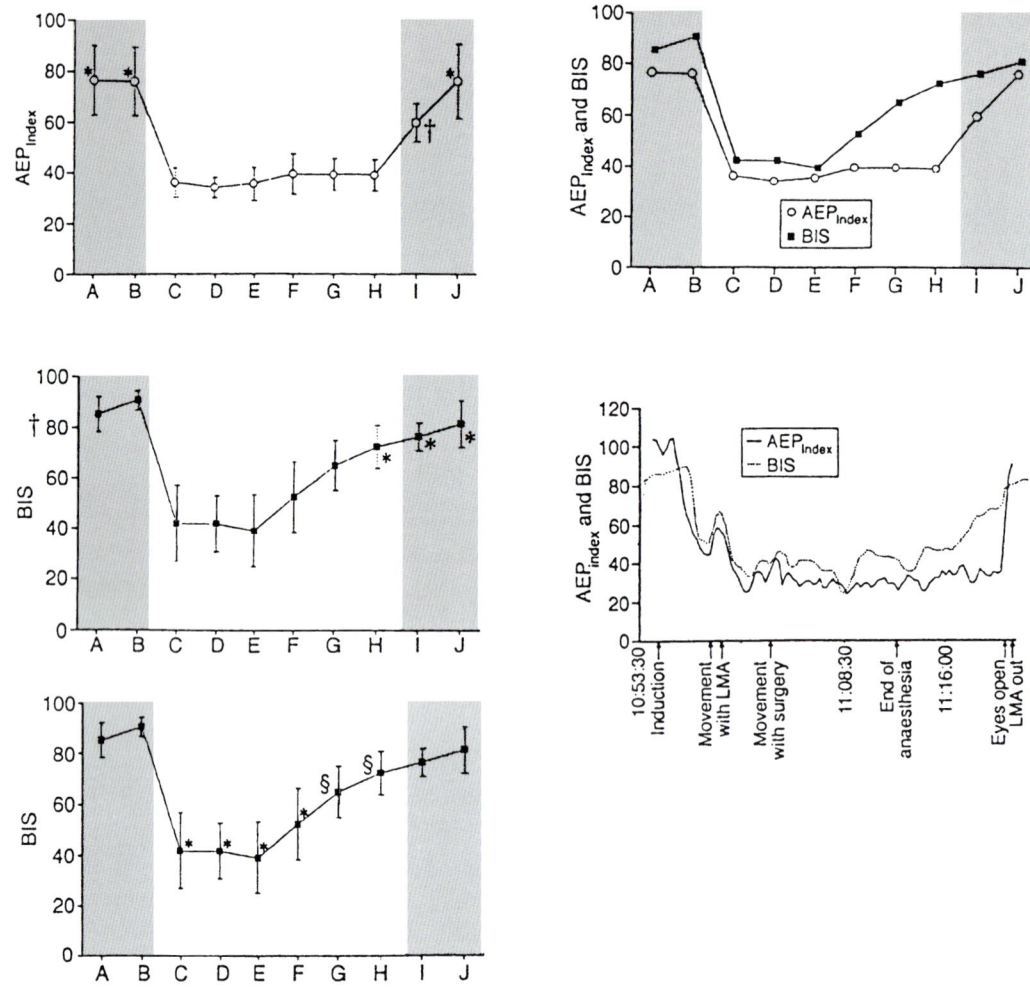

Fig. 5.20 Comparison between bispectral analysis of the EEG (BIS) and AEPs during surgery under propofol anaesthesia. Mean values for 20 patients with observation periods as follows: (A) baseline awake value, (B) before induction, (C) 30 s before skin incision, (D) 30 s after skin incision, (E) 5 min after skin incision, (F) at end of surgery and anaesthesia, (G) 3 min before eye opening, (H) 1 min before eye opening, (I) at eye opening, (J) at removal of the laryngeal mask. Shading marks the periods when patients were conscious. Plots (a)–(c) give mean and SD values for (a) AEP index, and (b) and (c) BIS. Statistical differences are indicated as follows: in plot (a) mean conscious AEP index values at point I are significantly lower than those at points A, B and J; in plot (b) unconscious BIS values at point H are not significantly different from those at points I and J; conscious BIS values at B were significantly higher than those at I; in plot (c) unconscious BIS values at points C to F are significantly lower than those at points G and H. Plots (d) and (e) compare group and individual data for the two indices. The authors conclude that whilst 'BIS may be able to predict recovery of consciousness during emergence from anaesthesia, the AEP was more able to detect the transition from unconsciousness to consciousness'. (From Gajraj RJ, Doi M, Mantzaridis H, Kenny GNC 1999 Comparison of bispectral EEG analysis and auditory evoked potentials for monitoring depth of anaesthesia during propofol anaesthesia. *British Journal of Anaesthesia* 82:672–678. © The Board of Management and Trustees of the *British Journal of Anaesthesia*. Reproduced by permission of Oxford University Press/*British Journal of Anaesthesia*.)

It is not only discomfort or pain that may lead to neurophysiological evidence of arousal phenomena during anaesthesia, but also the relatively loud noise levels encountered during some surgical procedures. These may approach those experienced when walking past a pneumatic drill being used to dig up the highway (see *Table 5.1*); this makes clarity in spoken communications important and increases fatigue of the surgical and monitoring teams (Jacobson and Balzer 1992).

Such noise levels are well above the typical levels for perceived or actual noise recorded in hospital wards by Soutar and Wilson (1986). However, even sounds such as a surgeon whistling quietly may elicit clear EEG responses during monitoring (*Fig. 5.25*). Words may be recalled postoperatively if anaesthesia is light. During a study with event-related potentials to frequent and infrequent tones of different pitch and to words, Van Hooff et al (1995) demonstrated that three of their twelve patients undergo-

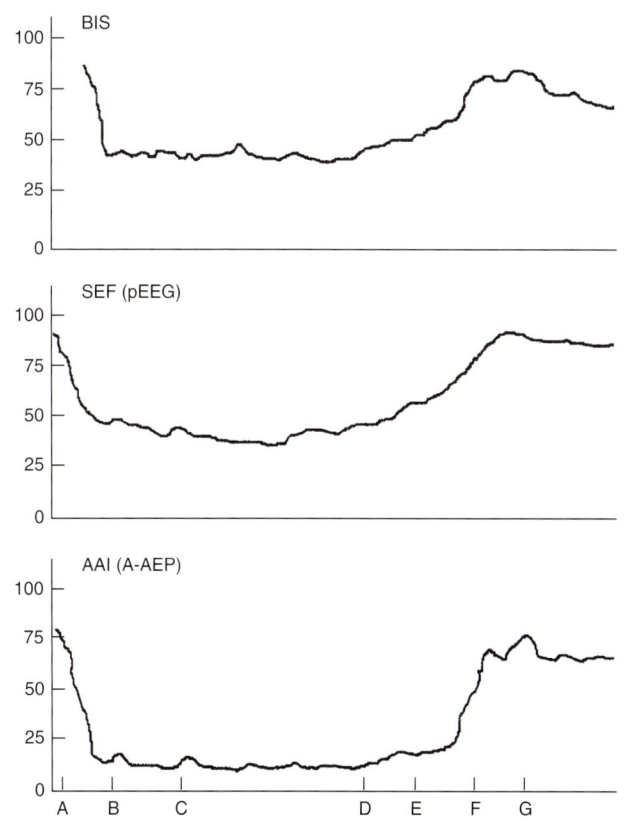

Fig. 5.21 Comparison between BIS (*n* = 30), processed EEG (pEEG, 90% spectral edge frequency (SEF) (*n* = 30) and Alaris MLAEPs (A-AEP, *n* = 30) in three groups of female patients during propofol and fentanyl induction and addition of nitrous oxide to maintain anaesthesia for mastectomy. A, start of anaesthesia; B, insertion of laryngeal mask airway; C, start of surgery; D, end of propofool infusion; E, end of surgery and nitrous oxide inhalation; F, respond to verbal command to open eyes; G, removal of laryngeal mask. (From Nishiyama T, Matsukawa T, Hanaoka K 2004 A comparison of the clinical usefulness of three different electroencephalogram monitors: Bispectral Index, processed electroencephalogram, and Alaris auditory evoked potentials. *Anesthesia and Analgesia* 98:1341–1345, by permission.)

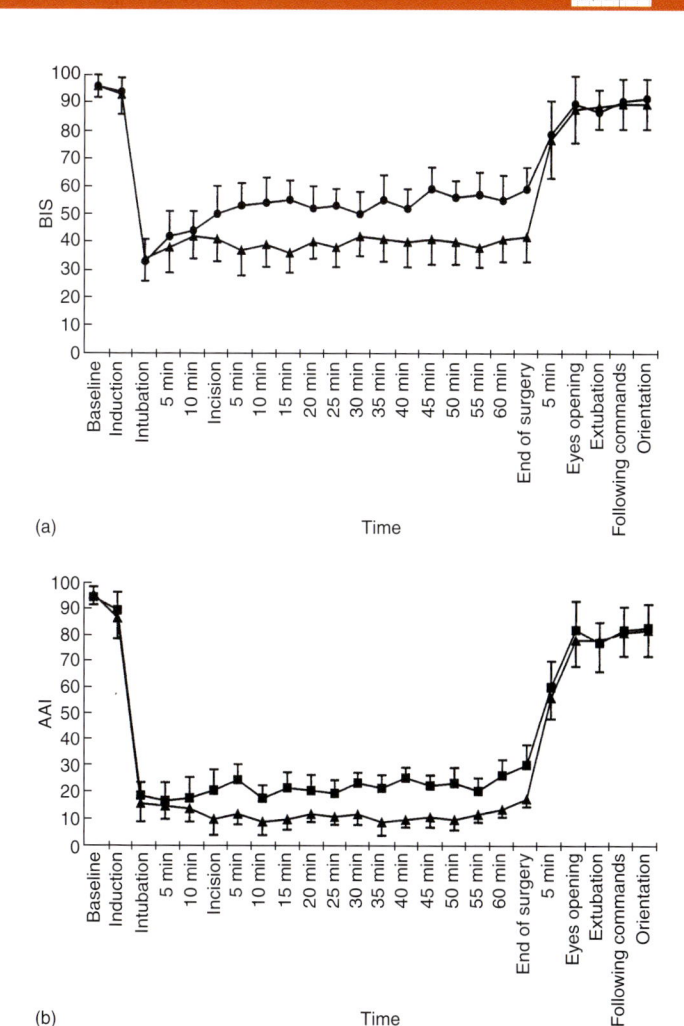

Fig. 5.22 (a) Perioperative BIS values in the control (triangles) and BIS-guided (circles) groups. Values are presented as mean ± SD; *$p \leq 0.05$ versus control group. (b) Perioperative auditory evoked potential index (AAI) values in the control (triangles) and AAI-guided (circles) groups. Values are presented as mean ± SD; *$p \leq 0.05$ versus baseline values. (White PF, Ma H, Tang J, Wender RH, Sloninsky A, Kariger R 2004 Does the use of electroencephalographic Bispectral Index or auditory evoked potential index monitoring facilitate recovery after desflurane anesthesia in the ambulatory setting? *Anesthesiology* 100:811–817, by permission.)

ing cardiac surgery under propofol–alfentanyl anaesthesia showed evidence of postoperative responses to words presented intraoperatively.

There have been reports linking 'awareness', judged by motor responses to commands using the isolated forearm technique (Tunstall 1977) to features such as the SEF 90% of the EEG (Gaitini et al 1995). However, Rodriguez et al (1993) monitoring relative power distribution, median frequency and fast/slow ratio serendipitously at the time of an episode which was followed by conscious recall, found no significant differences in these simple spectral parameters that could have differentiated the recall and non-recall periods. Similarly when SEF methods were compared with autonomic stress responses (Lindgren et al 1993) during prostatic surgery in a study of 89 patients, similar conclusions were reached (Ghouri et al 1993). These findings are somewhat comparable to those in prediction of eye opening during emergence from anaesthe-

Table 5.1 Sound levels generated by surgical tools in the operating theatre

Surgical tool	Monitored sound (dBA)		
	By surgeon's ear	At tool	By monitoring person
PD	83	85	78
CUSA	73	78	73
CO$_2$ laser	71	79	74
PD + CUSA	85	–	79
CUSA + laser	86	–	80

Data from Jacobson and Balzer (1992)
CUSA, Cavitron ultrasonic surgical aspirator; PD, pneumatic drill

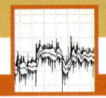

sia (Traast and Kalkman 1995) and support the view that such simple spectral measures are not adequate for such a task. It is perhaps rather more surprising that more advanced algorithms such as BIS do not always succeed either. Kerssens et al (2003) reported that with standard anaesthesia with BIS scores of 60–70, studies of 1082 commands in 56 patients showed no response from the isolated forearm in 82%, equivocal responses in 5% and an unequivocal response in 13% of commands. Of this 13% a quarter had conscious recall, leading the authors to conclude that EEG parameters did not distinguish between those with and without recall.

A major randomized controlled trial in 2463 patients in Australia (Myles et al 2004) showed that BIS monitoring reduced the risk of awareness during anaesthesia by 82% (at a cost of BIS monitoring of US $16 per use and a cost of preventing one case of awareness of $2200). Ekman et al (2004) report similar significant reduction of the incidence of awareness in a Swedish study of 4945 consecutive surgical patients compared with an historical control group of 7826 similar cases in whom no cerebral monitoring had been used.

Whether EEG arousal responses or event-related potentials predict or even equate with intraoperative 'awareness' and/or postoperative recall is quite another question (Lunn and Rosen 1987), but undoubtedly they suggest the risk of such a state. It is therefore important to notify the anaesthetic team when there is evidence of significant alteration in response to surgical or other events consisting either of a transient (and often reproducible) EEG event or of a sustained change in the overall pattern to that typical of a higher level of arousal. Comparable neurophysiological phenomena occur in comatose patients (see pp. 118 and 193). When seen, a trial of the efficacy of increased analgesia or anaesthesia in abolishing such responses to pain or noise should be undertaken to avoid such EEG events being followed by awareness and postoperative recall. Personal experience suggests that such observations often lead to beneficial modifications of an established anaesthetic protocol.

Medico-legal aspects of awareness during anaesthesia

The evidence that some patients undoubtedly have been aware during procedures under general anaesthesia, and have suffered severe and protracted psychological sequelae, has given rise to much anxiety as well as substantial litigation (Liu et al 1991, Moerman et al 1993, Sandin and Nordström 1993, Jones 1994, Payne 1994, Sebel et al 2004). It is considered that whilst many of these events are due to individual human failure, some reflect the extraordinary difficulties in clinical assessment of the level of

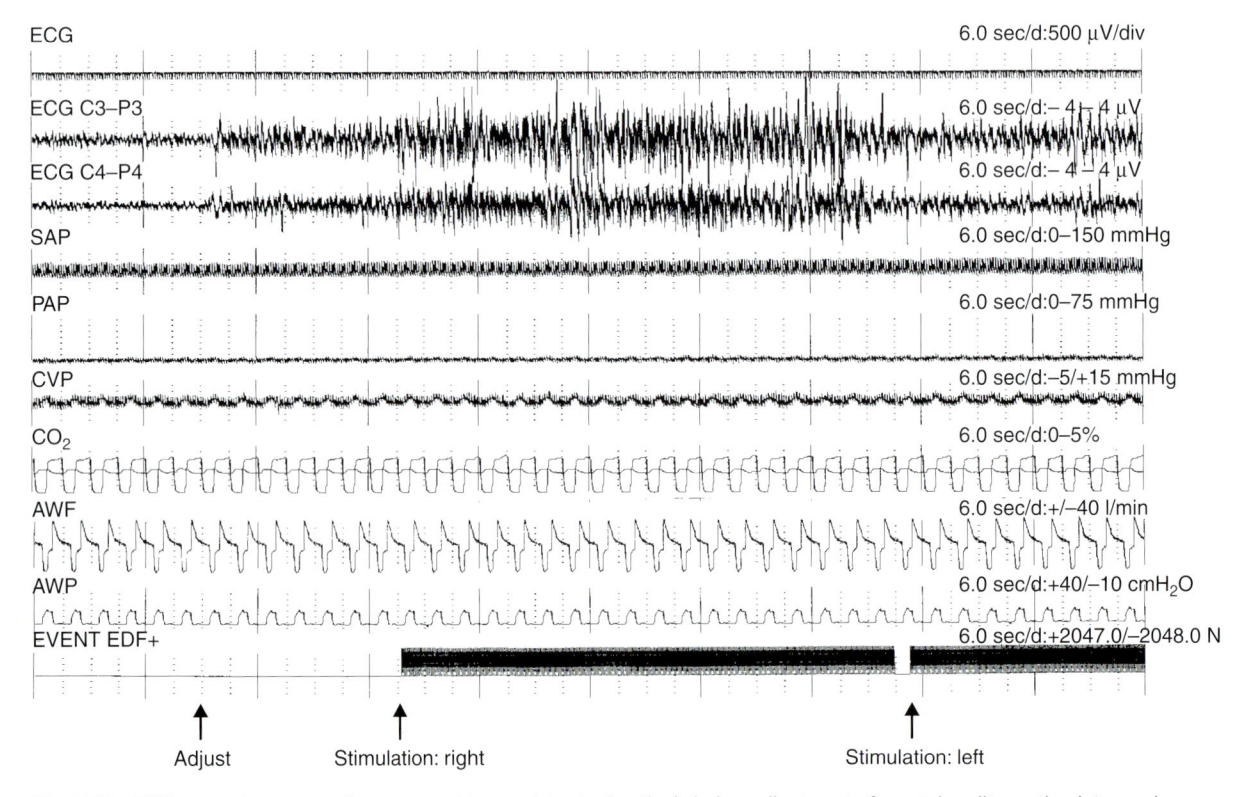

Fig. 5.23 EEG arousal response (compressed trace of 4 min duration) during adjustment of meatal auditory stimulator and right- and left-sided BAEP stimulation (event EDF) in an anaesthetized patient during cardiac surgery. SAP, PAP and CVP indicate systemic arterial, pulmonary artery and central venous pressures; CO_2, end-tidal CO_2 concentration; AWF and AWP, airway flow and pressure. Note abrupt and prolonged increase of EEG activity during adjustment. (Data from the EU-Biomed-2 project IBIS, (Carson et al 2000), from Binnie et al (2003), by permission.)

Awareness during anaesthesia

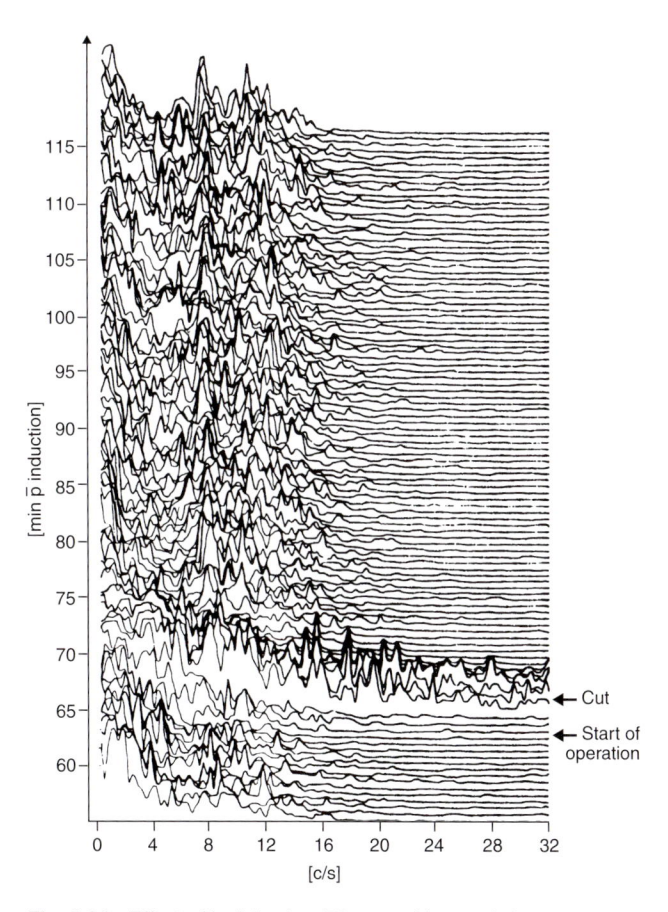

Fig. 5.24 Effect of incision in a 75-year-old man during surgery under 'neurolept-analgesia', CSA display of EEG from Cz-Al; other details as in previous figure. Sudden increase in fast activity at incision suggests that EMG components might be present. (From Pichlmayr I, Lips U, Künkel H 1984 *The Electroencephalogram in Anesthesia. Fundamentals, Practical Applications, Examples*, English edition (trans. E Bonatz, T Masyk-Iversen). © Springer-Verlag, Berlin, with kind permission of Springer Science and Business Media.)

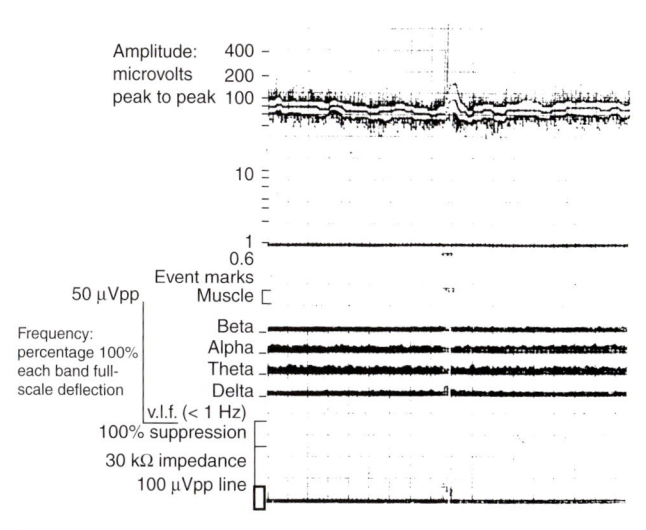

Fig. 5.25 Arousal response when surgeon whistled briefly during a prolonged spinal operation. EEG monitoring with CFAM during spinal monitoring with cortical SEPs. Note the initial artefact and brief flurry of scalp muscle potentials suggesting a behavioural response. (From Binnie et al (2003), by permission.)

consciousness with modern anaesthetic techniques. One role of the neurophysiologist may be to assist in the training of anaesthetists by providing some form of interactive human or neurophysiological feedback of the quality of anaesthetic depth control, reinforced when new techniques or agents are introduced; this could well enhance the role of simulators in training.

The close correlations between quantified EEG changes and concentrations of various anaesthetic agents have been known for many years (Faulconer and Bickford 1960, Hudson et al 1983, Schwilden et al 1985, Lunn and Rosen 1987). Discussion now centres upon comparisons of the relative merits of various types of EEG signal processing used in anaesthetic monitors with the aim of guiding the anaesthetist as to the adequacy of anaesthesia for the particular surgical procedure in hand. Such methods focus on quantitative measures of the ongoing pattern of the EEG but are rarely universally applicable because of intersubject and interanaesthetic agent variability (Dwyer et al 1994,

Thomsen and Prior 1996, Barr et al 2001, Daunderer and Schwender 2001, O'Connor et al 2001, Schneider et al 2002). It has been less common to consider what appears to be a more fruitful area for early warning of possible arousal, i.e. the specific EEG alterations following more or less noxious stimuli in subjects (see the example in *Fig. 5.24*) who are not protected by adequate anaesthesia and analgesia (e.g. Prior 1987, Kearse et al 1994, Kochs et al 1994, Wilder-Smith et al 1995, De Beer et al 1996, Mi et al 1998). The use of middle- and long-latency AEPs has indicated a possible (but complex) tool that may overcome problems of agent specificity with certain EEG measures (Heneghan et al 1987, Thornton et al 1989, Plourde and Picton 1990, Plourde et al 1993, Schwender et al 1996, Gajraj et al 1999).

Heier and Steen (1996a,b) after an extensive appraisal of the potential contributions of neurophysiological (EEG and AEP) monitoring techniques in detecting the possibility of awareness during anaesthesia, from which they concluded that because of insufficient sensitivity or specificity, 'at present a reliable monitor for consciousness is not available'. Now, we can be more optimistic, but the anaesthetist utilizing the information provided by modern 'awareness' monitors should always use his or her common sense and learn to read the raw signal. There may be clues such as the presence of muscle potentials or brief arousal responses in the EEG to suggest the possibility of arousals and hence awareness. In addition, the possibility of analysis of a corrupted or inadequate signal can be tested, much as Professor Yli-Hankala suggests in his Foreword. Golden rules for EEG and EP monitoring should include quality control of the validity of signals to help prevent erroneous conclusions being drawn on the basis of corrupted signals.

REFERENCES

Abrahamian HA, Allison T, Goff WR, et al 1963 Effects of thiopental on human cerebral evoked responses. Anesthesiology 24:650–657.

Algotsson L, Messeter K, Rosén I, et al 1992 Effects of nitrous oxide on cerebral haemodynamics and metabolism during isoflurane anaesthesia in man. Acta Anaesthesiol Scand 36:46–52.

Ali HH, Utting JE, Gray TC 1971 Quantitative assessment of residual antidepolarizing block (Part I). Br J Anaesth 43:473–476.

Arén C, Badr G, Feddersen K, et al 1985 Somatosensory evoked potentials and cerebral metabolism during cardiopulmonary bypass with special reference to hypotension induced by prostacyclin infusion. J Thorac Cardiovasc Surg 90:73–79.

Arfel G, Walter ST 1977 Aspects EEG au stade chirurgical d'anesthésies du longue durée en chirurgie cardiaque: effets des variation thermiques. Rev EEG Neurophysiol Clin 7:45–61.

Avramov MN, White PF 1995 Methods for monitoring the level of sedation. Crit Care Clin 11:803–826.

Baines DB, Whittle IR, Chaseling RW, et al 1985 Effect of halothane on spinal somatosensory evoked potentials in sheep. Br J Anaesth 57:896–899.

Banoczi W 1994 Anesthetic and metabolic effects during EEG and EP intraoperative monitoring. Am J EEG Technol 34:129–137.

Banoub M, Tetzlaff JE, Schubert A 2003 Pharmacologic and physiologic influences affecting sensory evoked potentials: implications for perioperative monitoring. Anesthesiology 99:716–737.

Barker J, Harper AM, McDowall DG, et al 1968 Cerebral blood flow, cerebral spinal fluid pressure and EEG activity during neuroleptanalgesia induced with dehydrobenzperidol and phenoperidine. Br J Anaesth 40:143–144.

Barr G, Anderson RE, Owall A, et al 2001 Being awake intermittently during propofol-induced hypnosis: a study of BIS, explicit and implicit memory. Acta Anaesthesiol Scand 45:834–838.

Bauerle K, Greim CA, Schroth M, et al 2004 Prediction of depth of sedation and anaesthesia by the Narcotrend EEG monitor. Br J Anaesth 92:841–845.

Beck A 1890 Localising brain and spinal cord function by electrical findings [in German]. Z Physiol 4:473–476.

Beecher HK, McDonagh FK 1939 Cortical action potentials during anesthesia. J Neurophysiol 2:289–307.

Bhargava AK, Setlur R, Sreevastava D 2004 Correlation of Bispectral Index and Guedel's stages of ether anesthesia. Anesth Analg 98:132–134.

Bickford RG 1950 Automatic electroencephalographic control of general anaesthesia. Electroencephalogr Clin Neurophysiol 1:93–96.

Bickford RG 1951 Use of frequency discrimination in the automatic electroencephalographic control of anaesthesia (servo-anaesthesia). Electroencephalogr Clin Neurophysiol 3:83–86.

Bingham RM, Procaccio F, Prior PF, et al 1985 Cerebral electrical activity influences the effects of etomidate on cerebral perfusion pressure in traumatic coma. Br J Anaesth 57:843–848.

Binnie CD, Cooper R, Mauguière F, et al (eds) 2003 *Clinical Neurophysiology*, Volume 2: *EEG, Paediatric Neurophysiology, Special Techniques and Applications*. Elsevier, Amsterdam.

Bischoff P, Kochs E, Haferkorn D, et al 1996 Intraoperative EEG changes in relation to the surgical procedure during isoflurane–nitrous oxide anesthesia: hysterectomy versus mastectomy. J Clin Anesth 8:36–43.

Blunnie WP, McIlroy PDA, Merrett JD, et al 1983 Cardiovascular and biochemical evidence of stress during major surgery associated with different techniques of anaesthesia. Br J Anaesth 55:611–618.

Bobbin RP, May JG, Lemoine RL 1979 Effects of pentobarbital and ketamine on brain-stem auditory potentials. Arch Otolaryngol 105:467–470.

Bogetz MS, Katz JA 1984 Recall of surgery for major trauma. Anesthesiology, 61:6–9.

Borgeat A, Dessibourg C, Popovic V, et al 1991 Propofol and spontaneous movements: an EEG study. Anesthesiology, 74:24–27.

Bovill JG, Sebel PS, Wauquier A, et al 1982 Electroencephalographic effects of sufentanil anaesthesia in man. Br J Anaesth 54:45–52.

Brain Trauma Foundation 1996 Guidelines for the management of severe head injury. American Association of Neurological Surgeons Joint Section on Neurotrauma and Critical Care. J Neurotrauma 13:641–734.

Brazier MAB 1970 Effect of anesthesia on visually evoked responses. Int Anesthesiol Clin 8:103–128.

Brazier MAB, Finesinger JE 1945 Action of barbiturates on the cerebral cortex. Arch Neurol Psychiatry (Chicago) 53:51–58.

Brierley JB, Graham DI 1984 Hypoxia and vascular disorders of the central nervous system. In: *Greenfield's Neuropathology*, 4th edition (eds J Hume Adams, JAN Corsellis, LW Duchan). Arnold, London, pp. 125–207.

Brocas E, Dupont H, Paugam-Burtz C, et al 2002 Bispectral index variations during tracheal suction in mechanically ventilated critically ill patients: effect of an alfentanil bolus. Intensive Care Med 28:211–213.

Buchwald JS, Hinman C, Norman RJ, et al 1981 Middle and long-latency evoked responses recorded from the vertex of normal and chronically lesioned cats. Brain Res 205:91–109.

Bulger EM, Nathens AB, Rivara FP, et al 2002 Management of severe head injury: institutional variations in care and effect on outcome. Crit Care Med 30:1870–1876.

Burchiel KS, Stockard JJ, Myers RR, et al 1975 Visual and auditory responses during enflurane anesthesia in man and cats. Electroencephalogr Clin Neurophysiol 39:434P.

Cabral RJ, Prior PF, Scott DF, et al 1977 Transient depression of EEG in hyperthermia. Electroencephalogr Clin Neurophysiol 42:697–701.

Carson ER, Van Gils M, Saranummi N (eds) 2000 IBIS: improved monitoring for brain dysfunction in intensive care and surgery. Comput Methods Programs Biomed 63:157–235.

Celesia G, Puletti F 1971 Auditory input in the human cortex during states of drowsiness and surgical anesthesia. Electroencephalogr Clin Neurophysiol 31:603–609.

Chabal CH, Jacobson L, Little J 1988 Effects of intrathecal fentanyl and lidocaine on somatosensory evoked potentials, the H-reflex and clinical responses. Anesth Analg 67:509–513.

Chi OZ, Field C 1990 Effects of enflurane on visual evoked potentials in humans. Br J Anaesth 64:163–166.

Chi OZ, Sommer W, Jasaitis D 1991 Power spectral analysis of EEG during sufentanil infusion in humans. Can J Anaesth 38:275–280.

Clark DL, Rosner BS 1973 Neurophysiologic effects of general anesthetics. I. The electroencephalogram and sensory evoked responses in man. Anesthesiology 38:564–582.

Clark DL, Hosick EC, Rosner BS 1971 Neurophysiological effects of different anesthetics in conscious man. J Appl Physiol 31:892–898.

Constant I, Leport Y, Richard P, et al 2004 Agitation and changes of Bispectral Index™ and electroencephalographic-derived variables during sevoflurane induction in children: clonidine premedication reduces agitation compared with midazolam. Br J Anaesth, 92:504–511.

Courtin RF, Bickford RG, Faulconer Jr A 1950 Classification and significance of electro-encephalographic patterns produced by nitrous oxide–ether anesthesia during surgical operations. Mayo Clinic Proc 25:197–206.

References

Dahaba AA 2005 Different conditions that could result in the bispectral index indicating an incorrect hypnotic state. Anesthesiology 101:765–773.

Daunderer M, Schwender D 2001 Depth of anesthesia, awareness and EEG [in German]. Anaesthesist 50:231–241.

De Beer NA, Van Hooff JC, Cluitmans PJ, et al 1996 Haemodynamic responses to incision and sternotomy in relation to the auditory evoked potential and spontaneous EEG. Br J Anaesth 76:685–693.

De Deyne C, Struys M, Decruyenaere J, et al 1998 Use of continuous bispectral EEG monitoring to assess depth of sedation in ICU patients. Intensive Care Med 24:1294–1298.

Deogaonkar A, Gupta R, DeGeorgia M, et al 2004 Bispectral Index monitoring correlates with sedation scales in brain-injured patients. Crit Care Med 32:2403–2406.

Desiderio DP, Thorne AC 1990 Awareness and general anaesthesia. Acta Anaesthesiol Scand 92(Suppl):48–50, 78 [discussion].

Doenicke A, Kugler J, Laub M 1967 Evaluation of recovery and 'street fitness' by EEG and psychodiagnostic tests after anaesthesia. Can Anaesth Soc J 14:567–583.

Doi M, Gajraj RJ, Mantzaridis H, et al 1999 Prediction of movement at laryngeal mask airway insertion: comparison of auditory evoked potential index, Bispectral Index, spectral edge frequency and median frequency. Br J Anaesth 82:203–207.

Doi M, Gajraj RJ, Mantzaridis H, et al 1997a Relationship between calculated blood concentration of propofol and electrophysiological variables during emergence from anaesthesia: comparison of Bispectral Index, spectral edge frequency, median frequency and auditory evoked potential index. Br J Anaesth 78:180–184.

Doi M, Gajraj RJ, Mantzaridis H, et al 1997b Effects of cardiopulmonary bypass and hypothermia on electroencephalographic variables. Anaesthesia 52:1048–1055.

Domino EF, Corssen G, Sweet RB 1963 Effects of various general anesthetics on the visually evoked response in man. Anesth Analg 42:735–747.

Drake ME, Pakalis A, Hietter SA, et al 1988 Antiepileptic drug effects on somatosensory evoked potentials. Epilepsia 29:265.

Driessen JJ, Harbers JB, van Egmond J, et al 1999 Evaluation of the electroencephalographic Bispectral Index during fentanyl–midazolam anaesthesia for cardiac surgery. Does it predict haemodynamic responses during endotracheal intubation and sternotomy? Eur J Anaesthesiol 16:622–627.

Drummond JC 1993 Brain protection during anesthesia. A reader's guide. Anesthesiology 79:877–880.

Drummond JC, Todd MM, U HS 1985 The effect of high dose sodium thiopental on brain stem auditory and median nerve somatosensory evoked responses in humans. Anesthesiology 63:249–254.

Dubois M, Sato S, Lees DE, et al 1980 Electroencephalographic changes during whole body hyperthermia in humans. Electroencephalogr Clin Neurophysiol 50:486–495.

Duncan PG, Sanders RA, McCullough DW 1979 Preservation of auditory-evoked brainstem responses in anesthetized children. Can Anesth Soc J 26:492–495.

Dutton RC, Smith WD, Smith NT 1996 EEG predicts movement response to surgical stimuli during general anesthesia with combinations of isoflurane, 70% N_2O, and fentanyl. J Clin Monit 12:127–139.

Dwyer RC, Rampil IJ, Eger EI II, et al 1994 The electroencephalogram does not predict depth of isoflurane anesthesia. Anesthesiology 81:403–409.

Eckert O, Werry C, Neulinger A, et al 1997 Intraoperative EEG monitoring using a neural network [in German]. Biomed Technik (Berlin) 42:78–84.

Eger EI, Saidman LJ, Brandstater B 1965 Minimum alveolar anesthetic concentration: a standard of anesthetic potency. Anesthesiology 26:756–763.

Eisenberg HM, Frankowski RF, Contant CF, et al 1988 High-dose barbiturate control of elevated intracranial pressure in patients with severe head injury. 69:15–23.

Ekman A, Lindholm ML, Lennmarken C, et al 2004 Reduction in the incidence of awareness using BIS monitoring. Acta Anaesthesiol Scand 48:20–26.

Ely EW, Truman B, Manzi DJ, et al 2004 Consciousness monitoring in ventilated patients: bispectral EEG monitor arousal not delirium. Intensive Care Med 30:1537–1543.

Engler GL, Speilholz NI, Bernhard WN, et al 1978 Somatosensory evoked potentials during Harrington instrumentation for scoliosis. J Bone Joint Surg Am 60A:528–532.

Faulconer A, Bickford RG 1960 *Electroencephalography in Anesthesiology. American Lectures in Anesthesiology* (ed. J Adriani), Publication No. 395, American Lecture Series. CC Thomas, Springfield, IL. [Some of the material will be found in Martin JT, Faulconer Jr A, Bickford RG 1959 Electroencephalography in anesthesiology. Anesthesiology 20:359–376.]

Fiol ME, Boening JA, Cruz-Rodriguez R, et al 1993 Effect of isoflurane (Forane) on intraoperative electro-corticogram. Epilepsia 34:897–900.

Forrest FC, Tooley MA, Saunders PR, et al 1994 Propofol infusion and the suppression of consciousness: the EEG and dose requirements. Br J Anaesth 72:35–41.

Frank M, Savege TM, Leigh M, et al 1982 Comparison of the cerebral function monitor and plasma concentrations of thiopentone and alphaxalone during total intravenous anaesthesia with repeated bolus doses of thiopentone and althesin. Br J Anaesth 54:609–616.

Frank M, Maynard DE, Tsanaclis LM, et al 1984 Changes in cerebral electrical activity measured by the Cerebral Function Analysing Monitor following bolus injections of thiopentone. Br J Anaesth 56:1075–1081.

Frenzel D, Greim CA, Sommer C, et al 2002 Is the Bispectral Index appropriate for monitoring the sedation level of mechanically ventilated surgical ICU patients? Intensive Care Med 28:178–183.

Gade J, Rosenfalck A, van Gils M, et al 1996 Modelling techniques and their application for monitoring in high dependency environments – learning models. Comput Methods Programs Biomed 51:75–84.

Gaitini L, Vaida S, Collins G, et al 1995 Awareness detection during caesarean section under general anaesthesia using EEG spectrum analysis. Can J Anaesth 42:377–381.

Gajraj RJ, Doi M, Mantzaridis H, et al 1999 Comparison of bispectral EEG analysis and auditory evoked potentials for monitoring depth of anaesthesia during propofol anaesthesia. Br J Anaesth 82:672–678.

Ganes T, Lundar T 1983 The effect of thiopentone on somatosensory evoked responses and EEGs in comatose patients. J Neurol Neurosurg Psychiatry 46:509–514.

García-Larrea L, Bertrand O, Artru F, et al 1987 Brainstem monitoring in coma. II: Dynamic interpretation of preterminal BAEP changes observed until brain death in deeply comatose patients. Electroencephalogr Clin Neurophysiol 68:446–457.

García-Larrea L, Artru F, Bertrand O, et al 1988 Transient drug-induced BAEP abolition in coma. Neurology 38:1487–1489.

García-Larrea L, Fischer C, Artru F 1993 Effet des anésthesiques sur les potentiels évoqués sensoriels. Neurophysiol Clin 23:141–162.

Ghouri AF, Monk TG, White PF 1993 Electroencephalogram spectral edge frequency, lower esophageal contractility, and autonomic responsiveness during general anesthesia. J Clin Monit 9:176–185.

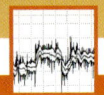

Gilbert TT, Wagner MR, Halukurike V, et al 2001 Use of bispectral electro-encephalogram monitoring to assess neurologic status in unsedated, critically ill patients. Crit Care Med 29:1996–2000.

Glaria AP, Murray A 1990 Monitoring brain function during cardiothoracic surgery in children and adults at two levels of hypothermia. Electroencephalogr Clin Neurophysiol 76:268–270.

Gloor P 1969 *Hans Berger on the Electroencephalogram of Man*. Electroencephalogr Clin Neurophysiol (Suppl 28).

Green JB, Walcoff MR, Lucke JF 1982 Phenytoin prolongs far-field somatosensory and auditory evoked potential interpeak latencies. Neurology 32:85–88.

Grote CL, Shanahan PT, Salmon P, et al 1992 Cognitive outcome after cardiac operations. Relationship to intraoperative computerized electro-encephalographic data. J Thorac Cardiovasc Surg 104:1405–1409.

Grundy BL, Brown RH, Greenberg PS 1979 Diazepam alters cortical evoked potentials. Anesthesiology 51P:S38.

Grundy BL, Brown RH, Berilla JA 1980 Fentanyl alters somatosensory cortical evoked potentials. Anesth Analg 59:544–545.

Guedel AE 1937 *Inhalational Anaesthesia. A Fundamental Guide*, 1st edn. MacMillan London.

Guérit JM 1998 Neuromonitoring in the operating room: why, when and how to monitor. Electroencephalogr Clin Neurophysiol 106:1–21.

Habibi S, Coursin DB 1996 Assessment of sedation, analgesia, and neuro-muscular blockade in the perioperative period. Int Anesthesiol Clin 34:215–241.

Hall JW 1985 The effects of high dose barbiturates on the acoustic reflex and auditory evoked responses. Acta Otolaryngol 100:387–398.

Hall JW, Mackey-Hargadine JR, Allen SJ 1985 Monitoring neurological status of comatose patients in the intensive care unit. In: *The Auditory Brainstem Response* (ed J Jacobson). College Hill, San Diego, CA, pp. 253–283.

Hall KD, Talton IH, Fox E, et al 1986 Brain function and level of consciousness in fentanyl anesthesia in heart surgery [in German]. Anaesthesist 35:226–230.

Harada K, Ichikawa G, Koh M, et al 1980 Effects of anesthesia on the middle latency components in cat [in Japanese]. Audiology 23:150–158.

Hartikainen KM, Rorarius M, Peräkylä JJ, et al 1995 Cortical reactivity during isoflurane burst suppression anesthesia. Anesth Analg 81:1223–1228.

Heier T, Steen PA 1996a Awareness in anaesthesia: incidence, consequences and prevention. Acta Anaesthesiol Scand 40:1073–1086.

Heier T, Steen PA 1996b Assessment of anaesthesia depth. Acta Anaesthesiol Scand 40:1087–1100.

Heneghan CPH, Thornton C, Navaratnarajah M, et al 1987 Effect of isoflurane on the auditory evoked response in man. Br J Anaesth 59:277–282.

Herkes GK, Wszolek ZK, Westmoreland BF, et al 1992 Effects of midazolam on electroencephalograms of seriously ill patients. Mayo Clinic Proc 67:334–338.

Herrick IA, Craen RA, Gelb AW, et al 1997 Propofol sedation during awake craniotomy for seizures: patient-controlled administration versus neurolept analgesia. Anesth Analg 84:1285–1291.

Hewitt PB, Chu DL, Polkey CE, et al 1999 Effect of propofol on the electrocorticogram in epileptic patients undergoing cortical resection. Br J Anaesth 82:199–202.

Hirose G, Kitagawa Y, Chujo T, et al 1986 Acute effects of phenytoin on brainstem auditory evoked potentials: clinical and experimental study. Neurology 36:1521–1524.

Hoffman WE, Edelman G 1995 Comparison of isoflurane and desflurane anesthetic depth using burst suppression of the electroencephalogram in neurosurgical patients. Anesth Analg 81:811–816.

Homer TD, Stanski DR 1985 The effect of increasing age on thiopental disposition and anesthetic requirement. Anesthesiology 62:714–724.

Hudson RJ, Stanski DR, Saidman LJ, et al 1983 A model for studying depth of anesthesia and acute tolerance to thiopental. Anesthesiology 59:301–308.

Hume AL, Cant BR 1981 Central somatosensory conduction after head injury. Ann Neural 10:411–419.

Hume AL, Durkin MA 1986 Central and spinal somatosensory conduction times during hypothermic cardiopulmonary bypass and some observations on the effects of fentanyl and isoflurane anesthesia. Electroencephalogr Clin Neurophysiol 65:46–58.

Hung OR, Varvel JR, Shafer SL, et al 1992 Thiopental pharmacodynamics. II. Quantitation of clinical and electroencephalographic depth of anesthesia. Anesthesiology 77:237–244.

Huotari AM, Koskinen M, Suominen K, et al 2004 Evoked EEG patterns during burst suppression with propofol. Br J Anaesth 92:18–24.

Ishikawa T, McDowall DG 1981 Electrical activity of the cerebral cortex during induced hypotension with sodium nitroprusside and trimetaphan in the cat. Br J Anaesth 53:605–611.

Ishikawa T, Funatsu N, Okamoto K, et al 1982 Cerebral and systemic effects of hypotension induced by trimetaphan or nitroprusside in dogs. Acta Anaesthesiol Scand 26:643–648.

Jacobson GP, Balzer GK 1992 Basic considerations in intra-operative monitoring: working in the operating room. In: *Neuromonitoring in Otology and Head and Neck Surgery* (eds JM Kartush, KR Bouchard). Raven Press, New York, pp. 21–60.

Jäntti V, Yli-Hankala A 2002 Neurophysiology of anaesthesia. In: *Clinical Neurophysiology at the Beginning of the 21st Century* (Clin Neurophysiol Suppl 53) (eds Z Ambler, S Nevšímalová, Z Kadaňka, PM Rossini). Elsevier, Amsterdam, pp. 84–88.

Jäntti V, Yli-Hankala A, Baer GA, et al 1993 Slow potentials of EEG burst suppression pattern during anaesthesia. Acta Anaesthesiol Scand 37:121–123.

Javel E, Mouney DF, McGee JA, et al 1982 Auditory brain-stem responses during systemic infusion of lidocaine. Arch Otolaryngol 108:71–76.

Jones JG 1989 *Depth of Anaesthesia*. Baillière's Clinical Anaesthesiology vol. 3, No. 3. Baillière Tindall, London.

Jones JG 1994 Perception and memory during general anaesthesia. Br J Anaesth 73:31–37.

Kalkman CJ, Van Rheineck AT, Hesselink EM, et al 1986 Effects of etomidate or midazolam on median nerve somatosensory evoked potentials [abstract]. Anesthesiology 65:A356.

Kálmánchey R, Avila A, Symon L 1986 The use of brainstem auditory evoked potentials during posterior fossa surgery as a monitor of brainstem function. Acta Neurochirurgica (Vienna), 82:128–136.

Kassell NF, Hitchon PW, Gerk MK, et al 1980 Alterations in cerebral blood flow, oxygen metabolism and electrical activity produced by high dose sodium thiopental. Neurosurgery 7:598–608.

Kearse Jr LA, Manberg P, DeBros F, et al 1994 Bispectral analysis of the electroencephalogram during induction of anesthesia may predict hemodynamic responses to laryngoscopy and intubation. Electroencephalogr Clin Neurophysiol 90:194–200.

Kerssens C, Klein J, Bonke B 2003 Awareness: Monitoring versus remembering what happened. Anesthesiology 99:570–575.

References

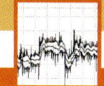

Kiyama S, Takeda J 1997 Effect of extradural analgesia on the paradoxical arousal response of the electroencephalogram. Br J Anaesth 79:750–753.

Kochs E, Treede RD, Roewer N, et al 1986 Alterations of somatosensory evoked potentials by etomidate and Diprivan. Anesthesiology 65, A353.

Kochs E, Bischoff P, Pichlmeier U, et al 1994 Surgical stimulation induces changes in brain electrical activity during isoflurane/nitrous oxide anesthesia. A topographic electroencephalographic analysis. Anesthesiology 80:1026–1034.

Kochs E, Stockmanns G, Thornton C, et al 2001 Wavelet analysis of middle latency auditory evoked responses: calculation of an index for detection of awareness during propofol administration. Anesthesiology, 95:1141–1150.

Kopf GS, Hume AL, Durkin MA, et al 1985 Measurement of central somatosensory conduction time in patients undergoing cardiopulmonary bypass: an index of neurologic function. Am J Surg 149:445–448.

Kovarik WD, Mayberg TS, Lam AM, et al 1994 Succinylcholine does not change intracranial pressure, cerebral blood flow velocity, or the electroencephalogram in patients with neurologic injury. Anesth Analg 78:469–473.

Kreuer S, Wilhelm W, Grundmann U, et al 2004 Narcotrend index versus Bispectral Index as electroencephalogram measures of anesthetic drug effect during propofol anesthesia. Anesth Analg 98:692–697.

Kumar A, Bhattacharya A, Makhija N 2000 Evoked potential monitoring in anaesthesia and analgesia. Anaesthesia 55:225–241.

Lam AM, Matta BF, Mayberg TS, et al 1995 Change in cerebral blood flow velocity with onset of EEG silence during inhalation anesthesia in humans: evidence of flow-metabolism coupling? J Cereb Blood Flow Metab 15:714–717.

Lancman ME, Marks S, Mahmood K, et al 1997 Atypical triphasic waves associated with the use of pentobarbital. Electroencephalogr Clin Neurophysiol 102:175–177.

Levy WJ 1984a Quantitative analysis of EEG changes during hypothermia. Anesthesiology 60:291–297.

Levy WJ 1984b Intraoperative EEG patterns: implications for EEG monitoring. Anesthesiology 60:430–434.

Lindgren L, Yli-Hankala A, Randell T, et al 1993 Haemodynamic and catecholamine responses to induction of anaesthesia and tracheal intubation: comparison between propofol and thiopentone. Br J Anaesth 70:306–310.

Litscher G, Schwarz G 1999 Is there paradoxical arousal reaction in the EEG subdelta range in patients during anesthesia? J Neurosurg Anesthesiol 11:49–52.

Liu J, Singh H, White PF 1997 Electroencephalographic Bispectral Index correlates with intraoperative recall and depth of propofol-induced sedation. Anesth Analg 84:185–189.

Liu WHD, Thorp TAS, Graham SG, et al 1991 Incidence of awareness with recall during general anaesthesia. Anaesthesia 46:435–437.

Lloyd-Thomas AR, Cole PV, et al 1990a Quantitative EEG and brainstem auditory evoked potentials: comparison of isoflurane with halothane using the cerebral function analysing monitor. Br J Anaesth 65:306–312.

Lloyd-Thomas AR, Cole PV, Prior PF 1990b Isoflurane prevents EEG depression during trimetaphan-induced hypotension in man. Br J Anaesth 65:313–318.

Loughnan BL, Sebel PS, Thomas D, et al 1987 Evoked potentials following diazepam or fentanyl. Anaesthesia 42:195–198.

Loughnan BA, Murdoch LJ, Hetreed MA, et al 1990 Effects of 2% lignocaine on somatosensory evoked potentials recorded in the extradural space. Br J Anaesth 65:643–647.

Lunn JN, Rosen M (eds) 1987 *Consciousness, Awareness, Pain and General Anaesthesia*. Butterworths, London.

Maas AI 2002 Guidelines for head injury: their use and limitations. Neurol Res 24:19–23.

Maas AI, Dearden M, Teasdale GM, et al 1997 European Brain Injury Consortium – guidelines for management of severe head injury in adults. Acta Neurochir (Wein) 139:286–294.

Maas AI, Dearden M, Servadei F, et al 2000 Current recommendations for neurotrauma. Curr Opin Crit Care 6:281–292.

Maas AI, Marmarou A, Murray GD, et al 2004 Clinical trials in traumatic brain injury: current problems and future solutions. Acta Neurochir (Suppl 89):113–118.

MacIver MB, Mandema JW, Stanski DR, et al 1996 Thiopental uncouples hippocampal and cortical synchronized electroencephalgraphic activity. Anesthesiology 84:1411–1424.

Madigan RR, Linton AE, Wallace SL, et al 1987 A new technique to improve cortical-evoked potentials in spinal cord monitoring: a ratio method of analysis. Spine 12:330–335.

Mäkelä K, Hartikainen K, Rorarius M, et al 1996 Suppression of F-VEP during isoflurane-induced EEG suppression. Electroenceph Clin Neurophysiol 100:269–272.

Manninen PH, Lam AM, Nicholas JF 1985 The effects of isoflurane and isoflurane–nitrous oxide anesthesia on brainstem auditory evoked potentials in humans. Anesth Analg 64:43–47.

Marcus DJ, Swift TR, McDonald TF 1981 Acute effects of phenytoin on peripheral nerve function in the rat. Muscle Nerve 4:48–50.

Markand ON, Lee BI, Warren C, et al 1987 Effects of hypothermia on BAEPs in humans. Ann Neurol 22:507–513.

Marsh RR, Frewent TC, Sutton LN, et al 1984 Resistance of the auditory brainstem responses to high barbiturate levels. Otolaryngol Head Neck Surg 92:685–688.

Marshall LF, Shapiro HM 1977 Barbiturate control of intracranial hypertension in head injury and other related conditions: iatrogenic coma. Acta Neurol Scand 56(Suppl 64):156–157.

Marshall LF, Smith RW, Shapiro HM 1979 The outcome with aggressive treatment in severe head injuries. Part II: Acute and chronic barbiturate administration in the management of head injury. J Neurosurg 50:26–30.

Martin JT, Faulconer Jr A, Bickford RG 1959 Electroencephalography in anesthesiology. Anesthesiology 20:359–376.

Matta BF, Lam AM 1995 Nitrous oxide increases cerebral blood flow velocity during pharmacologically induced EEG silence in humans. J Neurosurg Anesthesiol 1:89–93.

Matta BF, Lam AM, Strebel S, et al 1995 Cerebral pressure autoregulation and carbon dioxide reactivity during propofol-induced EEG suppression. Br J Anaesth 74:159–163.

Maurette P, Simeon F, Castagnera L, et al 1988 Propofol anaesthesia alters somatosensory evoked cortical potentials. Anaesthesia 43(Suppl):44–45.

Mavroudakis N, Brunko E, Nogueira MC, et al 1991 Acute effects of diphenylhydantoin on peripheral and central somatosensory conduction. Electroencephalogr Clin Neurophysiol 78:263–266.

McDowall DG, Okuda Y, Heuser D, et al 1978 Control of cerebral vascular smooth muscle during general anaesthesia. Ciba Found Symp 56:257–273.

McPherson RW, Mahla M, Johnson R, et al 1985 Effects of enflurane, isoflurane and nitrous oxide on somatosensory evoked potentials during fentanyl anesthesia. Anesthesiology 62:626–633.

185

Chapter 5 Neurophysiological Monitoring during Sedation and Anaesthesia

McPherson RW, Sell B, Traystman RJ 1986 Effects of thiopental, fentanyl, and etomidate on upper extremity somatosensory evoked potentials in humans. Anesthesiology 65:584–589.

Mi WD, Sakai T, Takahashi S, Matsuki A 1998 Haemodynamic and electro-encephalograph responses to intubation during induction with propofol or propofol/fentanyl. Can J Anaesth 45:19–22.

Mi WD, Sakai T, Singh H, et al 1999 Hypnotic endpoints vs. the Bispectral Index, 95% spectral edge frequency and median frequency during propofol infusion with or without fentanyl. Eur J Anaesthesiol 16:47–52.

Michenfelder JD 1974 The interdependency of cerebral functional and metabolic effects following massive doses of thiopental in the dog. Anesthesiology 41:231–236.

Michenfelder JD 1986 A valid demonstration of barbiturate-induced brain protection in man – at last. Anesthesiology 64:140–142.

Modica PA, Tempelhoff R, White PF 1990 Pro- and anticonvulsant effects of anesthetics. Anesth Analg 70:433–444.

Moerman N, Bonke B, Oosting J 1993 Awareness and recall during general anesthesia. Anesthesiology 79:445–464.

Munglani R, Andrade J, Sapsford DJ, et al 1993 A measure of consciousness and memory during isoflurane administration: the coherent frequency. Br J Anaesth 71:633–641.

Munson ES, Tucker WK, Ausinsch B, et al 1975 Etidocaine, bupivacaine and lidocaine seizure thresholds in monkeys. Anesthesiology 42:471–478.

Myles PS, Leslie K, McNeil J, et al 2004 Bispectral index monitoring to prevent awareness during anaesthesia: the B-Aware randomised controlled trial. Lancet 363(9423):1757–1763.

Narayan RK, Michel ME, Ansell B, et al 2002 Clinical trials in head injury. J Neurotrauma 19:503–557.

Newlon PG, Greenberg RP, Enas GG, et al 1983 Effects of therapeutic pentobarbital coma on multimodality evoked potentials recorded from severely head-injured patients. Neurosurgery 12:613–619.

Nilsson E, Ingvar DH 1967 EEG findings in neuroleptanalgesia. Acta Anaesthiol Scand 11:121–127.

Nishiyama T, Matsukawa T, Hanaoka K 2004 A comparison of the clinical usefulness of three different electroencephalogram monitors: Bispectral Index, processed electroencephalogram, and Alaris auditory evoked potentials. Anesth Analg 98:1341–1345.

Ochiai R, Sato K, Koitabashi T, et al 1992 Electroencephalographic change during induction of high-dose fentanyl anesthesia – evaluation by LIFESCAN EEG monitor [in Japanese, English summary]. Masui 41:799–804.

O'Connor MF, Daves SM, Tung A, et al 2001 BIS monitoring to prevent awareness during general anesthesia. Anesthesiology 94:520–522.

Patel H 1981 Experience with the cerebral function monitor during deliberate hypotension. Br J Anaesth 53:639–645.

Payne JP 1994 Awareness and its medicolegal implications. Br J Anaesth 73:38–45.

Peterson DO, Drummond JC, Todd MM 1986 Effects of halothane, enflurane, isoflurane and nitrous oxide on somatosensory evoked potentials in humans. Anesthesiology 65:35–40.

Pfurtscheller G, Schwarz G, Schroettner O, et al 1987 Continuous and simultaneous monitoring of EEG spectra and brainstem auditory and somatosensory evoked potentials in the intensive care unit and the operating room. J Clin Neurophysiol 4:389–396.

Pichlmayr I, Lips U, Künkel H 1984 *The Electroencephalogram in Anaesthesia. Fundamentals Practical Applications Examples*, English edition (trans. E Bonatz, T Masyk-Iversen). Springer-Verlag, Berlin.

Plourde G, Picton TW 1990 Human auditory steady-state response during general anesthesia. Anesth Analg 71:460–468.

Plourde G, Joffe D, Villemure C, et al 1993 The P3a wave of the auditory event-related potential reveals registration of pitch change during sufentanil anesthesia for cardiac surgery. Anesthesiology 78:498–509.

Ponte J 1995 Neuromuscular blockers during general anaesthesia. BMJ 310:1218–1219.

Prevec TS 1980 Effect of Valium on the somatosensory evoked potentials. In: *Clinical Uses of Cerebral Brainstem and Spinal Somatosensory Evoked Potentials* (ed. JE Desmedt). Karger, Basel, pp. 311–318.

Prichep LS, Gugino LD, John ER, et al 2004 The Patient State Index as an indicator of the level of hypnosis under general anaesthesia. Br J Anaesth 92:393–399.

Prior PF 1987 The EEG and detection of responsiveness during anaesthesia and coma. In: *Consciousness Awareness Pain and General Anaesthesia* (eds JN Lunn, M Rosen). Butterworths, London, pp. 34–45.

Prior PF, Maynard DE 1986 *Monitoring Cerebral Function. Long-term Monitoring of EEG and Evoked Potentials*. Elsevier, Amsterdam.

Procaccio F, Bingham RM, Hinds CJ, et al 1988 Continuous EEG and ICP monitoring as a guide to the administration of Althesin sedation in severe head injury. Intensive Care Med 14:148–155.

Rampil IJ 1998 A primer for EEG signal processing in anesthesia. Anesthesiology 89:980–1002.

Ramsay MAE, Savege TM, Simpson BRJ, et al 1974 Controlled sedation with alphaxalone–alphadolone. BMJ 2:656–659.

Reddy RV, Moorthy SS, Mattice T, et al 1992 An electroencephalographic comparison of effects of propofol and methohexitol. Electroencephalogr Clin Neurophysiol 83:162–168.

Reddy RV, Moorthy SS, Dierdorf SF, et al 1993 Excitatory effects and electroencephalographic correlation of etomidate, thiopental, methohexital, and propofol. Anesth Analg 77:1008–1011.

Reilly EL, Brunberg JA, Doty DB 1974 Effect of deep hypothermia and total circulatory arrest on the electroencephalogram in children. Electroencephalogr Clin Neurophysiol 36:661–667.

Reilly EL, Kondo C, Brunberg JA, et al 1978 Visual evoked potentials during hypothermia and prolonged circulatory arrest. Electroencephalogr Clin Neurophysiol 45:100–106.

Riker RR, Fraser GL, Wilkins ML 2003 Comparing the Bispectral Index and suppression ratio with burst suppression of the electroencephalogram during pentobarbital infusions in adult intensive care patients. Pharmacotherapy 23:1087–1093.

Rockoff MA, Marshall LF, Shapiro HM 1979 High-dose barbiturate therapy in humans: A clinical review of 60 patients. Ann Neurol 6:194–199.

Rodriguez RA, Edmonds Jr HL, Schroeder JA 1993 Quantitative EEG analysis during cardiac surgery in a case of spontaneous recall. Electroencephalogr Clin Neurophysiol 87:250–253.

Röpcke H, Schwilden H 1996 The interaction of nitrous oxide and enflurane on the EEG median of 2–3 Hz is additive, but weaker than at 1.0 MAC [in German]. Anaesthesist 45:819–825.

Rosén I, Hagerdal M 1976 Electroencephalographic study of children during ketamine anesthesia. Acta Anaesthiol Scand 20:32–39.

Rosner BS, Clark DL 1973 Neurophysiologic effects of general anesthetics. II: Sequential regional actions in the brain. Anesthesiology 39:59–81.

Russ W, Kling D, Loesevitz A, et al 1984 Effect of hypothermia on visual evoked potentials (VEP) in humans. Anesthesiology 61:207–210.

Russ W, Kling D, Sauerwein G, et al 1987a Spectral analysis of the EEG during hypothermic cardiopulmonary bypass. Acta Anaesthiol Scand 31:111–116.

Russ W, Sticher J, Scheld H, et al 1987b Effects of hypothermia on somatosensory evoked responses in man. Br J Anaesth 59:1484–1491.

Rytky S, Huotari AM, Alahuhta S, et al 1999 Tibial nerve somatosensory evoked potentials during EEG suppression in sevoflurane anaesthesia. Clin Neurophysiol 110:1655–1658.

Salzman SK, Beckman AL, Marks HG, et al 1986 Effects of halothane on intra-operative scalp recorded somatosensory evoked potentials to tibial nerve stimulation in man. Electroencephalogr Clin Neurophysiol 65:36–45.

Sandin R, Nordström O 1993 Awareness during total i.v. anaesthesia. Br J Anaesth 71:782–787.

Schmidt GN, Bischoff P, Standl T, et al 2003 Narcotrend and Bispectral Index monitor are superior to classic electroencephalographic parameters for the assessment of anesthetic states during propofol–remifentanil anesthesia. Anesthesiology 99:1072–1077.

Schneider G, Heglmeier S, Schneider J, et al 2004 Patient State Index (PSI) measures depth of sedation in intensive care patients. Intensive Care Med 30:213–216.

Schneider G, Wagner K, Reeker W, et al 2002 Bispectral Index (BIS) may not predict awareness reaction to intubation in surgical patients. J Neurosurg Anesthesiol 14:7–11.

Scholz J, Bischoff P, Szafarczyk W, et al 1996 Comparison of sevoflurane and isoflurane in ambulatory surgery. Results of a multicenter study [in German]. Anaesthesist 45(Suppl 1):S63–S70.

Schubert A, Licina MG, Lineberry PJ 1988 Ketamine enhances somatosensory evoked potentials in man but fails to prevent the depressant effect of nitrous oxide. Anesthesiology 69:A310.

Schubert A, Licina MG, Lineberry PJ 1990 The effect of ketamine on human somatosensory evoked potentials and its modification by nitrous oxide. Anesthesiology 72:33–39.

Schultz B, Schultz A, Grouven U, et al 1995 Changes with age in EEG during anesthesia [in German]. Anaesthesist 44:467–472.

Schultz A, Grouven U, Zander I, et al 2004 Age-related effects in the EEG during propofol anaesthesia. Acta Anaesthesiol Scand 48:27–34.

Schwartz MS, Virden S, Scott DF 1974 Effects of ketamine on the electroencephalograph. Anaesthesia 29:135–140.

Schwender D, Daunderer M, Klasing S, et al 1998a Power spectral analysis of the electroencephalogram during increasing end-expiratory concentrations of isoflurane, desflurane and sevoflurane. Anaesthesia 53:335–342.

Schwender D, Kunze-Kronawitter H, Dietrich P, et al 1998b Conscious awareness during general anaesthesia: patients' perceptions, emotions, cognition and reactions. Br J Anaesth 80:133–139.

Schwender D, Daunderer M, Klasing S, et al 1996 Monitoring intraoperative awareness. Vegetative signs, isolated forearm technique, electroencephalogram, and acute evoked potentials [in German]. Anaesthesist 45:708–721.

Schwilden H, Stoeckel H 1987 Quantitative EEC analysis during anaesthesia with isoflurane in nitrous oxide at 1.3 and 1.5 MAC. Br J Anaesth 59:738–745.

Schwilden H, Stoeckel H 1993 Closed-loop feedback controlled administration of alfentanil during alfentanil–nitrous oxide anaesthesia. Br J Anaesth 70:389–393.

Schwilden H, Schüttler J, Stoeckel H 1985 Quantitation of the EEG and pharmacodynamic modelling of hypnotic drugs: etomidate as an example. Eur J Anaesthesiol 2:121–131.

Schwilden H, Stoeckel H, Schüttler J 1989 Closed-loop feedback control of propofol anaesthesia by quantitative EEG analysis in humans. Br J Anaesth 62:290–296.

Schwilden H, Ropcke H, Drosler S 1995 Clinical potency of nitrous oxide – is MAC the gold standard? [in German]. Anaesth Intensivmed Notfallmed Schmerzther 30:337–340.

Sebel PS, Flynn PJ, Ingram DA 1984 Effect of nitrous oxide on visual, auditory and somatosensory evoked potentials. Br J Anaesth 56:1403–1407.

Sebel PS, Erwin CW, Neville WK 1987 Effects of halothane and enflurane on far and near field somatosensory evoked potentials. Br J Anaesth 59:1492–1496.

Sebel PS, Lang E, Rampil IJ, et al 1997 A multicenter study of bispectral electroencephalogram analysis for monitoring anesthetic effect. Anesth Analg 84:891–899.

Sebel PS, Dowdle TA, Ghoneim MM, et al 2004 The incidence of awareness during anesthesia: a multicenter United States study. Anesth Analg 99:833–839.

Shapira Y, Gurman G, Artru AA, et al 1997 Combined hemodilution and hypotension monitored with jugular bulb oxygen saturation EEG, and ECG decreases transfusion volume and length of ICU stay for major orthopedic surgery. J Clin Anesth 9:643–649.

Shapiro SM, Moller A, Shiu GK 1984 Brain-stem auditory evoked potentials in rats with high-dose pentobarbital. Electroencephalogr Clin Neurophysiol 58:266–276.

Sharpe RM, Nathwani D, Pal SK, et al 1997 Auditory evoked response, median frequency and 95% spectral edge during anaesthesia with desflurane and nitrous oxide. Br J Anaesth 78:282–285.

Shin IJ, Peinkofer JF, Paroski MW, et al 1988 The effect of non-toxic levels of phenytoin on brainstem auditory potentials in normal subjects. Neurology 38 1(Suppl):337.

Sidi A, Halimi P, Cotev S 1990 Estimating anesthetic depth by electroencephalography during anesthetic induction and intubation in patients undergoing cardiac surgery. J Clin Anesth 2:101–107.

Sloan TB 1998 Anesthetic effects on electrophysiologic recordings. J Clin Neurophysiol 15:217–226.

Sloan TB, Heyer EJ 2002 Anesthesia for intraoperative neurophysiologic monitoring of the spinal cord. J Clin Neurophysiol 19:430–443.

Sloan TB, Koht A 1985 Depression of cortical somatosensory evoked potentials by nitrous oxide. Br J Anaesth 57:849–852.

Sloan TB, Ronai AK, Toleikis JR, et al 1988 Improvement of intraoperative somatosensory evoked potentials by etomidate. Anesth Analg 67:582–585.

Smith B 1975 Anesthesia for orbital surgery: observed changes in the visually evoked response at low blood pressures. Modern Prob Ophthalmol 14:457–459.

Smith M, Smith SJ, Scott CA, et al 1996 Activation of the electrocorticogram by propofol during surgery for epilepsy. Br J Anaesth 76:499–502.

Smith NT, Dec-Silver H, Sanford Jr TJ, et al 1984 EEGs during high-dose fentanyl–, sufentanil–, or morphine–oxygen anesthesia. Anesth Analg 63:386–393.

Smith NT, Westover Jr CJ, Quinn M, et al 1985 An electroencephalographic comparison of alfentanil with other narcotics and with thiopental. J Clin Monit 1:236–244.

Sohmer H, Gafni M, Chisin R 1978 Auditory nerve and brainstem responses: comparison in awake and unconscious subjects. Arch Neurol 35:228–230.

Soliman HM, Mélot C, Vincent JL 2001 Sedative and analgesic practice in the intensive care unit: the results of a European survey. Br J Anaesth 87:186–192.

Soutar RL, Wilson JA 1986 Does hospital noise disturb patients? BMJ 292:305.

Spencer EM, Green JL, Willatts SM 1994 Continuous monitoring of depth of sedation by EEG spectral analysis in patients requiring mechanical ventilation. Br J Anaesth 73:649–654.

Steen PA, Michenfelder JD 1979 Neurotoxicity of anesthetics. Anesthesiology 50:437–453.

Stockard JJ, Werner SS, Aalbers JA, et al 1976 Electroencephalographic findings in phencyclidine intoxication. Arch. Neural 33:200–203.

Stockard JJ, Sharbrough FW, Tinker JA 1978 Effects of hypothermia on the human brainstem auditory response. Ann Neurol 3:368–370.

Stoeckel H, Schwilden H 1986 Methoden der automatischen Feedback-Regelung fur die Narkose. Konzepte und klinische Anwendung. Anasthesiol Intensivmed Notfallmed Schmerzther 21:60–67.

Sutton LN, Frewen T, Marsh R, et al 1982 The effects of deep barbiturate coma on multimodality evoked potentials. J Neurosurg 57:178–185.

Theilen HJ, Ragaller M, Tscho U, et al 2000 Electroencephalogram silence ratio for early outcome prognosis in severe head trauma. Crit Care Med 28:3522–3529.

Thomas WA, Cole PV, Etherington NJ, et al 1985 Electrical activity of the cerebral cortex during induced hypotension in man; a comparison of sodium nitroprusside and trimetaphan. Br J Anaesth 57:134–141.

Thomsen CE, Prior PF 1996 Quantitative EEG in assessment of anaesthetic depth: comparative study of methodology. Br J Anaesth 77:172–178.

Thornton C, Newton DEF 1989 The auditory evoked response: a measure of depth of anaesthesia. In: *Depth of Anaesthesia*, Baillière's Clinical Anaesthesiology, vol. 3, No. 3 (ed JG Jones). Baillière Tindall, London, pp. 559–585.

Thornton C, Sharpe RM 1998 Evoked responses in anaesthesia. Br J Anaesth 81:771–781.

Thornton C, Heneghan CPH, James MFM, et al 1984 Effects of halothane or enflurane with controlled ventilation on auditory evoked potentials. Br J Anaesth 56:315–323.

Thornton C, Barrowcliffe MP, Konieczko KM, et al 1989 The auditory evoked response as an indicator of awareness. Br J Anaesth 63:113–115.

Traast HS, Kalkman CJ 1995 Electroencephalographic characteristics of emergence from propofol/sufentanil total intravenous anaesthesia. Anesth Analg 81:366–371.

Trauner DA 1986 Barbiturate therapy in acute brain injury. J Pediatr 109:742–746.

Tunstall ME 1977 Detecting wakefulness during general anaesthesia for caesarian section. BMJ i:316–319.

Uhl RR, Squires KC, Bruce DL, et al 1980 Effect of halothane anesthesia on the human cortical visual evoked response. Anesthesiology 53:273–276.

Vakkuri A, Yli-Hankala A, Talja P, et al 2004 Time-frequency balanced spectral entropy as a measure of anesthetic drug effect in central nervous system during sevoflurane, propofol, and thiopental anesthesia. Acta Anesthesiol Scand 48:145–153.

Vakkuri A, Yli-Hankala A, Sandin R, et al 2005 Spectral entropy monitoring is associated with reduced propofol use and faster emergence in propofol–nitrous oxide–alfentanyl anaesthesia. *Anesthesiology* 103:274–279.

Van de Velde M, van Erp G, Cluitmans PJ 1998 Detection of muscle artefact in the normal human awake EEG. Electroencephalogr Clin Neurophysiol 107:149–158.

Van de Velde M, Ghosh IR, Cluitmans PJM 1999 Context related artefact detection in prolonged EEG recordings. Comput Method Program Biomed 60:183–196.

Van der Kouwe AJW, Burgess RC 2003 Neurointensive care unit system for continuous electrophysiological monitoring with remote web-based review. IEEE Trans Inf Technol Biomed 7:130–140.

Van Gils M, Rosenfalck A, White S, et al 1997 Signal processing in prolonged EEG recordings during intensive care. Methods for analyzing and displaying EEG signals. IEEE Eng Med Biol Mag 16:56–63.

Van Gils M, Korhonen I, Yli-Hankala A 2002 Methods for assessing adequacy of anesthesia. Crit Rev Biomed Eng 30:99–130.

Van Hooff JC, De Beer NA, Brunia CH, et al 1995 Information processing during cardiac surgery: an event related potential study. Electroencephalogr Clin Neurophysiol 96:433–452.

Verma A, Bedlack RS, Radtke RA, et al 1999 Succinylcholine induced hyperkalemia and cardiac arrest: death related to an EEG study. J Clin Neurophysiol 16:46–50.

Veselis RA, Reinsel R, Sommer S, et al 1991 Use of neural network analysis to classify electroencephalographic patterns against depth of midazolam sedation in intensive care unit patients. J Clin Monit 7:259–267.

Veselis RA, Reinsel R, Wronski M 1993 Analytical methods to differentiate similar electroencephalographic spectra: neural network and discriminant analysis. J Clin Monit 9:257–267.

Von Marxow EF 1890 Mittheilung, betreffend die Physiologie der Hirnrinde. Z Physiol 4:537–540.

Walder B, Suter PM, Romand JA 2001 Evaluation of two processed EEG analyzers for assessment of sedation after coronary artery bypass grafting. Intensive Care Med 27:107–114.

Walter WG 1936 The location of cerebral tumours by electro-encephalography. Lancet ii:305–308.

Wang AD, Costa E, Silva I, et al 1985 The effects of halothane on somatosensory and flash visual evoked potentials during operations. Neurol Res 7:58–62.

Wärsterström S-A 1985 Auditory brainstem evoked response after single-dose injection of lidocaine and tocainide. Scand Audiol 14:41–45.

White PF, Ma H, Tang J, et al 2004 Does the use of electroencephalographic Bispectral Index or auditory evoked potential index monitoring facilitate recovery after desflurane anesthesia in the ambulatory setting? Anesthesiology 100:811–817.

White PF, Schuttler J, Shafer A, et al 1985 Comparative pharmacology of the ketamine isomers. Studies in volunteers. Br J Anaesth 57:197–203.

Whittle IR, Johnston IH, Besser M 1986 Recording of spinal somatosensory evoked potentials for intraoperative spinal cord monitoring. J Neurosurg 64:601–612.

Wilder-Smith OH, Hagon O, Tassonyi E 1995 EEG arousal during laryngoscopy and intubation: comparison of thiopentone or propofol supplemented with nitrous oxide. Br J Anaesth 75:441–446.

Winer JW, Rosenwasser RH, Jimenez F 1991 Electroencephalographic activity and serum and cerebrospinal fluid pentobarbital levels in determining the therapeutic end point during barbiturate coma. Neurosurgery 29:739–741.

Yli-Hankala A, Heikkilä H, Värri A, et al 1990 Correlation between EEG and heart rate variation in deep enflurane anaesthesia. Acta Anaesthesiol Scand 34:138–143.

Yli-Hankala A, Lindgren L, Porkkala T, et al 1993 Nitrous oxide-mediated activation of the EEG during isoflurane anaesthesia in patients. Br J Anaesth 70:54–57.

Yli-Hankala A, Vakkuri A, Annila P, et al 1999a EEG Bispectral Index monitoring in sevoflurane or propofol anaesthesia: analysis of direct costs and immediate recovery. Acta Anaesthesiol Scand 43:545–549.

Yli-Hankala A, Vakkuri A, Särkelä M, et al 1999b Epileptiform electroencephalogram during mask induction of anesthesia with sevoflurane. Anesthesiology 91:1596–1603.

Chapter 6

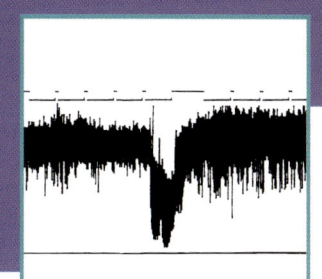

Neurophysiological Work in the ICU

INTRODUCTION

Neurophysiological parameters to be monitored and procedures

In this chapter we reflect the methodology for clinical neurophysiology appropriate to ICU work, with different diagnostic and monitoring tools and the mixture of electroencephalographic (EEG), evoked potential (EP) and electromyographic (EMG) techniques – using a problem-orientated approach. Information from neurophysiological techniques contributes to the diagnosis, prognosis, and monitoring the condition of patients undergoing intensive care; the accompanying *Table 6.1* summarizes the neurophysiological parameters to be monitored and procedures for handling the signals to optimize their clinical value (based on Guérit et al (1999a)).

Sleep in the ICU

Seriously ill patients undergoing intensive care are subject not only to the discomforts of their disease but also to environmental factors including noise, medical and nursing interventions, and effects of medications which often preclude normal restorative functions such as sleep. Disrupted sleep, and the effects of internal and external noxious stimuli, can be reflected in unexpected or unusual changes which may confuse interpretation in various neurophysiological parameters, particularly the EEG. Formal sleep studies have demonstrated that factors such as ventilator mode have significant effects in disrupting sleep (Cooper et al 2000, Parthasarathy and Tobin 2004) (*Fig. 6.1*); these last authors provide an excellent review. Continuous neurophysiological monitoring of EEG and of EPs is not only of value in guiding patient management to allow 'quiet resting periods' when possible, but also demonstrates how changes in physiological cyclical changes in signal waveforms may provide useful prognostic information.

CLINICAL CONDITIONS AFFECTING THE CENTRAL NERVOUS SYSTEM ENCOUNTERED IN THE ICU

COMA AND RELATED STATES

EEG recording and interpretation in coma and related states
Introduction

Pathological phenomena in the EEGs of patients who are stuporous or comatose present interesting and, at times, complex problems for proper characterization and inter-

Table 6.1 Neurophysiological parameters to be monitored and procedures

Steps	Procedure for EEG	Procedure for EPs
1. Quality screening, followed by automatic feature extraction	(a) Frequency analysis of steady EEG segments (b) Time analysis of transients	(a) Automatic identification of peaks
2. Selecting and combining parameters to define patterns	(a) Recombination of Fourier parameters to classical spectral measures (absolute or relative power, median frequencies, spectral edges, etc.) and user-defined indices (asymmetry index, power ratios, etc.) (b) Time analyses to identify specific patterns such as burst suppression, complex arousal patterns, epileptiform discharge, EP and EEG encephalopathy grades, ECS, etc. (c) Recombination of neurophysiological tools to reflect overall CNS function (BAEPs and SEPs to differentiate pontine from midbrain involvement – for early detection of brainstem involvement in transtentorial herniation)	
3. Integration of neurophysiological features into the clinical context	Reference to evidence-based data libraries showing clusters or patterns of high specificity and sensitivity	

Based on Guérit et al (1999).

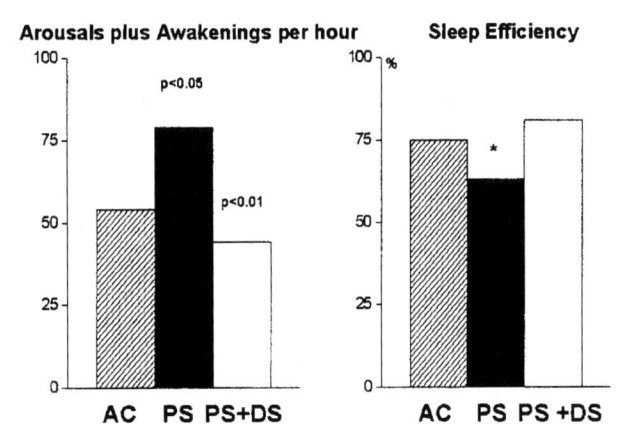

Fig. 6.1 Sleep fragmentation (left) and sleep efficiency (right) during assist-controlled ventilation and pressure support with and without dead space in 11 critically ill patients. Sleep fragmentation, measured as the number of arousals and awakenings, was greater during pressure support (solid columns) than during assist-control ventilation (hatched columns) or pressure support with dead space (open columns). Sleep efficiency (right) was also lower during pressure support (solid columns) than during assist-control ventilation (hatched columns) or pressure support with dead space (open columns). (From Parthasarathy S, Tobin MJ 2004 Sleep in the intensive care unit. *Intensive Care Medicine* 30:197–206. © 2004 Springer-Verlag, with kind permission of Springer Science and Business Media.)

pretation. There is probably a greater need for contextual information and immediate dialogue with the referring physician than in most other circumstances.

Clinical definitions and rating scales for stupor and coma abound. The typical stages, as outlined by the Medical Research Council Brain Injuries Committee (1941), with respect to traumatic coma include:

- *Coma* – 'A state of absolute unconsciousness, as judged by the absence of any psychologically understandable response (including, e.g., change of expression) to external stimuli or inner need. Note: If a patient is moving it is important to record the state of activity at levels of nervous integration lower than that which is the substratum of consciousness. Such activites are reflex (e.g. swallowing, pupillary and corneal reflexes, tendon jerks, plantar responses).'
- *Semicoma* – 'A state in which psychologically understandable responses are elicited only by painful or other disagreeable stimuli.'
- *Confusion* – 'Disturbance of consciousness characterized by impaired capacity to think clearly and with customary repetition and to perceive, respond to and remember current stimuli; there is also disorientation.'
- *Delirium* – 'A state of much disturbed consciousness (confusion) with motor restlessness, transient hallucinations, disorientation and perhaps delusions.'
- *Traumatic stupor* – 'A state in which the patient, though not unconscious, exhibits little or no spontaneous activity.'

The classic work on the comatose patient is *The Diagnosis of Stupor and Coma* by Plum and Posner (1966, 1980); their detailed accounts of clinical states and their pathophysiological bases are hard to surpass. The range of causes of disordered consciousness is considerable, including structural brain disease, trauma, encephalitides, encephalopathies, non-convulsive status epilepticus and psychogenic problems. For clinical grading of depth of coma, the Glasgow Coma Scale (GCS) (Teasdale and Jennett 1974) provides a widely used tool which can be reliably reproduced by different observers for communication. This, together with the Glasgow Outcome Score (Jennett and Bond 1975), has provided the basis for most comparisons with neurophysiological features. Overall, the depth of coma itself is more relevant to the nature of the EEG abnormalities than the underlying disease process. Specific EEG patterns associated with particular pathologies are few in number, being virtually limited to the triphasic waves of hepatic encephalopathy and the asynchronous periodic features of herpes simplex encephalitis. Some findings, for instance continuous epileptiform discharges suggesting non-convulsive status epilepticus, may have implications for urgent treatment.

The general approach to recording from patients with disorders of consciousness will be governed by the severity of the patient's condition and the presence of life-threatening factors requiring nursing in a high dependency unit or ICU. The rapid evolution of many conditions leading to coma necessitates more than a snapshot view of the EEG to identify various specific diagnostic phenomena. A modern integrated neurophysiological approach including EEG–polygraphy, EPs and long-term monitoring generally provides the most satisfactory documentation of pathophysiological processes and effects of treatment.

Optimal recording technique is based upon the principles set out in Chapter 2; it requires a highly organized systematic approach, even a degree of obsessionality and the instincts of a Sherlock Holmes! EEG phenomena have to be elicited and evaluated in the context of the state of the patient and the historical evidence regarding the circumstances of onset of coma. The possibility of causation by drugs, toxic states or unsuspected trauma always exists. It must also be considered that the patient may not be in coma but in fact 'locked-in' or in a psychogenic state of pseudocoma or of depressive stupor. It is good practice always to assume that apparently comatose patients are in fact conscious, and to talk and explain what is going to happen at each stage throughout the recording.

It is essential that the investigation does not make the patient worse – care should be taken to avoid undue head movement or excessive stimulation that might interfere with cervical spine injuries, the airway and vascular lines and drug infusions – or induce problems with venous return, or stimulus-induced reflex changes in arterial or intracranial pressure (ICP). The safety issues are given in the section beginning on page 55, which details matters regarding infection control, electrical safety and the special problems of artefact and interference that may be found in the ICU environment.

Arousal patterns in stuporous and comatose patients

The functioning human brain signals a response to stimuli by EEG alterations, which may or may not be accompanied by autonomic or motor changes suggesting arousal of the organism. With disease, the EEG responses may be modified or absent. The nature of abnormal responses may provide an indication of dysfunction in arousal mechanisms in the reticular activating system and its projection pathways.

Procedure A systematic protocol for presentation of stimuli to check for arousal responses is helpful. Event marks should be carefully annotated so that the effects of graded stimuli can be compared with responses; extraneous stimuli that might lead to misinterpretation should also be noted. Stimulation should start with quietly calling the patient's name, first by one ear, then by the other. In the absence of a response, clapping the hands should follow. Thereafter, discrete touch stimulation (not stroking) should be applied at intervals to the forehead or face and then to each of the four extremities. This should be followed by pain stimulation (supraorbital pressure and nail bed stimulation by pressing an object such as a pen or small screwdriver horizontally across the proximal edge of the finger or toe nail). The EEG should be allowed to return to its resting state between stimuli. If responses are elicited, care should be taken to avoid major arousals with reflex changes which might affect arterial or ICP. Some patients with brain injuries are highly sensitive to very minor stimuli.

Evidence that EEG changes are more than coincidental should be sought by checking reproducibility and looking for other evidence of arousal, for example heart rate changes or muscle potentials (unless the patient is receiving medications that would prevent such responses). Exhaustive testing should be used with caution if there are major reflex alterations in arterial or ICP associated with the EEG response. However, arousal responses can provide a simple and useful means of testing the integrity of pathways from peripheral limb stimulation, thus providing some information normally only available from somatosensory evoked potentials (SEPs).

Nature of responses Arousal patterns depend on the prestimulus state of the subject. In the alert normal subject, attenuation or 'blocking' of posterior rhythms is seen on eye opening and with visual attention (see *Fig. 4.1*). In drowsiness an alerting stimulus such as a sound or touch will produce an attenuation of theta activity with return to a waking EEG, sometimes with a paradoxical appearance of alpha rhythm on eye opening. In light sleep, such a stimulus produces a K-complex (see *Fig. 4.4(b)*). In deeper sleep responses become much less conspicuous, in part owing to the higher voltage of the ongoing activity, or are only elicited by repeated or stronger stimuli, as the families of most heavy sleepers will know! Patients who are stuporous or in light coma may show either an instantaneous brief response akin to a K-complex, or a longer lasting change in the pattern of the EEG, often consisting of a reduction in slow components suggesting sustained arousal (see *Fig. 6.2(a–c)*). Long-lasting changes of EEG state may be induced by quite minor stimuli, for example those relating to EP studies (see *Fig. 5.23*). Pfurtscheller et al (1986) investigated 20 patients with severe head injury and a GCS of 4–6 during a multimodality EP study. In 50% of patients, there was a long-lasting change in the spectral theta/beta ratio of the EEG compared with the prestimulus state and stimulus-induced cardiac alterations were also found. Those in whom long-term EEG monitoring is undertaken may show clusters of arousals following nursing or other disturbing stimuli, separated by long periods without responses. Careful observation of the timing of these phases often shows patterns resembling circadian cycles of the normal subject.

The more pathological arousal responses may be divided into those with 'positive' features, such as the paradoxical slow wave arousal pattern (*Figs 6.2(d)* and *4.10*), and 'negative' ones, such as attenuation of frontal intermittent rhythmic data (FIRDA) or periodic lateralized epileptiform discharges (PLEDs), decreased variability, short periods of suppression sometimes associated with di- or triphasic transients, change (increase or decrease) in periodic phenomena or the severity of the burst suppression pattern, or a total absence of any response

EEG reactivity should always be tested in the stuporous or comatose patient since it can provide valuable prognostic information, regardless of the morphology of the response; in addition it provides a degree of standardization for quantitative measurements (Claassen et al 2004a). Early appearance of EEG reactivity and the central conduction time (CCT) of the SEPs correctly classified 92% and 82%, respectively, of 50 severely head injured patients into good or bad outcome groups at 1.5 year follow-up (Gütling et al 1995a). Initial GCS allowed a correct classification in only 72% and did not improve prognostication, even when combined with EEG reactivity or CCT. The combination of the two neurophysiological measures achieved a prognostic accuracy of 98%, better than that of any other reported method.

Paradoxical slow wave arousal responses A common type of abnormal arousal response in light to moderate coma is the paradoxical slow wave arousal response (see *Fig. 6.2(d)* or *4.10*). It is most often encountered in subjects in the first few decades of life and may be the only abnormality in the EEG following head injury in a child or teenager. The response to a minor auditory stimulus, such as whispering the subject's name, may be a dramatic and prolonged run of slow delta waves with associated polygraphic evidence of arousal. Once the response dies away, often only after many minutes, the ongoing rhythms may appear virtually normal in frequency, amplitude, symmetry and distribution.

The response may be mistaken for ongoing activity if protracted arousal occurs from disturbance of the patient whilst adjusting electrodes. If long-duration, frontally accentuated, bisynchronous delta activity is dominant, it is important to continue the recording without applying any stimulus to see if the EEG changes to a quieter pattern without delta waves.

Reproducible paradoxical slow wave arousal patterns may be readily elicited in the child, often being more dramatic with a familiar auditory stimulus such as the voice of a parent than with a noxious stimulus. Especial caution should be exercised regarding repeated stimulation in patients with vulnerability to changes in cerebral perfusion; those with this pattern of abnormal response may fall into such a category.

Periodicity of arousal patterns in comatose patients

Cycles of reactivity in the comatose patient (see p. 113) are often separated by quieter periods with timing similar to that of sleeping (Aserinsky and Kleitman 1955) and waking patterns described as 'basic rest and activity cycles' (BRAC) (Webb 1969, Simon et al 1977) in the normal subject. Similar periodicities have long been recognized in the occurrence of epileptic seizures (Langdon-Down and Brain 1929, Daly 1973, Broughton et al 1986).

As well as these long-duration patterns, periodicities affecting EEG and autonomic variables occur about every 15 s in drowsiness and sleep (Evans 1992). Designated the cyclical alternating pattern (CAP), they can be detected in some comatose patients with a cycle of 0.5–2 min duration (Evans 1976, 2002, Evans and Bartlett 1995). They

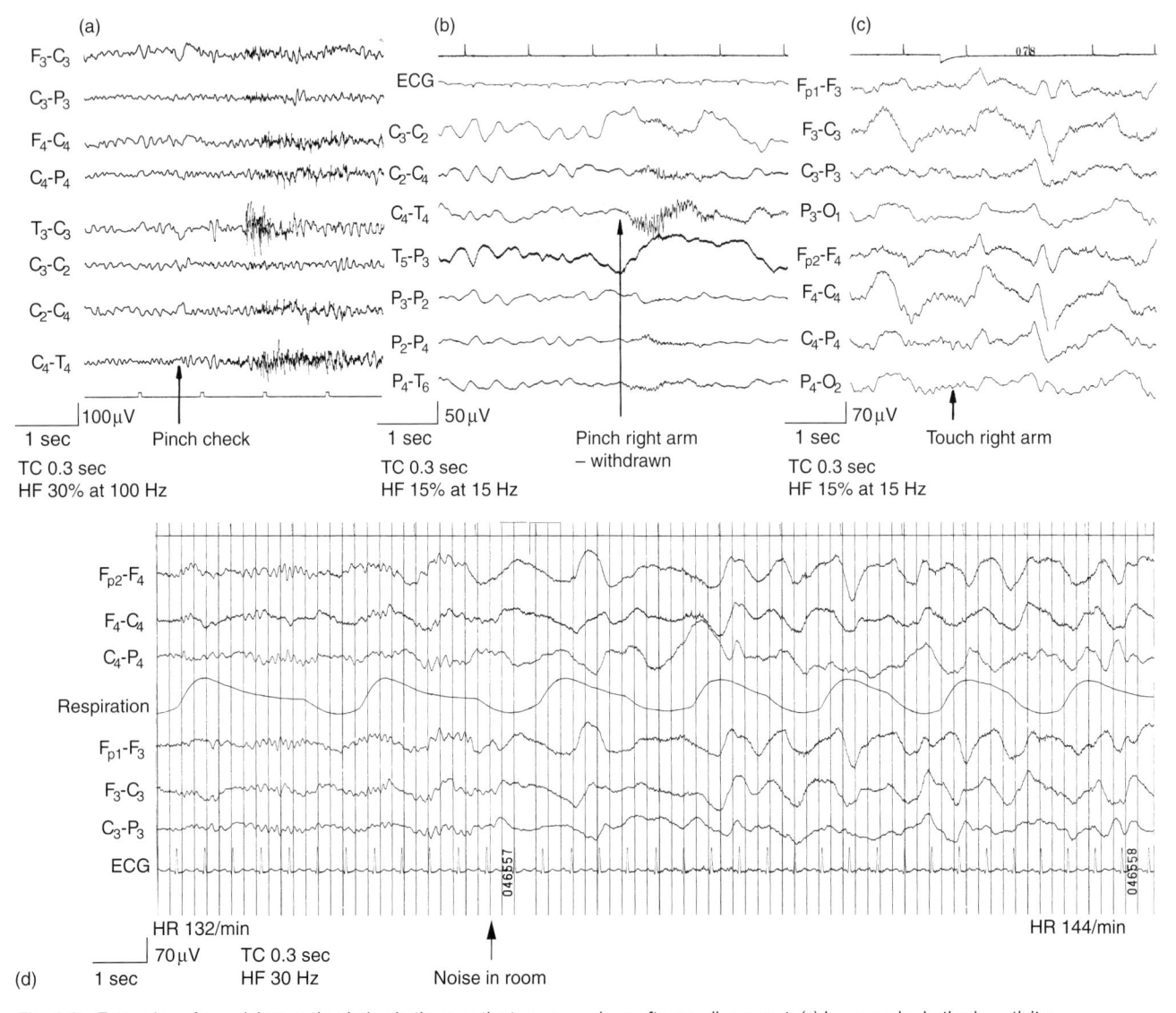

Fig. 6.2 Examples of reactivity to stimulation in three patients unconscious after cardiac arrest. (a) Increase in rhythmic activity; (b) attenuation of slow wave activity; (c) appearance of arousal response suggesting a primitive K-complex. Note different montages and sensitivities. (From Prior (1973), by permission.) (d) Paradoxical slow wave arousal pattern induced by noise in a man comatose 3 days after an acute episode of profound arterial hypotension prior to surgery for bowel obstruction. Note the increase in respiratory rate and heart rate (from 132/min to 144/min) following stimulus at arrow. (From Binnie et al (2003), by permission.)

are not evident when deep coma reflects severe brain damage and outcome is death or a vegetative state.

EEG features in comatose patients

Depth of coma is defined clinically according to reactivity to commands and to noxious stimuli. Such reactivity can be objectively documented, and indeed quantified, with EEG monitoring. Reactivity may well be intermittent in relation to cyclical changes in level of arousal; this means that if the patient's potential to respond is reduced at the time of a formal 'coma scoring' assessment, unrepresentative values may be obtained. Repetition of the assessment when monitoring suggests that EEG reactivity is present may provide more reliable results (*Figs 6.3–6.5*).

Failure to respond clinically or in the EEG to external stimulation following withdrawal of sedation may not imply that the patient is comatose from encephalopathic or other cerebral disorders. Non-convulsive status epilepticus (see *Fig. 4.12(a)* and p. 218) may be identified with continuous EEG monitoring, providing a possible explanation of the patient's obtunded state.

Traditional scales for EEG changes with increasing depth of stupor and coma are based on the observation that there is a gradual shift from reactive posterior rhythms in the alpha frequency range to a record dominated by diffuse, unreactive, delta activity (Romano and Engel 1944). Such scales show a clear overall relation to clinical staging of coma, provided reactivity is checked. However, these simple scales do not permit EEG categorization in as many as one-third of unconscious patients because many other abnormal patterns, such as periodic transients or 'alpha coma', are encountered which do not fit into a simple classification based on progressive slowing of the EEG. This means that a broader approach is required (Binnie et al 1970).

Seven key concepts underpin our understanding of the pathological features of the EEG in comatose patients:

1. *There are both 'positive' and 'negative' abnormalities.* When discussing the phenomenology of the EEG in the comatose patient, it is important to consider two general categories of abnormality which may be considered as 'positive' and 'negative' phenomena. The term 'positive' is used to imply active appearance of abnormal waveforms, such as triphasic waves, PLEDs or diffuse periodic transients. The term 'negative' is used to indicate abnormalities involving loss of expected activity, ranging from the 0.5–4 s duration 'episodic low-amplitude events' (ELAEs) of Rae-Grant et al (1991a) (see p. 115), to burst suppression pattern (see p. 123) and eventually a low amplitude, featureless and unreactive EEG and electrocerebral silence (ECS) (see p. 124).

2. *Stimulus-induced arousal patterns must be checked.* The whole basis of both clinical and EEG coma scoring systems hinges on the question of reactivity to external stimuli. A protocol for systematic stimulation (which may conveniently utilize a check-sheet) is needed together with a system for classifying EEG arousal patterns. The integrated nervous system in the patient capable of a full recovery from coma will be evidenced by a responsive, changeable EEG–polygraphic trace. Neurophysiological evidence of reactivity has an advantage over neurological examination when clinical signs of responses to stimulation may be masked, distorted or unobtainable because of drugs such as muscle relaxants, or physical problems such as limbs in plaster or traction.

3. *Abnormalities may be modified by medication and metabolic disturbance in patients undergoing intensive care.* The

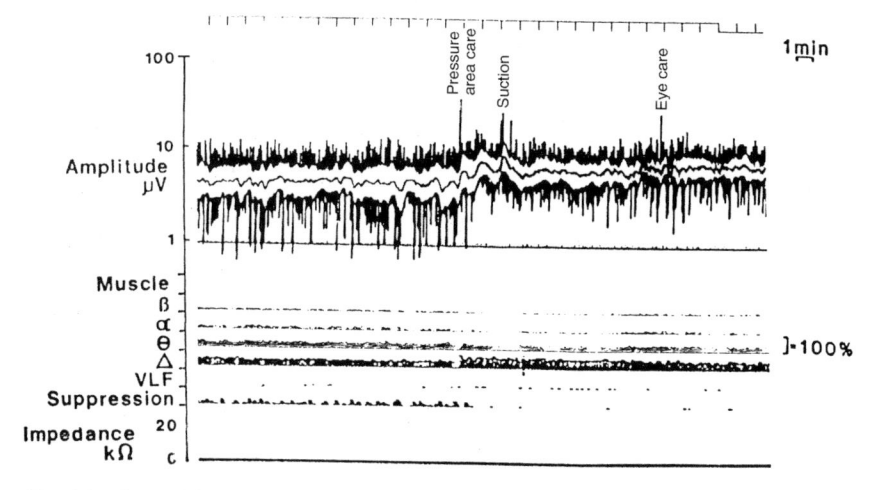

Fig. 6.3 Arousal features in the Cerebral Function Analysing Monitor (CFAM1) (RDM Consultants, Uckfield, Sussex, UK, http://www.cfams.com) trace of a woman in the ICU with posthypoxic coma. She was undergoing nursing care to pressure areas followed by tracheal suction. Note the change from a burst suppression pattern (see both minimum amplitude and suppression displays) to a higher amplitude trace. Such EEG changes in response to stimuli may precede clinical signs of responsiveness by several hours or days. (From Frank and Prior (1987), by permission.)

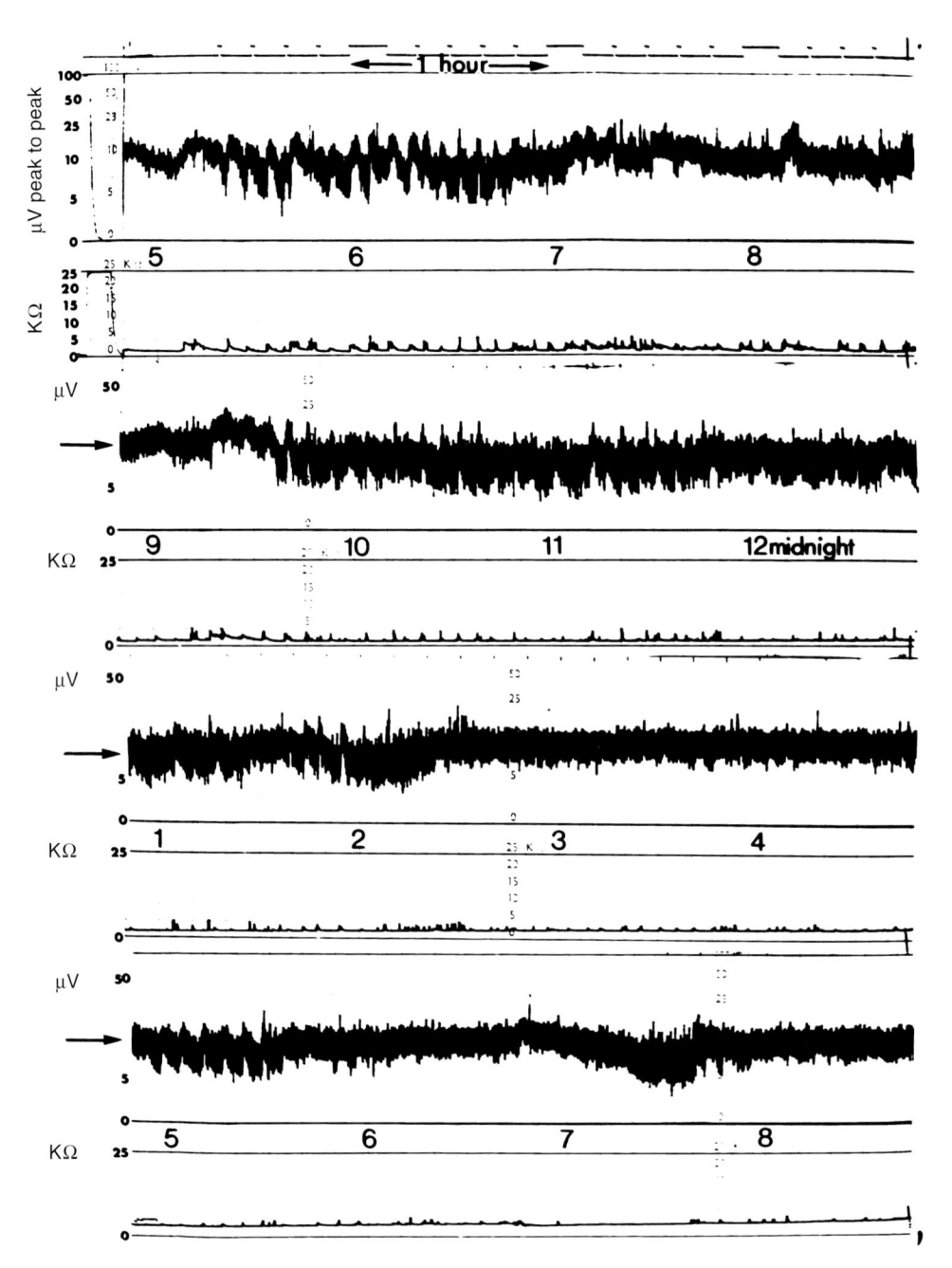

Fig. 6.4 CFM trace from a 40-year-old woman with postinfluenzal encephalitis showing cyclical changes in the degree of variability of the processed EEG and in the amount of movement and scalp muscle artefact on the electrode impedance trace. Clock times are marked. Note the 'quiet' trace from 02.20 h to 04.50 h with reduction in artefacts. The cyclical changes evolved towards a physiological sleep–wake pattern with some semblance of sleep phases over the subsequent few weeks as the patient began to emerge from coma. (From Prior and Maynard (1986), by permission.)

character of the ongoing activity in the EEG and the responses to stimuli may be markedly altered by drugs (e.g. sedatives or the lingering effects of overdose or toxic agents such as alcohol) and metabolic disturbance (e.g. in multisystem failure) in comatose patients. Such potentially transient factors may produce a false impression of unduly severe EEG abnormality, which may continue for several days, for instance if metabolic clearance of drugs is impaired by hepatic or renal problems. Major sedation will not affect short latency components of auditory brainstem or SEPs; EPs can provide a useful tool in interpreting the significance of an EEG with marked suppression from induced barbiturate narcosis (Ganes and Lundar 1983, Lundar et al 1983). Of course, local injury or pre-existing deficits such as deafness or other sensory impairment may impair our ability to demonstrate reactivity to external stimuli with either EEG or EP techniques.

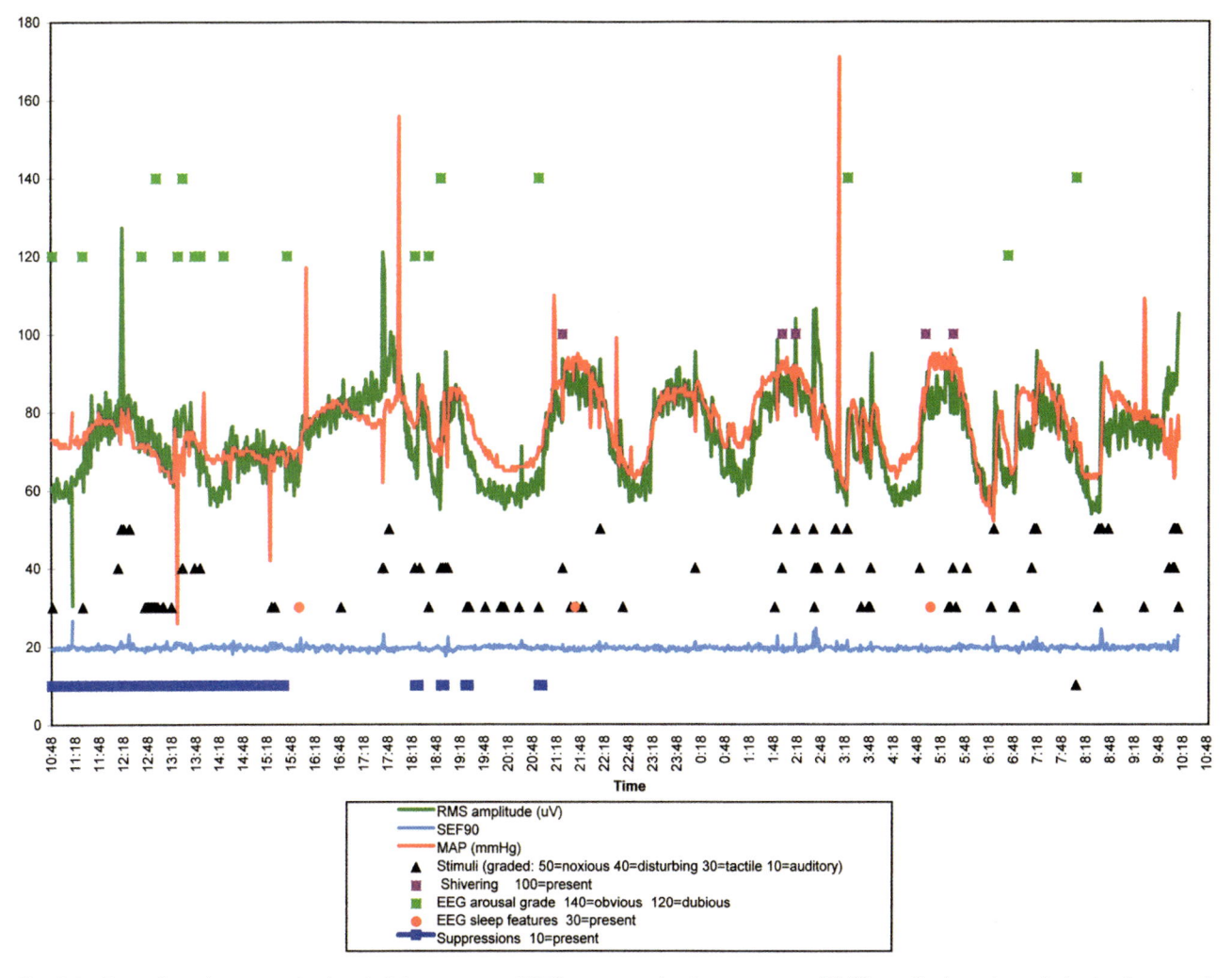

Clinical conditions affecting the central nervous system encountered in the ICU

Fig. 6.5 Recording of mean systemic arterial pressure and EEG measures (root mean square (RMS) amplitude and spectral edge frequency) to show the relationship with external stimuli due to ICU procedures (graded according to severity), in a 24 h recording from an 18-year-old man who exhibited unusual cyclical oscillations in multisystem parameters including EEG following high dose thiopental therapy for status epilepticus secondary to herpes zoster encephalitis (see also *Fig. 3.11*). Episodes when the patient shivered, and periods with EEG features of sleep, arousal and burst suppression, were all scored by two independent observers. (Data from the patient reported by Ghosh et al (2000), from the IMPROVE project (Cerutti and Saranummi 1997), courtesy of Professor N Saranummi. From Binnie et al (2003), by permission.)

4. *Sleep features may be present in comatose patients.* Although there is a clear distinction between sleep and coma, there is no doubt that phenomena such as sleep spindles, vertex sharp waves and K-complexes, may occur in the EEG of stuporous and comatose patients. This most often happens during the phase of emergence from coma and represents an encouraging prognostic sign. Rarely, abnormal sleep-like states such as 'spindle coma' are also encountered.

5. *Short- and long-term variability must be assessed.* Complex short- and long-duration cyclical variations in the presence or degree of reactivity of the EEG occur in comatose patients and may be of prognostic value (Bricolo and Turella 1973, Prior and Maynard 1986, Evans 1992, 2002, Evans and Bartlett 1995). These are best seen in low-speed recordings. Some relate to innate biological periodicities such as those affecting the cardiovascular system or ICP (Cooper and Hulme 1966, Lundberg 1969) or to sleep–wake or sleep cycles. Others may be due to external environmental events such as changes in level of background noise, or nursing and other attention to the patient. This has two consequences for interpretation of the EEG in comatose patients. First, attention must be paid to extraneous events (which should be monitored by annotation on the trace or electronically), or internal stimuli (such as pain) in the genesis of possible EEG responses such as prolonged paradoxical slow wave arousals which might otherwise be mistaken for ongoing activity. Second, the limited sampling time of a routine recording may not provide valid evidence of the absence of reactivity because either the sample falls during a period of prolonged arousal (e.g. following

the disturbance of electrode placement) or because the patient is in a cyclical period of decreased reactivity.

6. *EEG phenomena must be evaluated in the context of patient state.* A key concept in assessing the EEG in coma is that of appropriateness of the phenomena present to the clinical state of the patient. Examples are: (a) an appearance suggestive of an awake patient with responsive alpha rhythm and muscle artefacts or eye movements in response to stimuli suggesting simulated coma or even 'locked-in' syndrome; or (b) widespread unreactive alpha frequency activity or beta spindles (as opposed to the physiological alpha rhythm or sleep spindles) dominating the EEG in deeply comatose patients. Visual rating systems or quantitative analysis protocols for EEG monitoring must distinguish these as special cases – 'alpha coma' or 'spindle coma' (see p. 198).

7. *EEG scoring systems help standardize interpretation.* The EEG phenomenology of the stuporous or comatose patient has been studied since the early days of clinical neurophysiology. A clear account of sequential EEG changes with increasing disturbance of consciousness was provided by Romano and Engel (1944). Based on serial studies in 53 patients, they proposed a 5-point EEG classification which included simple quantification (a form of spectral analysis by manual baseline crossing counts), characterized by decreasing frequency and organization with loss of reactivity to eye opening or closure. The scale represented a continuum, which correlated with changes in level of consciousness independent of diagnosis in a wide range of clinical conditions. In subsequent work they demonstrated the reversibility of the EEG changes with treatment. Many refined classifications have been proposed over the years (Fischgold and Mathis 1959, Silverman 1963, Bricolo and Turella 1973, Prior 1973, Synek 1990) and are comprehensively reviewed by Chatrian (1990). One problem has been that no univariate scale (i.e. frequency content alone) addresses the real complexity of the EEG in stupor or coma. Approaches utilizing computerized pattern recognition methods to evaluate multivariate EEG phenomena have been applied both to visually graded variables and to the results of spectral analysis. These have proved far more successful both in providing widely applicable descriptions of the EEG and in terms of markedly increased accuracy of prognostic information (Binnie et al 1970, Thatcher et al 1989, Rae-Grant et al 1991b). Significant EEG features related to outcome from coma, whatever its cause, have proved very similar and tend to be far removed from the simple deviations from the normal frequency distribution that might be expected from some early coma rating scales. Apart from total ECS, adverse features include periodicity of any particular EEG phenomenon, stereotyped transients of simple mono- or diphasic waveforms, epileptiform discharges, low amplitude overall or episodic flattening, reduced alpha power posteriorly or altered anterior–posterior power ratios, localized abnormalities and, importantly, absence of variability and responses to stimulation. Most scales include consideration of the appropriateness of EEG features, for example deep coma with unreactive alpha frequency activity would suggest the highly adverse state of 'alpha coma'. Comparisons of EEG scoring with clinical coma scoring show that the neurophysiologist can provide useful supplementary information to the clinician and aid communication about depth of coma (Matousek et al 1996). It was of interest in this study, that semiquantitative visual scoring was comparable to commercially available quantitative EEG (qEEG) spectrum analysis systems in this regard, but that the addition of multivariate statistical processing of the qEEG gave it an advantage over the visual grading methods. Other recent developments utilize computerized modelling to define optimal algorithms for automated pattern recognition with artificial neural networks, knowledge-based systems and related techniques (Jansen 1986, Kemp et al 1987, Veselis et al 1991, Webber et al 1994).

Locked-in syndrome

The first modern clinical reports of this condition began to appear in the 1960s. Plum and Posner (1966) provided a clear description of locked-in syndrome in the first edition of *The Diagnosis of Stupor and Coma* indicating that patients are alert and can communicate intelligently by the voluntary vertical movements of the eyes, which are the only motor features to survive the typical bilateral lesions of the ventral part of the pons. Pharmacological misadventures may give rise to a functionally similar state, as described by those unfortunate patients given muscle relaxants but inadequate anaesthesia (Moerman et al 1993). The locked-in syndrome provides a strong reason for always approaching a comatose patient with the assumption that they might be able to hear, comprehend and feel pain – a state movingly described by Bauby (1997).

The older term, 'akinetic mutism', may give rise to confusion in the context of the locked-in syndrome. Cairns et al (1941) used the term 'akinetic mutism' (which equates to general usage of the French term *coma vigile*) to describe a patient with a third ventricle cyst who neither spoke nor moved but whose eyes followed the observer. Whilst this state had some clinical features in common with that subsequently described as locked-in syndrome, there are in fact clear differences; patients with this latter condition are totally unable to move or speak, whereas those with akinetic mutism, because of dysfunction at a higher level, cannot be persuaded to do so (Nordgren et al 1971).

EEG features of the locked-in syndrome (*Fig. 6.6(a)*) were mentioned by Nordgren et al (1971) who described a responsive 8–9 Hz alpha rhythm in one of their seven patients. Hawkes and Bryan-Smith (1974) detailed the EEG findings in a series of seven patients, all considered to be conscious, although some exhibited periods of drowsiness and arousal. The EEGs reflected this by showing a dominant alpha rhythm at some stage in all patients but with slowing to the theta range during drowsiness. Reactivity of the alpha rhythm to eye opening was observed in four patients and, of the alerting stimuli, noise elicited the most consistent responses in the five patients in whom it was tested. Other abnormalities were present, varying between diffuse posterior or localized temporal slowing. The presence of normal sleep patterns was also noted.

Comparable clinical and EEG states may occasionally follow cardiac arrest or severe head injury (Boisen and Siemkowicz 1976, Turazzi and Bricolo 1977). The latter authors described 13 patients, from a series of 1000 following severe head injury, in whom the clinical picture of a pontine lesion developed; EEGs were normal or borderline with reactivity present in four and sleep features in six. Turazzi and Bricolo (1977) pointed out the danger of mistaking a non-reacting, but conscious, patient for one who is deeply comatose, and indicated the value of repeated, careful clinical examinations and repeated EEGs which typically show tracings similar to a normal pattern.

It should be emphasized that reactivity of the alpha activity cannot be demonstrated in all patients who are locked-in; virtually all reports indicate this. Three patients of Jacome and Morilla-Pastor (1990) with acute brainstem strokes had repeated unresponsive EEGs more akin to those of alpha coma. They noted the importance of this finding since preservation of consciousness is often difficult to ascertain clinically in locked-in syndrome and also that the presence of unreactive alpha activity in the EEG of unresponsive patients is normally equated with vegetative states and thought to exclude locked-in syndrome. Strenuous attempts to document vertical eye movements should always be made when unresponsive patients exhibit an EEG with alpha activity, even if it is non-reactive. Gütling et al (1995b) used a combination of EEG and SEPs in five patients in whom neuroradiological and neuropathological findings were also reported. EEG reactivity was present in two patients and absent in three. They found no specific pattern of associated SEP abnormality, demonstrating a range of findings varying from unilateral normality to bilateral absence of potentials. They concluded with the caution that EEG reactivity cannot be taken as a sole measure of consciousness.

Locked-in syndrome can present diagnostic problems even when those responsible have considered this possibility. The neurophysiology (both EPs and EEG) can be very variable because of the variable extent of the lesion(s) and the fluctuating state of the patient who may not respond or be able to cooperate in the attempt to elicit classical eye movement responses because he or she is too ill or exhausted and drowsy (hence slow wave abnormality in the EEG) or terrified (hence a low voltage fast EEG which can appear almost isoelectric!) at the time of the examination. It may take considerable ingenuity to confirm

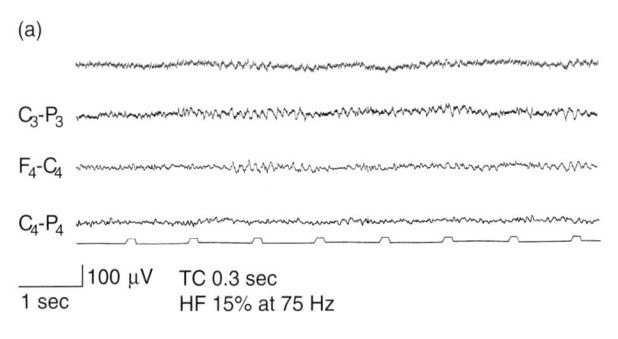

Fig. 6.6 (a) Alpha activity fluctuating with conversation near the bed in a 21-year-old man who appeared to be unconscious 3.5 h after cardiac arrest during induction of anaesthesia for repair of an atrial septal defect. The final diagnosis was of a drug-induced locked-in syndrome due to suxamethonium sensitivity. Examples of (b) alpha, (c) theta and (d) spindle coma in three patients after cardiac arrest. See text regarding the importance of testing EEG reactivity in such patients. Note different montages and recording sensitivities. (From Prior (1973), with permission.)

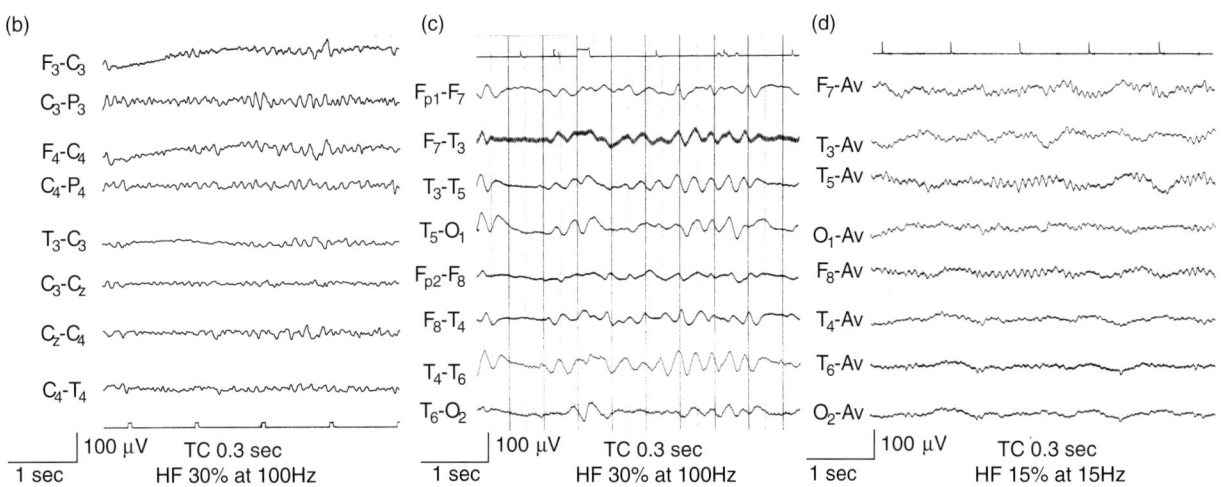

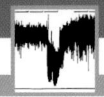

the diagnosis in reality. In practice this is where prolonged EEG recordings, or preferably EEG–polygraphic monitoring, are valuable because they open the possibility of documenting cyclical changes in state relating to some retention of 'sleep–wake' and 'sleep' cycles.

Sleep patterns on polygraphic recording may show little abnormality in locked-in syndrome. Oksenberg et al (1991) described such a patient with a typical clinical picture and with evidence of ischaemic infarction involving the ventral portion of the upper part of the pons bilaterally and extending posteromedially into the tegmentum. There was probable involvement of reticular structures, such as the median raphé nuclei, thus raising doubts about their regulatory role in sleep.

EPs may be intact but findings are somewhat variable, depending on the extent of the causative lesion; patients may be both de-efferented and de-afferented. Findings are discussed on page 208.

Alpha, theta and spindle coma

These relatively rare conditions are examples of EEGs inappropriate to the patient's clinical state (deep coma) in which records are dominated by superficially 'normal' rhythms but fail to show reactivity to external stimuli (see Fig. 6.6(b)). They usually represent a transient stage in an evolving EEG abnormality (e.g. to burst suppression and then ECS), and are associated with a poor outcome. Confusion arises when definitions are not strictly adhered to. Many lightly comatose patients show some activity of alpha or theta frequency, or sleep spindles, mixed with other components, particularly during emergence from coma. This is quite different from such waveforms being the only activity present and failing to fulfil the criteria normally applied (Chatrian et al 1974, 1983).

Alpha coma began to be reported in the 1950s. Most of the original patients had brainstem lesions and the alpha rhythm was widespread, often with frontocentral enhancement, being monotonous and unreactive to stimuli. This is quite distinct from physiological alpha rhythm. Alpha-like rhythms were described by Chokroverty (1975) in 12 patients, comatose after cardiac arrest and were contrasted with the alpha rhythm in three patients with locked-in syndrome from localized ventral pontine lesions who showed quite different clinical features. These locked-in patients with brainstem lesions had alpha rhythms of lower frequency (8–9 Hz versus 9–12 Hz), more posterior distribution, with more frequent reactivity and more frequently associated with sleep patterns. Neuropathological examinations in three of the postanoxic patients showed typical neuronal loss with no evidence of brainstem lesions. A clue to the pathophysiology may be provided by the observation that repeatable brief episodes akin to alpha coma have been produced pharmacologically by accidental injections of amytal into the vertebral arteries (Amin and Binnie 1987). The outcome of alpha coma is closely related to the cause, with mortality ranging from 88–90% after cardiac arrest or stroke, to 8% when it was induced by drugs, However, regardless of cause, the single most important feature for prognostic purposes in patients with alpha coma is the presence of EEG reactivity to noxious stimuli which predicts survival, albeit often without meaningful recovery (Kaplan et al 1999).

An important quantitative methodological approach to differentiation between the EEG activity of alpha coma from the awake alpha activity using non-linear dynamics of EEG has been reported by Kim et al (1996). This work makes use of the quantitative analytical techniques of spectral density function and phase space plotting so that measures could be compared statistically in seven alpha coma patients and 10 controls (seven patients whose EEGs were read as normal and three healthy volunteers). Both alpha coma patients and controls showed a broadband spectral density without distinguishing features in phase space plots. However, alpha coma patients showed greater segment-to-segment variation in correlation dimension than controls; this was ascribed to the greater amount of low frequency alpha components, albeit not visually appreciated. McKeown and Young (1997) demonstrated a significant difference in interhemispheric coherence in their five alpha coma patients compared with normal awake controls, suggestive of a significant thalamocortical disruption.

Theta coma is a related state that may occur, usually transiently, with hypoxic brain damage, brainstem lesions, and in some drug intoxications. Of 115 patients studied during coma after cardiac arrest 33 had initial EEGs dominated by monotonous theta activity; of these 33% recovered fully, 24% showed initial recovery but died from unrelated causes, 6% survived with brain damage and 36% died without regaining consciousness (Prior 1973). In another study, Young et al (1994) describe a clinical outcome study of 50 patients meeting criteria for alpha, theta or alpha–theta coma who they followed with serial EEG recordings. These represented 13% of the patients recorded in their ICU; 40 failed but 10 succeeded in recovering conscious awareness. Most of those with poor outcome had suffered hypoxic–ischaemic insults whilst other diagnoses were held responsible for the coma in one third of those recovering consciousness. Recordings were of long duration and included auditory stimulation (shouting in the patient's ear) and somatosensory stimulation (nail bed and supraorbital pressure) as well as passive eye opening.

These alpha, theta or mixed alpha–theta coma patterns occur as transient clinico-EEG patterns which do not differ from each other in aetiology or outcome; all are indicative of severe disturbance in thalamocortical physiology. Although most patients have a poor outlook, the patterns are not reliably predictive of outcome, regardless of the underlying aetiology. The 3 patterns characteristically last up to 5 days, being superseded by EEGs of more definite prognostic value. Change to a reactive pattern is favourable, whereas change to an unreactive burst suppression pattern is unfavourable in patients with anoxic–ischaemic pathology. Neuropathological examination in five post-cardiac-arrest patients showed considerable interindividual variation in severity and distribution of ischaemic cell damage.

Spindle coma provides a good example of the need for circumspection in the use of terminology. Chatrian et al (1963, 1964) described patterns resembling those of sleep in patients after head injury, noting that brainstem lesions incompatible with life could be associated with an EEG showing a 'normal' pattern of either alpha rhythm or sleep spindles. A similar picture occurs occasionally after cardiac arrest (Binnie et al 1970). In these and subsequent reports, the key diagnostic factor is the typical absence of EEG reactivity to noxious stimuli in true spindle coma and the adverse prognostic implications that implies (Kaplan et al 2000). These authors reviewed a total of 242 patients from personal data and published reports and found that EEG reactivity to noxious stimuli best predicted outcome (but neurological examination did not). Outcome depended on the underlying cause of coma, being worst with hypoxic–ischaemic insults or infarction or structural lesions affecting the brainstem or cerebrum: overall 23% of patients with unreactive spindle coma died or remained in a vegetative state. Thus the picture differs from (a) those patients emerging from coma in whom 'normal' sleep features appear cyclically as they move towards recovery, and (b) those patients in persistent vegetative states in whom preservation of sleep features of subcortical origin may be observed during eyes-closed, quiescent, 'sleep-like' periods.

EPs in comatose patients
Introduction
Electrophysiological testing in coma aims at reflecting brain dysfunction, and complements the anatomical information provided by radiological investigations. Since the value of clinical examination is limited in patients who are curarized or heavily sedated, EPs are being increasingly used in the ICU for the evaluation of central nervous system (CNS) function. Although preliminary studies were published, for instance by Larson et al (1973), the widespread acceptance of EPs as an aid to the assessment of comatose patients was promoted mainly by the work of Greenberg and his colleagues, first reported in 1977 (Greenberg et al 1977). Early research favoured a 'multimodality' approach that implied recording of short and long latency auditory evoked potentials (LLAEPs), SEPs and visual evoked potentials (VEPs) in coma, in an attempt to investigate as many sensory pathways as possible. However, later studies suggested that the different EP modalities were not of equal value, and that almost the same information could be obtained with the use of a limited number of EP types. Now, over 25 years later, brainstem auditory, and short and middle latency somatosensory potentials are widely recognized as useful contributors to diagnosis and prognosis in coma.

Most published works on coma outcome prediction are based on electrophysiology, which is a non-invasive approach, offering several possibilities to assess CNS function, all of which are technically easy to perform at bedside and relatively inexpensive (Chiappa and Hill 1998). The use of multimodality EPs as a prognostic tool has been steadily increasing over the past 25 years (Narayan et al 1981, Walser et al 1985, Cant et al 1986, Ganes and Lundar 1988, Lindsay et al 1990, Barelli et al 1991, Cusumano et al 1992, Zentner and Rohde 1992, Facco et al 1993, Guérit et al 1993, Madl et al 1996, Pohlmann-Eden et al 1997, Facco et al 1998, Sleigh et al 1999, Madl et al 2000, Rothstein 2000, Sherman et al 2000). Most of these studies combine early SEP with brainstem auditory evoked potential (BAEP) recordings; only a few more recent ones have been published on the recording of middle latency auditory evoked potentials (MLAEPs) in comatose patients (Litscher 1995, Morlet et al 1997) with special regard to outcome prediction (Fischer et al 1999).

The standard EP strategy in comatose patients has generally been a 'static' one, based on a single recording session at a given moment in the evolution of the patient's condition. Prognostic conclusions are derived from this information, which may be complemented by further recordings over subsequent days. For prognostic assessment, it is recommended that EPs should be recorded when their contribution to prognosis is expected to be the most reliable. In the hyperacute stage of coma there are too many neurological and non-neurological factors that may affect the outcome. Concerning SEPs it has been reported that after 24–48 hours no significant changes in peak latency occur in most cases (Haupt et al 2000, Gendo et al 2001, Nakabayashi et al 2001). However it is often safer to wait for a few days before recording evoked responses for prognostic assessment, at least until all sedative drugs have been discontinued and once it has been established that no metabolic disturbance could cause major changes in cortical EPs.

Since it is increasingly evident that EPs can offer the first sign of deterioration or improvement, repeated or serial testing is advisable when early recordings have shown preserved responses in a patient who has not regained consciousness. The obvious consequence of this is frequent, repeated EP testing or 'dynamic' EP monitoring, which is now becoming a complementary (but not yet widespread) adjunct to the standard use of EPs in comatose patients (see p. 208 *et seq.*). The International Federation of Clinical Neurophysiology (IFCN) has recently updated standards of clinical practice of EEG and EP in comatose and other unresponsive states (Guérit et al 1999a).

Non-pathological causes of evoked potential abnormalities in the ICU
The effects of sedatives, anaesthetics and temperature on EPs are covered in Chapter 5 (see pp. 157, 172 and 176). It is important to note that temperature has not been rigorously controlled in most published studies of EPs in coma, and this may be the reason for some inconsistency in results, particularly increases in central latencies.

Clinical use of evoked potentials during coma

The usefulness of EP recording in comatose patients is now considered incontrovertible. Single or repeated (often daily) EP recordings are widely used in ICUs either for diagnosis or prognosis, or as an ancillary technique for determination of brain death. It is well known that gross abnormalities or abolition of brainstem auditory potentials (BAEPs) in a comatose patient carry an ominous prognosis, provided that lesions to the peripheral auditory system have been ruled out. A similar adverse significance has been demonstrated for bilateral loss of cortical somatosensory N20 and/or auditory Pa potentials, although the greater susceptibility of the cortical components of evoked responses to the effects of anaesthetic or major sedative drugs may render their interpretation in ICU patients more problematic.

Almost every head-injured patient who survives has BAEPs within normal limits, as well as bilaterally present auditory Pa and somatosensory N20 components (Hume and Cant 1981, Greenberg et al 1982, Mauguière et al 1982, Facco et al 1985, Cant et al 1986, Ottaviani et al 1986, Cenzato et al 1988, Fischer et al 1988). Conversely, depending on the cause of coma, patients with absent SEPs in the first 24–48 h may later produce normal SEPs and go on to make a reasonable recovery, so a pessimistic early assessment can be unreliable (Guérit et al 1993, Schwartz et al 1999, Gendo et al 2001), and this is particularly so in children (Carter et al 1999). The following sections cover the prognostic information to be gained from single recordings of BAEPs and SEPs in comatose patients. The continuous monitoring of EPs is described on page 208 *et seq*.

Somatosensory evoked potentials

Central somatosensory conduction time   Hume and Cant (1978) introduced the concept of somatosensory 'central conduction time', or CCT, which they calculated by subtracting the latency of the main negativity over the cervical spine (N13 or 'N14') from that of the first cortical potential (N20). There is an anatomical ambiguity in this measure, since N13 reflects a true spinal component only when recorded with a non-cephalic reference, whereas brainstem P14 dominates the waveform when a cervico-scalp montage is used, as has been the case in most published studies (Hume and Cant 1981). However, this ambiguity does not present a real problem in clinical ICU practice, provided that a standardized recording technique is used and electrode position always specified when reporting results. Whatever the technique employed, CCT has proved to be a robust and reproducible measure which can provide a quantitative index for follow-up in sequential records. Measuring CCT usually implies two recording channels (cervical and parietal), but it can also be calculated with a single parietal to ear lobe derivation, which allows the simultaneous recording of P14 and N20 potentials.

Middle latency components   In general, SEP recording in coma has been limited to short latency components. Cortical responses beyond 30 ms are still poorly characterized in normals (*Fig. 6.7*) but qualitative studies of SEP components in the 30–200 ms range have proved prognostically relevant in several reports and may indeed have superior prognostic accuracy (*Table 6.2*). It must be remembered, however, that although their presence is of considerable prognostic value, their great variability in normals and their susceptibility to drugs make their absence virtually uninterpretable in comatose patients.

SEP features encountered during coma and their clinical significance   EPs are highly sensitive to brain damage, but quite unspecific as to its aetiology. Identical abnormal patterns can be encountered in a variety of pathological processes, although certain combinations of abnormalities may point to the pathophysiological mechanisms involved. *Table 6.2* summarizes the results of 690 SEP studies in 16 series published in or before 1998 where traumatic or cerebrovascular coma were predominant. Logi et al (2003) have recently reviewed literature data published since 1984 on the prognostic value of SEPs in 1818 patients (*Table 6.3*). In this latter table the positive predictive value of SEP parameters have been evaluated with regard to a bad outcome characterized by the absence of return to consciousness (Glasgow Outcome Scale scores 1 (death) and 2 (persistent vegetative state (PVS)), or to a favourable life outcome defined as a return to consciousness with or without disability (Glasgow Outcome Scale scores 3–5). Since neither SEP methodology nor criteria for estimation of clinical outcome are uniform across studies from different authors, the overall results presented are to be taken as a general indication. However the data are so convergent that the prognostic value of SEP recordings can be considered as firmly established.

Absence of cortical SEPs   When there is bilateral absence of cortical SEPs, this finding is of highly adverse significance. Failure to record any reproducible cortical potential in spite of a low residual noise level and in the presence of peripheral and cervical components is associated with death or PVS in about 90% of head-injured patients (see *Tables 6.2* and *6.3*), and in almost 100% of postanoxic patients. In the rare cases followed by recovery, this is usually achieved at the expense of severe disabilities (Anderson et al 1984); less than 1% of patients make a good recovery or have only moderate disabilities according to the classification of the Glasgow Outcome Scale (Jennett et al 1981b). The above comments are pertinent only for lesions involving the hemispheres; primary brainstem lesions may abolish cortical potentials by interrupting ascending pathways, but the outcome is usually good if the brainstem lesions are reversible. Thus moderate or good recovery may follow bilateral N20 abolition in such patients (Rumpl et al 1983). Furthermore, comatose patients may have SEP abnormalities related to pre-existing pathological conditions, including degenerative diseases, multiple sclerosis and Arnold–Chiari malformation. Such conditions may themselves lead to absence of P14 and N20 (*Fig. 6.8*) and loss of brainstem reflexes (Ringel et al 1988), thus hampering the reliability of prognosis.

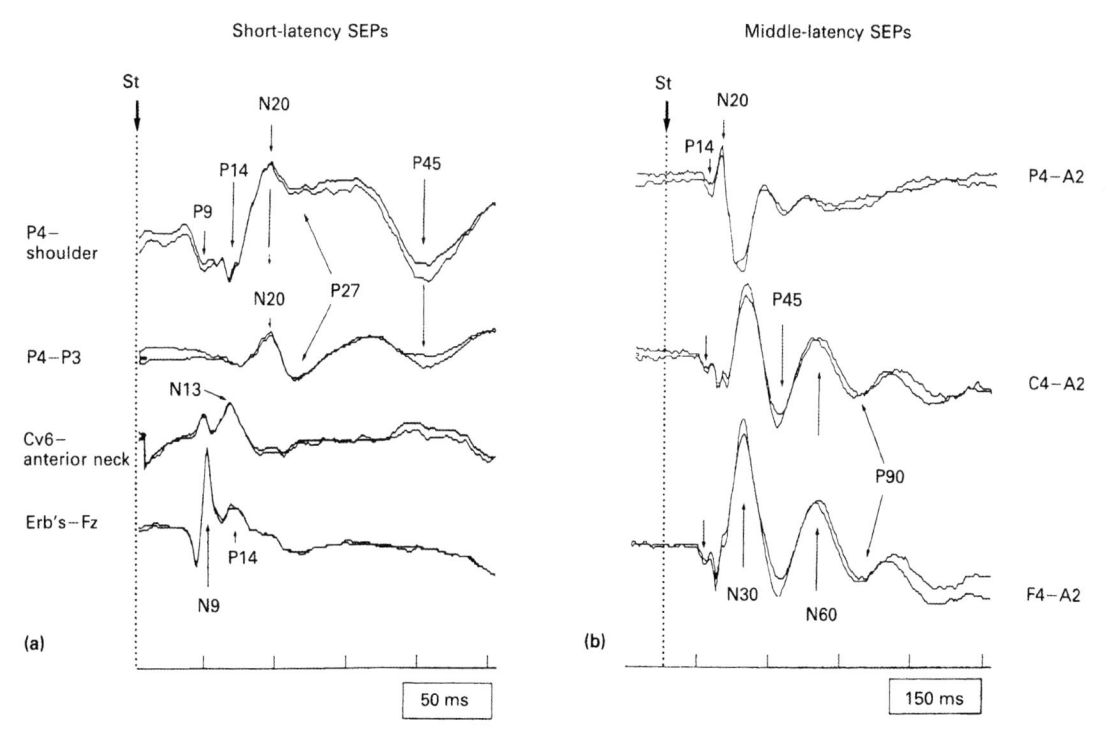

Fig. 6.7 Normal SEPs to left median nerve stimulation. (a) Short latency SEPs within a 50 ms time window. Brachial plexus N9 is recorded at Erb's point, while its counterpart far-field P9 may be observed anywhere on the scalp if a non-cephalic reference is used (uppermost trace). N13 reflects activation of spinal dorsal horn and is best recorded with a transverse montage between posterior (Cv6) and anterior neck (second trace from bottom) which avoids contamination by far-field potentials. Cortical N20 and P25 are recorded with a parietal electrode, either with a non-cephalic or a scalp reference, but the preceding far-fields P9, P11 and P14 are visible only if an extracephalic reference is used (uppermost trace). (b) Middle latency SEPs have been obtained from three scalp sites referred to the ear lobe; brainstem P14 is the only visible far-field. Middle latency components are quite dissimilar at different scalp locations and are usually more conspicuous at precentral recording sites (C4 and 4). (From Binnie et al (2004), by permission.)

Thus the value of SEPs for predicting a bad outcome varies according to aetiologies and must be analysed separately in anoxia, which causes mostly diffuse cortical lesions, and in traumatic and vascular lesions that can affect directly the sensory pathways at the brainstem or subcortical levels.

In postanoxic coma, all studies converge to a 100% specificity of bilateral N20 absence in predicting poor outcome in postanoxic comatose adult patients (Walser et al 1985, Brunko and Zegers de Beyl 1987, Rothstein et al 1991, Berek et al 1995, Bassetti et al 1996, Chen et al 1996, Madl et al 2000, Sherman et al 2000, Logi et al 2003). Furthermore, Logi et al (2003) reported that a peak-to-peak N20–P24 amplitude value below 1.2 µV on both sides had the same prognostic value as bilateral N20 abolition in this group of patients.

In head-injured patients or those with vascular lesions, bilateral absence of N20 also carries a very adverse prognosis with a predictive value for no return to consciousness in 90–100% of patients (see *Tables 6.2* and *6.3*). Conversely, unilateral absence of N20 with preserved P14, indicating a focal lesion involving the parietal cortex and/or the thalamoparietal radiations, is much more likely to occur in these conditions.

Unilateral N20 loss has been described as occurring in about 5–10% of patients in several series of traumatic coma, although matching groups of patients for severity of injury is notoriously difficult. The relationship of unilateral N20 loss with outcome is far less straightforward than with bilateral loss. However, there is a high incidence of poor outcome with unilateral N20 loss (about 75%, see *Table 6.2*) in most series. Outcome is more probably related to the responses from the less damaged hemisphere; if these are strictly normal, recovery is not infrequent. Indeed, unilateral loss of cortical components was by far the most frequent SEP abnormality encountered in a cohort of 500 surviving Vietnam veterans 20 years after penetrating head injury (Jabbari et al 1987).

Several authors have reported motor and sensory deficits associated with unilateral loss of N20, or significant unilateral attenuation of the N20–P24 (or P27) complex (Hume and Cant 1981, Cenzato et al 1988, Rumpl et al 1988b), the probability of permanent neurological sequelae increasing with persistence of SEP abnormalities. In the late chronic phase of head injury, unilateral N20 abolition is strongly associated with contralateral sensory loss, especially astereognosis, and also with motor deficits with hemiplegia (Jabbari et al 1987).

Table 6.2 SEPs in coma*

Study	Bilateral N20 loss G/MD	Bilateral N20 loss SD	Bilateral N20 loss PVS/D	Unilateral N20 loss G/MD	Unilateral N20 loss SD/PVS/D	CCT increase G/MD	CCT increase SD/PVS/D	Normal early SEPs (< 30 ms) G/MD	Normal early SEPs (< 30 ms) SD/PVS/D	Normal mid/late SEPs (> 50 ms) G/MD	Normal mid/late SEPs (> 50 ms) SD/PVS/D
De la Torre et al (1978)	–	–	–	–	–	–	–	–	–	8	0
Hume et al (1979)	–	–	–	–	–	3	2	3	2	ND	ND
Goldie et al (1981)	0	0	12	2	6	–	–	–	–	ND	ND
Lindsay et al (1981)	1	1	10	–	–	–	–	4	8	5	3
Hume and Cant (1981)	0	0	8	–	–	6	9	14	5	ND	ND
Greenberg et al (1982)	0	4	16	–	–	–	–	69	28	36	5
Rumpl et al (1983)	1	1	3	2	0	6	5	–	–	ND	ND
Pfurtscheller et al (1983)	–	–	–	–	–	–	–	–	–	8	1
Anderson et al (1984)	0	2	6	0	4	6	1	3	1	ND	ND
Marcus and Stone (1984)[†]	0	0	18	0	5	9	7	5	5	ND	ND
Frank et al (1985)[†]	0	0	12	3	6	–	–	–	–	ND	ND
Cant et al (1986)	0	1	7	1	5	2	2	17	0	ND	ND
Cenzato et al (1988)	0	0	17	–	–	7	42	4	0	ND	ND
Ganes and Lundar (1988)	0	0	57	–	–	–	–	–	–	ND	ND
Mauguière et al (1988)	0	3	18	–	–	–	–	9	22	ND	ND
Rumpl et al (1988b)	0	7	10	3	6	24	21	11	4	ND	ND
Global (n = 690)	2 (1%)	19 (9%)	194 (90%)	11 (26%)	32 (74%)	21 (41.5%)	89 (58.5%)	139 (65%)	75 (35%)	57 (86%)	9 (14%)

–, EP pattern or clinical outcome not unequivocally stated; G/MD, good or moderate disability; SD, severe disability; PVS/D, persistent vegetative state or death; ND, test not done.

*This table summarizes the results of early SEP studies in coma of predominantly traumatic or cerebrovascular origin. Patients from 16 series are included in whom SEPs were recorded during the acute phase of coma. SEP studies in series of anoxic, infectious or metabolic coma are not presented herein. An effort has been made to harmonize the description of SEP abnormalities and clinical outcome. The results are quite consistent and provide an overall view based on 690 recorded patients, but differences in recording methodology and definition of normal limits are unavoidable, and this summary should therefore be considered as purely orientative.

[†]Patients with cardiovascular (anoxic) coma in these two series have been excluded.

Abnormal central conduction time Increased CCT may result from damage to the somatosensory pathways at any level between the brainstem and the parietal cortex, and thus has little localizing value *per se*. Left–right comparisons, consideration of the clinical context and recording of other EP modalities often help in interpretation of CCT measurements.

Unilateral or asymmetrical increases of CCT are more likely to arise with hemispheric than with primary brainstem lesions. It must be borne in mind that progressive unilateral CCT increase in a head-injured patient may be one of the earliest signs of impending transtentorial herniation. Unfortunately, detection of this type of change requires continuous monitoring of SEPs, which is not yet available in most ICUs.

Increased CCT was considered by early investigators as a reliable predictor of bad outcome. Indeed, when groups of patients are compared, mean CCT has repeatedly been found to increase with worsening of outcome (Hume and Cant 1981, Rumpl et al 1983, 1988a). However, such epidemiological data are of little value when evaluating EP results in a single patient, and predictions of outcome based solely on CCTs are quite unreliable and incongruent across series (see *Table 6.2*). CCT is a continuously distributed variable; definition of its normal limits is statistical and derived from recordings in normal controls. A statistically 'normal' CCT may actually be increased with respect to the precoma value in the same individual, thus giving rise to falsely optimistic predictions. On the other hand, premature and falsely pessimistic conclusions may be derived from abnormal CCTs, which can improve rapidly during post-traumatic coma (Hume and Cant 1981) or occur with reversible metabolic or toxic conditions (Rumpl et al 1988b), or during barbiturate therapy (Drummond et al 1985). This might partly explain the high incidence of good or moderate recovery reported in patients with abnormal CCT during the acute phase of coma. This figure is about 40% when data from 17 series are grouped together (see *Table 6.2*).

Increased CCT was found as an isolated feature in only 1% of 500 head-injured patients who had SEPs

Table 6.3 SEP predictive value of bad outcome in coma

Study	Aetiology	N	SEP parameters	Sensitivity	Specificity	Positive predictive value
Claassen and Hansen (2001)	Traumatic brain injury	31	N20 abs	75%*	100%*	100%*
			N20–P25 < 0.9 µV	33%*	72%*	25%*
Madl et al (2000)	Anoxic–ischaemic	162	N20 abs	31%	100%	100%
			N70 > 130 ms	94%	97%	98%
			CCT > 7.0 ms	83%	79%	94%
Sherman et al (2000)	Anoxic–ischaemic	72	N20 abs or < 0.3 µV	55%	100%	100%
			N70 abs or latency > 176 ms	67%	100%	100%
Rothstein (2000)	Anoxic–ischaemic	50	N20 abs	68%	100%	100%
Sleigh et al (1999)	Traumatic brain injury	105	N20 abs	43%	99%	91%
Facco et al (1998)	Haemorrhagic stroke	70	N20 abs	73%	95%	97%
Pohlmann-Eden et al (1997)	Traumatic brain injury, stroke	42	N20 abs	70%	87%	90%
Madl et al (1996)	Anoxic–ischaemic, stroke, neurosurgery, encephalitis, metabolic and others	441	N20 abs	29%*	100%*	100%*
Chen et al (1996)	Anoxic–ischaemic	34	N20 < 0.5 µV or abs	66%	100%	100%
			N20 latency > 22.1 ms	59%	71%	89%
Bassetti et al (1996)	Anoxic–ischaemic	56	N20 abs	51%*	100%*	100%*
Berek et al (1995)	Anoxic–ischaemic	25	N20/N13 < 0.5	44%*	100%*	100%*
			N20 abs	40%	100%	100%
Gütling et al (1994)	Traumatic brain injury	50	N20 abs	64%*	97%*	90%*
Zentner and Rohde (1992)	Traumatic brain injury, stroke, anoxic–ischaemic, toxic	213	N20 abs	47%	98%	96%
Rothstein et al (1991)	Anoxic–ischaemic	40	N20 abs	73%	100%*	100%
			N9–N20 > 11.2 ms	33%	44%*	43%
Hutchinson et al (1991)	Traumatic brain injury	90	N20 abs	65%*	98%*	96%*
Barelli et al (1991)	Traumatic brain injury	24	N20 abs	43%*	100%*	100%*
Brunko and Zegers de Beyl (1987)	Anoxic–ischaemic	50	N20 abs	67%*	100%*	100%*
Walser et al (1985)	Anoxic–ischaemic	26	N20 abs	60%*	100%*	100%*
Anderson et al (1984)	Traumatic brain injury	23	N20 abs	86%*	100%*	100%*

Modified from Logi et al (2003), by permission.

Sensitivity, specificity and positive predictive value refer to the capability of SEP parameters to predict a bad outcome defined as a no-return to consciousness (Glasgow Outcome Scale scores 1 and 2). Sensitivity refers to the capability of excluding false negatives, specificity to the capability of excluding false positives and positive predictive value refers to the capability of excluding false predictions.

CCT, central conduction time; abs, absent bilaterally.

*Values calculated from raw data reported in the study.

recorded 20 years after the acute episode; in another 5% it was associated with other abnormalities (Jabbari et al 1987). This very low incidence suggests that, when the patient survives, abnormal CCT has an intrinsic potential for recovery that could serve to monitor the patient's status by means of serial recordings. Thus, several investigators have reported that sequential assessment of CCT during coma accurately reflects the patient's evolution and allows more reliable prediction of clinical outcome than does a single CCT measurement obtained during the acute phase of coma (Hume et al 1979, Hume and Cant 1981, Newlon et al 1982, Rumpl et al 1988b). Attempts to perform automatic CCT monitoring in the ICU have been described (Pfurtscheller et al 1987), but unfortunately such systems have not yet come into routine use. Concerning the prediction of recovery of consciousness, the measurement of

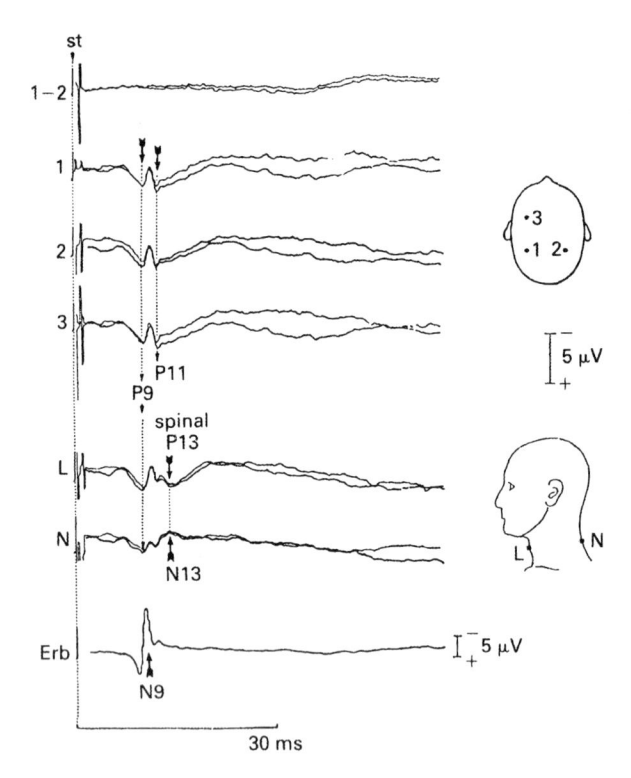

Fig. 6.8 Absence of SEP N20 and P14 in a patient with multiple sclerosis. This SEP pattern with abolition of cortical and brainstem components is almost indistinguishable from that obtained in brain death (see *Fig. 6.17*), but in this case was recorded in a conscious subject suffering from multiple sclerosis, in whom loss of reproducible potentials was related to extreme desynchronization of afferent impulses at the upper cervical cord (note that cervical N13 is normal). (From Binnie et al (2004), by permission.)

CCT in addition to that of N20–P24 voltage, on a single recording performed when the clinical status of the patient can be considered as stable, does not add any reliable information (Logi et al 2003).

Normal SEPs Most SEP studies in comatose patients have dealt only with subcortical and primary cortical components, results being considered as 'normal' on the basis of early cortical responses N20 and P24 (or P27). Given that primary inputs to any cortical region represent less than 2% of its total afferents (Braitenberg 1978), it is clear that interpreting a normal early SEP as a sign of good hemispheric function is far too optimistic; not surprisingly, death or severe disability occurs in about 30% of comatose patients in spite of bilaterally normal N20 (see *Table 6.2*). However, when later SEP components also are included in the analysis, the predictive value of a normal waveform is greatly enhanced, the incidence of poor outcome decreasing to about 15% (Greenberg et al 1982, Pfurtscheller et al 1983), or even lower if cases with secondary, non-neurological complications such as septicaemia are eliminated. Although currently available data come from only a few series, predictions of outcome based on normal SEPs are substantially more accurate when based on middle latency, rather than primary cortical components. Thus, to improve the predictive power of 'normal' SEP recordings, it will probably be necessary to employ analysis times longer than those used in most reported studies. It must be remembered, however, that cortical SEPs beyond 50 ms are quite variable in normal subjects, and are much more affected by CNS depressant medication than the early components. Presence of these later components indicates good cortical functioning, but their absence provides little information, especially if the patient is anaesthetized.

Auditory evoked potentials The frequent occurrence of peripheral auditory dysfunction in ICU patients represents one of the major drawbacks to routine use of AEPs in coma. Peripheral auditory deficits can result from traumatic injury to the cochlear nerve in cases of basal skull fracture, or more frequently from middle-ear dysfunction (Hall et al 1982), secondary to intubation and loss of swallowing movements. Recording of tympanograms is a simple and useful technique for assessing this latter point and ideally should be added to BAEP examination every time a peripheral auditory problem is suspected. Tympanography reflects the dynamic response of the tympanic membrane to pressure changes in the external auditory canal; this technique is easily performed at the bedside with commercially available devices. A normal and a 'flat' tympanic curve (this latter indicating middle-ear dysfunction) are illustrated in *Fig. 6.9*. Aminoglycosides such as gentamicin or kanamycin, the diuretic frusemide, and other drugs, may also induce peripheral hearing loss by damaging the inner-ear cells. This usually appears later in coma evolution than middle-ear problems. It should be suspected if BAEP abnormalities of 'peripheral' type (affecting wave I latency and amplitude, with preservation of the I–V IPI) develop in a patient in whom a middle-ear problem cannot be demonstrated. Of course, the two levels of peripheral auditory dysfunction (middle and inner ear) may coexist.

Short latency AEPs The brainstem components of BAEPs (waves III–V) are reported as being almost invariably normal when recorded during the first hours of traumatic coma (Cenzato et al 1988), suggesting that most abnormalities encountered later are secondary to hemispheric lesions and reflect an ongoing process of deterioration. The universal finding of an adverse prognostic significance of severely abnormal BAEPs in coma corroborates this view.

Most investigators have recorded BAEPs in comatose patients using a conventional Cz (lead 2) versus ear (lead 1) derivation, with stimulus rates ranging from 10 to 20 Hz. Some controversy surrounds the possible use of higher stimulation rates (up to 70 Hz) in order to increase BAEP sensitivity to minimal brainstem dysfunction. While some authors find it useful in post-traumatic patients (Pratt et al 1981, Cenzato et al 1988), experimental evidence in animals has suggested that it may be less efficient in detecting synaptic dysfunction than conventional 10 Hz stimulation (Sohmer and Goitein 1988).

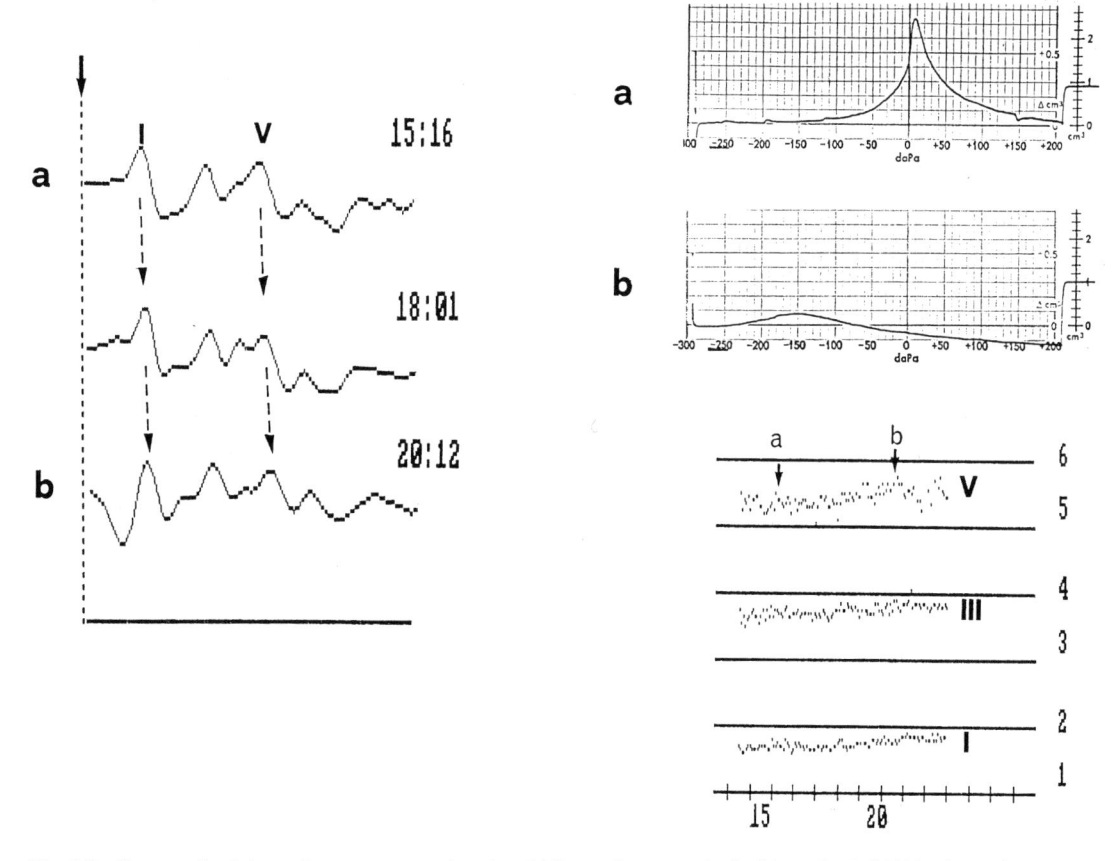

Fig. 6.9 Progressive latency increase secondary to middle-ear hypoacusis. In this patient, BAEPs showed a progressive latency increase affecting all waves to a similar degree (upper traces), reflected in latency trends by downward-pointing arrows with similar slopes for waves I, III and V. These BAEP changes are accounted for by development of middle-ear hearing loss between 15.16 h and 20.12 h, as shown by tympanograms (top right). (From Binnie et al (2004), by permission.)

Disparity in BAEP classification is one of the main difficulties that arise in comparing the results obtained by different groups. Although a number of authors differentiate between 'mild', 'moderate' and 'marked' abnormalities, the criteria used to include traces in such categories are quite inconsistent across series. For instance, loss of wave V was considered by Tsubokawa et al (1980) as a 'mild' dysfunction, but it represented a 'moderate' abnormality for Hall et al (1982) and a 'markedly abnormal' pattern for Karnaze et al (1985). To overcome this difficulty, the results of 14 series published between 1977 and 1986 are summarized and pooled in *Table 6.4*. Most published BAEP results can be placed in three simple categories, namely normal, abnormal and absent potentials, this last group also including BAEPs with only the peripheral wave I. Only series in which traumatic or cerebrovascular aetiologies predominate have been considered, as the relationship between EP and outcome may be slightly different for other types of coma (Hecox and Cone 1981, Jain and Maseshwari 1984, Rosenberg et al 1984).

Absent AEPs Absent BAEPs, or those consisting of an isolated wave I, are almost invariably followed by death whatever the origin of coma, provided that peripheral auditory problems, primary brainstem lesions or drug effects are ruled out. As was the case with SEPs, the existence of previous pathological conditions capable of altering waveforms, such as multiple sclerosis or hereditary degenerative diseases, must be carefully assessed before interpretation.

Every effort should be made to record a reproducible wave I; otherwise the absence of BAEPs may be indistinguishable from that due to peripheral lesions. Transtympanic ECochG may be used to check the presence of an auditory nerve response when a reasonable doubt exists. Isolated wave I or waves I and II have comparable adverse prognostic significance.

Present, but abnormal, AEPs The prognostic value of abnormal BAEPs is far less straightforward than that of absent potentials, since about 35% of patients with this pattern nevertheless have a favourable outcome (see *Table 6.4*). Such a high incidence of good and moderate outcomes may be surprising if abnormal BAEPs reflect actual brainstem damage. It may be explained by the very different degrees of abnormality that are grouped in this category. Indeed, isolated bilateral abnormality of the I–V interval or I/V amplitude ratio may, in many instances, not reflect significant brainstem pathology, but only hypothermia or high drug levels. These variables are not controlled

Table 6.4 AEPs in coma*

Study	Absent BAEPs[†] G/MD	SD	PVS/D	Abnormal BAEPs G/MD	SD/PVS/D	Normal BAEPs G/MD	SD/PVS/D	Normal MLAEPs G/MD	SD/PVS/D
Greenberg et al (1977)	–	–	–	–	–	27	7	ND	ND
Seales et al (1979)	0	0	3	3	2	9	0	ND	ND
Tsubokawa et al (1980)	0	0	23	8	5	25	3	ND	ND
Goldie et al (1981)	–	–	–	–	–	12	14	ND	ND
Uziel et al (1982)	0	0	3	3	5	14	5	ND	ND
Lütschg et al (1983)	0	0	5	4	8	6	3	ND	ND
Anderson et al (1984)	0	2	6	6	14	10	1	ND	ND
Brewer and Resnick (1984)	0	0	26	0	7	11	8	ND	ND
Marcus and Stone (1984)[‡]	0	0	5	3	16	11	13	ND	ND
Kaga et al (1985)	0	0	13	1	9	16	16	10	5
Karnaze et al (1985)	0	0	3	6	7	25	4	ND	ND
De Weerd and Groeneveld (1985)	–	–	–	0	5	7	6	ND	ND
Cant et al (1986)	–	–	–	–	–	19	9	ND	ND
Ottaviani et al (1986)	0	0	23	17	16	10	0	8	2
Totals	0	2	110	51	94	202	89	18	7
	(0%)	(2%)	(98%)	(35%)	(65%)	(69%)	(31%)	(72%)	(28%)

–, Either BAEP pattern or outcome not unequivocally stated.
*AEPs obtained during the first week of traumatic or cerebrovascular coma. Series in which EP or outcome classification were insufficient are not summarized. This includes those in which abnormal and absent BAEPs were pooled together, or in which outcome was only evaluated as survival/death. Since determination of normal limits or outcome classification are not uniform across series, these data should be considered only as orientative.
[†]Or isolated wave 1.
[‡]Patients in anoxic coma excluded.

in most published series. In addition, setting of upper normal latency limits at unusually low values (2 standard deviations (SDs) above the mean) in some series may have induced falsely pessimistic results in patients who would have been considered as having normal BAEPs with more conventional criteria.

Loss or extreme desynchronization of wave V carries a much more serious prognosis for life than isolated latency abnormalities. The outcome in these cases is very similar to that of absent BAEPs, with no false pessimistic results reported in studies where this pattern is specifically noted (Anderson et al 1984, Karnaze et al 1985, Ottaviani et al 1986).

Some authors have claimed that survivors and non-survivors from traumatic coma could be discriminated on the sole basis of the I–V interpeak interval (IPI), the limit being empirically determined in one study at 4.48 ms (Facco et al 1985). Such an approach seems untenable, especially since normal upper limits for I–V IPI (mean ± 3 SD) are remarkably similar between laboratories and vary around 4.5–4.7 ms – well above the proposed 'survival threshold'.

Normal BAEPs during acute coma Another fact evident from *Table 6.4* is that normal BAEPs are by no means a guarantee of a good outcome from coma. In patients with bilaterally normal BAEPs during the acute phase of traumatic or cerebrovascular coma, the average incidence of unfavourable evolution (severe disabilities, persistent vegetative state or death) is about 30%. This is not entirely unexpected, since (a) this evoked response, in contrast with SEPs, does not reflect the extent of hemispheric dysfunction; and (b) progressively deteriorating BAEPs may still remain within normal limits for a long time. This latter point is illustrated by *Fig. 6.10*, showing changes observed in I–V IPI between the first and second weeks of coma in a series of 20 patients. Although latencies had increased significantly by the time of the second examination in those who subsequently died, values remained well within statistical normal limits. This stresses the limitations of statistical concepts of 'normality' and the need to look at trends with time, in which each patient acts as his own control, rather than absolute values.

'Late' normal BAEPs Several studies have suggested that normal BAEPs recorded beyond the acute phase of coma ('late' normal BAEPs) indicate that the patient is capable of survival (Greenberg et al 1977, Seales et al 1979, De Weerd and Groeneveld 1985). This view is based on data from large series of traumatic or vascular coma, in

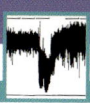

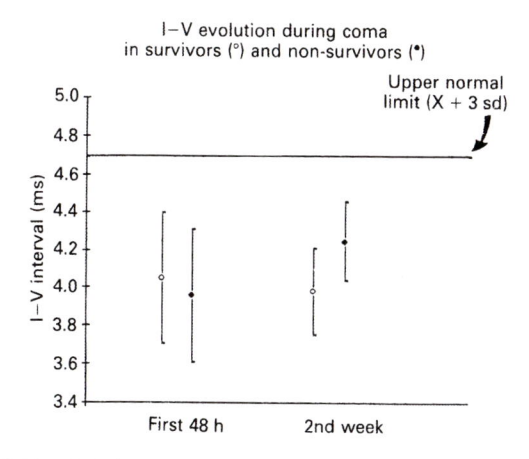

Fig. 6.10 Variations of the BAEP I–V interval between the first 48 hours and the second week of coma. Mean ± SD are plotted. Data from 20 comatose patients. Open circles represent patients who survived, and black squares those who subsequently died. Mean I–V IPI was not different between groups in the first recording session. IPIs increased significantly in the second week for the group of bad-outcome patients (filled circles), while remaining stable in survivors. Note that in both groups the absolute values of the I–V interval remained well within normal limits. This illustrates the limitations of single EP recordings, which do not account for evolution in time. A single normal BAEP does not give *per se* a good estimation of the chances of survival, but sequential recordings provide clues as to the evolution of brainstem function and are relevant for early detection of brainstem damage. (From Binnie et al (2004), by permission.)

which death occurred mainly during the first week, and especially within 48 h of admission (Carlsson et al 1968, Clifton et al 1980, Plum and Posner 1980). Since rostrocaudal brainstem deterioration secondary to raised ICP is the main mechanism of brain death in these patients, it is reasonable to assume that preserved brainstem function beyond that first, most dangerous period of one week should indicate an increased probability of survival. However, the nature of the lesion causing coma is relevant to the prognostic value of these 'late' BAEPs. In patients with traumatic or cerebrovascular lesions remaining in deep coma (GCS score < 8), the presence of strictly normal BAEPs during the second week after the insult is only associated with a good prognosis if CT scan shows no evidence of intracerebral haemorrhagic lesions (García-Larrea et al 1992). This finding is in accordance with the known potential for intracerebral lesions to produce late neurological deterioration, supervening one week or more after admission (Clifton et al 1980).

A more sensible way of improving the prognostic value of brainstem responses recorded at different moments of coma is to assess their evolution over time. As shown in *Fig. 6.10*, the comparison of sequential BAEPs (obtained one week apart) may provide prognostic information that is largely independent of the categorization of each response separately. In that way, EPs have been proved useful in monitoring neurological recovery or deterioration following secondary insults before they became evident on clinical grounds (Newlon et al 1982, Newlon and Greenberg 1984).

Coma due to metabolic disorders, hypoxia or inflammatory disease has been less extensively studied with BAEPs than coma of post-traumatic and cerebrovascular origin. Absent or severely abnormal waveforms are indicative of a bad prognosis whatever the underlying cause of coma (Hecox and Cone 1981, Jain and Maseshwari 1984, Rosenberg et al 1984). In postanoxic coma, consistently normal BAEPs with absence of cortical somatosensory or auditory responses frequently predict subsequent PVS (Frank et al 1985). Patterns of rostrocaudal BAEP deterioration in serial recordings have been reported after anoxia (Ganes and Lundar 1988), suggesting that intracranial hypertension may be one mechanism of deterioration in some of these patients.

The resistance of BAEP components to CNS depressants (except in extremely high dosage) has proved useful clinically in cases of coma due to drug overdose. A discrepancy between clinical signs of severe brainstem dysfunction, such as apnoea and absent cephalic reflexes, and persistence of normal BAEPs suggests a toxic or metabolic origin of coma, rather than a structural brainstem lesion (Starr and Achor 1975, Stockard et al 1980, Rumpl et al 1988a). This principle should, however, be applied exclusively in the presence of clinical signs of global brainstem dysfunction, and then with extreme caution, since normal responses may also be obtained with discrete tegmental lesions leading to irreversible coma but sparing the auditory pathways (Stockard et al 1980).

Middle and long latency AEPs Since the late 1980s MLAEPs have been introduced into the examination of comatose patients. They are more difficult to record than BAEPs and less resistant to CNS depressants; therefore a bilateral absence of the Pa component is less reliable than that of bilateral absence of the N20 SEP as factor of bad prognosis (Logi et al 2003). Conversely, unlike BAEPs, the persistence of MLAEPs during traumatic coma may be a reliable predictor of good outcome (Hall et al 1985a, Kaga et al 1985, Ottaviani et al 1986, Fischer et al 1988). In particular, MLAEPs can help in predicting outcome in the presence of a normal BAEP by identifying patients with a high risk of deterioration.

Late auditory potentials have been recorded in coma (Greenberg et al 1977, Kaga et al 1985, Karnaze et al 1985), but they have long been considered as too variable to permit a reliable assessment. When they are present, outcome is generally good (Karnaze et al 1985), but as the GCS score is usually also high, the additional information provided by the EPs is questionable. Moreover, due to recording difficulties, their absence has been considered as virtually uninterpretable. Conversely more recent studies point to the conclusion that preservation of long-latency EPs, including the mismatch negativity, are reliable predictors of awakening (Gott et al 1991, De Giorgio et al 1993, Kane et al 1996, Fischer et al 1999, Guérit et al 1999b, Mazzini et al 2001).

Combination of several EP modalities Early work on multimodality EPs involved recording short and long latency SEPs and AEPs, as well as flash VEPs at the same session (Greenberg et al 1977, Lindsay et al 1981, Narayan et al 1981, Newlon et al 1982). This can be very time consuming (about 3 h per session) and is hardly acceptable in everyday practice. Moreover, it was soon realized that the diagnostic and prognostic value of different EP modalities was not equal, and notably that flash VEPs and LLAEPs did not add any substantial information to that obtained by means of BAEPs and SEPs. With respect to prognostic value, Greenberg et al (1982) remarked that SEPs alone provided the same information as the combination of all EP modalities. Lindsay et al (1981) suggested, after recording multimodality potentials in 32 comatose patients, that SEPs should be considered the method of choice in establishing a prognostic baseline, and other modalities added subsequently, depending on the diagnosis or clinical course.

As a rule, combined recording of subcortical and cortical responses is desirable when testing comatose patients, even if it requires increasing the length of the recording. The acquisition of SEPs and BAEPs can be carried out in a reasonable time, and allows identification of certain combinations of abnormal patterns with important diagnostic and prognostic significance. The two extremes are:
(a) severely abnormal or absent BAEPs coexisting with bilaterally normal SEPs, which point to a primary brainstem lesion; and
(b) the reverse, i.e. absent SEPs with bilaterally normal BAEPs, consistent with widespread cortical dysfunction without brainstem involvement.

The benefits of combined use of cortical and subcortical EPs are greatest in serial records; thus, in traumatic coma, selective degradation of cortical EPs with preservation of brainstem responses points to progressive cortical injury, and the possibility of subsequent transtentorial herniation. Lemniscal fibres running superficially in the mesencephalon are likely to be distorted in impending herniation; thus, progressive abnormalities of CCT or a diminished N20 may provide the earliest signs of this complication.

The specific type of EP to record, the combinations and the appropriate protocol for repetition of tests vary in different patients. In some cases the combined recording of BAEPs and MLAEPs may be a better solution than the combination of BAEPs with SEPs, but the choice will generally depend on the clinicopathological context. For instance, it is self-evident that somatosensory responses rather than MLAEPs should be recorded whenever BAEPs are bilaterally absent and a peripheral auditory problem or a primary brainstem lesion is suspected.

EPs in comatose or pseudocomatose states caused by primary brainstem lesions The EP assessment of primary brainstem lesions causing coma or pseudocoma deserves a separate comment, since the clinical assessment of these patients may greatly benefit from accurate electrophysiological testing. Furthermore, some of the interpretative guidelines provided so far, which apply to coma due to supratentorial lesions, are not valid in the case of primary brainstem damage. In 'locked-in' syndrome (see p. 196), a condition induced by massive de-efferentation due to interruption of descending fibres at the basis pontis, the appearance of BAEPs and SEPs depends basically on the posterior extension of the lesion towards the tegmentum. If this is spared, EPs are likely to be unaffected. There does not seem to be a clear correlation between early EP recordings and clinical outcome in the relatively few documented patients with this syndrome. Thus, abnormal early BAEPs were recorded in the two surviving patients reported by Stern et al (1982), whereas the one patient who died had initially exhibited normal responses. Serial recordings are more likely to correlate with the clinical evolution (Seales et al 1981), but no attempt to monitor EPs systematically has yet been reported.

Abnormal EPs in clinically 'locked-in' patients demonstrate that in a number of cases they are not only de-efferented, but also de-afferented (Starr and Hamilton 1976, Gilroy et al 1977, Seales et al 1981, Stern et al 1982, Ferbert et al 1988). This possibility is hard to demonstrate clinically, but should be remembered during the evaluation of such patients since they may perceive somatosensory, but not auditory, inputs due to ischaemia of auditory structures (Ferbert et al 1988).

Patients in coma due to basilar artery occlusion most often show abnormal BAEPs, especially waves IV and V (Kaji et al 1985, Ferbert et al 1988). Normal BAEPs or an increased I–III IPI are mainly seen in patients with caudally located lesions which do not alter consciousness. Isolated alteration of waves IV and V in brainstem infarcts suggests a basilar occlusion distal to the anteroinferior cerebellar artery (AICA) (Gilroy et al 1977, Ferbert et al 1988). However, when the occlusion is proximal to, or involves, the AICA (which usually supplies both the cochlear nucleus and the peripheral auditory structures), early auditory components may be abolished, including wave I. This constitutes one of the major drawbacks in the assessment of brainstem function by BAEPs in these patients. Ferbert et al (1988) found an isolated wave I in 14% of 28 patients with basilar occlusion, but in only one case were clinical signs of brain death present. The utility of BAEPs for the diagnosis of brain death is therefore seriously reduced in patients with primary brainstem lesions; their value is much less than in the case of supratentorial conditions. In more extensive brainstem infarcts, both BAEPs and SEPs may be abolished. Thus when a primary brainstem lesion is suspected, demonstration of an isoelectric EEG is imperative before considering the absence of brainstem potentials as synonymous with brain death.

Evoked potential monitoring

Why EP monitoring? As explained in the previous section (p. 200 *et seq.*) a single normal BAEP does not provide adequate relevant information as to the final outcome of coma (Seales et al 1979, Lindsay et al 1981, Chiappa 1983,

Brewer and Resnick 1984, De Weerd and Groeneveld 1985, Cant et al 1986) and although normal SEPs and MLAEPs have proved to be of a higher predictive value for good outcome (Greenberg et al 1982, Anderson et al 1984, Rosenberg et al 1984, Kaga et al 1985, Cant et al 1986, Rothstein et al 1991, Ying et al 1992, Zentner et al 1992, Berek et al 1995, Chen et al 1996, Zandbergen et al 1998, 2000, Guérit 1999, 2000, Rothstein 2000, Carter and Butt 2001), there is still a figure of about 20% falsely optimistic predictions with these modalities (Anderson et al 1984, Bricolo et al 1987, Cant 1987, Pohlmann-Eden et al 1997). There are at least two main reasons for this: first, a sudden change in the patient's homeostasis such as a rapid increase of ICP may cause death shortly after normal EPs have been recorded; second, slow but steady degradation of EPs over several hours or days may go unnoticed if recordings take place when latencies and amplitudes, although deteriorating progressively, still remain within standard normal limits (García-Larrea et al 1987). On the other hand, absence of SEPs in the first 24–48 h does not necessarily predict a poor outcome (Guérit et al 1993, Schwartz et al 1999, Gendo et al 2001), particularly in children (Carter et al 1999). To overcome these limitations there is a need for systems capable of detecting both rapid and long-lasting EP changes, and this can only be done by frequent repetition of recordings, i.e. iterative recordings or continuous EP monitoring. Data provided by such monitoring systems should be immediately available to the physician in charge of the patient, and reliable enough to permit interactive interpretation in relation to therapeutic decisions. Rather than establishing a long-term prognosis, continuous feedback to the clinician is the ultimate goal of monitoring; accordingly, as stated by Hacke (1985), repeating examinations at long intervals, or storing continuously obtained data for later interpretation should not be regarded as monitoring.

BAEP monitoring in the ICU In this section we shall describe the utility, and also some of the limitations and uncertainties of BAEP monitoring in the ICU. The BAEP is chosen as an exemplar because of the vulnerability of the brainstem to homeostatic variations in patients with severe head injuries, and brainstem damage is potentially life-threatening. In addition BAEPs are particularly well adapted to monitoring given their stability in normals, even during physiological sleep (Bastuji et al 1988) and their resistance to most intravenous anaesthetics used in the ICU (Sohmer et al 1978, Bobbin et al 1979, Sutton et al 1982, Drummond et al 1985, Hall 1985). Several systems for automatic recording of BAEPs in the ICU have been proposed (Maresch and Pfurtscheller 1983, Boston and Deneault 1984, Maynard and Jenkinson 1984, Bertrand et al 1987), but their application to comatose patients has seldom been documented (García-Larrea et al 1987, 1988a,b, Pfurtscheller et al 1987, Kawahara et al 1989, Litscher et al 1990). This section summarizes results obtained with the system described by Bertrand et al (1987), which included an adaptive digital filter so that interpretable averages were available after 200–400 stimuli, reducing the recording time to 10–20 s for each BAEP.

Clinically relevant abnormal features

Irreversible changes With regard to the time course and characteristics of BAEP deterioration, two different patterns have been identified in patients whose BAEPs, ICP and mean arterial pressure (MAP) were continuously monitored up to brain death.

- *Simultaneous deterioration of waves I, III and V.* BAEP changes may occur either gradually over several hours (*Figs 6.11* and *6.12*) or suddenly, in a few minutes (*Fig. 6.13*). Low cerebral perfusion pressure (CPP < 40 mmHg) is a finding common to patients presenting with this deteriorating pattern; this can be due to a fall of MAP, an increased ICP or the combination of both. The time course of CPP decrease and that of BAEP changes correlate closely; in patients with a progressive latency increase of all BAEPs there was a prolonged and steady reduction of mean CPP values, the onset of BAEP degradation being time-locked to CPP decrease (see *Fig. 6.11*). Conversely, when death occurred after abrupt CPP fall there was also a sudden disappearance of BAEPs.

  The cause of simultaneous deterioration of peripheral and central BAEP components is most likely to be reduced perfusion in the vertebrobasilar system and its branches which supply the cochlea, cochlear nerve and brainstem (Brodal 1969). With gradual decrease of CPP, BAEP waves III and V usually disappear before wave I (Sohmer et al 1984), but a sudden fall in CPP, following cardiac arrest for example, can obliterate all BAEP components because the cochlea is infarcted (Brunko et al 1985, Kaga et al 1985). Autoregulatory mechanisms usually keep cerebral blood flow roughly constant for CPP values higher than 50 mmHg (Harper 1966, Nilsson 1969, Purves 1972, Plum and Posner 1980). However, in patients with severe head injury, BAEP changes can occur while CPP, though progressively decreasing, still remains within normal values. This may be explained by impaired vascular autoregulatory mechanisms after severe head injury, when reduction of blood flow, oxygen consumption and metabolic rate have been reported even in the presence of a CPP higher than 50 mmHg (Frewen et al 1985, Gray and Rosner 1987). This suggests that in such patients monitoring the activity of neural structures by EPs would be more useful than indirect assessment by CPP recording.

- *Deterioration of waves III and V with preserved wave I.* In this second terminal pattern, also described by Buchner et al (1986), BAEP changes are limited to the brainstem components (III–V) with preservation of wave I (*Figs 6.14* and *6.15*). Despite their latency increase, ultimate abolition of brainstem components

Chapter 6 Neurophysiological Work in the ICU

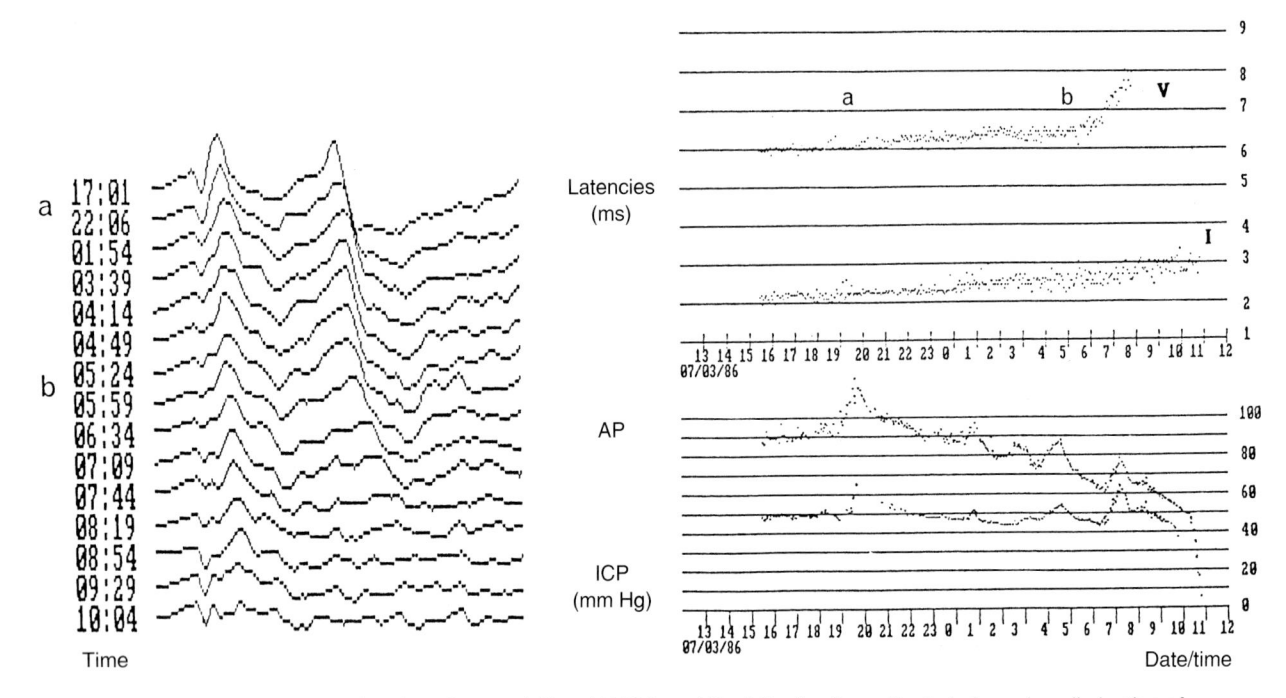

Fig. 6.11 Slow simultaneous deterioration of waves I, III and V. This and the following figure illustrate how slow diminution of CPP, with or without increased ICP values, is paralleled by changes of BAEPs. Responses selected at different recording times are displayed with timing in hours and minutes on the left side; trend lines of BAEP latencies, ICP and AP values are plotted in the right diagrams. MAP, mean arterial pressure; SAP, systolic arterial pressure; ICP, intracranial pressure. The rate of latency increase and of CPP decrease was constant between a and b. At point b (05.00 h), when CPP was 23 mmHg, wave V latency began to increase at a much faster rate up to complete abolition (see raw traces at 07.44 h). Wave I still persisted for 3 h more with steadily increasing latency, and finally disappeared when CPP reached 8 mmHg. (From García-Larrea et al (1987), by permission.)

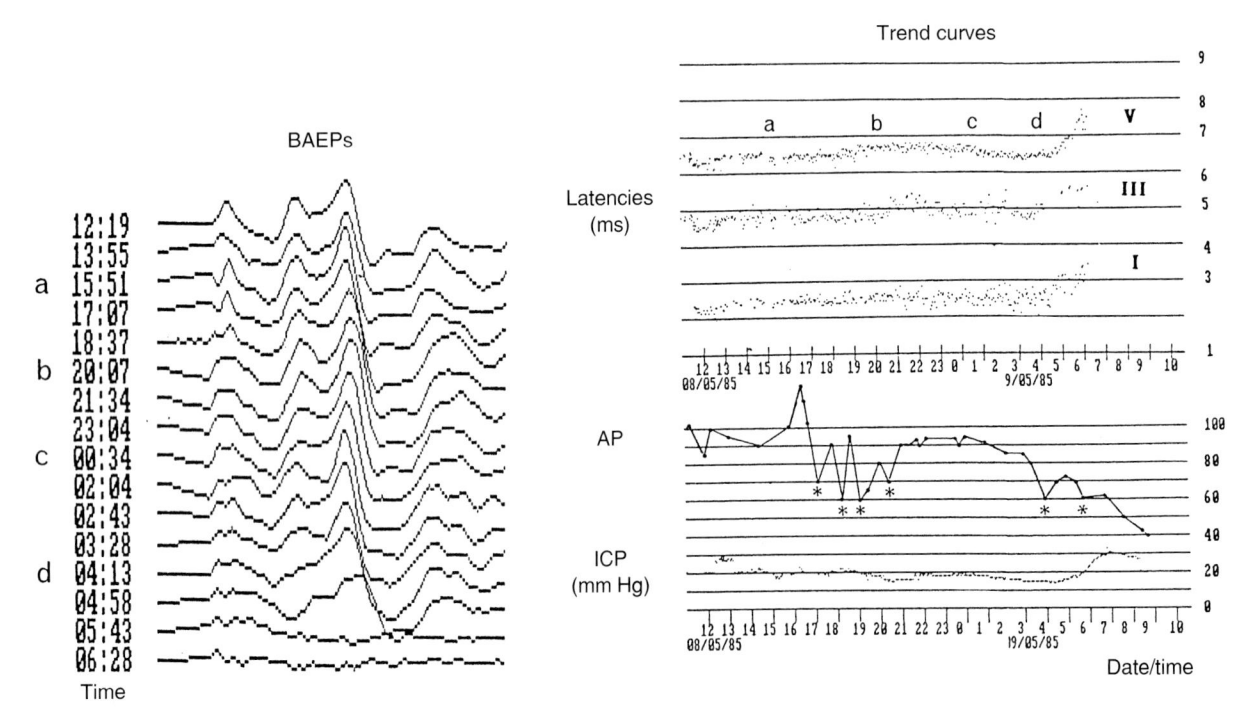

Fig. 6.12 Same presentation as *Fig. 6.11*. Multiple episodes of arterial pressure fall, rapidly reversed by dopamine injections (asterisks), were detected in this patient between points a and b; simultaneously, there was a small progressive increase of all BAEP latencies. From b to d the CPP remained constant at 70 mmHg and previous BAEP changes were found to be reversible (from c to d). At this point a sudden drop in AP, only partially and transiently corrected by dopamine (asterisks), was immediately followed by a rapid latency increase of waves I and V. Two hours later BAEPs were totally lost and only then was a rise of ICP detected. (From García-Larrea et al (1987), by permission.)

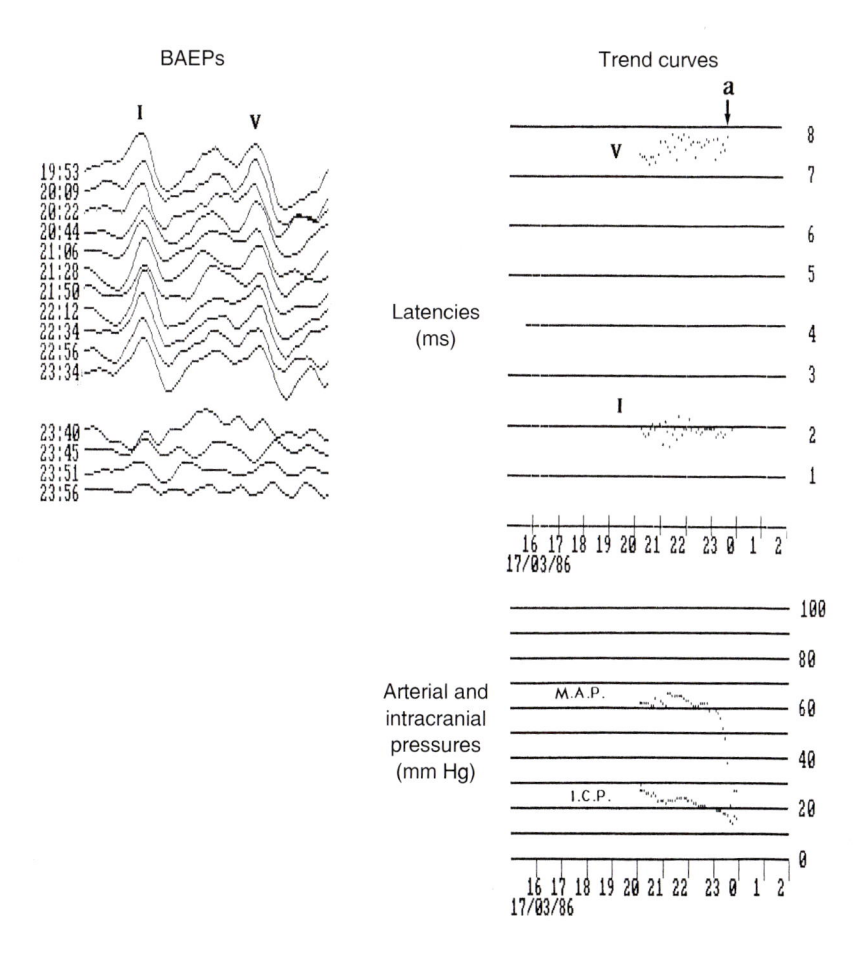

Clinical conditions affecting the central nervous system encountered in the ICU

Fig. 6.13 Sudden degradation of all BAEP waves in hypotensive shock. This patient underwent a rapid decrease of CPP due to a fall of arterial pressure (same presentation as in *Figs 6.11* and *6.12*). BAEP latencies had been abnormal since the beginning of monitoring, probably because of continuous lidocaine/ thiopental infusion (see raw traces on the left), and disappeared at 23.40 h for CPP values of about 20 mmHg without any previous significant changes of latency or amplitude (a, arrow). This kind of evolution has always been associated with sudden hypotension. (From García-Larrea et al (1987), by permission.)

often occurs before the I–V IPI has increased to statistically abnormal values. This terminal pattern does not correlate with any consistent pattern of modification of ICP/CPP, at least during its very terminal phase. When components III–V disappear, ICP and CPP values are usually still normal. Wave I persists several hours after the loss of all other components, and in most patients there is an increase of wave I amplitude concomitant with the progressive deterioration of brainstem waves (see *Fig. 6.14*).

Experimentally, this BAEP pattern has been shown to be associated with rostrocaudal deterioration of brainstem function, related to increased ICP (Nagao et al 1979, Nagata et al 1984). On the basis of these animal studies it has been claimed that latency increase of wave V could be an early sign of impending transtentorial herniation, while loss of wave V has been thought to correlate with pupillary dilatation – in patients not undergoing EP monitoring (Anderson et al 1984). In our experience pupillary dilatation has always preceded abolition of wave V. BAEPs are still present after loss of pupillary reactivity occurs, often with latencies and amplitudes within normal limits if single traces are considered (see *Figs 6.15* and *6.16*). However, continuous BAEP monitoring frequently reveals a progressive latency increase which is evident before pupillary abnormalities are detected (see below). The rapid and ultimate BAEP degradation leading to loss of waves III–V after a latency increase greater than 0.2 ms/h (see *Figs 6.14* and *6.16*) seems to occur only with terminal brainstem dysfunction, rather than being the first sign of transtentorial herniation. This is suggested by the frequent lack of ICP increase during development of this BAEP pattern, in spite of unequivocal electrophysiological evidence of brainstem deterioration.

Since the terminal phase of selective III–V loss reflects a late stage of brainstem dysfunction, it is conceivable that other, more subtle BAEP changes may bear witness to early brainstem distress. Indeed, in patients whose monitoring extended for at least 10 h before brain death, a slow latency increase of waves III and V was evident before terminal deterioration occurred (see *Fig. 6.16*), and it was during this period that loss of pupillary reactivity occurred. In some of the patients these long-lasting changes were preceded by a transient, reversible I–V latency increase which coincided with acute intracranial hypertension, although this pattern was not evident in all cases. Therefore, long-lasting, progressive BAEP changes may represent a more reliable sign of early brainstem distress, and reversible latency increases could mark the onset of such a progressive dysfunction that may lead ultimately to brain death. The significance of these transient latency increases will be discussed in the next section.

Enhanced wave I has already been reported in brainstem damage (Starr 1976, Hall et al 1985) and also during barbiturate coma in man and animals (Marsh et al 1984, Hall 1985b, Shaw 1986). Early explanatory hypotheses included inhibition of the acoustic reflex because of the brainstem lesion, but this can be excluded since curarized patients do not exhibit this phenomenon (Hall 1985, and personal observations). Interruption of inhibitory brainstem centrifugal pathways such as the olivo-cochlear bundle (Warr 1980) has also been proposed and is consistent with our monitoring results. Since enhancement of wave I occurs during the rapid terminal phase of BAEP deterioration, both seem to reflect a late, probably pontine, level of brainstem dysfunction.

Transient BAEP changes Transient, self-limited BAEP changes lasting from some minutes to hours are frequent in comatose patients. Many of these changes are, as discussed in previous paragraphs, non-pathological in origin and mainly related to alterations in temperature or drug levels. However, prompt detection of clinically relevant transient EP changes is crucial, for they may represent the earliest signs of brainstem dysfunction, especially if associated with critical alterations of other markers such as ICP or CPP. Several criteria may be proposed to distinguish changes related to CNS dysfunction from those of non-pathological origin. First, changes must occur in the absence of body temperature variations capable of accounting for them, or of blood levels of anaesthetic drugs known to affect BAEPs. This latter restriction should also apply to drug combinations whose action on BAEP has not been specifically tested. Moreover, changes should be progressive in nature and occur in at least 5 BAEPs in the ascending/descending slope of the trend plots (assuming a recording rate of 1 EP/5 min). In our experience, sudden latency variations exceeding 0.2 ms between two consecutive BAEPs are likely to be artefactual in origin and should be disregarded (see *Fig. 3.22*). It is useful to calculate linear regression coefficients on trend plots during phases of latency variation. If correlation coefficients are less than 0.6, the dispersion of individual values is probably too big and the original traces should be rechecked. Although latency changes are usually better detected on trend curves than on individual BAEPs (see *Figs 6.11*, *6.12* and *6.14*), frequent visual inspection of actual traces is imperative to check the quality of waveforms used to derive latency curves.

Applying the restrictions specified above, 25 episodes of transient, reversible latency BAEP changes have been observed so far in 18 patients of our series (*Figs 6.17* and *6.18*) (García-Larrea et al 1992). Only latencies were taken into account since amplitudes are still, in our experience, too susceptible to uncontrolled, non-pathological sources of variation.

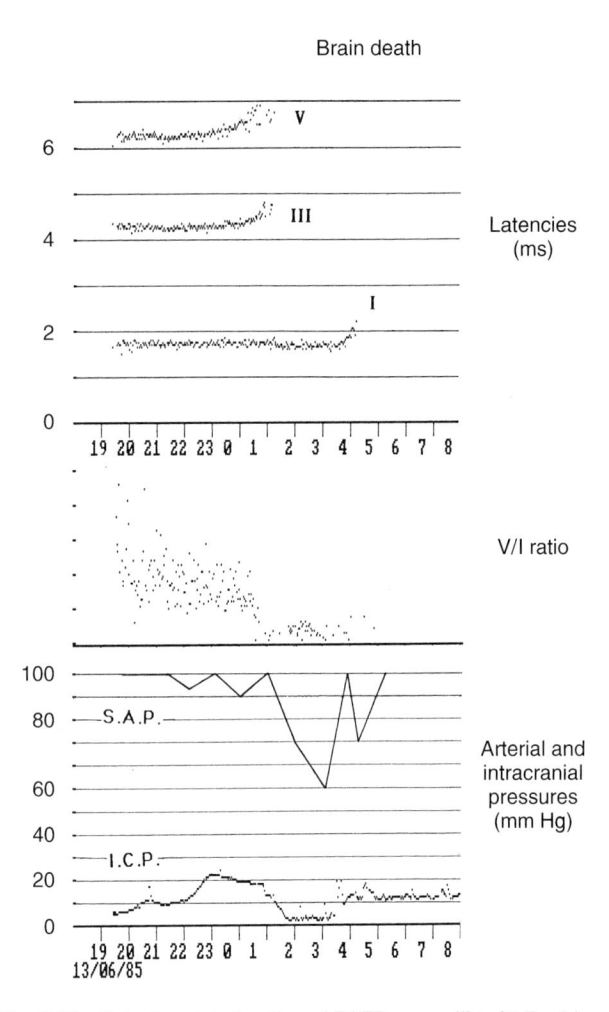

Fig. 6.14 Selective deterioration of BAEP waves III to V. Rapid latency increase of waves III and V at 22.30 h coincided with moderate ICP increase from 10 to 20 mmHg. This episode was preceded by loss of pupillary reflexes and progressive wave V amplitude loss demonstrated by fall of V/I amplitude ratio (middle traces). Wave I persisted for 4 h with enhanced amplitude until its final degradation (see *Fig. 6.22*). (From Binnie et al (2003), by permission.)

Latency changes fulfilling our quite restrictive criteria do not necessarily cause the BAEP to exceed statistical latency norms established in a control population. Indeed, in most patients BAEP latencies remained within normal limits during the whole episode, thus emphasizing that relative changes from baseline are more important during monitoring than comparisons with statistical norms.

Transient BAEP latency changes, in our experience, are significantly associated with poor outcome, even if other clinical or paraclinical parameters are not taken into consideration (*Fig. 6.19*). However, the significance of transient latency increase is best assessed when these are studied in association with ICP and CPP, since EPs can provide physiological evidence of the adverse effects of changes in arterial or ICP on brainstem function. Only 42% of acute CPP or ICP abnormalities in our series were accompanied by an increase of I–V IPI, suggesting that not all 'significant' pressure changes lead to functional

AMPLITUDE INCREASE OF WAVE I IN BRAIN DEATH

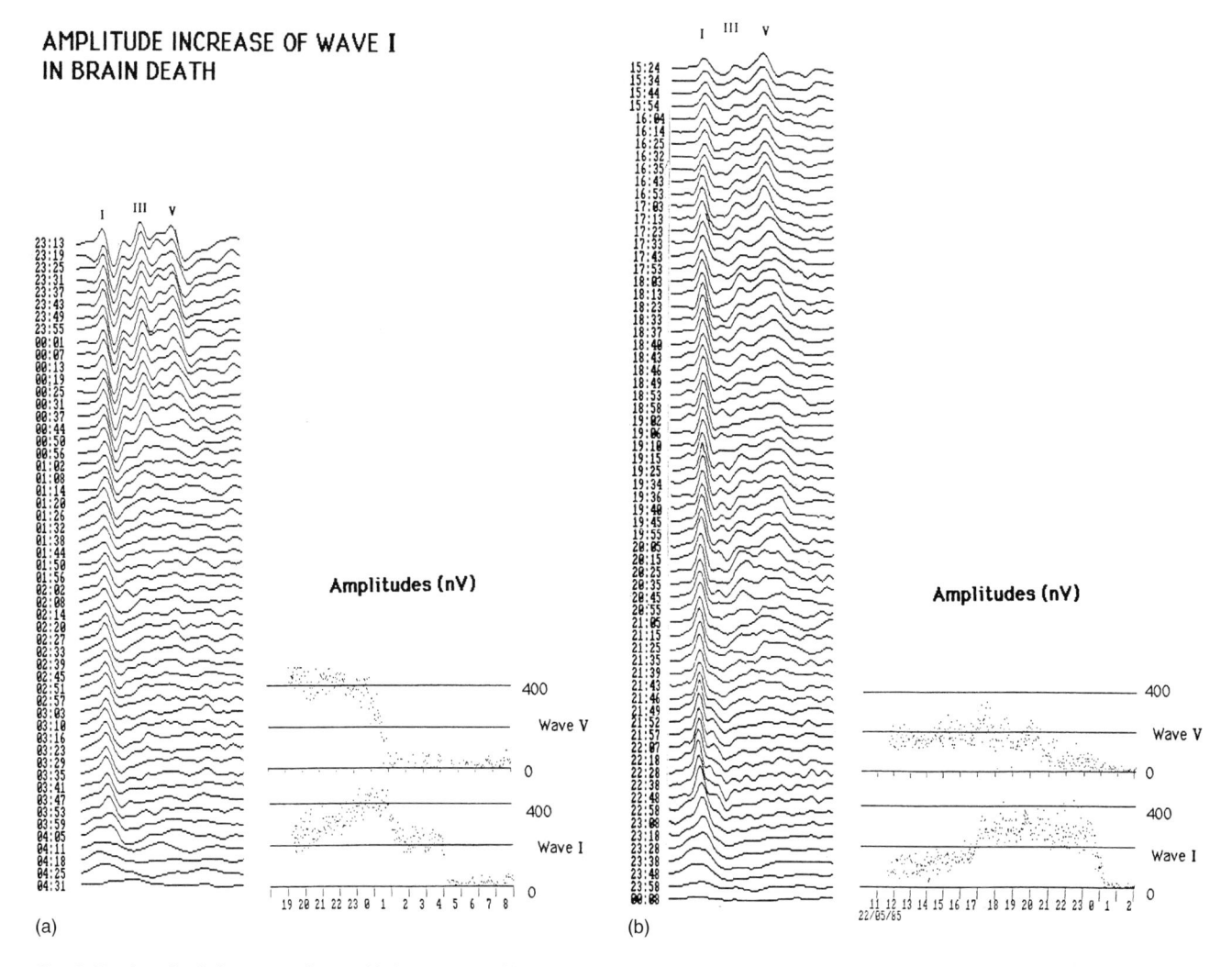

Fig. 6.15 Amplitude increase of wave I in brain death, (a) A case where wave I amplitude enhancement coincided with progressive deterioration of all brainstem components, (b) In this case progressive enhancement of wave I began 5 h before the terminal decrease of waves III and V. This BAEP pattern was preceded in both cases by pupillary areactivity and was followed by a rapid deterioration of clinical status and brain death. (From García-Larrea et al (1987), by permission.)

abnormalities of the brainstem. When acute ICP changes were accompanied by transient wave I–V IPI increase (see *Fig. 6.19*), subsequent clinical evolution was almost always bad, with more than 90% followed by brain death in the following hours or days. Outcome was significantly better (about 20% of brain death or PVS) in patients whose episodes of ICP increase did not produce BAEP changes (*Fig. 6.20*).

The reason why acute intracranial hypertension does not systematically lead to BAEP abnormalities is difficult to ascertain by EP recordings alone. When mass lesions impede or prevent communication between supra- and infratentorial compartments, ICP increases may create pressure gradients which accelerate the development of diencephalic herniation (Kaufmann and Clark 1970, Plum and Posner 1980). BAEP changes appearing concomitantly with an ICP increase could therefore indicate the existence of such a pressure gradient leading to mechanical distortion of brainstem structures. Conversely, waveform stability during and after ICP increase should indicate a good cerebrospinal fluid (CSF) communication between supra- and infratentorial compartments, which lessens the probability of mechanical distortion of the brainstem.

Development of intraparenchymal haemorrhages secondary to transient brainstem distortion is a well known phenomenon (Johnson and Yates 1956, Hassler 1967, Klintworth 1968, Plum and Posner 1980) that could perhaps explain the slow latency increase observed after acute episodes of I–V alteration, and which preceded terminal BAEP changes in some of our patients. At the time when our recordings were obtained these changes did not lead systematically to a modification of patient's management; therefore we cannot yet ascertain whether they are the first sign of a potentially reversible process, or on the contrary implicate an irreversible progression to brain death. What seems certain is that BAEP monitoring provides important information as to whether a particular episode of ICP increase is affecting brainstem function or not. The asso-

Chapter 6 Neurophysiological Work in the ICU

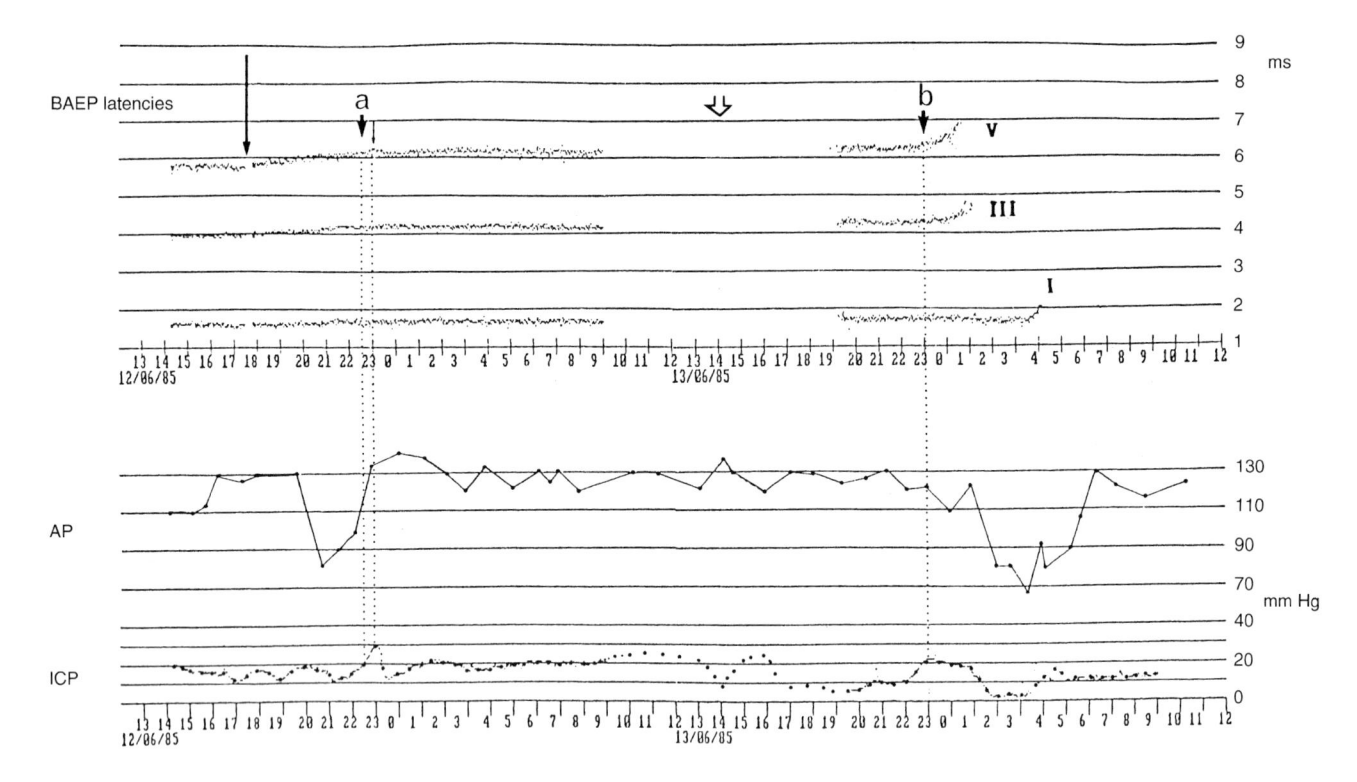

Fig. 6.16 Forty hours' preterminal BAEP evolution in a comatose patient. A slow progressive latency increase of waves III and V could be recorded from 17.30 h (left arrow). At 19.30 h a sudden fall of arterial pressure did not affect BAEP latencies. Less than 30 min later there was a small latency shift of wave V, time-locked to a 20 mmHg ICP increase (a, left arrow). BAEP normalized immediately when the ICP fell after mannitol injection (a, right small arrow) and remained stable until monitoring was discontinued at 09.00 h the next day. When BAEP monitoring was recommenced at 19.15 h on the same day, I–V IPI had increased slightly from 4.3 to 4.5 ms, and at 23.00 h (arrow b) a small increase of ICP from 10 to 20 mmHg was paralleled by rapid latency increase and ultimate abolition of waves III and V. Pupillary areactivity occurred at 14.00 h (open arrow), suggesting that the rapid BAEP latency increase at b represents a late, irreversible stage of brainstem damage. (From Binnie et al (2003), by permission.)

ciation of acute ICP increases with BAEP modifications, even if both are transient and totally reversible, is a reliable sign of alarm. The high mortality rate observed under such conditions (90%) probably justifies the adoption of aggressive therapeutic measures. This applies even to pressure modifications which do not reach standard 'significant' values, as with those shown in *Fig. 6.21*, for BAEP latency changes in that context seem a more reliable warning than absolute pressure levels.

Conclusions on BAEP monitoring BAEP monitoring is a relatively new technique that has proved useful in following both rapid and long-lasting BAEP changes, but presently with some limitations and uncertainties. In the case of sudden haemodynamic deterioration due to acute arterial hypotension or cardiac arrest, BAEP monitoring can only provide evidence of sudden disappearance of all components, which is mainly related to peripheral auditory damage. Conversely, in patients with slower preterminal BAEP deterioration, monitoring is a reliable indicator of brainstem dysfunction even when other parameters such as CPP or ICP remain within the normal limits. Even when baseline responses are abnormal – i.e.

those obtained at the beginning of the monitoring session – tendencies towards deterioration or improvement can still be detected.

Progressive increase in all BAEP latencies is also observed during hypothermia; consequently temperature variations should be systematically controlled when monitoring BAEPs. Similar changes can also be recorded in patients treated by a combination of lidocaine and sodium thiopental, thus emphasizing the additive effects of sedative drug combinations on BAEPs. Since in most studies the effect of drugs on EPs have been tested with single agents, caution must be exercised when extrapolating these results to patients under sedative polytherapy.

The two distinct BAEP preterminal modifications identified so far undoubtedly coexist in many patients. However our results show that they can also occur separately and that they are associated with different pathophysiological mechanisms, namely (a) ischaemia of all posterior fossa auditory structures, including their peripheral section and (b) intrinsic brainstem dysfunction (usually, but not always, secondary to intracranial hypertension). Early identification of these modalities of deterioration is a prerequisite for a rational therapeutic

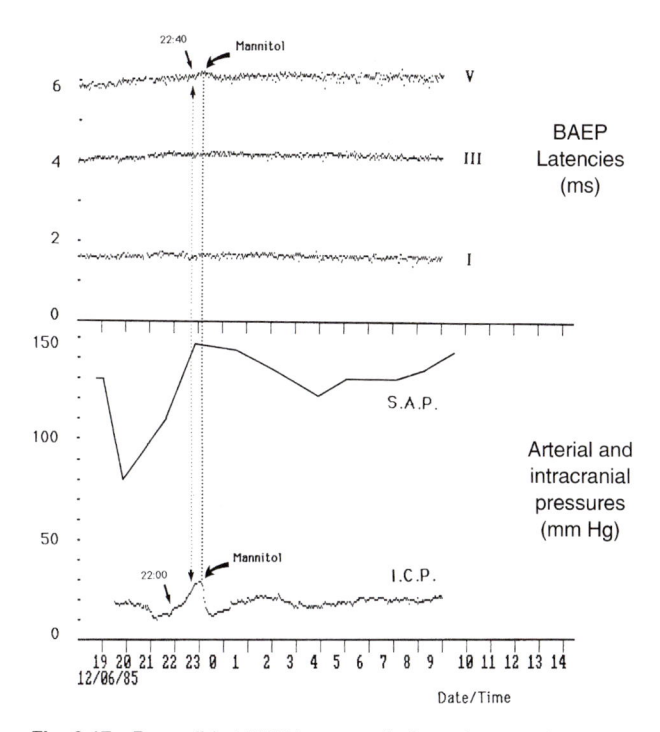

Fig. 6.17 Reversible I–V IPI increase (I). Acute intracranial hypertension at 20 h (from 10 to 30 mmHg) was paralleled by a small latency increase of wave V, which reversed when ICP was diminished after a mannitol injection (arrows). This transient episode was superimposed on a long-lasting, progressive increase of the I–V interval which had started 3 h before and pursued its course after the episode was over. The patient died 30 h later. (From Binnie et al (2003), by permission.)

approach in such patients. Whereas the former pattern is always well correlated with haemodynamic changes, the latter is not accompanied by systematic ICP or CPP alterations; in such cases (and perhaps also in the former) the rapid terminal phase probably reflects a late and irreversible stage of brainstem deterioration, the first manifestation of which must be looked for several hours before, possibly in the form of transient BAEP changes followed by progressive latency increase. Indeed, I–V IPI increase, even if totally reversible, has proved to be highly adverse when concomitant with episodes of intracranial hypertension, as brain death has followed within a few days in more than 90% of our patients. We suggest that this is one of the first electrophysiological signs of brainstem dysfunction, which merits energetic therapeutic efforts even if ICP, CPP or clinical changes do not reach levels considered as significant.

EPs combined with other variables

In addition to CNS pathology, EP changes during monitoring can result from variations in stimulus characteristics, body temperature, blood levels of anaesthetic drugs and possibly other, not yet identified variables. Simultaneous monitoring of a number of such parameters is often indispensable, since EP abnormalities resulting from their variation can mimic changes seen under pathological conditions (see below) and lead to unnecessary alarms.

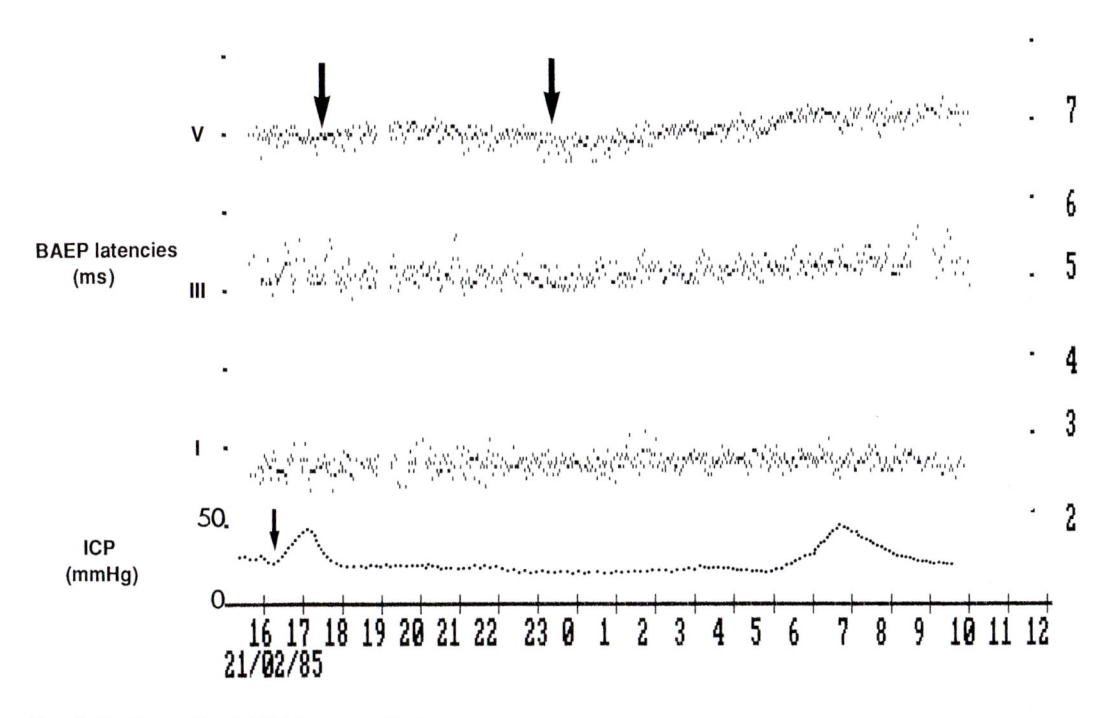

Fig. 6.18 Reversible I–V IPI increase (II). The onset of transient BAEP latency increase was roughly coincident with acute intracranial hypertension (left arrows), but the changes persisted several hours after ICP had regained basal values. The end of the episode (right arrow) was followed by a long-lasting latency increase of waves III and V which still persisted when monitoring was discontinued at 10.00 h. Note that I–V IPI remained within normal limits (between 4 and 4.3 ms) throughout the episode. The patient died 15 h after discontinuation of monitoring. (From Binnie et al (2003), by permission.)

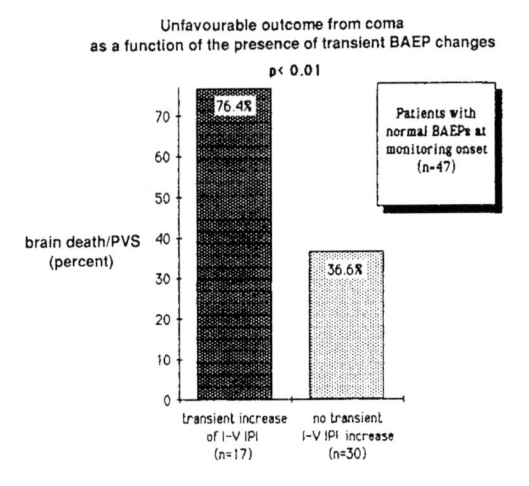

Fig. 6.19 Mortality and PVS incidence are significantly higher in patients with episodes of I–V IPI increase (excluded those linked to hypothermia or drugs). Data obtained from 47 comatose patients with normal BAEPs at monitoring onset. (From Binnie et al (2003), by permission.)

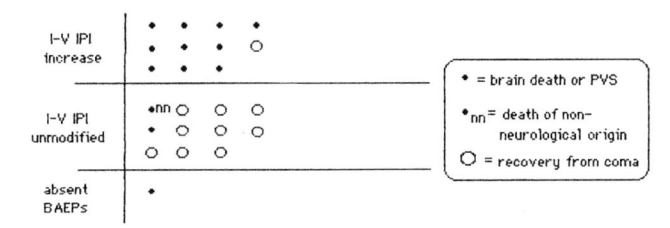

Fig. 6.20 Outcome in patients with episodes of acute abnormal ICP or CPP, as a function of the appearance of central BAEP latency increase. Note that concomitant pressure and BAEP changes were followed by death in 10 out of 11 patients, whereas mortality was low when BAEP remained stable during ICP/CPP episodes. In three other patients with transient ICP increase high levels of lidocaine prevented reliable interpretation of BAEPs. (From Binnie et al (2003), by permission.)

On the other hand, pathological EP changes during monitoring are best assessed when correlated with abnormalities of other physical or physiological markers, such as ICP, CPP, heart rate, etc. Several monitoring systems have already been described that allow simultaneous acquisition of analogue physiological signals, which are digitized and averaged simultaneously with EPs (Maresch and Pfurtscheller 1983, Bertrand et al 1987, Bricolo et al 1987). The combination of EEG and EP monitoring is increasingly used to achieve the separate but parallel tasks of identifying changes in cortical function (especially the occurrence of epileptiform discharges, and the suppression produced by drugs or inadequate cerebral perfusion or oxygenation) and the integrity of neural pathways with EPs. Monitor displays combining bilateral compressed spectral arrays (CSAs) and various EP modalities have been proposed as a convenient way of assessing such monitoring (see *Figs 3.23* and *8.33*) (Pfurtscheller et al 1987). These have been used in a range of intraoperative applications during neurosurgery, often in combination with display of the signals more conventionally monitored in these circumstances.

CARDIAC ARREST AND HYPOXIC–ISCHAEMIC ENCEPHALOPATHIES

At the time of cardiac arrest there are clear EEG indicators of the cerebral effects of abrupt cessation of cerebral perfusion (Young and Ornstein 1985). When prompt resuscitation restores an adequate circulation to the brain, the EEG returns quickly and the patient recovers without evidence of damage (*Fig. 6.22*). Conversely, if resuscitation is either delayed or is ineffective in restoring an adequate supply of oxygenated blood to the brain, then the neurological and neurophysiological restoration is delayed and permanent sequelae may ensue.

Cerebral outcome after resuscitation from cardiac arrest has been related to the rate of recovery of functions dependent on brainstem reflexes; predictive accuracy is increased by including variables based on EEG and SEPs (Edgren et al 1987, Bassetti et al 1996, Chen et al 1996), although some studies suggest that prediction of minor degrees of damage is difficult (Rothstein et al 1991). *Table 6.5* gives the longest times for ECS compatible with various grades of functional recovery based upon detailed studies of 613 patients who were examined intensively from the time of resuscitation (Jørgensen and Malchow-Møller 1978, 1981). In a subsequent study of 231 patients in whom EEG recordings had begun immediately after resuscitation, Jørgensen and Holm (1998) reported that the presence of any EEG activity was associated with survival at one year in 33% of patients, compared with survival in only 16% of those who showed ECS. Apart from research studies, sequential testing of brainstem reflexes is generally fairly impractical in the first minutes or hours after resuscitation, although the rate of return of intermittent and then continuous EEG activity (*Fig. 6.23*) may be more readily determined if an EEG monitor is available in the ICU. Considerable predictive accuracy about the presence or absence of hypoxic brain damage was achieved with a method used to develop discriminant scores based on a cluster of EEG variables, coma score and interval since resuscitation in a study of 93 EEGs from patients after cardiac arrest (Binnie et al 1970). When the rate of improvement of sequential scores was taken into account in a larger series of 916 EEGs from 371 patients, a clearer indication of the degree of brain damage was found (*Fig. 6.24*). All these studies give clues as to possible approaches that might usefully be developed in algorithms for more advanced monitoring equipment where learning systems compare rates of improvement with that found in material from appropriate databases.

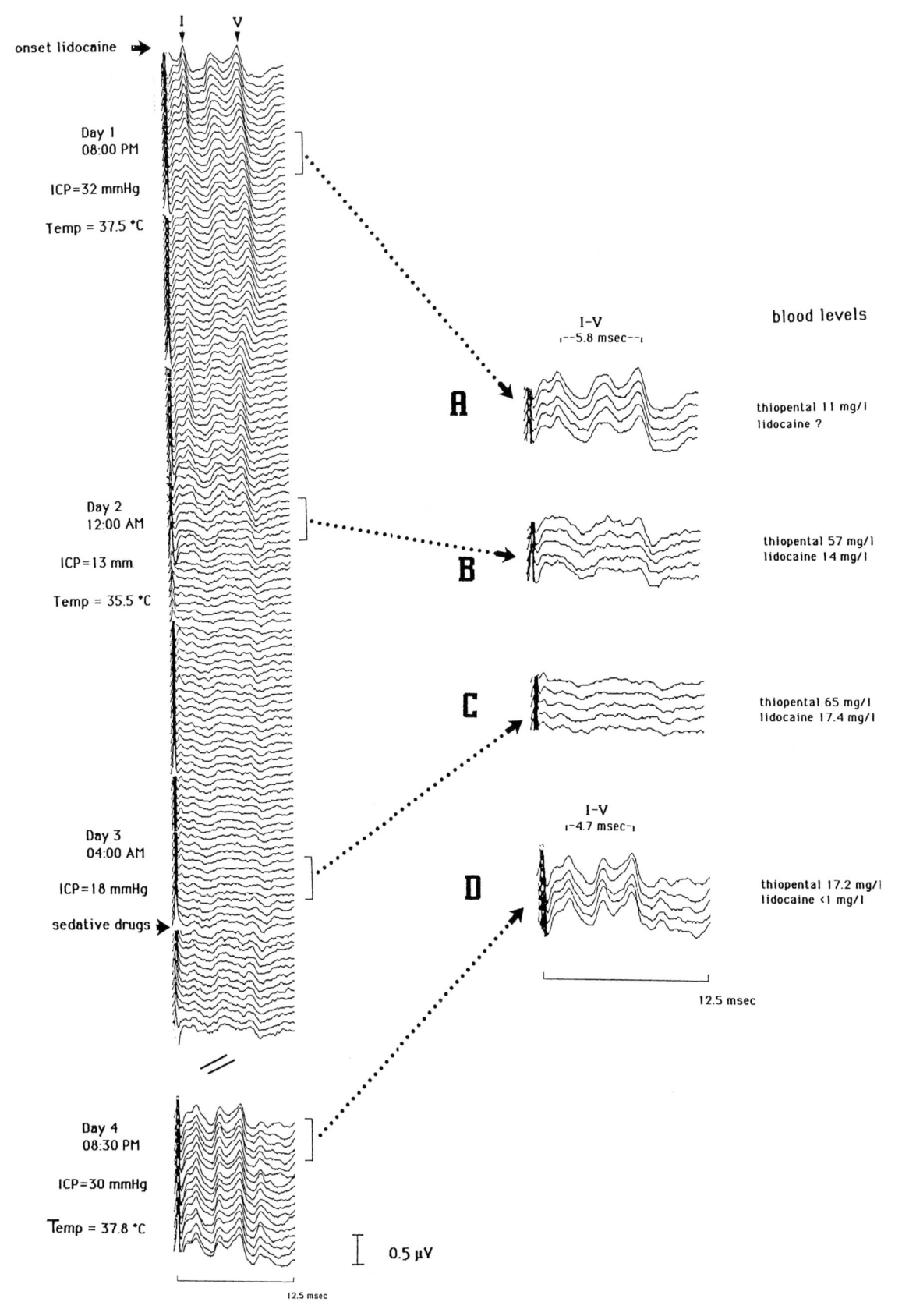

Clinical conditions affecting the central nervous system encountered in the ICU

Fig. 6.21 Major BAEP deterioration under high doses of thiopental and lidocaine in a 6-year-old boy after head injury. Increasing blood levels of lidocaine and sodium thiopental, given in continuous intravenous infusion, were associated with a progressive deterioration of BAEPs manifested by latency increase and amplitude decrease of all components (B). BAEPs were almost abolished at 65 mg/l of thiopental and 17.4 mg/l of lidocaine (C). Interruption of drug infusion was followed by complete recovery of normal BAEP waveforms (D). The patient regained full and independent normal life. (From Binnie et al (2003), by permission.)

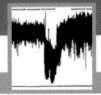

OTHER METABOLIC AND TOXIC ENCEPHALOPATHIES AND MULTIPLE ORGAN FAILURE

The need for objective, reproducible, measurements to indicate change in the state of encephalopathic patients led to an early application of spectral analysis, most importantly in patients with hepatic disease.

The reduction in inter-operator variability, combined with reliability of parameters correlated with mental status, allows detection of early encephalopathy (Amodio et al 1999), replicating the classification based upon visual scoring of Van der Rijt et al (1984). In renal encephalopathies, values from syntactometric analysis of frequency spectra have been related to clinical state and levels of creatinine and urea in renal encephalopathies (Hernandez Sande and Arias Rodriguez 1985).

Sepsis-related encephalopathies have been examined by Young et al (1992) who classified EEGs from 62 patients into five groups matched according to severity. Features of importance were increasing slow components, appearance of triphasic waves and suppression or burst suppression patterns. It was found that the EEG was more sensitive than clinical criteria for encephalopathy, showed identifiable features compatible with reversibility, and well-defined features correlating with mortality. The sequence of EEG changes, in common with that in many types of encephalopathy encountered in the ICU, has considerable implications for the design of algorithms for monitoring.

When multiple organ failure occurs, encephalopathic changes are additive in their effect on monitor tracings (as they are on mortality), with increasing deterioration in activity that cannot be adequately explained by accumulation of sedative drugs. Typical changes are monotonous slowing, loss of reactivity and cyclical patterns leading to a marked reduction in total EEG power, suppressions, and eventual EEG silence in irreversible cases.

ENCEPHALITIS

Typical EEG monitor findings in encephalitis are consistent with the depth of coma. Exaggerated cyclical changes have already been mentioned on page 192 (see *Figs 3.11* and *6.5*); florid epileptiform discharges may also be evident. Periodic discharges may be present in herpes simplex encephalitis, subacute sclerosing panencephalitis and classical Creutzfeldt–Jakob disease. This phenomenon may be evident in EEG monitor recordings in the time domain, depending on the relative amplitudes of periodic complexes and intervening activity and the degree of time compression in the display. In terminal states with residual complexes on an isoelectric background, the monitor trace may be indistinguishable from a burst suppression pattern.

EPILEPTIFORM DISCHARGES AND STATUS EPILEPTICUS

EEG confirmation of the nature of epileptiform discharges in status epilepticus is always an emergency procedure and requires skilled recording technique and immediate interpretation by the clinical neurophysiologist at the bedside. The ensuing discussion with the ICU team will help planning concerning the appropriateness of a trial of intravenous therapy under continuing EEG control and of subsequent EEG–polygraphic monitoring to guide therapeutic management.

In the ICU, status epilepticus will be encountered in two quite different clinical settings (see review by Walker

Table 6.5 Outcome after cardiac arrest in 613 patients

Clinical and EEG features	Longest delays compatible with		
	Any functional recovery	Recovery of consciousness	Full recovery at 1 year
Respiratory movements	15 min	7 min	
Pupillary light reflex	28 min	12 min	
Coughing/swallowing reflex	58 min	23 min	
Caloric vestibular reflex	15 h	2.75 h	
Decorticate posturing	9 h	3 h	
Stereotyped reactivity	7.5 h	3 h	
Intermittent EEG	7.5 h	3.3 h	3 h
Continuous EEG	17 h	10.5 h	
Consciousness			3 days
Speech			6.5 days
Coping with personal needs			2 weeks

Based on Jørgensen and Malchow-Møller (1978, 1981), by permission.

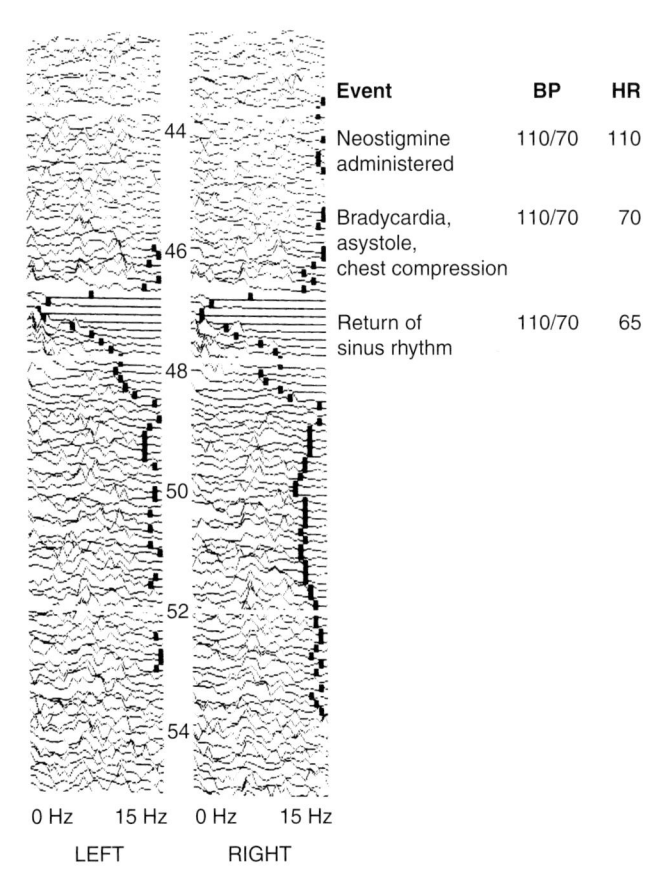

Fig. 6.22 Cardiac arrest and resuscitation at the end of surgery for left carotid artery stenosis. EEG monitoring over a period of 12 min with left- and right-sided CSAs (8 s analysis epochs, display up to 15 Hz) and spectral edge frequency markers. Note abrupt decrease in higher frequencies and loss of virtually all EEG components following asystole, followed by gradual recovery over the following 2 min. No postoperative neurological deficit was observed. (From Young WL, Ornstein E 1985 Compressed EEG monitoring during cardiac arrest and resuscitation. *Anesthesiology* 62:535–538, by permission.)

2003). The first are patients with a history of epilepsy, or epilepsy presenting *de novo*, with status epilepticus. The second are patients requiring intensive care in the context of a major medical or surgical emergency who develop seizure discharges secondary to that condition, including those with severe encephalitis or post-traumatic coma with major cerebral lesions or sequelae following evacuation of haematomata. In many of these, seizures may prove slow to respond to anticonvulsant medication and the EEG may show localized origin for them; an example is given in a child with cerebral malaria (*Fig. 6.25*). A series of studies from the Kilifi group in Kenya showed that almost two-thirds admitted to hospital with cerebral malaria had seizures, half with periods of status epilepticus; 84% were complex seizures, and 47% of these were partial and over 70% repetitive; they were considered to be a significant cause of longer term morbidity in malaria endemic areas (Crawley et al 1996, 2001, Waruiru et al 1996).

The EEGs of comatose patients with acute cerebral disorders (in the context of cardiac arrest, infection, metabolic disorder, etc.) may have three distinct patterns of epileptiform discharge, regardless of aetiology (Lowenstein and Aminoff 1992). These are: (1) repetitive electrographic seizure discharges or bilateral continuous spike-and-wave activity typical of status epilepticus; (2) irregular slowing with frequent spikes and sharp waves or irregular mixed frequency background activity with episodic accentuation, or diffuse slowing, or intermittent burst suppression pattern, or (3), in patients without any clinical evidence of seizures, repetitive electrographic seizures or continuous spike-and-wave activity or repetitive sharp waves. This latter group are often described as having non-convulsive status epilepticus or electroclinical dissociation and they represent a high proportion of patients in the ICU in whom electrographic seizures are found (Claassen et al 2004b); of course, in ICU patients, this may be due to masking of clinical expression of seizures by major sedative, anaesthetic or muscle relaxant drugs (*Fig. 6.26*). Seizures in postanoxic coma are often myoclonic and may be stimulus sensitive; typically the bursts of generalized spike and polyspike discharge appear on a background devoid of any other form of EEG activity (Van Cott et al 1996). The wide, and often changing, range of patterns of discharge mean that conventional multichannel diagnostic EEG recording is essential for initial assessment and for regular surveillance. Any form of automatic monitoring used to quantify discharges to assess response to treatment, must be demonstrably able to reliably distinguish degraded or idiosyncratic discharges in the individual patient.

Whether status epilepticus is in the context of a seizure disorder or an acute illness, it is essential that the patient is treated urgently until all discharges are controlled to avoid the serious consequences of brain damage or death. Apart from factors such as age and aetiology, the response to treatment and the outcome in neuro-ICU patients depends on the delay in identification and the duration of seizure discharges, mortality figures between 33% and 57% being reported (Young et al 1996). The control of intravenous therapy is normally by EEG monitoring (*Fig. 6.27*). In their reviews of the genesis and management of major status epilepticus in the context of critical care, Payne and Bleck (1997) and Walker (2003) emphasize the risk of brain damage and its mechanisms. Hypoxia, hypotension and hyperthermia are significant factors, along with the high neuronal metabolic burden of the massive epileptiform discharges, in contributing to the genesis of ischaemic brain damage and death in patients with uncontrolled status epilepticus.

Some controversy has arisen as to whether treatment regimens involving depression of cortical activity to the level of EEG burst suppression by anaesthetic agents equates with a true anticonvulsant effect. Ancillary treatment to ensure adequate blood levels of appropriate anticonvulsant drugs usually runs in parallel with the emergency control of status epilepticus with regimens of major sedation or anaesthesia, artificial ventilation and possibly muscular paralysis. Once burst suppression has

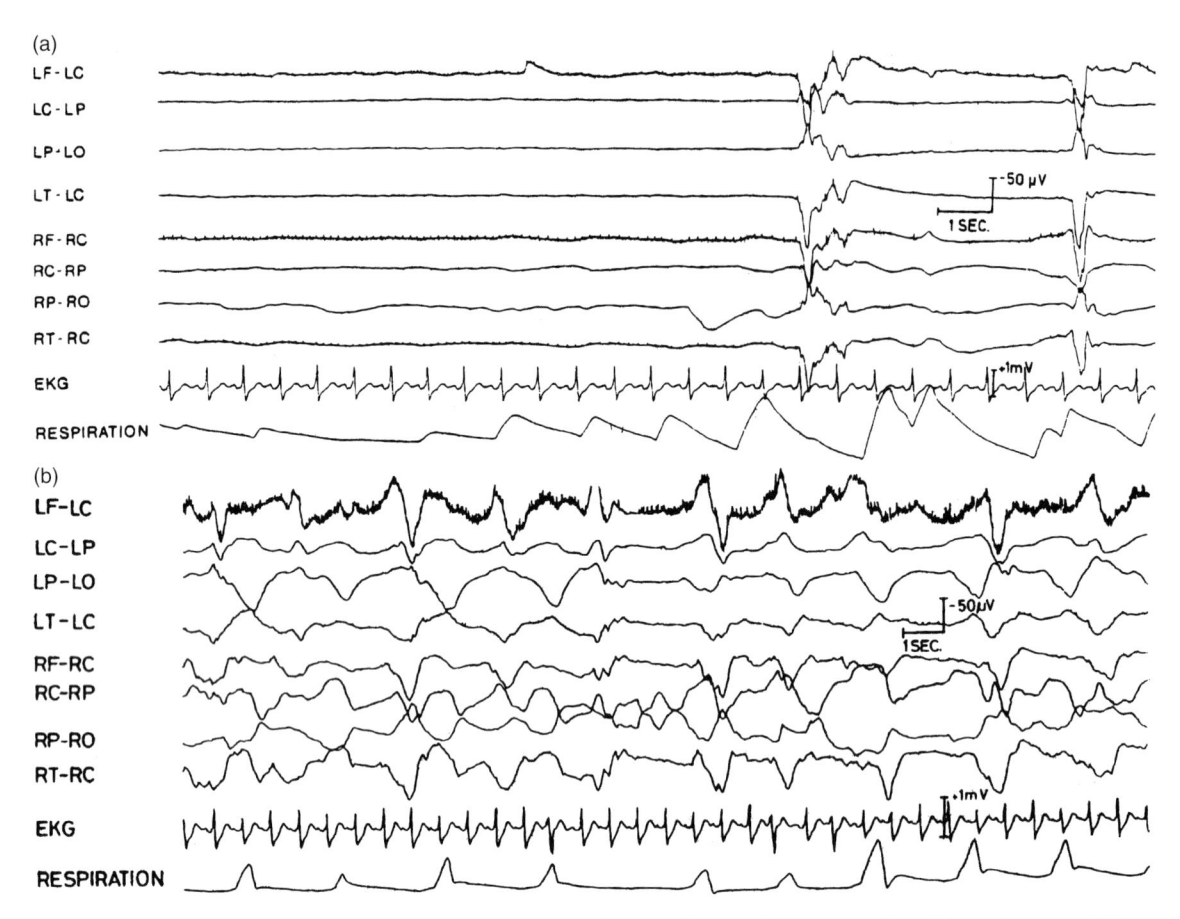

Fig. 6.23(a,b) Evolution of EEG changes after resuscitation from cardiac arrest. Two recordings taken after resuscitation from circulatory arrest in a 69-year-old woman with myocardial infarction. (a) At 1 h the EEG showed intermittent delta activity with spikes and sharp waves at a time when the patient had Cheyne–Stokes' respiration; all cranial nerve reflexes except the caloric vestibular ones could be elicited and decorticate posturing occurred spontaneously or following cutaneous stimulation. (b) After 13 h the EEG showed continuous delta activity, the caloric vestibular reflex was still absent and somatosensory stimulation elicited stereotyped motor responses. The patient deteriorated and lost her electrocortical activity during a second, reversible, cardiac arrest; spontaneous breathing, coughing and swallowing movements and decerebrate postural responses continued until final asystole 88 h after the primary resuscitation. (From Jørgensen and Holm (1998), by permission.)

been achieved, some clarification may be required as to the role and interpretation of EEG monitoring; it does, however, regain its importance during weaning from the emergency regimen and the establishment of the functional adequacy of anticonvulsants in controlling discharges (van Ness 1990, Jäntti et al 1994, Payne and Bleck 1997).

The considerable body of experience in the world of epilepsy monitoring has shown that detection of seizure discharges and warning of clinical seizures can be assisted by use of automatic systems based on signal processing of the EEG and other variables. However, in the context of the general or neuro-ICU even the simple application of trend analysis using EEG envelope detection (*Fig. 6.28*) can be helpful. Their application to patients in status epilepticus is germane to management of this serious condition in the ICU. The impact of and methodology for EEG monitoring has been assessed in both paediatric and adult practice for patients with either convulsive or non-convulsive status epilepticus (Tasker et al 1989, Altafullah et al 1991, Jordan 1994, 1999, Young et al 1996). Methods for monitoring and automatic seizure detection in the neonate require special consideration because of the more subtle phenomenology of the EEG discharges (Gotman et al 1997a,b).

In spite of training schemes to help non-expert staff to recognize seizure discharges, Leira et al (2004) reported that their recognition by bedside caregivers was disturbingly low. They asked 50 staff (neurology and neurosurgery residents, intensive care fellows, critcal care and neurological floor nurses, and EEG technicians) to answer questions on EEG samples before and after undertaking a PowerPoint training programme. Analysis of 2398 EEG responses showed that pretraining responses were correct in 46–94%, depending on the subgroup of participants; EEG technologists and neurology residents having the highest scores. Training improved performance to give a range of correct answers of 52–93%. From experience in neonatal seizure detection, the work of Gotman et al

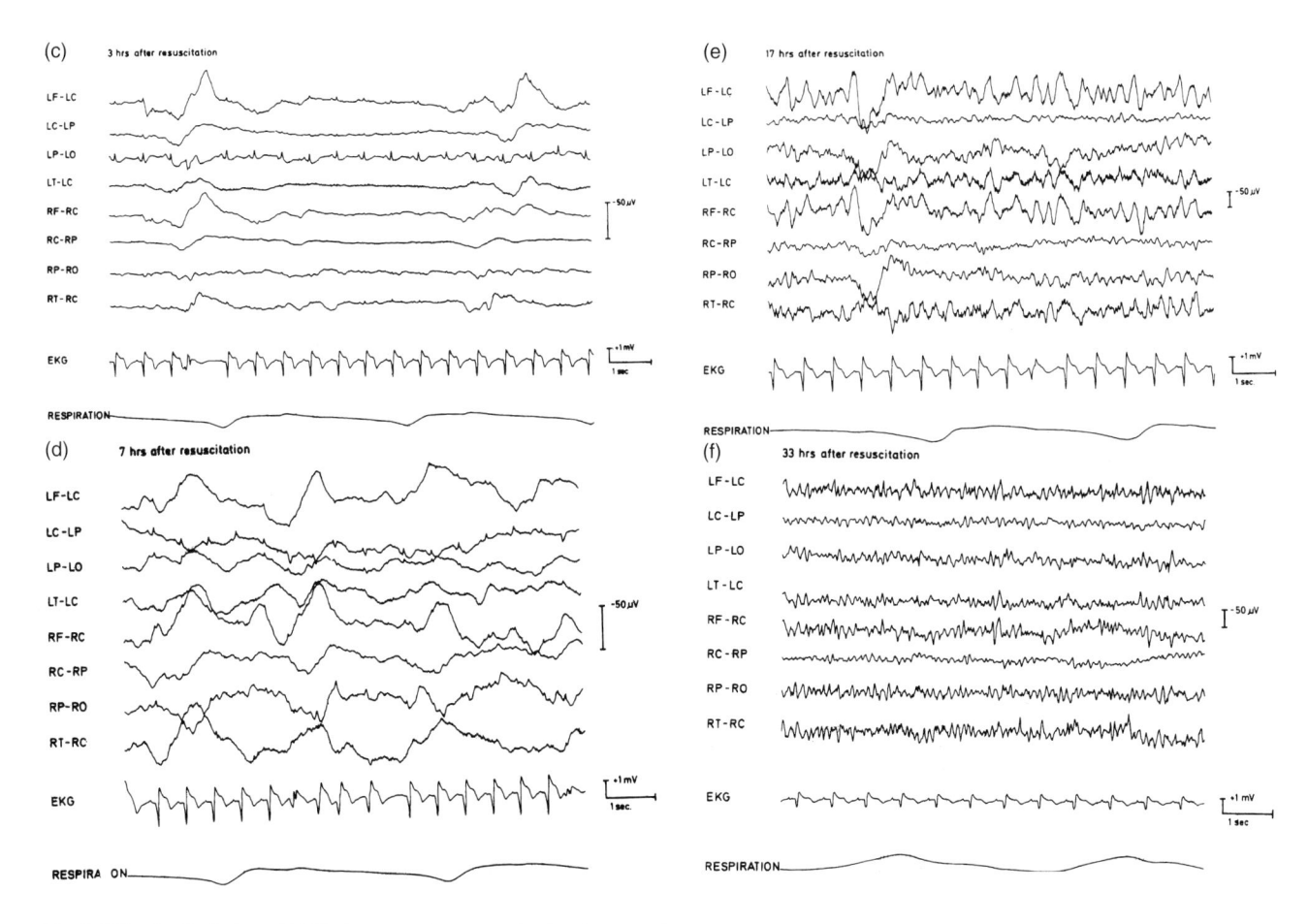

Fig. 6.23(c–f) Evolution of EEG changes after resuscitation from cardiac arrest. Sequence of four recordings in a 59-year-old man after resuscitation from circulatory arrest following myocardial infarction; initially there was no detectable cortical activity in the EEG but all cranial nerve reflexes and decerebrate posturing could be elicited. (c) At 3 h, intermittent delta activity appeared with large amplitude in the frontal leads and cutaneous stimulation elicited decorticate posturing. (d) After 7 h the EEG showed continuous delta activity whilst clinical examination revealed stereotyped reactivity. (e) At 17 h continuous theta activity with largest amplitude frontally. (f) At 33 h continuous alpha activity was widespread. The patient recovered consciousness after 29 days and gradually began to utter syllables but remained unable to sit, stand or walk, and required assistance with all personal needs until final asystole 79 days after resuscitation. (From Jørgensen and Holm (1998), by permission.)

(1997a,b) suggests that there is a place for automated detection of EEG seizure discharges in providing more consistent detection.

HEAD INJURY AND OTHER NEUROSURGICAL APPLICATIONS

Modern guidelines for the management of severe head injuries are now fairly standardized (Bullock et al 1996, Maas et al 1997), with emphasis on proper control of raised ICP and maintenance of adequate CPP (above 60–70 mmHg). Sedation remains an important tool, its management being aided by EEG monitoring.

Secondary brain injury remains one of the most important determinants of outcome in those surviving severe trauma (e.g. Miller et al 1992, Chesnut et al 1993). This governs the principles of early resuscitation and subsequent management (Miller et al 1994, O'Sullivan et al 1994, Kirkpatrick et al 1995, Robertson et al 1995), including exploration of pharmacological therapies (Marshall and Marshall 1995, Miller and Marshall 1996). Factors strongly associated with ischaemic brain damage and eventual outcome include hypotension and hypoxia, raised ICP, intracranial haematomas and multiple injuries (Aldrich et al 1992), with problems being compounded by failure of autoregulation and loss of carbon dioxide sensitivity of cerebral blood flow (Marion and Bouma 1991). Extracranial complications include pneumonia, hypotension, coagulopathy and sepsis (Piek et al 1992).

Identification of transient or sustained episodes with failure of cerebral perfusion and of understanding the cascade of precipitating events can be assisted by polygraphic monitoring of cardiorespiratory, temperature and intracranial parameters. Computer-supported multimodal systems at the bedside have performed well in research studies and, apart from simple recording of time trends of monitored variables, have calculated key parameters for

Chapter 6 Neurophysiological Work in the ICU

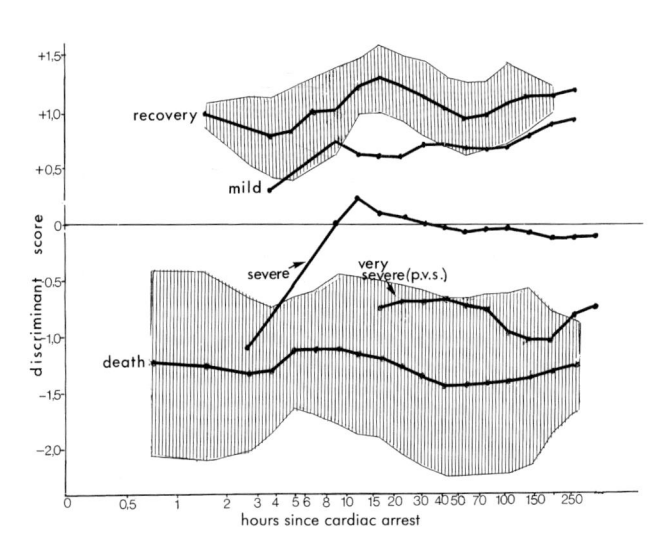

Fig. 6.24 Time course of group mean serial discriminant scores in 371 patients (916 EEGs) following cardiac arrest, grouped according to outcome (P Prior, unpublished data, 1991). Scoring based on non-linear discriminant function analysis of 11 EEG and two clinical variables (Prior 1973). Group outcomes, from above down, are: 19% recovered fully, 6% survived with mild brain damage, 4% with moderate or severe brain damage, 1.5% with PVS, and 64% died without regaining consciousness. In 5.5% of patients accurate information about outcome was not available. Shaded areas represent ± 1 SD from the mean for recovery and death groups. Note that time from cardiac arrest is shown on a logarithmic scale. The scoring is such that the initial isoelectric state, which would inevitably occur at the time of cardiac arrest, would theoretically attract a strongly negative discriminant score. (From Prior P 1993 Cerebral anoxia: clinical aspects. In: *Electro-encephalography. Basic Principles, Clinical Applications and Related Fields*, 3rd edn (eds E Niedermeyer, FH Lopes da Silva), Williams and Wilkins, Baltimore, pp. 431–444, by permission.)

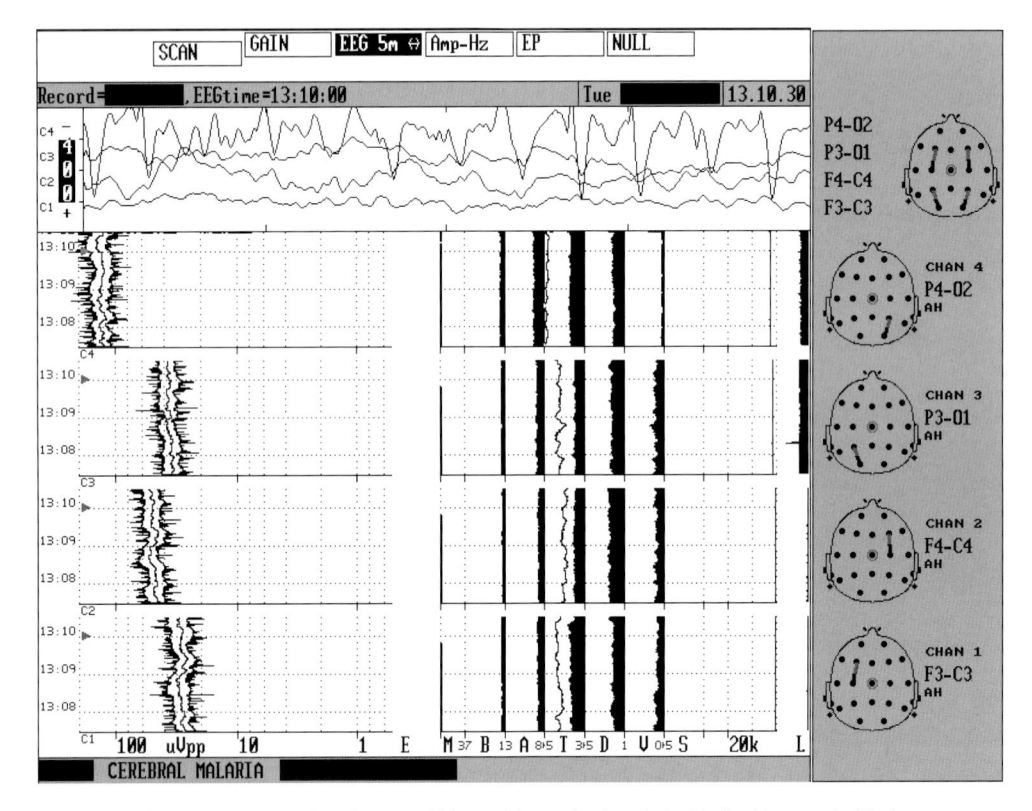

Fig. 6.25 Status epilepticus in a 3-year-old boy with cerebral malaria. He had been admitted unconscious following a 5-day history of fever, with two generalized convulsions within 24 h of admission. He was unconscious (unable to localize a painful stimulus), but without localizing or lateralizing signs and within 2 h of admission (12.08 h), he had another generalized convulsion lasting 5 min. This 3 min extract from a 4-channel CFAM3 monitor starts an hour later and shows, from above downward, simultaneous 4 s EEG samples from alternate right and left sided parieto-occipital and frontocentral regions, followed by simultaneous CFAM amplitude and frequency analysis outputs from the same derivations (mean frequency superimposed on frequency analysis), the timing of the EEG samples being marked on each analysis trace by an arrow. Note the continuous repetitive sharp wave discharge arising from the right parieto-occipital region and the markedly increased amplitude analysis output from the same area. The child was able to localize pain after 8 h and had one further short fit at 18 h, his recovery thereafter was uneventful. Although this 4-channel monitor recording highlights localization and the nature of the localized activity, a full diagnostic EEG is also essential. (Courtesy of Dr CRJC Newton. From Binnie et al (2003), by permission.)

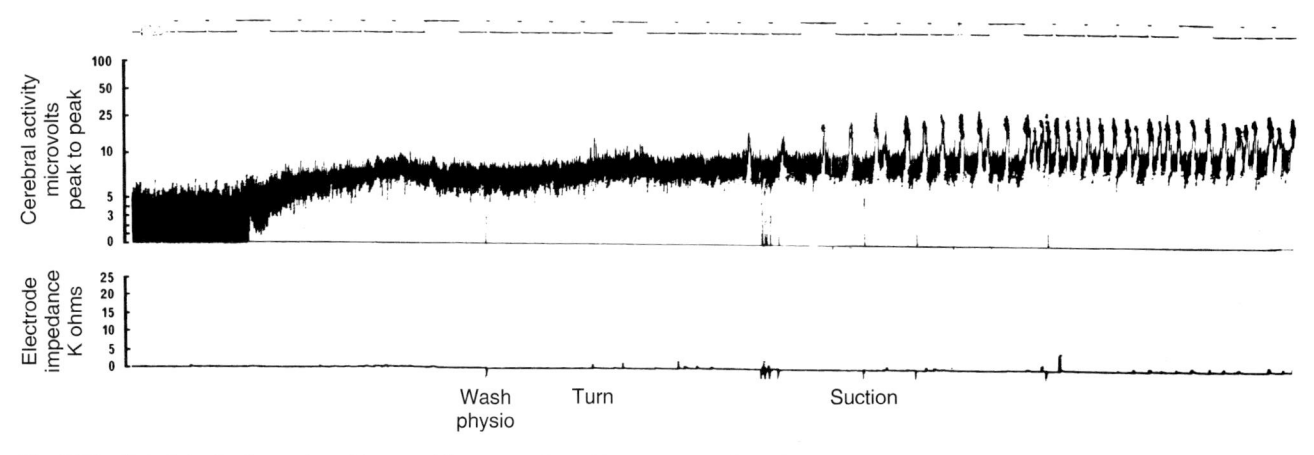

Fig. 6.26 Subclinical seizures in a 32-year-old man monitored in a neurosurgical ICU following surgical evacuation of a subdural haematoma and partial temporal lobectomy. CFM monitoring starts on arrival in the unit after surgery and continues for 6 h (large and small time marks at top of recording indicate hourly and 10 min intervals). The initial tracing represents a burst suppression pattern; as anaesthesia wears off the amplitude level rises. After 3 h, abrupt peaks of high amplitude occur, initially at intervals of 10 min, then increasing in rate and becoming associated with brief bursts of scalp muscle potentials (on the electrode impedance trace) before any clinical evidence of seizures was seen. The discharges were rapidly controlled by anticonvulsants and the patient made a full recovery. (From Prior and Maynard (1986), by permission.)

identification and continuous assessment of cerebral haemodynamic and compensatory reserves. In one 2 year study in a neuro-ICU, these included transcranial Doppler pulsatility indices, the dependence between blood flow velocity or laser Doppler flux and CPP, and the correlation coefficient between pulse amplitude and mean ICP (Czosnyka et al 1994).

In patients with raised ICP, multimodal studies have helped to identify factors such as optimal head position, which has implications for the placement of EEG electrodes. Excessive changes in positioning, including elevation and neck flexion (which may impede venous return), may lead to adverse effects on cerebral perfusion in addition to those associated with arousal due to discomfort. In 22 head-injured patients it was found that when elevation of the head to 30° was compared with a horizontal position, ICP was significantly reduced in the majority without reducing CPP or blood flow (Feldman et al 1992). The EEG was not monitored in that study, but evidence from using the Cerebral Function Monitor (CFM) (RDM Consultants, Uckfield, Sussex, UK, http://www.cfams.com) monitoring in the general ICU has long suggested that impairment of brain perfusion with postural changes during routine nursing procedures can occur in patients with reduced cardiac output (e.g. Prior et al 1978).

There are three reasons for using EEG modules to supplement multimodal neuro-ICU monitoring systems: first, to provide more direct evidence of the onset of cortical ischaemic changes in the individual patient (*Fig. 6.29*); second, to identify factors such as subclinical seizure discharges (Vespa et al 1999) (*Fig. 6.26*) and excessive responses to noxious stimuli, both of which increase cerebral metabolic demand and may be accompanied by abrupt elevation of ICP associated with the increased cerebral blood flow; and third, to assist in guiding management of

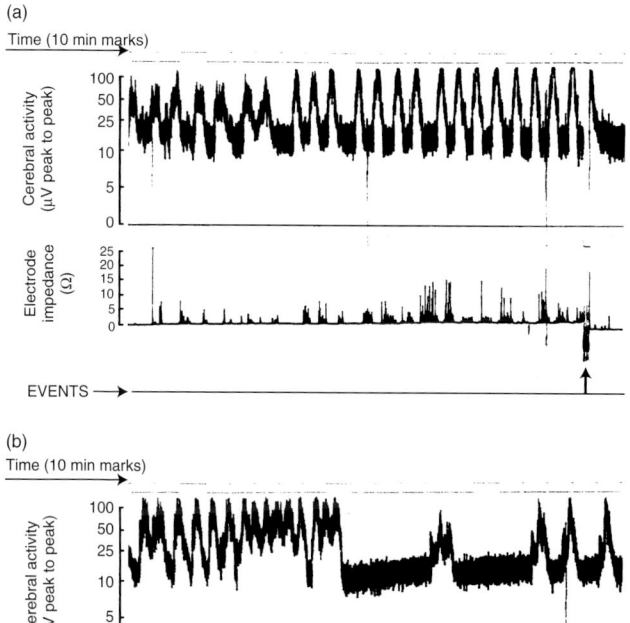

Fig. 6.27 CFM monitoring of intractable status epilepticus and attempts at its control in an 8-year-old child in a paediatric ICU with rubella encephalitis. (a) Note the repeated build up of high peaks in the CFM trace and the associated scalp muscle potentials on the electrode impedance trace. 5 ml intramuscular paraldehyde given at arrow achieves (short-lived) control. (b) Subsequent partial control is achieved with bolus increments of midazolam which cause the level of the trace to fall abruptly, but discharges soon reappear (Modified from Prior and Maynard (1986), by permission.)

Chapter 6 Neurophysiological Work in the ICU

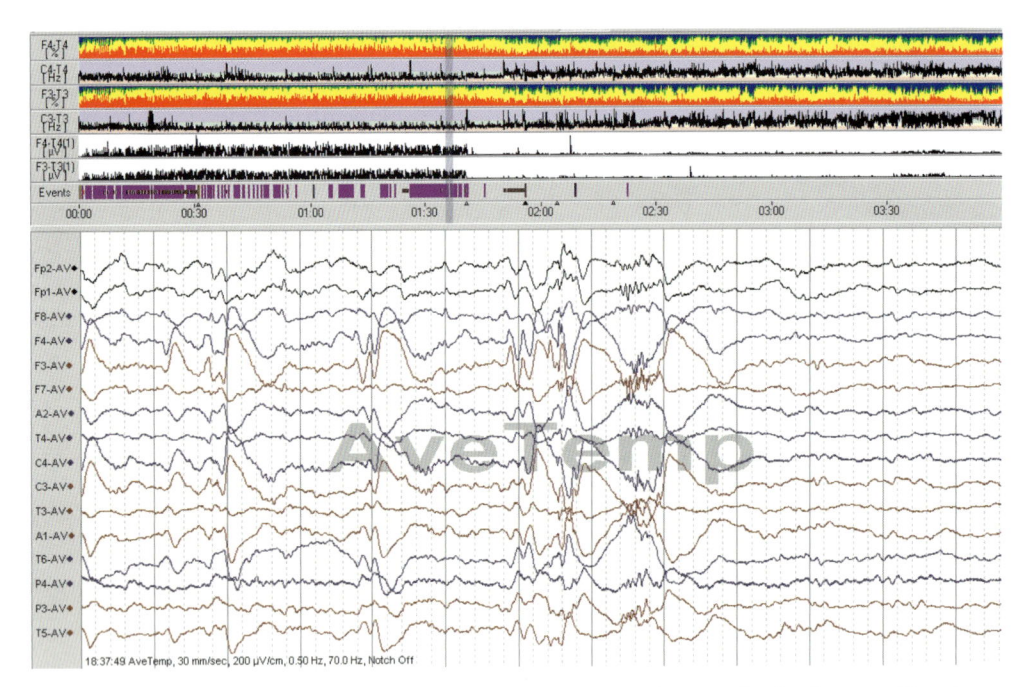

Fig. 6.28 Status epilepticus in an 11-year-old boy with symptomatic generalized epilepsy and learning disability since one year of age; complex partial, atonic, myoclonic and atypical absence seizures occurred. He presented with stupor and increased frequency of myoclonus for two days. This 16-channel digital registration is combined with a multichannel trend analysis. The user can freely select the measures used for trend displays, their parameters and the derivations used. The shaded vertical bar across the 4 h trend display indicates the timing of the section of recording currently displayed in the 12.5 s EEG window. Trends: channels one and three, relative power; channels two and four, mean frequency; channels five and six, EEG envelope. Event markers: mainly myoclonicas. Note the dramatic change in the middle of the display with cessation of discharges subsequent to an increase in mean frequency during recovery after the period of status epilepticus. (From Binnie et al (2003), by permission.)

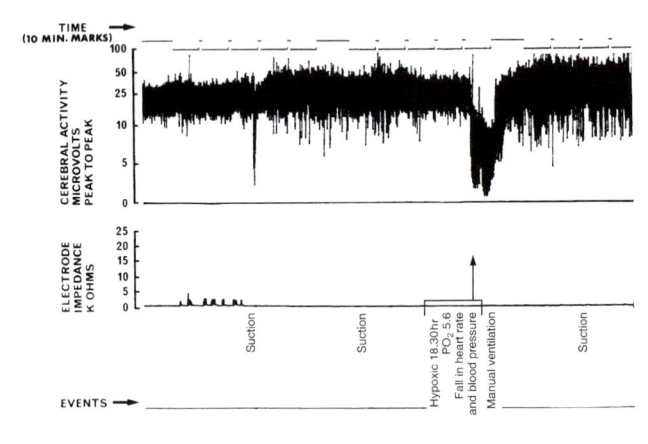

Fig. 6.29 Effect of hypoxic events during CFM monitoring in a 4-year-old child unconscious in an ICU following a road traffic accident in which he sustained a head injury and a cardiac arrest. The trace shows a brief fall in amplitude following suction and a more protracted fall during a hypoxic episode with bradycardia and arterial hypotension (at arrow). Note the scalp muscle potentials and electrode movement artefacts in the early part of the impedance trace. The CFM trace did not return to its previous state after the second event and later fell to zero level. A multichannel EEG was isoelectric and the patient died 3 days after injury. Contrast these appearances with falls in the level of the CFM trace due to bolus injections of an anaesthetic agent in another child in post-traumatic coma (see *Fig. 6.30*). (From Binnie et al (2003), by permission.)

elevated ICP by major sedative regimens. From the point of view of prognostication, EEG signs such as impaired percentage alpha variability are indicative of a poor outcome (Vespa et al 2002) and there is some evidence that short latency SEPs are more powerful than EEG reactivity and GCS (Amantini et al 2005, Moulton et al 1998).

The general aim in the use of EEG monitoring of sedation has been to identify a chosen degree of burst suppression as an index of reduction of cerebral oxygen metabolism to a minimum compatible with cerebral neuronal viability (Michenfelder 1974) (see *Fig. 5.7*). Work on advanced algorithms for use of EEG control to assist in titrating intravenous infusions of barbiturates has already been outlined on page 158. Practical guidelines for application to patients in post-traumatic coma have been reported by Procaccio et al (1988), who reported the use of visually assessed continuous EEG and ICP monitoring to guide control of sedation (*Fig. 6.30*). This work also showed that such monitoring could be used to predict which patients would benefit from treatment with intravenous sedative boluses. The effects of 142 boluses of the short-acting steroid anaesthetic and sedative agent alphaxalone–alphadolone ('althesin', no longer in use) were analysed after subdivision into those with ICP above or below 20 mmHg and into those with minimum peak-to-peak

EEG voltages (measured from a CFM trace) above or below 5 μV. When the effects of althesin on ICP were balanced against those on arterial blood pressure, it was evident that beneficial effects on CPP were largely confined to those patients with ICP > 20 mmHg and minimum EEG activity > 5 μV. If bolus administration was restricted to such patients, 90% of adverse effects on cerebral perfusion due to greater depression of arterial than of ICP (albeit minor) would have been avoided.

Evidence from work examining the effects of noxious stimulation indicates that routine ICU procedures such as tracheal suction may lead to a rise in ICP (Fisher et al 1982). When such procedures are investigated with the addition of EEG monitoring, it is evident that, in all except irreversibly brain-damaged patients, changes associated with external stimuli graded according to intensity could be detected non-invasively. This helped identify clinically relevant effects on cerebral perfusion in comatose and sedated patients.

Neurophysiological work in neurosurgical patients in the ICU is not confined to those with head injuries or raised ICP. Subarachnoid haemorrhage may be followed by vasospasm of cerebral vessels leading to an adverse outcome due to secondary ischaemic changes. Surgery to clip bleeding cerebral aneurysms in the presence of vasospasm carries a greater risk of a worse prognosis. There is evidence that qEEG monitoring may demonstrate alterations that antedate clinical evidence of vasospasm (Labar et al 1991, Vespa et al 1997) and, further, that trend analysis of single variables or their variability contributes to more accurate prediction than CSA. In one study of 11 patients with ischaemic events after subarachnoid haemorrhage, the most sensitive trends were found in the changes in total power (91%), alpha ratio (64%), frequency centroid (55%) and percent delta (45%) (Labar et al 1991). In another, a range of quantitative continuous EEG parameters measured after a standardized stimulus to produce maximum arousal showed that a decrease in alpha/delta power ratio provided a sensitive measure which allowed earlier detection of delayed cerebral ischaemia and initiation of interventions at a reversible stage, thus preventing infarction and neurological morbidity (Claassen et al 2004a).

VEGETATIVE STATES AND BRAINSTEM DEATH (BRAIN DEATH)

Vegetative states

There is a range of conditions in which the state of patients with severe brain damage gradually evolves from one of coma into a *vegetative state*. Prognosis depends on the cause of the damage as well as its extent; some degree of late improvement may occur following severe head injury, but is much less common in patients whose pathology includes severe neocortical damage following cardiac arrest. Recent consideration suggests that such a state should be described as a persistent vegetative state if it continues for more than 4 weeks and as a permanent vegetative state (PVS) when

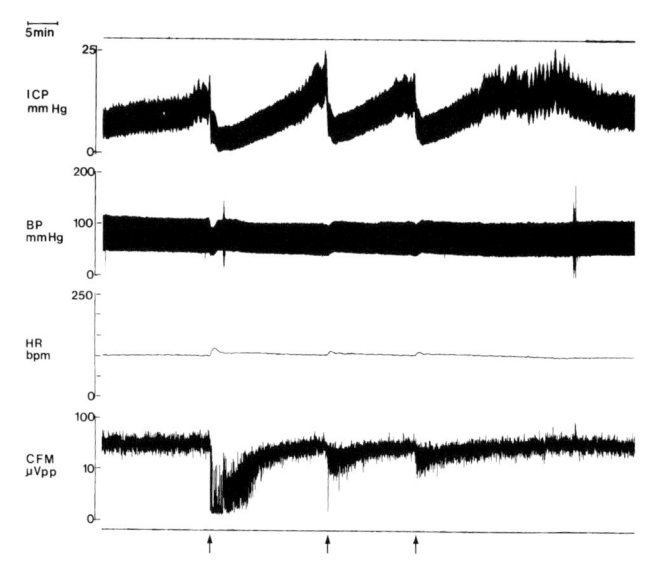

Fig. 6.30 The value of polygraphic recording in monitoring the pharmacological management of raised ICP in an 8-year-old child after head injury. The three abrupt falls in CFM amplitude were due to boluses of the short-acting anaesthetic agent althesin (arrows indicate 2 ml added to infusion at 6 ml/h with fentanyl 1 ml/h). They were accompanied by a fall in ICP, but no significant alteration in arterial blood pressure, and resulted in an overall increase in cerebral perfusion pressure. Note the 1–2/min waves in ICP in the later part of the recording; these are often a positive prognostic sign in post-traumatic coma. The patient made a full recovery. (From Procaccio F, Bingham RM, Hinds CJ, Prior PF 1988 Continuous EEG and ICP monitoring as a guide to the administration of Althesin sedation in severe head injury. *Intensive Care Medicine* 14:148–155. © 1988 Springer-Verlag, with kind permission of Springer Science and Business Media.)

the diagnosis of irreversibility can be established with a high degree of accuracy, usually only when a patient has been in a continuing vegetative state following head injury for more than 12 months or following other causes of brain damage for more than 6 months (Royal College of Physicians Working Party 2003).

The classic description of PVS after ischaemic or traumatic brain damage is that of Jennett and Plum (1972). They described it as 'a syndrome in search of a name' in which patients with severe brain damage may survive indefinitely and 'never regain recognizable mental function, but recover from sleep-like coma in that they have periods of wakefulness when their eyes are open and move; their responsiveness is limited to primitive postural and reflex movements of the limbs and they never speak.' The lesions may be in any or all of the following sites – the cerebral cortex, the subcortical structures of the cerebral hemispheres or the brainstem. Jennett and Plum (1972) devoted considerable attention to difficulties of nomenclature, carefully distinguishing PVS from brain death or *coma dépassé* (Mollaret and Goulon 1959), from the apallic syndrome (Kretschmer 1940) and from akinetic mutism (*coma vigile*). Recently, Jennett (2002) has provided a masterly review of the development of concepts since his original report with Plum in 1972. He gives a detailed account of the ethical and legal dilemmas associated with this condition and sets this in the context of present practice in different countries.

The onset of periods of eye opening provides crucial differentiation from 'eyes closed coma' after major insults to the brain, which rarely lasts more than a matter of 2–4 weeks before patients either die, recover fully or lapse into a vegetative state (Levy et al 1978, 1981). Indeed, Dougherty et al (1981) reported a study of 10 patients with hypoxic–ischaemic brain damage following cardiac arrest in eight and subarachnoid haemorrhage in two, all of whom regained brainstem function and 'awoke' within days of the insult, and opened their eyes within 2 weeks. None showed evidence of cognitive awareness. All the patients died after 2–8 weeks in a vegetative state and all showed widespread ischaemic neuronal damage in the cerebral hemispheres, only one having any brainstem abnormality. In patients with traumatic coma (Jennett et al 1979) the eventual outcome from vegetative states may be less clearly predictable, with occasional late improvement reported (Arts et al 1985), probably in relation to the more variable pattern of damage compared to that of neocortical death after cardiac arrest. There is not yet adequate evidence as to whether early neurophysiological features can reliably predict these rare patients with late improvement.

The EEG in patients with PVS shows a wide range of patterns, often consistent with the nature and distribution of neuropathology. Patients may exhibit sustained isoelectric EEGs over weeks, months or years, whilst retaining some reflex and short latency EP evidence of brainstem function, in the context of brain damage which overwhelmingly affects the neocortex with variable degrees of deep grey matter damage, as in the patients after cardiac arrest reported by Brierley et al (1971), Cole and Cowie (1987) and Wytrzes et al (1989). Of the 10 patients with PVS due to hypoxic–ischaemic brain injury of Dougherty et al (1981) mentioned above, seven had EEGs, six showing 'severe bilateral cerebral dysfunction' (diffuse low-amplitude irregular delta activity), the seventh showing paroxysmal activity on a background of disorganized slow waves. In patients in a vegetative state as a result of cardiac arrest, there is generally widespread damage in both superficial and deep grey matter and it is the particular combination of these two that is reflected in the clinical and neurophysiological findings in any individual. Patients following blunt head injury, such as the 25 who remained vegetative or severely disabled for more than 1 month described by McLellan et al (1986), show that the structural basis is a combination of diffuse axonal injury (present in 20) and moderate to severe ischaemic brain damage (also present in 10 of them), with secondary damage in the brainstem (in eight patients). Reviews by Kinney and Samuels (1994) and by Graham and Gennarelli (1997) conclude that vegetative states after acute brain damage show three patterns of damage – widespread bilateral neocortical damage, diffuse damage to white matter and bilateral damage to the thalami. The picture has been clarified by the detailed studies of Adams et al (2000) in 49 patients surviving from 1 month to 8 years in a vegetative state, 35 following blunt head injury and 14 some form of non-traumatic brain damage. Essentially, this work confirmed differences in patterns of damage according to causation, the head-injured patients showed most diffuse axonal and thalamic injury, whilst those with non-traumatic causes showed a greater proportion of neocortical ischaemic damage. Such variations in distribution of pathology in vegetative patients help explain the range of abnormal EEG and EP patterns. The EEG phenomenology may range from diffuse slow waves or unresponsive higher frequency activities to the transient alpha or beta coma patterns comparable to those in some patients with pontine lesions.

Generally, in patients with an established vegetative state, the EEG varies considerably according to the state of the patient at the time of recording. This makes it essential to record in both the eyes-open 'awake' and the eyes-closed 'asleep' states; in the latter there may be preservation of 'sleep' spindles, which are generated subcortically (*Fig. 6.31*). Thus, in PVS patients, an EEG that appears isoelectric in a routine recording does not necessarily remain so throughout prolonged recordings or 24 h monitoring, but may reflect cyclical behavioural patterns. Such patients may show periods with eye-opening, yawning and associated movements and certain reflex activities during which sleep-like features disappear from the EEG. These features may be superimposed upon the long-term trend towards decline of all EEG activities in the presence of major cortical death (*Fig. 6.32*). Such findings point to the dangers of drawing overenthusiastic prognostic conclusions from serial EEG changes based upon short 'snapshot' recordings, and are one of the main reasons for our recommendation of prolonged polygraphic studies in possible PVS patients.

Prolonged EEG–polygraphic recording should be used to look for and document such cyclical patterns and must be of sufficient duration to capture the eyes-open and the 'sleep-like' states that characterize vegetative patients. Systematic use of stimulation (calling the patient's name, clapping, touching and applying pain to both cranial nerve territories and to all four limbs) is essential to distinguish PVS from coma, and also from locked-in syndrome and other causes of unresponsiveness such as depressive stupor. Consideration should always be given to the use of somatosensory and auditory brainstem EP studies to rule out any unexpected drug-induced effects from earlier overdose or from excessive therapeutic zeal.

To summarize, recording techniques and neurophysiological reports concerning studies in patients with suspected or definite PVS should not be restricted to a bland description of the frequency content of any ongoing EEG activity, but should include precise documentation, in as quantitative terms as possible, of:

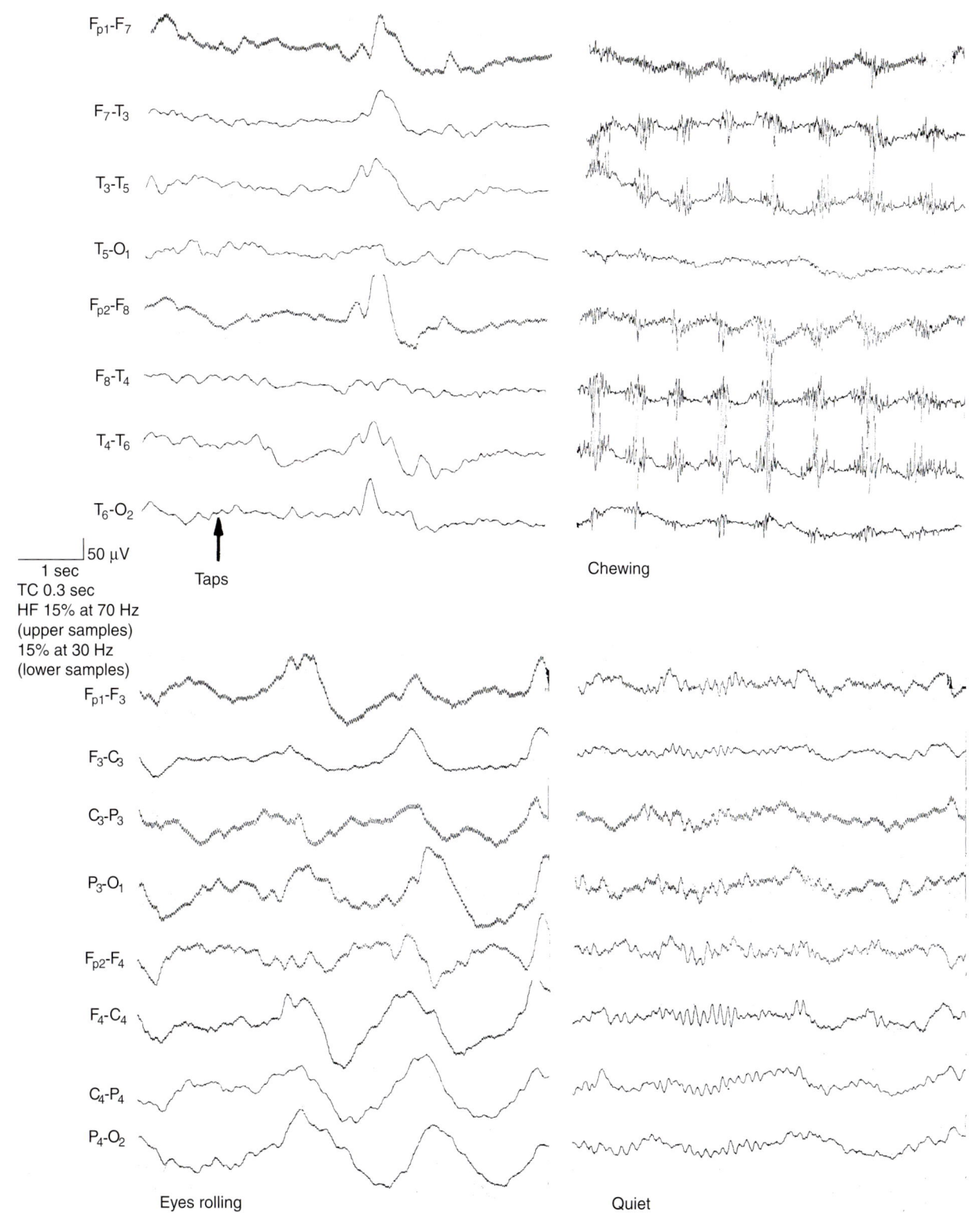

Clinical conditions affecting the central nervous system encountered in the ICU

Fig. 6.31 Short-term EEG variability during a prolonged recording in a woman of 20 years in a persistent vegetative state following cardiac arrest 7 months earlier. (From Prior (1973), with permission.)

1. the EEG and heart rate responses to systematic stimulation in cranial nerve territories and to all four limbs;

2. well-annotated EEG–polygraphic observation of sufficient duration and in appropriate recording modalities (i.e. eye movements, heart rate,

227

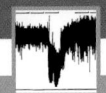

respiration, etc.) to characterize the nature of the cyclical changes in state which typify the 24 h patterns of behaviour in these patients; and
3. sufficiently rigorous early short-latency somatosensory and auditory brainstem EP studies to exclude any question of functional depression by drugs and to identify functioning pathways ascending though the brainstem to primary cortical projection areas.

Unfortunately, documentation of the original clinical and neurophysiological features of the vegetative state has not always been available or adequate in some of the more recent controversial cases with medico-legal or other ramifications relating to lawful withdrawal of life-support. PVS patients are not a homogeneous group; some follow hypoxic–ischaemic insults such as cardiac arrest, others follow head injury and have quite different patterns of brain damage. Different aetiologies may be associated with a similar clinical syndrome but not necessarily with the same prognosis. The occasional report of a patient with some degree of late improvement can be no surprise; late recovery is considered more likely to occur in patients with predominately diffuse axonal injury than with diffuse hypoxic damage, but this has not yet been established (DI Graham, personal communication, 1996). Reports on individual cases with prolonged survival have highlighted the diagnostic and ethical dilemmas of the persistent vegetative state and that of Karen Ann Quinlan (Angell 1994, Kinney et al 1994), who survived for 10 years without regaining consciousness, has been of considerable interest. It included descriptions of the two phases of EEG coincident with the patient's 'awake' and 'asleep' states and of the neuropathological findings. Machado (2005) reviews investigation of residual cortical processing. Recent guidelines on the medical aspects and ethical issues are available, but some lack of clarity remains regarding the contribution of neurophysiology (American Academy of Neurology Committee on Ethical Affairs 1993, Multi-Society Task Force on PVS 1994, Giacino et al 2002, Royal College of Physicians Working Party 2003).

Brainstem death (brain death)
Historical introduction

Developments in our understanding of the clinical and electrophysiological background of the diagnosis of brain death are detailed by Chatrian (1986, 1990), whilst much of the medico-legal, cultural and religious argument is given by Pallis (1990); both authors describe the claims for and against neurophysiological components in codes of practice in some detail. Nonetheless, by the close of the 20th century, an increasing number of national and international codes of practice now state that the diagnosis of brain death is basically clinical (areactive coma, loss of brainstem reflexes, apnoea) providing the origin of coma is known as sufficient to cause brain death; electrophysiological testing (to demonstrate an isoelectric EEG and loss

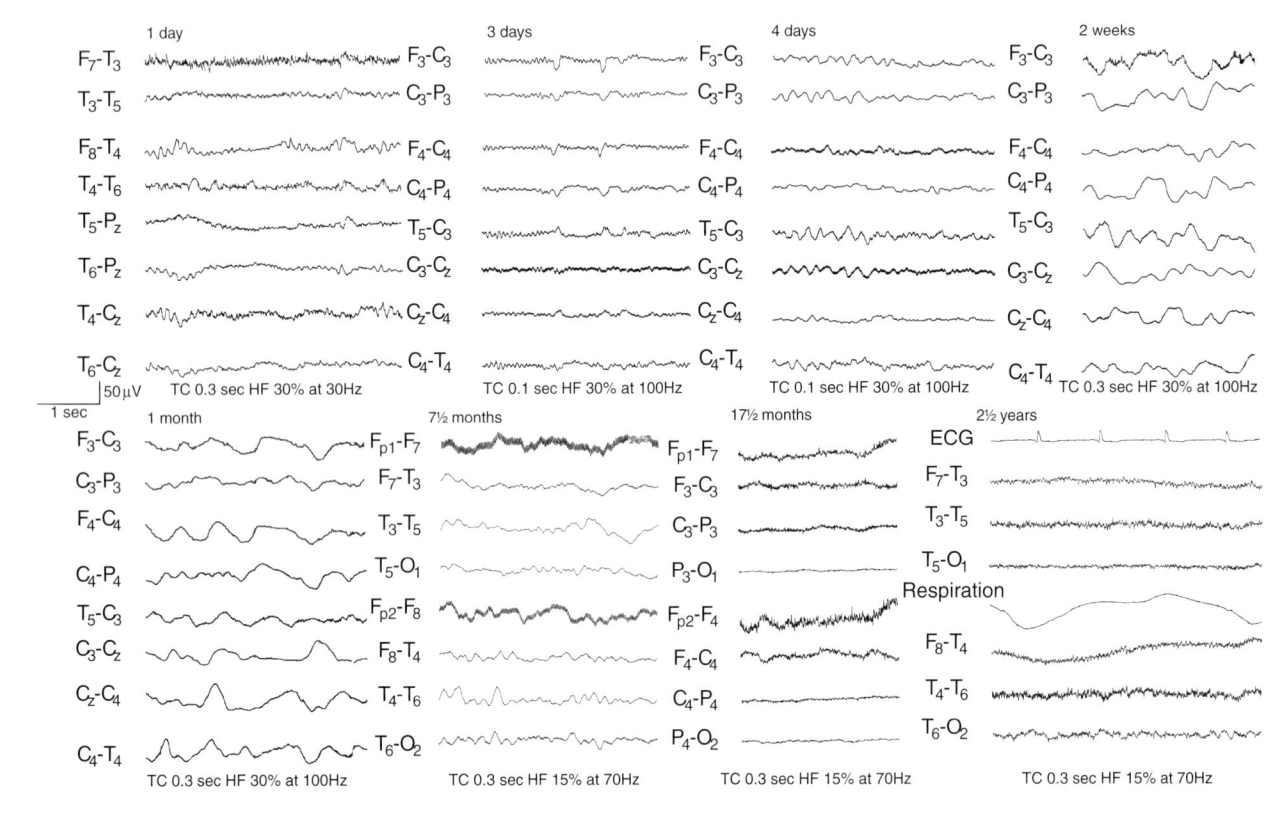

Fig. 6.32 Long-term evolution of EEG findings in the patient described in *Fig. 6.31* who survived for almost 8 years. Samples at 1, 3 and 4 days, 2 weeks, 1, 7.5 and 17.5 months, and 2.5 years after cardiac arrest. All recorded at same sensitivity and paper speed (From Prior (1973), with permission.)

of all EPs of intracranial origin) is only considered a confirmatory tool (Guérit et al 1999a).

Because of continuing confusion about the term or concept of brainstem death, there is a need to underline explicitly the nature of the essential preconditions:
1. irreparable, massive, cerebral hemisphere damage, confidently diagnosed, which has led to the terminal demise of the brainstem, rendering the whole process irretrievable; *or*
2. massive, irreparable, primary brainstem lesions.

Although the irreversible failure of brainstem function is the key to diagnosing what was previously called brain death, in clinical practice including trauma and neurosurgical patients it is more often the end result of massive hemispheric disease or damage leading to herniation (coning) and demise of the brainstem from compression of its vascular supply etc., than due to primary brainstem lesions.

The strict emphasis on this combination of factors is because: (1) irreversible cerebral mass lesions and/or swelling lead to *secondary death of the brainstem*, as opposed to *isolated irreversible primary brainstem lesions* which do so directly; and (2) cardiac standstill (asystole) always follows brainstem death, usually in a matter of days (Jennett et al 1981a, Pallis and Harley 1996). The same cannot be said for the consequences of electrocerebral inactivity.

Neurophysiology is, of course, of very great importance in patients in whom brainstem death may be considered. Its role is at the stage when it can be used in a very positive way, both to identify those who may have reversible conditions, and to help characterize the extent and severity of major brain damage at a time when treatment may still be feasible. It is thus in relation to the question of the *precondition* of a definite diagnosis of irreversible, massive brain damage which the EEG, EPs and other investigations may help characterize before it necessarily proceeds to the demise of the brainstem. Most authorities are agreed that EEG and EPs have no place in the final testing of brainstem death (American Academy of Pediatrics Task Force 1987, British Paediatric Association 1991, Lynch and Eldadah 1992, American Academy of Neurology Quality Standards Subcommittee 1995, Royal College of Physicians Working Group 1995, Wijdicks 1995, Chatrian et al 1996, Guérit et al 1999a). A survey by Pallis and Harley (1996) showed that 12 of the 37 countries for whom information was available still required EEG, 6 required it or other confirmatory studies and 18 depended entirely on clinical testing. When EEG is required as part of a confirmatory procedure, technique should follow the latest IFCN Recommendations (Guérit et al 1999a), but always with the understanding that persistence of EEG activity does not exclude the diagnosis of brain death (Chatrian et al 1996) – a view fully endorsed by the IFCN panel of experts.

The IFCN recommendations (Guérit et al 1999a) set out the meticulous technique that is required when the *absence* of potentials is a criterion of abnormality and the burden rests on the neurophysiological team to prove that the lack of signals (or the presence of a few dubious waveforms) does not have a technological or artefactual basis. Protocols for neurophysiological assessment in patients with suspected irreversible brain damage should include documentation of other biological activities, careful checking of responses to a systematic sequence of external stimuli and interpretation in the light of the whole clinical picture. A high standard of documentation of EEG and EP recordings is required and the use of a checklist is helpful. It is important that the clinical neurophysiologist is present for at least part of the recording to be personally satisfied as to the quality of the recordings, to assess the data in the light of all the relevant medical circumstances and to discuss the findings with the clinical team. A preliminary written opinion (signed and with the date and time of the observations) should always be entered into the patient's case notes at the end of recording, followed promptly by a full conventional report. This should indicate any limitations and, as with any assessment in the evolving course of a disease process, should indicate the possible significance of any potentially reversible elements or confusing factors. These include high levels of CNS depressant drugs, profound hypothermia, major metabolic encephalopathies, which can affect both EEG and EPs (especially their cortical components), and also disease of, or damage to, peripheral receptors or pathways involved in sensory EP testing.

EEG and brainstem death

It is important not to equate electrocerebral inactivity with brain or brainstem death (for recording methodology see p. 124). Whilst sustained EEG silence not explicable by potentially reversible factors may be an important indication of irreversible loss of cortical function (at least in the areas accessible to recording with scalp electrodes), this only provides part of the diagnostic criteria of irrecoverable cerebral hemisphere cortical damage and does not provide any indication of the functional state of the vital centres of the brainstem. Overwhelming, irrecoverable cerebral hemisphere damage may occur, for example, in neocortical death – a consequence of selective hypoxic brain damage after cardiac arrest in which function of the brainstem may be fully preserved with the possibility of prolonged survival in a vegetative state. It is also evident from frequent serial recordings that some transient return of activity may be seen even in patients sustaining severe hypoxic damage (Prior 1973) (see *Fig. 4.16(b)*). In a series of 231 patients observed from the time of resuscitation after cardiorespiratory arrest, the 106 with some initial EEG activity had a 33% 1-year survival rate, whilst only 16% of the 125 with initial ECS survived 1 year, many succumbing to brain death (Jørgensen and Holm 1999). Bassetti et al (1996), on the basis of their study of 60 postcardiac arrest patients, considered that a combination of GCS score (Teasdale and Jennett 1974) at 48 h, SEP, and if these are non-conclusive, EEG (combined prognostic accuracy 82% with no falsely pessimistic predictions), permitted a more reliable prediction of outcome than clinical examination alone. An overview by Sandroni et al (1995) of the 'best test' to predict outcome after prolonged cardiac arrest in 62 patients with ECS, ranked a combination of GCS and multimodality EPs the strongest predictor of outcome at 6 months. This was in part due to the value of the EPs, with their 100% specificity, in indicating effects of sedatives and muscle relaxants which impair the accuracy of GCS and EEG predictions.

EPs and brainstem death

'Brain death' and 'brainstem death' are relatively recent terms (Adams and Jequier 1969, Pallis 1990) that have now largely replaced the more picturesque label of *coma dépassé* ('coma beyond coma'), originally coined by Mollaret and Goulon (1959). All three terms in their proper usage imply irreversible loss of function in cerebrum and brainstem; consequently, the only EP pattern consistent with them is one in which no evidence of CNS activity is detected rostrally to the foramen magnum (*Fig. 6.33*). In the case of SEPs, this is represented by the persistence of peripheral and spinal potentials (N9 and N13) with bilateral absence of brainstem P14 and of all subsequent cortical waves (Anziska and Cracco 1980, Belsh and Chokroverty 1987, Facco et al 1990). The equivalent BAEP pattern consists of an isolated wave I, or waves I and II, without subsequent brainstem potentials (Starr and Hamilton 1976, Goldie et al 1981, Hall et al 1985b, Machado et al 1991). BAEP abolition is also a pattern seen in brain-dead patients, but its interpretation requires exclusion of any direct damage to or antecedent dysfunction in peripheral auditory structures. In general, recording of reproducible peripheral responses (N9 or wave I) is of utmost importance before interpreting an EP pattern (absence of potentials) as consistent with brain death.

Brainstem death is not the sole condition leading to abolished central EPs and this is one reason for strict assessment of patients in the context of a thorough clinical history and examination (Pallis 1990). Besides the existence of previous diseases that may severely depress waveforms (*Fig. 6.33*), bilateral BAEP and SEP abolition may be observed in lesions localized to the brainstem (see above). In this rare eventuality, the diagnosis is given by the persistence of EEG and VEPs (Trojaborg and Jørgensen 1973, Ferbert et al 1986).

Except for cases complicated by spinal injury, cervical somatosensory responses (P11, N11, N13) remain normal in brain death (see *Fig. 6.17*). Unfortunately, some confusion about this issue has arisen due to the common practice of recording cervical components with a cephalic reference, which injects rostral far-field responses into the cervical waveform. In these circumstances, loss or attenuation of brainstem P14 during coma induces a spurious decrease of 'cervical' components (Ferbert et al 1988, Ganes and Lundar 1988) which may be erroneously interpreted as a sign of spinal damage. This was illustrated ele-

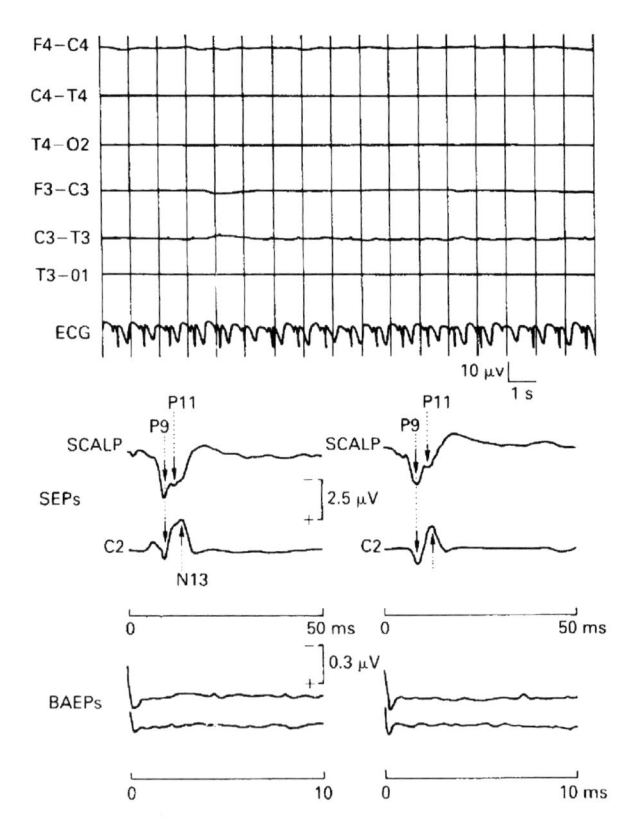

Fig. 6.33 Electrophysiological study in a brain-dead patient. EEG and EP patterns consistent with a diagnosis of brain death, recorded from a 20-year-old woman. At the time of recordings, all cephalic reflexes were clinically absent. Calibration mark for EEG, 10 μV, 1 s. The EEG is isoelectric even under maximal amplification. BAEPs show absence of all components, including wave I. Somatosensory stimulation elicits only peripheral (P9) and spinal (P11, N13) components, while brainstem P14 and cortical N20/P27 are bilaterally absent. (From Binnie et al (2004), by permission.)

gantly by Buchner et al (1988), who monitored SEPs during developing brain death and showed that cervical N13 remained stable if recorded with a non-cephalic reference, but appeared to decrease due to P14 loss if a frontal reference was employed.

A few patients have been described in whom persistence of EPs coexisted with abolition of all brainstem reflexes and clinical 'brain death' (Hall et al 1985a, Ferbert et al 1986). In at least one case, in the former report, the patient eventually recovered good cognitive function, thus making electrophysiological testing a crucial step before such a diagnosis is accepted.

Present status of neurophysiological testing in brainstem death

The reasons for the current view that there is 'no evidence at present that the EEG or EPs improve upon clinical diagnosis of brainstem death' and phrases such as 'should not presently form part of the diagnostic requirements' may be summarized as follows.

The EEG may: (1) give falsely pessimistic information (e.g. be isoelectric because of unsuspected but potentially recoverable drug overdose); (2) give falsely optimistic information (e.g. show some short-lived residual waves) because it is recording residual traces of activity arising in the cerebral cortex, i.e. away from the site of the critical damage that governs the irreversibility of the comatose state (i.e. the brainstem) – indeed the EEG can provide information only about those areas of the (superficial convolutions of the) cerebral cortex that are accessible to scalp electrodes.

Evoked potentials (1) These record from only a few selected pathways through the brainstem; (2) rigorous studies make it clear that correct identification of potentials is a problem in the face of major lesions (i.e. there may be confusion between those of subcortical and cortical origin); and (3) the techniques are not sufficiently widely available to a sufficiently high standard at the bedside in the ICU to be part of what must be a highly reliable and reproducible test.

Combined EEG + EP monitoring There are several factors which, either singly or in combination, can produce a transient isoelectric EEG, or loss of cortical EP components, from which the patient may make a full recovery. In spite of its crucial importance in assessment and monitoring of the functional state of the nervous system to assist in diagnosis of the cause of an unresponsive comatose state with absence of brainstem reflexes, it is unwise to overstate the role of clinical neurophysiology in the actual determination of brain death.

CLINICAL CONDITIONS AFFECTING THE PERIPHERAL NERVOUS SYSTEM AND MUSCLES ENCOUNTERED IN THE ICU

NEUROMUSCULAR SYNDROMES OF CRITICAL ILLNESS (CRITICAL ILLNESS NEUROPATHY)

There are certain primary neurological disorders which require management in the ICU because the illness has become complicated by respiratory difficulties or failure. These include myasthenia gravis, inflammatory demyelinating neuropathies and other disorders such as botulism, poliomyelitis and severe myopathies. In addition, it is now recognized that there are also various peripheral neurological problems which arise *de novo* during, or possibly as a result of treatment in the ICU for other, primarily non-neurological, conditions. Significant muscle weakness following critical illness is more common than was previously thought. A number of pathologically different syndromes have now been reported (Van Marie and Woods 1980, Bolton et al 1984, Op de Coul et al 1985, Knox et al 1986, Kupfer et al 1987, Coakley et al 1992, Zochodne et al 1994). The dramatic increase in the frequency of such reports during the last two decades may be related to one or more of the following:

1. The increasingly sophisticated management of patients in the ICU has resulted in the ability of physicians successfully to manage much more severely ill patients than ever before.
2. Since such patients may be kept alive for much longer periods in the ICU, they may be exposed for longer periods to potentially harmful agents as well as adverse (deficient) nutritional factors that could, theoretically, give rise to the development of neuropathies.
3. The more frequent use of complex drug regimens containing ever-increasing ranges of drugs which may include steroids, major sedatives, several antibiotic and antifungal preparations, and other potentially toxic drugs.
4. The now commonplace ICU regimen to assist respiration in which mechanical ventilation is accompanied by heavy intravenous sedation coupled with prolonged use of continuous intravenous muscle relaxants; this is widely used, but the potentially adverse effects on the human peripheral nervous system are not known.

The various clinical syndromes described by different groups have many similarities. Often, the patients have suffered from multiple organ failure or respiratory failure from bronchial asthma, chronic obstructive airway disease or some other form of respiratory illness and required mechanical ventilation. Most would appear to have received sedatives, neuromuscular blocking agents by continuous intravenous infusion, a range of antimicrobials and parenteral nutrition. Most patients have respiratory weakness, which may be either of central or peripheral origin, the distinction between the two often being difficult. Reflexes may be lost, diminished, normal or even exaggerated. Clinically the most important and the most striking feature is diffuse muscle wasting and weakness, involving both proximal and distal musculature. Extraocular and facial muscle involvement is rare. Sensory loss in these patients is often difficult to assess, but in those who can cooperate this appears to be the least affected.

Bolton et al (1984) were the first to describe in detail the clinical and electrophysiological features of 'polyneuropathy in critically ill patients'. In all 5 of their patients, collected over a 4-year period, there was difficulty in weaning from the ventilator and areflexic, flaccid limb weakness. In addition, two had some facial weakness. CSF findings were unremarkable. All the patients were reported as having sepsis and multiple organ dysfunction. All required ventilatory assistance and received total parenteral nutrition, multivitamin therapy and mineral supplements. No details are available on muscle relaxant administration. Two patients made spontaneous gradual recoveries; the remaining three died of other, non-neurological, causes.

The neurophysiological findings in the original 1984 report from Bolton's group were of absent or abnormally low amplitude SAPs or compound motor action potentials (CMAPs), without evidence of slowing of conduction velocity. EMG of both proximal and distal muscles showed fibrillation, positive sharp waves and reduced motor unit activity, suggesting widespread axonal degeneration. Morphological studies carried out from autopsy specimens showed features consistent with moderate to severe sensory motor polyneuropathy. In one patient there was also evidence of primary muscle disease. A specific search for both virus infection and heavy metal poisoning was negative. The authors excluded Guillain–Barré syndrome (GBS) on the basis of the lack of evidence of demyelination in electrical and pathological studies and the normal CSF findings. Bolton and his colleagues interpreted their results as showing a severe form of axonal sensory motor polyneuropathy and hypothesized (Bolton and Young 1989) that the primary problem might be involvement of the axonal transport system by a cause yet to be identified.

Op de Coul et al (1985) reported 12 critically ill patients managed in the ICU with sedation and mechanical ventilation who were clinically similar to those described by the Bolton group. They exhibited severe areflexic tetraparesis and diffuse muscle atrophy, the latter being more prominent distally. A few also had facial, extraocular and pharyngeal muscle involvement. Most had multiple organ failure and sepsis and there was universal respiratory involvement. Of the 12 patients, eight had chest injuries, eight had renal failure and 1 had asthma. All had received frequently repeated neuromuscular blockade with pancuronium bromide; biochemically this is a biquaternary amino-steroid which produces non-depolarizing postsynaptic neuromuscular blockade. The main difference in the patients reported by Op de Coul's group, when compared to those of Bolton et al (1984), was that they had not had respiratory difficulties when weaned from the ventilator and none had had any sensory disturbance.

The neurophysiological findings reported by Op de Coul et al (1985) were of widespread denervation, but no data were given on SAP or CMAP amplitudes, nor was there information on any studies specifically assessing the possibility of a neuromuscular junction disorder. Histological examination of muscle biopsies from four patients showed neurogenic changes in two and myopathic changes in two. Three patients died of non-neurological causes but nearly all the others rapidly improved; in seven, complete recovery was achieved within 2–5 months. Op de Coul's group attributed this syndrome to prolonged neuromuscular blockade resulting from long-term muscle relaxant therapy. Although these drugs are short-acting, there have been reports that drug interactions and metabolic derangements may prolong their duration of action (e.g. Fogdall and Miller 1974). The widespread denervation observed may well have resulted from continuous neuromuscular blockade simulating axonotomesis.

There have also been some reports, with varying neurophysiological findings, in which flaccid quadriparesis has been attributed to pancuronium therapy administered in the management of status asthmaticus. Kupfer et al (1987) described a patient, apparently showing a normal neurophysiological study, in whom the weakness was simply explained on the basis of disuse atrophy. Bachmann et al (1987) reported four cases of acute 'myopathy' in patients with status asthmaticus. All had been treated by mechanical ventilation and had received prolonged neuromuscular blockade and high doses of corticosteroids. Neurophysiological studies were carried out in two of the four, both apparently showing normal conduction velocities, although in one fibrillations had been noted at rest. Histologically these two patients showed diffuse atrophy, and in one a vacuolar myopathy was demonstrated.

Steroid therapy has been implicitly or explicitly implicated in a few reports of patients in a similar clinical setting, i.e. of some form of respiratory failure, often status asthmaticus or chronic obstructive airway disease, without evidence of multiple organ failure. Knox et al (1986) described an asthmatic patient with proximal and distal wasting but preserved reflexes. Neurophysiological information was limited but suggestive of myopathy. The patient improved over 1 week and it was concluded that the disorder was due to steroid therapy. Reports from MacFarlane and Rosenthal (1977) and Van Marie and Woods (1980) describe a nearly identical picture and reach similar conclusions. It is worth noting that all these patients had been mechanically ventilated.

Coakley et al (1992) reported three patients with chronic obstructive airway disease managed in the ICU with muscle relaxants, sedation and mechanical ventilation. Clinically these patients had both proximal and distal weakness with diminished or absent reflexes. Electrophysiologically, the most striking abnormality was marked reduction of the CMAP amplitudes, in some instances to less than 5% of control data. Upper and lower limb muscles tested were equally affected. One patient, studied serially, showed marked improvement of the CMAP recorded from abductor digiti minimi, with amplitudes increasing from 0.2 mV to 4.0 mV over about 4 months after the initial study. Needle EMG revealed widespread, profuse fibrillations in proximal as well as distal muscles. Sensory studies were less abnormal or even within the normal range. None of the patients showed any slowing of conduction or evidence of a demyelinating peripheral nerve disorder; all three improved rapidly, achieving normal mobility within 4–6 months.

It is clear that limb weakness is the most striking feature in all the patients in each of the reports described above. Difficulty in weaning the patient from the ventilator may well have been the earliest event which drew attention to the generalized weakness in the majority, but it should be noted that respiratory weakness is not a constant feature of the condition. There are also other notable differences. Multiple organ failure and sepsis were not invariably present. Electrophysiological sensory abnormalities have been reported in only a few studies (Bolton et al 1984, Zochodne et al 1987, Coakley et al 1993) and whilst most patients have been described as being areflexic, this has not always been the case. The time taken for full functional recovery also seems to have varied from one week to several months.

It is possible that the patients studied by Bolton et al (1984) were suffering from a severe peripheral neuropathy, but there were features to suggest that this was not the sole problem. Firstly, close inspection of the electrophysiological data suggests that at least one patient had marked upper limb involvement and one other had disproportionately severe motor involvement. These features do not follow the general pattern of neurophysiological findings in the toxic or metabolic polyneuropathies of primarily axonal, symmetrical, sensorimotor type. In addition, Bolton's first two patients, in spite of showing marked proximal involvement (including tongue and facial muscles, suggesting a very severe polyneuropathy) and widespread severe denervation, had subsequently made rapid progress, being able to walk independently within 10–24 months. It is unlikely that such satisfactory progress could take place in peripheral neuropathy with axonal degeneration given the degree and extent of denervation.

A neuromuscular junction abnormality of either post- or presynaptic type cannot be entirely ruled out as the basis of these syndromes of weakness in critically ill patients. Although many studies have failed to demonstrate a decremental response to repetitive stimulation, there are occasional reports which do give some indication of electrophysiological abnormality at the neuromuscular junction (Subramony et al 1991). It is important to realize that appropriate studies are rather cumbersome to perform in the ICU because of limited access to the usual sites and the problem of lack of patient cooperation for measuring jitter. This latter problem may be overcome by choosing the method of stimulated single fibre EMG (Trontelj et al 1986).

Most authors have reported myopathy as occurring, but only in a relatively small proportion of their patients. Indeed, myopathy alone cannot explain the full syndrome as it has been clearly shown by all groups that the most common EMG finding is the presence of neurogenic changes. Nonetheless, it is again important to realize that the number of myopathies confirmed neurophysiologically is probably an underestimate, as most patients would not be in position to cooperate for a detailed EMG study. Disuse atrophy and nutritional factors may well have a contributory role in producing weakness in these critically ill patients, but could not explain the denervation and myopathic EMG changes.

The only logical conclusion that can be made from the literature to date is that a multifactorial aetiological basis must exist for the generalized muscle weakness presenting in critically ill patients undergoing intensive care.

In a prospective study, the author's group has examined more than 40 patients using a multidisciplinary approach (Coakley et al 1993, 1998, Nagendran et al 1993). All patients who had been managed in the ICU for 7 days or more were included and clinical, neurophysiological and neuropathological assessments made serially. Seventeen of the patients were collected consecutively to assess the incidence of the various abnormalities. Not all the patients showed clinical evidence of muscle weakness, but neurophysiological and neuropathological abnormalities were nonetheless present in the majority (76%). Patients could be categorized in four groups according to the EMG findings, as follows:

1. *Normal* (24%) – Patients had normal median and sural SAPs, normal motor conduction velocity, normal CMAPs from abductor pollicis brevis and extensor digitorum brevis, and normal needle EMG findings, where this examination had proved possible.
2. *Primarily axonal and predominantly sensory polyneuropathies* (28%) – Patients had reduced SAP amplitudes, particularly those of the sural SAPs, but the studies were otherwise unremarkable.
3. *Peripheral motor syndrome of uncertain aetiology* (24%) – Patients had marked reduction of CMAPs without any sensory involvement. Clinical weakness was observable in this group, particularly when the reduction in CMAP was most severe. Concentric needle (CN-EMG) studies often showed widespread fibrillations.
4. *A combined picture of (2) and (3)* with both sensory as well as motor abnormalities with reduction of CMAPs as well as SAPs without any slowing or conduction block (24%).

Repetitive stimulation carried out in four of our group 3 patients showed no abnormality. Some patients who had extremely small CMAPs (of the order of ≤ 5% of normal) did not show evidence of post-tetanic potentiation. Acute denervation (fibrillation) was evident in patients who showed marked reduction of CMAP. Some of these patients also had myopathic (questionably pseudomyopathic – see below) changes in the proximal muscles sampled. Only four patients showed EMG evidence of possible myopathy. Sequential follow-up studies during clinical recovery of group 3 and 4 patients showed rapid improvement of CMAP amplitude during a period of 4–12 weeks that cannot be explained purely on the basis of axonal regeneration following a severe, extensive axonal neuropathy (*Fig. 6.34*).

Some patients during this recovery period show pseudomyopathic EMG changes. In one such case careful motor unit studies confirmed this impression. It is possible this might have been the case in the many patients described in various studies who reportedly showed 'myopathic' EMG. In a more recent report, Zochodne et al (1994) now recognize a group of patients identical to Group 3 patients with 'myopathic' EMG changes.

Thus the clinical neurophysiological and neuropathological findings of this ongoing prospective study clearly suggest a multifactorial problem. The exact aetiology of the neurophysiological and neuropathological abnormalities, even in clinically unaffected patients who have spent at least a week under intensive care, remains elusive. One possible factor, in at least some patients, may well prove to be an as yet unexplained side-effect of prolonged neuromuscular blockade (Hinds et al 1993). The rapid clinical recovery observed would favour this hypothesis.

In this study, the outcome for weakness in those who survived the underlying illness was good. Recovery often began within weeks after weaning from the ventilator and continued for several months. Thus it may be concluded that the syndrome of weakness in the ICU is multifactorial in origin, and that as yet the exact aetiological factors remain hypothetical. It appears most likely that the so-called 'critical illness neuropathy' reported by Bolton's group (Bolton et al 1984, Zochodne et al 1987) does in fact reflect a combination of various different pathophysiological mechanisms. These appear to include an unusual form of pure or predominantly motor syndrome with wide-spread denervation, axonal peripheral neuropathy, myopathy, neuromuscular junction disorder and possibly also simple atrophy. It is even possible that the pure motor type of the syndrome is a form of toxic or metabolic channelopathy. We suggest that at present it is prudent to group them simply as *critical illness neuromuscular syndromes* rather than designating them as specific neuropathies or myopathies. Since the presence of weakness complicating the course of critically ill patients in the ICU is often difficult to assess by purely clinical methods, more systematic neurophysiological assessment may well be particularly helpful both in planning management and in prognosis.

ACUTE ONSET NEUROPATHIES

Guillain–Barré syndrome (GBS)

The syndrome is named after Guillain, Barré and Strohl (1916), who described a benign, rapidly progressive acute flaccid paralytic illness associated with areflexia and albuminocytologic dissociation in the CSF. Following the detailed autopsy study by Asbury (1981) which established the immune-mediated inflammatory nature of the syndrome, the term GBS was used synonymously with 'acute inflammatory demyelinating polyneuropathy' (AIDP). Although an acute axonal form of GBS was reported over two decades ago (Feasby et al 1986), the concept of an 'axonal' form was not widely accepted. However, this situation changed dramatically about 12 years ago with the collaborative work between members of John Hopkins University, USA, and the Second Teaching Hospital of Hebei Province in Northern China, who had the unique opportunity to study systematically a large number of patients with clinically defined GBS. In a series of joint reports, they firmly established evidence for the existence of many different forms of GBS, including two distinct axonal variants. Their studies have also provided further insights into the ultrastructural immunopathological mechanisms involved in this group of disorders. The present view is that the different variants of GBS result from 'molecular mimicry' whereby an antecedent infection generates an immune response to specific antigenic targets in the organism which are also shared by normal host tissue components.

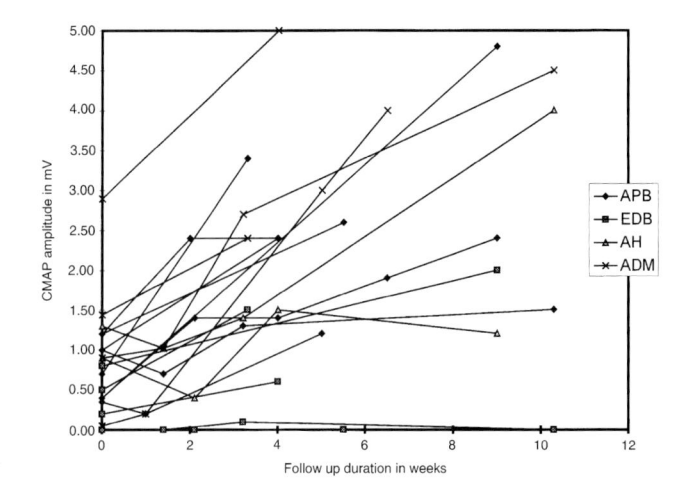

Fig. 6.34 Neuromuscular syndromes of critical illness. Compound muscle action potential (CMAP) amplitude change during recovery in 6 patients belonging to group 3 or 4 (see text) who had sequential studies. A total of 19 muscle groups was studied 9 from the abductor pollicis brevis (APB), 2 from abductor digiti minimi (ADM), 2 from the abductor hallucis (AH) and 4 from the extensor digitorum brevis (EDB). Note markedly abnormal CMAP amplitude in the initial study with varying degrees of improvement over a period of up to 11 weeks. Note also that APB and ADM show more striking improvement; EDB shows least or no improvement. (From Binnie et al (2003), by permission.)

The following is a classification of the different subtypes of GBS currently recognized:

1. *Acute inflammatory demyelinating polyneuropathy (AIDP)*. This is by far the commonest form of GBS in the west and accounts for 85–90% of cases of GBS (Govoni and Granieri 2001). Recent studies have shown binding of complement-fixing antibodies directed against the components of the outer Schwann cell membrane, leading to compliment activation followed by macrophage-mediated segmental demyelination and stripping of compact myelin (Griffin et al 1996).

2. *Axonal forms*.
   - *Acute motor axonal neuropathy (AMAN)*. This unique, almost purely motor form of GBS is recognized as the most frequent cause of acute flaccid paralysis in rural areas of northern China (McKhann et al 1993). Antibodies to the target antigen, GM1 ganglioside (a constituent of axonal membrane in the node of Ranvier), GD1a, Ga1NAc-GD1a and GD1b have been demonstrated in a large number of patients (Ogawara et al 2000). Antibody binding resulting in impairment of nodal conduction is believed to be the cause of inexcitable nerves seen in this condition. Neuropathological studies show no significant inflammation or demyelination of the compact myelin.
   - *Acute motor sensory axonal neuropathy (AMSAN)*. This form, where severe motor and sensory axonal

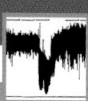

degeneration is seen, appears to result from a primary macrophage-mediated immune attack on axons (Griffin et al 1996). Again there is minimal inflammation or demyelination. It is likely that the cases originally reported by Feasby et al (1986) belonged to this group.

The following two clinical presentations may also be considered as other subtypes:

3. *Miller–Fisher syndrome (MFS)*. This is characterized by ophthalmoplegia, areflexia and ataxia. Affected cases show significantly raised antibodies against GQ1b glycolipid antigen. Cases of ataxic variant with anti-GQ1b antibody is also described suggesting that the ataxic GBS and MFS have a common autoimmune mechanism and form a continuous spectrum.

4. *Sensory GBS*. A purely sensory, acute relapsing sensory neuropathy has been described (Oh et al 2001) with antibodies directed against the GD1b antigenic components of dorsal root ganglia. In the strict clinical sense, this is not a form of GBS as there is no motor weakness; however, its inclusion cannot be disallowed in classifications which are now based primarily on the properties of antigenic determinants and their pathological effect. A pure pandysautonomia has also been described as a variant of GBS.

Clinical and electrodiagnostic features of the AIDP variant of GBS

The AIDP variant is the most prevalent form in Europe, the USA and Australia. It is now well established as an immunologically mediated acute neuropathy involving both cell-mediated and humoral mechanisms. Characteristically there is multifocal inflammatory demyelination and a variable degree of secondary axonal loss. Approximately 50–65% of the cases have been preceded by infections or some other antecedent event such as surgery or vaccination during the preceding weeks (Arnason 1984, Emilia-Romagna Study Group 1997). Several studies have implicated recent infection by *Campylobacter jejuni* (CJ), cytomegalovirus, Epstein–Barr virus and *Mycoplasma pneumoniae* as being of probable significance, but CJ is the most important agent associated with axonal degeneration, lack of sensory disturbances, severe disability, slow recovery and poor outcome (Winer et al 1988a, Govoni and Granieri 2001).

Conduction block is the main cause of the muscle weakness in AIDP. Limb weakness of acute onset is the most common symptom. In the majority of cases this weakness starts symmetrically in the legs and ascends, but in a small number of patients the onset of weakness may begin in the cranial nerves or proximally in the upper limbs and descend. The interval from onset to greatest weakness can vary between a few hours to some weeks. By definition, the severity of weakness must reach a plateau by 4 weeks from the onset of symptoms. Sensory symptoms are less prominent and more variable; a patient can suffer a complete flaccid tetraparesis without any sensory deficit, or have demonstrable and subjective sensory loss. At the nadir of the condition, assisted ventilation may be necessary. Cardiac arrhythmias and other symptoms of dysautonomia which necessitate monitoring can occur; their severity is unrelated to the extent of sensory or motor involvement (Tuck and McLeod 1981).

There is a mortality of 5–13%, with death occurring in the most severe phase, usually due to autonomic disturbances. For those patients who survive the most severe phase of the illness, the prognosis is good, with return to normal function in 60–67%. The remaining 20% or so may be left with a persisting disability (Wiederholt et al 1964, McLeod et al 1976, Winer et al 1988b) but are usually able to walk with assistance. In the latter study, the major features in the initial assessment which were associated with persistent disability were the time taken to become bed-bound, requirement for ventilation, age greater than 40 years, and small or absent compound abductor pollicis brevis muscle action potentials elicited by stimulation of the median nerve at the wrist.

Neurophysiological studies provide the basis for making the diagnosis of a primarily demyelinating peripheral nerve disorder, in assessing the anatomical extent of the lesions and the degree of secondary axonopathy. This latter factor can determine the final outcome. They have also been used to monitor the course of the disease or to describe specific patterns of abnormality.

Electrophysiological abnormalities represent changes resulting from patchy, but extensive, multifocal demyelinating peripheral nerve lesions with a variable degree of secondary axonopathy. Motor abnormalities are consistently more severe than sensory changes. Common findings include prolonged distal motor latencies – particularly those of median nerve at wrist and common peroneal nerve at ankle – and temporal dispersion of motor responses (*Fig. 6.35*). There may or may not be conduction block in the routinely tested elbow–wrist and knee–ankle segments. Van der Meché et al (1988) have pointed to a pattern of length-dependent reduction of compound muscle action potentials (CMAPs), with the response decreasing on moving the stimulating site proximally. Motor conduction velocities are usually reduced to a range of 30–40 m/s, but severe reduction is unusual. Neurophysiological findings may occasionally appear to deteriorate despite clinical improvement (*Fig. 6.36*) and it is not unusual to find one nerve showing improvement while another shows apparent deterioration (*Fig. 6.37*). Unlike classical axonal neuropathies, lower limb studies are no more likely to be abnormal than upper limb responses.

SAP abnormalities are common but variable. They include some slowing and dispersion with reduction of amplitude. These changes probably result from a combination of secondary axonopathy and desynchronized conduction. Complete sparing of SAPs may occur in patients with predominantly radicular presentation and in the early stages of the disease. A curious sparing of the sural nerve has been observed. In GBS, in contrast with most peripheral neuropathies, the median SAP is more often and more severely abnormal than the sural SAP (Murray and Wade 1980).

Minimum F wave latencies are sometimes slightly increased but may occasionally be normal. F waves are more often absent, or show reduced persistence, than significantly delayed. Kimura and Butzer (1975) were the first to measure F wave conduction velocity in GBS and show it to be slowed in proximal segments. Recording the F response can increase the yield of abnormality; Lachman et al (1980) found abnormalities of F wave latency in the median nerve in 10 out of 11 patients who had GBS and in whom peripheral nerve conduction studies were normal.

An important observation is that in all published series of patients examined within 3 weeks of onset of the illness 12–20% had no demonstrable abnormality of nerve conduction (Lambert and Mulder 1964, Eisen and Humphreys 1974, McLeod et al 1976, Asbury 1981, Walsh et al 1984), although McLeod (1981) considered that the likelihood of detecting abnormality increases with the number of nerves studied. The experience of examining a patient acutely disabled by presumed peripheral nerve disease, and being unable to demonstrate a conduction defect, can be disconcerting for the neurophysiologist. However, if followed sequentially, the majority of patients do develop abnormalities (Ropper and Shahani

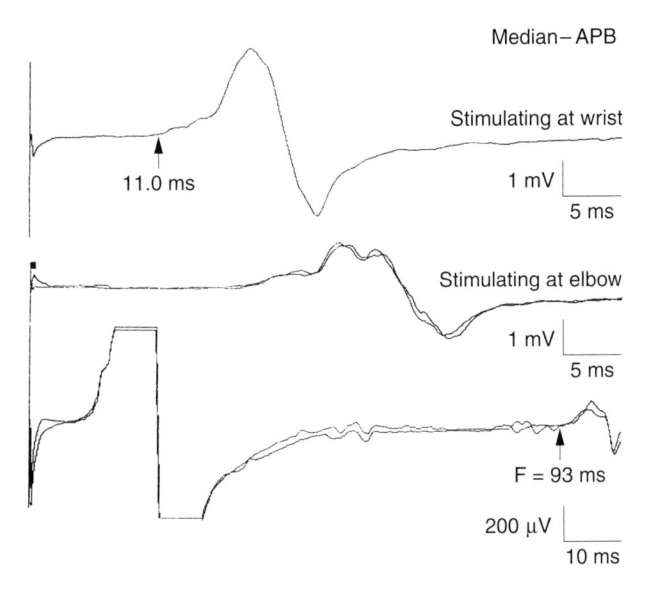

Fig. 6.35 Typical nerve conduction abnormalities in a 36-year-old patient with Guillain–Barré syndrome presenting with acute generalized weakness. Note prolonged distal motor latency to abductor pollicis brevis at 11 ms (arrow, top trace), dispersed and reduced amplitude response to stimulation at elbow (middle trace), with a calculated motor conduction velocity of 34 m/s and markedly delayed F-response at 93 ms (bottom trace). (From Binnie et al (2003), by permission.)

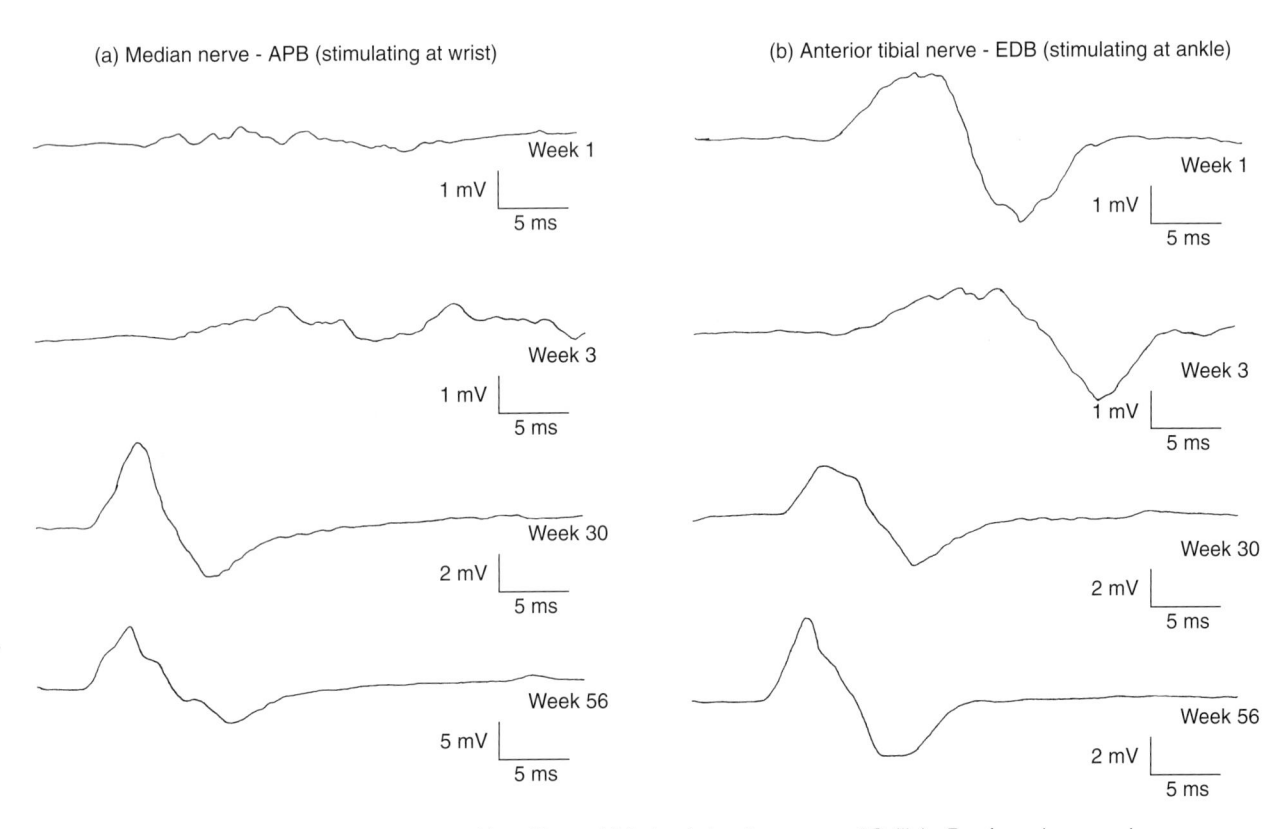

Fig. 6.36 CMAP changes recorded from upper (a) and lower (b) limbs during the course of Guillain–Barré syndrome and recovery period. Note prolonged distal motor latency (DML) and dispersion of waveforms at initial clinical presentation (top traces), and subsequent electrophysiological deterioration at 3 weeks, with further dispersion of the CMAP despite clinical improvement (second trace). Further follow-up studies at 30 and 56 weeks show progressive reduction in DML with increasing amplitude and an improvement in synchronization of conduction (third and fourth traces). (From Binnie et al (2003), by permission.)

(a) Median nerve – APB (stimulating at wrist)

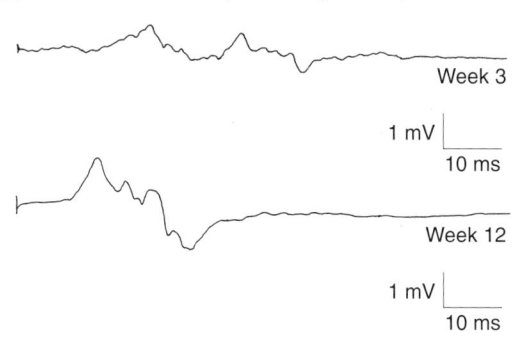

Week 3

1 mV
10 ms

Week 12

1 mV
10 ms

(b) Anterior tibial nerve – EDB (stimulating at ankle)

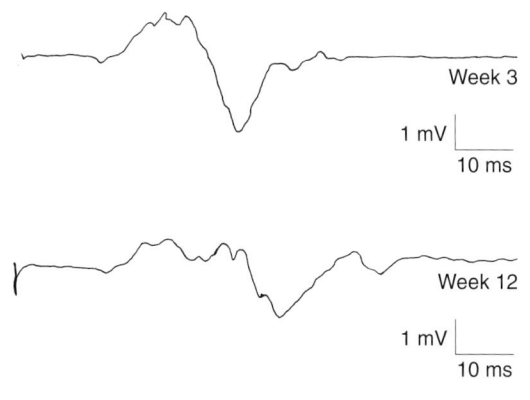

Week 3

1 mV
10 ms

Week 12

1 mV
10 ms

Fig. 6.37 Electrophysiological improvement in one nerve concurrent with deterioration in another during the course of illness in a patient with Guillain–Barré syndrome. (a) Median nerve recording following stimulation at the wrist from abductor pollicis brevis shows marked improvement in CMAP amplitude, duration and waveform between weeks 3 and 12. (b) Anterior tibial nerve study following stimulation at the ankle, in the same patient carried out at the same times. The follow-up at 12 weeks shows deterioration in DML and further desynchronization of the CMAP recorded from extensor digitorum brevis. (From Binnie et al (2003), by permission.)

1984, Albers et al 1985) but these may only become apparent after a week to 10 days. It is worth remembering that most routine nerve conduction studies test only distal segments of major nerves and only a small proportion of functioning nerve fibres. Studies which have included vertebral stimulation have clearly shown evidence of proximal conduction block, which would explain why some patients who otherwise clinically have GBS fail to show abnormalities in routine distal segment studies (Brown and Feasby 1984a, Mills and Murray 1986). It also explains why routine nerve conduction studies may appear to show deterioration at a time when the patient is clinically improving.

Studies of SEPs have added further information about conduction in proximal segments (Brown and Feasby 1984b, Walsh et al 1984, Gilmore and Nelson 1989) and the demonstration of an abnormal median SEP adds weight to the clinical significance of absent F waves, which may otherwise be the only abnormality in the early stages of paresis. However, both the study by Walsh et al (1984), and that by Gilmore and Nelson (1989), showed it was possible to have normal F wave latencies and an abnormal median SEP, or vice versa.

Clinical and electrodiagnostic features of the AMAN variant of GBS

This form of GBS has been recognized for more than 20 years in Northern China (McKhann et al 1993, Griffin et al 1995). It has recently been shown to occur in Mexico and probably also occurs in other parts of the world. In Northern China it occurs mainly in rural areas and shows a peak incidence during the late summer months, predominantly affecting children. Clinically, the motor syndrome is indistinguishable from the AIDP form. There is, however, no sensory loss. Electrophysiologically, it is characterized by normal SAPs and a lack of demyelinating changes. Motor conduction velocities and distal motor latencies are normal. There is marked reduction of CMAP amplitudes, with affected muscles showing evidence of acute denervation but, interestingly, the speed of recovery is comparable to the demyelinating forms of GBS. It has now been shown that the pathology in this condition is a specific immunological attack by cross-reacting, anti-GM1, IgG antibodies against lipopolysaccharide epitopes of motor axon membrane in the region of the node of Ranvier. Pathologically there is minimal cellular reaction and no internodal demyelination. It is believed that the weakness is caused by antibodies blocking axonal conduction and/or by axonal degeneration affecting the exposed and vulnerable parts of most distal intramuscular nerve branches. Serological studies carried out within 10 days of weakness show antecedent *Campylobactor jejuni* infection in about 75% of cases.

Clinical and electrodiagnostic features of other GBS variants

The pathological features of the acute motor–sensory axonal (AMSAN) type are essentially the same as those of AMAN except for the following:
1. sensory fibres are also involved;
2. there is more severe axonal damage with prominent Wallerian degeneration; and
3. prognosis is worse than for the AIDP and AMAN types.

Miller–Fisher syndrome (MFS) accounts for 3–7% of cases of GBS in western literature. Serum IgG anti GQ1b antibody has been found in the acute phase of 80–100% of cases. Although controversy has arisen since the introduction of the term 'brainstem encephalitis' by Bickerstaff, presently there is ample evidence to indicate that MFS is a variant of GBS. Nerve conduction studies in the pure form may show reduced sensory action potential amplitudes. Conventional motor conduction studies are often normal.

Acute intermittent porphyria

Peripheral neuropathy can occur in several of the forms of inherited porphyria: acute intermittent porphyria

Table 6.6 Channelopathies

Type of channelopathy	Inheritance: gene and chromosomal location	Ion channel abnormality	Main clinical features	Precipitants of attacks	Investigations	Treatment
Inherited channelopathies presenting with myotonia or muscle stiffness						
Myotonia congenita	AR (Becker type) more common, more severe; AD (Thomsen) mild form. CLCN1, 7q35	Defective inward chloride current	Onset in early (Thomsen) or late (Becker) childhood. Non-progressive. Generalized and painless muscle stiffness. Lower limbs often affected. Paralysis prominent after a period of rest. Improves with continued activity ('warm-up' phenomenon). Muscle hypertrophy common but wasting and some fixed weakness may also occur; no periodic paralysis. Some distal forearm weakness and atrophy in recessive form, may be confused with myotonic dystrophy. Heterozygous parents of Becker type may show myotonia	Prolonged rest maintaining the same posture. Not influenced by cold	No ECG changes. Repetitive stimulation may show a decremental response. EMG: prominent myotonia, normal motor units. Muscle biopsy: type 2B fibres may be absent. Cause not known. No myopathic changes	Exercise and avoidance of prolonged rest. Mexiletine (a sodium channel blocker) to prevent myotonia. In most cases, no treatment required
Myotonic dystrophy (Charlet et al 2002, Mankodi et al 2002)	Type 1: 19q13.3 Type 2: 3q abnormal splicing of CLC-1 RNA	Defective chloride channel function	Type 1: classical features including muscle wasting, myotonia, subcapsular cataracts, cardiac conduction abnormalities, hearing defects, gonadal atrophy and cognitive deficits. Expansion of an unstable CTG repeat in myotonic dystrophy protein kinase gene (*DMPK*). Type 2: more proximal pattern (PROMM), other features variable. Expansion of a CCTG tetranucleotide repeats in zinc finger protein 9 (*ZNF9*) gene	Not episodic	EMG: typical dive-bomber myotonic discharges. Genetic testing	Symptomatic
Potassium aggravated myotonia	AD: sodium channel alpha subunit SCN4A, 17q23-q25	Mildly reduced fast sodium channel inactivation	Includes myotonia fluctuans (mild form), acetazolamide responsive myotonia and myotonia permanens (severe form). Clinical severity depends on degree of channel function disruption. Onset in childhood. Face, especially eyelids, hands and paraspinal muscles mainly affected. Stiffness only, often painful but no weakness and not temperature dependent. Some cases worsened by cold or exercise. In myotonia fluctuans there is marked variability in the severity of muscle stiffness, fluctuating daily. Myotonia is constant in myotonia permanens. A painful form is also reported	Ingestion of K+ rich food, fasting, exertion, exposure to cold, infection	ECG changes of hyperkalaemia. EMG: myotonia	Mexiletine or acetazolamide. Monitor during surgery for rigidity and rhabdomyolysis

Clinical conditions affecting the peripheral nervous system and muscles encountered in the ICU

Paramyotonia congenita	AD: sodium channel alpha subunit SCN4A, 17q23-q25	Reduced fast sodium channel inactivation	Onset in early childhood. Myotonia most obvious in facial, eyelid, forearm and hand muscles. Muscle stiffness is usually painless, aggravated by cold or exercise (paradoxical myotonia). Further continued cooling or exercise cause attacks of weakness which may last hours. Condition often non-progressive. (There may also be coexisting hyperPP.) No hypertrophy or atrophy of muscles	Cold exposure followed by exercise causes focal paralysis. (When associated with hyperPP there are also other precipitating factors)	Myotonia at rest but prominent fibrillations with cooling; with further cooling fibrillations disappear. CMAP amplitude is reduced after cooling of the limb. Serum K^+ may be low, normal or raised (raised in the presence of coexisting hyperPP). CK slightly elevated. Muscle biopsy myopathic	Mexiletine prevents myotonia as well as cold-provoked paralysis. Acetazolamide often provokes weakness, but some patients may benefit
Malignant hyperthermia (MH) and central core disease (CCD) (Quane et al 1993, Greenberg 1999)	MH: commonly AD, RYR1 or CACNL1AS calcium channel. Others include CACNL2A and SCN4A CCD: RYR1	MH: defective ryanodine receptor causing increased Ca^{2+} release from sarcoplastic reticulum CCD: defective Ca^{2+} release	MH: attacks of skeletal muscle rigidity, pain and weakness and rhabdomyolysis associated with fever, sympathetic hyperactivity and hypermetabolism. Myoglobinurea may cause renal shutdown. CCD: caused by different mutations of the same gene RYR1. Presents during infancy with hypotonia, proximal weakness and skeletal abnormalities. Both dominant and recessive forms exist. Susceptible to MH	MH: inhalational anaesthetic agents and succinylcholine. Less severe episodes by alcohol, exercise, neuroleptic drugs and infections	MH: high CK, acidosis and myoglobinurea during an episode. Halothane contracture test. Gene testing. CCD: typical muscle histochemistry with amorphous central cores in type 1 fibres	MH: during an episode dantrolene, cooling, bicarbonate and maintenance of urine output. Discontinuation of anaesthesia, avoidance of Ca^{2+}, Ca^{2+} antagonists and beta blockers

Channelopathies presenting with periodic paralysis

Hyperkalaemic periodic paralysis with myotonia (hyperPP + myotonia)	AD, sodium alpha subunit SCN4A, 17q23-q25	Reduced inactivation fast and slow sodium channels	Symptoms first noticed during childhood, episodic flaccid weakness, rapid onset, affecting mainly proximal limb muscles, may be focal when followed by specific exercise or generalized following sleep. Respiratory and bulbar muscles spared. Attacks begin within minutes, usually in the morning, last 20 min to 2 h, and can be aborted or delayed by mild exercise. Weakness milder than with hypoPP. No residual weakness. Reflexes reduced or absent. Occasional myotonia between paralytic attacks affecting facial, eyelid, tongue and hand muscles. Frequency of attacks lessens with time but repeated attacks over years may lead to persistent proximal muscle weakness	Rest after exercise, fasting, exposure to cold, sleep, anaesthesia, K^+ rich food	ECG changes of hyperkalaemia, CK increased during attacks. EMG: fibrillations and myotonia often precede an attack, affected muscles inexcitable or less excitable during paralysis. During an attack CMAP amplitude is reduced. Muscle biopsy myopathic. Provocative tests using potassium potentially dangerous	During an attack hydrochloro-thiazide or beta adrenergic inhalation. Mexiletine to prevent further attacks of myotonia. Also avoidance of strenuous exercise, prolonged fasting and exposure to cold. CHO rich and low potassium diet beneficial

Continued

Table 6.6 Continued

Type of channelopathy	Inheritance: gene and chromosomal location	Ion channel abnormality	Main clinical features attacks	Precipitants of	Investigations	Treatment
Hypokalaemic periodic paralysis (Hypo)	AD, calcium alpha subunit CACNL1AS, 1q31-q32, poorly expressed in females. M:F = 3:1, commonest type of HypoPP	Calcium channel inactive, leading to defect in control of resting membrane potential	Onset in adolescence (or 5–35 years); M:F = 3:1; may spontaneously improve thereafter. Episodic proximal weakness of limbs, lasts hours or days, hypotonic and often areflexic limbs during attack, rapid recovery. Weakness may be severe. Frequency: several/week to one every few months. Hypokalaemia during an attack. Residual, often slowly progressive, proximal leg weakness present, occasionally permanent. Eyelid myotonia may be present	CHO or salt rich food, after sleep, rest following exercise, alcohol, emotional changes, etc. Light physical activity may prevent an attack	Hypokalemia during, but not between, attacks. CK increased during attack. EMG: no myotonia, muscles inexcitable during an attack. CMAP amplitude reduced during attack. Exercise test to demonstrate defective repolarization. Biopsy myopathic, with vacuoles and tubular aggregates. Provocative tests: oral glucose with or without insulin	Oral potassium solution for an attack. Low CHO and low Na diet, acetazolamide and other K+ sparing drugs for prevention
Thyrotoxic periodic paralysis	Sporadic or autosomal dominant, males more often affected	Not known	Affects particularly patients of Asian ethnic origin. Common in summer months. Onset 18–40 years. M:F = 20:1. Hypokalaemia during paralysis. Weakness lasts hours or days and involves proximal > distal, legs > arms. Respiratory and bulbar muscles may also be affected	CHO rich food, rest after exercise, muscle cooling, thyroxine ingestion	Hypokalemia during attack. CMAP reduced during attack. Exercise test to demonstrate defective repolarization. Myopathy uncommon. No myotonia	As for HypoPP, careful potassium replacement and correction of thyroid function. Not acetazolamide |//

Other channelopathies causing periodic paralysis (PP)

Hypokalaemic periodic paralysis (HypoPP2), AD (Sternberg et al 2001)	AD, sodium alpha subunit SCN4A, 17q23-q25	? Reduced number of excitable Na+ channels	Most features similar to calcium channel HypoPP. Rare. Single family reported. Onset 9 years. Muscle pain common after attacks. Permanent muscle weakness in some cases	Rest following exercise, sleep	ECG-changes of hypokalaemia. EMG: usually no myotonia. Muscle biopsy: myopathic with tubular aggregates	Not acetazolamide, often deleterious

HypoPP3 (Abbott et al 2001)	Potassium channel, *KCNE3*, 11q13	Not known	Rare. Paralysis lasts hours to days. No myotonia, not precipitated by high CHO meals. Myopathy with tubular aggregates. Both hypo- and hyperkalaemic forms reported	Strenuous exercise	Serum K^+ may be high, low or normal. ECG changes	CHO loading
Anderson syndrome (Plaster et al 2001, Tristani-Firouzi et al 2001)	AD, potassium channel *KCNJ2* Kir 2.1, 17q23	Reduced inward rectifying K^+ current	Features other than PP include skeletal and facial dysmorphism, cardiac conduction abnormalities causing syncope and sudden death, prolonged QT interval. Onset 2–18 years. Paralytic episodes last 1–36 h	Exercise and K^+	Serum K^+ may be high, low or normal. ECG changes	Amiodorone. or acetazolamide
Distal renal tubular acidosis (Nicoletta and Schwartz 2004)	AS or AR, *SLC4A1*, 17q21	Defective chloride–bicarbonate ion exchange	Endemic in north-east Thailand. Paralysis lasts hours. Other features include osteomalacia and pathological fractures. No myotonia	Not known	Hypokalaemia and hyperchloraemic acidosis in the presence of alkaline urine	K^+ and bicarbonate. Not acetazolamide (worsens acidosis)
X-linked form of periodic paralysis (Ryan et al 1999)	Xp 22.3, recessive	Not known	Onset in infancy and childhood, respiratory and bulbar muscles may be involved. Severity variable. Episodic prolonged weakness lasting days to months. Myalgia and cramps may be present	Not known	EM: dilatation and proliferation of sarcoplasmic reticulum. Basal ganglia calcification	Not known

CHO, carbohydrate; hyperPP, hyperkalaemic periodic paralysis.

(mostly found in Sweden), porphyria variegata (most common in the Afrikaans population of South Africa), and in some of the hereditary coproporphyrias. Only in porphyria variegata are there the associated skin lesions with characteristic light-sensitive bullous eruptions. The change of urine colour on standing to a 'port-wine colour' reflects the abnormally high levels of excretion of porphobilinogen and is seen in all types of inherited porphyria. The onset of neuropathy may be precipitated by the administration of barbiturates, given as part of an anaesthetic. Up-to-date lists of porphyrinogenic agents can be found in the major pharmacopoeias (e.g. the *British National Formulary* and its equivalents in other countries); common precipitants include sulphonamides (e.g. co-trimoxazole, erythromycin), anticonvulsants such as barbiturates, carbamazepine and phenytoin, benzodiazepines, antihistamines, diuretics, oral contraceptives, tamoxifen. Alcohol, especially red wine and whisky, and starvation are also known to precipitate attacks of acute porphyria.

Patients rarely present before adolescence. Features of the neuropathy are similar, whatever the associated metabolic defect, and in the acute phase the patient can present with a flaccid quadriparesis. Characterization of the neuropathy has shown it to be predominantly an axonal motor neuropathy with involvement of proximal muscles, although fibrillations may be found in both proximal and distal muscles. Cranial nerves may be involved. There is relative sparing of sensory fibres (Albers et al 1978). Autonomic neuropathy is considered to be the cause of acute features such as abdominal pain, vomiting, tachycardia and hypertension.

The exact mechanism of causation of the neuropsychiatric manifestations is not known. In most cases of acute porphyria abdominal or psychiatric symptoms have preceded the onset of neuropathy, but the diagnosis should be considered in patients with acute onset neuropathy who do not show slowing of conduction velocity or temporal dispersion of CMAPs. Unfortunately the diagnosis is easy to miss due to the rarity of the condition and the low sensitivity of the screening tests, if not requested early, during the acute phase.

Other causes of acute or subacute peripheral neuropathy
Other causes of acute or subacute neuropathy are much less common than GBS but should be considered if there are atypical features of the illness.

ACUTE WEAKNESS DUE TO DISORDERS OF NEUROMUSCULAR TRANSMISSION

Botulism
Though rare, botulism is an important treatable condition, which often requires management in the ICU for weeks or months. It is caused by one of the most potent toxins known to man. The classical food-borne botulism is now rare. Recently reported cases are due to wound botulism (e.g. intravenous drug abusers). Other subtypes include infant botulism (usually infants of less than 6 months of age due to their inability to resist colonization of bacterial spores from a contaminated source, e.g. honey products), inadvertent botulism (due to iatrogenic use of botulinum toxin for movement disorders, etc.) and hidden botulism (without obvious wound or source of contamination of food). Generalized weakness comes on 12–36 h after the ingestion of botulinum toxin, accompanied by respiratory, pharyngeal and extraocular paralysis and blurred vision (Cherington 1974). The weakness may continue to worsen for 4 or 5 days.

Botulinum toxin has a presynaptic action of reducing the number of quanta of acetylcholine released by a depolarizing potential (Kao et al 1976). The electrophysiological findings therefore reflect blocked neuromuscular transmission with preserved nerve conduction (Oh 1977). Sensory conduction studies are normal. CMAPs can be markedly diminished in amplitude with normal or near-normal motor conduction velocities, particularly in limbs which are clinically weak (Cherington 1982). With slow rates of repetitive stimulation (2 Hz), a decremental response may be seen, whereas with rapid rates (50 Hz), an increase in the amplitude of the responses may occur (Cherington 1974, Oh 1977). This phenomenon of post-tetanic potentiation can also be elicited by demonstration of the baseline CMAP amplitude to a single supramaximal stimulus and again repeating it after about 15–20 s of maximum voluntary contraction. An increase of more than 50% supports the diagnosis. This is not an invariable finding and may be more common in some types of botulinum intoxication than others. Any increment that does occur is less pronounced than in patients with Lambert–Eaton syndrome (Valli et al 1983). Single fibre EMG shows increased jitter and blocking (Schiller and Stålberg 1978) which may improve with continued contraction of the muscle or with higher rates of stimulation. EMG sampling reveals small, short-duration motor units; spontaneous fibrillations are a variable finding (Oh 1977). Management is mainly supportive. Recovery may take several weeks or months, but is usually complete.

Myasthenia gravis
The clinical distinction between myasthenia and GBS, both of which can present with acute tetraplegia and ophthalmoplegia, can be difficult, although Ropper and Shahani (1984) have pointed to the relative preservation of jaw power compared with facial muscles in GBS, the jaw sometimes hanging loose in myasthenic patients who have facial weakness. Responses to repetitive stimulation are clearly important in these circumstances.

Familial periodic paralysis and the channelopathies
Recent developments in molecular genetics (Hoffman et al 1995, Davies and Hanna 2001) have resulted in identi-

fication of genes of many ion channels that are vital for proper functioning of the cell, particularly nerve and muscle. Consequently the classification of this group of disorders is now based on molecular mechanisms rather than on phenotypes; they are becoming known as 'channelopathies'. Disorders with varied presentation have now been reclassified under a new title, a classical example being the mutations of the skeletal muscle sodium channel alpha subunit gene, now called *SCN4A*. Different mutations of the same gene are now known to cause *hyperkalaemic periodic paralysis*, *paramyotonia congenita* and *potassium aggravated myotonia*. More recently muscle disorders such as malignant hyperthermia (ryanodine calcium channel dysfunction), Anderson syndrome (periodic paralysis, ventricular arrhythmia and dysmorphism resulting from potassium channel disorder) and central core disease have also been attributed as ion channel disorders.

Acute episodic paralysis or muscle stiffness is often the clinical presentation in these rare inherited ion channel disorders, particularly those associated with hypo- or hyperkalaemia. Serum potassium during a paralytic episode can be raised, low or normal (Chesson et al 1979). Attacks may be precipitated by a high carbohydrate intake, alcohol, heavy exercise, rest following exercise, sleep, stress or cold; they may occur nocturnally. The precipitating cause varies depending on the ion channel abnormality.

During attacks of periodic paralysis, muscle is inexcitable either voluntarily or by electrical stimulation; motor units can be observed to decrease in amplitude as the paralysis progresses (Layzer et al 1967, Gordon et al 1970). Sensory conduction studies are normal. Motor studies may show inexcitable nerves or reduced amplitude CMAP. *Table 6.6* briefly summarizes the genetic abnormalities, salient clinical features and diagnostic methods in the periodic paralyses and related disorders.

Non-peripheral causes of acute onset generalized weakness

An alternative, which is not infrequently considered in the differential diagnosis of the acute onset of weakness, is a high cervical cord lesion. This may present as an ascending paresis during the acute phase of spinal shock. Tendon reflexes may be lost. Likewise ophthalmoplegia and quadraparesis due to basilar occlusion can present a clinical picture virtually indistinguishable from GBS (Ropper and Shahani 1984).

CONCLUSIONS ON PRESENT STATUS OF NEUROPHYSIOLOGICAL WORK IN THE ICU

EEG monitors, and even EP recorders, may eventually become a routine module of general patient monitoring systems in the ICU, as their potential contribution becomes increasingly regarded in clinical decision-making. At present their role is primarily in the management of major sedative regimens and in patients with cerebral disorders where information from conventional neurological assessment is inadequate or inaccessible. EMG and nerve conduction studies have important contributions to make in the management of neuromuscular conditions commonly leading to admission to an ICU.

REFERENCES

Abbott GW, Butler MH, Bendahhou S, et al 2001 MiRP2 forms potassium channels in skeletal muscle with Kv 3.4 and is associated with periodic paralysis. Cell 104:217–231.

Adams JH, Graham DI, Jennett B 2000 The neuropathology of the vegetative state after an acute brain insult. Brain 123:1327–1338.

Adams RD, Jequier M 1969 The brain death syndrome: hypoxemic panencephalopathy. Schweiz Med Wochen 99:65–69.

Albers JW, Roberston WC, Daube JR 1978 Electrodiagnostic findings in acute porphyric neuropathy. Muscle Nerve 1:292–296.

Albers JW, Donofrio PD, McGonagle TK 1985 Sequential electrodiagnostic abnormalities in acute inflammatory demyelinating polyradiculoneuropathy. Muscle Nerve 8:528–539.

Aldrich EF, Eisenberg HM, Saydjari C, et al 1992 Predictors of mortality in severely head-injured patients with civilian gunshot wounds: a report from the NIH Traumatic Coma Data Bank. Surg Neural 38:418–423.

Altafullah I, Asaikar S, Torres F 1991 Status epilepticus: clinical experience with two special devices for continuous cerebral monitoring. Acta Neural Scand 84:374–381.

Amantini A, Grippo A, Fossi S, et al 2005 Prediction of 'awakening' and outcome in prolonged acute coma from severe traumatic brain injury: evidence for validity of short-latency SEPs. Clin Neurophysiol 116:229–235.

American Academy of Neurology Committee on Ethical Affairs 1993 Persistent vegetative state: report of the American Neurological Association Committee on Ethical Affairs. Ann Neurol 33:386–390.

American Academy of Neurology Quality Standards Subcommittee 1995 Practice parameters for determining brain death in adults. Neurology 45:1012–1014.

American Academy of Pediatrics Task Force 1987 Report of a special task force: guidelines for the determination of brain death in children. Pediatrics 80:298–300.

Amin D, Binnie CD 1987 Pharmacologically-induced alpha coma. Electroencephalogr Clin Neurophysiol 67:50P.

Amodio P, Marchetti P, Del Piccolo F, et al 1999 Spectral versus visual EEG analysis in mild hepatic encephalopathy. Clin Neurophysiol 110:1334–1344.

Anderson DC, Bundlie S, Rockswold GL 1984 Multimodality evoked potentials in closed head trauma. Arch Neurol 41:369–374.

Angell M 1994 After Quinlan: the dilemma of the persistent vegetative state. N Engl J Med 330:1524–1525.

Arnason BG 1984 Acute inflammatory demyelinating polyradiculoneuropathies. In: *Peripheral Neuropathy* (eds PJ Dyck, PK Thomas, EH Lambert, R Bunge). WB Saunders, Philadelphia, pp. 2050–2100.

Anziska BJ, Cracco RQ 1980 Short-latency somatosensory evoked potentials: studies in patients with focal neurological disease. Electroencephalogr Clin Neurophysiol 49:227–239.

Arts WFM, van Dongen HR, van Hof-van Duin J, et al 1985 Unexpected improvement after prolonged posttraumatic vegetative state. J Neurol Neurosurg Psychiatry 48:1300–1303.

Chapter 6 Neurophysiological Work in the ICU

Asbury AK 1981 Diagnostic considerations in Guillain–Barré syndrome. Ann Neurol 9(Suppl):1–5.

Aserinsky E, Kleitman N 1955 Regular periods of eye motility and concomitant phenomena during sleep. Science 118:273–274.

Bachmann P, Gaussorgues P, Piperno D, et al 1987 Acute myopathy after status asthmaticus [in French]. Presse Méd 16:1486.

Barelli A, Valente MR, Clemente A, et al 1991 Serial multimodality-evoked potentials in severely head-injured patients: diagnostic and prognostic implications. Crit Care Med 19:1374–1381.

Bassetti C, Bomio F, Mathis J, et al 1996 Early prognosis in coma after cardiac arrest: a prospective clinical, electrophysiological, and biochemical study of 60 patients. J Neurol Neurosurg Psychiatry 61:610–615.

Bastuji H, García-Larrea L, Bertrand O, et al 1988 BAEP latency changes during nocturnal sleep are correlated with body temperature variations, but not with sleep stages. Electroencephalogr Clin Neurophysiol 70:9–15.

Bauby J-D 1997 *The Diving-Bell and the Butterfly*. Fourth Estate, London.

Belsh JM, Chokroverty S 1987 Short-latency somatosensory evoked potentials in brain-death patients. Electroencephalogr Clin Neurophysiol 68:75–78.

Berek K, Lechleitner P, Luef G, et al 1995 Early determination of neurological outcome after prehospital cardiopulmonary arrest. Stroke 26:543–549.

Bertrand O, García-Larrea L, Artru F, et al 1987 Brainstem monitoring. I: A system for high rate BAEP sequential monitoring and feature extraction. Electroencephalogr Clin Neurophysiol 68:433–145.

Binnie CD, Prior PF, Lloyd DSL, et al 1970 Electroencephalographic prediction of fatal anoxic brain damage after resuscitation from cardiac arrest. BMJ 4:265–268.

Binnie CD, Cooper R, Mauguière F, et al 2003 *Clinical Neurophysiology*, volume 2, *EEG, Paediatric Neurophysiology, Special Techniques and Applications*. Elsevier, Amsterdam.

Binnie CD, Cooper R, Mauguière F, et al 2004 *Clinical Neurophysiology*, volume 1, revised and enlarged edition, *EMG, Nerve Conduction and Evoked Potentials*. Elsevier, Amsterdam.

Bobbin RP, May JG, Lemoine RL 1979 Effects of pentobarbital and ketamine on brain-stem auditory potentials. Arch Otolaryngol 105:467–470.

Boisen E, Siemkowicz E 1976 Six cases of cerebromedullospinal disconnection after cardiac arrest. Lancet 1:1381–383.

Bolton CF, Young GB 1989 Neurological complications in critically ill patients. In: *Neurology and General Medicine* (ed MJ Aminoff). Churchill Livingstone, New York, pp. 713–720.

Bolton CF, Gilbert JJ, Hahn AF, et al 1984 Polyneuropathy in critically ill patients. J Neurol Neurosurg Psychiatry 47:1223–1231.

Boston JR, Deneault LG 1984 Sensory evoked potentials: a system for clinical testing and patient monitoring. Int J Clin Monit Comput 1:13–19.

Braitenberg V 1978 Cortical architectonics: general and areal. In: *Architectonics of the Cerebral Cortex* (eds MAB Brazier, H Petsche). Raven Press, New York, pp. 443–465.

Brewer CC, Resnick DM 1984 The value of BAEPs in assessment of the comatose patient. In: *Evoked Potentials II* (eds RH Nodar, C Barber). Butterworth, Boston, MA, pp. 578–581.

Bricolo A, Turella G 1973 Electroencephalographic patterns of acute post-traumatic coma: diagnostic and prognostic value. J Neurosurg Sci 17:278–285.

Bricolo A, Faccioli F, Grosslercher JC, et al 1987 Electrophysiological monitoring in the intensive care unit. In: *The London Symposia* (Electroencephalogr Clin Neurophysiol Suppl 39) (eds RJ Ellingson, NMF Murray, AM Halliday). Elsevier, Amsterdam, pp. 255–263.

Brierley JB, Adams JH, Graham DI, et al 1971 Neocortical death after cardiac arrest. Lancet ii:560–565.

British Paediatric Association 1991 *Diagnosis of Brain Stem Death in Infants and Children*. A Working Party Report. British Paediatric Association, London, pp. 1–6.

Brodal A 1969 Neurological Anatomy. Oxford University Press, New York.

Broughton R, Stampi C, Romano F, et al 1986 A case of petit mal epilepsy with discharges recurring with BRAC rate ultradian rhythmicity. Electroencephalogr Clin Neurophysiol 64:95P.

Brown WF, Feasby TE 1984a Conduction block and denervation in Guillain–Barré polyneuropathy. Brain 107:219–239.

Brown WF, Feasby TE 1984b Sensory evoked potentials in Guillain–Barrè polyneuropathy. J Neurol Neurosurg Psychiatry 47:288–291.

Brunko E, Zegers de Beyl D 1987 Prognostic value of early cortical somatosensory evoked potentials after resuscitation from cardiac arrest. Electroencephalogr Clin Neurophysiol 66:15–24.

Brunko E, Delecluse F, Herbaut AG, et al 1985 Unusual pattern of somatosensory and brainstem auditory evoked potentials after cardio-respiratory arrest. Electroencephalogr Clin Neurophysiol 62:338–342.

Buchner H, Ferbert A, Scherg M, et al 1986 Evoked potentials monitoring in brain death. Generators of BAEPs and spinal SEP. In: *Clinical Problems of Brainstem Disorders* (eds K Krunze, WH Zangemneister, A Arlt). Springer-Verlag, Vienna, pp. 130–133.

Buchner H, Ferbert A, Hacke W 1988 Serial recording of median nerve stimulated subcortical somatosensory evoked potentials (SEPs) in developing brain death. Electroencephalogr Clin Neurophysiol 69:14–23.

Bullock R, Chesnut RM, Clifton G, et al 1996 Guidelines for the management of severe head injury. J Neurotrauma 13:639–731.

Cairns H, Oldfield RC, Pennybacker JB, et al 1941 Akinetic mutism with an epidermoid cyst of the third ventricle. Brain 64:273–290.

Cant BR 1987 Evoked potential monitoring of post-traumatic coma and its relation to outcome. In: *The London Symposia* (Electroencephalogr Clin Neurophysiol Suppl 39) (eds RJ Ellingson, NMF Murray, AM Halliday). Elsevier, Amsterdam, pp. 250–254.

Cant BR, Hume AL, Judson JA, et al 1986 The assessment of severe head injury by short latency somatosensory and brain-stem auditory evoked potentials. Electroencephalogr Clin Neurophysiol 65:188–195.

Carlsson CA, Von Essen C, Löfgren J 1968 Factors affecting the clinical course of patients with severe head injuries. I: Influence of biological factors. II. Significance of posttraumatic coma. J Neurosurg 29:242–251.

Carter BG, Butt W 2001 Review of the use of somatosensory evoked potentials in the prediction of outcome after severe brain injury. Crit Care Med 29:178–186.

Carter BG, Taylor A, Butt W 1999 Severe brain injury in children: long-term outcome and its prediction using somatosensory evoked potentials (SEPs). Intensive Care Med 25:722–728.

Cenzato M, Ducati A, Fava E, et al 1988 Evaluation of traumatic coma by means of multimodality evoked potentials. In: *Evoked Potentials. ICU, Surgical Monitoring* (eds BL Grundy, RM Villani). Springer-Verlag, Berlin, pp. 153–164.

Cerutti S, Saranummi N (eds) 1997 Improving control of patient status in critical care. IEEE Eng Med Biol Mag 16:19–79.

Charlet B, Singh GN, Philips AV, et al 2002 Loss of the muscle-specific chloride channel in type 1 myotonic dystrophy due to misregulated alternative splicing. Mol Cell 10:45–53.

Chatrian GE 1986 Electrophysiologic evaluation of brain death: a critical appraisal. In: *Electrodiagnosis in Clinical Neurology*, 2nd edition (ed. MJ Aminoff). Churchill Livingstone, New York, pp. 669–736.

References

Chatrian GE 1990 Coma, other states of altered responsiveness, and brain death. In: *Current Practice of Clinical Electroencephalography*, 2nd edition (eds DD Daly, TA Pedley). Raven Press, New York, pp. 425–487.

Chatrian GE, White Jr LE, Daly D 1963 Electroencephalographic patterns resembling those of sleep in certain comatose states after injuries to the head. Electroencephalogr Clin Neurophysiol 15:272–280.

Chatrian GE, Shaw CM, Leffman H 1964 The significance of periodic lateralized epileptiform discharges in EEG: an electrographic, clinical and pathological study. Electroencephalogr Clin Neurophysiol 17:177–193.

Chatrian GE, Bergamini L, Dondey M, et al 1983 A glossary of terms most commonly used by clinical electroencephalographers. In: *Recommendations for the Practice of Clinical Neurophysiology*. Elsevier, Amsterdam, pp. 11–27.

Chatrian CE, Bergamini L, Dondey M, et al 1974 A glossary of terms used by clinical electroencephalographers. Electroencephalogr Clin Neurophysiol 37:538–548.

Chatrian G-E, Bergamasco B, Bricolo A, et al 1996 IFCN Recommended standards for electrophysiologic monitoring in coma and other unresponsive states. Report of an IFCN committee. Electroencephalogr Clin Neurophysiol 99:103–122.

Chen R, Bolton CF, Young B 1996 Prediction of outcome in patients with anoxic coma: a clinical and electrophysiologic study. Crit Care Med 24:672–678.

Cherington M 1974 Botulism: ten year experience. Arch Neurol 30:432–437.

Cherington M 1982 Electrophysiologic methods as an aid in diagnosis of botulism: a review. Muscle Nerve 5(9 Suppl):S28–S29.

Chesnut RM, Marshall LF, Klauber MR, et al 1993 The role of secondary brain injury in determining outcome from severe head injury. J Trauma 34:216–222.

Chesson AL, Schochet SS, Peters BH 1979 Biphasic periodic paralysis. Arch Neural 36:700–704.

Chiappa KH 1983 *Evoked Potentials in Clinical Medicine*. Raven Press, New York, pp. 145–189.

Chiappa KH, Hill RA 1998 Evaluation and prognostication in coma. Electroencephalogr Clin Neurophysiol 106:149–155.

Chokroverty S 1975 'Alpha-like' rhythms in electroencephalograms in coma after cardiac arrest. Neurology 25:655–663.

Claassen J, Hansen HC 2001 Early recovery after closed traumatic head injury: somatosensory evoked potentials and clinical findings. Crit Care Med 29:494–502.

Claassen J, Hirsch LJ, Kreiter KT, et al 2004a Quantitative continuous EEG for detecting delayed cerebral ischemia in patients with poor-grade subarachnoid hemorrhage. Clin Neurophysiol 115:2699–2710.

Claassen J, Mayer SA, Kowalski RG, et al 2004b Detection of elctrographic seizures with continuous EEG monitoring in critically ill patients. Neurology 62:1743–1748.

Clifton GL, Grossman RG, Makela ME, et al 1980 Neurological course and computerised tomography findings after severe closed head injury. J Neurosurg 52:611–624.

Coakley JH, Nagendran K, Ormerod IEC, et al 1992 Prolonged neurogenic weakness in patients requiring mechanical ventilation for acute airflow limitation. Chest 101:1413–1416.

Coakley JH, Nagendran K, Honavar M, et al 1993 Preliminary observations on the neuromuscular abnormalities in patients with organ failure and sepsis. Intensive Care Med 19:323–328.

Coakley JH, Nagendran K, Yarwood GD, et al 1998 Patterns of neuro-physiological abnormality in prolonged critical illness. Intensive Care Med 24:801–807.

Cole G, Cowie VA 1987 Long survival after cardiac arrest: case report and neuropathological findings. Clin Neuropathol 6:104–109.

Cooper AB, Thornley KS, Young GB, et al 2000 Sleep in critically ill patients requiring mechanical ventilation. Chest 117:809–818.

Cooper R, Hulme A 1966 Intracranial pressure and related phenomena during sleep. J Neurol Neurosurg Psychiatry 29:564–570.

Crawley J, Smith S, Kirkham F, et al 1996 Seizures and status epilepticus in childhood cerebral malaria. Q J Med 89:591–597.

Crawley J, Smith S, Muthinji P, et al 2001 Electroencephalographic and clinical features of cerebral malaria. Arch Dis Child 84:247–253.

Cusumano S, Paolin A, Di Paola F, et al 1992 Assessing brain function in post-traumatic coma by means of bit-mapped SEPs BAEPs CT, SPET and clinical scores. Prognostic implications. Electroencephalogr Clin Neurophysiol 84:499–514.

Czosnyka M, Whitehouse H, Smielewski P, et al 1994 Computer supported multimodal bed-side monitoring for neuro intensive care. Int J Clin Monit Comput 11:223–232.

Daly DD 1973 Circadian cycles and seizures. In: *Epilepsy: Its Phenomena in Man* (ed MAB Brazier). UCLA Forum in Medical Sciences No. 17. Academic Press, New York, pp. 173–190.

Davies NP, Hanna GH 2001 The skeletal muscle channelopathies: basic science, clinical genetics and treatment. Curr Opin Neurol 14:539–551.

De Giorgio CM, Rabinowicz AL, Gott PS 1993 Predictive value of P300 event-related potentials compared with EEG and somatosensory evoked potentials in non-traumatic coma. Acta Neurol Scand 87:423–427.

De la Torre JC, Trimble JL, Beard RT, et al 1978 Somatosensory evoked potentials for the prognosis of coma in humans. Exp Neurol 60:304–317.

De Weerd AW, Groeneveld C 1985 The use of evoked potentials in the management of patients with severe cerebral trauma. Acta Neurol Scand 72:489–494.

Dougherty JH, Rawlinson DG, Levy DE, et al 1981 Hypoxic–ischemic brain injury and the vegetative state: clinical and neuropathologic correlation. Neurology 31:993–997.

Drummond JC, Todd MM, U HS 1985 The effect of high dose sodium thiopental on brainstem auditory and median nerve somatosensory evoked responses in humans. Anesthesiology 63:249–254.

Edgren E, Hedstand U, Nordin M, et al 1987 Prediction of outcome after cardiac arrest. Crit Care Med 15:820–825.

Eisen A, Humphreys P 1974 The Guillain–Barré syndrome. Arch Neurol 30:438–443.

Emilia-Romagna Study Group on Clinical and Epidemiological Problems in Neurology 1997 A prospective study on the incidence and prognosis of Guillain–Barré syndrome in Emilia-Romagna region, Italy (1992–1993). Neurology 48:214–221.

Evans BM 1976 Patterns of arousal in comatose patients. J Neurol Neurosurg Psychiatry 39:392–402.

Evans BM 1992 Periodic activity in cerebral arousal mechanisms – the relationship to sleep and brain damage. Electroencephalogr Clin Neurophysiol 83:130–137.

Evans BM 2002 What does brain damage tell us about the mechanisms of sleep? J R Soc Med 95:591–597.

Evans BM, Bartlett JR 1995 Prediction of outcome in severe head injury based on recognition of sleep related activity in the polygraphic electro-encephalogram. J Neurol Neurosurg Psychiatry 59:17–25.

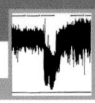

Facco E, Martini A, Zuccarello M, et al 1985 Is the auditory brain-stem response (ABR) effective in the assessment of post-traumatic coma? Electroencephalogr Clin Neurophysiol 62:332–337.

Facco E, Casartelli Livero M, Munari M, et al 1990 Short-latency evoked potentials: new criteria for brain death? J Neurol Neurosurg Psychiatry 53:351–353.

Facco E, Munari M, Baratto F, et al 1993 Multimodality evoked potentials (auditory, somatosensory and motor) in coma. Neurophysiol Clin 23:237–258.

Facco E, Behr AU, Munari M, et al 1998 Auditory and somatosensory evoked potentials in coma following spontaneous cerebral hemorrhage: early prognosis and outcome. Electroencephalogr Clin Neurophysiol 107:332–338.

Feasby TE, Gilbert JJ, Brown WF, et al 1986 An acute axonal form of Guillain–Barré polyneuropathy. Brain 109:1115–1126.

Feldman Z, Kanter MJ, Robertson CS, et al 1992 Effect of head elevation on intracranial pressure, cerebral perfusion pressure, and cerebral blood flow in head-injured patients. J Neurosurg 76:207–211.

Ferbert A, Buchner H, Ringelstein EB, et al 1986 Isolated brainstem death. Case report with demonstration of preserved visual evoked potentials (VEPs). Electroencephalogr Clin Neurophysiol 65:157–160.

Ferbert A, Buchner H, Brüickmann H, et al 1988 Evoked potentials in basilar artery thrombosis: correlation with clinical and angiographic findings. Electroencephalogr Clin Neurophysiol 69:136–147.

Fischer C, Ibañez V, Jourdan C, et al 1988 Potentiels évoqués auditifs précoces (PEAP), auditifs de latence moyenne (PALM) et somesthésiques (PES) dans le pronostic vital et fonctionnel des traumatismes craniens graves en réanimation. Agressologie 29:359–363.

Fischer C, Morlet D, Bouchet P, et al 1999 Mismatch negativity and late auditory evoked potentials in comatose patients. Clin Neurophysiol 110:1601–1610.

Fisher DM, Frewin T, Swedlow DB 1982 Increase in intracranial pressure during suctioning – stimulation vs rise in $PaCO_2$. Anesthesiology 57:416–417.

Fischgold H, Mathis P 1959 Obnubilations, comas et stupeurs. Electroencephalogr Clin Neurophysiol Suppl 11:13–26.

Fogdall RP, Miller RD 1974 Prolongation of a pancuronium induced neuromuscular blockade by polymyxin. Anesthesiology 40:84–87.

Frank LM, Furgiville TL, Etheridge JE 1985 Prediction of chronic vegetative state in children using evoked potentials. Neurology 35:931–934.

Frank M, Prior P 1987 The CFAM, the principle of the method and its potential use. In: *Consciousness Awareness Pain and General Anaesthesia* (eds JN Lunn, M Rosen). Butterworths, London, pp. 61–71.

Frewen TC, Sumabat WO, Del Maestro RF 1985 Cerebral blood flow, metabolic rate and cross-brain oxygen consumption in brain injury. J Pediatr 107:510–513.

Ganes T, Lundar T 1983 The effect of thiopentone on somatosensory evoked responses and EEGs in comatose patients. J Neurol Neurosurg Psychiatry 46:509–514.

Ganes T, Lundar T 1988 EEG and evoked potentials in comatose patients with severe brain damage. Electroencephalogr Clin Neurophysiol 69:6–13.

García-Larrea L, Bertrand O, Artru F, et al 1987 Brainstem monitoring in coma II: dynamic interpretation of preterminal BAEP changes observed until brain death in deeply comatose patients. Electroencephalogr Clin Neurophysiol 68:446–457.

García-Larrea L, Artru F, Bertrand O, et al 1988a Transient drug-induced BAEP abolition in coma. Neurology 38:1487–1489.

García-Larrea L, Artru F, Bertrand O, et al 1988b Monitorage des potentiels évoqués auditifs du tronc cérébral lors des altérations aiguës de la pression intracrânienne. Agressologie 29:329–332.

García-Larrea L, Artru F, Bertrand O, et al 1992 The combined use of brainstem auditory potentials and intracranial pressure monitoring in coma: a study of 57 patients. J Neurol Neurosurg Psychiatry 55:792–798.

Gendo A, Kramer L, Häfner M, et al 2001 Time-dependency of sensory evoked potentials in comatose cardiac arrest survivors. Intensive Care Med 27:1305–1311.

Ghosh IR, Langford RM, Nieminen K, et al 2000 Repetitive cyclical oscillations of multisystem parameters subsequent to high-dose thiopental therapy for status epilepticus secondary to herpes encephalitis. Br J Anaesth 85:471–473.

Giacino JT, Ashwal S, Childs N, et al 2002 The minimally unconscious state: definition and diagnostic criteria. Neurology 58:349–353.

Gilmore RL, Nelson KR 1989 SSEP, F-wave studies in acute inflammatory demyelinating polyradiculoneuropathy. Muscle Nerve 12:538–543.

Gilroy J, Lynn GE, Ristow GE, et al 1977 Auditory evoked brain stem potentials in a case of 'locked-in' syndrome. Arch Neurol 34:492–495.

Goldie WD, Chiappa KH, Young RR, et al 1981 Brainstem auditory and short-latency somatosensory evoked responses in brain death. Neurology 31:248–256.

Gordon EM, Green JR, Lagunoff D 1970 Studies on a patient with hypokalemic familial periodic paralysis. Am J Med 48:185–195.

Gotman J, Flanagan D, Zhang J, et al 1997a Automatic seizure detection in the newborn: methods and initial evaluation. Electroencephalogr Clin Neurophysiol 103:356–362.

Gotman J, Flanagan D, Rosenblatt B, et al 1997b Evaluation of an automatic seizure detection method for the newborn EEG. Electroencephalogr Clin Neurophysiol 103:363–369.

Gott PS, Rabinowicz AL, DeGiorgio CM 1991 P300 auditory event-related potentials in nontraumatic coma. Association with Glasgow Coma Score and awakening. Arch Neurol 48:1267–1270.

Govoni V, Granieri E 2001 Epidemiology of the Guillain–Barré syndrome. Curr Opin Neurol 14:605–613.

Graham DI, Gennarelli TA 1997 Trauma. In: *Greenfield's Neuropathology*, 6th edition (eds DI Graham, PL Lantos). Arnold, London, pp. 197–262.

Gray WJ, Rosner MJ 1987 Pressure–volume index as a function of cerebral perfusion pressure. Part 2: The effects of low cerebral perfusion pressure and autoregulation. J Neurosurg 67:377–380.

Greenberg DA 1999 Neuromuscular disease and calcium channels. Muscle Nerve 22:1341–1349.

Greenberg RP, Becker DP, Miller JD, et al 1977 Evaluation of brain function in severe human head trauma with multimodality evoked potentials. Part 2. Localization of brain dysfunction and correlation with posttraumatic neurological conditions. J Neurosurg 47:163–177.

Greenberg RP, Newlon PG, Becker DP 1982 The somatosensory evoked potential in patients with severe head injury: outcome prediction and monitoring of brain function. Ann NY Acad Sci 388:683–688.

Griffin JW, Li CY, Ho TW, et al 1995 Guillain–Barré syndrome in Northern China. The spectrum of neuropathological changes in clinically defined cases. Brain 118:577–595.

Griffin JW, Li CY, Ho TW, et al 1996 Pathology of the motor–sensory axonal Guillain–Barré. Ann Neurol 39:17–28.

Guérit JM 1999 EEG and evoked potentials in the intensive care unit. Neurophysiol Clin 29:301–317.

Guérit JM 2000 The usefulness of EEG, exogenous evoked potentials, and cognitive evoked potentials in the acute stage of post-anoxic and post-traumatic coma. Acta Neurol Belg 100:229–236.

Guérit JM, de Tourtchaninoff M, Soveges L, et al 1993 The prognostic value of three-modality evoked potentials (TMEPs) in anoxic and traumatic comas. Neurophysiol Clin 23:209–226.

Guérit JM, Fischer C, Facco E, et al 1999a Standards of clinical practice of EEG, EPs in comatose and other unresponsive states. The International Federation of Clinical Neurophysiology. Electroencephalogr Clin Neurophysiol 52(Suppl):117–131.

Guérit JM, Verougstraete D, de Tourtchaninoff M, et al 1999b ERPs obtained with the auditory oddball paradigm in coma and altered states of consciousness: clinical relationships, prognostic value, and origin of components. Clin Neurophysiol 110:1260–1269.

Guillain G, Barré JA, Strohl A 1916 Sur un syndrome de radiculo-névrite avec hyperalbuminose du liquide céphalo-rachidien sans réaction cellulaire. Remarques sur les caractères cliniques et graphiqéues des réflexes tendinaux. Bull Mém Soc Méd Hôp 40:1462–1470.

Gütling E, Gonser A, Imhof HG, et al 1994 Prognostic value of frontal and parietal somatosensory evoked potentials in severe head injury: a long term follow-up study. Electroencephalogr Clin Neurophysiol 92:568–570.

Gütling E, Gonser A, Imhof HG, et al 1995a EEG reactivity in the prognosis of severe head injury. Neurology 45:915–918.

Gütling E, Isermann S, Wichmann W 1995b Electrophysiology in the locked-in syndrome. Neurology 46:1092–1101.

Hacke W 1985 Neuromonitoring. J Neurol 232:125–133.

Hall JW 1985 The effects of high-dose barbiturates on the acoustic reflex and auditory evoked responses. Acta Otolaryngol (Stockholm) 100:387–398.

Hall JW, Huang-Fu M, Gennarelli TA 1982 Auditory function in acute severe head injury. 92:883–890.

Hall JW, Mackey-Hargadine JR, Allen SJ 1985a Monitoring neurological status of comatose patients in the intensive care unit. In: *The Auditory Brainstem Response* (ed J Jacobson). College Hill, San Diego, CA, pp. 253–283.

Hall JW, Mackey-Hargadine JR, Kim EE 1985b Auditory brain-stem response in determination of brain death. Arch Otolaryngol 111:613–620.

Harper AM 1966 Autoregulation of cerebral blood flow: influence of the arterial blood pressure on the blood flow through the cerebral cortex. J Neurol Neurosurg Psychiatry 29:398–403.

Hassler O 1967 Arterial pattern of human brain-stem. Normal appearance and deformation in expanding supratentorial conditions. Neurology 17:368–375

Haupt WF, Birkmann C, Halber M 2000 Serial evoked potentials and outcome in cerebrovascular critical care patients. J Clin Neurophysiol 17:326–330.

Hawkes CH, Bryan-Smyth L 1974 The electroencephalogram in the 'locked-in' syndrome. Neurology 24:1015–1018.

Hecox KE, Cone B 1981 Prognostic importance of brainstem auditory evoked responses after asphyxia. Neurology 31:1429–1433.

Hernandez Sande C, Arias Rodriguez JE 1985 Monitoring kidney patients by syntactometric EEG analysis. J Biomed Eng 7:334–336.

Hinds CJ, Nagendran K, Honavar M, Coakley JH 1993 Muscle relaxants in intensive care patients [letter]. Crit Care Med 21:1403–1404.

Hoffman EP, Lehmann-Horn F, Rudel R 1995 Overexcited or inactive: ion channels in muscle disease. Cell 80:681–686

Hume AL, Cant BR 1978 Conduction time in central somatosensory pathways in man. Electroencephalogr Clin Neurophysiol 45:361–375.

Hume AL, Cant BR 1981 Central somatosensory conduction after head injury. Ann Neural 10:411–419.

Hume AL, Cant BR, Shaw NA 1979 Central somatosensory conduction time in comatose patients. Ann Neural 5:379–384.

Hutchinson DO, Frith RW, Shaw NA, et al 1991 A comparison between electroencephalography and somatosensory evoked potentials for outcome prediction following severe head injury. Electroencephalogr Clin Neurophysiol 78:228–233.

Jabbari B, Vance SC, Harper MG, et al 1987 Clinical and radiological correlates of somatosensory evoked potentials in the late phase of head injury: a study of 500 Vietnam veterans. Electroencephalogr Clin Neurophysiol 67:289–297.

Jacome DE, Morilla-Pastor D 1990 Unreactive EEG: pattern in locked-in syndrome. Clin Electroencephalogr 21:31–36.

Jain S, Maseshwari MC 1984 Brainstem auditory evoked responses in coma due to meningoencephalitis. Acta Neurol Scand 69:163–167.

Jansen BH 1986 Quantitative EEG analysis in renal disease. In: *Clinical Applications of Computer Analysis of EEG and other Neurophysiological Signals* (eds FH Lopes da Silva, W Storm van Leeuwen, A Rémond). *Handbook of Electroencephalography and Clinical Neurophysiology*, revised series, volume 2. Elsevier, Amsterdam, pp. 239–257.

Jäntti V, Eriksson K, Hartiainen K, et al 1994 Epileptic EEG discharges during burst-suppression. NeuroPediatrics 25:271–273.

Jennett B 2002 *The Vegetative State. Medical Facts, Ethical and Legal Dilemmas*. Cambridge University Press, Cambridge.

Jennett B, Bond M 1975 Assessment of outcome after severe brain damage. A practical scale. Lancet i:480–484.

Jennett B, Plum F 1972 The persistent vegetative state; a syndrome in search of a name. Lancet i:734–737.

Jennett B, Teasdale G, Braakman R, et al 1979 Prognosis of patients with severe head injury. Neurosurgery 4:283–289.

Jennett B, Gleave J, Wilson P 1981a Brain deaths in three neurosurgical units. BMJ 282:533–539.

Jennett B, Snoek J, Bond MR, et al 1981b Disability after severe head injury: observations on the use of the Glasgow Outcome Scale. J Neurol Neurosurg Psychiatry 44:285–293.

Johnson RT, Yates PO 1956 Clinico-pathological aspects of pressure changes at the tentorium. Acta Radiol 46:242–249.

Jordan KG 1994 Status epilepticus. A perspective from the neuroscience intensive care unit. Neurosurg Clin North Am 5:671–686.

Jordan KG 1999 Continuous EEG monitoring in the neuroscience intensive care unit and emergency department. J Clin Neurophysiol 16:14–39.

Jørgensen EO, Holm S 1998 The natural course of neurological recovery following cardiopulmonary resuscitation. Resuscitation 36:111–122.

Jørgensen EO, Malchow-Møller A 1978 Cerebral prognostic signs during cardiopulmonary resuscitation. Resuscitation 6:217–225.

Jørgensen EO, Malchow-Møller A 1981 Natural history of global and critical brain ischaemia. Part III: Cerebral prognostic signs after cardio-pulmonary resuscitation. Cerebral recovery course and rate during the first year after global and critical ischaemia monitored and predicted by EEG and neurological signs. Resuscitation 9:155–174.

Kaga K, Nagai T, Takamori A 1985 Auditory short-, middle- and long-latency responses in acutely comatose patients. Laryngoscope 95:321–325.

Kaji R, McCormick F, Kameyama M, et al 1985 Brainstem auditory evoked potentials in early diagnosis of basilar artery occlusion. Neurology 35:240–243.

Kane NM, Curry SH, Rowlands CA, et al 1996 Event related potentials. Neurophysiological tools for predicting emergence and early outcome from traumatic coma. Intensive Care Med 22:39–46.

Kao I, Drachman DB, Price DL 1976 Botulinum toxin: mechanism of presynaptic blockade. Science 193:1256–1258.

Kaplan PW, Genoud D, Ho TW, et al 1999 Etiology, neurlgic correlations, and prognosis in alpha coma. Clin Neurophysiol 110:205–213.

Kaplan PW, Genoud D, Ho TW, et al 2000 Clinical correlates and prognosis in early spindle coma. Clin Neurophysiol 111:584–590.

Karnaze DS, Weiner JM, Marshall LF 1985 Auditory evoked potentials in coma after closed head-injury. Neurology 35:1122–1126.

Kaufmann GE, Clark K 1970 Continuous simultaneous monitoring of intraventricular and cervical subarachnoid cerebrospinal fluid pressure to indicate development of cerebral or tonsilar herniation. J Neurosurg 33:145–150.

Kawahara N, Sasaki Mii K, Tsuzuki M, et al 1989 Sequential changes of auditory brain-stem responses in relation to intracranial and cerebral perfusion pressures and initiation of secondary brain-stem damage. Acta Neurochir 100:142–149.

Kemp B, Gröneveld EW, Janssen AJ, et al 1987 A model-based monitor of human sleep stages. Biol Cybern (Berlin) 57:365–378.

Kim YW, Krieble KK, Kim CB, et al 1996 Differentiation of alpha coma from awake alpha by non-linear dynamics of electroencephalography. Clin Electroencephalogr 98:35–41.

Kimura J, Butzer JF 1975 F wave conduction velocity in Guillain–Barré syndrome. Arch Neurol 32:524–529.

Kinney HC, Samuels MA 1994 Neuropathology of the persistent vegetative state. A review. J Neuropathol Exp Neurol 53:548–558.

Kinney HC, Korein J, Panigrahy A, et al 1994 Neuropathological findings in the brain of Karen Ann Quinlan. The role of the thalamus in the persistent vegetative state. N Engl J Med 330:1469–1475.

Kirkpatrick PJ, Smielewski P, Czosnyka M, et al 1995 Near-infrared spectroscopy use in patients with head injury. J Neurosurg 83:963–70.

Klintworth KG 1968 Paratentorial growing of human brains with particular reference to transtentorial herniation and the pathogenesis of secondary brainstem hemorrhages. Am J Pathol 53:391–399.

Knox AJ, Mascie-Taylor BH, Muers MF 1986 Acute hydrocortisone myopathy in acute severe asthma. Thorax 41:411–412.

Kretschmer E 1940 Das apallische Syndrom. Z Neurol Psychiatric 169:576–579.

Kupfer Y, Okrent DG, Twersky RA, et al 1987 Disuse atrophy in a ventilated patient with status asthmaticus receiving neuromuscular blockade. Crit Care Med 15:795–796.

Labar DR, Fisch BJ, Pedley TA, et al 1991 Quantitative EEG monitoring for patients with subarachnoid hemorrhage. Electroencephalogr Clin Neurophysiol 78:325–332.

Lachman T, Shahani BT, Young RR 1980 Late responses as aids to diagnosis in peripheral neuropathy. J Neurol Neurosurg Psychiatry 43:156–162.

Lambert EH, Mulder DW 1964 Nerve conduction in the Guillain–Barré syndrome [abstract]. Electroencephalogr Clin Neurophysiol 17:86.

Langdon-Down M, Brain WR 1929 Time of day in relation to convulsions in epilepsy. Lancet i:1029–1032.

Larson SJ, Sances A, Ackman J 1973 Noninvasive evaluation of head trauma patients. Surgery 74:34–40.

Layzer RB, Lovelace RE, Rowland LP 1967 Hyperkalemic periodic paralysis. Arch Neural 16:455–472.

Leira EC, Bertrand ME, Hogan RE, et al 2004 Continuous or emergent EEG: can bedside caregivers recognize epileptiform discharges? Intensive Care Med 30:207–212.

Levy DE, Knill-Jones RP, Plum F 1978 The vegetative state and its prognosis following non-traumatic coma. Ann NY Acad Sci 315:293–306.

Levy DE, Bates D, Caronna JJ, et al 1981 Prognosis in non-traumatic coma. Ann Intern Med 94:293–301.

Lindsay KW, Carlin J, Kennedy I, et al 1981 Evoked potentials in severe head injury. Analysis and relation to outcome. J Neurol Neurosurg Psychiatry 44:796–802.

Lindsay K, Pasaoglu A, Hirst D, et al 1990 Somatosensory and auditory brain stem conduction after head injury: a comparison with clinical features in prediction of outcome. Neurosurgery 26:278–285.

Litscher G 1995 Middle latency auditory evoked potentials in intensive care patients and normal controls. Int J Neurosci 83:253–267.

Litscher G, Frichs G, Maresch H, et al 1990 Electro-encephalographic and evoked potentials monitoring in hyperbaric environment. J Clin Monit 6:10–17.

Logi F, Fischer C, Murri L, et al 2003 The prognostic value of evoked responses from primary somatosensory and auditory cortex in comatose patients. Clin Neurophysiol 114:1615–1627

Lowenstein DH, Aminoff MJ 1992 Clinical and EEG features of status epilepticus in comatose patients. Neurology 42:100–104.

Lundar T, Ganes T, Lindegaard KF 1983 Induced barbiturate coma: methods for evaluation of patients. Crit Care Med 11:559–562.

Lundberg N 1969 Continuous recording and control of ventricular fluid pressure in neurosurgical practice. Acta Psychiatr Neurol Scand 149(Suppl):1–193.

Lütschg J, Pfenninger J, Ludin HP, et al 1983 Brain stem auditory evoked potentials and early somatosensory evoked potentials in neurointensively treated comatose children. Am J Dis Child 137:421–426.

Lynch J, Eldadah MK 1992 Brain-death criteria currently used by pediatric intensivists. Clin Pediatr 31:457–160.

Machada C 2005 Can vegetative state patients retain cortical processing? Clin Neurophysiologr 116:2253–2254.

Maas AIR, Teasdale GM, Braakman R, et al 1997 EBIC guidelines for the management of severe head injury in adults. European Brain Consortium. Acta Neurochirurgia (Wien) 139:286–294.

MacFarlane IA, Rosenthal FD 1977 Severe myopathy after status asthmaticus. Lancet ii:615.

Machado C, Valdés P, García-Tijera J, et al 1991 Brain-stem auditory evoked potentials and brain death. Electroencephalogr Clin Neurophysiol 80:392–398.

Madl C, Kramer L, Yeganehfar W, et al 1996 Detection of nontraumatic comatose patients with no benefit of intensive care treatment by recording of sensory evoked potentials. Arch Neurol 53:512–516.

Madl C, Kramer L, Domanovits H, et al 2000 Improved outcome prediction in unconscious cardiac arrest survivors with sensory evoked potentials compared with clinical assessment. Crit Care Med 28:721–726.

Mankodi A, Takahashi MP, Jiang H, et al 2002 Expanded CUG repeats trigger aberrant splicing of CIC-1 chloride channel pre-mRNA and hyperexcitability of skeletal muscle in myotonic dystrophy. Mol Cell 10:35–44.

References

Marcus EM, Stone B 1984 Short-latency median nerve somatosensory evoked potentials in coma. Relationship to BAEPs, etiology, age and outcome. In: *Evoked Potentials II* (eds H Nodar, C Barber). Butterworths, Boston, MA, pp. 357–362.

Maresch H, Pfurtscheller G 1983 Simultaneous measurements of auditory brainstem potentials and EEG spectra. Electroencephalogr Clin Neurophysiol 56:531–533.

Marion DW, Bouma GJ 1991 The use of stable xenon-enhanced computed tomographic studies of cerebral blood flow to define changes in cerebral carbon dioxide vasoresponsivity caused by a severe head injury. Neurosurgery 29:869–873.

Marsh RR, Frewent TC, Sutton LN, et al 1984 Resistance of the auditory brainstem responses to high barbiturate levels. Otolaryngol Head Neck Surg 92:685–688.

Marshall LF, Marshall SB 1995 Pharmacologic therapy: promising clinical investigations. New Horizons 3:573–580.

Matousek M, Takeuchi E, Starmark JE, et al 1996 Quantitative EEG analysis as a supplement to the clinical coma scale RLS85. Acta Anaesthesiol Scand 40:824–831.

Mauguière F, Grand C, Fischer C, et al 1982 Aspects des potentiels evoques auditifs et somesthésiques précoces dans les comas neurologiques et la mort cerebrale. Rev EEG Neurophysiol Clin 12:280–286.

Mauguière F, García-Larrea L, Bertrand O 1988 Utility and uncertainties of evoked potentials monitoring in the intensive care unit. In: *Evoked Potentials: ICU, Surgical Monitoring* (eds BL Grundy, RM Villani). Springer-Verlag, Vienna, pp. 153–167.

Maynard DE, Jenkinson JL 1984 The cerebral function analysing monitor. Initial clinical experience, application and further development. Anaesthesia 39:678–690.

Mazzini L, Zaccala M, Gareri F, et al 2001 Long-latency auditory-evoked potentials in severe traumatic brain injury. Arch Phys Med Rehab 82:57–65.

McKeown MJ, Young GB 1997 Comparison between the alpha pattern in normal subjects and in alpha pattern coma. J Clin Neurophysiol 14:414–418.

McKhann GM, Cornblath DR, Griffin JW, et al 1993 Acute motor axonal neuropathy: a frequent cause of acute flaccid paralysis in China. Ann Neural 33:333–342.

McLellan DR, Adams JH, Graham DI, et al 1986 The structural basis of the vegetative state and prolonged coma after non-missile head injury. In: *Le Coma Traumatique* (eds I Papo, F Cohadon, M Massarotti). Liviana Editrice, Padova, pp. 165–185.

McLeod JG 1981 Electrophysiologic studies in the Guillain–Barré syndrome. Ann Neurol 9:20–27.

McLeod JG, Walsh JC, Prineas JW, et al 1976 Acute idiopathic polyneuritis. A clinical and electrophysiological follow-up study. J Neurol Sci 27:145–162.

Medical Research Council Brain Injuries Committee 1941 *A Glossary of Psychological Terms Commonly used in Cases of Head Injury*. MRC War Memorandum No. 4. HMSO, London.

Michenfelder JD 1974 The interdependency of cerebral functional and metabolic effects following massive doses of thiopental in the dog. Anesthesiology 41:231–236.

Miller JD, Marshall LF 1996 Are steroids useful in the treatment of head-injured patients? Surg Neurol 45:296.

Miller JD, Jones PA, Dearden NM, et al 1992 Progress in the management of head injury. Br J Surg 79:60–64.

Miller JD, Piper IR, Jones PA 1994 Integrated multimodality monitoring in the neurosurgical intensive care unit. Neurosurg Clin North Am 5:661–670.

Mills KR, Murray NMF 1986 Proximal conduction block in early Guillain Barré syndrome. Lancet i:105–106.

Moerman N, Bonke B, Oosting J 1993 Awareness and recall during general anesthesia. Facts and feelings. Anesthesiology 79:454–464.

Mollaret P, Goulon M 1959 Le coma dépassé (Mémoir préliminaire). Rev Neurol 101:3–15.

Morlet D, Bertrand O, Salord F, et al 1997 Dynamics of MLAEP changes in midazolam induced sedation. Electroencephalogr Clin Neurophysiol 104:437–446.

Moulton RJ, Brown JI, Konasiewicz SJ 1998 Monitoring severe head injury: a comparison of EEG and somatosensory evoked potentials. Can J Neurol Sci 25, S7–S11.

Multi-Society Task Force on PVS 1994 Medical aspects of the persistent vegetative state [in two parts]. N Engl J Med 330:1499–1508, 1572–1579.

Murray NMF, Wade DT 1980 The sural sensory action potential in Guillain–Barré syndrome [letter]. Muscle Nerve 3:444.

Nagao S, Roccaforte P, Moody RA 1979 Acute intracranial hypertension and auditory brain-stem responses (Parts I, II). J Neurosurg 51:669–676.

Nagata K, Tazawa T, Mizukami M, et al 1984 Application of brainstem auditory evoked potentials to evaluation of cerebral herniation. In: *Evoked Potentials II* (eds RH Nodard, C Barber). Butterworth, Boston, pp. 183–193.

Nagendran K, Honavar M, Yarwood G, et al 1993 EMG and muscle biopsy findings in critically ill patients in the intensive therapy unit (ITU) [abstract]. Electroencephalogr Clin Neurophysiol 87:S65.

Nakabayashi M, Kurokawa A, Yamamoto Y 2001 Immediate prediction of recovery of consciousness after cardiac arrest. Intensive Care Med 27:1210–1214.

Narayan RK, Greenberg RP, Miller DJ, et al 1981 Improved confidence of outcome prediction in severe head injury. J Neurosurg 54:751–762.

Newlon PG, Greenberg RP 1984 Evoked potentials in severe head injury. J Trauma 24:61–66.

Newlon PG, Greenberg RP, Hyatt MS, et al 1982 The dynamics of neuronal dysfunction and recovery following severe head injury, assessed with serial multimodality evoked potentials. J Neurosurg 57:168–177.

Nicoletta JA, Schwartz GJ 2004 Distal renal tubular acidosis. Curr Opin Pediatr 16(2):194–198.

Nilsson J 1969 Influence of induced changes in CSF pressure on the cerebral blood flow. In: *Third International Symposium on Cerebral Circulation Research* (ed JS Meyer). CC Thomas, Springfield, IL.

Nordgren RE, Markesbery WR, Fukada K, et al 1971 Seven cases of cerebromedullospinal disconnection: the 'locked-in' syndrome. Neurology 21:1140–1148.

Ogawara K, Kuwabara S, Mori M, et al 2000 Axonal Guillain–Barré syndrome: relation to anti-ganglioside antibodies and *Campylobacter jejuni* infection in Japan. Ann Neural 48:624–631.

Oh SJ 1977 Botulism: electrophysiological studies. Ann Neurol 1:481–485.

Oh SJ, LaGanke C, Claussen GC 2001 Sensory Guillain–Barré syndrome. Neurology 56:82–86.

Oksenberg A, Soroker N, Solzi P, et al 1991 Polysomnography in locked-in syndrome. Electroencephalogr Clin Neurophysiol 78:314–317.

Op de Coul AAW, Lambregts PCLA, Koeman J, et al 1985 Neuromuscular complications in patients given Pavulon (pancuronium bromide) during artificial ventilation. Clin Neurol Neurosurg 87:17–22.

O'Sullivan MG, Statham PF, Jones PA, et al 1994 Role of intracranial pressure monitoring in severely head-injured patients without signs of intracranial hypertension on initial computerized tomography. J Neurosurg 80:46–50.

Ottaviani F, Almadori G, Calderazzo AB, et al 1986 Auditory brain-stem and middle-latency auditory responses in the prognosis of severely head-injured patients. Electroencephalogr Clin Neurophysiol 65:196–202.

Pallis C 1990 Brainstem death. In: *Head Injury. Handbook of Clinical Neurology*, volume 13, revised series (ed R Braakman). Elsevier, Amsterdam, pp. 441–496.

Pallis C, Harley DH 1996 ABC of Brainstem Death, 2nd edition. British Medical Journal, London.

Parthasarathy S, Tobin MJ 2004 Sleep in the intensive care unit. Intensive Care Med 30:197–206.

Payne TA, Bleck TP 1997 Status epilepticus. In: *Critical Care Clinics*. Update on Neurologic Critical Care, volume 13 (eds RW Carlson, MA Geheb). WB Saunders, Philadelphia, pp. 17–38.

Pfurtscheller G, Schwartz G, Gravenstein N 1983 Clinical relevance of long-latency SEPs and VEPs during coma. Electroencephalogr Clin Neurophysiol 62:88–98.

Pfurtscheller G, Schwarz G, List W 1986 Long-lasting EEG reactions in comatose patients after repetitive stimulation. Electroencephalogr Clin Neurophysiol 64:402–410.

Pfurtscheller G, Schwarz G, Schroettner O, et al 1987 Continuous and simultaneous monitoring of EEG spectra and brainstem auditory and somatosensory evoked potentials in the intensive care unit and the operating room. J Clin Neurophysiol 4:389–396.

Piek J, Chesnut RM, Marshall LF, et al 1992 Extracranial complications of severe head injury. J Neurosurg 77:901–907.

Plaster NM, Tawil R, Tristani-Firouzi M, et al 2001 Mutations in Kir 2.1 cause the developmental and episodic electrical phenotypes of Anderson's syndrome. Cell 105:511–519.

Plum F, Posner J 1966 *Diagnosis of Stupor and Coma*, 1st edition. Davis, Philadelphia.

Plum F, Posner JB 1980 *Diagnosis of Stupor and Coma*, 3rd edition. FA Davis, Philadelphia, PA.

Pohlmann-Eden B, Dingethal K, Bender H-J, et al 1997 How reliable is the predictive value of SEP (somatosensory evoked potentials) patterns in severe brain damage with special regard to the bilateral loss of cortical responses? Intensive Care Med 23:301–308.

Pratt H, Ben David Y, Peled R, et al 1981 Auditory brain-stem evoked potentials: clinical promise of increasing stimulus rate. Electroencephalogr Clin Neurophysiol 51:80–90.

Prior PF 1973 The EEG in Acute Cerebral Anoxia. Excerpta Medica, Amsterdam.

Prior P 1993 Cerebral anoxia: clinical aspects. In: *Electroencephalography. Basic Principles Clinical Applications and Related Fields*, 3rd edition (eds E Niedermeyer, FH Lopes da Silva). Williams and Wilkins, Baltimore, OH, pp. 431–444.

Prior PF, Maynard DE 1986 *Monitoring Cerebral Function*, 2nd edition. Elsevier, Amsterdam.

Prior PF, Brigden W, Maynard DE, et al 1978 Nursing procedures and cardiovascular status. Lancet i:938–939.

Procaccio F, Bingham RM, Hinds CJ, et al 1988 Continuous EEG, ICP monitoring as a guide to the administration of Althesin sedation in severe head injury. Intensive Care Med 14:148–155.

Purves MJ 1972 *The Physiology of Cerebral Circulation*. Monographs of the Physiological Society. Cambridge University Press, London, pp. 161–165.

Quane KA, Healy JMS, Keating KE, et al 1993 Mutations in the ryanodine receptor gene in central core disease and malignant hyperthermia. Nature Genet 5:51–55.

Rae-Grant AD, Strapple C, Barbour PJ 1991a Episodic low-amplitude events: an under-recognized phenomenon in clinical electroencephalography. J Clin Neurophysiol 8:203–211.

Rae-Grant AD, Barbour PJ, Reed J 1991b Development of a novel EEG rating scale for head injury using dichotomous variables. Electroencephalogr Clin Neurophysiol 79:349–357.

Ringel RA, Riggs JE, Brick JF 1988 Reversible coma with prolonged absence of pupillary and brainstem reflexes: an unusual response to a hypoxic–ischemic event in multiple sclerosis. Neurology 38:1275–1278.

Robertson CS, Gopinath SP, Goodman JC, et al 1995 SjvO$_2$ monitoring in head-injured patients. J Neurotrauma 12:891–896.

Romano J, Engel GL 1944 Delirium. I. Electroencephalographic data. Arch Neurol Psychiatry 51:356–377.

Ropper AH, Shahani BT 1984 Pain in Guillain–Barré syndrome. Arch Neurol 41:511–514.

Rosenberg C, Wogensen K, Starr A 1984 Auditory brain-stem and middle and long-latency evoked potentials in coma. Arch Neurol 41:835–838.

Rothstein TL 2000 The role of evoked potentials in anoxic–ischemic coma and severe brain trauma. J Clin Neurophysiol 17:486–497.

Rothstein TL, Thomas EM, Sumi SM 1991 Predicting outcome in hypoxic–ischemic coma. A prospective clinical and electrophysiological study. Electroencephalogr Clin Neurophysiol 79:101–107.

Royal College of Physicians Working Group 1995 Criteria for the diagnosis of brain stem death. J R Coll Physicians London 29:381–382.

Royal College of Physicians Working Party 2003 The vegetative state: guidance on diagnosis and management. Clin Med 3:249–254.

Rumpl E, Prugger M, Gerstenbrand F, et al 1983 Central somatosensory conduction time and short-latency somatosensory evoked potentials in post-traumatic coma. Electroencephalogr Clin Neurophysiol 56:583–596.

Rumpl E, Prugger M, Battista HJ, et al 1988a Short latency somatosensory evoked potentials and brain-stem auditory evoked potentials in coma due to CNS depressant drug poisoning. Preliminary observations. Electroencephalogr Clin Neurophysiol 70:482–489.

Rumpl E, Prugger M, Gerstenbrand F, et al 1988b Central somatosensory conduction time and acoustic brainstem transmission time in post-traumatic coma. J Clin Neurophysiol 5:237–360.

Ryan MM, Taylor P, Donald JA, et al 1999 A novel syndrome of episodic muscle weakness maps to xp22.3. Am J Hum Genet 65(4):1104–1113.

Sandroni C, Barelli A, Piazza O, et al 1995 What is the best test to predict outcome after prolonged cardiac arrest? Eur J Emerg Med 2:33–37.

Schiller HH, Stålberg E 1978 Human botulism studied with single-fiber electromyography. Arch Neurol 35:346–349.

Schwartz S, Schwab S, Aschoff A, et al 1999 Favorable recovery from bilateral loss of somatosensory evoked potentials. Crit Care Med 27:182–187.

Seales DM, Rossiter VS, Weinstein ME 1979 Brainstem auditory evoked responses in patients comatose as a result of blunt head trauma. J Trauma 19:347–352.

Seales DM, Torkelson RD, Shuman RM, et al 1981 Abnormal brainstem auditory potentials and neuropathology in 'locked-in' syndrome. Neurology 31:893–896.

Shaw NA 1986 The effect of pentobarbital on the auditory evoked response in the brainstem of the rat. Neuropharmacology 25:63–69.

Sherman AL, Tirschwell DL, Micklesen PJ, et al 2000 Somatosensory potentials CSF creatine kinase BB activity, and awakening after cardiac arrest. Neurology 54:889–894.

Silverman D 1963 Retrospective study of the EEG in coma. Electroencephalogr Clin Neurophysiol 15:486–503.

Simon O, Schulz H, Rasmann W 1977 The definition of waking stages on the basis of continuous polygraphic recordings in normal subjects. Electroencephalogr Clin Neurophysiol 42:48–56.

Sleigh JW, Havill JH, Frith R, et al 1999 Somatosensory evoked potentials in severe traumatic brain injury: a blinded study. J Neurosurg 91:577–580.

Sohmer H, Gafni M, Havatselet G 1984 Persistence of auditory nerve response and absence of brain-stem response in severe cerebral ischaemia. Electroencephalogr Clin Neurophysiol 58:66–72.

Sohmer H, Goitein K 1988 Auditory brainstem and somatosensory evoked potentials in an animal model of synaptic lesions: elevated plasma barbiturate levels. Electroencephalogr Clin Neurophysiol 71:382–388.

Sohmer H, Gafni M, Chisin R 1978 Auditory nerve and brainstem responses: comparison in awake and unconscious subjects. Arch Neurol 35:228–230.

Starr A 1976 Auditory brainstem responses in brain death. Brain 99:543–544.

Starr A, Achor J 1975 Auditory brainstem responses in neurological disease. Arch Neurol 32:761–768.

Starr A, Hamilton AE 1976 Correlation between confirmed sites of neurological lesions and abnormalities of far-field auditory brainstem responses. Electroencephalogr Clin Neurophysiol 41:595–608.

Stern BJ, Krumholz A, Weiss HD, et al 1982 Evaluation of brainstem stroke using brainstem auditory evoked potentials. Stroke 13:705–711.

Sternberg D, Maisonobe T, Jurkat-Rott K, et al 2001 Hypokalemic periodic paralysis type 2 caused by mutations at codon 672 in the muscle sodium channel gene SCN4A. Brain 124:1091–1099.

Stockard JJ, Stockard JE, Sharbrough FW 1980 Brainstem auditory evoked potentials in neurology: methodology, interpretation and clinical application. In: *Electrodiagnosis in Clinical Neurology* (ed. MJ Aminoff). Churchill Livingstone, New York, pp. 370–400.

Subramony SH, Carpenter DE, Raju S, et al 1991 Pancuronium induced prolonged neuromuscular blockade. Crit Care Med 19:1583–1587.

Sutton LN, Frewen T, Marsh R, et al 1982 The effects of deep barbiturate coma on multimodality evoked potentials. J Neurosurg 57:178–185.

Synek VM 1990 Revised EEG coma scale in diffuse acute head injuries in adults. Clin Exp Neurol 27:99–111.

Tasker RC, Boyd SG, Harden A, et al 1989 EEG monitoring of prolonged thiopentone administration for intractable seizures and status epilepticus in infants and young children. Neuropediatrics 20:147–153.

Teasdale G, Jennett B 1974 Assessment of coma and impaired consciousness. A practical scale. Lancet ii:7872:81–84.

Thatcher RW, Walker RA, Gerson I, et al 1989 EEG discriminant analyses of mild head trauma. Electroencephalogr Clin Neurophysiol 73:94–106.

Tristani-Firouzi M, Chen J, Mitcheson JJ, et al 2001 Molecular biology of K^+ channels and their role in cardiac arrhythmias. Am J Med 110:50–59.

Trojaborg W, Jørgensen EO 1973 Evoked cortical potentials in patients with 'isoelectric' EEGs. Electroencephalogr Clin Neurophysiol 35:301–309.

Trontelj JV, Mihelin M, Fernandez JM, et al 1986 Axonal stimulation for end-plate jitter studies. J Neurol Neurosurg Psychiatry 49:677–685.

Tsubokawa T, Nishimoto H, Yamamoto T, et al 1980 Assessment of brainstem damage by the auditory brainstem responses in acute severe head injury. J Neurol Neurosurg Psychiatry 43:1005–1011.

Tuck RR, McLeod JG 1981 Autonomic dysfunction in Guillain–Barré syndrome. J Neurol Neurosurg Psychiatry 44:983–990.

Turazzi S, Bricolo A 1977 Acute pontine syndromes following head injury. Lancet ii:62–64.

Uziel A, Benezech J, Lorenzo S, et al 1982 Clinical application of brainstem auditory evoked potentials in comatose patients. In: *Clinical Applications of Evoked Potentials in Neurology* (eds J Courjon, F Mauguière, M Revol). Raven Press, New York, pp. 195–202.

Valli G, Barbieri S, Scarlato G 1983 Neurophysiological tests in human botulism. Electromyogr Clin Neurophysiol 23:3–11.

Van Cott AC, Blatt I, Brenner RP 1996 Stimulus-sensitive seizures in postanoxic coma. Epilepsia 37:868–874.

Van der Meché FG, Meulstee J, Vermeulen M, et al 1988 Patterns of conduction failure in the Guillain–Barré syndrome. Brain 111:405–416.

Van der Rijt CDC, Scalm S, De Groot GH, et al 1984 Objective measurement of hepatic encephalopathy in an unselected population of patients with liver cirrhosis in general practice. Electroencephalogr Clin Neurophysiol 57:423–426.

Van Marie W, Woods KL 1980 Acute hydrocortisone myopathy. BMJ 281:271–272.

Van Ness PC 1990 Pentobarbital and EEG burst-suppression in treatment of status epilepticus refractory to benzodiazepines and phenytoin. Epilepsia 31:61–67.

Veselis RA, Reinsel R, Sommer S, et al 1991 Use of neural network analysis to classify electroencephalographic patterns against depth of midazolam sedation in intensive care unit patients. J Clin Monit 7:259–267.

Vespa P, Nuwer MR, Juhasz C, et al 1997 Early detection of vasospasm after acute subarachnoid hemorrhage using continuous EEG ICU monitoring. Electroencephalogr Clin Neurophysiol 103:607–615.

Vespa PM, Nuwer MR, Nenov V, et al 1999 Increased incidence and impact of nonconvulsive and convulsive seizures after traumatic brain injury as detected by continuous electroencephalographic monitoring. J Neurosurg 91:750–760.

Vespa PM, Boscardin WJ, Hovda DA, et al 2002 Early and persistent impaired percent alpha variability on continuous electroencephalography monitoring as predictive of poor outcome after traumatic brain injury. J Neurosurg 97:84–92.

Walker MC 2003 Status epilepticus on the intensive care unit. J Neurol 250:401–406.

Walser H, Mattle H, Keller HM, et al 1985 Early cortical median nerve somatosensory evoked potentials. Prognostic value in anoxic coma. Arch Neurol 42:32–38.

Walsh JC, Yiannikas C, McLeod JG 1984 Abnormalities of proximal in acute idiopathic polyneuritis: comparison of short latency evoked potentials and F-waves. J Neurol Neurosurg Psychiatry 47:197–200.

Warr B 1980 Efferent components of the auditory system. Ann Otol Rhinol Laryngol 89:114–120.

Waruiru CM, Newton CR, Forster D, et al 1996 Epileptic seizures and malaria in Kenyan children. Trans R Sch Trop Med Hygiene 90:152–155.

Webb W 1969 Twenty-four hour sleep cycling. In: *Sleep, Physiology and Pathology* (ed A Kales). Lippincott, Philadelphia, pp. 53–66.

Webber WR, Litt B, Wilson K, et al 1994 Practical detection of epileptiform discharges (EDs) in the EEG using an artificial neural network: a comparison of raw and parameterized EEG data. Electroencephalogr Clin Neurophysiol 91:194–204.

Wiederholt WC, Mulder DW, Lambert EH 1964 The Landry–Guillain–Barré–Strohl syndrome or polyradiculoneuropathy: historical review, report on 97 patients and present concepts. Mayo Clin Proc 39:427–51.

Wijdicks EFM 1995 Determining brain death in adults. Neurology 45:1003–1011.

Winer JB, Hughes RA, Anderson MJ, et al 1988a A prospective study of acute idiopathic neuropathy. II: Antecedent events. J Neurol Neurosurg Psychiatry 51:613–618.

Winer JB, Hughes RA, Osmond C 1988b A prospective study of acute idiopathic neuropathy. I. Clinical features and their prognostic value. J Neurol Neurosurg Psychiatry 51:605–612.

Wytrzes LM, Chatrian G-E, Shaw C-M, et al 1989 Acute failure of forebrain with sparing of brain-stem function. Electroencephalographic, multimodality evoked potential, and pathologic findings. Arch Neurol 46:93–97.

Ying Z, Schmid UD, Schmid J, et al 1992 Motor and somatosensory evoked potentials in coma analysis and relation to clinical status and outcome. J Neurol Neurosurg Psychiatry 55:470–474.

Young GB, Bolton CF, Archibald YM, et al 1992 The electroencephalogram in sepsis-associated encephalopathy. J Clin Neurophysiol 9:145–152.

Young GB, Blume WT, Campbell VM, et al 1994 Alpha, theta and alpha–theta coma: a clinical outcome study utilizing serial recordings. Electroencephalogr Clin Neurophysiol 91:93–99.

Young GB, Jordan KG, Doig GS 1996 An assessment of nonconvulsive seizures in the intensive care unit using continuous EEG monitoring: an investigation of variables associated with mortality. Neurology 47:83–89.

Young WL, Ornstein E 1985 Compressed EEG monitoring during cardiac arrest and resuscitation. Anesthesiology 62:535–538.

Zandbergen EGJ, de Haan RJ, Stoutenbeek CP, et al 1998 Systematic review of early prediction of poor outcome in anoxic–ischaemic coma. Lancet 352:1808–1812.

Zandbergen EGJ, de Haan RJ, Koelman JHTM, et al 2000 Prediction of poor outcome in anoxic–ischemic coma. J Clin Neurophysiol 17:498–501.

Zentner J, Rohde V 1992 The prognostic value of somatosensory and motor evoked potentials in comatose patients. Neurosurgery 31:429–434.

Zentner J, Rohde V, Friedman WA, et al 1992 The prognostic value of somatosensory and motor evoked potentials in comatose patients. Neurosurgery 31:429–434.

Zochodne DW, Bolton CF, Wells GA, et al 1987 Critical illness polyneuropathy: a complication of sepsis and multiple organ failure. Brain 110:819–842.

Zochodne DW, Ramsay DA, Saly V, et al 1994 Acute necrotising myopathy of intensive care: electrophysiological studies. Muscle Nerve 17:285–292.

Chapter 7

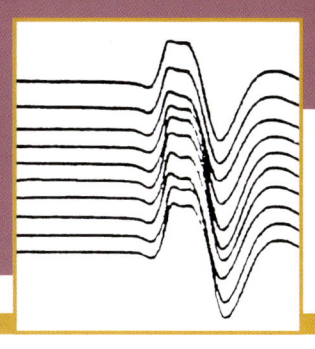

Neurophysiological Monitoring during Surgical Operations

INTRODUCTION

This chapter introduces neurophysiological monitoring work during surgical operations and explains the choice and the practical use of tools for this purpose (*Table 7.1*).

INTRACRANIAL SURGERY

EEG monitoring in intracranial surgery
Historical background
Monitoring during neurosurgical operations has been undertaken since the time of the early demonstrations of the clinical utility of the electroencephalogram (EEG) (Walter 1936). Comparisons of preoperative and intraoperative localization of cerebral mass lesions, demonstrations of the effects of raised intracranial pressure in masking localization of tumours and of the changes wrought by anaesthesia, highlighted technical expertise at a time when few other methods existed for detection of areas of altered function (*Fig. 7.1*). Since then, in addition to electrocorticography and related techniques used in the surgical treatment of epilepsy, a wide range of neurosurgical applications has been developed.

Procedures
Because of the high incidence of abnormalities due to the presenting condition, preoperative baseline recordings are required using a conventional multichannel recording and include evaluation of responses to external stimuli of graded severity. In stuporous or comatose patients it may be possible to elicit abnormal, high amplitude, slow wave responses such as frontal intermittent rhythmic delta activity (FIRDA) or paradoxical slow wave arousal responses described in Chapter 3. These could lead to misinterpretation in computerized monitoring systems (Kochs et al 1994, Bischoff 1996, Dahaba 2005).

Monitoring during neurosurgical procedures may involve recording from scalp electrodes outside the surgical field, but equally may require the use of various forms of specialized electrodes similar to those for electrocorticography and related procedures.

Subarachnoid haemorrhage and vasospasm
Early prediction of ischaemia in patients with subarachnoid haemorrhage may assist in planning the appropriateness and timing of surgery. Quantitative EEG monitoring with compressed spectral arrays (CSAs) and trend analysis was correlated with clinical and radiographic findings in 11 patients by Labar et al (1991). The most sensitive parameters were found on trend analysis; these appeared before clinical changes in 4 out of the 11 ischaemic events and comprised changes in total power (in 91% of ischaemic events), changes in alpha ratio (64%), frequency centroid (55%) and per cent delta (45%). Comparable CSA features were changes in power (44%) and slowing (39%). Total power and frequency were noted to vary independently, an important characteristic with implications for the design of appropriate monitoring algorithms. In a more recent study, Vespa et al (1997) showed that vasospasm could be predicted on the basis of reduced variability of relative (per cent) alpha in 19 patients out of a total of 32 studied after subarachnoid haemorrhage. Positive and negative predictive values were 76% and 100%, respectively. The EEG changes in 10 of the patients preceded clinical diagnosis of vasospasm by a mean of 2.9 (SD ± 1.73) days; resolution of vasospasm was accompanied by recovery of the EEG.

Intracranial aneurysms and vascular malformations
Evaluation of localized brain ischaemia by EEG monitoring during test occlusion of the internal carotid artery and during treatment of intracranial aneurysms and vascular malformations has been advocated on the basis that prompt warning of consequent cortical ischaemic changes can be provided (Andrews et al 1989, Dabbagh et al 1990). In one case report, EEG monitoring demonstrated short duration ischaemic changes, not evident on somatosensory evoked potential (SEP) testing, when short periods of asystole occurred on a background of induced hypotension and mild hypothermia during clipping of a basilar aneurysm. Since the ischaemic episodes each lasted for a matter of seconds, it is hardly surprising that averaged evoked potentials (EPs) did not demonstrate them (Groff et al 1999). EEG monitoring may also be helpful during interventional neuroradiology for testing the adequacy of collateral circulation during balloon occlusions, and as an adjunct to sodium amytal testing, for instance prior to embolization (Hacke 1985, Paiva et al 1995), whether with scalp electrodes or even with intra-arterial recording (Nakase et al 1995, Boniface and

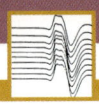

Antoun 1997). Since this latter electrode site provided an EEG of higher amplitude than scalp recordings and showed interictal spikes comparable to those from subdural strip electrodes over the medial temporal lobe, Nakase and colleagues also considered it useful in patients with complex partial seizures.

Successful surgical treatment of intracranial aneurysms or vascular malformations must not be at the price of production of ischaemic damage. Following subarachnoid haemorrhage, such patients may have impaired autoregulation of cerebral blood flow (CBF) and this increases the risk of hypoperfusion during periods of accidental or controlled hypotension during surgery. Risk of ischaemic damage is increased by surgical events such as retraction and interference with local blood supply by clamping, particularly when there is already a degree of localized vasospasm. Retraction of the brain for surgical access may lead to contusion or infarction in an estimated 5% of patients during intracranial aneurysm procedures and in 10% of cranial base procedures. Policies for minimizing such injuries include a combination of judicious retraction, appropriate anaesthetic and pharmacological management, and aggressive intraoperative monitoring, including the use of EEG to monitor drug-induced burst suppression activity and lateralized or localized abnormalities (Muizelaar 1989, Andrews et al 1993). Localized ischaemia resulting in new neurological deficits due to vasospasm could be predicted on the basis of intraoperative EEG changes consisting of marked ipsilateral attenuation or marked ipsilateral delta, theta and alpha components in a study of 27 patients undergoing surgery for anterior circulation aneurysms (Tempelhoff et al 1989).

Table 7.1 Procedures

Surgical procedure	MEP	VEP	BAEP	SEP	EMG	EEG	Other techniques	Comments
Removal of cerebral tumour	•	•		•				Cortical localization as well as prevention of damage
Clipping intracranial aneurysm				•				SEP measurement of central conduction time
Carotid endarterectomy				•		•	Near infrared spectroscopy; transcranial Doppler; stump pressure	
Orbital surgery, removal of optic nerve tumour		•						
Removal of vestibular schwannoma (acoustic neuroma) facial	•		•	•	•			ECochG; 8NAP; facial muscle response to seventh nerve stimulation; spontaneous EMG of muscles
Cranial nerve surgery other than vestibular schwannoma	•				•			Spontaneous and evoked EMG of appropriate muscles
Cervical spine surgery	•			•				
Correction of spinal deformity	•			•				
Spinal tumour	•			•				MEP: spinal D wave to transcranial stimulation
Dorsal rhizotomy	•				•			Multichannel EMG responses to root stimulation
Spinal dysraphism and myelomeningocoele	•			•	•			SEP; MEP; EMG responses to root stimulation
Insertion of pedicle screw					•			EMG responses to pedicle screw stimulation
Spinal root, plexus or peripheral nerve	•				•		Nerve action potential	
Cardiac surgery				•		•		
Aortic coarctation and aneurysm				•				

Intracranial surgery

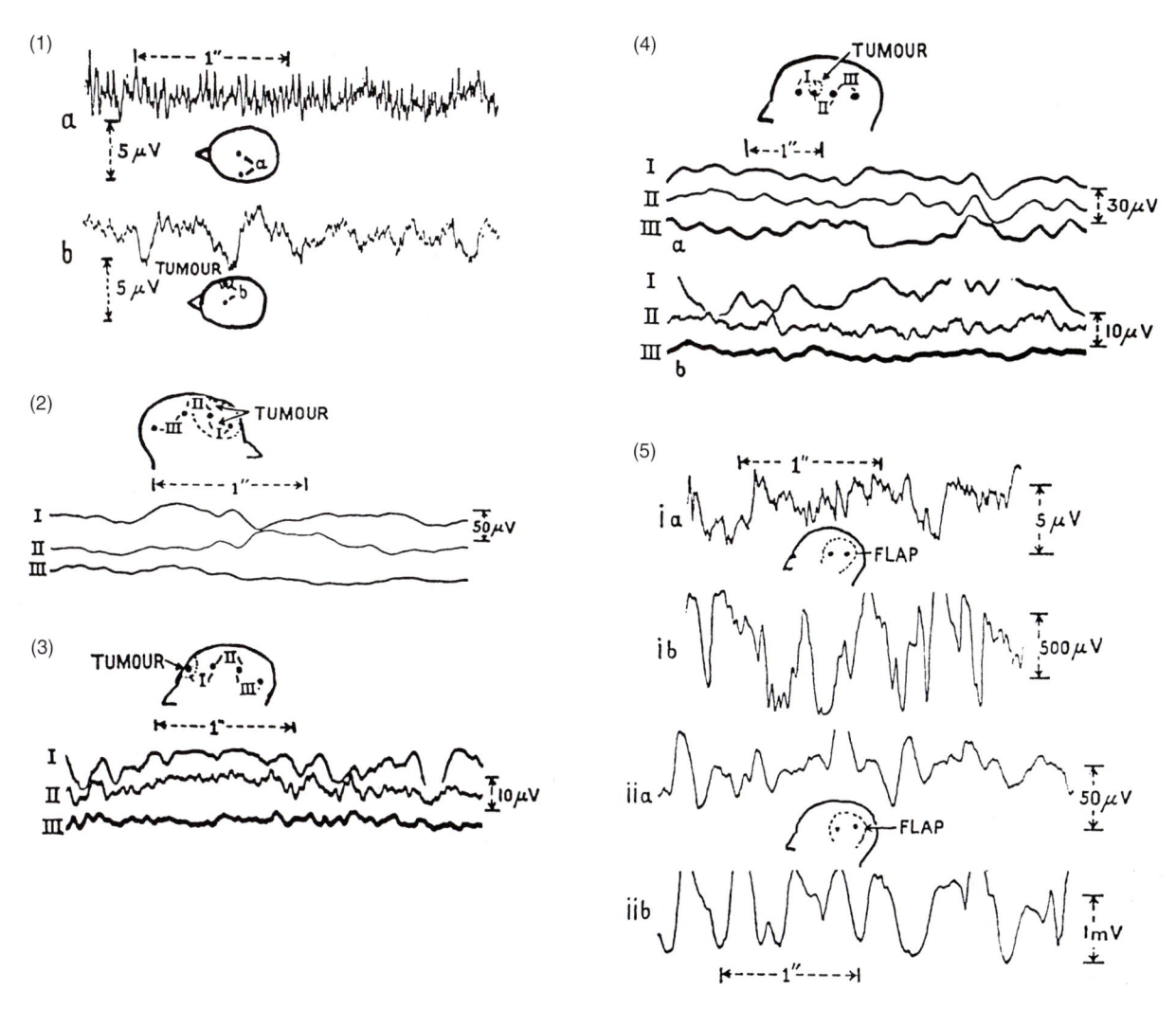

Fig. 7.1 Monitoring during neurosurgery. Montage of figures from Grey Walter's 1936 report on 'The location of cerebral tumours by electro-encephalography'. Extracts from his original legends still provide an excellent basis for understanding the general principles of monitoring work in neurosurgery, far beyond that described with cerebral tumours. (1) The first case in which the tumour was located by EEG. (a) From the left side, showing almost continuous beta waves at a frequency of about 25/s. (b) From the right side, showing the slow delta waves and smaller and less regular beta waves. (2) The amplification had to be kept low because the patient was liable to move. The varying phase-relations of the delta waves in leads I and II suggested that the area enclosed by the dotted line was abnormal. This was verified by numerous records taken with other electrode arrangements. (3) The delta waves originate in the extreme frontal region. The area was further delimited by recording with more elaborate electrode patterns. (4) (a) Taken while the intracranial pressure was high. The slow waves involve the whole cranium. No location is possible, (b) Taken with the same electrode arrangement but after some reduction in the intracranial pressure. The slow waves are now restricted to the area of the tumour. The amplification in (b) is about three times that in (a). (5) (i) Patient A: (a) from the skull with high amplification, (b) during operation. From the corresponding area on the cortex with much lower amplification. There was only a local anaesthetic and morphine. The eyes were closed in both records. (ii) Patient B: (a) from the skull, (b) from the underlying brain at operation, both when the patient was under nitrous oxide and ether. (From Walter (1936), by permission.)

In patients with 'unclippable' aneurysms, permanent occlusion of the carotid artery by ligation may be used to encourage thrombosis of the aneurysm (Swearingen and Heros 1987). The effects of a test occlusion are checked during surgery by awake examination, EEG monitoring, and carotid stump pressure measurements. Additional extracranial–intracranial bypass is only required when preoperative carotid cross-compression studies suggest inadequate collateral circulation or when intraoperative testing demonstrates new neurological deficits or EEG abnormalities during test occlusion.

Monitoring brain protection regimens in neurosurgery

The brain can be protected from the effects of local ischaemia during periods of arterial occlusion during treatment of aneurysms by anaesthetic-induced suppression of the metabolic requirement for oxygen (Belopavlovic et al 1985, Dabbagh et al 1990, Ravussin and de Tribolet 1993). Various agents have been used successfully with EEG monitoring for detection of the desired end-point of burst suppression. Hypothermia has been an additional source of protection in some circumstances (Stone et al 1993,

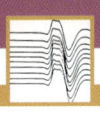

1996). In one study of 38 patients during middle cerebral aneurysm surgery, in those given thiopental to produce burst suppression there was a reduction of SEP deterioration and no postoperative neurological deficits. Deficits in those without burst suppression were attributed to clamping and to brain retraction (Parenti et al 1996).

Visual evoked potential monitoring

Visual evoked potentials (VEPs) have been monitored intraoperatively during orbital and pituitary surgery to prevent damage to optic nerves or chiasm, and to help identify these structures if they are embedded in, or infiltrated by, tumour. Intraoperative visual potentials have also been investigated as a possible adjunct to other neurophysiological methods for detecting cerebral ischaemia (Reilly et al 1978, Russ et al 1984, Keenan et al 1987). Although VEP recording was the first form of intraoperative EP monitoring to be reported (Wright et al 1973), it has not yet found widespread application.

Methods of VEP monitoring

The response to flash stimulation is monitored, since responses to pattern reversal cannot be recorded in unconscious patients. Unfortunately the presence of a response to flash indicates only that light perception should be possible, but this may represent a very considerable reduction in visual acuity (Harding 1982). An attempt has been made to record the action potential from the optic nerve directly, but the responses are of low amplitude and are not recordable in all subjects (Møller et al 1987).

Practical aspects of VEP monitoring

Stimulation The commonly used stroboscopic stimulator is not suitable for operative use because of its size and the need to position it in front of the patient's face; furthermore, the repeated bright flashes would seriously impede the surgeon. Stimulators for surgical use fall into two types, goggles and contact lenses. An array of light emitting diodes (LEDs) mounted in goggles has been used by some workers, but goggles have the disadvantage of being unsuitable for orbital surgery because of their size and position, and because the commercially available goggles are only supplied with red LEDs. Two different contact lens stimulators have been described: a scleral shell upon which LEDs are mounted (Feinsod et al 1976, Møller 1988, 1995), and a haptic lens to which one end of a fibre-optic cable is connected, the other end being connected to a stroboscope (Wright et al 1973). An improved contact lens stimulator has been developed so that the fibre-optic cable does not interfere with the surgeon's movements, and does not tend to pull the lens away from the pupil (*Fig. 7.2*) (Harding et al 1987). The lens must be carefully positioned over the pupil and held in place by stitches in the eyelids.

Recording The response is recorded from an electrode at either Oz or 5 cm above the inion with a reference at either Fz, Cz or 12 cm above the nasion (Costa e Silva et al 1985). The filter settings usually chosen are low frequency (LF) 5 Hz and high frequency (HF) 100 Hz. A notch filter (50 or 60 Hz) cannot be used during VEP recording, because the maximum spectral power in the response is close to these frequencies; such a filter may almost completely obliterate the response. In most operations only one optic nerve is at risk from surgical manipulation, so comparison of the responses from the two eyes will reduce any uncertainty arising from possible drug effects, hypothermia, etc. Alternatively, during intracranial operations a nerve action potential can be recorded directly from the optic nerve, but this is not possible in all patients (Møller et al 1987). To confirm that an adequate stimulus has been given simultaneous electroretinogram (ERG) recording is desirable (Wright et al 1973).

Clinical applications and assessment of results VEP monitoring has been carried out in the very young (Reilly et al 1978, Albright and Sclabassi 1985), and the elderly (Russ et al 1984). However, it is important to recognize that the waveform changes markedly at different ages, and the latencies of different components alter with age (Harding 1982). Two studies have demonstrated that the VEP remains stable under anaesthesia during operations which do not involve interference with the visual pathways (Uhl et al 1980, Russ et al 1984). However, during operations for tumours at various sites in the visual pathways the response to LED stimulation often deteriorates before tumour dissection begins, and disappears in some patients, either spontaneously or as a result of pressure on the optic nerve, without any postoperative deterioration of vision (Smith 1975, Cedzich et al 1987). This variability of the response has led many workers to regard intraoperative VEP monitoring as being unreliable. Others have found

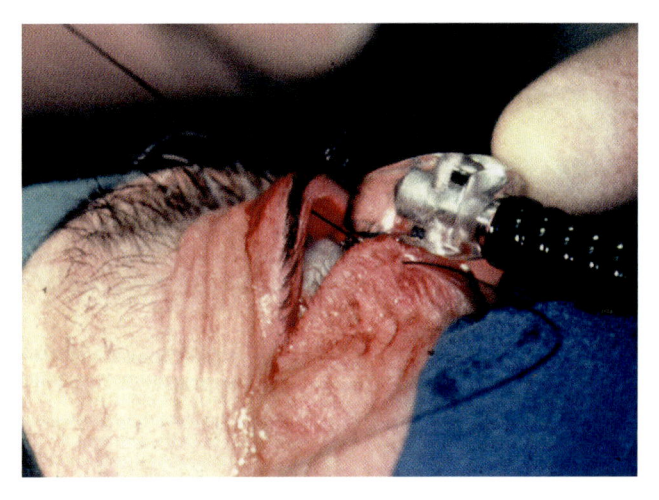

Fig. 7.2 Contact lens stimulator. The outer surface of the haptic contact lens is fitted with a small prism. In the surface of the prism, at right angles to the axis of the contact lens (which is secured with sutures), is a small socket suitable for a 5 mm fibre-optic cable. There is a small set screw to lock the fibre-optic cable in place. (Courtesy of Professor GFA Harding. From Binnie et al (2003), by permission.)

that intraoperative VEP deterioration predicts impairment of vision postoperatively, and that improvement in the response following decompression of optic pathways is often associated with improved vision (*Fig. 7.3*) (Feinsod et al 1976, Allen et al 1981, Raudzens 1982). A grading system for VEP abnormalities has been proposed, ranging from normal (Grade 1) to absence of response (Grade 5) (Costa e Silva et al 1985). Results obtained with the fibre-optic contact lens stimulator suggest that transient disappearance of the response is not necessarily associated with impairment of vision, but that absence of response for 4 min suggests significant damage to the optic pathways (Harding et al 1990). Flash VEP monitoring carries a significant risk of false negatives, since the response may still be recordable in patients with gross impairment of acuity and large visual field defects (Harding 1982).

Operations on the skull base and paranasal sinuses carry a risk of damaging the optic nerves, and some surgical teams have used VEP monitoring to prevent permanent visual impairment (Herzon and Zelear 1994, Hussain et al 1996), but this form of monitoring is regarded as being too cumbersome, and the results too variable, to justify widespread adoption at present (Jones 1997).

Intraoperative VEP recording has been used to identify the optic pathways if they are either embedded in, or infiltrated by, tumour (Albright and Sclabassi 1985, Costa e Silva et al 1985). VEP monitoring in these circumstances permits a more complete tumour removal without sacrificing the optic nerves or chiasm; the volume of tumour remaining to be treated by radiotherapy is thereby reduced. This may prove to be a more important application of the technique than visual pathway protection.

SEP monitoring in intracranial surgery
Experimental studies

Studies of experimental ischaemia in animals have shown the existence of several threshold levels of regional CBF (Hossmann 1994, 1999), two of which are important electrophysiologically (Symon 1985). Electrical activity, both spontaneous and evoked, ceases when blood flow falls to about 15 ml/100 g/min (Branston et al 1974, Heiss et al 1976) and a similar value has been identified in man (Trojaborg and Boysen 1973, Sundt et al 1974). Disappearance of the cortical SEP due to reduced blood flow is preceded by increase in latency (Hargadine et al 1980), and in the cat thalamocortical conduction time is prolonged when cortical blood flow is reduced to 45% of its normal value (Lesnick et al 1984).

If blood flow falls to 10 ml/100g/min the membrane sodium–potassium pump fails and potassium begins to leak out of neurones; when a critical extracellular potassium concentration is reached, calcium begins to enter cells with damaging effects (Astrup et al 1977, Branston et al 1977, Harris and Symon 1984). If blood flow is maintained below this ionic threshold then infarction rapidly follows (Morawetz et al 1978); longer periods of blood flow above this level, but below the electrical threshold (i.e. between 10 and 15 ml/100 g/min), will also result in infarction (Bell et al 1985). Electrophysiological monitoring can therefore

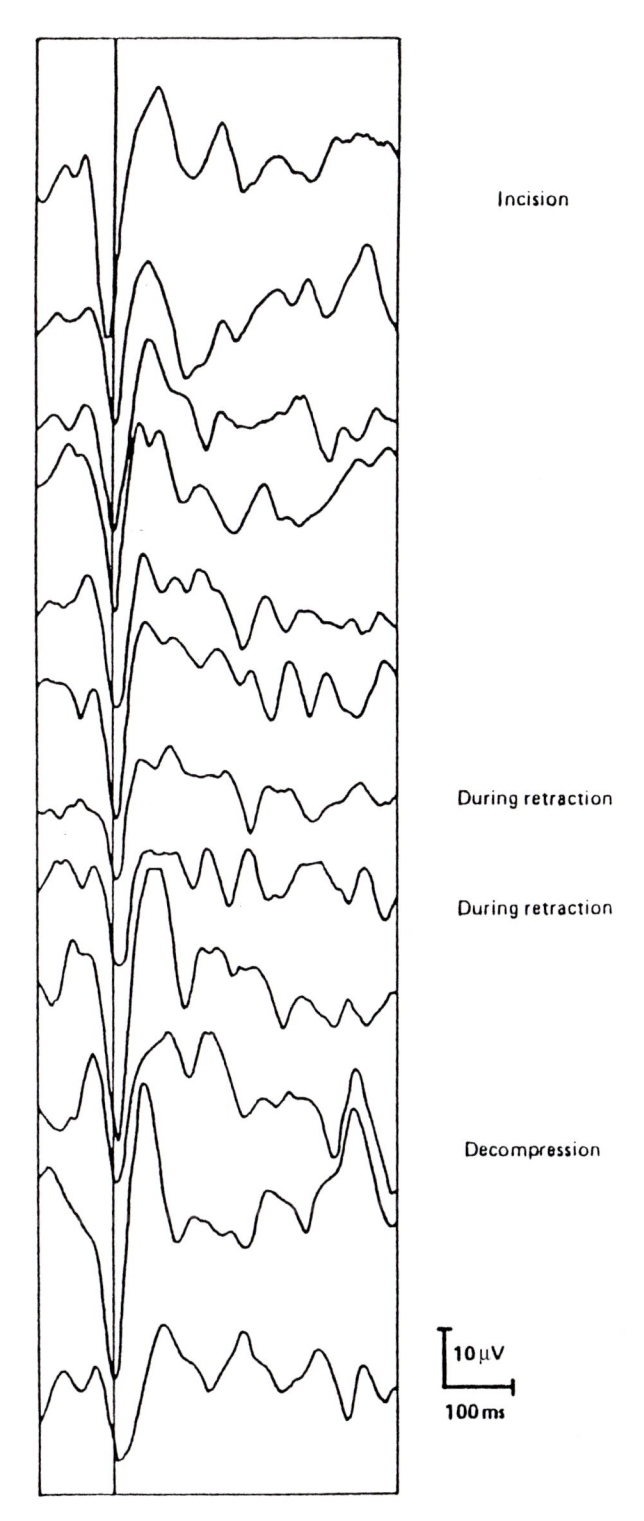

Fig. 7.3 VEPs recorded during bimedial decompression for dysthyroid exophthalmos in a 41-year-old woman. Stimuli were delivered at 1.6 Hz through a contact lens. Each trace is the average of 32 responses from Oz referred to Fz. The amplitude is reduced by retraction. (From Harding et al 1990 Visual evoked potential monitoring of optic nerve function during surgery. *Journal of Neurology, Neurosurgery and Psychiatry* 53:890–895, by permission of the BMJ Publishing Group.)

identify a level of blood flow which is close to the threshold for rapid infarction; if flow is maintained just above this threshold, infarction will eventually develop (*Fig. 7.4*). Following occlusion of a cerebral artery, blood flow in the centre of its territory falls below the ionic threshold, and infarction can only be prevented by rapid restoration of perfusion. The size of the ischaemic area depends on systemic blood pressure and the adequacy of collateral circulation (Jones et al 1981). Blood flow to a surrounding border of tissue may be maintained by collaterals at a level between the two thresholds; this 'ischaemic penumbra' (Astrup et al 1981) can be easily identified, since it is electrically silent. Although the penumbra can recover if blood flow is rapidly restored, delay in re-establishing normal blood flow may result in loss of viability (Oberenovitch 1995, Back 1998).

The effects of ischaemia upon the SEP to median nerve stimulation, and particularly its initial negative component (N20), have been studied in detail. The amplitude of this potential is reduced by ischaemia, but this is not a specific indicator of early ischaemia during operations because the amplitude is sensitive to fluctuating concentrations of halogenated anaesthetics (see p. 172). Ischaemia is more reliably detected by increasing N20 latency, particularly if the variability due to arm length and temperature effects on peripheral nerve conduction are eliminated by calculating the central conduction time (CCT), which is the interval between cervical N13 and cortical N20 (Hume and Cant 1978). In normal subjects between the ages of 15 and 50 years the mean CCT is 5.6 ms, standard deviation 0.5 ms; over the age of 50 years the mean CCT increases by about 0.3 ms (Hume et al 1982). Studies relating CCT to CBF, determined by the xenon technique, have shown that CCT increases when blood flow falls below 30 ml/100g/min (*Fig. 7.5*). Further CCT prolongation accompanies flow reduction down to 15 ml/100g/min, at which level the cortical response disappears. The apparent threshold for change in CCT at 30 ml/100g/min, rather than 20 ml/100g/min as mentioned above, probably indicates that flow has fallen below the electrical threshold in small areas not detectable by the xenon technique, which averages values over a wide area (Rosenstein et al 1985); similar results have been obtained with flow measurement by the microsphere technique which is similarly insensitive to localized ischaemia (Kochs and Schulte am Esch 1991).

SEP monitoring for ischaemia

Aneurysm surgery During operations upon intracranial aneurysms several factors may contribute to ischaemia: retractors compress vessels, and ischaemia is then particularly likely if the systemic arterial pressure is low (Ducati et al 1988); bleeding induces arterial spasm, both before and during the operation (Dorsch 1995); the surgeon may

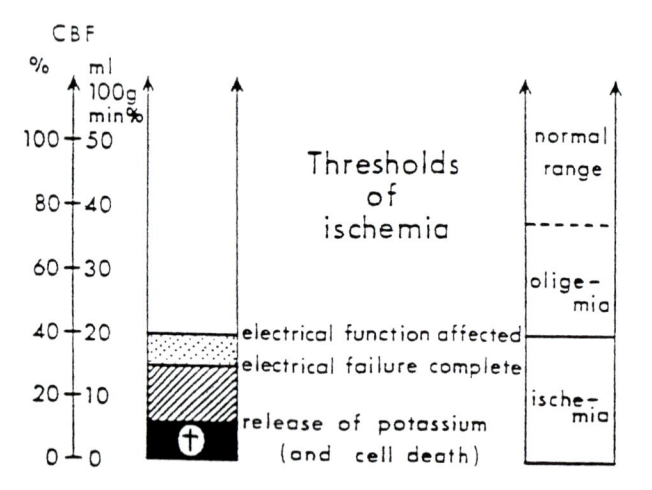

Fig. 7.4 Diagram illustrating the four thresholds (change in electrical activity, failure of activity, potassium leakage, and cell death) associated with reduction in CBF. (From Astrup J, Symon L, Branston NM, Lassen NA 1977 Cortical evoked potential and extracellular K^+ and H^+ at critical levels of brain ischaemia. *Stroke* 8:51–57, by permission.)

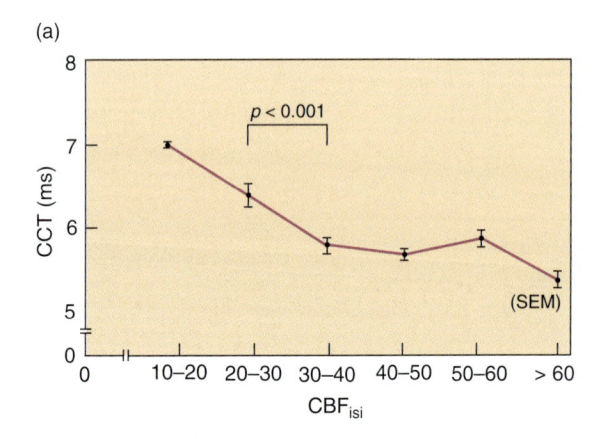

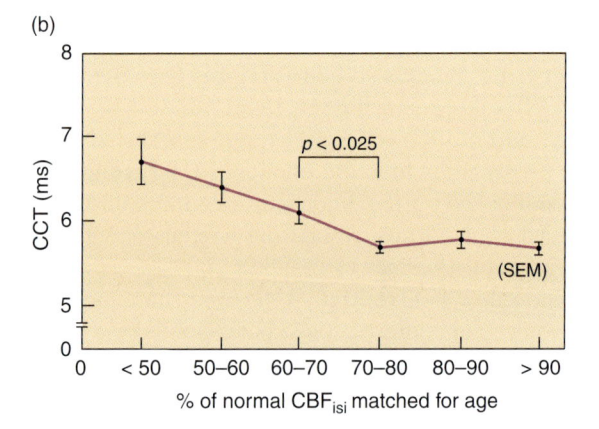

Fig. 7.5 (a) Relation between central conduction time (CCT) and initial-slope index of the CBF_{isi} measured in 120 hemispheres. A threshold phenomenon is noted at flow values below 30 where CCT becomes significantly prolonged. (b) Relation between CCT and CBF_{isi} taken as a percentage of normal CBF_{isi} matched for age. A threshold phenomenon is noted at flow values below 70% of normal, where CCT becomes significantly prolonged. SEM, standard error of mean. (From Rosenstein et al (1985), by permission of the *Journal of Neurosurgery*.)

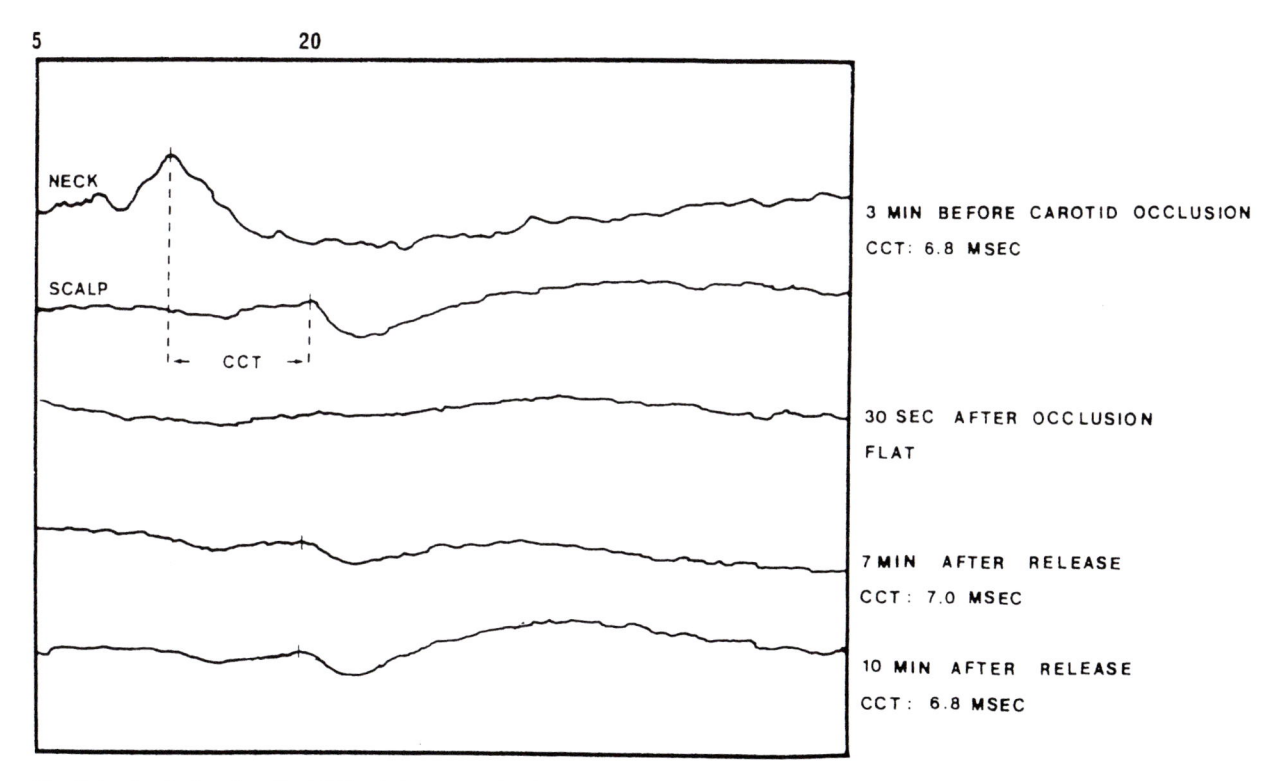

Fig. 7.6 Cervical and cortical SEPs during operation for giant aneurysm at the origin of the ophthalmic artery. The cervical potential (top trace) remained constant throughout the operation, but carotid artery occlusion led to loss of the cortical N20. The artery was occluded for 12 min, and N20 reappeared 20 min after release of occlusion. (From Momma et al (1987), by permission.)

apply a temporary clip to a major artery to obtain a bloodless field (Momma et al 1987, Mooij et al 1987, Schramm et al 1994); and, finally, permanent carotid ligation is sometimes used to reduce the risk of haemorrhage from aneurysms which cannot be clipped (Annotation 1987, Swearingen and Heros 1987).

The choice of EP recording for the detection of cerebral ischaemia depends upon which vascular territory is most at risk. The greatest experience has been acquired in operations upon the internal carotid or middle cerebral arteries, where the vascular territory includes the upper limb primary sensory area which can be monitored by the SEP to median nerve stimulation (*Fig. 7.6*). Complete disappearance of the cortical N20 is associated with a high risk of postoperative deficit, even if the potential subsequently reappears (McPherson et al 1983, Friedman et al 1987). Two other factors have predictive value: the duration of loss of response and the rapidity with which the potential disappears after any surgical manoeuvre, such as clipping a vessel, which might be expected to induce ischaemia (Momma et al 1987). Adverse factors are disappearance within 4 min, or loss of the response for more than 12 min in the case of middle cerebral artery occlusion, or 16 min for the terminal carotid (Momma et al 1987, Symon et al 1988). Careful application of median nerve SEP monitoring can prevent the appearance of new neurological deficits in patients undergoing operation for middle cerebral artery aneurysms (Schramm et al 1990, Friedman et al 1991).

CCT prolongation precedes disappearance of N20 and is a more sensitive indicator of ischaemia. However, significant increase in CCT is associated with opening the skull, brain retraction and aneurysm clipping, even in uneventful operations, after which no deficit can be detected (Symon et al 1984). A prolonged CCT is therefore not specific for ischaemia, and unacceptably high rates of both false positives and false negatives result if an arbitrary CCT limit is used throughout the operation (Friedman et al 1987). Greater reliability depends upon defining confidence limits for each stage of the operation. A CCT prolongation of less than 10 ms can usually be tolerated without postoperative deficit, provided that the cause of ischaemia is removed within 30 min (Symon et al 1984). SEP monitoring has been found to have a high correlation with postoperative deficit in aneurysm surgery provided the area likely to become ischaemic can be monitored, and SEP changes are often reversed if the surgeon can remove temporary arterial occlusion, reposition retractors or reapply the aneurysm clip (Schramm et al 1990, Friedman et al 1991, Lopéz et al 1999). Although SEP monitoring might be expected to predict only sensory deficits, the correlation with motor function appears good; only one example of a pure motor hemiparesis has been reported, following no significant change in SEP latency or amplitude during operation for intracranial aneurysm (Krieger et al 1992).

For the anterior communicating and anterior cerebral arteries, the vascular territory does not include the primary sensory area of the upper limb, so the SEP produced by median nerve stimulation does not provide a reliable indication of ischaemia in this territory (Symon et al 1984). The lower limb sensory area is in the relevant vascular territory, but only limited experience has been obtained with tibial nerve SEPs as a monitor of ischaemia (Grundy et al 1982a, Sako et al 1998). Subcortical strokes have occasionally developed in spite of unchanged SEPs during operations on the anterior circulation, hence methods need further refinement for anterior cerebral aneurysms (Holland 1998).

Monitoring operations for basilar artery aneurysms was initially less satisfactory. Neither SEP monitoring alone, nor a combination of brainstem auditory evoked potentials (BAEPs) and SEPs, were able to reliably detect ischaemic damage to the brainstem (Friedman et al 1987, Little et al 1987), but more recent reports suggest that a combination of BAEP and SEP monitoring may provide adequate sensitivity (Manninen et al 1994, Lopéz et al 1999). SEP monitoring is sometimes useful when the basilar artery is temporarily occluded (Symon et al 1988), and cortical ischaemia has been detected following unintentional compression of the middle cerebral artery by a retractor during operation for basilar aneurysm (Buchthal and Belopavlovic 1992).

Intraoperative SEPs for cortical localization

Wide resection of tumours or epileptic foci without postoperative sensory or motor deficit depends upon accurate identification of sensorimotor cortex; this may be difficult and unreliable during operations, particularly if the anatomy is distorted by the presence of tumour or cerebral oedema (Ebeling and Huber 1992). If the operation is performed under local anaesthetic, with the patient awake, then cortical areas can be identified by the response to direct stimulation, as was done originally by Penfield and Boldrey (1937); renewed interest has been shown in this method more recently (Ojemann et al 1996, Taylor and Bernstein 1999). Localization of sensory cortex by direct stimulation is impossible if general anaesthesia is to be used, but EPs can be recorded from the exposed cortex under general anaesthesia, and they have a distribution similar to the cortical somatic representation demonstrated by direct stimulation (Jasper et al 1960, Woolsey et al 1979); similarly, spinal or EMG responses to stimulation of the exposed cortex can be recorded under anaesthesia to identify motor cortex (Cedzich et al 1996, Yingling et al 2000, Kombos et al 2001). In the presence of a large tumour, or where anatomy is grossly distorted, some surgeons have found a combination of sensory and motor methods to be essential for localization (Horikoshi et al 2000, Neuloh and Schramm 2002, Romstock et al 2002).

The EP to contralateral median nerve stimulation recorded by an electrode on the cortex, with respect to a reference outside the wound, has a large amplitude (over 20 mV and sometimes up to 200 mV), and the area of maximum amplitude is sharply localized to a few square centimetres on either side of the central sulcus, corresponding to the hand area (Gregorie and Goldring 1984, Lüders et al 1986, Wood et al 1988). Localization is further aided by the fact that the response shows phase reversal across the sulcus; precentrally the initial deflection is positive followed by negative, while postcentrally the sequence is negative followed by positive (*Fig. 7.7*). The peak latency of the postcentral negative potential is almost identical to that of the much smaller N20 recorded from a central or parietal scalp electrode, which is sometimes useful if the postcentral cortex is not fully exposed. Correct localization can be achieved by exploring the exposed cortex with a single electrode, but a much less time consuming method is to use a specially constructed array of electrodes mounted in a holder, from which several channels can be recorded simultaneously. Some workers have used bipolar chains of electrodes, but this method does not always permit a distinction between pre- and postcentral gyri (Gregorie and Goldring 1984, King and Schell 1987). A possible source of error is the presence of a prominent positive–negative potential (P25, N35) which is maximal over the postcentral gyrus, but medial to the postcentral negative–positive response (N20, P30) (Wood et al 1988).

However, the largest negative–positive waveform is always postcentral and, if this potential is identified, incorrect localization of the postcentral gyrus will be avoided. In some patients the central sulcus takes an anterior–posterior course, in which case potentials must be mapped in two directions at right angles (Legatt and Kader 2000).

Attempts have been made to identify the sensory areas for the lip and the foot, but this has not proved reliable (Lüders et al 1986, Wood et al 1988). Identification of auditory cortex has been attempted using subdural electrodes, but the responses are small (Lee et al 1984). Visual cortex has been satisfactorily identified by EP recording during operations to remove epileptogenic foci (Curatolo et al 2000).

Posterior fossa surgery

The aims of EP monitoring during operations in the posterior fossa are to preserve hearing and to prevent damage to the brainstem and cranial nerves. The operations which have been monitored are those upon the auditory nerve or its immediate surroundings (removal of acoustic neuromas and other cerebellopontine angle tumours, retrolabyrinthine vestibular neurectomy and microvascular decompression of cranial nerves), those for brainstem lesions such as tumours, aneurysms and vascular malformations, and fracture dislocations of the first cervical vertebra. Some form of EP from the auditory system, usually the BAEP, is monitored for hearing preservation, but there are distinct advantages to recording the electrocochleogram (ECochG) and eighth nerve action potential (8NAP) in addition. The BAEP is also used for brainstem protection, sometimes with the addition of somatosensory potentials. Monitoring posterior fossa surgery can be very demanding, during many operations

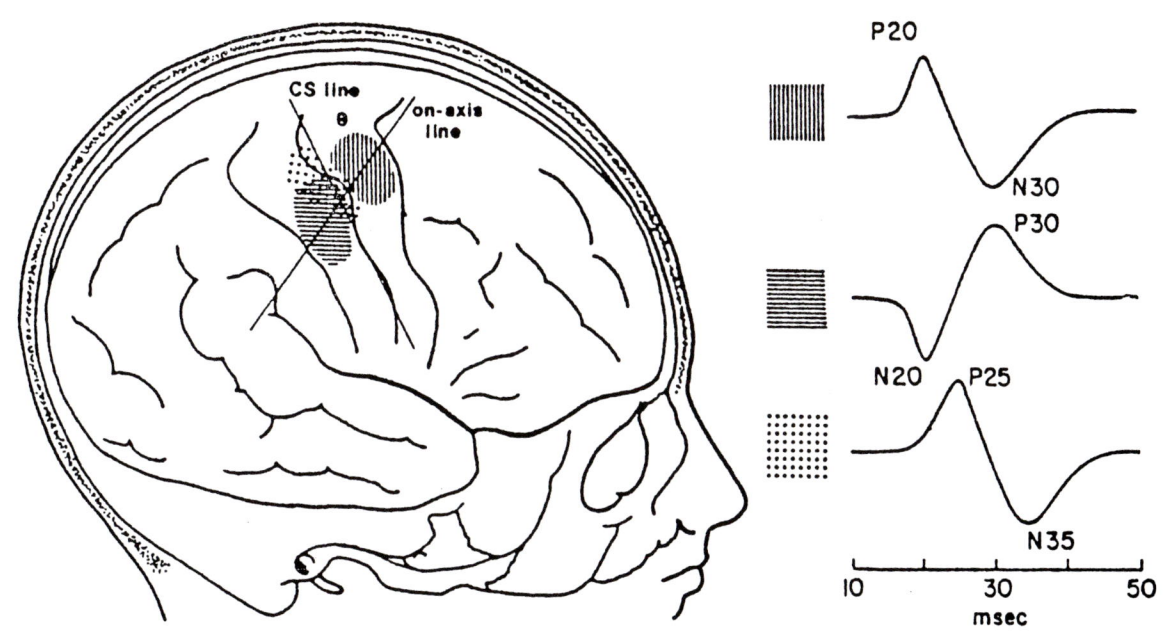

Fig. 7.7 Localization of the sensorimotor cortex by means of SEPs to median nerve stimulation. The on-axis line joins the precentral and postcentral amplitude maxima, and makes an angle (θ) of c. 70° with the line indicating the overall course of the central sulcus (CS line). Maximum P20–N30 amplitude is recorded in the area marked with vertical lines, maximum N20–P30 amplitude in the area marked with horizontal lines. The stippled area indicates the maximum P25 amplitude. (From Wood et al (1988), by permission of the *Journal of Neurosurgery*.)

several different auditory potentials must be recorded, together with SEPs, and EMG monitoring of the facial nerve (and possibly other nerves). Precisely which potentials are monitored, and at what stage of the operation, must be determined by discussion with the surgeon. This complex subject is reviewed by Møller (1988, 1995), Kartush and Bouchard (1992), Yingling and Gardi (1992), Cheek (1993), Beck (1994) and Kartush (1998).

Experimental studies

Two important factors have been identified as contributing to changes in auditory potentials during operations:

Manipulation of the eighth cranial nerve The intracranial auditory nerve has neither perineurium nor endoneurium, and the Schwann–glial junction is just inside the porus acousticus (Berthold et al 1984); the nerve is therefore very sensitive to compression and stretching, particularly at its exit from the bony cochlea into the internal auditory canal (Sekiya et al 1988). Cerebellar retraction, particularly in a lateral to medial direction, stretches the nerve, and produces initially a reversible decrease in amplitude of the 8NAP and wave V of the BAEP, and an increase in latency of wave V (Møller and Jannetta 1983). Animal studies have shown that this conduction block is associated with fibrin deposition and small haemorrhages around nerve fibres at their exit from the cochlea (Sekiya and Møller 1987, Sekiya et al 1988). Further stretching severs some nerve fibres as they leave the basal turn of the cochlea, and haemorrhages appear at the Schwann–glial junction. This results in disappearance of the 8NAP and all components of the BAEP. Early BAEP changes due to conduction block may be reversible for 30–60 min (Silverstein et al 1985a), but, if the cause cannot be corrected, axons will be damaged, resulting in a permanent deficit. Unfortunately the degree of BAEP change that indicates the progression of a physiological conduction block to permanent axonal damage has not yet been defined (Sekiya et al 1985). The 8NAP is much more sensitive than the BAEP to the presence of a partial conduction block because the later BAEP waves are not affected by desynchronization of impulses to the same degree as a nerve action potential (Møller and Jannetta 1983). A complete conduction block in the distal 8th nerve results in disappearance of the 8NAP negative peak and all waves of the BAEP, but the ECochG is not affected (*Fig. 7.8*).

Interruption of cochlear blood supply Several factors may interfere with cochlear blood supply. Surgical manipulation sometimes causes spasm of the anterior inferior cerebellar or internal auditory artery. If the latter vessel is embedded in tumour it may be cut inadvertently (Jannetta et al 1984), and occasionally coagulation of apparently insignificant vessels in the meatus results in loss of all auditory responses, hence such vessels must be assumed to be contributing to the arterial supply to the cochlea (Symon et al 1986). Animal studies have shown that the internal auditory artery can be damaged at the fundus of the internal meatus (*area cribrosa*) by traction on the 8th nerve (Sekiya and Møller 1987). When cochlear blood supply is interrupted, the ECochG (including cochlear microphonic and summating potential), 8NAP and all waves of the

BAEP, including wave I, are lost irreversibly (Levine et al 1984). In practice an alternative explanation for a total loss of response would be failure of stimulation, either because the transducer has been displaced, or fluid has accumulated in the external meatus. If the stimulator has been accidentally displaced or disconnected the stimulus artefact will disappear, but the artefact will not be affected if fluid has accumulated in the meatus, or if there has been a vascular lesion of the cochlea.

Methods of monitoring

The method of monitoring depends upon the type of operation and which structure (auditory nerve, facial nerve or brainstem) is regarded as being at greatest risk. For hearing preservation, most workers have used the BAEP, combined with either the ECochG or 8NAP (Ojemann et al 1984, Møller 1988, 1995, Mullatti et al 1999). Brainstem function is best monitored by the BAEP from the contralateral ear, supplemented by somatosensory responses from the limbs. Acoustic neuroma surgery therefore ideally requires a combination of monitoring methods; in patients with small tumours hearing preservation will be a priority, while brainstem protection is more important with large tumours; the facial nerve is endangered in all operations. Operative recording of auditory potentials (BAEP, ECochG and 8NAP) is reviewed by Lüders (1988), Møller (1988, 1992, 1995), Schwaber and Hall (1992), Yingling and Gardi (1992), Cheek (1993), Beck (1994) and Møller (1995) who give extensive detail on the technical aspects of recording.

Practical aspects of auditory monitoring

Stimulation Large earphones are unsuitable for operations in which the incision is close to the ear. Some small earphones, such as those used with portable radios, have been shown to be suitable acoustically (Erwin and Gulevich 1985), but may still be large enough to impede the surgeon. If small earphones are used, they should be fixed firmly so that the sound radiating area lies over the meatus rather than the pinna. Alternatively, a plastic tube can be connected to a transducer and held in the external meatus with either foam rubber or moulded wax. The advantages of the latter method are that the meatus is sealed both acoustically and against seepage of fluids, which could spuriously alter responses, and the stimulus artefact is reduced by the separation between transducer and ear. The length of tubing will introduce a latency delay; with a tube of 10–15 cm the delay is 0.5–1.0 ms, and the extra separation between stimulus artefact and response usually aids interpretation. Care must be taken to ensure that the tube is not kinked when the head is turned or positioned on a head rest. Hearing aid transducers are not suitable for click stimulation because of their limited high frequency output, but may be used if the stimulus is a tone burst (Møller 1988, 1995).

The rate of stimulation is usually about 10 Hz, but increasing the rate to 11 Hz will reduce interference from the mains supply if this cannot be eliminated in any other way. Higher rates are sometimes attempted for BAEP recording, but while this shortens the time required for acquisition of each average, response components are less easily recognized since the amplitude is diminished at rates over 20 Hz. The stimulus intensity should be at least 60 dB above threshold in each ear, with masking noise in the opposite ear to ensure unilateral stimulation. If the rate of stimulation is increased to 500 Hz, action potentials are produced in the cochlear nerve with the same frequency. If these are recorded with a bipolar probe, an audible 500 Hz tone is produced after amplification, which alerts the surgeon to the presence of acoustic nerve fibres (Butler et al 1995).

Stimulus polarity may affect the results. Some workers have used alternating clicks, while others have chosen the polarity which gives responses of highest amplitude. Lesions may produce different effects upon the responses to condensation and rarefaction clicks, so ideally alternate

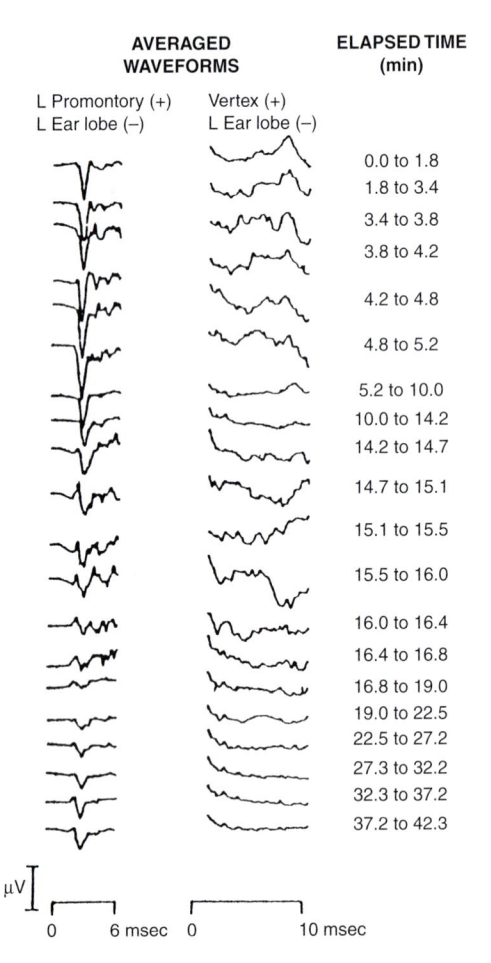

Fig. 7.8 ECochG and BAEP recorded to 44 Hz condensation clicks at 78 dB nHL during dissection of acoustic neuroma from the underlying eighth nerve. Wave V disappears before the negative component (NI) of the ECochG, suggesting localized conduction block in the auditory nerve. (From Levine et al 1984 Monitoring of auditory evoked potentials during acoustic neuroma surgery. Insight into mechanisms of the hearing loss. *Annals of Otology, Rhinology and Laryngology* 93:116–123, by permission.)

averages should be acquired with opposite stimulus polarities. If a single polarity is used, an average should be acquired to stimuli of the opposite polarity following the appearance of any change (Mokrusch et al 1988).

Recording

BAEP The BAEP is usually recorded between electrodes placed over the ipsilateral mastoid and the vertex. For operative recording the filter settings usually have to be adjusted from those normally used for BAEP recording in the clinic so as to reduce interference. A low frequency setting of 100 Hz is often necessary, but 150 Hz can be used, and some workers have reduced the high frequency setting to 1000 Hz. Such a restricted pass band will inevitably distort the response somewhat, but this is justified if troublesome noise is thereby reduced (Lüders 1988). A high frequency filter with increased roll-off (18 or 24 dB/octave), which provides better attenuation of high frequency interference, is particularly valuable in operative BAEP recording (Møller 1988, 1995). A serious difficulty with the BAEP is the time taken to produce an average acceptably free from noise, since at least 1000, and often more, responses will be required. Depending on the rate of stimulation, a new average can only be acquired every 1–2 min or even longer, which may not provide the surgeon with sufficiently rapid indication of changes at critical stages of the operation. Ideally BAEP monitoring equipment should include digital filtering or some other method of reducing the time taken to acquire each average (see p. 66).

ECochG The ECochG is recorded by an electrode inserted through the tympanic membrane with its tip resting on the medial wall of the middle ear. Two different types of electrode have been used, either a needle (Ojemann et al 1984) or a small silver ball similar to a nasopharygeal EEG electrode (Symon et al 1986); the advantage of the latter is its lower impedance. The transtympanic electrode is better secured if it is inserted through the tragus so that it pierces the anterior wall of the external meatus before passing through the membrane (*Fig. 7.9*) (Prass et al 1987). The same stimulus and amplifier settings are used as for BAEP recording. The response consists of three components: the cochlear microphonic, which is a small short latency deflection resembling the stimulus waveform; a summating potential, which is a negative deflection, only recorded with a high stimulus intensity; and a larger negative potential (N1). The cochlear microphonic is best recorded with a rarefaction stimulus and is often difficult to identify if condensation or alternating clicks are used. The summating potential is not phase dependent, and its amplitude may be greater than that of the microphonic with high intensity stimuli. N1 has a larger amplitude, and is more commonly used for surgical monitoring. Although N1 is thought to arise from the distal eighth nerve, its peak latency is not identical with wave I of the BAEP; the peak of N1 is often 0.5 ms later than wave I, and the two may vary independently during operations (Silverstein et al 1985a, Prass et al 1987).

Intracranial 8th nerve action potential After the auditory nerve has been exposed, direct recordings can be made from the nerve using a specially prepared wire electrode with a piece of cotton wick sutured to the tip (*Fig. 7.10*) (Møller et al 1981). The reference is either a needle inserted in muscle at the edge of the wound, or a disc on the earlobe. The high amplitude of the 8NAP compensates for the extra time taken in placing an electrode on the nerve, since an acceptable average can often be obtained from less than 10 responses; changes are therefore recognized much more rapidly than would be the case with the BAEP. During most operations only the distal portion of the nerve is accessible initially, but as more of the nerve is exposed the recording electrode should be moved proximally.

The normal 8NAP consists of a triphasic potential, having a major negative component, preceded and followed by smaller positivities (*Fig. 7.11*). The latency of the negative peak corresponds to wave II of the BAEP (Møller et al 1981, Silverstein et al 1985a). Following a conduction block distal to the recording site, the latency of the negative peak increases and, when the conduction block is complete, the negative peak and succeeding positivity disappear, leaving only the first positive peak, which represents the flow of current into the active region of the nerve distal to the block. The 8NAP is much more sensitive than the BAEP to the presence of such a block because the later BAEP waves are postsynaptic potentials, and are not therefore affected by desynchronization of impulses to the same degree as a nerve action potential (Møller and Jannetta 1983).

The cochlear nucleus When the brainstem is exposed at operation an electrode can be placed in the lateral recess of the fourth ventricle (Møller et al 1994a,b, Kuroki and Møller 1995). In this position the electrode is close to the cochlear nucleus and records a potential in response to auditory stimulation which has an amplitude much higher than the BAEP – comparable to that of the 8NAP. Only a few responses are required for a satisfactory average, so that repeated averages can be acquired much more quickly than with the BAEP.

Factors affecting auditory responses during operations

Several factors affect auditory responses during operations, in addition to anaesthetics, hypotension and temperature which are mentioned on pages 172–176 (Colletti et al 1997b, Mullatti et al 1999). Draining cerebrospinal fluid (CSF) either by lumbar puncture or by opening the dura increases the latency and decreases the amplitude of wave V of the BAEP (Kálmánchey et al 1986). Irrigating the surgical field with electrolyte solutions can also affect the response (Daspit et al 1982), either because of the temperature of the solution or the effects of its constituents. Positioning of the head and neck for retromastoid craniectomy, regardless of the presence of tumour, may attenuate or obliterate the BAEP. The explanation for this is not entirely clear, although transient ischaemia has been proposed, since the EP may recover after several hours, without detectable postoperative hearing impairment (Grundy et al 1982b,c). Much more serious is loss of the BAEP while positioning the head of a patient with fracture dislocation of the cervical spine, since the cause may be vertebral artery occlusion (Lüders 1988). During drilling of the walls of the internal auditory canal the response may be obliterated. This may be due to two different mechanisms: first, masking of stimuli by the bone conducted noise of the drill; secondly, thermal injury of the auditory nerve associated with the heating effect of the drill (Møller 1988, 1995).

Definition of a significant change in auditory monitoring

Since so many different factors can affect the BAEP, and transient changes are common, the definition of a significant change is very difficult. 'Significant' here means having a high probability of predicting a postoperative deficit if not corrected; this is quite distinct from a statistically significant change, which may have no surgical significance during a particular operation. A further difficulty is that the mechanisms producing changes in the BAEP are not the same in all types of operation, so that it may be dangerous to apply the definition of significant change established in one operation to a completely different procedure. Nevertheless, there is no doubt that an abrupt

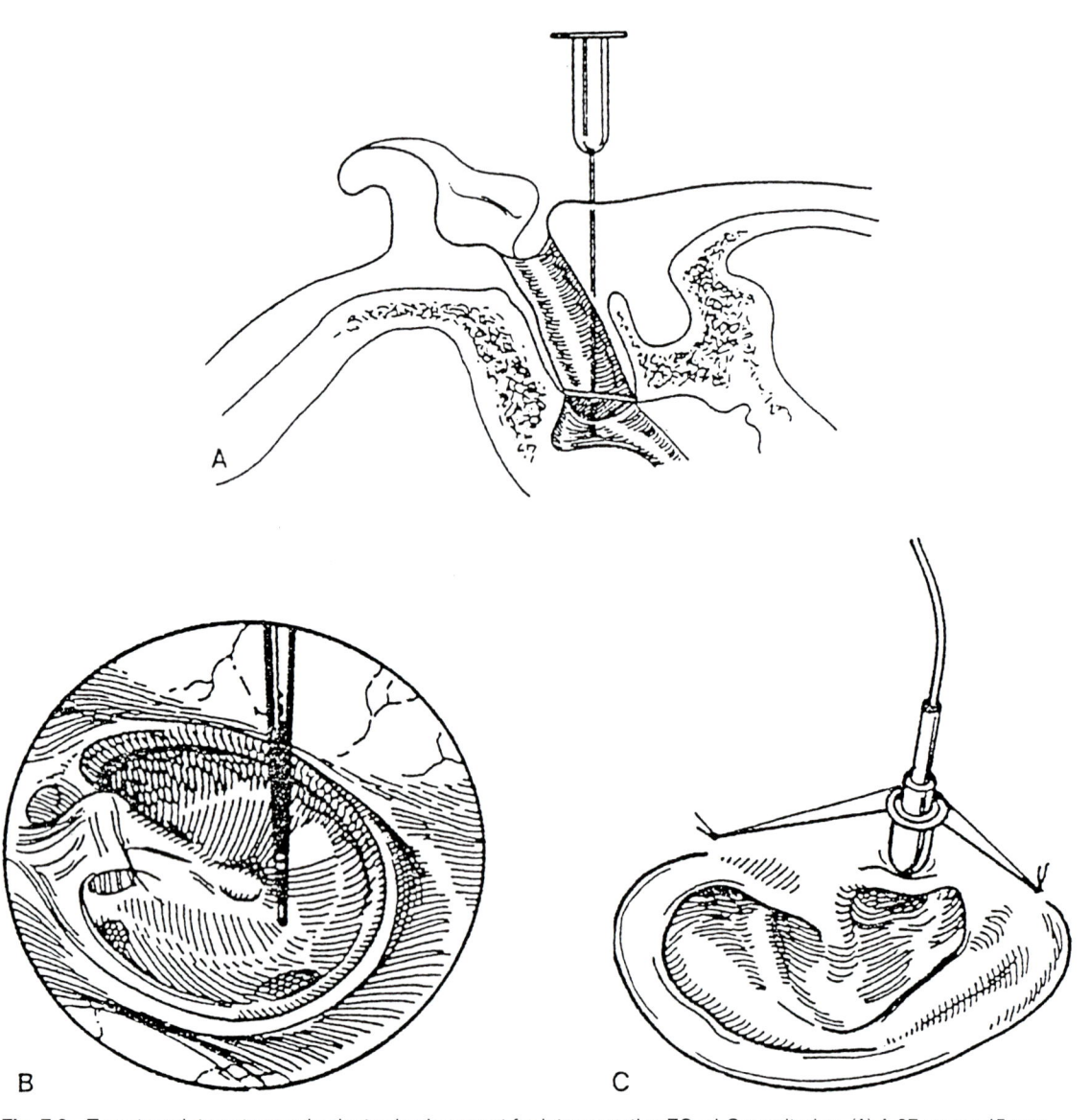

Fig. 7.9 Transtragal, transtympanic electrode placement for intraoperative ECochG monitoring. (A) A 27 gauge 45 mm Teflon-coated monopolar needle electrode is passed through a pretragal crease such that the needle exits into the external auditory canal along its anterior wall. (B) The needle is then advanced through the posteroinferior quadrant of the tympanic membrane until the bone of the cochlear promontory is encountered. (C) The electrode is secured via two sutures and is placed in a preauricular crease above and below the external auditory canal. (From Lüders 1988 Surgical monitoring with auditory evoked potentials. *Annals of Otology, Rhinology and Laryngology* 5:261–285, by permission.)

Intracranial surgery

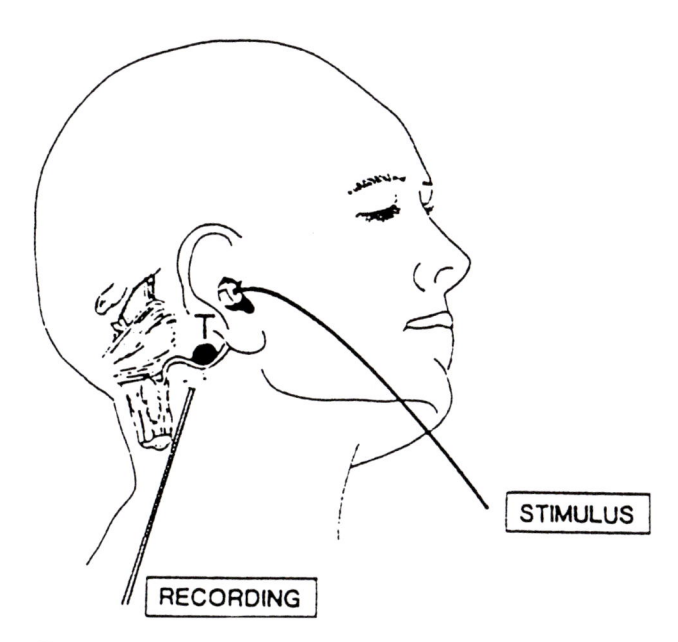

Fig. 7.10 Intraoperative monitoring of the cochlear nerve action potential from the eighth nerve during removal of an eighth nerve tumour (T). Auditory stimuli are delivered through an ear insert and the responses recorded with a cotton-wick electrode placed on the exposed cochlear nerve. (From Linden et al 1988 Electrophysiological monitoring during acoustic neuroma and other posterior fossa surgery. *Canadian Journal of Neurological Sciences* 15:73–81, by permission.)

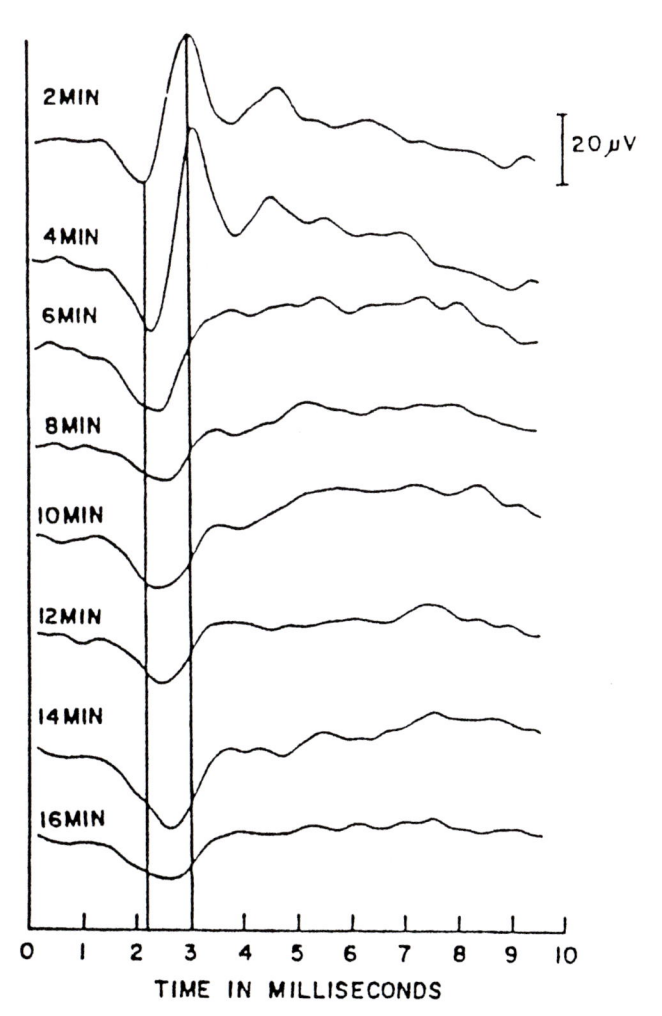

Fig. 7.11 Intracranial recordings from the distal eighth nerve in a patient undergoing microvascular decompression to relieve trigeminal neuralgia. (From Møller and Jannetta (1983), by permission of the *Journal of Neurosurgery*.)

permanent loss of all responses (ECochG, 8NAP and BAEP) is usually associated with severe postoperative deafness (Raudzens and Shetter 1982, Ojemann et al 1984), although a more gradual loss of wave V may persist for as long as 3 h without any postoperative hearing impairment (Grundy et al 1982b,c). In microvascular decompression operations, Friedman et al (1985) found that latency changes were not well correlated with postoperative hearing impairment, and Schramm et al (1988) reached the same conclusion in a series consisting predominantly of tumours. Several authors have noted that the number of operations in which there is a latency increase of more than 1 ms far exceeds the rate of postoperative deafness. There is therefore great uncertainty about the importance of latency changes, and about the level at which the surgeon should be notified of a slow progressive change in amplitude. Until more information is available on this point, it would be prudent to follow a graded system, such as that suggested by Nuwer (1986). The surgeon should initially be warned when the latency increases by 1.0 ms, or if the amplitude falls by 50%, and possible causes other than surgical manipulation should be sought. If the latency increases by 1.5 ms, or amplitude reduction progresses to 80%, then tension or pressure on the 8th nerve should be released, and any other adverse factors, such as hypotension, should if possible be corrected. This approach will inevitably lead to some false positives, but the number of false negatives should be minimized. Any change which persists for less than 15 min after removing the cause is unlikely to be associated with a postoperative deficit. A flexible attitude should be maintained, since the importance of any change depends on the stage of the procedure at which it occurs, and different levels of significance may be appropriate during some operations.

Contribution of monitoring to preservation of hearing

Many patients with acoustic neuroma, or other cerebellopontine angle tumours, cannot benefit from the possibility of hearing preservation, because there is no useful hearing on the affected side preoperatively. In patients with small acoustic neuromas hearing can now be preserved in many cases. With the help of BAEP and ECochG monitoring, Ojemann et al (1984) achieved postoperative speech discrimination scores over 35% in 75% of patients with tumours less than 1 cm in diameter, 25% or more with tumours between 1.5 cm and 2.5 cm, but only rarely if the tumour was larger than 3 cm. Similar results have been obtained by others (e.g. Fischer et al 1992, Brookes and Woo 1994, Torrens et al 1995, Neu et al 1999, Tonn et al 2000, Schlake et al 2001b), although some have found the 8NAP to predict postoperative hearing better than either the EcochG or BAEP (Colletti et al 1997a,b). Retrospective comparison has shown that the rate of hearing preservation with monitoring was superior to that obtained in patients matched for tumour size and preoperative hearing, whose operations were performed without benefit of BAEP monitoring (Slavit et al 1991, Harper et al 1992). The precise contribution of monitoring to hearing preservation in acoustic neuroma surgery is difficult to determine, since hearing preservation depends upon earlier diagnosis, increased willingness of surgeons to operate before all hearing is lost and developments in microsurgical technique which occurred at about the same time as intraoperative monitoring was introduced. However, some of the changes in operative procedure were adopted as a result of information obtained from monitoring (Jannetta et al 1984). Monitoring may be helpful in some operations by informing the surgeon that his original aim of hearing preservation is unattainable, and he may concentrate on removing the tumour totally without further attempts to save the 8th nerve, although intraoperative loss of BAEP wave V does not completely exclude the possibility of hearing preservation (Harner et al 1996). While many have concluded that BAEP monitoring contributed to hearing preservation, some have found that the only contribution was from the 8NAP (Jackson and Robertson 2000).

Two reports (Møller and Møller 1989, Radtke et al 1989) have retrospectively compared results of microvascular decompression operations performed with and without BAEP monitoring; both have shown a statistically significant reduction in the incidence of hearing impairment. In the absence of a prospective controlled trial, this is the best evidence yet available for reduction of postoperative morbidity by BAEP monitoring in this type of operation. By using BAEP monitoring in microvascular decompression operations it is suggested that hearing preservation is more likely if cerebellar retraction is reduced to a level at which BAEP wave I–V interpeak latency is not increased by more than 1.5 ms (Rizvi et al 1999).

During vestibular neurectomy BAEP and 8NAP monitoring has been used successfully to confirm that the vestibular nerve has been correctly identified, and that the auditory nerve is not divided (Silverstein et al 1985b).

Monitoring brainstem function

In patients with tumours or vascular lesions (aneurysms, malformations) in the posterior fossa, there is serious risk of operative damage to the brainstem. Early impairment of brainstem function has traditionally been detected by cardiac monitoring, since interference with the medullary cardiovascular centres produces slowing and irregularity of the heart rate. Wave III and subsequent waves of the BAEP depend on the integrity of pontine structures; BAEP monitoring should therefore be complementary to cardiac monitoring. Early reports indicated that loss of BAEP waves III–V from the contralateral ear was correlated with evidence of a severe brainstem disturbance postoperatively (Hashimoto et al 1980, Raudzens and Shetter 1982). Further work has shown that increase in latency of waves III or V often gives an earlier warning of impaired brainstem function than cardiac monitoring (Møller 1988, 1995), but some workers have found the BAEP to be excessively sensitive to brainstem manipulation during tumour operations (Kálmánchey et al 1986, Angelo and Møller 1996). During vascular procedures in the posterior fossa prominent BAEP deterioration has predicted a poor outcome, but only small series have been reported so far (Little et al 1983, 1987, Lam et al 1985).

An important source of error in intraoperative BAEP recordings is that the auditory pathways form only a relatively small part of the pons, so BAEP monitoring may be insensitive to serious brainstem disturbances which happen to spare these pathways (Piatt et al 1985). In those cases where brainstem function is the primary concern, a more complete indication of brainstem function is obtained by monitoring somatosensory responses from the limbs as well as the BAEP (McPherson et al 1984, Gentili et al 1985, Little et al 1987).

Monitoring cranial nerve motor function

Facial palsy is a serious complication of operations to remove tumours in the cerebellopontine angle. The surgeon may have great difficulty in identifying the facial nerve when the anatomy is distorted by the presence of a large tumour, the nerve often being compressed, stretched or embedded in the tumour. The first attempt to identify the nerve consisted of electrical stimulation of the exposed nerve, the facial muscle response being detected either by palpation or by the light of a flashlight under the surgical drapes (Parsons 1966). This technique is extremely inconvenient, and has a serious disadvantage that a contraction of the masseter resulting from stimulation of the motor branch of the 5th nerve might be mistaken for a facial response. Greater reliability is provided by intraoperative recording of EMG responses from facial muscles (*Fig. 7.12*) (Delgado et al 1979). This consists of displaying facial muscle compound action potentials on an oscilloscope, their presence being reported to the surgeon by a member of the monitoring team. An alternative method is to record the facial muscle responses by accelerometer (Sugita and Kobayashi 1982) or pressure sensor (Shibuya et al 1993, Silverstein et al 1993), rather than electrically; the surgeon can then hear the responses clearly if the amplified output from the accelerometer or pressure sensor is fed into a loudspeaker. Similar audible indication of a response is possible with EMG activity, provided the stimulus artefact can be suppressed so as not to be confused with a muscle response; the necessary circuit was described by Møller and Jannetta (1984), and has been incorporated into commercial monitoring devices. In addition to evoked activity, spontaneous potentials (neurotonic discharges) in facial muscles give a warning of

Intracranial surgery

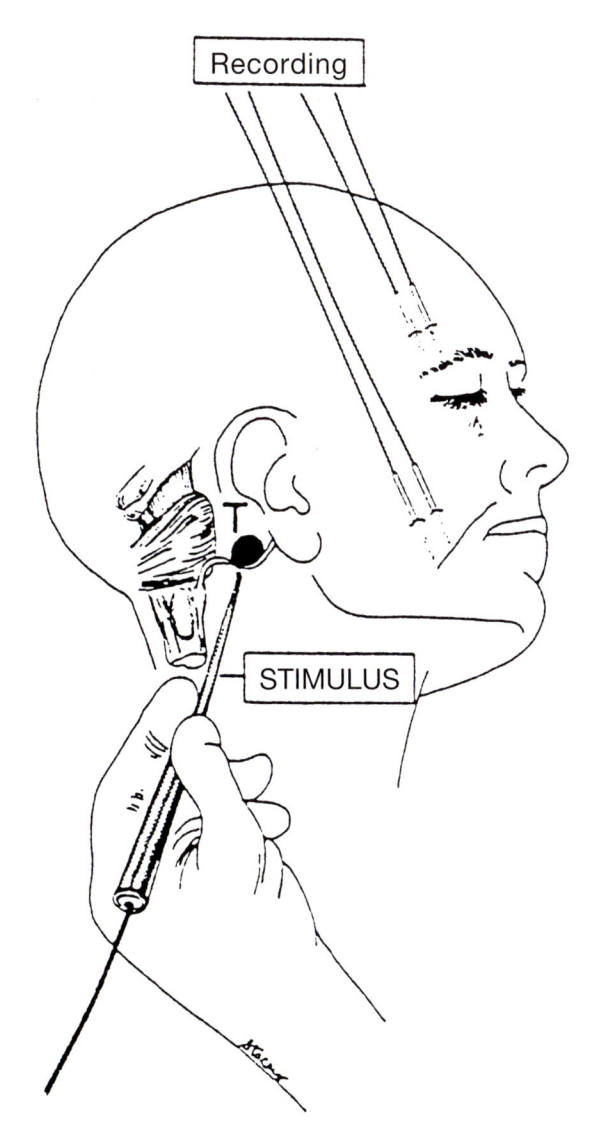

Fig. 7.12 Intraoperative monitoring of facial nerve function. EMG activity is recorded from the orbicularis oris (or mentalis) and orbicularis oculi muscles. A hand-held stimulator is used to identify the facial nerve displaced by a tumour (T). (From Linden et al (1988) Electrophysiological monitoring during acoustic neuroma and other posterior fossa surgery. *Canadian Journal of Neurological Sciences* 15:73–81, by permission.)

nerve irritation resulting from manipulation or excessive heating from diathermy (Prass and Lüders 1986, Harner et al 1987). For any of these methods to be successful, the patient must not be completely paralysed, and the anaesthetist should be made aware of this before the operation so that long-acting paralysing agents can be avoided; although responses can often be recorded from facial muscles, even with a high degree of neuromuscular paralysis, as shown by the conventional method of a train of stimuli applied to the ulnar nerve at the wrist (Blair et al 1994, Brauer et al 1996). In an attempt to avoid any possible interference with monitoring from neuromuscular blockade, some workers have recorded antidromic impulses in the facial nerve, evoked by peripheral stimulation (Colletti et al 1996, 1997a), but this is not a widely used method.

Although conventional EMG equipment is capable of recording facial nerve activity, both spontaneous and evoked, a stimulator intended for transcutaneous stimulation is unlikely to be satisfactory for direct nerve stimulation. This is because such a stimulator will be capable of delivering a pulse of several hundred volts (constant voltage) or up to 100 mA (constant current) (Smith 1997). A shock of this magnitude would be almost certain to damage the nerve if applied directly. For intraoperative use the output must be limited to less than 1 V (constant voltage) or 1 mA (constant current); few standard stimulators are capable of delivering such a small pulse accurately. However, suitable stimulators are available, and equipment designed specifically for facial monitoring includes a stimulator with the appropriate output. There has been considerable debate as to whether the facial nerve should be stimulated by a constant current or a constant voltage pulse, and whether the stimulating electrode should be monopolar or bipolar. Møller (1988) recommends a low impedance constant voltage stimulator, connected to a monopolar probe, because constant voltage stimulation is less affected by variable shunting due to changes in the amount of fluid in the operative field; in addition, a monopolar probe is smaller and the stimulus does not depend upon its orientation with respect to the nerve. Special probes must be used with a completely insulated shaft, so that the stimulating current emerges only at the tip, thus minimizing the risk of current spread to the facial nerve when the surgeon applies the probe to some other structure. With this type of monopolar probe a constant current stimulator gives satisfactory results in tumour surgery (Prass and Lüders 1985), where large areas can be quickly demonstrated to be remote from the facial nerve, but a bipolar probe may be preferable for operations such as vestibular neurectomy, since bipolar stimulation permits more accurate localization of the stimulus (Kartush et al 1985).

Harner et al (1987) have shown a definite contribution of monitoring to preservation of facial nerve function compared with a series of patients matched for age and size of tumour whose operations were performed without facial monitoring. There was no difference between the groups immediately after the operation, but the monitored group showed a significant improvement in facial nerve function at 3 months. Retrospective comparison has shown that EMG monitoring is followed by a higher rate of facial nerve preservation than facial muscle contraction monitoring by the anaesthetist (Jellinek et al 1991b).

Other authors have reported improved preservation of facial nerve function after operations performed with monitoring (e.g. Nadol et al 1992, Lalwani et al 1994, Torrens et al 1995, Samii and Matthies 1997, Slattery et al 1997, Tonn et al 2000), but it is important to remember that intraoperative monitoring is only one of the factors determining facial nerve function postoperatively; age of the patient, size of the tumour, surgical approach, and completeness of tumour removal are also important (Grey et al 1996, Nissen et al 1997). Some surgeons (e.g. Prasad et al 1993, Fenton et al 1999), have found that the threshold to proximal nerve stimulation at the end of the operation predicts postoperative facial nerve function, while others (e.g. Sobottka et al 1998, Goldbrunner et al 2000) consider that the response amplitude gives a better indication of the postoperative state of the nerve. Occasionally, facial palsy develops some time after an operation to remove an acoustic neuroma, even though intraoperative monitoring detected no nerve damage, and the nerve appeared to have been preserved immediately after surgery (Lalwani et al 1995). Delayed facial palsy should not be confused with false negative monitoring.

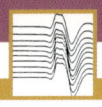

The methods used for facial nerve monitoring are applicable to operations involving other cranial motor nerves, such as the trigeminal (Kartush and Bouchard 1992, Schlake et al 2001a), and similar methods have been used to identify cranial nerve nuclei during operation for brainstem cavernous malformation (Chang et al 1999).

SPINAL CORD FUNCTION MONITORING

Spinal cord function has been monitored during orthopaedic operations for correction of kyphoscoliosis and stabilization of fractures, during aortic surgery for coarctation and aneurysm, and during some neurosurgical operations for spinal tumour. A considerable literature has accumulated on the different spinal monitoring techniques and their clinical application, including several symposia (Nash 1979, Homma and Tamaki 1984, Schramm and Jones 1985, Ducker and Brown 1988, Salzman 1990, Shimoji et al 1991, Jones et al 1994, Stålberg et al 1998), together with recent authoritative reviews of monitoring sensory EPs (Nuwer 1998), motor responses (Burke and Hicks 1998, Isley et al 2001) or a combination of the two (Burke and Hicks 1997, Emerson et al 1997, Nuwer 1999, Owen 1999, Padberg and Bridwell 1999, Calancie et al 2001, Luk et al 2001). In addition, many of the general works on operative monitoring, to which reference was made on page 59 *et seq*, include detailed discussion of applications to spinal surgery.

Experimental studies and mechanisms of spinal cord damage

Compression and distraction damage the spinal cord by two distinct mechanisms. First, the external force may be sufficient to cause either conduction block or axonal degeneration similar to the changes produced in peripheral nerve by compression (Gelfan and Tarlov 1956). However, with external force only sufficient to affect a minority of neurones directly, a more important mechanism is the secondary damage arising predominantly from ischaemia and haemorrhage (Ducker 1976, Tator and Fehlings 1991) and the associated metabolic effects (Schwab and Bartholdi 1996, Li and Tator 1998, Winkler et al 1998).

Several models of spinal cord injury have been studied: weight-drop, graded compression and distraction; the electrophysiological findings depend on the model used. When the cord is crushed by dropping a weight onto it, cortical potentials evoked by hindlimb stimulation disappear if the injury is severe enough to render the animal paraplegic, indicating that spinal conduction has failed (Gossman et al 1968, Ducker 1976). With slightly less severe trauma from weight-drop, where minimal motor function persists, cortical EPs from the hindlimbs are initially attenuated for up to 2 hours but later recover in spite of a severe motor deficit. Occasionally, following early improvement, the EP disappears a few days after injury (Blight and Young 1990). It is important to recognize that the weight-drop method, when applied *in vivo* through a laminectomy, produces greatest damage in the dorsal cord, and may have little relevance to the mechanism of human cord injury arising during operations.

When graded compression is applied to the cord, disappearance of sensory EPs predicts subsequent motor dysfunction (Croft et al 1972, Kojima et al 1979, Bennett 1983); motor EPs may be more sensitive to early changes, but sensory responses distinguish more accurately between mild and moderate injury (Fehlings et al 1989a).

Experimental spinal distraction until sensory EPs disappear is associated with total abolition of spinal cord blood flow (Dolan et al 1980). When spinal EPs are recorded from vertebral bone, the amplitude must be reduced to less than 50% before paresis appears (*Fig. 7.13*) (Nordwall et al 1979), and reduction in EP amplitude of more than 50% is usually followed by total loss of motor function at 24 hours (Maiman et al 1989). Although these findings indicate that sensory EPs should provide satisfactory early warning of cord damage from distraction, motor EPs are more sensitive to the early effects of distraction (Machida et al 1989, Owen et al 1990a,b). If rotation is combined with distraction, spinal cord damage is more severe than with distraction alone.

In addition to compression and distraction, which reduce cord blood flow, the spinal cord can be damaged by ischaemia resulting directly from interruption of arterial blood supply. Blood reaches the cord by two different routes; the anterior cord is supplied by the anterior spinal artery which usually depends upon a single large vessel, the *arteria radicularis magna* of Adamkiewicz, arising from a segmental branch of the descending aorta in the lower thoracic or upper lumbar area, while the dorsal columns are supplied by the posterior spinal artery which is formed from an anastomosis between branches of the vertebral arteries and segmental vessels at different levels. There is, however, great variation in the level at which the artery of Adamkiewicz arises, and in the contribution made to both anterior and posterior spinal arteries by segmental vessels. Nevertheless, interruption of aortic flow above the artery of Adamkiewicz renders the anterior part of the cord ischaemic, with relative sparing of the dorsal columns. Although ischaemia is greater in the anterior cord, experimental aortic occlusion undoubtedly affects sensory EPs (Larson et al 1980, Coles et al 1982, Laschinger et al 1984). EPs are likely to disappear if the distal aortic pressure falls below 70 mmHg, and the risk of paraplegia is related to the duration of loss of potentials (Laschinger et al 1983, 1987a,b). If occlusion is maintained until EPs are completely lost, recovery of sensory and motor function is possible, provided blood flow is restored within a few minutes. Return of sensory EPs usually precedes recovery of motor function, but a severe motor deficit may persist in spite of reappearance of sensory potentials. Some studies suggest that motor potentials are more sensitive to ischaemia from experimental aortic occlusion (Kai et al 1995, Gonzalez-Fajardo et al 1998), while others report no significant difference between motor and sensory EPs (Osenbach et al 1993). The precise method of stimulation and recording appears to be important in detecting effects of experimental arterial occlusion; for sensory pathways the potentials evoked by spinal stimulation are more sensitive than those produced by peripheral nerve stimulation (Grossi et al 1988), whereas muscle and peripheral nerve responses to motor stimulation are more sensitive than potentials recorded from the spinal cord (Konrad et al 1987, Machida et al 1990).

It seems reasonable to conclude from experimental data that either sensory or motor EPs can detect a disturbance of cord function due to compression, distraction or arterial occlusion, at a stage before permanent loss of function has occurred. However, an important question remains: how can we explain the high sensitivity of SEPs for motor deficit, when sensory responses depend on impulses ascending in the dorsal columns? The answer is that compression, distraction and arterial insufficiency produce widespread changes in the cord: reduction in blood flow, oedema from damage to capillaries, tissue electrolyte disturbance, enzyme activation and metabolite accumulation, release of excitotoxins, and haemorrhage beginning in the centre of the cord and spreading into white matter. These processes do not spare the dorsal columns (Schwab and Bartholdi 1996, Li and Tator 1998). This is confirmed by the only human pathological material obtained after disappearance of sensory EPs during an operation to correct scoliosis, which revealed an intramedullary haemorrhage disrupting the dorsal horn and extending into the dorsal column on one side; in spite of apparent sparing of the opposite dorsal column, responses from that leg disappeared, presumably due to oedema (Shukla et al 1988). Indeed, clinical evidence of dorsal column sparing is rare in spinal cord trauma and ischaemia, with the exception of anterior cordotomy and isolated occlusion of the anterior spinal artery, after both of which SEPs from the legs may be preserved (Owen et al 1989, Zornow et al 1990). An additional factor is that most ascending fibres from the legs make a detour out of the dorsal columns into the lateral columns (Webster 1977); furthermore, following mixed nerve stimulation in man, some impulses ascend in high velocity lateral tracts (Jones et al 1982, Halonen et al 1989). SEPs may, therefore, be attenuated by lesions outside the dorsal columns.

The importance of cord blood flow is emphasized by the finding that the effects of experimental cord compression and systemic hypotension are additive and the clinical observation that induced hypertension ameliorates the effects of cord compression, as shown by improvement in sensory EPs (Brodkey et al 1972, Hardy et al 1972, Grundy et al 1981, Fehlings et al 1989b). Cord blood flow is clearly the critical factor when the aorta is clamped, and an additional factor contributing to the risk of paraplegia during aortic surgery is that clamping the aorta increases intrathecal pressure; if intrathecal pressure exceeds pressure in the aorta distal to the clamp, the cord cannot be perfused (Blaisdell and Cooley 1962). Both in aortic surgery, and in orthopaedic procedures employing the anterior approach to the spine, paraplegia may result from ligation of segmental arteries which are critical for maintenance of cord perfusion (Cunningham et al 1982, Apel et al 1991).

Methods of monitoring cord function

The earliest attempt to detect intraoperative spinal cord damage was the 'wake up' test (Vauzelle et al 1973). In this procedure, after neuromuscular blockade has been reversed, the patient is awakened and told to move his feet. This has the disadvantage that information is only

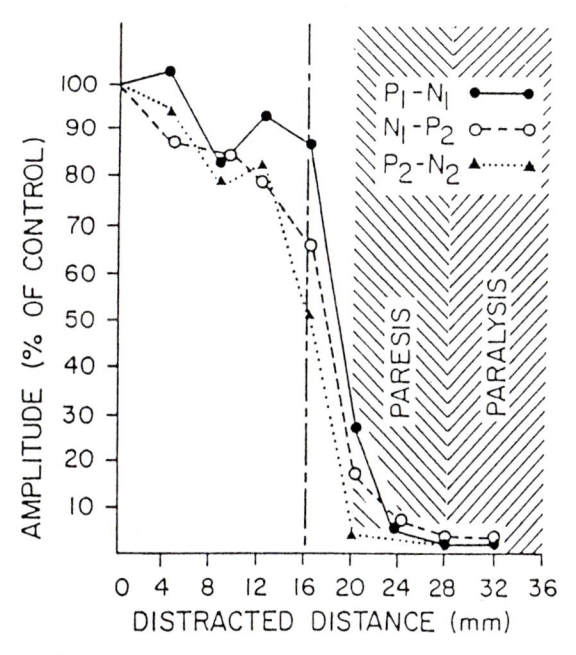

Fig. 7.13 Amplitudes of EP components (plotted as percentages of predistraction control values) recorded from the spine of T9, during experimental spinal cord distraction in the cat. The amplitude is reduced by more than 50% before paresis develops, and marked reduction in amplitude precedes total paralysis. (From Nordwall A, Axelgaard J, Harada Y, Valencia P, McNeal DR, Brown JC 1979 Spinal cord monitoring using evoked potentials recorded from the feline vertebral bone. *Spine* 4:486–494, by permission.)

available at a single instant, and the test cannot be performed repeatedly; it is therefore only useful in those operations in which there is a clearly defined time of maximum danger to the cord (Owen 1999). In addition, the 'wake up' test carries a risk of extubation, and of surgical trauma since the patient is not paralysed and has a large open wound. There is also the possibility that the experience may be recollected postoperatively, with distressing results. Dissatisfaction with the 'wake up' test led to the search for a safer method of detecting cord damage which would be available throughout the operation.

Sensory monitoring
Following the suggestion of Croft et al (1972) that SEPs might be used to monitor cord function, reports of clinical applications began to appear (McCallum and Bennett 1975, Nash et al 1977, Engler et al 1978), although Japanese surgeons had been monitoring for some time with invasive techniques (Kurokawa 1972, Tamaki 1998). The first workers to attempt operative spinal cord monitoring with SEPs recorded responses in a conventional manner through electrodes applied to the scalp (*Fig. 7.14(a)*). Subsequently electrode positions and amplifier passband were modified to reduce electrical interference and allow interpretable averages to be acquired more quickly (Nuwer and Dawson 1984a,b). Because of the inherent variability of scalp-recorded SEPs from the lower limbs, and the tendency of halogenated anaesthetic agents to attenuate cortical responses, subcortical potentials proved attractive, and they have been recorded through skin electrodes over the occiput or cervical spine (Lüders et al 1982, Abel et al 1990, Hurlbert et al 1995, Bernard et al 1996, Ecker et al 1996, Papastefanou et al 2000), or from electrodes inserted into interspinous ligaments (Lüders et al 1982), spinous processes (Maccabee et al 1983) or the spinal subdural space (Whittle et al 1984) (*Fig. 7.14(b)*). A popular method in Britain involves recording from an electrode positioned in the spinal epidural space at the rostral end of the wound, using a remote reference (Jones et al 1983), while others have used a second epidural electrode as reference (Macon and Poletti 1982, Anderson et al 1990, Erwin and Erwin 1993).

Motor monitoring
The methods mentioned so far rely upon stimulating a peripheral nerve at ankle or knee, but in Japan some workers use epidural electrodes for both stimulation and recording, so that either ascending impulses (Grabitz et al 1993, Fujioka et al 1994), or a combination of ascending and descending impulses can be monitored (Tsuyama et al 1978, Koyanagi et al 1993), while others record descending impulses from electrodes in the subarachnoid space after stimulating the cord through epidural electrodes (Tamaki et al 1981, 1984, Tamaki 1998) (*Fig. 7.14(c)*). This approach has been extended to recording peripheral nerve action potentials evoked by stimulating the cord, either through needle electrodes inserted into spinous processes at the rostral end of the wound (Owen et al 1991, Owen 1993, Phillips et al 1995, Bernard et al 1996, Péréon et al 1998), or by electrodes introduced into the epidural space

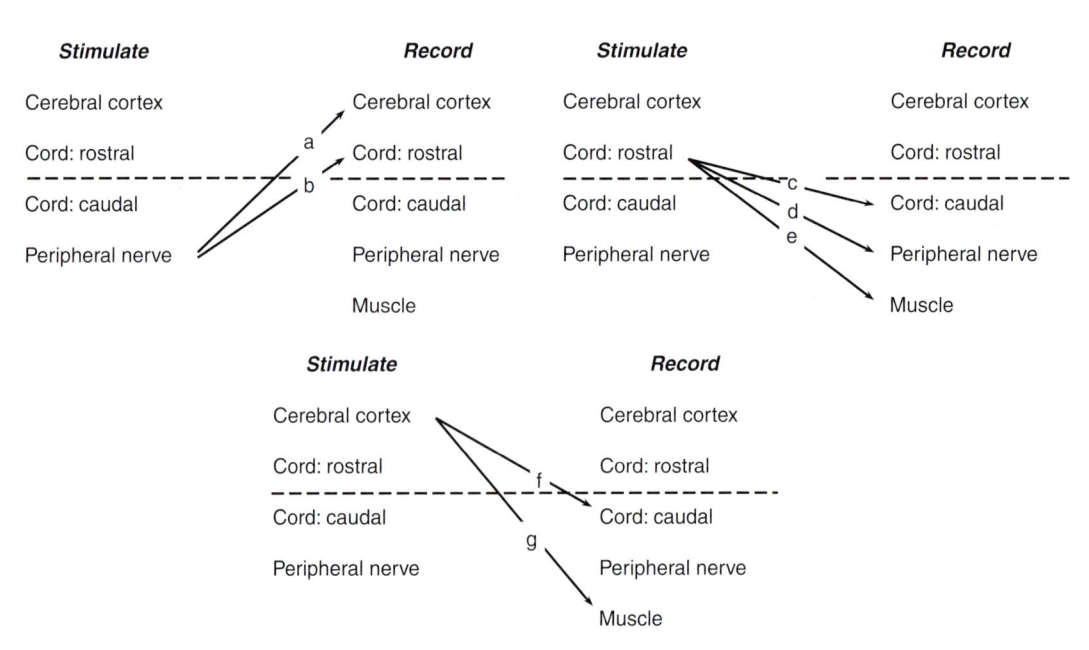

Fig. 7.14 Methods of monitoring spinal cord function during an operation at the level indicated by the horizontal dashed line. Ascending impulses from peripheral nerve stimulation can be recorded form the cerebral cortex, or from the cord rostral to the operation site (a and b). Descending impulses from stimulation of the rostral cord can be recorded from caudal cord, peripheral nerve or muscle (c, d and e), and cortical stimulation produces descending impulses which can be recorded from the cord (f), or muscle responses can be monitored (g). (From Binnie et al (2003), by permission.)

percutaneously (Komanetsky et al 1998) (*Fig. 7.14(d)*). Although often referred to as 'neurogenic motor monitoring', methods of recording peripheral nerve responses to spinal cord stimulation are better described as monitoring descending impulses, because the recorded activity probably depends largely, if not exclusively, upon antidromic conduction in sensory pathways (Su et al 1992, Toleikis et al 2000, Deletis 2001, Minahan et al 2001).

An alternative approach, which avoids confusion with antidromic sensory conduction, is to record the muscle response to spinal cord stimulation (Machida et al 1988, Adams et al 1993) (*Fig. 7.14(e)*). Unfortunately most anaesthetics reduce the responsiveness of anterior horn cells to descending impulses (Kalkman et al 1995, Woodforth et al 1996), and neuromuscular blockade eliminates the muscle response to nerve action potentials. However, provided that neuromuscular blockade is not complete, temporal summation from pairs of stimuli (Taylor et al 1993), or trains of stimuli (Mochida et al 1997), with an interstimulus interval of 2 ms, greatly increases the amplitude of the myogenic potentials produced by spinal stimulation. This technique permits reliable recording of muscle responses under anaesthesia.

Motor pathways can also be monitored by stimulating the motor cortex non-invasively (*Fig. 7.14(f, g)*). Three methods are available: constant current stimulation with the anode on the scalp and the cathode pressed against the hard palate (Levy 1987, Kitagawa et al 1989); a high voltage capacitor discharge with both electrodes on the scalp (Boyd et al 1986, Zentner 1991, Burke et al 1992, Tabaraud et al 1993, Lang et al 1996, Stephen et al 1996, Cioni et al 1999); or transcranial magnetic stimulation (Edmonds et al 1989, Shields et al 1990, Deletis 1993, Herdmann et al 1993, Anderson et al 1994, Glassman et al 1995, Lee et al 1995). The descending volley in the cord may be recorded through epidural electrodes (Boyd et al 1986, Kitagawa et al 1989, Burke et al 1992, Stephen et al 1996, Gokaslan et al 1997, Cioni et al 1999) (*Fig. 7.14(f)*), or muscle action potentials can be recorded if neuromuscular block is not complete (Levy 1987, Zentner 1991, Herdmann et al 1993, Tabaraud et al 1993) (*Fig. 7.14(g)*). Volatile anaesthetics severely attenuate responses to cortical stimulation, leading some workers to adopt total intravenous anaesthesia (Jellinek et al 1991a). Fortunately, however, temporal summation partially reverses anaesthetic-induced inhibition, and a train of 3–6 stimuli at intervals of 2 ms is much more effective than a single stimulus (Kalkman et al 1995, Jones et al 1996, Calancie et al 1998, Kawaguchi et al 1998, Van Dongen et al 1999), but repetitive transcranial stimulation may be harmful and should be used with care (Wassermann 1998). The recent finding that, under anaesthesia, a train of stimuli to the sole of the foot facilitates the response to a single transcranial stimulus may reduce the need for repetitive transcranial stimulation (Andersson and Ohlin 1999).

The various methods of monitoring motor pathways are reviewed by Burke and Hicks (1998), Isley et al (2001) and Deletis (2002).

Comparison of methods

The different methods (ascending, descending; invasive, non-invasive) each have their own advantages and disadvantages. Non-invasive methods have the great advantage that skin electrodes can be applied quickly before the patient is anaesthetized, and do not impede the surgeon; monitoring by these methods is available throughout the operation and can continue postoperatively if necessary. Invasive techniques, involving insertion of electrodes into either a spinous process, the interspinous ligament or the epidural or subarachnoid space, produce responses of higher amplitude than those obtained from surface electrodes, combined usually with a higher signal-to-noise ratio resulting in less variability of amplitude and latency during uneventful operations (Dinner et al 1986, 1990). The principal disadvantage of invasive techniques, particularly those involving stimulation of the spinal cord or motor cortex, is that these procedures are potentially dangerous, and may be blamed for a postoperative deficit. Further disadvantages are that if electrodes are inserted through the surgical incision they may be dislodged or require repositioning as the operation proceeds, and percutaneous insertion before the operation begins can be time-consuming. Whether monitored invasively or non-invasively, spinal cord responses have the advantage of being relatively free from anaesthetic effects, unlike cortical responses. On the other hand, motor pathway monitoring is often applicable in patients in whom sensory responses cannot be recorded because of pre-existing impairment of spinal cord function.

Although considerable experience has been accumulated with each technique, no formal comparison has been carried out to determine which is the most sensitive and specific, or whether a particular method is preferable in certain types of operation. Several groups now use a combination of ascending and descending methods (Machida et al 1985, Katayama et al 1986, Burke et al 1992, Erwin and Erwin 1993, Segal 1995, Stephen et al 1996, Burke and Hicks 1997, Emerson et al 1997, Calancie et al 1998, Padberg et al 1998, Tamaki 1998, MacDonald et al 2003) (*Fig. 7.15*). While it is commonly assumed that monitoring both sensory and motor pathways should be more sensitive than monitoring either alone, this assumption has not been tested by randomized controlled trial, and the optimal combination of sensory and motor methods has yet to be established.

For the sake of completeness, before concluding this account of methods of spinal cord monitoring, a relatively recent development must be mentioned, although it is clinical rather than electrophysiological. This is the ankle clonus test, which relies upon the fact that, during recovery from anaesthesia, dorsiflexion of the ankle is followed by a few beats of clonus if spinal cord function is normal. However, presence of clonus depends upon performing the test at the appropriate stage of recovery from anaesthesia; absence of clonus cannot, therefore, be taken as unequivocal evidence of spinal cord damage (Owen 1999).

High sensitivity and specificity have been reported in a large series of operations, with accuracy superior to either the 'wake up' test or SEP monitoring (Hoppenfeld et al 1997), but the clonus test is only applicable while anaesthesia is wearing off, rather than being available continuously throughout the operation.

Practical aspects of spinal cord monitoring

Monitoring ascending impulses by SEP recording is still the most widely used technique, so the following section emphasizes the practical points that must be remembered if reliable results are to be obtained with this method.

Stimulation

If peripheral nerve stimulation is used, responses from the right and left legs must be monitored separately. A severe disturbance of function may be confined to one side of the cord, and may not be detected by monitoring the response from one leg, or both legs simultaneously (Molaie 1986, Friedman and Richards 1988). Either the peroneal or tibial nerve may be stimulated; there is no significant difference in amplitude between the responses produced (Nuwer and Dawson 1984a,b). If spinal recording is used, stimulation of the tibial nerve at the knee produces a component in the response due to impulses conducted in dorsal spinocerebellar tracts; this component is missing if the nerve is stimulated at the ankle (Halonen et al 1989). Choice of stimulation rate depends on whether spinal or cortical responses are to be recorded. Rates of stimulation above 3 Hz attenuate cortical responses, particularly if there is any abnormality of the cord preoperatively (Schubert et al 1987). There is thus a conflict between the need to increase the rate of stimulation so as to acquire averages more quickly, and amplitude reduction at higher rates. In most patients a rate of about 5 Hz provides the best compromise for cortical responses (Nuwer and Dawson 1984a,b). Spinal responses are not affected by increasing the stimulation rate to 20 Hz (Lüders et al 1982, Ryan and Britt 1986, Hu et al 2001).

Recording

Noise is reduced by avoiding wide separation between active and reference electrodes. For the cortical response from the legs, satisfactory results are obtained with the active electrode at Cz or 2 cm behind Cz, referred to Fz or midway between Fz and Fpz (Nuwer and Dawson 1984a,b), while responses from the arms are best recorded from an active electrode over the parietal region (P3 or P4) contralateral to the stimulated arm, referred to the same frontal reference as used for the legs. In some patients optimal cortical responses can only be obtained by individually selecting electrode positions, and they may have to be readjusted in the course of the operation (MacDonald 2001, MacDonald et al 2004).

Early reports of monitoring using cortical evoked responses employed a wide passband, but greater reproducibility is obtained with a passband from 30 Hz to 1000 Hz (Nuwer and Dawson 1984a,b). The cortical response is attenuated by low frequency filter settings above 30 Hz, but a setting of 100 Hz can be used to record spinal responses.

If a multichannel recorder is available, reliability is improved by recording several channels, at least one above and one below the level of operation, to confirm the adequacy of stimulation (*Fig. 7.16*) (Lüders et al 1982, Ryan and Britt 1986, Guérit et al 1996, Burke and Hicks 1997). In addition, if a peripheral nerve is stimulated, it is advisable to record the nerve action potential to confirm that the stimulus is adequate, and to identify the effect of cooling on nerve conduction velocity, which can lead to a latency increase of several milliseconds. This must be correctly identified, because a marked change in response due to cooling might be erroneously attributed to a surgical complication (Guérit et al 1994, Nuwer 1998). If cortical SEPs are used to monitor operations on the thoracic or lumbar spine, a supplementary channel recording cortical respons-

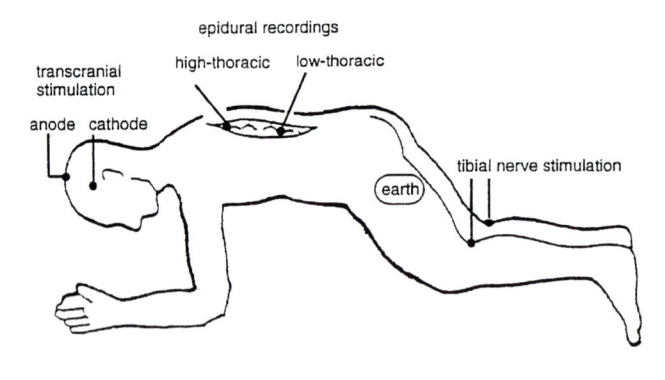

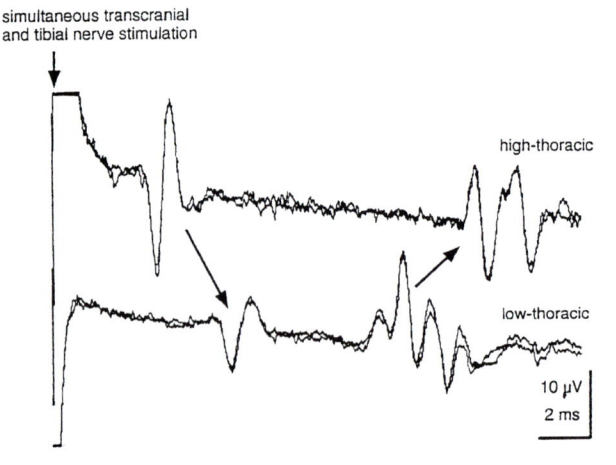

Fig. 7.15 Simultaneous assessment of corticospinal and somatosensory conduction during scoliosis surgery in a 14-year-old girl. Because bipolar electrodes were used, the negative deflections of descending motor and ascending sensory volleys are in opposite directions (downwards for the motor volley and upwards for the sensory). The motor volley consists of a single D wave at both recording sites (24 μV at the rostral site, 10 μV at the caudal), occurring at relatively short latencies: 12 and 16.2 ms at the lower and higher recording sites respectively. (From Burke et al (1992), by permission.)

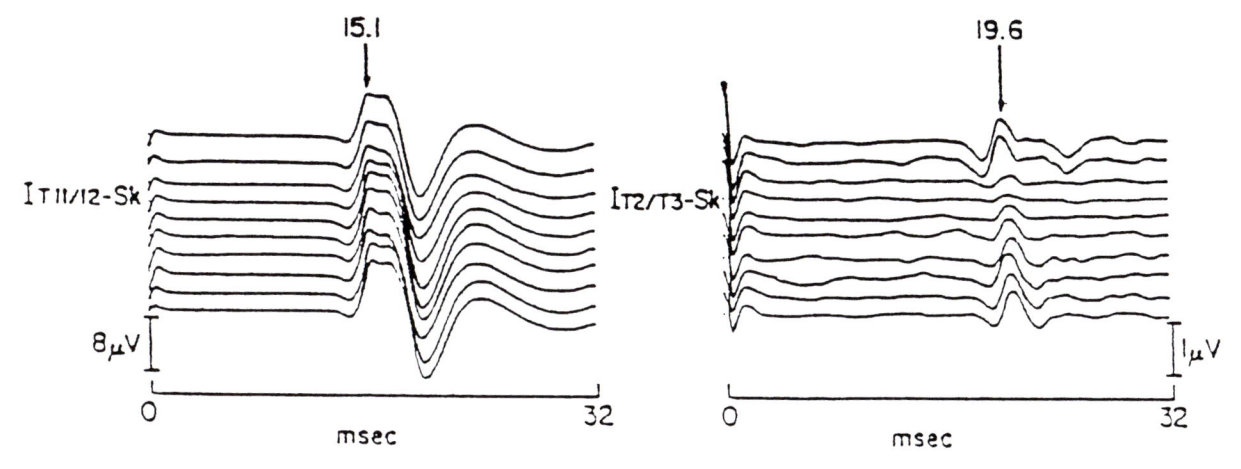

Fig. 7.16 Spinal EPs from interspinous ligament after tibial nerve stimulation recorded from two levels, above (right) and below (left) the level of operation during correction of scoliosis with Harrington rods. Pressure was applied to the cord after trace 2, and released after trace 4. Only potentials from above the level of operation are affected. (From Lüders H, Gurd A, Hahn J, Andrish J, Weiker G, Klem G (1982) A new technique for intraoperative monitoring of spinal cord function. Multichannel recording of spinal cord and subcortical evoked potentials. *Spine* 7:110–115, by permission.)

es from an arm will help to identify anaesthetic effects, which may otherwise be confusing (Madigan et al 1987). If only a limited number of recording channels can be used, the most stable response should be monitored and, in the event of any deterioration, the other responses should be checked to eliminate the possibility of interference by extraneous factors such as anaesthetics, or loss of stimulation.

Criteria for identification of a significant change

Regardless of the method of monitoring, response amplitude and latency do not remain absolutely constant, even in uneventful operations. The magnitude of amplitude and latency variation depends upon the precise stimulating and recording technique, and on anaesthetic factors; even with a standardized procedure greater variation is encountered in some patients than others, and variation often increases at certain stages of the operation (Schramm et al 1985, York et al 1987, Lubicky et al 1989, Spielholz and Engler 1990, Kalkman et al 1991, Luk et al 1999, Van Dongen et al 1999). Most workers have concluded that a fall in amplitude of 50% or more is likely to predict a postoperative deficit (Schramm et al 1985), although some have warned the surgeon only if the amplitude is reduced by 60% (Jones et al 1985). However, a transient postoperative deficit has occasionally followed a fall in EP amplitude of less than 50% (Brown and Nash 1985, Jones et al 1988), so a warning after amplitude attenuation of 40% has been recommended, at least for cortical responses (More et al 1988). Further work is needed to determine the precise level of amplitude reduction of each response which will reliably predict a postoperative deficit, and thus the need for corrective action during the operation. Perhaps rather than concentrating on quantification of variability by conventional statistics, particularly the coefficient of variation, attention should be turned to the method of statistical process control (see p. 354), where a significant change is defined with reference to the mean consecutive difference (Wheeler and Chambers 1992, Wheeler 1995). Until more precise criteria can be established, the surgeon should be informed when the amplitude falls to 50% of its original value, or if a surgical manoeuvre is followed by either a sharp amplitude reduction of 40% or more, or a sustained increase in latency is observed.

False positives, false negatives and the predictive value of spinal monitoring

Although the aim of monitoring is to prevent a neurological deficit developing intraoperatively, occasionally new symptoms and signs are identified postoperatively. If this occurs without a significant change in monitoring having been identified it is a false negative result. False negative results are inextricably linked with the criteria for issuing a warning to the surgeon; more stringent criteria reduce the risk of overlooking a significant change (false negative), but increase the risk of giving a false alarm (false positive) (More et al 1988). Many false negative results have been reported, both with the 'wake up' test (Ben-David et al 1987, Bar-On et al 1995) and with SEPs (Tamaki et al 1984, Wilber et al 1984, Ginsburg et al 1985, Johnston et al 1986, Lesser et al 1986, Molaie 1986, Ben-David et al 1987, Chatrian et al 1988, Harper et al 1988, More et al 1988, Roy et al 1988, Ashkenaze et al 1993, Wagner et al 1994, Noordeen et al 1997, Manninen 1998, Pelosi et al 1999, Deutsch et al 2000, Minahan et al 2001), and a survey of American and European scoliosis surgeons identified 96 cases of postoperative deficit not predicted by SEP monitoring (Dawson et al 1991). Some of these reports might be dismissed on the grounds that the recorded traces were of questionable technical quality, or a period of marked attenuation of responses had been ignored (possibly due to inappropriate criteria for issuing a warning), or the deficit was due to a lesion in a pathway that was not monitored, or the deficit appeared after monitoring had been discontinued (Friedman and Grundy 1987). The first two explanations for a false negative result (poor quality traces, and failure to recognize a significant change) should be minimized by ensuring that monitoring is only undertaken by personnel with appropriate training and experience. Deficits arising in pathways not monitored should be eliminated by using one of the combined ascending and descending methods mentioned above (see p. 269). The possibility of delayed paraplegia after surgery for spinal deformity (Johnston et al 1986 Taylor et al 1994, Mineiro and Weinstein 1997, Pelosi et al 1999) and after operations to repair aortic aneurysms (Crawford et al 1988, Schepens et al 1994, Guérit et al 1996, 1999) should not be confused with false negative results of intraoperative monitoring, because monitoring which ceases at the end of the operation cannot be blamed for problems that do not begin until later. Finally, it must be accepted that false negative cases may arise occasionally even when the appropriate pathways have been monitored by competent staff using the best available technique; however, the risk of paraplegia is lower with monitoring than without it (Nuwer et al 1995) (see *Is the risk eliminated by monitoring?*, p. 13).

When a significant change in monitoring is identified without any neurological manifestations postoperatively there are two possibilities: either monitoring correctly identified a change in spinal cord function which was reversed by action taken after the warning was given, or monitoring indicated a lesion when none was present. In the former instance monitoring has fulfilled its primary function by preventing damage to the spinal cord, and the monitoring result can safely be regarded as a true positive (Papastefanou et al 2000, Luk et al 2001); the latter is a false positive result. In many cases a warning is given at a stage in the operation when no manoeuvre likely to damage the cord has been performed and, if the outcome is favourable, the warning must be regarded as a false alarm. On the other hand, if a warning leads to corrective action following a potentially dangerous manoeuvre, the favourable outcome may have been achieved because of, or in spite of, the warning provided by monitoring. In some cases it is possible to confirm the electrophysiological findings by performing a 'wake up' test, and many surgeons regard this as essential before removing instrumentation, but a recent report suggests that both the sensitivity and the specificity of monitoring are now so high that the 'wake up' test is no longer needed, at least for scoliosis surgery (Padberg et al 1998).

If there are no false negatives, then the sensitivity of monitoring is 100%, and the probability of no deficit after negative monitoring is 100%; this is the negative predictive value of monitoring. The positive predictive value is the probability that a significant change in monitoring will either be followed by a new deficit, or requires corrective action. This is the ratio of true positives to the total number of positive results (see *Clinical significance of changes*, p. 8). For example, the report quoted above (Padberg et al 1998) included two true positives and seven false positives,

so the positive predictive value is 2/9, or 22.2%. Lower positive predictive values have been reported by others, leading some surgeons to regard current methods as interrupting uneventful operations too frequently with false positives (Annotation 1991); others accept that this is the price that must be paid for eliminating false negatives.

Clinical applications of spinal cord monitoring
Surgery for spinal deformity (kyphosis and scoliosis)

The risk of paraplegia following operations for kyphoscoliosis is about 1% (MacEwen et al 1975, Dove 1989), although the risk is higher in patients with severe deformities, congenital scoliosis, prominent kyphosis or scoliosis secondary to neuromuscular disease, and in operations involving either osteotomy or the insertion of sublaminar wires or following preoperative traction. About 30% of postoperative paraplegias are permanent, but the risk of persistence is reduced if distraction can be released as soon as possible after paraplegia is detected. Corrective action is therefore likely to be successful provided that an early warning of paresis can be given. Many surgeons have relied upon a 'wake up' test immediately after correction of deformity (Vauzelle et al 1973, Hall et al 1978, Dorgan et al 1984), but this policy has several disadvantages (Owen 1999) (see *Methods of monitoring cord function*, p. 269). Electrophysiological methods, both invasive and noninvasive, have been used to monitor operations for scoliosis in many centres since the mid-1970s. In a series of 1168 consecutive operations monitored with a somatosensory technique there were no false negatives and 52 false positives (Forbes et al 1991), while 500 operations monitored with a combined somatosensory and motor technique produced only two false positives, again without false negatives (Padberg et al 1998). The authors of the latter report considered that their method made the 'wake up' test unnecessary.

Where data are available for operations performed with and without monitoring, the incidence of serious neurological deficits is lower following operations performed with monitoring (Dawson et al 1991, Forbes et al 1991, Nuwer et al 1995). However, in the absence of a randomized controlled trial, such comparisons are usually retrospective, and as such can never provide conclusive evidence for the benefit of monitoring, since improvements in surgical and anaesthetic technique may have been partly responsible for the reduction in morbidity. On the other hand, the contribution of monitoring is quite clear in those operations where monitoring identifies a significant disturbance of cord function, a 'wake up' test confirms the presence of paraplegia or severe paresis, but appropriate surgical intervention prevents the appearance of a deficit postoperatively; cases like this are mentioned in many reports (*Fig. 7.17*).

Scoliosis due to neuromuscular disease presents a particular problem with respect to monitoring. The risk of

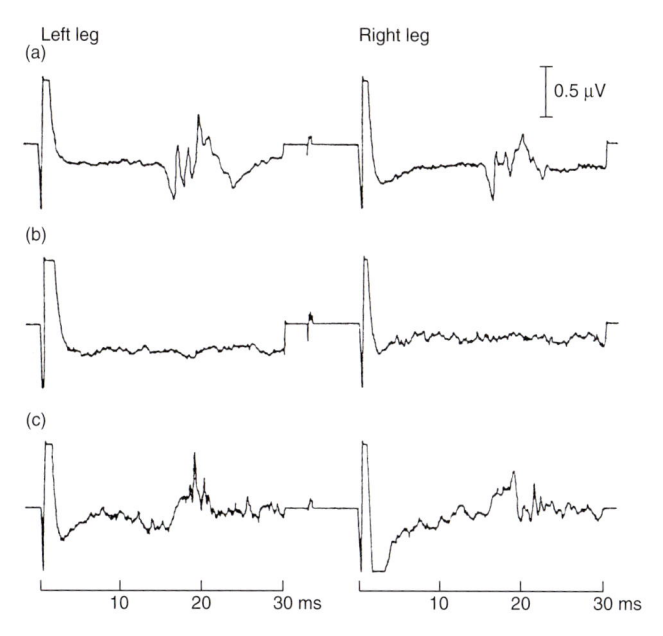

Fig. 7.17 Spinal potentials recorded from the epidural space during scoliosis surgery. (a) Initial responses recorded at the C7 level, showing several negative components (upward deflections). (b) Absent responses recorded about 20 min after spinal distraction. At 'wake up' test there was no movement of either foot. (c) Partially restored responses with abnormal waveform recorded after release of distraction and laminectomy. On recovery from anaesthetic the patient could move the feet, but had paraesthesiae in the right leg and a band of numbness on the trunk. (From Jones et al 1983 A system for the electrophysiological monitoring of the spinal cord during operations for scoliosis. *Journal of Bone and Joint Surgery (British)* 65B:134–139.)

paraplegia is greater in operations for symptomatic scoliosis, but in many cases the condition causing the scoliosis (e.g. cerebral palsy or Friedreich's ataxia) also severely attenuates SEPs. Cortical responses are likely to be unreliable, and may be unobtainable, in such patients (Ashkenaze et al 1993), whereas other causes of scoliosis, particularly Duchenne dystrophy and spinal muscular atrophy, do not affect somatosensory conduction, so that satisfactory monitoring can be achieved, particularly if spinal responses are monitored rather than cortical ones (Williamson and Galasko 1992, Noordeen et al 1997). Even in children with severe disabilities, subcortical responses can usually be monitored (Ecker et al 1996, Tucker et al 2001). A descending (motor) technique often permits monitoring in scoliosis secondary to neuromuscular disease, when it would have been impossible by a somatosensory technique (Owen et al 1995, Langeloo et al 2001), but in children with spinal cord lesions monitoring shows greater variability, and the 'wake up' test is often needed to confirm changes identified by electrophysiological methods (Wilson-Holden et al 1999).

When correction of deformity involves anterior vertebrectomy, the risk of a false negative result from somatosensory monitoring techniques may be particularly high (Deutsch et al 2000), and it is probably advisable to recommend motor monitoring for anterior vertebrectomy if at all possible.

Spinal fractures

Many patients with unstable spinal fractures are treated by operative fixation methods similar to those used for scoliosis. Reproducible EPs can often be recorded if paraplegia is not complete preoperatively, and operative monitoring is then possible (*Fig. 7.18*). In a series of operations performed without monitoring a new deficit appeared postoperatively in 6.9%, but when SEP monitoring was added new deficits were found in less than 1%, SEP deterioration having prompted a change in intraoperative management in 4% (Meyer et al 1988). This retrospective analysis suggests that SEP monitoring is capable of giving early warning of deteriorating cord function during fixation of unstable spinal fractures. In patients with spinal fractures, especially cervical, the cord is at particular risk during positioning on the table, and extra attention should be paid to monitoring at this stage of the procedure.

The possibility that spinal roots might be damaged during pedicle screw insertion has led some surgeons to monitor spontaneous EMG activity, and EMG responses to stimulation through the screw track (Weiss 2001, Toleikis 2002, Shi et al 2003), although results are conflicting (Rose et al 1997, Reidy et al 2001). When trauma affects the lumbar roots (T12 to L4), responses to femoral nerve stimulation may provide more information than stimulation of either tibial or peroneal nerves (Robinson et al 1993). For lower lumbar and sacral roots the bulbocavernous reflex can be monitored (Rodi and Vodušek 2001).

Cervical spine surgery

Operations on the cervical spine present particular problems, because the site of operation usually precludes the use of invasive electrodes above the incision, and sensory pathways from the arms must be monitored as well as those from the legs (*Fig. 7.19*). If the anterior approach is used, retraction may compress the carotid artery, leading to reduction in amplitude and increase in latency of the cortical SEP (Sloan et al 1986). In a variety of operations on the cervical spine, monitoring of cortical SEPs from the upper and lower limbs contributed to prevention of a postoperative deficit in several patients (Veilleux et al 1987, Smith, et al 1994a, Wagner et al 1994, Dennis et al 1996, May et al 1996). SEP monitoring has been performed during operations on the odontoid by the transoral approach (Spetzler et al 1979, Selman et al 1981), during excision of a cervical arteriovenous malformation (Owen et al 1979), and in operations for cervical myeloradiculopathy (Sebastian et al 1997). Retrospective comparison of operations on the cervical spine performed with and without SEP monitoring shows that monitoring contributed to a significant reduction in the combined rate of death and quadriplegia (Epstein et al 1993). In a few small series of operations on the cervical spine, the results of surgery have been predicted by changes in intraoperative motor monitoring (Kitagawa et al 1989, Darden et al 1996, Gokaslan et al 1997). Monitoring, both sensory and motor, for cervical spinal surgery has been recently reviewed (Weisz et al 1999).

Spinal EPs have been used to identify the level of the principal lesion in patients with cervical spondylosis, in whom imaging suggested involvement of multiple levels. Restricting operation to the level of maximum compression minimizes complications, especially for elderly patients (Tani et al 2000).

Electrophysiological monitoring by methods similar to those used for spinal deformity surgery and cervical spondylosis are now being advocated for lumbar stenosis operations (Norcross-Nechay et al 1999, Weiss 2001).

Spinal tumours and arteriovenous malformations

Although the risk of paraplegia following operations for spinal tumour, especially intramedullary, is 15–20% (Greenwood 1967), relatively few such operations have been monitored, because paraplegia developing in the course of operation for tumour is unlikely to be correctable. Furthermore, many patients with spinal tumour have sustained damage to the cord before surgery is undertaken, so monitoring may be either impossible or unreliable (Koyanagi et al 1993). Paraplegia has occasionally followed disappearance of EPs during essential manipulation of the tumour (McCallum and Bennett 1975, Allen et al 1981), but documentation of this phenomenon is not a justification for monitoring. Nevertheless, several surgeons have found that monitoring provided useful information during operation for spinal tumour, and reduced the risk of a new deficit (e.g. Lang et al 1996, Lorenzini and Schneider 1996, Kothbauer et al 1997, Morota et al 1997, Szekely et al 1998, Prestor and Golob 1999) (*Fig. 7.20*). In some cases the surgeon may be able to perform a more complete eradication of malignant tissue if he can be assured that he is not damaging the cord (*Fig. 7.21*); monitoring may be able to provide such assurance. This is particularly the case following loss of the muscle response to transcranial stimulation, when there is often postoperative weakness, but if the epidurally recorded D-wave is retained the deficit is unlikely to be permanent (Jallo et al 2001, Kothbauer 2002). Monitoring certainly seems to have a place in operations to remove spinal arteriovenous malformations, since feeding vessels can be sac-

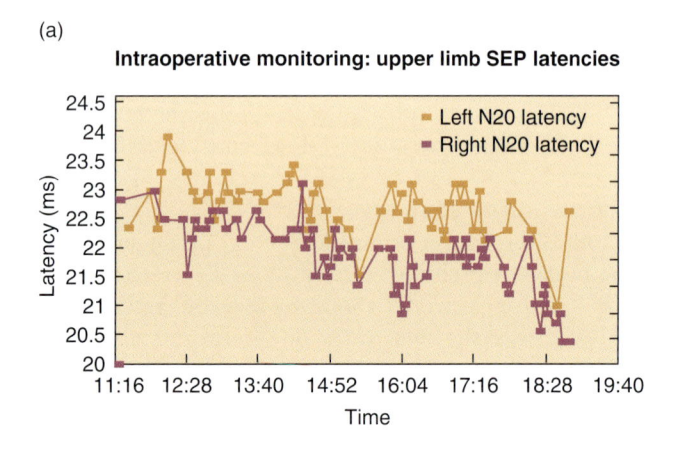

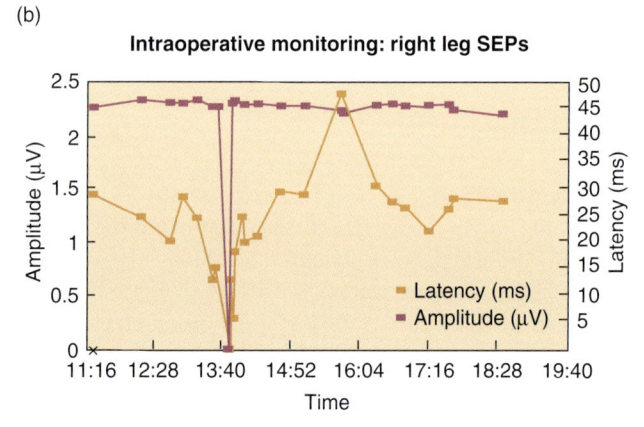

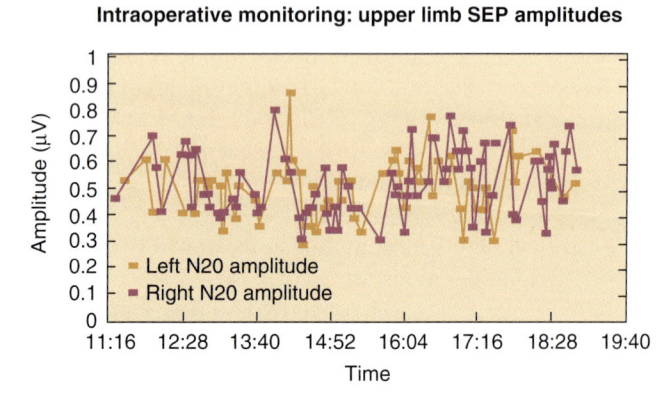

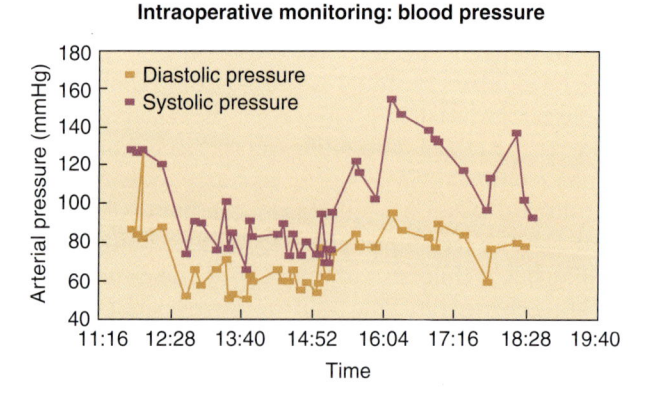

Fig. 7.18 Amplitudes and latencies of cortical SEPs from arms (a) and right leg (b), together with systolic and diastolic blood pressure, during operation for cervical fracture. During a period of hypotension whilst the surgeon was working at the step in the bone just before removing one of the loose fragments, the amplitude of the leg SEP decreased, but had recovered 2 min later when blood pressure was restored. The patient made a good recovery. (From Binnie et al (2003), by permission.)

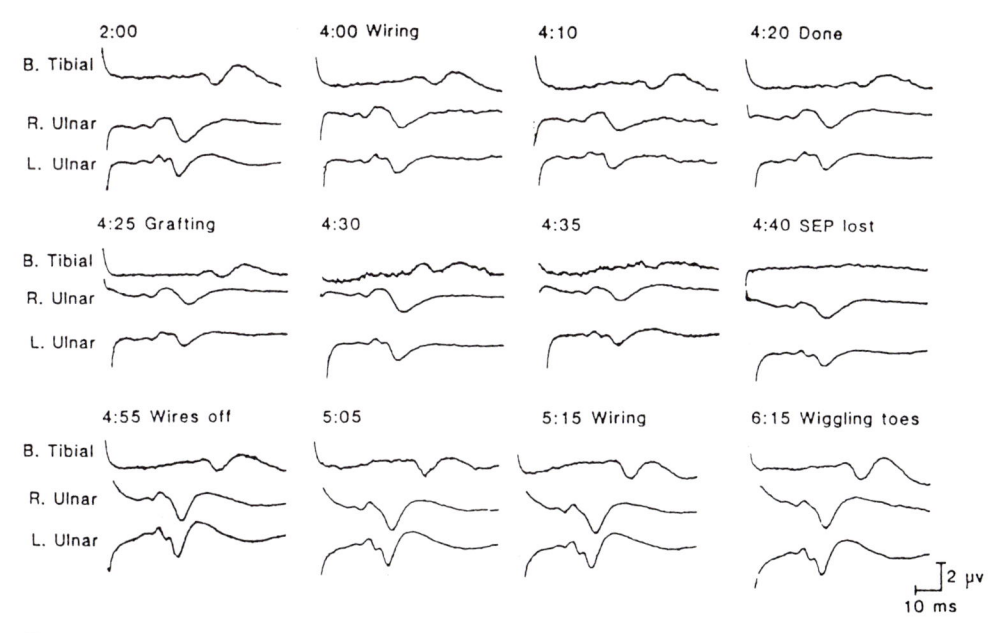

Fig. 7.19 Transient loss of ulnar and tibial cortical EPs in 60-year-old man who had clinical signs of cervical cord compression and unstable C6 fracture on x-ray of cervical spine at time of initial examination. During operation, ulnar and tibial cortical EPs progressively decreased in amplitude after C5–7 wire fixation of cervical spine, and tibial cortical EPs were eventually lost. All cortical EPs reappeared when wires were removed. Wire fixation of C5–7 vertebrae took place between 4.00 h and 4.20 h and was followed by bone grafting at 4.25 h. The concentration of isoflurane was maintained at 0.75% until 4.30 h, at which time it was decreased to 0.5% for the rest of the operation. At 4.35, tibial cortical EPs were appreciably reduced in amplitude, and were lost at 4.40 h. Wires were removed at 4.55 h and repositioned at 5.15 h. Anaesthesia was terminated at 6.15 h, when the patient could wiggle his toes. B, bilateral; L, left; R, right. (From Veilleux et al (1987), by permission.)

rificed with confidence if monitoring confirms that cord blood supply is not compromised (Owen et al 1979); the same applies to embolization of feeding vessels (Niimi and Berenstein 1999, Sala et al 2001).

Aortic surgery

Paraplegia may result from operations upon the aorta for aneurysm, coarctation or traumatic rupture. In the case of coarctation, the risk of paraplegia is between 0.3% and 0.4% (Brewer et al 1972, Keen 1987), whereas the risk of postoperative paraplegia associated with ruptured or dissecting thoracoabdominal aneurysm may be as high as 44%, depending on the site and size of the lesion (Jex et al 1986). Operation for all these conditions involves clamping the aorta, and whether this can be done without damaging the spinal cord depends on the adequacy of collateral circulation, because many segmental vessels may be occluded (Jacobs et al 2002). Since there is great anatomical variability in the blood vessels feeding the anterior spinal artery, the cord may become severely ischaemic within a few minutes of clamping the aorta at any level. If the distal pressure cannot be maintained above 60 mmHg, and intercostal vessels isolated between the clamps cannot be reattached, then the cord is likely to be damaged with resulting paraplegia. An additional factor increasing the risk of paraplegia in operation for abdominal aneurysm is failure of perfusion of the principal radicular artery (*arteria radicularis magna* of Adamkiewicz), which arises between T5 and L2 (most commonly T9 to T12), and is the main vessel supplying the anterior spinal artery in the thoracolumbar area (Connolly 1998, Wan et al 2001).

If spinal cord blood flow cannot be maintained then EPs are likely to be lost or grossly attenuated (*Fig. 7.22*) (Cunningham et al 1982). Several surgeons have monitored aortic operations with SEPs (e.g. Mizrahi and Crawford 1984, Cunningham et al 1987, Crawford et al 1988, Faberowski et al 1999, Galla et al 1999, Guérit et al 1999), and some have adopted descending (motor) methods (e.g. Laschinger et al 1988, De Haan et al 1998, Jacobs et al 1999, Van Dongen et al 1999, Jacobs et al 2000, Dong 2002, MacDonald and Janusz 2002), as well as techniques in which the cord is stimulated directly (North et al 1991, Yamamoto et al 1994, Grabitz et al 1996). Some workers have reported unacceptable false negative rates for SEP monitoring in aortic surgery (e.g. Cunningham et al 1987, Galla et al 1994, Schepens et al 1994), perhaps because SEPs depend on dorsal column conduction and may therefore be unaffected by ischaemia in the anterior spinal artery territory. However, in the hands of others SEP monitoring has achieved high sensitivity in aortic surgery (Guérit et al 1999).

Three important differences must be recognized between aortic and spinal surgery as far as monitoring is concerned:

Chapter 7 Neurophysiological Monitoring during Surgical Operations

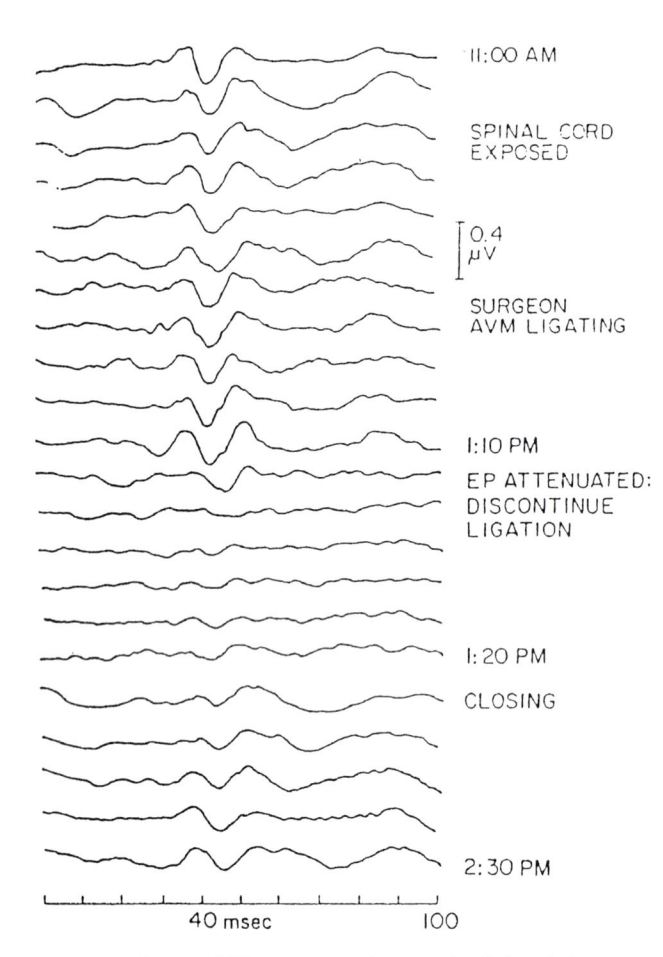

Fig. 7.20 Cortical SEPs to peroneal nerve stimulation during ligation of a spinal arteriovenous malformation. After the SEP disappeared no more ligation was undertaken. The SEP recovered after about 1 h, and the patient awoke without any new neurological impairment. (From Nuwer and Dawson (1984b), by permission.)

1. Clamping the aorta may render the lower limbs ischaemic, so that monitoring which relies entirely upon peripheral nerve stimulation will be associated with a high false positive rate (North et al 1991);
2. Different warning criteria are needed; the change in SEP amplitude or latency that should be regarded as significant is probably smaller for aortic operations than for spinal surgery (Stuhmeier et al 1993); and
3. Monitoring should ideally be continued postoperatively, because of the risk of delayed paraplegia due to hypotension or reperfusion injury after uneventful operations (Guérit et al 1999).

Comparison of SEP and motor monitoring in the same operations shows that SEP monitoring is slower to detect spinal cord ischaemia, and is associated with more false positives than motor monitoring (Meylaerts et al 1999). Nevertheless, it must be remembered that neurophysiological monitoring is only one of many strategies adopted to reduce the risk of paraplegia complicating aortic surgery, and that the need for monitoring is not accepted by all surgeons (Connolly 1998, Robertazzi and Cunningham 1998, Wan et al 2001).

SPINAL ROOT SURGERY AND PERIPHERAL NERVE SURGERY

Intraoperative EP recording has found several applications in operations on peripheral nerves and spinal nerve roots. During exploration of a damaged nerve the demonstration that nerve action potentials can be conducted across the lesion indicates that fibres are beginning to regenerate, so repair is unnecessary (*Fig. 7.23*) (Terzis et al 1980, Slimp 2000, Spinner and Kline 2000, Happel and Kline 2002). This type of intraoperative recording requires a meticulous technique to minimize stimulus artefact because of the short distance between stimulating and recording electrodes on the exposed nerve (Brown and Veitch 1994, Tiel et al 1996). Brachial plexus injuries often include avulsion of nerve roots from the cord, the so-called preganglionic lesion, following which there is no hope of recovery. If this has occurred, no cortical SEP can be recorded in response to stimulation of the proximal nerve stump, and repair of a distal transection is then not worthwhile (Landi et al 1980, Oberle et al 1998).

In most surgical operations there is a risk of nerve damage from compression or stretching while the patient is anaesthetized, but the risk is minimized by careful positioning and padding over areas where nerves may be externally compressed (Payan 1987, Dawson and Krarup 1989, Warner 1998). However, in some operations there is a particular risk of nerve damage because of the nature of the surgical procedure; impending nerve damage can then be identified intraoperatively by recording appropriate EPs (Brown and Veitch 1994). For example, operation to correct spondylolisthesis (i.e. forward displacement of one vertebra upon the vertebra below) carries a risk of compression or excessive stretching of spinal nerve roots (Cohen and Huizenga 1988); limb lengthening procedures may damage nerves (Wexler et al 1998); operations on the pelvis or acetabulum can damage the lumbosacral nerve roots or sciatic nerve (Helfet et al 1997, Arrington et al 2000); the brachial plexus may be at risk in operations performed in the lateral decubitus position (Mahla et al 1984) or during surgery for scoliosis in the prone position (O'Brien et al 1994, Schwartz et al 2000); the radial nerve is at risk during operative reduction of humeral fractures (Mills et al 2000). The ulnar nerve is frequently damaged, the risk being particularly high in certain types of surgery (Prielipp et al 1999). In view of the medico-legal consequences of intraoperative nerve damage, monitoring may become more important in this context, although care is essential because some have found SEP monitoring to be unreliable in detecting incipient nerve damage from faulty positioning on the operating table (Lorenzini and Poterack 1996).

A further use of electrophysiological recording intraoperatively is the selection of dorsal spinal rootlets for rhizotomy in the treatment of spasticity, according to the electromyographic response to root stimulation (Abbott 2002). The technique depends upon the observation that,

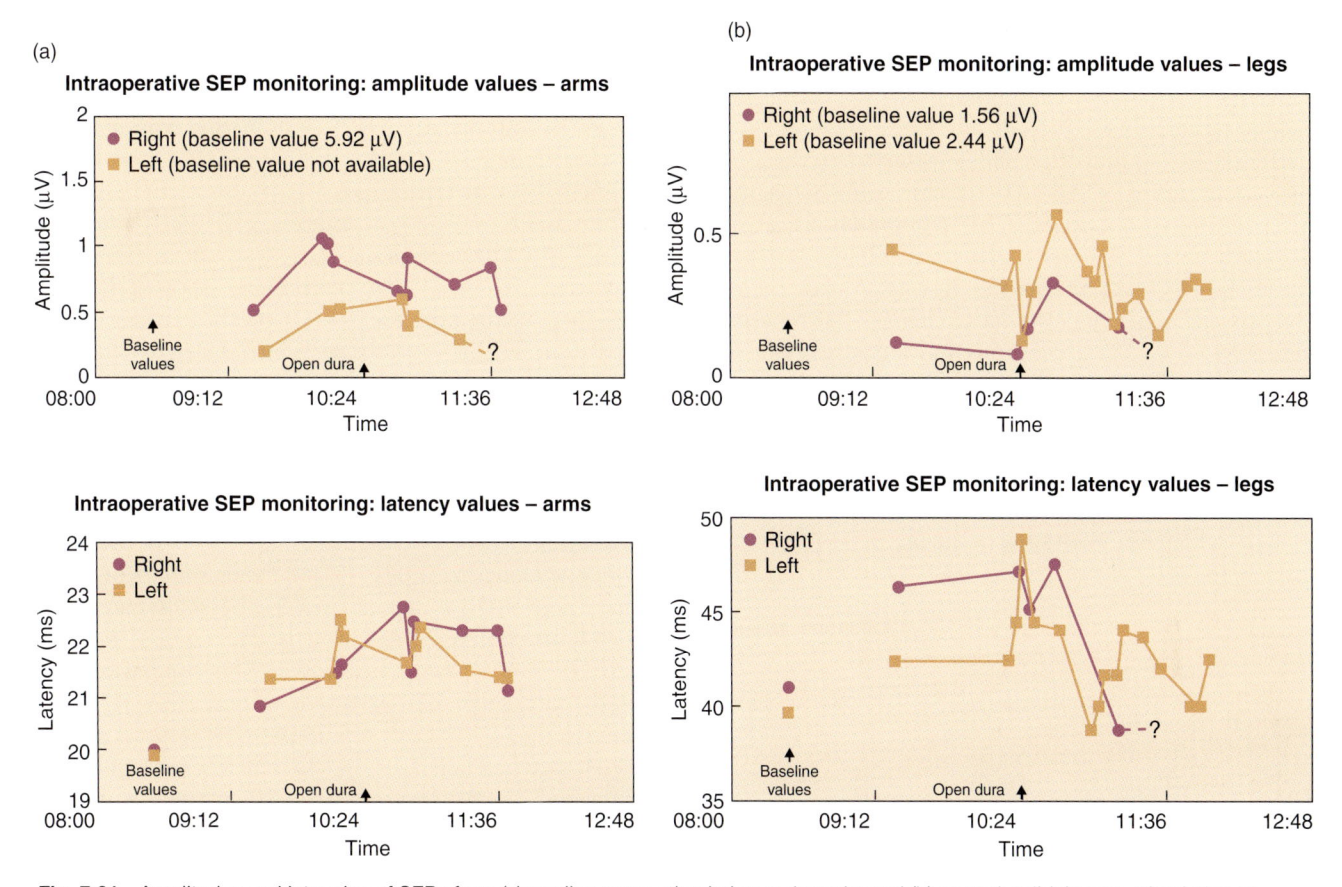

Fig. 7.21 Amplitudes and latencies of SEPs from (a) median nerve stimulation at the wrist and (b) posterior tibial nerve stimulation at the ankles during operation for a cervical ependymoma. Baseline values are 5 min pre-induction of anaesthesia Decompression following opening of the dura at 10.24 h resulted in a marked increase in SEP amplitude from the legs. The tumour was debulked and a biopsy was taken between 11.04 and 11.27 h without significant further reduction in amplitude or increase in latency. (From Binnie et al (2003), by permission.)

in patients with spasticity, electrical stimulation of some dorsal rootlets produces exaggerated and prolonged EMG responses which spread to other muscles, whereas only a brief localized response follows stimulation of other rootlets (Smyth and Peacock 2000). In selective dorsal rhizotomy, multichannel intraoperative recording is used to identify the involvement of additional muscles in response to stimulation of a single root. Some surgeons have reported reduction of spasticity after cutting rootlets responsible for exaggerated responses (e.g. Peacock and Staudt 1990), but the value of this technique is not universally accepted (Landau and Hunt 1990, Sacco et al 2000). Care must be taken to ensure that the intensity of stimuli applied to dorsal roots is kept below the threshold for ventral root stimulation, or else the results of intraoperative recordings will be misleading (Logigian et al 1996).

CEREBRAL ISCHAEMIA DURING NON-INTRACRANIAL SURGERY

Cardiac surgery
Historical background

With the advent of 'open' heart surgery using a 'heart–lung' machine to oxygenate and recirculate the blood (CPB), it soon became apparent that such procedures carried considerable risk of brain damage. Its incidence varied widely between reports using differing criteria over the early years, e.g. 53% (Javid et al 1969), 44% (Tufo et al 1970), 34% (Gilman 1965) and 19% (Branthwaite 1972) to 12% (Åberg 1974). Early neuropathological studies confirmed typical cerebral arterial 'boundary zone' (also described as 'watershed', see *Fig. 2.31*) distribution of ischaemic damage, whether from hypoperfusion or emboli (Brierley 1967, Aguilar et al 1971, Witoszka et al 1973). The ischaemic changes showed a predilection for the triple boundary zone between the territories of the anterior, middle and posterior cerebral arteries in the parietal regions, extending anteriorly along the anterior–middle cerebral arterial boundary zone in more severe cases. These findings are consistent with the clinical effects of predominantly posterior cerebral circulatory inadequacies producing curious clinical syndromes including visual agnosias (Ross Russell and Bharucha 1978, McAuley and Ross Russell 1979). Current views on the mechanisms and neuropathology of brain damage after cardiac surgery in adults, and suggestions for prevention or amelioration, are given by Ahonen and Salmenperä (2004).

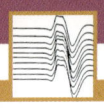

The early introduction of monitoring with conventional multichannel EEG equipment (Bellville et al 1955, Davenport et al 1959, Pampiglione and Waterston 1961) helped to identify the periods of greatest risk and led to a better understanding of possible measures to reduce cerebral complications. The EEG findings using conventional recordings on transportable equipment were often dramatic (*Fig. 7.24*) and merit study when considering optimal qEEG features for monitoring. A wide range of the classical appearances of EEG changes during cardiac surgery with CPB, hypothermia and the anaesthetic procedures then in common use are illustrated by Arfel et al (1961), Pampiglione and Waterston (1961), Fischer-Williams and Cooper (1964) and Arfel and Walter (1977).

As purpose-built EEG monitors became available, several studies showed a decrease in the incidence of brain damage which was attributed to the use of EEG monitoring. An incidence of postoperative neurological damage of 19.2% was reported from retrospective analysis of the notes of 417 patients submitted to open heart surgery during 1970 (Branthwaite 1972). Monitoring with an early Cerebral Function Monitor (CFM) (RDM Consultants, Uckfield, Sussex, UK, http://www.cfams.com) helped to identify the timing and, by inference, the likely causes of periods of EEG depression; these were most commonly associated with inadequate cerebral perfusion at onset of CPB or presumed microemboli from the extracorporeal circulation. When measures were introduced to eliminate or minimize the adverse effects of such factors, the incidence of neurological damage fell to 7.4% in 538 patients operated on during 1973 (Branthwaite 1975).

Incidence of neurological and neuropsychiatric sequelae

Evidence began to appear in the 1980s suggesting that minor neurological or neuropsychological damage was much more common after coronary artery bypass grafting (CABG) than previously supposed. This conclusion was based upon several prospective studies showing that, although the incidence of serious postoperative neurological damage had fallen, being less than 1% for fatal brain injury and 2% for severe strokes in an analysis of 421 patients at the Cleveland Clinic (Breuer et al 1983), minor

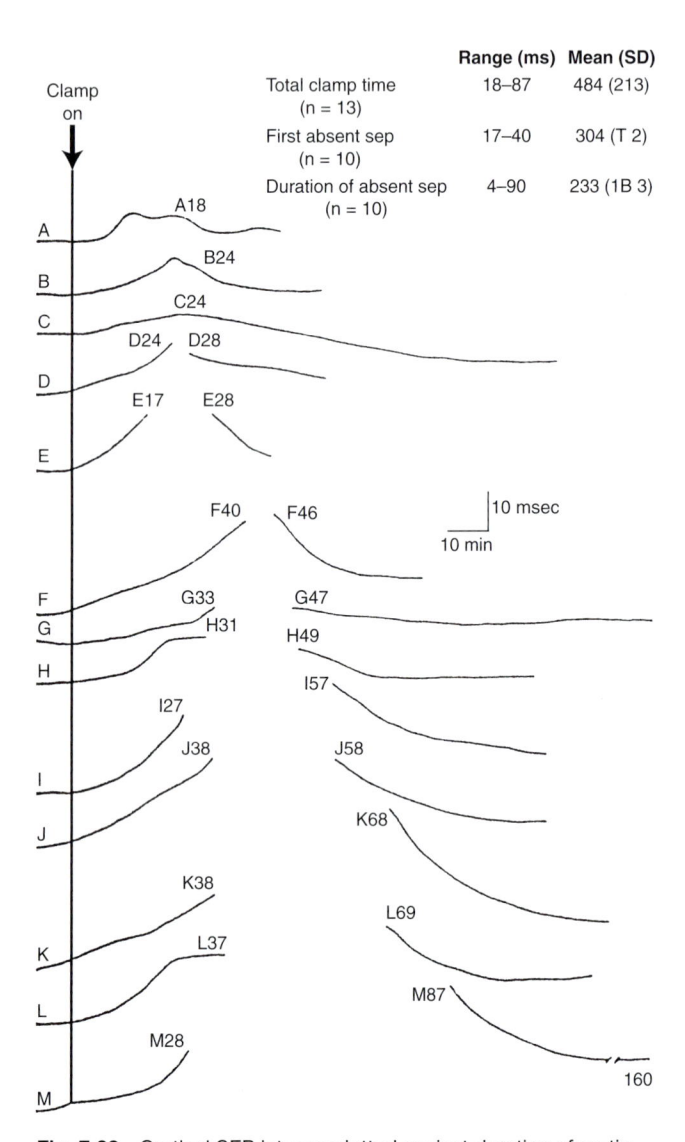

Fig. 7.22 Cortical SEP latency plotted against duration of aortic clamping during operation for aortic aneurysm in 13 patients (A–M). In 10 cases SEPs disappeared during aortic clamping (patients D–M) and patient M was paraplegic postoperatively. (From Mizrahi and Crawford (1984), by permission.)

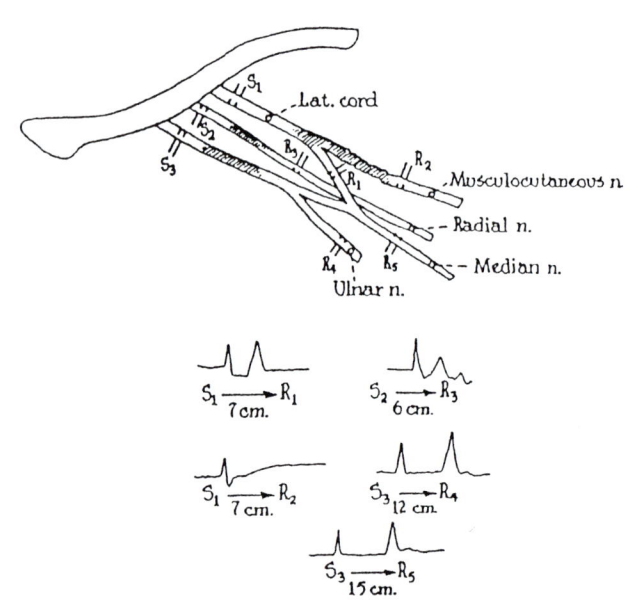

Fig. 7.23 A gunshot wound to the left shoulder resulted in complete lateral and medial and incomplete posterior cord palsies. Exploration 3 months after injury showed lesions in continuity. Since NAPs could be recorded from medial as well as posterior cords, only neurolysis of these elements was carried out. No NAPs could be recorded from the lateral cord to the musculocutaneous nerve, and there was no contraction of biceps in response to stimulation of the lateral cord. Since NAPs were recorded from the lateral cord to the median nerve, this was split away from the musculocutaneous portion of the lateral cord, and the latter replaced by grafts. Biceps function returned to a Grade 4 level within 14 months. (From Kline and Judice (1983), by permission of the *Journal of Neurosurgery*.)

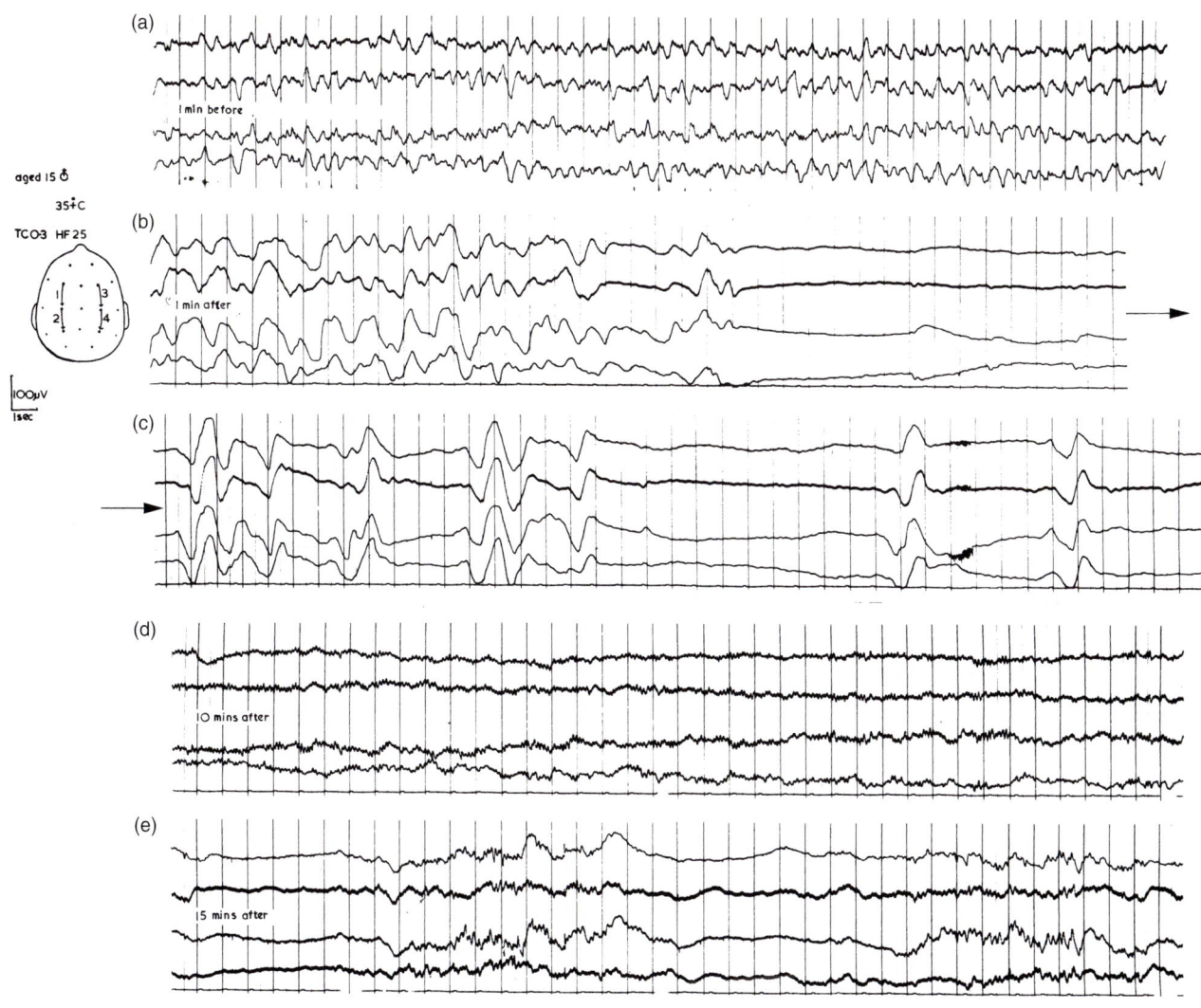

Fig. 7.24 Historical recording from a 15-year-old boy undergoing surgery for congenital heart disease (Fallot's tetralogy) in the early days of such procedures. (a) EEG during CPB after rewarming from 12.5 to 35°C; (b, c) slowing and burst-suppression develop abruptly following rupture of the pulmonary artery; (d, e) partial EEG recovery occurs 10 and 15 min later. (from Binnie et al (2003), by permission.)

deficits could be detected by fairly detailed neurological examination in almost two-thirds of patients in a study in Newcastle upon Tyne, UK (Shaw 1986). In the Newcastle study 312 patients were followed for 6 months after CABG. Early neurological disorders were found in 61% but only one resulted in death (0.3%) and four in severe disability (1.3%), the rest being of a lesser degree (Shaw et al 1985). Six-month follow-up examinations showed that only 10 patients had persisting neurological disability which was severe in the four who had suffered major perioperative strokes. Out of 139 patients of working age who had not returned to work by 6 months, in only four was this because of neurological disability (Shaw et al 1986). Several preoperative and intraoperative risk factors for adverse cerebral outcomes were defined in a large American multicentre study in which 24 surgical teams surveyed outcomes in 2108 patients undergoing CABG; adverse outcomes were found in 6%, half being major neurological problems including death from cerebral injury or stroke, and half deterioration of intellectual function or seizures (Roach et al 1996). Risk factors included proximal aortic atherosclerosis, arterial hypertension, previous neurological disorders, excessive consumption of alcohol and older age (which of itself increases the chances of postoperative cognitive dysfunction in several types of non-cardiac operation) (Moller et al 1998). Postoperative neuropsychiatric deficits attributed to ischaemia are not confined to patients undergoing cardiac surgery, but may also occur in those undergoing other major thoracic or vascular operations, albeit with less persistent disabilities (Smith et al 1986).

Abnormalities in the preoperative EEG

Localized abnormalities in the preoperative EEG, suggestive of previous, often clinically silent, cerebral insults, appear to be associated with an increased likelihood of postoperative sequelae in the adult; in infants with congenital cardiac abnormalities, preoperative epileptiform discharges in the EEG predict the likelihood of postoperative seizures (Helmers et al 1997). Global preoperative qEEG measures also differ between patients who do and do not develop postoperative cognitive changes, suggesting that preoperative EEG deficits indicate a vulnerability of patients with already compromised cerebral circulation to damage during CPB (Hofste et al 1997, Toner et al 1998). Very low amplitude tracings which did not improve with CPB, were most commonly found in patients with long-standing severe cardiac disease and predicted adverse neurological outcomes in an earlier study (Schwartz et al 1973); differences between patients with stenotic and incompetent aortic and mitral valve disease were also shown. Even patients without subsequent permanent neurological damage may show both intraoperative EEG monitor abnormalities (Schwartz et al 1973, Malone et al 1981, Nevin et al 1989) and evidence of short-lived early postoperative brain swelling on MRI imaging (Harris et al 1993), suggesting that insults to the brain have to reach a critical threshold before clinical or neuropathological sequelae are clinically evident.

Characteristics of intraoperative EEG changes

Distinctive EEG changes occur during routine cardiac surgery with CPB and moderate hypothermia; they are illustrated in an EEG–polygraphic CSA monitor display in *Fig. 3.20* and, using spectral measures from the synchronized outputs of bandpass filters for alpha, theta and delta frequencies and the amplitude analysis of the original CFM (*Fig. 7.25*). It is instructive to compare how similar events at the time of changes in arterial pressure during CPB and with hypothermia are characterized in the different displays.

An early, but still relevant, classification of abnormalities was proposed by Schwartz et al (1973) as a result of their outcome studies in 100 patients undergoing cardiac surgery with CPB at the London Hospital in 1970–1971. Simple measurements of CFM traces were made independently of assessment of neurological outcome (*Table 7.2*).

Comparison with actual outcome showed a predictive accuracy of 83% and suggested a tendency to overestimate probable severity of outcome. Monitoring with the Cerebral Function Analysing Monitor (CFAM) (RDM Consultants, Uckfield, Sussex, UK, http://www.cfams.com) has led to similar predictive indicators of neurological damage after CPB based on abrupt falls in amplitude and reversal of slow/fast frequency ratios (*Fig. 7.26*) (Nevin et al 1989).

A separate neuropathological investigation examined the brains of all patients in the London Hospital necropsy records who had died at any stage following cardiac surgery with CPB and EEG monitoring with the CFM between 1970 and 1977. Twenty patients were found out of the 1022 who had been operated on during this 8 year period (Malone et al 1981). The cause of death was attributable to cardiac disease or its complications in 11 without clinical evidence of postoperative brain damage who had died between 1 day and 3 years (mean 5.5 months) after surgery. The nine with clinical evidence of brain damage had survived between 28 h and 1 month (mean 19 days) except for two with mild damage who survived for 14 and

Table 7.2 Classification of EEG abnormalities during cardiopulmonary bypass

Type of recording	Nature of changes	Incidence	Predicted outcome
Uneventful	'Slow' changes related to anaesthesia, start of CPB, or hypothermia	68%	Uncomplicated with consciousness regained within 12 h of operation
Mild changes	(a) 1–2 'quick' changes in level of cerebral activity with prompt return to the previous level (b) One 'quick' change with return to the previous level within 3–10 min (c) A constant low level recording (2–4 µV peak-to-peak) throughout	24%	Return of full consciousness delayed beyond 12 h *or* minor and probably reversible neurological changes
Moderate changes	> 2 'quick' changes followed by a partial return to the previous level taking 3–10 min	3%	Irreversible neurological changes
Severe changes	(a) Several episodes of 'quick' change with incomplete recovery to the previous level (b) No recovery of EEG activity or with a low level of activity (< 1–2 µV peak-to-peak) sustained for more than 20 min following the acute change	5%	Death during or immediately after surgery *or* consciousness never regained

Data from Schwartz et al (1978).

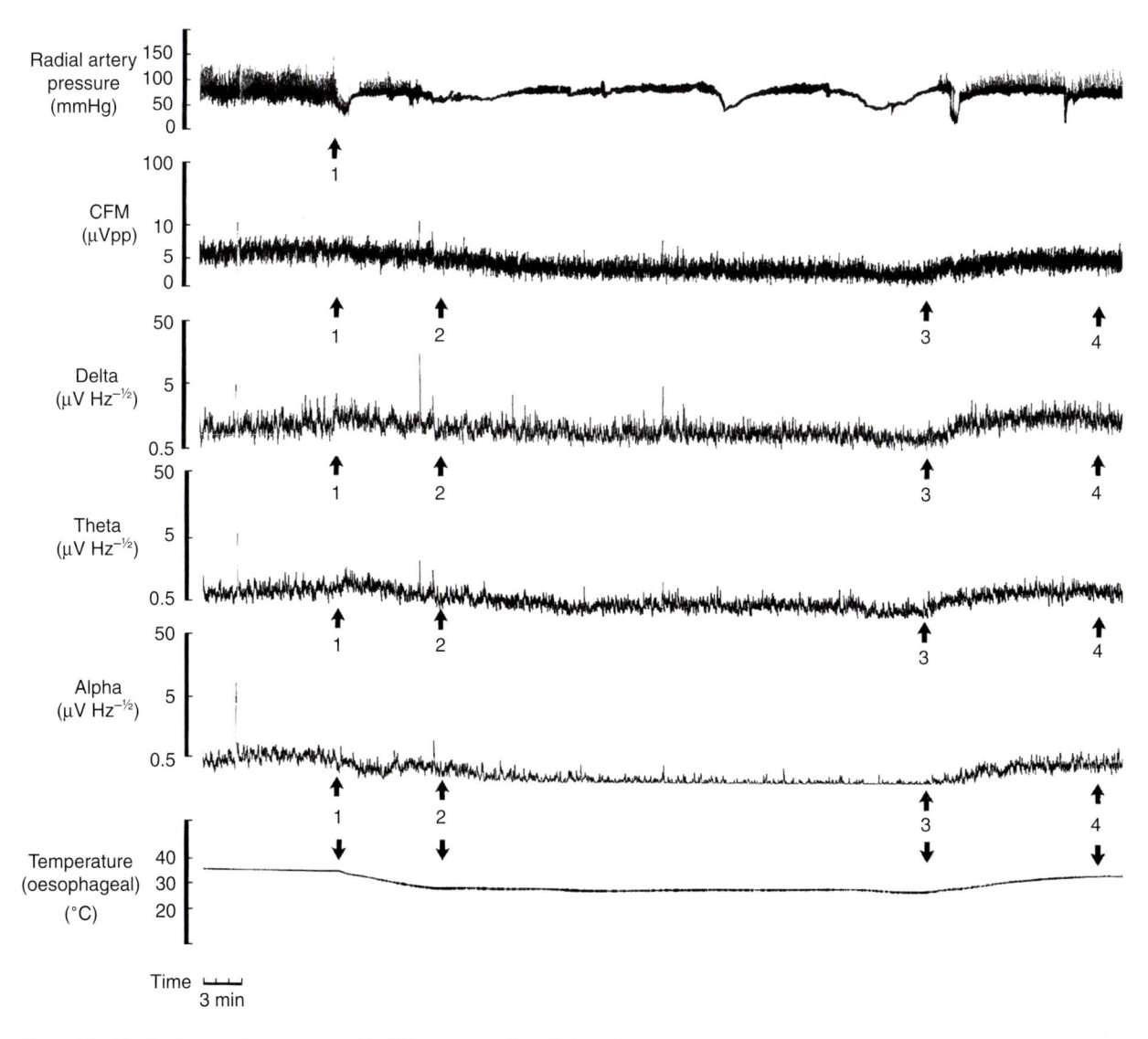

Fig. 7.25 Monitoring cardiac surgery with CPB and hypothermia in a 48-year-old man undergoing his second coronary artery vein graft. Display of radial artery pressure, CFM, outputs of Chebyshev bandpass filters (corner frequencies: delta 0.5–4 Hz, theta 4.5–8 Hz, alpha 8.5–13 Hz) and oesophageal temperature. Arrows mark (1) onset of CPB and start of cooling, (2) temperature reached 30°C, (3) start of rewarming, (4) temperature reached 37°C. CFM trace shows amplitude changes which mirror temperature; also minor amplitude changes at onset of CPB. Spectral outputs show an increase in slow components accompanied by a decrease in the alpha band, reflected in a fall in the r.m.s. power spectral density slow/fast ratio [(^{1}delta + ^{1}theta)/(^{1}alpha + ^{1}beta)] from 6.5 at 1 min before onset of CPB to 3.5 at time of minimum arterial pressure. (From Etherington (1981) and Binnie et al (2003), by permission.)

18 months respectively; their deaths were due to a combination of factors with cardiac disease complicated by renal and hepatic failure. Ischaemic lesions along the cerebral arterial boundary zones were present in the nine patients dying with neurological deficits, their extent correlating with the severity of deficit and of the intraoperative CFM abnormalities. In the 11 patients without deficits, the brains were macroscopically normal and their intraoperative CFMs were classified as normal ($n = 7$), or as showing minor changes or low amplitude ($n = 4$), or a depression of less than 7 min duration ($n = 2$). This confirms the view that relatively brief loss of EEG amplitude, usually short of ECS, with rapid recovery may not be associated with either clinical or neuropathological evidence of ischaemic brain damage. Experimental studies support this view, suggesting that reduction in total EEG power or increase in the duration of suppressions has to reach a critical threshold before it is quantitatively associated with severity of neurological deficit (Mezrow et al 1995) of the typical cerebral arterial boundary zone areas of infarction that follow hypoperfusion during cardiac surgery (Brierley 1967, Brierley et al 1980, Brierley and Graham 1984).

Intraoperative factors affecting the EEG and incidence of brain damage

Arterial hypotension during CPB was considered to be the direct cause of EEG change in 15 patients in the study of Stockard et al (1974); eight of these patients subsequently

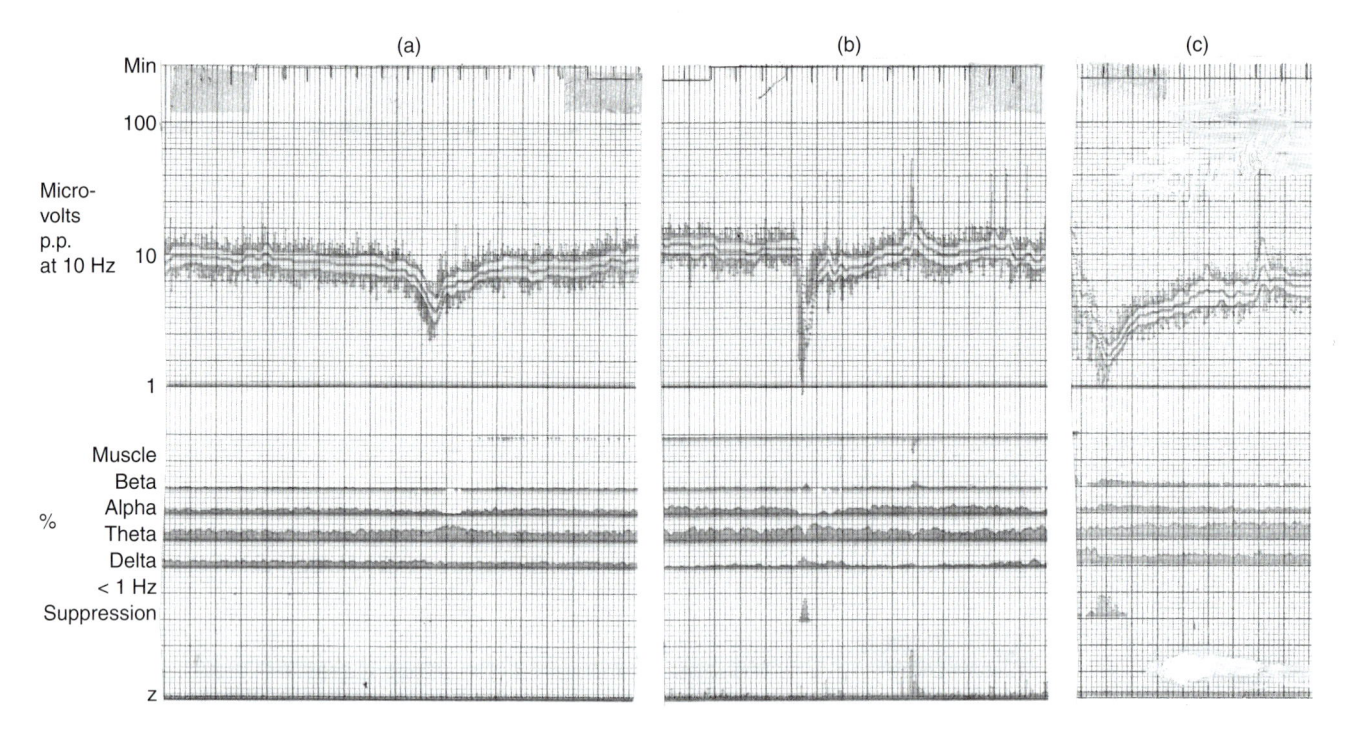

Fig. 7.26 CFAM1 display during cardiac surgery in a 55-year-old man undergoing coronary artery vein grafting and repair of a left ventricular aneurysm. Extracts show: (a) fall in CFAM amplitude with associated decrease in percent alpha and theta activities, and increase in per cent delta activity as attempts were made to wean the patient off CPB; (b) more marked CFAM changes with an episode of ventricular fibrillation and DC shock just after sternal wires were placed during chest closure; (c) a further event after reopening the chest followed by gradual recovery of the CFAM trace with direct cardiac massage and reconnection to CPB. Note movement artefacts in (b) and the registration of suppressions at the time of maximal EEG depression in (b) and (c). The patient exhibited transient, mild, neurological problems postoperatively. (From Binnie et al (2003), by permission.)

showed postoperative neurological deficits accompanied by persisting EEG abnormalities of proportionate severity. Advanced age and previous cerebrovascular insufficiency were considered risk factors. The relative merits of non-pulsatile and pulsatile flow during CPB in terms of cerebral perfusion have been much discussed; there is evidence that CBF is increased during non-pulsatile flow in a majority of high-risk patients but is pressure-dependent, indicating an absence of autoregulation of CBF which requires careful EEG monitoring (Lundar et al 1985).

Arterial carbon dioxide concentration requires careful management to avoid hypocapnia and hypoperfusion; this has been advocated as a means of optimizing CBF, hence reducing EEG abnormalities during CPB together with postoperative neurological deficits. In one study of 65 patients undergoing CABG with cooling to 32°C, routine procedures were adopted in 35 patients and strict measures to ensure normocapnia (35–45 mmHg 4.7–6.0 kPa) in the remaining 30 patients (Nevin et al 1987). Neurological deficits on the third day after surgery were found in 46% of the former and 27% of the latter groups whilst psychometric deficits were present in 71% and 40% respectively. On review of the operative data it was clear that more than half of the 'routine' group had been hypocapnic immediately before onset of CPB; those with deficits had greater changes in $PaCO_2$ and lower cerebral perfusion pressures (usually due to a rise in central venous pressure) in the early part of CPB. In spite of several studies, the matter remained controversial (Bashein et al 1990, Prough et al 1990) until discussion shifted to reduction of neuropsychological impairment by careful choice of mechanisms for acid–base regulation (Patel et al 1996).

Hypothermia, with or without addition of high-dose barbiturates has a clearer record for reduction of the risk of ischaemic brain damage. Management by EEG control with an end-point of ECS before prolonged circulatory arrest has been considered a safe and reliable procedure, permitting definition of appropriate temperatures for brain protection in individual patients (Coselli et al 1988, Mizrahi et al 1989). There has been some suggestion that avoidance of hypothermia might reduce adverse effects on the myocardium. A large randomized study compared normothermic (33–37°C) and hypothermic (25–30°C) perfusion in 1732 patients undergoing isolated coronary bypass surgery in three adult cardiac surgery centres (Warm Heart Investigators 1994). Although there were some differences in cardiac factors, there were none in the rates of 'stroke'.

SEPs in cardiac surgery

SEPs to median nerve stimulation have been monitored during cardiopulmonary bypass in adults (Arén et al 1985, Hume and Durkin 1986) and children (Coles et al 1984). However, the possibility of preventing cerebral ischaemia by monitoring SEPs during cardiac surgery depends upon reliable identification of temperature effects, because the operations are often performed under hypothermia. Some have reported a linear relationship between temperature and SEP latency during both cooling and rewarming (Kopf et al 1985, Yang et al 1995), whereas others find that latency changes during rewarming are not simply the reverse of those produced by cooling, and that there is evidence of hysteresis (Markand et al 1990, Zeitlhofer et al 1990). Whether SEP monitoring will contribute to reducing neurological complications associated with cardiac surgery has yet to be determined.

Carotid endarterectomy
Historical background

Since the 1970s, EEG monitoring has been considered a valuable monitoring tool during carotid endarterectomy, guiding decisions on the need for placement of a shunt to provide a bypass for supply of oxygenated blood to the ipsilateral cerebral hemisphere during carotid clamping (Sundt et al 1977, 1981, Chiappa et al 1979, Rampil et al 1983). EEG abnormalities appear when carotid clamping reduces regional cerebral blood flow (rCBF) to 18–24 ml/100 g/min or below and are considered a more accurate indicator of regional cerebral ischaemia than carotid stump pressure (McKay et al 1976). The cost benefits of EEG monitoring have been assessed by Green et al (1985) on the basis of evidence from 562 carotid endarterectomies. They found that the benefit of relatively expensive EEG monitoring may be mainly in the avoidance of the more expensive risks of stroke from technical problems with shunting in those who do not show signs of ischaemia on placement of a carotid clamp. In more recent years there have been some new explorations of the performance of measures for monitoring during carotid surgery, such as stump pressure (Cherry et al 1991, Harada et al 1995), transcranial Doppler for warning of microembolism (Jorgensen and Schroeder 1992, Jansen et al 1993) and near-infrared spectroscopy (De Letter et al 1998, Kirkpatrick et al 1998) compared with that of the 'gold standard' of EEG. Contemporary practice has been reviewed by Ackerstaff and Van de Vlasakker (1998).

Surgical approach and place of monitoring

Since endarterectomy involves clamping the carotid artery, there is a risk of cerebral infarction if blood flow to the ipsilateral hemisphere is not maintained through the circle of Willis. The overall complication rate is about 3.5%, with a rate of about 10% in high risk patients (Sundt et al 1981), although some surgeons have reported rates as high as 20% (Easton and Sherman 1977). However, not all the complications are associated with carotid clamping, since a proportion of strokes are produced by either carotid thrombosis, cerebral haemorrhage resulting from hyperperfusion after endarterectomy or emboli which may be dislodged from an atheromatous artery at any stage of the procedure (Steed et al 1982). Myocardial infarction also contributes to the perioperative morbidity and mortality, related to the high prevalence of ischaemic heart disease among patients with carotid atheroma (Riles et al 1979, Sundt et al 1981). Inadequate perfusion during carotid clamping is therefore only one of several possible sources of operative morbidity, but it differs from the others in being preventable by using a shunt while the artery is clamped.

There are three different policies with respect to shunt insertion during carotid endarterectomy (Sandmann et al 1993a,b, Murie et al 1994); each has enthusiastic followers, but none has been shown to be superior (Counsell et al 2000). Many surgeons routinely insert a shunt (e.g. Browse and Ross Russell 1984), but this has the disadvantage of increasing both the difficulty of operation and the risk of embolization, and may therefore cause strokes in some patients with adequate flow through the circle of Willis (Salvian et al 1997). Others never use a shunt, in the belief that most intraoperative strokes are embolic (Whitney et al 1980, Ferguson 1982, Boontje 1994). However, some have modified their 'no shunt' policy because occasionally patients develop hemiparesis after having shown neurophysiological evidence of ischaemia while the artery was clamped (Ferguson 1986, Spetzler et al 1986). A widely accepted view is, therefore, that although only a minority of strokes during carotid endarterectomy are due to inadequate blood flow, strokes with this aetiology should be preventable if perfusion failure can be detected at an early stage (Sundt 1983, Ojemann and Heros 1986). It is important to note, however, that only two randomized controlled trials of shunting have been performed (Gumerlock and Neuwelt 1988, Sandmann et al 1993a,b). Both studies compared routine shunting with no shunt, but the results were inadequate to determine which approach gave better results (Counsell et al 2000). Selective shunting has not yet been compared in a randomized controlled trial with either routine shunting, or with no shunt.

If a policy of selective shunting is adopted, inadequate cerebral perfusion must be identified during carotid occlusion. Several methods have been used (reviewed by Loftus and Quest 1995), including: operating under local anaesthesia with assessment of mental performance and contralateral motor function (Lee et al 1988, Benjamin et al 1993); measurement of CBF by an isotope method (Sundt et al 1974); pressure measurement in the carotid artery distal to the clamp ('stump pressure'); determining flow velocity in the middle cerebral artery by transcranial Doppler ultrasonography (Gaunt 1998); assessing cerebral oxygenation by near-infrared spectroscopy (Kirkpatrick 1997); and finally recording either EEG (Nuwer 1993, McGrail 1996) or SEPs. Many surgeons use two or more monitoring methods simultaneously. Only one randomized trial has compared results obtained by different monitoring policies; this demonstrated that there was no difference in stroke rate between patients selected for shunting on the basis of stump pressure plus EEG, compared with stump pressure alone; the possible contribution of the SEP was not investigated (Fletcher et al 1988, Counsell et al 2000). On the other hand, a retrospective review found that shunting selectively on the basis of EEG monitoring gave better results than routine shunting (Salvian et al 1997).

Value of EEG monitoring

The main role of EEG monitoring during carotid endarterectomy is to increase the safety of the procedure by providing early warning of ischaemic changes, particularly at the time of clamping, in order that a shunt may be placed to prevent postoperative stroke. Thus, it is necessary to know the sensitivity and specificity of the EEG as a monitoring method and the overall effect on outcome when it is used. Several large series, both retrospective and prospective, have addressed these questions and provided data for sensitivity of the EEG, often in comparison with other monitoring techniques. The majority provide clear evidence in support of the role of EEG in preventing adverse outcomes (Zampella et al 1991, Facco et al 1992, Redekop and Ferguson 1993, Deriu et al 1994, Plestis et al 1997, Salvian et al 1997, Julia et al 1998). Some reports suggest limitations of varying degrees and causation, including false negatives, possibly due to inadequate EEG monitoring techniques (Rosenthal et al 1981, Silbert et al 1989, Kresowik et al 1991, McCarthy et al 1996, McKinsey et al 1996). There are a few randomized trials of EEG monitoring during carotid surgery, but Fletcher et al (1988) report a study in which 131 patients undergoing 142 endarterectomies were randomly allocated as to whether or not operative EEG monitoring was used. They found significantly fewer neurological defects in the monitored group and concluded that EEG monitoring is 'useful in identifying patients requiring shunting during carotid endarterectomy'. They recommended use of a shunt if there was EEG change regardless of stump pressure and, conversely, if stump pressure was low but without EEG change, that it would appear safe to proceed without shunting. In a multicentre audit of 3328 carotid endarterectomies performed in 1981, when patients who were not monitored during surgery were compared with those who had EEG monitoring, there was found to be a significant statistical difference in favour of the EEG group (Fode et al 1986). A review of 1661 carotid endarterectomies using the EEG as a criterion for the need for a shunt also endorsed the significant role of the EEG in stroke prevention (Pinkerton 2002). However, when the European Carotid Surgery Trialists Group (Bond et al 2002) made a survey of surgical practice, they were unable to demonstrate a statistically significant association between operative risk and intraoperative EEG, or any other form of monitoring. However, they considered that the considerable variation between surgeons, and between countries, in the use of ancillary techniques was in-keeping with the lack of convincing data from randomized controlled trials, and suggested that there was still sufficient uncertainty to make large pragmatic trials possible

The requirement for neurophysiological monitoring during carotid endarterectomy has increased substantially since the major American and European clinical trials (North American Symptomatic Carotid Endarterectomy Trial Collaborators (NAS-CET) 1991, European Carotid Surgery Trialists' Collaborative Group (ECST) 1991, Executive Committee for the Asymptomatic Carotid Atherosclerosis Study (ACAS) 1995) demonstrated the value of such surgery in selected symptomatic and asymptomatic patients with severe internal carotid artery stenosis. The efficacy of both EEG and SEP monitoring have been reviewed by Fisher et al (1995) on the basis of a meta-analysis of 3028 patients undergoing carotid endarterectomy. This work was undertaken against a background of previous data suggesting a perioperative stroke rate of 2% and death rate of 0.8–1.5%. Although placement of a shunt during surgery on the carotid artery reduces risk of impaired blood flow, it also has adverse effects; thus an accurate method for predicting the need for a shunt is required. They concluded that the EEG was a better predictor of the need for a shunt than stump pressure and that more patients might have been damaged had the carotid blood supply not been maintained by shunting. Whilst regretting the lack of properly controlled prospective studies of outcome, Fisher and his colleagues also concluded that the weight of evidence suggests that loss of spontaneous EEG and SEP correlates well with critical reduction in CBF and that these are useful indicators to monitor.

A subsidiary role for the EEG concerns preoperative screening which, although strictly beyond the remit of this section, deserves brief mention because of its value as a functional marker of ischaemic changes during carotid compression balloon testing (Chiesa et al 1992, Cloughesy et al 1993, Herkes et al 1993). The preoperative multichannel EEG may also provide evidence suggestive of pre-existing localized vascular disorders; these were found in one-third of the 303 patients screened by Illig et al (1998). Whilst they were highly predictive of changes with anaesthesia, they did not predict adverse vascular events during surgery.

Intraoperative factors affecting the EEG and incidence of stroke

During carotid endarterectomy, the position of EEG monitoring may be influenced by considerable variation in anaesthetic protocols. Some protagonists take the view that it is desirable to avoid general anaesthesia to permit clinical evaluation of impending neurological deficits (Benjamin et al 1993, Fiorani et al 1997). This has implications for intraoperative EEG monitoring; indeed Stoughton et al (1998) found that EEG monitoring yielded a significant number of false positive (6.7%) and false negative (4.5%) results in the detection of neurological deficits when compared with mental state examination in awake patients managed with regional anaesthesia. This experience was shared by Wellman et al (1998) who concluded that EEG monitoring could be insensitive and might fail to detect ischaemia in patients under regional anaesthesia whilst the presence of general anaesthetics might alter the character of the EEG findings, increasing the sensitivity of EEG monitoring to ischaemic events. However, the consensus view of anaesthetists in the USA is that general anaesthesia should be used (85% of respondents) rather than regional anaesthesia (15%). Intraoperative monitoring was reported by 90%, with EEG the favoured modality in 61%. Specific neuroprotective measures were used by 22%, barbiturates being chosen by half of the anaesthetists, and with elective intraoperative hypertension to preoperative baseline plus 20% to ensure adequate CBF (Cheng et al 1997). Preoperative blood volume expansion combined with barbiturate-induced burst suppression has also been used and in one study this reduced lateralized EEG abnormalities with clamping, an effect which persisted when blood pressure was raised, in 63% (Gross et al 1998).

The type of anaesthetic agent chosen may affect the relationship between stump pressure, rCBF and EEG. Thus McKay et al (1976) found that cerebral vasodilatation with halothane and enflurane, but vasoconstriction with neuroleptic anaesthesia, led to differences in rCBF and EEG with and without ischaemic changes. With isoflurane, the critical level of rCBF for onset of EEG signs of ischaemia was less than with halothane (Messick et al 1987); and the incidence of ischaemia was reduced when isoflurane was compared with enflurane and halothane (Michenfelder et al 1987). Using 0.6–1.2% sevoflurane in 50% nitrous oxide and a logistic regression analysis in 52 patients, Grady et al (1998) were able to determine $rCBF_{50}$ as 11.5 ± 1.4 ml/100 g/min; they reported that with this anaesthetic regimen they were able to accurately and rapidly detect ischaemic EEG changes due to carotid occlusion. Two different anaesthetic regimens (isoflurane *versus* propofol) resulted in different patterns of EEG change with carotid clamping in 152 patients by Laman et al (2001); for both regimens, information from the anterior head regions was the most informative in this particular study (*Fig. 7.27*).

Cerebral protection regimens, which have been reviewed on page 255, have some specific implications for EEG monitoring during carotid surgery, since some authors suggest that high-dose thiopental may avoid the need for shunting in high-risk patients (Hicks et al 1986, Frawley et al 1997). The latter group found that the correlation between stump pressure and EEG signs of ischaemia fell from the usually accepted 100% to 27%, presumably because the expected ischaemic EEG changes were 'prevented by thiopental' *pari passu* with the prevention of strokes.

The occurrence of late changes due to, for example, occluded shunts (Artru and Strandness 1989) or early postoperative thrombosis, means that monitoring should continue in the recovery room, at least overnight.

Assessment of significant changes in the EEG

Methodology for analysis during EEG monitoring for carotid surgery has varied from multichannel paper recordings to advanced quantitative EEG systems. Essential features are separate recording and displays from the vulnerable areas of each cerebral hemisphere and methods suitable for detection of significant asymmetries, decrease in total power, burst suppression or ECS and intermittent localized features such as periodic lateralized epileptiform discharges or zeta waves. Some groups have advocated computerized topographic mapping to give adequate cover nearer to their 'gold standard' of a multichannel conventional EEG (Ahn et al 1988) and most emphasize that appropriate qEEG methods are more efficient at identifying significant changes than visual assessment of a multichannel EEG. It is, however, clear that

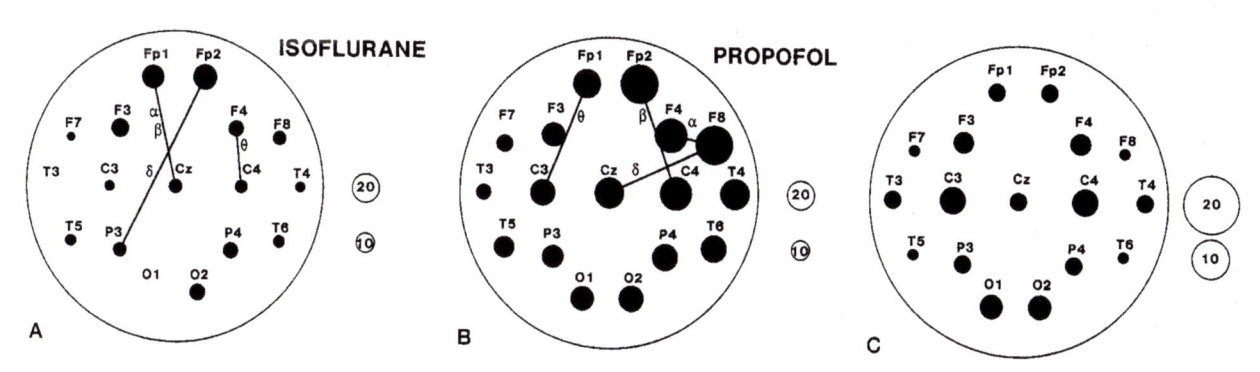

Fig. 7.27 Optimal electrode positions found by receiver operating characteristics curves for isoflurane (A) and propofol (B) anaesthesia. (C) Electrode positions found in the literature. The area of the circle on each electrode position is proportional to the frequency in which that electrode position was involved in the 20% highest ranking derivations in each frequency band. Lines in A and B are the highest ranking derivations for the four indicated frequency bands. Note the different scaling in C compared with A and B. (From Laman DM, van der Reijden CS, Wieneke GH, van Duijn H, van Huffelen AC 2001 EEG evidence for shunt requirement during carotid endarterectomy. Optimal EEG derivations with respect to frequency bands and anesthetic regimen. *Journal of Clinical Neurophysiology* 18:353–363. Lippincott, Williams and Wilkins Inc., Philadelphia. © 2001 American Clinical Neurophysiology Society, by permission.)

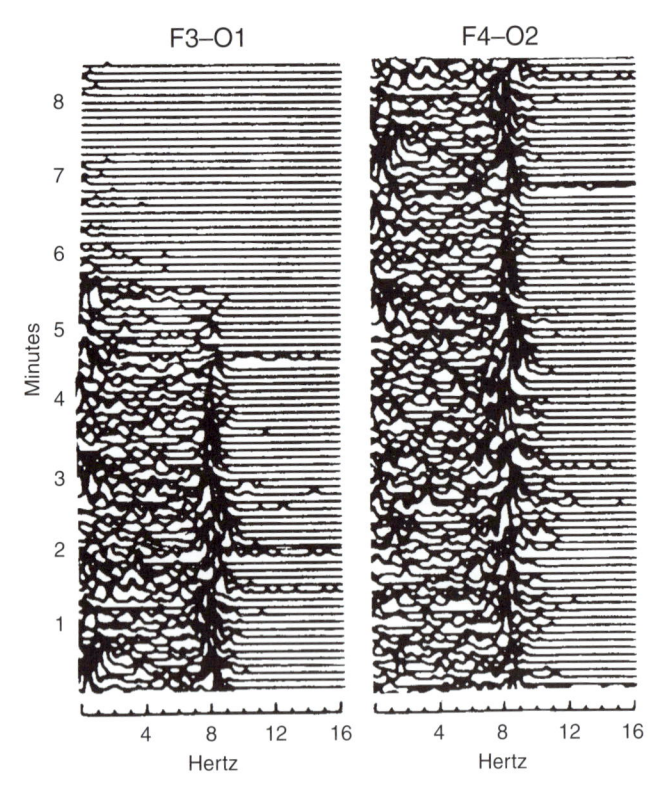

Fig. 7.28 Carotid artery surgery. CSA from wide-spaced left (F3–Ol) and right (F4–O2) derivations during test clamping (at arrow) of the left internal carotid artery prior to endarterectomy. Deterioration of left-sided EEG activity, comprising loss of faster components and overall power, led to placement of a vascular shunt to maintain internal carotid flow during surgery. (From Myers RR, Stockard JJ, Saidman LJ 1977 Monitoring of cerebral perfusion during anesthesia by time-compressed Fourier analysis of the electroencephalogram. *Stroke* 8:331–337, by permission.)

even simple two-channel systems can display readily understandable signs of unilateral deterioration of the EEG (*Fig. 7.28*) and also that a qEEG-based brain symmetry index can assist in detecting subtle ischaemic changes not apparent on visual inspection of the EEG (van Putten et al 2004).

Methods chosen for qEEG monitoring must be demonstrably able to:

1. provide immediate detection of any significant deviation from the steady state;
2. be able to detect changes in the presence of bilateral carotid or cerebrovascular disease (note that the use of monopolar derivations referred to a single midline electrode such as Cz is not suitable for detection of subtle asymmetries);
3. reliably detect the changes that are likely to be early warning EEG signs of cerebral ischaemia (i.e. asymmetries, localized changes including reduction of fast and increase in slow components);
4. detect changes that specifically predict a high probability of an adverse neurological outcome (i.e. lateralized burst suppression or major reduction in total EEG power); and
5. match the gold standard of multichannel EEG interpreted by experts.

Definitive standards are outlined in the international recommendations concerning methodology for neurophysiological monitoring that have been set out by the International Federation for Clinical Neurophysiology (IFCN) (Burke et al 1999); the criteria for significant change during monitoring to detect cerebral ischaemia are detailed in *Table 7.3*. Although changes are most commonly ipsilateral to the carotid artery clamped, it is essential to recognize that bilateral changes may follow

Table 7.3 IFCN criteria of changes in EEG suggestive of impaired brain perfusion and severe ischaemia during carotid artery surgery

Feature	Significance
Greater than 50% loss of fast activity	Suspicious
Unilateral loss of fast activity	Alarming
Greater than 50% increase in background slow activity	Suspicious
Unilateral or focal increase in slow activity	Sign of impairment
Development of new asymmetry or localized change	Sign of impairment (providing electrode contact problems excluded)
Flattening (localized or generalized); Loss of overall amplitude (localized or generalized) Appearance of burst suppression pattern (localized or generalized)	Often the principal changes with impaired perfusion
Abrupt loss of EEG activity across all frequency bands	Ominous sign often associated with severe ischaemia

Information tabulated from Burke et al (1999).
Note: EEG may change bilaterally even after unilateral vascular clamping; this should not be overlooked, especially by persons who try to compare the EEG from the two hemispheres as the principal measure of monitoring.

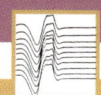

unilateral clamping (with or without contralateral stenosis). False negatives are unusual, providing strict criteria are used (Ballotta et al 1997).

There is little evidence that the easily detected early warning signs of decrease in higher frequencies and increase in slower frequencies alone correlate with actual neurological damage – indeed experimental work indicates that we should expect complete extinction of the EEG and the SEP in the region of local ischaemia which, unless reversed, will be associated with neurological damage (Astrup et al 1977, Morawetz et al 1979, Jones et al 1981). Clinical work confirms that patients do not experience new neurological deficits after carotid endarterectomy unless there is marked voltage attenuation (Chiappa et al 1979) that persists for at least 10 min (Rampil et al 1983). This time delay before permanent damage occurs provides the window of opportunity for remedial action which forms the whole basis of the use of intraoperative monitoring. It will thus be evident that early warning signs (e.g. van Putten et al 2004) are useful just as a warning that the major change of marked depression or extinction of the EEG indicates that damage is imminent, and that its reversibility will herald the likelihood of successful prevention of permanent neurological sequelae.

Choice of methods for qEEG monitoring during carotid artery surgery

In essence, the qEEG design requirement is for rapidly responding tracking (trend detection) methods which can indicate a significant change within a matter of seconds when the carotid clamp is applied. This is where the time domain has an advantage over frequency domain methods with their inherent delay for averaging – a disadvantage that also applies to SEP monitoring. Barlow (1984) compared elective analogue filtering, matched inverse filtering and automatic adaptive segmentation for automated

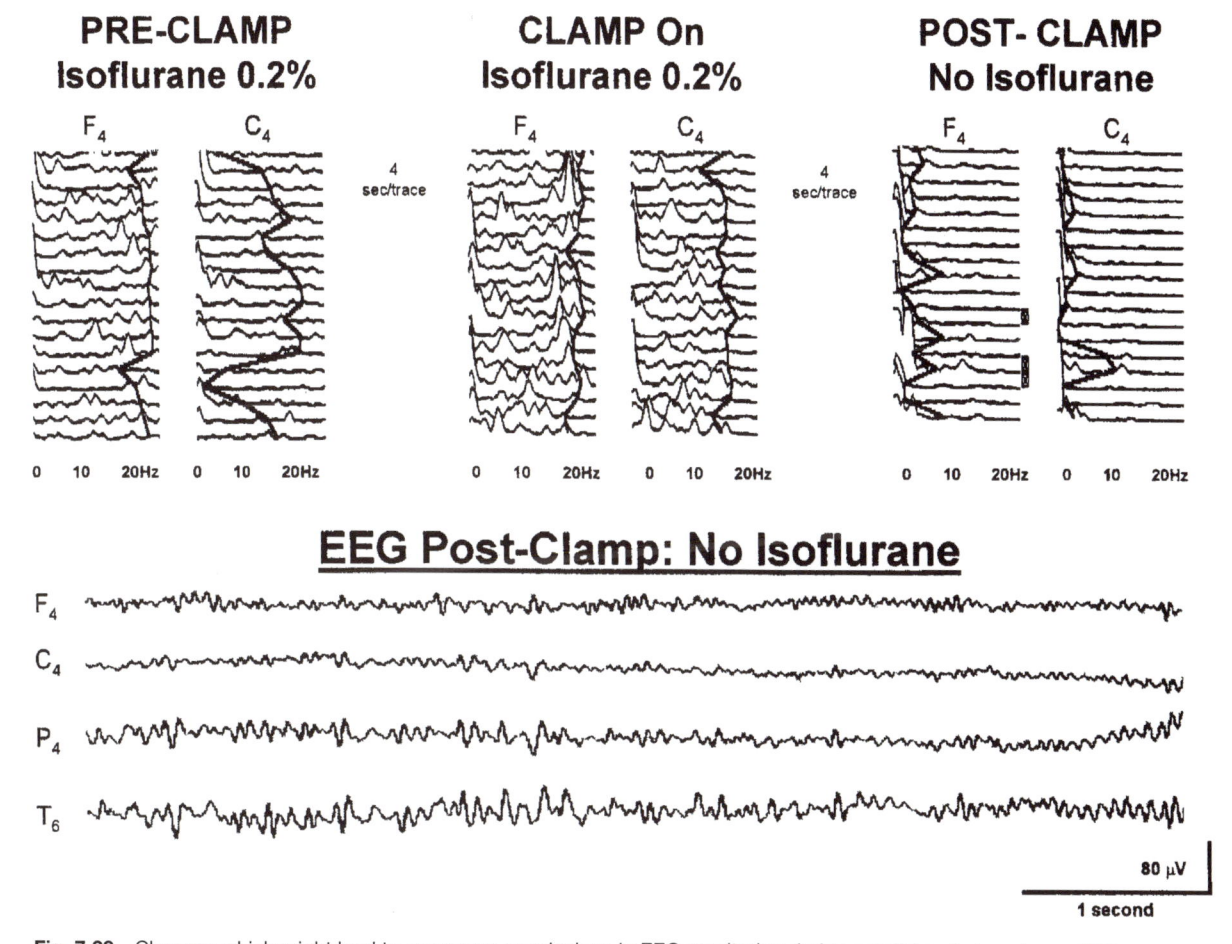

Fig. 7.29 Changes which might lead to erroneous conclusions in EEG monitoring during carotid endarterectomy. CSAs pre-, on and postclamping of the right internal carotid artery and EEG in the postclamping period during wound closure. Monopolar derivations with Cz reference. Note that isoflurane, already at a low end-tidal concentration, had been withdrawn before the post-clamp sample – the patient was probably in a drowsy state near to wakening. The drop in bilateral EEG power on the CSA post-clamping is consistent with this state and, in view of the lack of initial warning signs or asymmetry, would be an unlikely finding in localized postendarterectomy cerebral ischaemia. (From Heyer EJ, Adams DC, Moses C, Quest DO, Connolly ES 2000 Erroneous conclusion from processed electroencephalogram with changing anesthetic depth. *Anesthesiology* 92:603–607, by permission.)

evaluation and signalling of significant EEG changes at the time of carotid clamping and showed a clear advantage for automatic adaptive segmentation for this purpose. He also remarked on his incidental finding that a low-speed write-out of the analogue filter outputs provided a good visual indication of EEG changes.

Comparative studies between qEEG methods and visual assessment of the EEG show that power changes tend to be more sensitive than spectral values, although the latter show wide ranges of sensitivity and specificity depending on the observer (Young et al 1988, Hanowell et al 1992, Kearse et al 1993). In an important study, Visser et al (1999) have examined spectral EEG changes in 94 patients monitored during carotid surgery to define indicators for shunting by best reflecting changes in blood pressure and flow velocity in the middle cerebral artery. These represented (a) a change in power in the alpha and beta frequency ranges in combination with a less pronounced opposite change in power in the delta frequency range, and (b) a change in power restricted to the delta and theta frequency ranges. The first factor distinguished two types of spectral EEG change indicative of cerebral ischaemia (decrease in fast activity and increase in slow activity) and arousal (the opposite changes). With the two factors combined, the changes indicative of minor ischaemia (decrease in fast activity only) could also be distinguished; their material has not yet been adequate to permit definition of more severe ischaemic changes.

Whatever method is chosen, it is important that interpretation is based on sound experience of clinical neurophysiology. Heyer et al (2000) have pointed out that if a patient wakes up during monitoring (in their example the anaesthetic had finished), the monitored changes of 'attenuation' (*Fig. 7.29*) could be misinterpreted as evidence of major depression due to cerebral ischaemia by the naive observer.

Recent developments include more advanced methods capable of incorporating decision systems derived from

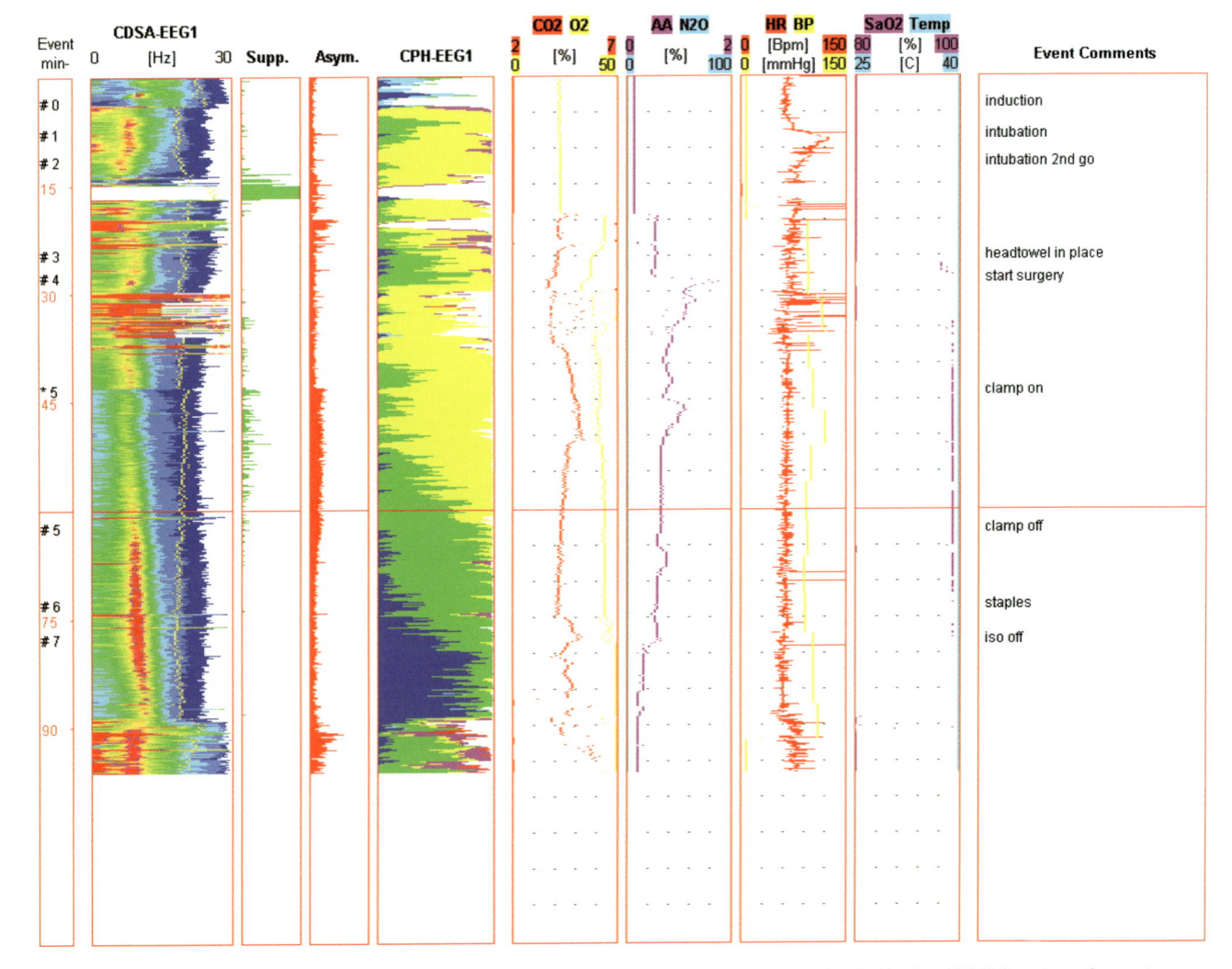

Fig. 7.30 Monitoring during left carotid endarterectomy using the Advanced Depth of Anaesthesia Monitor (ADAM) system. General features of display as described in *Fig. 3.13*. Note appearance of EEG suppressions and increase in pre-existing asymmetry at the time of decrease in alpha–beta activity during carotid clamping. (Courtesy of Dr RM Langford, Dr CE Thomsen and Professor J Lumley. From Binnie et al (2003), by permission)

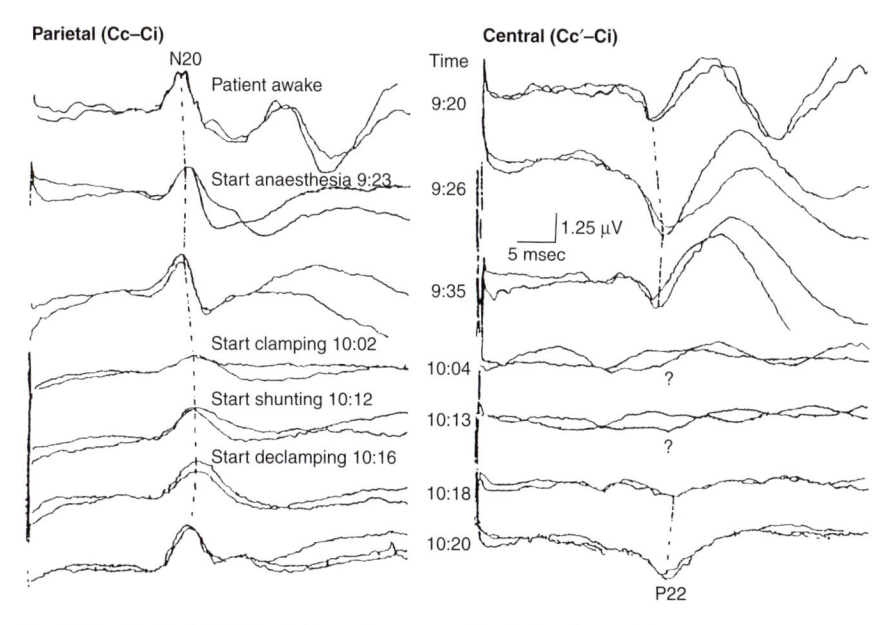

Fig. 7.31 Cortical SEPs to right median nerve stimulation during left carotid endarterectomy. The traces on the left were recorded from a position 2 cm behind C3, reference 2 cm behind C4, and those on the right from 2 cm in front of C3, reference 2 cm in front of C4. Clamping the carotid artery is followed by a reduction in amplitude and increase in latency of the N20 and P22 components. After inserting a shunt, N20 partially recovered, but P22 recovered more slowly. Postoperatively the patient had a right hemiparesis and mild motor aphasia, both of which recovered in the following 3 days. (From Gigli et al (1987), by permission.)

knowledge bases (*Fig. 7.30*); further advances are likely in this area.

Finally, as will be evident from perusal of *SEP monitoring for ischaemia* (p. 258), it is necessary to consider the use of simultaneous EEG and SEPs for monitoring during carotid endarterectomy because (a) neither is able to detect all ischaemic changes in subcortical and cortical areas around the vulnerable arterial boundary zone areas of the middle cerebral artery territory, (b) the EEG may give the earliest warning because of the time taken to average each EP run (a continuously moving average, updated by relatively small stimulus numbers, will of course respond more quickly – see *Intraoperative and ICU EP recording*, p. 65), and (c) they are complementary during deep anaesthesia or cerebral protective regimens based on high-dose barbiturates and other agents which depress the EEG to a marked burst suppression pattern (Pozzessere et al 1987, Lam et al 1991, Fava et al 1992, Pistolese et al 1993, Haupt et al 1994, Fiori and Parenti 1995, Prokop et al 1996, Arnold et al 1997).

SEPs in carotid endarterectomy

The sensitivity of the median nerve SEP to cerebral ischaemia is well recognized (see *SEP monitoring for ischaemia*, p. 258), so this method has also been used to detect ischaemia during carotid endarterectomy. Initial results in small series show that marked attenuation of the SEP developing after application of the clamp, and persisting after its removal, indicates a strong probability of postoperative deficit (*Fig. 7.31*) (Jacobs et al 1983, Gigli et al 1987). Marked changes in the response (reduction in amplitude and increase in latency) during clamping can be reversed by shunt insertion (Markand et al 1984). In some patients the CCT is prolonged by any reduction in systemic blood pressure while the artery is occluded, suggesting a failure of cerebral autoregulation (Hargadine 1985). There is some evidence that the earliest sign of inadequate perfusion is a reduction in amplitude and increase in latency of the later components of the response (N30, etc.), rather than the CCT or N20 latency (Jacobs et al 1983, Amantini et al 1987), but the precise levels have yet to be determined at which amplitude reduction or latency increase of any component will reliably predict a postoperative deficit. While some authors report no false negatives with SEP monitoring in large series (e.g. Amantini et al 1992, Beese et al 1998, Pedrini et al 1998), others have encountered patients in whom a stroke developed in spite of no change in SEP during the operation (e.g. Prokop et al 1996, Linstedt et al 1998, Wober et al 1998). Clearly, detection of a significant change depends upon the criteria adopted (Isley et al 1998). Some workers have used complete disappearance of N20 (e.g. Schweiger et al 1991), but most use a 50% reduction in N20 amplitude, often combined with a latency criterion, although one group found that more patients who needed a shunt were identified by qualitative criteria (Guérit et al 1997, Witdoeckt et al 1997). There is sometimes a discrepancy between changes identified by EEG and SEP. For example, a 50% amplitude reduction of N20 detects some ischaemic episodes during carotid clamping not detected by EEG recording (Lam et al 1991, Fiori and Parenti 1995). On the other hand, some presumed ischaemic events are demonstrated by the EEG when the SEP shows no significant change (Kearse et al 1992). The lack of agreement arises because the SEP is affected by ischaemia confined to subcortical areas, which will not affect the EEG (Fava et al 1990), whereas the EEG covers a wider area of cortex and may detect ischaemia in watershed areas to which the SEP is insensitive (Prokop et al 1996). In any event, the two techniques (EEG and SEP) are often used together during carotid endarterectomy, and they may eventually prove to be complementary (Smith et al 1994b). The various methods of monitoring for carotid endarterectomy are reviewed by Ackerstaff and Van de Vlasakker (1998) and by Isley et al (1998).

REFERENCES

Abbott R 2002 Sensory rhizotomy for the treatment of childhood spasticity. In: *Neurophysiology in Neurosurgery. A Modern Intraoperative Approach* (eds V Deletis, JL Shils). Academic Press, Amsterdam, pp. 219–230.

Abel MF, Mubarak SJ, Wenger DR, et al 1990 Brainstem evoked potentials for scoliosis surgery: a reliable method allowing use of halogenated anesthetic agents. J Pediatr Orthop 10:208–213.

Åberg T 1974 Effect of open heart surgery on intellectual function. Scandinavian J Thorac Cardiovasc Surg 15(Suppl):1–63.

Ackerstaff RG, van de Vlasakker CJ 1998 Monitoring of brain function during carotid endarterectomy: an analysis of contemporary methods. J Cardiothorac Vasc Anesth 12:341–347.

Adams DC, Emerson RG, Heyer EJ, et al 1993 Monitoring of intraoperative motor-evoked potentials under conditions of controlled neuromuscular blockade. Anesth Analg 77:913–918.

Aguilar J, Gerbode F, Hill JD 1971 Neuropathological complications of cardiac surgery. J Thorac Cardiovasc Surg 61:676–685.

Ahn SS, Jordan SE, Nuwer MR, et al 1988 Computed electroencephalographic topographic brain mapping. A new and accurate monitor of cerebral circulation and function for patients having carotid endarterectomy. J Vasc Surg 8:247–254.

Ahonen J, Salmenperä M 2004 Brain injury after adult cardiac surgery. Acta Anaesthesiol Scand 48:4–19.

Albright AL, Sclabassi RJ 1985 Cavitron ultrasonic surgical aspirator and visual evoked potential monitoring for chiasmal gliomas in children. J Neurosurg 63:138–140.

Allen A, Starr A, Nudleman K 1981 Assessment of sensory function in the operating room utilizing cerebral evoked potentials: a study of fifty-six surgically anesthetized patients. Clin Neurosurg 28:457–481.

Amantini A, De Scisciolo G, Bartelli M, et al 1987 Selective shunting based on somatosensory evoked potential monitoring during carotid endarterectomy. Int Angiol 6:387–390.

Amantini A, Bartelli M, de Scisciolo G, et al 1992 Monitoring of somatosensory evoked potentials during carotid endarterectomy. J Neurol 239:241–247.

Anderson SK, Loughnan BA, Hetreed MA 1990 A technique for monitoring evoked potentials during scoliosis and brachial plexus surgery. Ann R Coll Surgeons Engl 72:321–323.

Anderson LC, Hemler DE, Luethke JM, et al 1994 Transcranial magnetic evoked potentials used to monitor the spinal cord during neuroradiologic angiography of the spine. Spine 19:613–616.

Andersson G, Ohlin A 1999 Spatial facilitation of motor evoked responses in monitoring during spinal surgery. Clin Neurophysiol 110:720–724.

Andrews JC, Valavanis A, Fisch U 1989 Management of the internal carotid artery in surgery of the skull base. Laryngoscope 99:1224–1229.

Andrews RJ, Bringas JR, Muizelaar JP, et al 1993 A review of brain retraction and recommendations for minimizing intraoperative brain injury [discussion]. Neurosurgery 33:1052–1063, 1063–1064.

Angelo R, Møller AR 1996 Contralateral evoked brainstem auditory potentials as an indicator of intraoperative brainstem manipulation in cerebellopontine angle tumors. Neurol Res 18:528–540.

Annotation 1987 Common carotid ligation for intracranial aneurysm. Lancet i:77–78.

Annotation 1991 Scoliosis: spinal cord monitoring during surgery. Lancet 338:219–220.

Apel DM, Marrero G, King J, et al 1991 Avoiding paraplegia during anterior spinal surgery. The role of somatosensory evoked potential monitoring with temporary occlusion of segmental spinal arteries. Spine 16(Suppl):S365–S370.

Arén C, Badr G, Feddersen K, et al 1985 Somatosensory evoked potentials and cerebral metabolism during cardiopulmonary bypass with special reference to hypotension induced by prostacyclin infusion. J Thorac Cardiovasc Surg 90:73–79.

Arfel G, Walter ST 1977 Aspects EEG au stade chirurgical d'anésthesies du longue durée en chirurgie cardiaque: effets des variation thermiques. Rev EEG Neurophysiol Clin 7:45–61.

Arfel G, Weiss J, DuBouchet N 1961 EEG findings during open-heart surgery with extra-corporeal circulation. In: *Cerebral Anoxia and the Electro-encephalogram* (eds H Gastaut, JS Meyer). CC Thomas, Springfield, IL, pp. 231–249.

Arnold M, Sturzenegger M, Schaffler L, et al 1997 Continuous intraoperative monitoring of middle cerebral artery blood flow velocities and electroencephalography during carotid endarterectomy. A comparison of the two methods to detect cerebral ischemia. Stroke 28:1345–1350.

Arrington ED, Hochschild DP, Steinagle TJ, et al 2000 Monitoring of somatosensory and motor evoked potentials during open reduction and internal fixation of pelvis and acetabular fractures. Orthopedics 23:1081–1083.

Artru AA, Strandness Jr DE 1989 Delayed carotid shunt occlusion detected by electroencephalographic monitoring. J Clin Monit 5:119–122.

Ashkenaze D, Mudiyam R, Boachie-Adjei O, et al 1993 Efficacy of spinal cord monitoring in neuromuscular scoliosis. Spine 18:1627–1633.

Astrup J, Symon L, Branston NM, et al 1977 Cortical evoked potential and extracellular K^+ and H^+ at critical levels of brain ischemia. Stroke 8:51–57.

Astrup J, Siesjö BK, Symon L 1981 Thresholds in cerebral ischemia – the ischemic penumbra. Stroke 12:723–725.

Back T 1998 Pathophysiology of the ischemic penumbra – revision of a concept. Cell Mol Neurobiol 18:621–638.

Ballotta E, Dagiau G, Saladini M, et al 1997 Results of electroencephalographic monitoring during 369 consecutive carotid artery revascularizations. Eur Neurol 37:43–47.

Barlow JS 1984 Analysis of EEG changes with carotid clamping by selective analog filtering, matched inverse filtering and automatic adaptive segmentation: a comparative study. Electroencephalogr Clin Neurophysiol 58:193–204.

Bar-On Z, Zeilig G, Blumen N, et al 1995 Paraplegia following surgical correction of scoliosis with Cotrel-Dubousset instrumentation. Bull Hosp Jt Dis 54:32–34.

References

Bashein G, Townes BD, Nessly ML, et al 1990 A randomized study of carbon dioxide management during hypothermic cardiopulmonary bypass. Anesthesiology 72:7–15.

Beck DL (ed.) 1994 *Handbook of Intraoperative Monitoring*. Singular, San Diego, CA.

Beese U, Langer H, Lang W, et al 1998 Comparison of near-infrared spectroscopy and somatosensory evoked potentials for the detection of cerebral ischaemia during carotid endarterectomy. Stroke 29:2032–2037.

Bell BA, Symon L, Branston NM 1985 CBF, time threshold for the formation of ischemic cerebral edema and effect of reperfusion in baboons. J Neurosurg 62:31–41.

Bellville JW, Artusio JF Jr, Glenn F 1955 The electroencephalogram in cardiac manipulation. Surgery 38:259–271.

Belopavlovic M, Buchthal A, Beks JW 1985 Barbiturates for cerebral aneurysm surgery. A review of preliminary results. Acta Neurochir (Wien) 76:73–81.

Ben-David B, Haller G, Taylor P 1987 Anterior spinal fusion complicated by paraplegia. A case report of a false-negative somatosensory evoked potential. Spine 12:536–539.

Benjamin ME, Silva MB, Watt C, et al 1993 Awake patient monitoring to determine the need for shunting during carotid endarterectomy. Surgery 114:673–681.

Bennett MH 1983 Effects of compression and ischemia on spinal cord evoked potentials. Exp Neurol 80:508–519.

Bernard JM, Péréon Y, Fayet G, Guiheneuc P 1996 Effects of isoflurane and desflurane on neurogenic motor- and somatosensory-evoked potential monitoring for scoliosis surgery. Anesthesiology 85:1013–1019.

Berthold C-H, Carlstedt T, Corneliuson O 1984 Anatomy of the nerve root at the central–peripheral transitional region. In: *Peripheral Neuropathy*, 2nd edition, vol. 1 (eds PJ Dyck, PK Thomas, EH Lambert, R Bunge). WB Saunders, Philadelphia, pp. 156–170.

Binnie CD, Cooper R, Mauguière F, et al (eds) 2003 *Clinical Neurophysiology*. Volume 2: *EEG, Paediatric Neurophysiology, Special Techniques and Applications*. Elsevier, Amsterdam.

Bischoff P, Kochs E, Haferkorn D, et al 1996 Intraoperative EEG changes in relation to the surgical procedure during isoflurane–nitrous oxide anesthesia: hysterectomy versus mastectomy. J Clin Anesth 8:36–43.

Blair EA, Teeple E, Sutherland RM, et al 1994 Effect of neuromuscular blockade on facial nerve monitoring. Am J Otol 15:161–167.

Blaisdell FW, Cooley DA 1962 The mechanism of paraplegia after temporary thoracic aortic occlusion and its relationship to spinal fluid pressure. Surgery 51:351–355.

Blight A, Young W 1990 Axonal and morphometric correlates of evoked potentials in experimental spinal cord injury. In: *Neural Monitoring* (ed SK Salzman). Humana Press, Clifton, NJ, pp. 87–113.

Bond R, Warlow CP, Naylor AR, et al 2002 European Carotid Surgery Trialists' Collaborative Group. 2002 Variation in surgical and anaesthetic technique and associations with operative risk in the European carotid surgery trial: implications for trials of ancillary techniques. Eur J Vasc Endovasc Surg 23:117–126.

Boniface SJ, Antoun N. 1997 Endovascular electroencephalography: the technique and its application during carotid amytal assessment. J Neurol Neurosurg Psychiatry 62:193–195.

Boontje AH 1994 Carotid endarterectomy without a temporary indwelling shunt: results and analysis of back pressure measurements. Cardiovasc Surg 2:549–554.

Boyd SG, Rothwell JC, Cowan JMA, et al 1986 A method of monitoring function in corticospinal pathways during scoliosis surgery with a note on motor conduction velocities. J Neurol Neurosurg Psychiatry 49:251–257.

Branston NM, Symon L, Crockard HA, et al 1974 Relationship between the cortical evoked potential and local cortical blood flow following acute middle cerebral artery occlusion in the baboon. Exp Neurol 45:195–208.

Branston NM, Strong AJ, Symon L 1977 Extracellular potassium activity, evoked potential and tissue blood flow. J Neurol Sci 32:305–321.

Branthwaite MA 1972 Neurological damage related to open-heart surgery. Thorax 27:748–753.

Branthwaite MA 1975 Prevention of neurological damage during open-heart surgery. Thorax 30:258–261.

Brauer M, Knuettgen D, Quester R, et al 1996 Electromyographic facial nerve monitoring during resection for acoustic neurinoma under moderate to profound levels of peripheral neuromuscular blockade. Eur J Anaesthesiol 13:612–615.

Breuer AC, Furlan AJ, Hanson MR, et al 1983 Central nervous system complications of coronary artery bypass graft surgery: prospective analysis of 421 patients. Stroke 14:682–687.

Brewer LA, Fosburg RG, Mulder GA, et al 1972 Spinal cord complications following surgery for coarctation of the aorta. J Thorac Cardiovasc Surg 64:368–379.

Brierley JB 1967 Brain damage complicating open-heart surgery: a neuropathological study of 46 patients. Proc R Soc Med 60:858–859.

Brierley JB, Graham DI 1984 Hypoxia and vascular disorders of the central nervous system. In: *Greenfield's Neuropathology*, 4th edition (eds J Hume Adams, JAN Corsellis, LW Duchan). Arnold, London, pp. 125–207.

Brierley JB, Prior PF, Calverley J, et al 1980 The pathogenesis of ischaemic neuronal damage along the cerebral arterial boundary zones. In: *Papio anubis*. Brain 103:929–965.

Brodkey JS, Richards DE, Blasingame JP, et al 1972 Reversible spinal cord trauma in cats. Additive effects of direct pressure and ischemia. J Neurosurg 37:591–593.

Brookes GB, Woo J 1994 Hearing preservation in acoustic neuroma surgery. Clin Otolaryngol 19:204–214.

Brown RH, Nash CL 1985 The 'grey zone' in intra operative SCEP monitoring. In: *Spinal Cord Monitoring* (eds J Schramm, SJ Jones). Springer-Verlag, Berlin, pp. 179–185.

Brown WF, Veitch JP 1994 AAEM Minimonograph no. 42. Intraoperative monitoring of peripheral and cranial nerves. Muscle Nerve 17:371–377.

Browse NL, Ross Russell R 1984 Carotid endarterectomy and the Javid shunt: the early results of 215 consecutive operations for transient ischaemic attacks. Br J Surg 71:53–57.

Buchthal A, Belopavlovic M 1992 Somatosensory evoked potentials in cerebral aneurysm surgery. Eur J Anaesthesiol 9:493–497.

Burke DJ, Hicks RG 1997 Intraopertive monitoring with motor and sensory evoked potentials. In: *Evoked Potentials in Clinical Medicine*, 3rd edition (ed KH Chiappa). Lippincott-Raven, Philadelphia, pp. 675–688.

Burke D, Hicks RG 1998 Surgical monitoring of motor pathways. J Clin Neurophysiol 15:194–205.

Burke D, Hicks RG, Stephen JPH, et al 1992 Assessment of corticospinal and somatosensory conduction simultaneously during scoliosis surgery. Electroencephalogr Clin Neurophysiol 85:388–396.

Burke D, Nuwer MR, Daube J, et al 1999 Intraoperative monitoring. In: *Recommendations for the Practice of Clinical Neurophysiology. Guidelines of the International Federation of Clinical Neurophysiology* (EEG Suppl 52) (eds G Deuschl, A Eisen). Elsevier Science, Amsterdam, pp. 133–148.

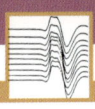

Butler S, Coakham H, Maw R, et al 1995 Physiological identification of the auditory nerve during surgery for acoustic neuroma. Clin Otolaryngol 20:312–317.

Calancie B, Harris W, Broton JG, et al 1998 'Threshold-level' multipulse transcranial electrical stimulation of motor cortex for intraoperative monitoring of spinal motor tracts: description of method and comparison to somato-sensory evoked potential monitoring. J Neurosurg 88:457–470.

Calancie B, Harris W, Brindle GF, et al 2001 Threshold-level repetitive transcranial electrical stimulation for intraoperative monitoring of central motor conduction. J Neurosurg 95(Suppl 2):161–168.

Cedzich C, Schramm J, Fahlbusch R 1987 Are flash-evoked visual potentials useful for intraoperative monitoring of visual pathway function? Neurosurgery 21:709–715.

Cedzich C, Taniguchi M, Schafer S, et al 1996 Somatosensory evoked potential phase reversal and direct motor cortex stimulation during surgery in and around the central region. Neurosurgery 38:962–970.

Chang SD, Lopez JR, Steinberg GK 1999 Intraoperative electrical stimulation for identification of cranial nerve nuclei. Muscle Nerve 22:1538–1543.

Chatrian GE, Berger MS, Wirch AL 1988 Discrepancy between intraoperative SSEPs and postoperative function. J Neurosurg 69:450–454.

Cheek JC 1993 Posterior fossa intra-operative monitoring. J Clin Neurophysiol 10:412–424.

Cheng MA, Theard MA, Tempelhoff R 1997 Anesthesia for carotid endarterectomy: a survey. J Neurosurg Anesthesiol 9:211–216.

Cherry Jr KJ, Roland CF, Hallett Jr JW, et al 1991 Stump pressure, the contralateral carotid artery, and electroencephalographic changes. Am J Surg 162:185–188.

Chiappa KH, Burke SR, Young RR 1979 Results of electroencephalographic monitoring during 367 carotid endarterectomies: use of a dedicated mini-computer. Stroke 10:381–388.

Chiesa R, Minicucci F, Melissano G, et al 1992 The role of transcranial Doppler in carotid artery surgery. Eur J Vasc Surg 6:211–216.

Cioni B, Meglio M, Rossi GF 1999 Intraoperative motor evoked potentials monitoring in spinal neurosurgery. Arch Ital Biol 137:115–126.

Cloughesy TF, Nuwer MR, Hoch D, et al 1993 Monitoring carotid test occlusions with continuous EEG, clinical examination. J Clin Neurophysiol 10:363–369.

Cohen BA, Huizenga BA 1988 Dermatomal monitoring for surgical correction of spondylolisthesis. Spine 13:1125–1128.

Coles JG, Wilson GJ, Sima F, et al 1982 Intraoperative detection of spinal cord ischemia using somatosensory cortical evoked potentials during thoracic aortic occlusion. Ann Thorac Surg 34:299–306.

Coles JG, Taylor MJ, Pearce JM, et al 1984 Cerebral monitoring of somatosensory evoked potentials during profoundly hypothermic circulatory arrest. Circulation 70(Suppl I):I96–I102.

Colletti V, Fiorino FG, Policante Z, et al 1996 New perspectives in intraoperative facial nerve monitoring with antidromic potentials. Am J Otol 17:755–762.

Colletti V, Fiorino F, Policante Z, Bruni L 1997a Intraoperative monitoring of facial nerve antidromic potentials during acoustic neuroma surgery. Acta Otolaryngol 117:663–669.

Colletti V, Fiorino FG, Carner M, et al 1997b Mechanisms of auditory impairment during acoustic neuroma surgery. Otolaryngol Head Neck Surg 117:596–605.

Connolly JE 1998 Prevention of spinal cord complications in aortic surgery. Am J Surg 176:92–101.

Coselli JS, Crawford ES, Beall Jr AC, et al 1988 Determination of brain temperatures for safe circulatory arrest during cardiovascular operation. Ann Thorac Surg 45:638–642.

Costa e Silva I, Wang AD-J, Symon L 1985 The application of flash visual evoked potentials during operations on the anterior visual pathways. Neurol Res 7:11–16.

Counsell C, Salinas R, Naylor R, et al 2000 Routine or selective carotid artery shunting for carotid endarterectomy (and different methods of monitoring in selective shunting). Cochrane Library of Systematic Reviews, Issue 2.

Crawford ES, Mizrahi EM, Hess KR, et al 1988 The impact of distal aortic perfusion and somatosensory evoked potential monitoring on prevention of paraplegia after aortic aneurysm operation. J Thorac Cardiovasc Surg 95:357–367.

Croft TJ, Brodkey JS, Nulsen FE 1972 Reversible spinal cord trauma: a model for electrical monitoring of spinal cord function. J Neurosurg 36:402–406.

Cunningham JN, Laschinger JC, Merkin HA, et al 1982 Measurement of spinal cord ischaemia during operations upon the thoracic aorta. Initial clinical experience. Ann Surg 196:285–293.

Cunningham JN, Laschinger JC, Spencer FC 1987 Monitoring of somatosensory evoked potentials during surgical procedures on the thoracoabdominal aorta. IV. Clinical observations and results. J Thorac Cardiovasc Surg 94:275–285.

Curatolo JM, Macdonell RA, Berkovic SF, et al 2000 Intraoperative monitoring to preserve central visual fields during occipital corticectomy for epilepsy. J Clin Neurosci 7:234–237.

Dabbagh E, Rosenthal I, Gavand E, et al 1990 Value of etomidate for the temporary clamping in early surgery in normotension, of ruptured intracranial aneurysms. Preliminary study of 8 cases [in French]. Agressologie 31:389–393.

Dahaba AA 2005 Different conditions that could result in the bispectral index indicating an incorrect hypnotic state. Anesthesiology 101:765–773.

Darden BV, Hatley MK, Owen JH 1996 Neurogenic motor evoked-potential monitoring in anterior cervical surgery. J Spinal Disord 9:485–493.

Daspit CP, Raudzens PA, Shetter AG 1982 Monitoring of intraoperative auditory brain stem responses. Otolaryngol Head Neck Surg 90:108–116.

Davenport HT, Arfel G, Sanchez FR 1959 The electroencephalogram in patients undergoing open-heart surgery with heart–lung bypass. Anesthesiology 20:674–684.

Dawson AM, Krarup C 1989 Perioperative nerve lesions. Arch Neurol 46:1355–1360.

Dawson EG, Sherman JE, Kanim LEA, et al 1991 Spinal cord monitoring; results of the Scoliosis Research Society and the European Spinal Deformity Society survey. Spine 16(Suppl):S361–S364.

De Haan P, Kalkman CJ, Jacobs MJ 1998 Spinal cord monitoring with myogenic motor evoked potentials: early detection of spinal cord ischemia as an integral part of spinal cord protective strategies during thoracoabdominal aneurysm surgery. Semin Thorac Cardiovasc Surg 10:19–24.

De Letter JA, Sie HT, Thomas BM, et al 1998 Near-infrared reflected spectroscopy and electroencephalography during carotid endarterectomy – in search of a new shunt criterion. Neurol Res 20(Suppl 1):S23–S27.

Deletis V 1993 Intraoperative monitoring of the functional integrity of the motor pathways. In: *Electrical and Magnetic Stimulation of the Brain and Spinal Cord* (eds O Devinsky, A Beric, M Dogali). Raven Press, New York, pp. 201–214.

Deletis V 2001 The 'motor' inaccuracy in neurogenic motor evoked potentials. Clin Neurophysiol 112:1365–1366.

Deletis V 2002 Intraoperative neurophysiology and methodologies used to monitor the functional integrity of the motor system. In: *Neurophysiology in Neurosurgery. A Modern Intraoperative Approach* (eds V Deletis, JL Shils). Academic Press, Amsterdam, pp. 25–51.

Delgado TE, Buchheit WA, Rosenholtz HR, et al 1979 Intraoperative monitoring of facial nerve evoked responses obtained by intracranial stimulation of the facial nerve: a more accurate technique for facial nerve dissection. Neurosurgery 4:418–421.

Dennis GC, Dehkordi O, Mills RM, et al 1996 Monitoring of median nerve somatosensory evoked potentials during cervical spinal cord decompression. J Clin Neurophysiol 13:51–59.

Deriu GP, Franceschi L, Milite D, et al 1994 Carotid artery endarterectomy in patients with contralateral carotid artery occlusion: perioperative hazards and late results. Ann Vasc Surg 8:337–342.

Deutsch H, Arginteanu M, Manhart K, et al 2000 Somatosensory evoked potential monitoring in anterior thoracic vertebrectomy. J Neurosurg 92(Suppl 2):155–161.

Dinner DS, Lüders H, Lesser RP, et al 1986 Invasive methods of somato-sensory evoked potential monitoring. J Clin Neurophysiol 3:113–130.

Dinner DS, Lüders H, Lesser RP, et al 1990 Invasive somatosensory-evoked potential monitoring. In: *Neural Monitoring* (ed SK Salzman). Humana Press, Clifton, NJ, pp. 179–196.

Dolan EJ, Transfeldt EE, Tator CH, et al 1980 The effect of spinal distraction on regional spinal cord blood flow in cats. J Neurosurg 53:756–764.

Dong CC, MacDonald DB, Janusz MT 2002 Intraoperative spinal cord monitoring during descending thoracic and thoracoabdominal aneurysm surgery. Ann Thorac Surg 74:S1873–S1876.

Dorgan JC, Abbott TR, Bentley G 1984 Intraoperative awakening to monitor spinal cord function during scoliosis surgery. J Bone Joint Surg Br 66B:716–719.

Dorsch NW 1995 Cerebral arterial spasm – a clinical review. Br J Neurosurg 9:403–412.

Dove J 1989 Segmental wiring for spinal deformity. A morbidity report. Spine 14:229–231.

Ducati A, Landi A, Cenzato M, et al 1988 Monitoring of brain function by means of evoked potentials in cerebral aneurysm surgery. Acta Neurochir 42(Suppl):8–13.

Ducker TB 1976 Experimental injury of the spinal cord. In: *Handbook of Clinical Neurology*, vol. 25. North Holland, Amsterdam, pp. 9–26.

Ducker TB, Brown RH 1988 *Neurophysiology and Standards of Spinal Cord Monitoring*. Springer-Verlag, New York.

Easton JD, Sherman DG 1977 Stroke and mortality rate in carotid endarterectomy: 228 consecutive operations. Stroke 8:565–568.

Ebeling U, Huber P 1992 Localization of central lesions by correlation of CT findings and neurological deficits. Acta Neurochir 119:17–22.

Ecker ML, Dormans JP, Schwartz DM, et al 1996 Efficacy of spinal cord monitoring in scoliosis surgery in patients with cerebral palsy. J Spinal Disord 9:159–164.

Edmonds HL, Paloheimo MPJ, Backman MH, et al 1989 Transcranial magnetic motor evoked potentials (tcMMEP) for functional monitoring of motor pathways during scoliosis surgery. Spine 14:683–686.

Emerson RG, Adams DC, Nagle KJ 1997 Monitoring of spinal cord function intraoperatively using motor and somatosensory evoked potentials. In: *Evoked Potentials in Clinical Medicine*, 3rd edition (ed KH Chiappa). Lippincott-Raven, Philadelphia, pp. 647–660.

Engler GL, Spielholz NI, Bernhard WN, et al 1978 Somatosensory evoked potentials during Harrington instrumentation for scoliosis. J Bone Joint Surg Am 60A:528–532.

Epstein NE, Danto J, Nardi D 1993 Evaluation of intraoperative somatosensory-evoked potential monitoring during 100 cervical operations. Spine 18:737–747.

Erwin CW, Erwin AC 1993 Up and down the spinal cord: intraoperative monitoring of sensory and motor spinal cord pathways. J Clin Neurophysiol 10:425–436.

Erwin CW, Gulevich SJ 1985 Evaluation of transducers for obtaining intraoperative short-latency auditory evoked potentials. Electroencephalogr Clin Neurophysiol 61:194–196.

Etherington NJ 1981 Continuous assessment of brain function: the design and evaluation of an electroencephalographic frequency band monitor. MSc thesis, University of London.

European Carotid Surgery Trialists' Collaborative Group 1991 MRC European Carotid Surgery Trial: interim results for symptomatic patients with severe (70–99%) or with mild (0–29%) carotid stenosis. Lancet 337:1235–1243.

Executive Committee for the Asymptomatic Carotid Atherosclerosis Study 1995 Endarterectomy for asymptomatic carotid artery stenosis. JAMA 273:1421–1428.

Faberowski LW, Black S, Trankina MF, et al 1999 Somatosensory-evoked potentials during aortic coarctation repair. J Cardiothorac Vasc Anesth 13:538–543.

Facco E, Deriu GP, Dona B, et al 1992 EEG monitoring of carotid endarterectomy with routine patch-graft angioplasty: an experience in a large series. Neurophysiol Clin 22:437–446.

Fava E, Ducati A, Bortolani E, et al 1990 Role of SEP monitoring in selection of patients requiring temporary shunting in carotid surgery. Neurophysiol Clin 20(Suppl):22.

Fava E, Bortolani E, Ducati A, et al 1992 Role of SEP in identifying patients requiring temporary shunt during carotid endarterectomy. Electroencephalogr Clin Neurophysiol 84:426–432.

Fehlings MG, Tator CH, Linden RD 1989a The relationships among the severity of spinal cord injury, motor and somatosensory evoked potentials and spinal cord blood flow. Electroencephalogr Clin Neurophysiol 74:241–259.

Fehlings MG, Tator CH, Linden RD 1989b The effect of nimodipine and dextran on axonal function and blood flow following experimental spinal cord injury. J Neurosurg 71:403–416.

Feinsod M, Selhorst JB, Hoyt WF, et al 1976 Monitoring optic nerve function during craniotomy. J Neurosurg 44:29–31.

Fenton JE, Chin RY, Shirazi A, et al 1999 Prediction of post-operative facial nerve function in acoustic neuroma surgery. Clin Otolaryngol 24:483–486.

Ferguson GG 1982 Intra-operative monitoring and internal shunts: are they necessary in carotid endarterectomy? Stroke 13:287–289.

Ferguson GG 1986 Carotid endarterectomy: to shunt or not to shunt? Arch Neurol 43:615–618.

Fiorani P, Sbarigia E, Speziale F, et al 1997 General anaesthesia versus cervical block and perioperative complications in carotid artery surgery. Eur J Vasc Endovasc Surg 13:37–42.

Fiori L, Parenti G 1995 Electrophysiological monitoring for selective shunting during carotid endarterectomy. J Neurosurg Anesthesial 7:168–173.

Fischer G, Fischer C, Remond J 1992 Hearing preservation in acoustic neurinoma surgery. J Neurosurg 76:910–917.

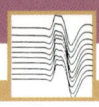

Fischer-Williams M, Cooper RA 1964 Some aspects of electroencephalographic changes during open-heart surgery. Neurology (Minneapolis) 14:472–482.

Fisher RS, Raudzens P, Nunemacher M 1995 Efficacy of intraoperative neurophysiological monitoring. J Clin Neurophysiol 12:97–109.

Fletcher JP, Morris JGL, Little JM, et al 1988 EEG monitoring during carotid endarterectomy. Aust NZ J Surg 58:285–288.

Fode NC, Sundt Jr TM, Robertson JT, et al 1986 Multicenter retrospective review of results and complications of carotid endarterectomy in 1981. Stroke 17:370–376.

Forbes HJ, Allen PW, Waller CS, et al 1991 Spinal cord monitoring in scoliosis surgery. Experience with 1168 cases. J Bone Joint Surg Br 73B:487–491.

Frawley JE, Hicks RG, Beaudoin M, et al 1997 Hemodynamic ischemic stroke during carotid endarterectomy: an appraisal of risk and cerebral protection. J Vasc Surg 25:611–619.

Friedman WA, Grundy BL 1987 Monitoring of sensory evoked potentials is highly reliable and helpful in the operating room. J Clin Monit 3:38–44.

Friedman WA, Richards R 1988 Somatosensory evoked potential monitoring accurately predicts hemi-spinal cord damage: a case report. Neurosurgery 22:140–142.

Friedman WA, Kaplan BJ, Gravestein D, et al 1985 Intraoperative brain-stem auditory evoked potentials during posterior fossa microvascular decompression. J Neurosurg 62:552–557.

Friedman WA, Kaplan BL, Day AL, et al 1987 Evoked potential monitoring during aneurysm operation: observations after fifty cases. Neurosurgery 20:678–687.

Friedman WA, Chadwick GM, Verhoeven FJ, et al 1991 Monitoring of somatosensory evoked potentials during surgery for middle cerebral artery aneurysms. Neurosurgery 29:83–88.

Fujioka H, Shimoji K, Tomita M, et al 1994 Spinal cord potential recordings from the extradural space during scoliosis surgery. Br J Anaesth 73:350–356.

Galla JD, Ergin MA, Sadeghi AM, et al 1994 A new technique using somatosensory evoked potential guidance during descending and thoraco-abdominal aortic repairs. J Card Surg 9:662–672.

Galla JD, Ergin MA, Lansman SL, et al 1999 Use of somatosensory evoked potentials for thoracic and thoracoabdominal aortic resections. Ann Thorac Surg 67:1947–1952.

Gaunt ME 1998 Transcranial Doppler: preventing stroke during carotid endarterectomy. Ann R Coll Surg Engl 80:377–387.

Gelfan S, Tarlov IM 1956 Physiology of spinal cord, nerve root and peripheral nerve compression. Am J Physiol 185:217–229.

Gentili F, Lougheed WM, Yamashiro K, et al 1985 Monitoring of sensory evoked potentials during surgery of skull base tumours. Can J Neurol Sci 12:336–340.

Gigli GL, Caramia M, Marciani MG, et al 1987 Monitoring of subcortical and cortical somatosensory evoked potentials during carotid endarterectomy: a comparison with stump pressures. Electroencephalogr Clin Neurophysiol 68:424–432.

Gilman S 1965 Cerebral disorders after open-heart operations. N Engl J Med 272:489–498.

Ginsburg HH, Shetter AG, Raudzens PA 1985 Postoperative paraplegia with preserved intraoperative somatosensory evoked potentials. J Neurosurg 63:296–300.

Glassman SD, Zhang YP, Shields CB, et al 1995 Transcranial magnetic motor-evoked potentials in scoliosis surgery. Orthopedics 18:1017–1023.

Gokaslan ZL, Samudrala S, Deletis V, et al 1997 Intraoperative monitoring of spinal cord function using motor evoked potentials via transcutaneous epidural electrode during anterior cervical spinal surgery. J Spinal Disord 10:299–303.

Goldbrunner RH, Schlake HP, Milewski C, et al 2000 Quantitative parameters of intraoperative electromyography predict facial nerve outcomes for vestibular schwannoma surgery. Neurosurgery 46:1140–1146.

Gonzalez-Fajardo JA, Toledano M, Alvarez T, et al 1998 Monitoring of evoked potentials during spinal cord ischaemia: experimental evaluation in a rabbit model. Eur J Vasc Endovasc Surg 16:320–328.

Gossman M, White R, Taslitz N, et al 1968 Electrophysiological responses immediately after experimental injury to the spinal cord. Anat Rec (New York) 160:473.

Grabitz K, Freye E, Stuhmeier K, et al 1993 Spinal evoked potential in patients undergoing thoracoabdominal aortic reconstruction: a prognostic indicator of postoperative motor deficit. J Clin Monit 9:186–190.

Grabitz K, Sandmann W, Stuhmeier K, et al 1996 The risk of ischemic spinal cord injury in patients undergoing graft replacement for thoracoabdominal aortic aneurysms. J Vasc Surg 23:230–240.

Grady RE, Weglinski MR, Sharbrough FW, et al 1998 Correlation of regional cerebral blood flow with ischemic electroencephalographic changes during sevoflurane–nitrous oxide anesthesia for carotid endarterectomy. Anesthesiology 88:892–897.

Green RM, Messick WJ, Ricotta JJ, et al 1985 Benefits, shortcomings, and costs of EEG monitoring. Ann Surg 201:785–792.

Greenwood J 1967 Surgical removal of intramedullary tumours. J Neurosurg 26:276–282.

Gregorie EM, Goldring S 1984 Localization of function in the excision of lesions from the sensorimotor region. J Neurosurg 61:1047–1054.

Grey PL, Moffat DA, Palmer CR, et al 1996 Factors which influence the facial nerve outcome in vestibular schwannoma Surgery Clin Otolaryngol 21:409–413.

Groff MW, Adams DC, Kahn RA, et al 1999 Adenosine-induced transient asystole for management of a basilar artery aneurysm. Case report. J Neurosurg 91:687–690.

Gross CE, Beditionar MM, Lew SM, et al 1998 Preoperative volume expansion improves tolerance to carotid artery cross-clamping during endarterectomy. Neurosurgery 43:222–226.

Grossi EA, Laschinger JC, Krieger KH, et al 1988 Epidural-evoked potentials: a more specific indicator of spinal cord ischemia. J Surg Res (New York) 44:224–228.

Grundy BL, Nash CL, Brown RH 1981 Arterial pressure manipulation alters spinal cord function during correction of scoliosis. Anesthesiology 54:249–253.

Grundy BL, Nelson PB, Lina A, et al 1982a Monitoring of somatosensory evoked potentials to determine the safety of sacrificing the anterior cerebral artery. Neurosurgery 11:64–67.

Grundy BL, Jannetta PJ, Procopio PT, et al 1982b Intraoperative monitoring of brain-stem auditory evoked potentials. J Neurosurg 57:674–681.

Grundy BL, Procopio PT, Jannetta PJ, et al 1982c Evoked potential changes produced by positioning for retromastoid craniectomy. Neurosurgery 10:766–770.

Guérit JM, Etienne PY, Dion R 1994 An explanation of the high 'false positive' rate in descending aorta surgery. In: *Handbook of Spinal Cord Monitoring* (eds SJ Jones, S Boyd, M Hetreed, NJ Smith). Kluwer, Dordrecht, pp. 135–145.

Guérit JM, Verhelst R, Rubay J, et al 1996 Multilevel somatosensory evoked potentials (SEPs) for spinal cord monitoring in descending thoracic and thoraco-abdominal aortic surgery. Eur J Cardiothorac Surg 10:93–103.

Guérit JM, Witdoeckt C, de Tourtchaninoff M, et al 1997 Somatosensory evoked potential monitoring in carotid surgery. I. Relationships between qualitative SEP alterations and intraoperative events. Electroencephalogr Clin Neurophysiol 104:459–69.

Guérit JM, Witdoekt C, Verhelst R, et al 1999 Sensitivity, specificity, and surgical impact of somatosensory evoked potentials in descending aorta surgery. Ann Thorac Surg 67:1943–1946.

Gumerlock MK, Neuwelt EA 1988 Carotid endarterectomy: to shunt or not to shunt. Stroke 19:1485–1490.

Hacke W 1985 Neuromonitoring during interventional neuro-radiology. Central Nervous System Trauma 2:123–136.

Hall JE, Levine CR, Sudhir KG 1978 Intraoperative awakening to monitor spinal cord function during Harrington instrumentation and spine fusion. J Bone Joint Surg Am 60A:533–536.

Halonen J-P, Jones SJ, Edgar MA, et al 1989 Wave form patterns of multiple level recordings of spinal SEPs during scoliosis surgery. Electroencephalogr Clin Neurophysiol 67:58P.

Hanowell LH, Soriano S, Bennett HL 1992 EEG power changes are more sensitive than spectral edge frequency variation for detection of cerebral ischemia during carotid artery surgery: a prospective assessment of processed EEG monitoring. J Cardiothorac Vasc Anesth 6:292–294.

Happel L, Kline D 2002 Intraoperative neurophysiology of the peripheral nervous system. In: *Neurophysiology in Neurosurgery. A Modern Intraoperative Approach* (eds V Deletis, JL Shils). Academic Press, Amsterdam, pp. 169–195.

Harada RN, Comerota AJ, Good GM, et al 1995 Stump pressure, electroencephalographic changes, and the contralateral carotid artery: another look at selective shunting. Am J Surg 170:148–153.

Harding GFA 1982 The flash evoked visual potential and its use in ocular conditions. The lost potential. J Electrophysiol Technol 8:63–78 [part 1], 8:110–130 [part 2].

Harding GFA, Smith VH, Yorke HC 1987 A contact lens photostimulator for surgical monitoring. Electroencephalogr Clin Neurophysiol 66:322–326.

Harding GFA, Bland JP, Smith VH 1990 Visual evoked potential monitoring of optic nerve function during surgery. J Neurol Neurosurg Psychiatry 53:890–895.

Hardy RH, Brodkey JS, Richards DE, et al 1972 Effects of systemic hypertension on compression block of spinal cord. Surgical Forum 23:434–435.

Hargadine JR 1985 Intraoperative monitoring of sensory evoked potentials. In: *MicroNeuroSurgery,* 3rd edition (ed RW Rand). CV Mosby, St. Louis, pp. 92–110.

Hargadine JR, Branston NM, Symon L 1980 Central conduction time in primate brain ischaemia – a study in baboons. Stroke 11:637–642.

Harner SG, Daube JR, Ebersold MJ, et al 1987 Improved preservation of facial nerve function with use of electrical monitoring during removal of acoustic neuromas. Mayo Clinic Proceedings 62:92–102.

Harner SG, Harper CM, Beatty CW, et al 1996 Far-field auditory brainstem response in neurotologic surgery. Am J Otol 17:150–153.

Harper CM, Daube JR, Litchy WJ, et al 1988 Lumbar radiculopathy after spinal fusion for scoliosis. Muscle Nerve 11:386–391.

Harper CM, Harner SG, Slavit DH, et al 1992 Effect of BAEP monitoring on hearing preservation during acoustic neuroma resection. Neurology 42:1551–1553.

Harris RJ, Symon L 1984 Extracellular pH, potassium and calcium activities in progressive ischemia in rat cortex. J Cereb Blood Flow Metab 4:178–186.

Harris DNF, Bailey SM, Smith PLC, et al 1993 Brain swelling in the first hour after coronary artery bypass surgery. Lancet 342:586–587.

Hashimoto I, Ishiyama Y, Totsuka G, et al 1980 Monitoring brainstem function during posterior fossa surgery with brainstem auditory potentials. In: *Evoked Potentials* (ed C Barber). MTP Press, Lancaster, pp. 377–390.

Haupt WF, Erasmi-Korber H, Lanfermann H 1994 Intraoperative recording of parietal SEP can miss hemodynamic infarction during carotid endarterectomy: a case study. Electroencephalogr Clin Neurophysiol 92:86–88.

Heiss WD, Hayakawa T, Waltz AG 1976 Cortical neuronal function during ischemia. Arch Neural 33:813–819.

Helfet DL, Anand N, Malkani ALL, et al 1997 Intraoperative monitoring of motor pathways during operative fixation of acute acetabular fractures. J Orthop Trauma 11:2–6.

Helmers SL, Wypij D, Constantinou JE, et al 1997 Perioperative electro-encephalographic seizures in infants undergoing repair of complex congenital cardiac defects. Electroencephalogr Clin Neurophysiol 102:27–36.

Herdmann J, Lumenta CB, Huse KOW 1993 Magnetic stimulation for monitoring motor pathways in spinal procedures. Spine 18:551–559.

Herkes GK, Morgan M, Grinnell V, et al 1993 EEG monitoring during angiographic balloon test carotid occlusion: experience in sixteen cases. Clin Exp Neurol 30:98–103.

Herzon GD, Zealear DL 1994 Intraoperative monitoring of the visual evoked potential during endoscopic sinus surgery. Otolaryngol Head Neck Surg 111:575–579.

Heyer EJ, Adams DC, Moses C, et al 2000 Erroneous conclusion from processed electroencephalogram with changing anesthetic depth. Anesthesiology 92:603–607.

Hicks RG, Kerr DR, Horton DA 1986 Thiopentone cerebral protection under EEG control during carotid endarterectomy. Anaesth Intensive Care 14:22–28.

Hofste WJ, Linssen CA, Boezeman EH, et al 1997 Delirium and cognitive disorders after cardiac operations: relationship to pre- and intra-operative quantitative electroencephalogram. Int J Clin Monit Comput 14:29–36.

Holland NR 1998 Subcortical strokes from intracranial aneurysm surgery: implications for intraoperative neuromonitoring. J Clin Neurophysiol 15:439–446.

Homma S, Tamaki T 1984 *Fundamentals and Clinical Application of Spinal Cord Monitoring*. Saikon, Tokyo.

Hoppenfeld S, Gross A, Andrews C, et al 1997 The ankle clonus test for assessment of the integrity of the spinal cord during operations for scoliosis. J Bone Joint Surg Am 79A:208–212.

Horikoshi T, Omata T, Uchida M, et al 2000 Usefulness and pitfalls of intraoperative spinal motor evoked potential recording by direct cortical electrical stimulation. Acta Neurochir 142:257–262.

Hossmann KA 1994 Viability thresholds and the penumbra of focal ischemia. Ann Neural 36:557–565.

Hossmann KA 1999 The hypoxic brain – insights from ischemia research. Adv Exp Med Biol 474:155–169.

Hu Y, Luk KD, Wong YW, et al 2001 Effect of stimulation parameters on intraoperative spinal cord evoked potential monitoring. J Spinal Disord 14:449–452.

Hume AL, Cant BR 1978 Conduction time in central somatosensory pathways in man. Electroencephalogr Clin Neurophysiol 45:361–375.

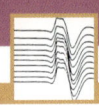

Hume AL, Durkin MA 1986 Central and spinal somatosensory conduction times during hypothermic cardiopulmonary bypass and some observations on the effects of fentanyl and isoflurane anesthesia. Electroencephalogr Clin Neurophysiol 65:46–58.

Hume AL, Cant BR, Shaw MA, et al 1982 Central somatosensory conduction time from 10 to 79 years. Electroencephalogr Clin Neurophysiol 54:49–54.

Hurlbert RJ, Fehlings MG, Moncada MS 1995 Use of sensory-evoked potentials recorded from the human occiput for intraoperative physiologic monitoring of the spinal cord. Spine 20:2318–2327.

Hussain SS, Laljee HCK, Horrocks JM, et al 1996 Monitoring of intra-operative visual evoked potentials during functional endoscopic sinus surgery (FESS) under general anaesthesia. J Laryngol Otol 110:31–36.

Illig KA, Burchfiel JL, Ouriel K, et al 1998 Value of preoperative EEG for carotid endarterectomy. Cardiovasc Surg 6:490–495.

Isley MR, Cohen MJ, Wadsworth JS, et al 1998 Multimodality neuro-monitoring for carotid endarterectomy: determination of critical cerebral ischemic thresholds. Am J Electroneurodiag Technol 38:65–122.

Isley MR, Balzer JR, Pearlman RC, et al 2001 Intraoperative motor evoked potentials. Am J Electroneurodiag Technol 41:266–338.

Jackson LE, Robertson JB 2000 Acoustic neuroma surgery: use of cochlear nerve action potential monitoring for hearing preservation. Am J Otol 21:249–259.

Jacobs LA, Brinkman SD, Morrell RM, et al 1983 Long-latency somatosensory evoked potentials during carotid endarterectomy. Am Surg 49:338–344.

Jacobs MJHM, Meylaerts SA, de Haan P, et al 1999 Strategies to prevent neurologic deficit based on motor-evoked potentials in type I, II thoraco-abdominal aortic aneurysm repair. J Vasc Surg 29:48–57.

Jacobs MJ, Meylaerts SA, de Haan P, et al 2000 Assessment of spinal cord ischemia by means of evoked potential monitoring during thoracoabdominal aortic surgery. Semin Vasc Surg 13:299–307.

Jacobs MJ, De Mol BA, Elenbaas T, et al 2002 Spinal cord blood supply in patients with thoracoabdominal aortic aneurysms. J Vasc Surg 35:30–37.

Jallo GI, Kothbauer KF, Epstein FJ 2001 Intrinsic spinal cord tumor resection. Neurosurgery 49:1124–1128.

Jannetta PJ, Møller AR, Møller MB 1984 Technique of hearing preservation in small acoustic neuromas. Ann Surg 200:513–522.

Jansen C, Moll FL, Vermeulen FE, et al 1993 Continuous transcranial Doppler ultra-sonography and electroencephalography during carotid endarterectomy: a multimodal monitoring system to detect intra-operative ischemia. Ann Vasc Surg 7:95–101.

Jasper H, Lende R, Rasmussen T 1960 Evoked potentials from the exposed somatosensory cortex in man. J Nerv Ment Dis 130:526–537.

Javid H, Tufo HM, Najafi H, et al 1969 Neurological abnormalities following open-heart surgery. J Thorac Cardiovasc Surg 58:502–509.

Jellinek D, Jewkes D, Symon L 1991a Noninvasive intra-operative monitoring of motor evoked potentials under propofol anesthesia: effects of spinal surgery on the amplitude and latency of motor evoked potentials. Neurosurgery 29:551–557.

Jellinek D, Tan LC, Symon L 1991b The impact of continuous electrophysio-logical monitoring on preservation of the facial nerve during acoustic tumour surgery. Br J Neurosurg 5:19–24.

Jex RK, Schaff HV, Piehler JM, et al 1986 Early and late results following repair of dissection of the descending thoracic aorta. J Vasc Surg 3:226–237.

Johnston CE, Happel LT, Norris R, et al 1986 Delayed paraplegia complicating sublaminar segmental spinal instrumentation. J Bone Joint Surg Am 68A:556–563.

Jones NS 1997 Visual evoked potentials in endoscopic and anterior skull base surgery: a review. J Laryngol Otol 111:513–516.

Jones SJ, Edgar MA, Ransford AO 1982 Sensory nerve conduction in the human spinal cord: epidural recordings made during scoliosis surgery. J Neural Neurosurg Psychiatry 45:446–451.

Jones SJ, Edgar MA, Ransford AO, et al 1983 A system for the electro-physiological monitoring of the spinal cord during operations for scoliosis. J Bone Joint Surg Br 65B:134–139.

Jones SJ, Carter L, Edgar MA, et al 1985 Experience of epidural spinal cord monitoring in 410 cases. In: *Spinal Cord Monitoring* (eds J Schramm, SJ Jones). Springer-Verlag, Berlin, pp. 215–220.

Jones SJ, Howard L, Shawkat F 1988 Criteria and pathological significance of response decrement during spinal cord monitoring. In: *Neurophysiology and Standards of Spinal Cord Monitoring* (eds TB Ducker, RH Brown). Springer-Verlag, New York, pp. 201–206.

Jones SJ, Boyd S, Hetreed M, et al 1994 *Handbook of Spinal Cord Monitoring*. Kluwer, Dordrecht.

Jones SJ, Harrison R, Koh KF, et al 1996 Motor evoked potential monitoring during spinal surgery: responses of distal limb muscles to transcranial cortical stimulation with pulse trains. Electroencephalogr Clin Neurophysiol 100:375–383.

Jones TH, Morawetz RB, Crowell RM, et al 1981 Thresholds of focal cerebral ischemia in awake monkeys. J Neurosurg 54:773–782.

Jorgensen LG, Schroeder TV 1992 Transcranial Doppler for detection of cerebral ischaemia during carotid endarterectomy. Eur J Vasc Surg 6:142–147.

Julia P, Chemla E, Mercier F, et al 1998 Influence of the status of the contralateral carotid artery on the outcome of carotid surgery. Ann Vasc Surg 12:566–571.

Kai Y, Owen JH, Allen BT, et al 1995 Relationship between evoked potentials and clinical status in spinal cord ischemia. Spine 20:291–296.

Kalkman CJ, ten Brink SA, Been HD, et al 1991 Variability of somatosensory cortical evoked potentials during spinal surgery. Effect of anesthetic technique and high-pass digital filtering. Spine 16:924–929.

Kalkman CJ, Ubags LH, Ben HD, et al 1995 Improved amplitude of myogenic motor evoked-responses after paired transcranial electrical-stimulation during sufentanil/nitrous oxide anesthesia. Anesthesiology 83:270–276.

Kálmánchey R, Avila A, Symon L 1986 The use of brainstem auditory evoked potentials during posterior fossa surgery as a monitor of brainstem function. Acta Neurochirurgica (Vienna) 82:128–136.

Kartush JM 1998 Intraoperative monitoring in acoustic neuroma surgery. Neural Res 20:593–596.

Kartush JM, Bouchard KR (eds) 1992 *Neuromonitoring in Head and Neck Surgery*. Raven Press, New York.

Kartush JM, Niparko JK, Bledsoe SC, et al 1985 Intraoperative facial nerve monitoring: a comparison of stimulating electrodes. Laryngoscope 95:1536–1540.

Katayama Y, Tsubokawa T, Sugitani M, et al 1986 Assessment of spinal cord injury with multimodality evoked spinal cord potentials. Part I. Localization of lesions in experimental spinal cord injury. Neuro-orthopedics 1:130–141.

Kawaguchi M, Inoue S, Kakimoto M, et al 1998 The effect of sevoflurane on myogenic motor-evoked potentials induced by single and paired transcranial electrical stimulation of the motor cortex during nitrous oxide. J Neurosurg Anesthesiol 10:131–136.

Kearse LA, Brown EM, McPeck K 1992 Somatosensory evoked potentials sensitivity relative to electroencephalography for cerebral ischaemia during carotid endarterectomy. Stroke 23:498–505.

Kearse Jr LA, Martin D, McPeck K, et al 1993 Computer-derived density spectral array in detection of mild analog electroencephalographic ischemic pattern changes during carotid endarterectomy. J Neurosurg 78:884–890.

Keen G 1987 Spinal cord damage and operations for coarctation of the aorta: aetiology, practice, and prospects. Thorax 42:11–18.

Keenan NK, Taylor MJ, Coles JG, et al 1987 The use of VEPs for CNS monitoring during continuous cardiopulmonary bypass and circulatory arrest. Electroencephalogr Clin Neurophysiol 68:241–246.

King RB, Schell GR 1987 Cortical localization and monitoring during cerebral operations. J Neurosurg 67:210–219.

Kirkpatrick PJ 1997 Use of near-infrared spectroscopy in the adult. Philos Trans R Soc Lond. Series B Biol Sci 352:701–705.

Kirkpatrick PJ, Lam J, Al-Rawi P, et al 1998 Defining thresholds for critical ischemia by using near-infrared spectroscopy in the adult brain. J Neurosurg 89:389–394.

Kitagawa H, Itoh T, Takano H, et al 1989 Motor evoked potential monitoring during upper cervical spine surgery. Spine 10:1078–1083.

Kline DG, Judice D 1983 Operative management of selected brachial plexus lesions. J Neurosurg 58:631–649.

Kochs E, Schulte am Esch J 1991 Somatosensory evoked responses during and after graded brain ischaemia in goats. Eur J Anaesthesial 8:257–265.

Kochs E, Bischoff P, Pichlmeier U, et al 1994 Surgical stimulation induces changes in brain electrical activity during isoflurane/nitrous oxide anesthesia. A topographic electroencephalographic analysis. Anesthesiology 80:1026–1034.

Kojima Y, Yamamoto T, Ogino H, et al 1979 Evoked spinal potentials as a monitor of spinal cord viability. Spine 4:471–477.

Komanetsky RM, Padberg AM, Lenke LG, et al 1998 Neurogenic motor evoked potentials: a prospective comparison of stimulation methods in spinal deformity surgery. J Spinal Disord 11:21–28.

Kombos T, Suess O, Ciklatekerlio O, et al 2001 Monitoring of intraoperative motor evoked potentials to increase the safety of surgery in and around the motor cortex. J Neurosurg 95:608–614.

Konrad PE, Tacker WA, Levy WJ, et al 1987 Motor evoked potentials in the dog: effects of global ischemia on spinal cord and peripheral nerve signals. Neurosurgery 20:117–124.

Kopf GS, Hume AL, Durkin MA, et al 1985 Measurement of central somatosensory conduction time in patients undergoing cardiopulmonary bypass: an index of neurologic function. Am J Surg 149:445–448.

Kothbauer K, Deletis V, Epstein FJ 1997 Intraoperative spinal cord monitoring for intramedullary surgery: an essential adjunct. Pediatr Neurosurg 26:247–254.

Kothbauer KF 2002 Motor evoked potential monitoring for intramedullary spinal cord tumor surgery. In: *Neurophysiology in Neurosurgery. A Modern Intraoperative Approach* (eds V Deletis, JL Shils). Academic Press, Amsterdam, pp. 73–92.

Koyanagi I, Iwasaki Y, Isu T, et al 1993 Spinal cord evoked potential monitoring after spinal cord stimulation during surgery of spinal cord tumors. Neurosurgery 33:451–459.

Kresowik TF, Worsey MJ, Khoury MD, et al 1991 Limitations of electro-encephalographic monitoring in the detection of cerebral ischemia accompanying carotid endarterectomy. J Vasc Surg 13:439–443.

Krieger D, Adams HP, Albert F, et al 1992 Pure motor hemiparesis with stable somatosensory evoked potential monitoring during aneurysm surgery: case report. Neurosurgery 31:145–150.

Kurokawa T 1972 Spinal cord action potentials evoked by epidural stimulation of cord – a report of human and animal record. Jpn J Electroencephalogr Electromyogr 1:64–66.

Kuroki A, Møller AR 1995 Microsurgical anatomy around the foramen of Luschka in relation to intraoperative recording of auditory evoked potentials from the cochlear nuclei. J Neurosurg 82:933–939.

Labar DR, Fisch BJ, Pedley TA, et al 1991 Quantitative EEG monitoring for patients with subarachnoid hemorrhage. Electroencephalogr Clin Neurophysiol 78:325–332.

Lalwani AK, Butt FY, Jackler RK, et al 1994 Facial nerve outcome after acoustic neuroma surgery: a study from the era of cranial nerve monitoring. Otolaryngol Head Neck Surg 111:561–570.

Lalwani AK, Butt FY, Jackler RK, et al 1995 Delayed onset facial nerve dysfunction following acoustic neuroma surgery. Am J Otol 16:758–764.

Lam AM, Keane JF, Manninen PH 1985 Monitoring of brainstem auditory evoked potentials during basilar artery occlusion in man. Br J Anaesth 57:924–928.

Lam AM, Manninen PH, Ferguson GF, et al 1991 Monitoring electrophysiologic function during carotid endarterectomy: a comparison of somatosensory evoked potentials and conventional electroencephalogram. Anesthesiology 75:15–21.

Laman DM, van der Reijden CS, Wieneke GH, et al 2001 EEG evidence for shunt requirement during carotid endarterectomy. Optimal EEG derivations with respect to frequency bands and anesthetic regimen. J Clin Neurophysiol 18:353–363.

Landau WM, Hunt CC 1990 Dorsal rhizotomy, a treatment of unproven efficacy. J Child Neural 5:174–178.

Landi A, Copeland SA, Wynn-Parry CB, et al 1980 The role of somatosensory evoked potentials and nerve conduction studies in the surgical management of brachial plexus injuries. J Bone Joint Surg Br 62B:492–496.

Lang EW, Beutler AS, Chesnut RM, et al 1996 Myogenic motor-evoked potential monitoring using partial neuro-muscular blockade in surgery of the spine. Spine 21:1676–1686.

Langeloo DD, Journee HL, Polak B, et al 2001 A new application of TCE–MEP: spinal cord monitoring in patients with severe neuromuscular weakness undergoing corrective spine surgery. J Spinal Disord 14:445–448.

Larson SJ, Walsh PR, Sances A, et al 1980 Evoked potentials in experimental myelopathy. Spine 5:299–302.

Laschinger JC, Cunningham JN, Nathan IM, et al 1983 Experimental and clinical assessment of the adequacy of partial bypass in maintenance of spinal cord blood flow during operation on the thoracic aorta. Ann Thorac Surg 36:417–426.

Laschinger JC, Cunningham JN, Cooper MM, et al 1984 Prevention of ischaemic spinal cord injury following aortic cross-clamping: use of corticosteroids. Ann Thorac Surg 38:500–507.

Laschinger JC, Cunningham JN, Cooper MM, et al 1987a Monitoring of somatosensory evoked potentials during surgical procedures on the thoracoabdominal aorta. 1. Relationship of aortic cross-clamp duration, changes in somatosensory evoked potentials, and incidence of neurologic dysfunction. J Thorac Cardiovasc Surg 94:260–265.

Laschinger JC, Cunningham JN, Baumann FG, et al 1987b Monitoring of somatosensory evoked potentials during surgical procedures on the thoracoabdominal aorta. 2. Use of somatosensory evoked potentials to assess adequacy of distal aortic bypass and perfusion after aortic cross-clamping. J Thorac Cardiovasc Surg 94:266–270.

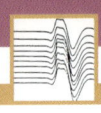

Laschinger JC, Owen JH, Rosenbloom M, et al 1988 Direct noninvasive monitoring of spinal cord motor function during thoracic aortic occlusion. J Vasc Surg 7:161–171.

Lee KS, Davis CH, McWhorter JM 1988 Low morbidity and mortality of carotid endarterectomy performed with regional anesthesia. J Neurosurg 69:483–487.

Lee YS, Lüders H, Dinner DS, et al 1984 Recording of auditory evoked potentials in man using chronic subdural electrodes. Brain 107:115–131.

Lee WY, Hou WY, Yang LH, et al 1995 Intraoperative monitoring of motor function by magnetic motor evoked potentials. Neurosurgery 36:493–500.

Legatt AD, Kader A. 2000 Topography of the initial cortical component of the median nerve somatosensory evoked potential. Relationship to central sulcus anatomy. J Clin Neurophysiol 17:321–325.

Lesnick JE, Michele JJ, Simeone FA, et al 1984 Alteration of somatosensory evoked potentials in response to global ischaemia. J Neurosurg 60:490–494.

Lesser RP, Raudzens P, Lüders H, et al 1986 Postoperative neurological deficits may occur despite unchanged intraoperative somatosensory evoked potentials. Ann Neurol 19:22–25.

Levine RA, Montgomery WM, Ojemann RG, et al 1984 Monitoring of auditory evoked potentials during acoustic neuroma surgery. Insight into mechanisms of the hearing loss. Ann Otol Rhinol Laryngol 93:116–123.

Levy WJ 1987 Clinical experience with motor and cerebellar evoked potential monitoring. Neurosurgery 20:169–182.

Li S, Tator CH 1998 Spinal cord blood flow and evoked potentials as outcome measures for experimental spinal cord injury. In: *Spinal Cord Monitoring. Basic Principles Regeneration Pathophysiology and Clinical Aspects* (eds E Stålberg, HS Sharma, Y Olsson). Springer-Verlag, Vienna, pp. 365–392.

Linden RD, Tator CH, Benedict C, et al 1988 Electrophysiological monitoring during acoustic neuroma and other posterior fossa surgery. Can J Neurol Sci 15:73–81.

Linstedt U, Maier C, Petry A 1998 Intraoperative monitoring with somatosensory evoked potentials in carotid artery surgery – less reliable in patients with preoperative neurologic deficiency? Acta Anaesthesiol Scand 42:13–16.

Little JR, Lesser RP, Lüders H, et al 1983 Brain stem auditory evoked potentials in posterior circulation surgery. Neurosurgery 12:496–502.

Little JR, Lesser RP, Lüders H 1987 Electrophysiological monitoring during basilar aneurysm operation. Neurosurgery 20:421–427.

Loftus CM, Quest DO 1995 Technical issues in carotid artery surgery. Neurosurgery 36:629–647.

Logigian EL, Shefner JM, Goumnerova L, et al 1996 The critical importance of stimulus intensity in intraoperative monitoring for partial dorsal rhizotomy. Muscle and Nerve 19:415–422.

Lopéz JR, Chang SD, Steinberg GK 1999 The use of electrophysiological monitoring in the intraoperative management of intracranial aneurysms. J Neurol Neurosurg Psychiatry 66:189–196.

Lorenzini NA, Poterack KA 1996 Somatosensory evoked potentials are not a sensitive indicator of potential positioning errors in the prone patient. J Clin Monit 12:171–176.

Lorenzini NA, Schneider JH 1996 Temporary loss of intraoperative motor-evoked potential and permanent loss of somato-sensory-evoked potentials associated with a post-operative sensory deficit. J Neurosurg Anesthesiol 8:142–147.

Lubicky JP, Spadaro JA, Yuan HA, et al 1989 Variability of somatosensory cortical evoked potential monitoring during spinal surgery. Spine 14:790–798.

Lüders H 1988 Surgical monitoring with auditory evoked potentials. J Clin Neurophysiol 5:261–285.

Lüders H, Gurd A, Hahn J, et al 1982 A new technique for intraoperative monitoring of spinal cord function. Multichannel recording of spinal cord and subcortical evoked potentials. Spine 1:110–115.

Lüders H, Dinner DS, Lesser RP, et al 1986 Evoked potentials in cortical localization. J Clin Neurophysiol 3:75–84.

Luk KDK, Hu Y, Wong YW, et al 1999 Variability of somatosensory-evoked potentials in different stages of scoliosis surgery. Spine 24:1799–1804.

Luk KD, Hu Y, Wong YW, et al 2001 Evaluation of various evoked potential techniques for spinal cord monitoring during scoliosis surgery. Spine 26:1772–1777.

Lundar T, Lindegaard KF, Froysaker T, et al 1985 Cerebral perfusion during nonpulsatile cardiopulmonary bypass. Ann Thorac Surg 40:144–150.

Maccabee PJ, Levine DB, Pinkhasov EI, et al 1983 Evoked potentials recorded from scalp and spinous process during spinal column surgery. Electroencephalogr Clin Neurophysiol 56:569–582.

MacDonald DB 2001 Individually optimizing posterior tibial somatosensory evoked potential P37 scalp derivations for intra-operative monitoring. J Clin Neurophysiol 18:364–371.

MacDonald DB, Janusz M 2002 An approach to intraoperative neurophysiologic monitoring of thoracoabdominal aneurysm surgery. J Clin Neurophysiol 19:43–54.

MacDonald DB, Al Zayed Z, Khoudeir I, et al 2003 Monitoring scoliosis surgery with combined multiple pulse transcranial electric motor and cortical somatosensory-evoked potentials from the lower and upper extremities. Spine 28:194–203.

MacDonald DB, Stigsby B, Al Zayed Z 2004 A comparison between derivation optimization and Cz′–FPz for posterior tibial P37 somatosensory evoked potential intraoperative monitoring. Clin Neurophysiol 115:1925–1930.

MacEwen GD, Bunnell WP, Sriram K 1975 Acute neurological complications in the treatment of scoliosis. A report of the Scoliosis Research Society. J Bone Joint Surg Am 57A:404–408.

Machida M, Weinstein SL, Yamada T, et al 1985 Spinal cord monitoring: electrophysiological measures of sensory and motor function during spinal surgery. Spine 10:407–413.

Machida M, Weinstein SL, Yamada T, et al 1988 Monitoring of motor action potentials after stimulation of the spinal cord. J Bone Joint Surg Am 70A:911–918.

Machida M, Weinstein SL, Imamura Y, et al 1989 Compound muscle action potentials and spinal evoked potentials in experimental spine maneuver. Spine 14:687–691.

Machida M, Yamada T, Ross M, et al 1990 Effect of spinal cord ischemia on compound muscle action potentials and spinal evoked potentials following spinal cord stimulation in the dog. J Spinal Disord 3:345–352.

Macon JB, Poletti CE 1982 Conducted somatosensory evoked potentials during spinal surgery. Part 1: Control conduction velocity measurements. J Neurosurg 57:349–353.

Madigan RR, Linton AE, Wallace SL, et al 1987 A new technique to improve cortical-evoked potentials in spinal cord monitoring: a ratio method of analysis. Spine 12:330–335.

Mahla ME, Long DM, McKennett J, et al 1984 Detection of brachial plexus dysfunction by somatosensory evoked potential monitoring – a report of two cases. Anesthesiology 60:248–252.

Maiman DJ, Mykebust JB, Ho KC, et al 1989 Experimental spinal cord injury produced by axial lesions. J Spinal Disord 2:6–13.

Malone M, Prior P, Scholtz CL 1981 Brain damage after cardiopulmonary bypass: correlations between neurophysiological and neuropathological findings. J Neurol Neurosurg Psychiatry 44:924–931.

Manninen PH 1998 Monitoring evoked potentials during spinal surgery in one institution. Can J Anaesth 45:460–465.

Manninen PH, Patterson S, Lam AM, et al 1994 Evoked potential monitoring during posterior fossa aneurysm surgery: a comparison of two modalities. Can J Anaesth 41:92–97.

Markand ON, Dilley RS, Moorthy SS, et al 1984 Monitoring of somatosensory evoked responses during carotid endarterectomy. Arch Neurol 41:375–378.

Markand ON, Warren C, Mallik GS, et al 1990 Temperature-dependent hysteresis in somatosensory and auditory evoked potentials. Electroencephalogr Clin Neurophysiol 77:425–435.

May DN, Jones SJ, Crockard HA 1996 Somatosensory evoked potential monitoring in cervical surgery: identification of pre- and intra-operative risk factors associated with neurological deterioration. J Neurosurg 85:566–573.

McAuley DL, Ross Russell RW 1979 Correlation of CAT scan and visual field defects in vascular lesions of the posterior visual pathways. J Neurol Neurosurg Psychiatry 40:298–311.

McCallum JE, Bennett MH 1975 Electrophysiologic monitoring of spinal cord function during intraspinal surgery. Surg Forum 26:469–471.

McCarthy WJ, Park AE, Koushanpour E, et al 1996 Carotid endarterectomy. Lessons from intra-operative monitoring – a decade of experience. Ann Surg 224:297–305.

McGrail KM 1996 Intraoperative use of electroencephalography as an assessment of cerebral bloodflow. Neurosurg Clin North Am 7:685–692.

McKay RD, Sundt TM, Michenfelder JD, et al 1976 Internal carotid artery stump pressure and cerebral blood flow during carotid endarterectomy: modification by halothane, enflurane and innovar. Anesthesiology 45:390–399.

McKinsey JF, Desai TR, Bassiouny HS, et al 1996 Mechanisms of neurologic deficits and mortality with carotid endarterectomy. Arch Surg 131:526–531.

McPherson RW, Niedermeyer EF, Otenasek RJ, et al 1983 Correlation of transient neurological deficit and somatosensory evoked potentials after intracranial aneurysm surgery. J Neurosurg 59:146–149.

McPherson RW, Szymanski J, Rogers MC 1984 Somatosensory evoked potential changes in position-related brain stem ischaemia. Anesthesiology 61:88–90.

Messick JM, Casement B, Sharbrough FW, et al 1987 Correlation of regional cerebral blood flow (rCBF) with EEG changes during isoflurane anesthesia for carotid endarterectomy: Critical rCBF. Anesthesiology 66:344–349.

Meyer PR, Coder HB, Gireesan GT 1988 Operative neurological complications resulting from thoracic and lumbar spine internal fixation. Clin Orthop 237:125–131.

Meylaerts SA, Jacobs MJ, van Iterson V, et al 1999 Comparison of transcranial motor evoked potentials and somatosensory evoked potentials during thoraco-abdominal aortic aneurysm repair. Ann Surg 230:742–749.

Mezrow CK, Midulla PS, Sadeghi AM, et al 1995 Quantitative electro-encephalography: a method to assess cerebral injury after hypothermic circulatory arrest. J Thorac Cardiovasc Surg 109:925–934.

Michenfelder JD, Sundt TM, Fode N, et al 1987 Isoflurane when compared to enflurane and halothane decreases the frequency of cerebral ischaemia during carotid endarterectomy. Anesthesiology 67:336–340.

Mills WJ, Chapman JR, Robinson LR, et al 2000 Somatosensory evoked potential monitoring during closed humeral nailing: a preliminary report. J Orthop Trauma 14:167–170.

Minahan RE, Sepkuty JP, Lesser RP, et al 2001 Anterior spinal cord injury with preserved neurogenic 'motor' evoked potentials. Clin Neurophysiol 112:1442–1450.

Mineiro J, Weinstein SL 1997 Delayed postoperative paraparesis in scoliosis surgery. A case report. Spine 22:1668–1672.

Mizrahi EM, Crawford ES 1984 Somatosensory evoked potentials during reversible spinal cord ischaemia in man. Electroencephalogr Clin Neurophysiol 58:120–126.

Mizrahi EM, Patel VM, Crawford ES, et al 1989 Hypothermic-induced electrocerebral silence, prolonged circulatory arrest, and cerebral protection during cardiovascular surgery. Electroencephalogr Clin Neurophysiol 72:81–85.

Mochida K, Komori H, Okawa A, et al 1997 Evaluation of motor function during thoracic and thoracolumbar spinal surgery based on motor-evoked potentials using train spinal stimulation. Spine 22:1385–1393.

Mokrusch T, Schramm J, Hochstetter A 1988 Effect of click polarity on abnormality of intra-operatively monitored brainstem acoustic evoked potentials. Neurosurg Rev 11:33–37.

Molaie M 1986 False negative intraoperative somatosensory evoked potentials with simultaneous bilateral stimulation. Clin Electroencephalogr 17:6–9.

Møller AR 1988 *Evoked Potentials in Intraoperative Monitoring*. Williams and Wilkins, Baltimore.

Møller AR 1992 Use of auditory evoked potentials in intraoperative neurophysiological monitoring. In: *Neuromonitoring in Head and Neck Surgery* (eds JM Kartush, KR Bouchard). Raven Press, New York, pp. 199–214.

Møller AR 1995 *Intraoperative Neurophysiologic Monitoring*. Harwood Academic, Luxembourg.

Møller AR, Jannetta PJ 1983 Monitoring auditory functions during cranial nerve microvascular decompression operations by direct recordings from the eighth nerve. J Neurosurg 59:493–499.

Møller AR, Jannetta PJ 1984 Preservation of facial function during removal of acoustic neuromas: use of monopolar constant-voltage stimulation and EMG. J Neurosurg 61:757–760.

Møller AR, Møller MB 1989 Does intra-operative monitoring of auditory evoked potentials reduce incidence of hearing loss as a complication of microvascular decompression of cranial nerves? Neurosurgery 24:257–263.

Møller AR, Jannetta PJ, Bennett M, et al 1981 Intracranially recorded responses from human auditory nerve: new insights into the origin of brainstem evoked potentials (BSEPs). Electroencephalogr Clin Neurophysiol 52:18–27.

Møller AR, Burgess JE, Sekhar LN 1987 Recording compound action potentials from the optic nerve in man and monkeys. Electroencephalogr Clin Neurophysiol 67:549–555.

Møller AR, Jannetta PJ, Jho HD 1994a Click-evoked responses from the cochlear nucleus: a study in human. Electroencephalogr Clin Neurophysiol 92:215–224.

Møller AR, Jho HD, Jannetta PJ 1994b Preservation of hearing in operations on acoustic tumors: an alternative to recording brainstem auditory evoked potentials. Neurosurgery 34:688–692.

Moller JT, Cluitmans P, Rasmussen LS, et al 1998 Long-term post-operative cognitive dysfunction in the elderly: ISPOCD1 study. Lancet 351:857–861.

Momma F, Wang A-D, Symon L 1987 Effects of temporary arterial occlusion on somatosensory evoked responses in aneurysm surgery. Surg Neurol 27:343–352.

Mooij JJA, Buchthal A, Belopavlovic M 1987 Somatosensory evoked potential monitoring of temporary middle cerebral artery occlusion during aneurysm operation. Neurosurgery 21:492–496.

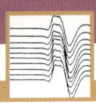

Morawetz RB, De Girolami U, Ojemann RG, et al 1978 Cerebral blood flow determined by hydrogen clearance during middle cerebral artery occlusion in unanesthetized monkeys. Stroke 9:143–149.

Morawetz RB, Crowell RH, De Girolami U, et al 1979 Regional cerebral blood flow thresholds during cerebral ischemia. Federation Proceedings 38:2493–2494.

More RC, Nuwer MR, Dawson EG 1988 Cortical evoked potential monitoring during spinal surgery: sensitivity, specificity, reliability, and criteria for alarm. J Spinal Disord 1:75–80.

Morota N, Deletis V, Constantini S, et al 1997 The role of motor evoked potentials during surgery for intramedullary spinal cord tumours. Neurosurgery 41:1327–1336.

Muizelaar JP 1989 The use of electroencephalography and brain protection during operation for basilar aneurysms. Neurosurgery 25:899–903.

Mullatti N, Coakham HB, Maw AR, et al 1999 Intraoperative monitoring during surgery for acoustic neuroma: benefits of an extratympanic intrameatal electrode. J Neurol Neurosurg Psychiatry 66:591–599.

Murie JA, John TG, Morris PJ 1994 Carotid endarterectomy in Great Britain and Ireland: practice between 1984 and 1992. Br J Surg 81:827–831.

Myers RR, Stockard JJ, Saidman LJ 1977 Monitoring of cerebral perfusion during anesthesia by time-compressed Fourier analysis of the electro-encephalogram. Stroke 8:331–337.

Nadol JB, Chiong CM, Ojemann RG, et al 1992 Preservation of hearing and facial nerve function in resection of acoustic neuroma. Laryngoscope 102:1153–1158.

Nakase H, Ohnishi H, Touho H, et al 1995 An intra-arterial electrode for intracranial electro-encephalogram recordings. Acta Neurochirugia (Wien) 136:103–105.

Nash CL 1979 Symposium: spinal cord monitoring. Spine 4:463–510.

Nash CL, Lorig RA, Schatzinger LA, et al 1977 Spinal cord monitoring during operative treatment of the spine. Clin Orthop 126:100–105.

Neu M, Strauss C, Romstock J, et al 1999 The prognostic value of intra-operative BAEP patterns in acoustic neurinoma surgery. Clin Neurophysiol 110:1935–1941.

Neuloh G, Schramm J 2002 Intraoperative neurophysiological mapping and monitoring for supratentorial procedures. In: *Neurophysiology in Neurosurgery. A Modern Intraoperative Approach* (eds V Deletis, JL Shils). Academic Press, Amsterdam, pp. 339–401.

Nevin M, Colchester ACF, Adams S, et al 1987 Evidence for involvement of hypocapnia and hypoperfusion in aetiology of neurological deficit after cardiopulmonary bypass. Lancet ii:1493–1495.

Nevin M, Colchester ACF, Adams S, et al 1989 Prediction of neurological damage after cardiopulmonary bypass surgery. Anaesthesia 44:725–729.

Niimi Y, Berenstein A 1999 Endovascular treatment of spinal vascular malformations. Neurosurg Clin North Am 10:47–71.

Nissen AJ, Sikand A, Welsh JE, et al 1997 A multifactorial analysis of facial nerve results in surgery for cerebellopontine angle tumors. Ear Nose Throat J 76:37–40.

Noordeen MH, Lee J, Gibbons CE, et al 1997 Spinal cord monitoring in operations for neuromuscular scoliosis. J Bone Joint Surg Br 79B:53–57.

Norcross-Nechay K, Mathew T, Simmons JW, et al 1999 Intraoperative somatosensory evoked potential findings in acute and chronic spinal canal compromise. Spine 24:1029–1033.

Nordwall A, Axelgaard J, Harada Y, et al 1979 Spinal cord monitoring using evoked potentials recorded from the feline vertebral bone. Spine 4:486–494.

North American Symptomatic Carotid Endarterectomy Trial Collaborators 1991 Beneficial effect of carotid endarterectomy in syptomatic patients with high-grade stenosis. N Engl J Med 325:445–453.

North RB, Drenger B, Beattie C, et al 1991 Monitoring of spinal cord stimulation evoked potentials during thoracoabdominal aneurysm surgery. Neurosurgery 28:325–330.

Nuwer MR 1986 *Evoked Potential Monitoring in the Operating Room*. Raven Press, New York.

Nuwer MR 1993 Intraoperative electroencephalography. J Clin Neurophysiol 10:437–444.

Nuwer MR 1998 Spinal cord monitoring with somatosensory techniques. J Clin Neurophysiol 15:183–193.

Nuwer MR 1999 Spinal cord monitoring. Muscle Nerve 22:1620–1630.

Nuwer MR, Dawson EG 1984a Intraoperative evoked potential monitoring of the spinal cord. A restricted filter, scalp method during Harrington instrumentation for scoliosis. Clin Orthop 183:42–50.

Nuwer MR, Dawson EG 1984b Intraoperative evoked potentials monitoring of the spinal cord: enhanced stability of cortical recordings. Electroencephalogr Clin Neurophysiol 59:318–327.

Nuwer MR, Dawson EG, Carlson LG, et al 1995 Somatosensory evoked potential spinal cord monitoring reduces neurologic deficits after scoliosis surgery: results of a large multicenter survey. Electroencephalogr Clin Neurophysiol 96:6–11.

Oberenovitch TP 1995 The ischemic penumbra – 20 years on. Cerebrovasc Brain Metab Rev 1:297–323.

Oberle J, Antoniadis G, Rath SA, et al 1998 Radiological investigations and intra-operative evoked potentials for the diagnosis of nerve root avulsion: evaluation of both modalities by intradural root inspection. Acta Neurochir 140:527–530.

O'Brien MF, Lenke LG, Bridwell KH, et al 1994 Evoked-potential monitoring of the upper extremities during thoracic and lumbar spinal deformity surgery – a prospective study. J Spinal Disord 7:277–284.

Ojemann RG, Heros RC 1986 Carotid endarterectomy. To shunt or not to shunt? Arch Neurol 43:617–618.

Ojemann RG, Levine RA, Montgomery WM, et al 1984 Use of intraoperative auditory evoked potentials to preserve hearing in unilateral acoustic neuroma removal. J Neurosurg 61:938–948.

Ojemann RG, Miller JW, Silbergeld DL 1996 Preserved function in brain invaded by tumor. Neurosurgery 39:253–258.

Osenbach RK, Hitchon PW, Mouw L, et al 1993 Effects of spinal cord ischemia on evoked potential recovery and postischemic regional spinal cord blood flow. J Spinal Disord 6:146–154.

Owen JH 1993 Intraoperative stimulation of the spinal cord for prevention of spinal cord injury. In: *Electrical and Magnetic Stimulation of the Brain and Spinal Cord* (eds O Devinsky, A Beric, M Dogali). Adv Neurol, vol. 63. Raven Press, New York, pp. 271–288.

Owen JH 1999 The application of intraoperative monitoring during surgery for spinal deformity. Spine 24:2649–2662.

Owen MP, Brown RH, Spetzler RF, et al 1979 Excision of intramedullary arteriovenous malformations using intraoperative spinal cord monitoring. Surg Neurol 12:271–276.

Owen JH, Jenny AB, Naito M, et al 1989 Effects of spinal cord lesioning on somatosensory and neurogenic-motor evoked potentials. Spine 14:673–682.

Owen JH, Naito M, Bridwell KH, et al 1990a Relationship between duration of spinal cord ischemia and postoperative neurologic deficits in animals. Spine 15:846–851.

Owen JH, Naito M, Bridwell KH 1990b Relationship among level of distraction, evoked potentials, spinal cord ischemia and integrity, and clinical status in animals. Spine 15:852–857.

Owen JH, Bridwell KH, Grubb R, et al 1991 The clinical application of neurogenic motor evoked potentials to monitor spinal cord function during surgery. Spine 16(Suppl):S385–S390.

Owen JH, Sponseller PD, Szymanski J, et al 1995 Efficacy of multimodality spinal cord monitoring during surgery for neuromuscular scoliosis. Spine 20:480–488.

Padberg AM, Bridwell KH 1999 Spinal cord monitoring: current state of the art. Orthopedic Clinics of North America 30:407–433.

Padberg AM, Wilson-Holden TJ, Lenke LG, et al 1998 Somatosensory- and motor-evoked potential monitoring without a wake-up test during idiopathic scoliosis surgery. An accepted standard of care. Spine 23:1392–1400.

Paiva T, Campos J, Baeta E, et al 1995 EEG monitoring during endovascular embolization of cerebral arteriovenous malformations. Electroencephalogr Clin Neurophysiol 95:3–13.

Pampiglione G, Waterston DJ 1961 Observations during changes in venous and arterial pressure. In: *Cerebral Anoxia and the Electroencephalogram* (eds H Gastaut, JS Meyer). Charles C Thomas, Springfield, Illinois, pp. 250–255.

Papastefanou SL, Henderson LM, Smith NJ, et al 2000 Surface electrode somatosensory-evoked potential in spinal surgery. Implications for indications and practice. Spine 25:2467–2472.

Parenti G, Marconi F, Fiori L 1996 Electrophysiological (EEG–SSEP) monitoring during middle cerebral aneurysm surgery. J Neurosurg Sci 140:195–205.

Parsons RC 1966 Electrical stimulation of the facial nerve. Laryngoscope 76:391–406.

Patel RL, Turtle MR, Chambers DJ, et al 1996 Alpha-stat acid–base regulation during cardiopulmonary bypass improves neuropsychologic outcome in patients undergoing coronary artery bypass grafting. J Thorac Cardiovasc Surg 111:1267–1279.

Payan J 1987 Nerve injury. In: *Hazards and Complications of Anaesthesia* (eds TH Taylor, E Major). Churchill Livingstone, Edinburgh, pp. 392–402.

Peacock WJ, Staudt LA 1990 Spasticity in cerebral palsy and the selective posterior rhizotomy procedure. J Child Neurol 5:179–185.

Pedrini L, Tarantini S, Cirelli MR, et al 1998 Intraoperative assessment of cerebral ischaemia during carotid surgery. Int Angiol 17:10–14.

Pelosi L, Jardine A, Webb JK 1999 Neurological complications of anterior spinal surgery for kyphosis. J Neurol Neurosurg Psychiatry 66:662–664.

Penfield W, Boldrey E 1937 Somatic motor and sensory representation in the cerebral cortex of man as studied by electrical stimulation. Brain 60:389–443.

Péréon Y, Bernard J-M, Fayet G, et al 1998 Usefulness of neurogenic motor evoked potentials for spinal cord monitoring: findings in 112 consecutive patients undergoing surgery for spinal deformity. Electroencephalogr Clin Neurophysiol 108:17–23.

Phillips LH, Blanco JS, Sussman MD 1995 Direct spinal stimulation for intraoperative monitoring during scoliosis surgery. Muscle and Nerve 18:319–325.

Piatt HR, Radtke RA, Erwin CW 1985 Limitations of brain stem auditory evoked potentials for intraoperative monitoring during posterior fossa operation: case report and technical note. Neurosurgery 16:818–821.

Pinkerton Jr JA 2002 EEG as a criterion for shunt need in carotid endarterectomy. Ann Vasc Surg 16:756–761.

Pistolese GR, Ippoliti A, Appolloni A, et al 1993 Cerebral haemodynamics during carotid cross-clamping. Eur J Vasc Surg 1(Suppl A):33–38.

Plestis KA, Loubser P, Mizrahi EM, et al 1997 Continuous electroencephalographic monitoring and selective shunting reduces neurologic morbidity rates in carotid endarterectomy. J Vasc Surg 25:620–628.

Pozzessere G, Valle E, Santoro A, et al 1987 Prognostic value of early somatosensory evoked potentials during carotid surgery: relationship with electroencephalogram, stump pressure and clinical outcome. Acta Neurochir (Wien) 89:28–33.

Prasad S, Hirsch BE, Kamerer DB, et al 1993 Facial nerve function following cerebellopontine angle surgery: prognostic value of intraoperative thresholds. Am J Otol 14:330–333.

Prass R, Lüders H 1985 Constant-current versus constant-voltage stimulation [letter]. J Neurosurg 62:622–623.

Prass RL, Lüders H 1986 Acoustic (loudspeaker) facial electromyographic monitoring. Part 1. Evoked electromyographic activity during acoustic neuroma resection. Neurosurgery 19:392–400.

Prass RL, Kinney SE, Lüders H 1987 Transtragal, transtympanic electrode placement for intraoperative electroco-chleographic monitoring. Otolaryngol Head Neck Surg 97:343–350.

Prestor B, Golub P 1999 Intra-operative spinal cord neuromonitoring in patients operated on for intramedullary tumours and syringmyelia. Neurol Res 21:125–129.

Prielipp RC, Morell RC, Walker FO, et al 1999 Ulnar nerve pressure – influence of arm position and relationship to somatosensory evoked potentials. Anesthesiology 91:345–354.

Prokop A, Meyer GP, Walter M, et al 1996 Validity of SEP monitoring in carotid surgery. Review and own results. J Cardiovasc Surg 37:337–342.

Prough DS, Stump DA, Troost BT 1990 $PaCO_2$ management during cardiopulmonary bypass: intriguing physiologic rationale, convincing clinical data, evolving hypothesis? Anesthesiology 72:3–6.

Radtke RA, Erwin CW, Wilkins RH 1989 Intraoperative brainstem auditory evoked potentials: significant decrease in postoperative morbidity. Neurology 39:187–191.

Rampil IJ, Holzer JA, Quest DO, et al 1983 Prognostic value of computerized EEG analysis during carotid endarterectomy. Anesth Analg 62:186–192.

Raudzens PA 1982 Intraoperative monitoring of evoked potentials. Ann NY Acad Sci 388:308–325.

Raudzens PA, Shetter AG 1982 Intraoperative monitoring of brain-stem auditory evoked potentials. J Neurosurg 57:341–348.

Ravussin P, de Tribolet N 1993 Total intravenous anesthesia with propofol for burst suppression in cerebral aneurysm surgery: preliminary report of 42 patients. Neurosurgery 32:236–240.

Redekop G, Ferguson G 1992 Correlation of contralateral stenosis and intraoperative electroencephalogram change with risk of stroke during carotid endarterectomy. Neurosurgery 30:191–194.

Reidy DP, Houlden D, Nolan PC, et al 2001 Evaluation of electromyographic monitoring during insertion of thoracic pedicle screws. J Bone Joint Surg Br 83:1009–1014.

Reilly EL, Kondo C, Brunberg JA, et al 1978 Visual evoked potentials during hypothermia and prolonged circulatory arrest. Electroencephalogr Clin Neurophysiol 45:100–106.

Riles TS, Kopelman I, Imparato AM 1979 Myocardial infarction following carotid endarterectomy: a review of 683 operations. Surgery 85:249–252.

Rizvi SS, Goyal RN, Calder HB 1999 Hearing preservation in microvascular decompression for trigeminal neuralgia. Laryngoscope 109:591–594.

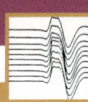

Roach GW, Kanchuger M, Mangano CM, et al 1996 Adverse cerebral outcomes after coronary bypass surgery. Multicenter Study of Perioperative Ischemia Research Group and the Ischemia Research and Education Foundation Investigators. N Engl J Med 335:1857–1863.

Robertazzi RR, Cunningham JN 1998 Intraoperative adjuncts of spinal cord protection. Semin Thorac Cardiovasc Surg 10:29–34.

Robinson LR, Slimp JC, Anderson PA, et al 1993 The efficacy of femoral nerve intraoperative somatosensory evoked potentials during surgical treatment of thoracolumbar fractures. Spine 18:1793–1797.

Rodi Z, Vodusek DB 2001 Intraoperative monitoring of the bulbocavernosus reflex: the method and its problems. Clin Neurophysiol 112:879–883.

Romstock J, Fahlbusch R, Ganslandt O, et al 2002 Localisation of the sensorimotor cortex during surgery for brain tumours: feasibility and waveform patterns of somatosensory evoked potentials. J Neurol Neurosurg Psychiatry 72:221–229.

Rose RD, Welch WC, Balzer JR, et al 1997 Persistently electrified pedicle stimulation instruments in spinal instrumentation. Technique and protocol development. Spine 22:334–343.

Rosenstein J, Wang AD-J, Symon L, et al 1985 Relationship between hemispheric cerebral blood flow, central conduction time, and clinical grade in aneurysmal subarachnoid haemorrhage. J Neurosurg 62:25–30.

Rosenthal D, Stanton Jr PE, Lamis PA 1981 Carotid endarterectomy. The unreliability of intraoperative monitoring in patients having had stroke or reversible ischemic neurologic deficit. Arch Surg 116:1569–1575.

Ross Russell RW, Bharucha N 1978 The recognition and prevention of border zone cerebral ischaemia during cardiac surgery. Q J Med 47:303–323.

Roy EP, Gutmann L, Riggs JE, et al 1988 Intraoperative somatosensory evoked potential monitoring in scoliosis. Clin Orthop 229:94–98.

Russ W, Kling D, Loesevitz A, et al 1984 Effect of hypothermia on visual evoked potentials (VEP) in humans. Anesthesiology 61:207–210.

Ryan TP, Britt RH 1986 Spinal and cortical somatosensory evoked potential monitoring during corrective spinal surgery with 108 patients. Spine 11:352–361.

Sacco DJ, Tylkowski CM, Warf BC 2000 Nonselective partial dorsal rhizotomy: a clinical experience with 1-year follow-up. Pediatr Neurosurg 32:114–118.

Sako K, Nakai H, Kawata Y, et al 1998 Temporary arterial occlusion during anterior communicating or anterior cerebral artery aneurysm operation under tibial nerve somatosensory evoked potential monitoring. Surg Neurol 19:316–322.

Sala F, Niimi Y, Berenstein A, et al 2001 Neuroprotective role of neurophysiological monitoring during endovascular procedures in the spinal cord. Ann NY Acad Sci 939:126–136.

Salvian AJ, Taylor DC, Hsiang YN, et al 1997 Selective shunting with EEG monitoring is safer than routine shunting for carotid endarterectomy. Cardiovasc Surg 5:481–485.

Salzman SK 1990 *Neural Monitoring*. Humana Press, Clifton, NJ.

Samii M, Matthies C 1997 Management of 1000 vestibular schwannomas (acoustic neuromas): the facial nerve – preservation and restitution of function. Neurosurgery 40:684–694.

Sandmann W, Kolvenbach R, Willeke F 1993a Risks and benefits of shunting in carotid endarterectomy. Stroke 24:1098.

Sandmann W, Willeke F, Kovenbach R, et al 1993b To shunt or not to shunt: the definite answer with a randomized study. In: *Current Critical Problems in Vascular Surgery*, volume 5 (ed FJ Veith). Quality Medical Publishing, St Louis, MI, pp. 434–440.

Schepens MA, Boezeman EH, Hamerlijnck RP, et al 1994 Somatosensory evoked potentials during exclusion and reperfusion of critical aortic segments in thoracoabdominal aortic aneurysm surgery. J Card Surg 9:692–702.

Schlake HP, Goldbrunner RH, Milewski C, et al 2001a Intraoperative electro-myographic monitoring of the lower cranial motor nerves (LCN IX–XII) in skull base surgery. Clin Neurol Neurosurg 103:72–82.

Schlake HP, Milewski C, Goldbrunner RH, et al 2001b Combined intra-operative monitoring of hearing by means of auditory brainstem responses (ABR) and transtympanic electrocochleography (ECochG) during surgery of intra- and extrameatal acoustic neurinomas. Acta Neurochir 143:985–995.

Schramm J, Jones SJ 1985 *Spinal Cord Monitoring*. Springer-Verlag, Berlin.

Schramm J, Romstöck J, Thurner F, et al 1985 Variance of latencies and amplitudes in SEP monitoring during operation with and without cord manipulation. In: *Spinal Cord Monitoring* (eds J Schramm, SJ Jones). Springer-Verlag, Berlin, pp. 29–34.

Schramm J, Mokrusch T, Fahlbusch R, et al 1988 Detailed analysis of intra-operative changes monitoring brain stem acoustic evoked potentials. Neurosurgery 22:694–702.

Schramm J, Koht A, Schmidt G, et al 1990 Surgical and electrophysiological observations during clipping of 134 aneurysms with evoked potential monitoring. Neurosurgery 26:61–70.

Schramm J, Zentner J, Pechstein U 1994 Intraoperative SEP monitoring in aneurysm surgery. Neurol Res 16:20–22.

Schubert A, Drummond JC, Garfin SR 1987 The influence of stimulus presentation rate on the cortical amplitude and latency of intraoperative somatosensory evoked potential recordings in patients with varying degrees of spinal cord injury. Spine 12:969–973.

Schwab ME, Bartholdi D 1996 Degeneration and regeneration of axons in the lesioned spinal cord. Physiol Rev 76:319–370.

Schwaber MK, Hall JW 1992 Intraoperative electrocochleography. In: *Neuromonitoring in Head and Neck Surgery* (eds JM Kartush, KR Bouchard). New York, Raven Press, pp. 215–228.

Schwartz MS, Colvin MP, Prior PF, et al 1973 The Cerebral Function Monitor: its value in predicting the neurological outcome in patients undergoing cardiopulmonary bypass. Anaesthesia 28:611–618.

Schwartz DM, Drummond DS, Hahn M, et al 2000 Prevention of positional brachial plexopathy during surgical correction for scoliosis. J Spinal Disord 13:178–182.

Schweiger H, Kamp H-D, Dinkel M 1991 Somatosensory-evoked potentials during carotid artery surgery: experience in 400 operations. Surgery 109:602–609.

Sebastian C, Raya JP, Ortega M, et al 1997 Intraoperative control by somatosensoery evoked potentials in the treatment of cervical myeloradiculopathy. Results in 210 cases. Eur Spine J 6:316–323.

Segal LS 1995 Combining somatosensory and motor evoked potentials for posterior spine fusion. In: *Primer of Intraoperative Monitoring* (eds GB Russell, LD Rodichok). Butterworth-Heinemann, Oxford, pp. 205–226.

Sekiya T, Møller AR 1987 Avulsion rupture of the internal auditory artery during operations in the cerebellopontine angle: a study in monkeys. Neurosurgery 21:631–637.

Sekiya T, Iwabuchi T, Kamata S, et al 1985 Deterioration of the auditory evoked potentials during cerebello-pontone angle manipulations: an interpretation based on an experimental model in dogs. J Neurosurg 63:598–607.

Sekiya T, Okabe S, Iwabuchi T 1988 Damage of the peripheral auditory system after operations in the cerebellopontine angle. A scanning electron-microscopic observation in dogs. Surg Neurol 30:117–124.

References

Selman WR, Spetzler RF, Brown R 1981 The use of intraoperative fluoroscopy and spinal cord monitoring for transoral microsurgical odontoid resection. Clin Orthoped 154:51–56.

Shaw PJ 1986 Neurological dysfunction following coronary artery bypass graft surgery. J R Soc Med 79:130–131.

Shaw PJ, Bates D, Cartlidge NEF, et al 1985 Early neurological complications of coronary artery bypass surgery. BMJ 291:1384–1387.

Shaw PJ, Bates D, Cartlidge NEF, et al 1986 Neurological complications of coronary artery bypass graft surgery: six month follow-up study. BMJ 293:165–167.

Shi YB, Binette M, Martin WH, et al 2003 Electrical stimulation for intraoperative evaluation of thoracic pedicle screw placement. Spine 28:595–601.

Shibuya M, Mutsuga N, Suzuki Y, et al 1993 A newly designed nerve monitor for microneurosurgery: bipolar constant current nerve stimulator and movement detector with a pressure sensor. Acta Neurochir 125:173–176.

Shields CB, Paloheimo LPJ, Backman MH, et al 1990 Intraoperative use of transcranial motor evoked potentials. In: *Magnetic Stimulation in Clinical Neurophysiology* (ed. S Chokroverty). Butterworth, Boston, pp. 173–184.

Shimoji K, Kurokawa T, Tamaki T, et al 1991 *Spinal Cord Monitoring and Electrodiagnosis*. Springer-Verlag, Berlin.

Shukla R, Docherty TB, Jackson RK, et al 1988 Loss of evoked potentials during spinal surgery due to spinal cord hemorrhage. Ann Neurol 24:272–275.

Silbert BS, Koumoundouros E, Davies MJ, et al 1989 The use of aperiodic analysis of the EEG during carotid artery surgery. Anaesth Intensive Care 17:16–23.

Silverstein H, McDaniel AB, Norrell H 1985a Hearing preservation after acoustic neuroma surgery using intraoperative direct eighth cranial nerve monitoring. Am J Otol 6(Suppl):99–106.

Silverstein H, McDaniel A, Wazen J, et al 1985b Retrolabyrinthine vestibular neurectomy with simultaneous monitoring of eighth nerve and brain stem auditory evoked potentials. Otolaryngol Head Neck Surg 93:736–742.

Silverstein H, Rosenberg SI, Flanzer J, et al 1993 Intraoperative facial nerve monitoring in acoustic neuroma surgery. Am J Otol 14:524–532.

Slattery WH, Brackmann DE, Hitselberger W 1997 Middle fossa approach for hearing preservation with acoustic neuromas. Am J Otol 18:596–601.

Slavit DH, Harner SG, Harper CM, et al 1991 Auditory monitoring during acoustic neuroma removal. Arch Otolaryngol Head Neck Surg 117:1153–1157.

Slimp JC 2000 Intraoperative monitoring of nerve repairs. Hand Clin 16:25–36.

Sloan TB, Ronai AK, Koht A 1986 Reversible loss of somatosensory evoked potentials during anterior cervical spinal fusion. Anesth Analg 65:96–99.

Smith B 1975 Anesthesia for orbital surgery: observed changes in the visually evoked response at low blood pressures. Modern Prob Ophthalmol 14:457–459.

Smith NJ 1997 Electric nerve stimulators: assessment and performance. J Electrophysiol Technol 23:4–17.

Smith NJ, Beer D, Clarke SA, et al 1994a Monitoring cortical evoked potentials (EPs) in operations on the cervical spine. In: *Handbook of Spinal Cord Monitoring* (eds SJ Jones S Boyd, M Hetreed, MNJ Smith). Kluwer, Dordrecht, pp. 216–221

Smith NJ, Henderson LM, Hope DT, et al 1994b Cerebral function analysing monitor (CFAM) and SSEP for detection of ischaemia during carotid endarterectomy. Electroenceph Clin Neurophysiol 91:37P.

Smith PL, Treasure T, Newman SP, et al 1986 Cerebral consequences of cardiopulmonary bypass. Lancet i(8485):823–825.

Smyth MD, Peacock WJ 2000 The surgical treatment of spasticity. Muscle and Nerve 23:153–163.

Sobottka SB, Schackert G, May SA, et al 1998 Intraoperative facial nerve monitoring (IFNM) predicts facial nerve outcome after resection of vestibular schwannoma. Acta Neurochirurgica 140:235–242.

Spetzler RF, Selman WR, Nash CL, et al 1979 Transoral microsurgical odontoid resection and spinal cord monitoring. Spine 4:506–510.

Spetzler RF, Martin N, Hadley MN, et al 1986 Microsurgical endarterectomy under barbiturate protection: a prospective study. J Neurosurg 65:63–73.

Spielholz NI, Engler GL 1990 The non-pathological variability of somatosensory-evoked potentials. In: *Neural Monitoring* (ed SK Salzman). Humana Press, Clifton, NJ, pp. 197–204.

Spinner RJ, Kline DG 2000 Surgery for peripheral nerve and brachial plexus injuries or other nerve lesions. Muscle and Nerve 23:680–695.

Stålberg A, Sharma HS, Olsson Y 1998 *Spinal Cord Monitoring. Basic Principles Regeneration Pathophysiology and Clinical Aspects*. Springer-Verlag, Vienna.

Steed DL, Peitzman AB, Grundy BL, Webster MW 1982 Causes of stroke in carotid endarterectomy. Surgery 92:634–639.

Stephen JP, Sullivan MR, Hicks RG, et al 1996 Cotrel-Dubousset instrumentation in children using simultaneous motor and somatosensory evoked potential monitoring. Spine 21:2450–2457.

Stockard JJ, Bickford RG, Myers RR, et al 1974 Hypotension-induced changes in cerebral function during cardiac surgery. Stroke 5:730–746.

Stone JG, Young WL, Marans ZS, et al 1993 Cardiac performance preserved despite thiopental loading. Anesthesiology 79:36–41.

Stone JG, Young WL, Marans ZS, et al 1996 Consequences of electroencephalographic-suppressive doses of propofol in conjunction with deep hypothermic circulatory arrest. Anesthesiology 85:497–501.

Stoughton J, Nath RL, Abbott WM 1998 Comparison of simultaneous electroencephalographic and mental status monitoring during carotid endarterectomy with regional anesthesia. J Vasc Surg 28:1014–1021.

Stuhmeier KD, Grabitz K, Mainzer B, et al 1993 Use of the electrospinogram for predicting harmful spinal cord ischemia during repair of thoracic or thoracoabdominal aortic aneurysms. Anesthesiology 79:1170–1176.

Su CF, Haghighi SS, Oro JJ, et al 1992 'Backfiring' in spinal cord monitoring. Spine 17:504–508.

Sugita K, Kobayashi S 1982 Technical and instrumental improvements in the surgical treatment of acoustic neurinomas. J Neurosurg 57:747–752.

Sundt TM 1983 The ischaemic tolerance of neural tissue and the need for monitoring and selective shunting during carotid endarterectomy. Stroke 14:93–98.

Sundt TM, Sharbrough FW, Anderson RE, et al 1974 Cerebral blood flow measurements and electroencephalograms during carotid endarterectomy. J Neurosurg 41:310–320.

Sundt TM, Houser OW, Sharbrough FW, et al 1977 Carotid endarterectomy: results, complications and monitoring techniques. In: *Advances in Neurology* vol. 16 (eds RA Thompson, JR Green). Raven Press, New York, pp. 97–119.

Sundt TM, Sharbrough FW, Piepgras DG, et al 1981 Correlation of cerebral blood flow and electroencephalographic changes during carotid endarterectomy, with results of surgery and hemodynamics of cerebral ischaemia. Mayo Clin Proc 56:533–543.

Swearingen B, Heros RC 1987 Common carotid occlusion for unclippable carotid aneurysms: an old but still effective operation. Neurosurgery 21:288–295.

Symon L 1985 Thresholds of ischaemia applied to aneurysm surgery. Acta Neurochir 77:1–7.

Symon L, Wang AD, Costa e Silva IE, et al 1984 Perioperative use of somatosensory evoked potentials in aneurysm surgery. J Neurosurg 60:269–275.

Symon L, Momma F, Schwerdtfeger P, et al 1986 Evoked potential monitoring in neurosurgical practice. Adv Tech Stand Neurosurg 14:25–70.

Symon L, Momma F, Murota T 1988 Assessment of reversible cerebral ischaemia in man: intraoperative monitoring of the somatosensory evoked response. Acta Neurochir 42(Suppl):3–7.

Szekely G, Csecsei GI, Miko L 1998 Somatosensory and motor evoked potentials in patients with tumours in the spinal canal. Acta Neurochir 140:533–538.

Tabaraud F, Boulesteix JM, Moulies D, et al 1993 Monitoring of the motor pathway during spinal surgery. Spine 18:546–550.

Tamaki T 1998 Intraoperative spinal cord monitoring-clinical overview. In: *Spinal Cord Monitoring. Basic Principles Regeneration Pathophysiology and Clinical Aspects* (eds E Stålberg, HS Sharma, Y Olsson). Springer-Verlag, Vienna, pp. 509–520.

Tamaki T, Tsuji H, Inoue S, et al 1981 The prevention of iatrogenic spinal cord injury utilizing the evoked spinal cord potential. Int Orthop 4:313–317.

Tamaki T, Noguchi T, Takano H, et al 1984 Spinal cord monitoring as a clinical utilization of the spinal evoked potential. Clin Orthop 184:58–64.

Tani T, Ishida K, Ushida T, et al 2000 Intraoperative electroneurography in the assessment of the level of operation for cervical spondylotic myelopathy in the elderly. J Bone Joint Surg Br 82:269–274.

Tator CH, Fehlings MG 1991 Review of the secondary injury theory of acute spinal cord trauma with emphasis on vascular mechanisms. J Neurosurg 75:15–26.

Taylor MD, Bernstein M 1999 Awake craniotomy with brain mapping as the routine surgical approach to treating patients with supratentorial intraaxial tumors: a prospective trial of 200 cases. J Neurosurg 90:35–41.

Taylor BA, Fennelly ME, Taylor A, et al 1993 Temporal summation the key to motor evoked potential spinal cord monitoring in humans. J Neurol Neurosurg Psychiatry 56:104–106.

Taylor BA, Webb PJ, Hetreed M, et al 1994 Delayed postoperative paraplegia with hypotension in adult revision scoliosis surgery. Spine 19:470–474.

Tempelhoff R, Modica PA, Rich KM, Grubb Jr RL 1989 Use of computerized electroencephalographic monitoring during aneurysm surgery. J Neurosurg 71:24–31.

Terzis JK, Daniel RK, Williams HB 1980 Intraoperative assessment of nerve lesions with fascicular dissection and electrophysiological recordings. In: *Management of Peripheral Nerve Problems* (eds GE Omer, M Spinner). WB Saunders, Philadelphia, pp. 462–474.

Tiel RL, Happel LT, Kline DG 1996 Nerve action potential recording method and equipment. Neurosurgery 39:103–108.

Toleikis JR 2002 Neurophysiological monitoring during pedicle screw placement. In: *Neurophysiology. A Modern Intraoperative Approach* (eds V Deletis, JL Shils). Academic Press, Amsterdam, pp. 231–264.

Toleikis JR, Skelly JP, Carlvin AO, et al 2000 Spinally elicited peripheral nerve responses are sensory rather than motor. Clin Neurophysiol 111:736–742.

Toner I, Taylor KM, Newman S, et al 1998 Cerebral functional changes following cardiac surgery: neuropsychological and EEG assessment. Eur J Cardiothorac Surg 13:13–20.

Tonn JC, Schlake HP, Goldbrunner R, et al 2000 Acoustic neuroma surgery as an interdisciplinary approach: a neurosurgical series of 508 patients. J Neurol Neurosurg Psychiatry 69:161–166.

Torrens M, Maw R, Coakham H, et al 1995 Facial and acoustic nerve preservation during excision of extracanalicular acoustic neuromas using the suboccipital approach. Br J Neurosurg 8:655–665.

Trojaborg W, Boysen G 1973 Relation between EEG, regional cerebral blood flow and internal carotid endarterectomy. Electroencephalogr Clin Neurophysiol 34:61–69.

Tsuyama N, Tsuzuki N, Kurokawa T, et al 1978 Clinical application of spinal cord action potential measurement. Int Orthop 2:39–46.

Tucker SK, Noordeen MH, Pitt MC 2001 Spinal cord monitoring in neuromuscular scoliosis. J Pediatr Orthop Part B 10:15.

Tufo HM, Ostfeld AM, Shekelle R 1970 Central nervous system dysfunction following open heart surgery. JAMA 212:1333–1340.

Uhl RR, Squires KC, Bruce DL, et al 1980 Effect of halothane anesthesia on the human cortical visual evoked response. Anesthesiology 53:273–276.

Van Dongen EP, Ter Beek HT, Schepens MA, et al 1999 Within-patient variability of myogenic motor-evoked potentials to multipulse transcranial electrical stimulation during two levels of partial neuromuscular blockade in aortic surgery. Anesth Analg 88:22–27.

van Putten MJ, Peters JM, Mulder SM, et al 2004 A brain symmetry index (BSI) for online EEG monitoring in carotid endarterectomy. Clin Neurophysiol 115:1189–1194.

Vauzelle C, Stagnara P, Jouvinroux P 1973 Functional monitoring of spinal cord activity during spinal surgery. Clin Orthop 93:173–178.

Veilleux M, Daube JR, Cucchiara RF 1987 Monitoring of cortical evoked potentials during surgical procedures on the cervical spine. Mayo Clin Proc 62:256–264.

Vespa PM, Nuwer MR, Juhasz C, et al 1997 Early detection of vasospasm after acute subarachnoid hemorrhage using continuous EEG ICU monitoring. Electroencephalogr Clin Neurophysiol 103:607–615.

Visser GH, Wieneke GH, van Huffelen AC 1999 Carotid endarterectomy monitoring: patterns of spectral EEG changes due to carotid artery clamping. Clin Neurophysiol 110:286–294.

Wagner W, Peghini-Halbig L, Maurer JC, et al 1994 Intraoperative SEP monitoring in neurosurgery around the brain stem and cervical spinal cord: differential recording of subcortical components. J Neurosurg 81:213–220.

Walter WG 1936 The location of cerebral tumours by electro-encephalography. Lancet ii:305–308.

Wan IY, Angelini GD, Bryan AJ, et al 2001 Prevention of spinal cord ischaemia during descending thoracic and thoracoabdominal aortic surgery. Eur J Cardiothorac Surg 19:203–213.

Warm Heart Investigators 1994 Randomised trial of normothermic versus hypothermic coronary bypass surgery. Lancet 343:559–563.

Warner MA 1998 Perioperative neuropathies. Mayo Clin Proc 73:567–574.

Wassermann EM 1998 Risk and safety of repetitive transcranial magnetic stimulation: report and suggested guidelines from the International Workshop on the Safety of Repetitive Transcranial Magnetic Stimulation June 5–7, 1996. Electroencephalogr Clin Neurophysiol 108:1–16.

Webster KE 1977 Somaesthetic pathways. Br Med Bull 33:113–120.

Weiss DS 2001 Spinal cord and nerve root monitoring during surgical treatment of lumbar stenosis. Clin Orthop Rel Res 384:82–100.

Weisz DJ, Yang BY, Fung K, et al 1999 Intraoperative neurophysiological monitoring of surgeries at the cervical spine. Tech Neurosurg 5:85–94.

Wellman BJ, Loftus CM, Kresowik TF, et al 1998 The differences in electro-encephalographic changes in patients undergoing carotid endarterectomies while under local versus general anesthesia. Neurosurgery 43:769–773.

Wexler I, Paley D, Herzenberg JE, et al 1998 Detection of nerve entrapment during limb lengthening by means of near-nerve recording. Electromyogr Clin Neurophysiol 38:161–167.

Wheeler DJ 1995 *Advanced Topics in Statistical Process Control*. SPC Press, Knoxville, TN.

Wheeler DJ, Chambers DS 1992 *Understanding Statistical Process Control*, 2nd edition. SPC Press, Knoxville, TN.

Whitney DW, Kahn EM, Estes JW, et al 1980 Carotid artery surgery without a temporary indwelling shunt. 1,917 consecutive procedures. Arch Surg 115:1393–1399.

Whittle IR, Johnston IH, Besser M 1984 Spinal cord monitoring during surgery by direct recording of somatosensory evoked potentials. Technical note. J Neurosurg 60:440–443.

Wilber RG, Thompson GH, Shaffer JW, et al 1984 Postoperative neurological deficits in segmental spinal instrumentation. A study using spinal cord monitoring. J Bone Joint Surg Am 66A:1178–1187.

Williamson JB, Galasko CS 1992 Spinal cord monitoring during operative correction of neuromuscular scoliosis. J Bone Joint Surg Br 74:870–872.

Wilson-Holden TJ, Padberg AM, Lenke LG, et al 1999 Efficacy of intraoperative monitoring for pediatric patients with spinal cord pathology undergoing spinal deformity surgery. Spine 24:1685–1692.

Winkler T, Sharma HS, Stålberg E, et al 1998 Spinal cord bioelectrical activity, edema and cell injury following a focal trauma to the rat spinal cord. An experimental study using pharmacological and morphological approaches. In: *Spinal Cord Monitoring. Basic Principles Regeneration Pathophysiology and Clinical Aspects* (eds E Stålberg, HS Sharma, Y Olsson). Springer-Verlag, Vienna, pp. 283–363.

Witdoeckt C, Ghariani S, Guérit J-M 1997 Somatosensory evoked potential monitoring in carotid artery surgery: II. Comparison between qualitative and quantitative scoring systems. Electroencephalogr Clin Neurophysiol 104:328–332.

Witoszka MM, Tamura H, Indeglia H, et al 1973 EEG changes and cerebral complications in open heart surgery. J Thorac Cardiovasc Surg 66:855–864.

Wober C, Zeitlhofer J, Asenbaum S, et al 1998 Monitoring of median nerve somatosensory evoked potentials in carotid surgery. J Clin Neurophysiol 15:429–438.

Wood CC, Spencer DD, Allison T, et al 1988 Localization of human sensorimotor cortex during surgery by cortical surface recording of somatosensory evoked potentials. J Neurosurg 68:99–111.

Woodforth IJ, Hicks RG, Crawford MR, et al 1996 Variability of motor-evoked potentials recorded during nitrous oxide anesthesia from the tibialis anterior muscle after transcranial electrical stimulation. Anesth Analg 82:744–749.

Woolsey CN, Erickson TC, Gilson WE 1979 Localization in somatic sensory and motor area of human cerebral cortex as determined by direct recording of evoked potentials and electrical stimulation. J Neurosurg 51:476–506.

Wright JE, Arden G, Jones BR 1973 Continuous monitoring of visually evoked responses during intraorbital surgery. Trans Ophthalmol Soc UK 93:311–314.

Yamamoto N, Takano H, Kitagawa H, et al 1994 Monitoring for spinal cord ischemia by use of the evoked spinal cord potentials during aortic aneurysm surgery. J Vasc Surg 20:826–833.

Yang LC, Jawan B, Chang KA, et al 1995 Effects of temperature on somatosensory evoked potentials during open heart surgery. Acta Anaesthesiol Scand 39:956–959.

Yingling CD, Gardi JN 1992 Intraoperative monitoring of facial and cochlear nerves during acoustic neuroma surgery. Otolaryngol Clin North Am 25:413–448.

Yingling CD, Ojemann S, Dodson B, et al 2000 Identification of motor pathways during tumor surgery facilitated by multichannel electromyographic recording. J Neurosurg 91:922–927.

York DH, Chabot RJ, Gaines RW 1987 Response variability of somatosensory evoked potential during scoliosis surgery. Spine 12:864–876.

Young WL, Moberg RS, Ornstein E, et al 1988 Electroencephalographic monitoring for ischemia during carotid endarterectomy: visual versus computer analysis. J Clin Monit 4:78–85.

Zampella E, Morawetz RB, McDowell HA, et al 1991 The importance of cerebral ischemia during carotid endarterectomy. Neurosurgery 29:727–730.

Zeitlhofer J, Steiner M, Bousek K, et al 1990 The influence of temperature on somatosensory-evoked potentials during cardiopulmonary bypass. Eur Neurol 30:284–290.

Zentner J 1991 Motor evoked potential monitoring during neurosurgical operations on the spinal cord. Neurosurg Rev 14:29–36.

Zornow MH, Grafe MR, Tybor C, et al 1990 Preservation of evoked potentials in a case of anterior spinal artery. Clin Neurophysiol 77:137–139.

Chapter 8

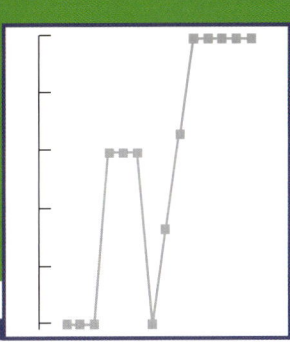

Further Signal Analysis

INTRODUCTION

In the preceding chapters many applications of neurophysiological monitoring have been discussed. We have tried to demonstrate how different states of the nervous system can be examined by using the appropriate measurement techniques. The 'input' of monitoring is typically the physiological system to be studied; the 'output' is a descriptor (or more usually a set of descriptors) that is meaningful to a human. Such descriptors come in very different forms: they can be a single rather 'low-level' feature of the recorded signal, like the root mean square (RMS) amplitude of the electroencephalogram (EEG) at a given moment, or the latency of peak Nb in an auditory evoked potential (AEP); they can be a series of numbers, maybe represented in a graph, such as a colour density spectral array (CDSA) or the RMS amplitude plotted as a function of time; and they can also be a 'high-level' descriptor like the statement 'patient is waking up' or 'high risk of ischaemia'. The conversion of the electric signals present at the measurement site into meaningful descriptors requires many steps, involving hardware (sensors, cables, filters, computers, displays) as well as signal processing methods implemented in software that act upon the signal samples to extract interpretable and useful information from it.

This chapter is meant both for the clinician who wants to understand how these systems work in order to make more efficient use of them but it is also meant for technologists and clinical scientists who will provide support and for the biomedical engineers involved in development and manufacture of new equipment.

Terms like 'hardware' and 'signal processing' may be perceived as belonging solely to the realms of electrical engineering, mathematics, software engineering and so forth. From a clinical user's perspective one can take the attitude that it is best to stay far away from these issues. To perform adequate monitoring one can manage by making sure that at the 'input side' everything is in good order (sensors applied properly, impedances satisfactory, calibrations done correctly, etc.) and then concentrate on interpretation of the monitor output as provided on screen or in files, leaving what happens in between as the province of engineers and equipment manufacturers. However, it does pay dividends to know at least something of the underlying principles of the methods that are used to generate the descriptors one relies upon to take action.

Having some knowledge of the basics of the underlying methods allows the clinician, for example:

- to understand what the limitations are of a spectral representation as provided by the ubiquitously available 'FFT-feature' on equipment; why does the spectrum shape sometimes look so noisy or smeared out?
- to appreciate what a signal processing system can or cannot do to remove artefacts from a recording;
- to get some idea of what underlies a 'magical number' that a Bispectral Index (BIS) or Spectral Entropy module in an EEG monitor outputs as a measure of the hypnotic component in anaesthesia;
- to participate in research and development (R&D) efforts together with biomedical engineers and equipment manufacturers to identify problems with current systems and, building upon current knowledge, initiate new research directions, which may lead to equipment that better fulfils clinical needs; and
- to assess merits of advertised features in new equipment from another point of view than only a clinical one.

The aim of this chapter is to give some insight into what happens 'behind the curtains' of the monitor display and provide the (non-technical) reader with basic knowledge on the usefulness of certain methods for certain applications.

A large part of the subject matter of this chapter could be categorized as traditionally belonging to the area of biomedical engineering. Biomedical engineering is a very wide field. The International Federation for Medical and Biological Engineering (IFMBE) defines it as:

Medical and Biological Engineering (MBE) integrates physical, mathematical and life sciences and engineering principles for the study of biology, medicine and health systems and for the application of technology to improving health and quality of life. It creates knowledge from the molecular to organ systems levels, develops materials, devices, systems, information approaches, technology management, and methods for assessment and evaluation of technology, for the prevention, diagnosis, and treatment of disease, for health care delivery and for patient care and rehabilitation.

Focusing on those parts of biomedical engineering most relevant to neurophysiological monitoring systems leads us to the area of 'biosignal processing' or 'biosignal interpretation and analysis' (BSI). It deals with methods for

Chapter 8 Further Signal Analysis

acquisition, processing, and analysis of signals that originate from living systems. We will limit the discussion to methods that are relevant to the application area of neurophysiological monitoring during intensive care and in the operating theatre. As the aim is to give an introduction to the methods, the text mainly concentrates on summarizing the principles of methods and their merits for their use in practice. Tables to allow easy and quick reference will be used in many sections. The aim is not to provide an exhaustive overview of all approaches and associated references, but rather to emphasize those that have been demonstrated to be useful in practical situations, together with the latest trends and ideas. The reader interested in more exhaustive treatments on the subject of biomedical signal processing in general is referred to in Devasahayam (2000), Bruce (2001) and Rangayyan (2002). Scientific journals that regularly publish articles in this area are, for example, *IEEE Transactions on Biomedical Engineering, Computer Methods and Programs in Biomedicine, Journal of Clinical Monitoring and Computing, IEEE Transactions on Information Technology in Biomedicine, Medical and Biological Engineering and Computing* and *EMB Magazine*.

The outline of this chapter is organized around the 'biosignal interpretation chain' as it is depicted in *Fig. 8.1*.

At the start of the chain is the (biological) system process under study. Sensors are used to register changes in physical quantities that are associated with this process. If the quantity is not electric in nature it typically is transformed to an electric signal using a transducer. As the resulting signal is typically very small it needs to be amplified for further processing. After amplification and filtering steps it can be digitized using an analogue-to-digital (AD) converter that takes samples of the signal at a given sampling rate. After digitization the samples are stored in the memory of the monitoring device (usually in the memory of the monitoring device). These steps together are often referred to as the data-acquisition phase; this is discussed in *Data acquisition* (p. 311).

Subsequently the recorded data enter the analysis phase. A crucial step before aiming to extract something useful from the data is making sure that the biological data at hand actually make sense, i.e. are valid for the clinical purpose at hand. The aim may have been to obtain data that are directly related to the process under study, but it may very well be that the sensors have registered signals originating from other sources as well. These other sources may be biological in nature (e.g. when we aim to study electrical activity originating from the brain but register the electrical effects of muscle activity as well) or they may originate from external, physical sources like electromagnetic interference caused by equipment in the vicinity of the measurement set-up or effects of movements of electrodes or problems with cables. Methods to deal with artefacts and noise are discussed in *Improving signal quality* (p. 317).

Once we are confident that the data thus obtained are of sufficient quality the next step is to actually analyse the data. The methods that will be used in that step are entirely dependent on the goal of the monitoring. For example, in long-term monitoring the goal may be to detect trends, whereas in other applications the aim may be to detect short-term events (e.g. epileptic seizures or a state of ischaemia), or to recognize a certain morphology (shape) of an underlying signal and classifying it, for example, as either 'normal' or 'abnormal' (e.g. in evoked potential (EP) analysis). This is thus an area in which many different signal processing, analysis and classification methods can be of use. Most of these methods can be considered as being 'general' analysis methods (that can in principle be applied on any signal) applied in a biomedical context, but some of them were developed specifically with the underlying signal/application in mind. *Signal processing and interpretation* (p. 327) aims to give an overview of those methods that are most prominent and relevant in the context of neurophysiological monitoring. The aim is to provide a description concentrating on their principle of operation and an outline of their advantages, disadvantages, and possible caveats related to their use in monitoring practice. References are provided for those wanting to explore the underlying technical details further.

The concluding section then 'takes a step backward' from the monitoring equipment box and discusses its role as a device that is supposed to help a user. The reader will have observed that mainly *digital* signal processing will be the subject of discussion, i.e. it is assumed that signals will be sampled and processed with digital equipment (typically based on personal computers (PCs)). This is not to imply that analogue methods are not relevant, but the fact is that most of today's (and tomorrow's) monitoring

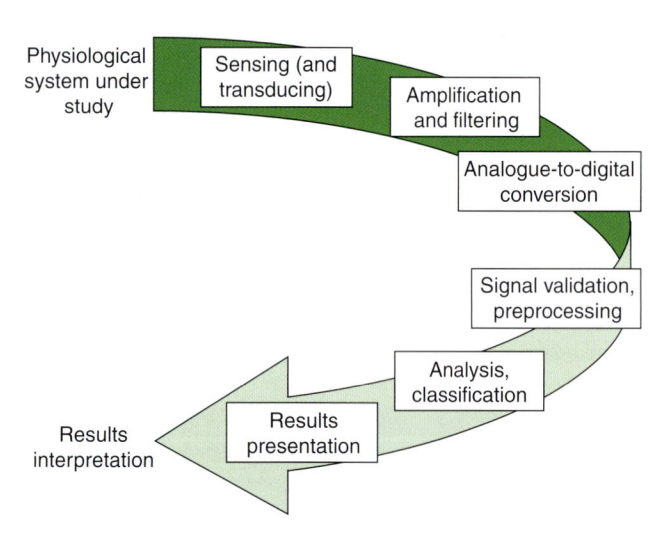

Fig. 8.1 The components of the biosignal interpretation chain. Starting with the subject under study at top left we encounter: the data-acquisition phase in which signals are picked up via sensors, amplified, filtered and subsequently digitized; the signal validation phase in which issues like artefacts and noise are dealt with; the analysis and classification phase in which information is extracted from the observed data and transformed into meaningful descriptors, that are finally presented to the user who will interpret them.

equipment is digital, and digital monitoring equipment is one part of the 'digital neurophysiology' environment (Binnie et al 2003, Ch. 7.8). Advantages of digital systems include the possibility for fast analysis, statistical postprocessing, comparison to reference values, and flexibility in methods of display, reporting and storage. The software for the recording system may be so flexible that layout of displays and some functions can be tailored for the user's needs. Great flexibility may, however, sometimes be demanding for the user and has usually not been used to its full extent. Questions arise about how to evaluate the usefulness of technical methods in clinical use, what kind of requirements to set upon the performance, on the usability aspects, how to decide which features are potentially useful and which not (both from a developer's perspective as well as from a user's perspective). This chapter ends by addressing these questions, thereby hoping to provide a practical guide for the potential user of new equipment.

DATA ACQUISITION

Technology

Traditional EEG machines recorded on a moving paper chart. 'Paperless' systems which display the signals on workstations and store them on digital media were an exciting new development in the 1990s, and now account for most sales of new equipment (even though in a few countries only analogue equipment is available at present). Modern machines use entirely digital storage and display and, apart from the amplifiers, may contain only a standard PC. Conventional EEG machines and digital workstations share many basic features. This text mainly concentrates on the data-acquisition process taking place within a digital system, noting that many of the underlying principles are equally valid for both traditional machines and digital systems.

If we 'zoom-in' in on the upper part of *Fig. 8.1* we can get a closer look at what happens in the first phase of a typical biosignal interpretation application; the data-acquisition phase. A more detailed representation of that phase is given *Fig. 8.2*.

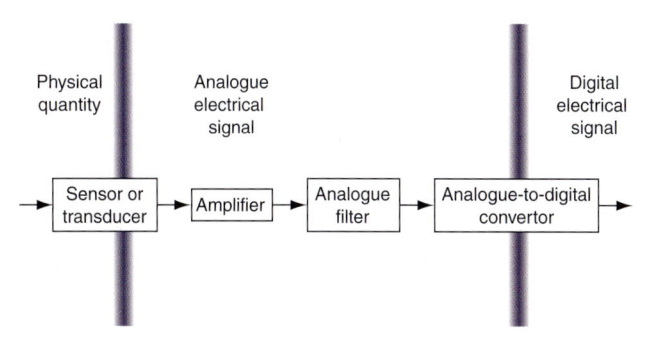

Fig. 8.2 The data-acquisition phase in more detail, depicting the transition from a physical quantity under study to a digital representation that can be further processed in a computer.

The recording process

Sensors and transducers The first step of measuring a physiological signal is to convert the physical quantity at hand into an electrical signal. In the case of electrophysiology, the quantity under study is already electrical in nature (differences between electrical potentials in the human body) and can be directly picked up by sensors (electrodes), but it is good to keep in mind that there are many applications in which we explicitly have to transform some other quantity (e.g. pressure, sound, light intensity) into an electrical signal using a transducer. The underlying technology of electrodes is the domain of biomedical instrumentation and we will not dwell on the details here, the interested reader is referred to *Electrodes for neurophysiological recording* (p. 18) in this volume as well as Sections 1.2.1.4, 1.2.1.5, 2.5.2.4 and 3.2.1.3 in Binnie et al (2004) and Section 4.2.1 in Binnie et al (2003) for details on electrodes for use in EEG, electromyography (EMG) and EP recordings. More treatments on biomedical instrumentation in general can be found in Webster (1992), Cobbold (1988), Bronzino (2000) and Cooper et al (2005).

Amplifiers EEG machines must be capable of recording electrical activity from a level of one or two microvolts up to one or two millivolts within a frequency range of about 0.1–100 Hz. For EP recordings the frequency range is considerably higher, ranging up to 3000 Hz when recording BAEPs. For EMG and nerve conduction studies they must be able to amplify signals with amplitudes of between a few microvolts to around 100 mV within a frequency range of 2 or 1 Hz to 20 kHz.

Exceptionally, direct current (DC) amplification (infinite time constant) should be available at the microvolt level. Current machines amplify and display 8, 16, 21 or more EEG channels with additional channels for polygraphic recording (respiration, electrocardiogram (ECG), etc.). For routine use in the diagnostic laboratory, 21 EEG channels may be considered the minimum standard as this allows all the electrodes of the 10–20 system to be included simultaneously in common reference montages. In special circumstances, for example in intracerebral recording (when many electrodes are usually available), more channels are required. For monitoring purposes in the intensive care unit (ICU) and surgical monitoring, many fewer channels (often 2–4) are used in order to reduce the data display to a valid minimum and allow recording of a range of other physiological signals such as ECG, respirations, body movements and surface EMG.

Machines having 16 or more channels with pen writers are transportable in that they can be wheeled but not carried; for recordings away from the base hospital 8-channel portable machines are available, but digital systems of up to 32 channels based on laptop computers are much to be preferred.

In most laboratories there are considerable electrical fields from mains (50/60 Hz) operated equipment including the EEG machine itself. These can cause interference in various ways, as any amplifier working at microvolt level is liable to be affected by mains-borne interference as well as artefact arising from movement of connecting wires and cables. The mechanisms and possible sources of mains interference are discussed in *Improving signal quality* (p. 317).

Most mains interference arises in the input circuits and the system is less susceptible if the leads between patient and amplifiers are kept as short as is practicable. Modern machines have preamplifiers in a headbox that can be placed close to the subject. The purpose of these preamplifiers is not so much for amplification of the signal but to convert the impedance of the signal source from several thousands of ohms at the electrodes to a few ohms, so that any subsequent transmission along cables to the main amplifiers in the EEG machine is done at low impedance, which reduces the possibility of pick-up and cable movement artefact.

The sockets on the headbox should provide not only for the standard electrode placements but also for additional EEG and polygraphy electrodes. A reasonable provision is for 21 standard placements, sphenoidal or nasopharyngeal electrodes, at least three earths (useful when connecting ancillary equipment), 5–12 non-standard microvolt inputs and two ECG sockets, possibly connected directly to an ECG amplifier. Sockets should be located on a head diagram, designated by the standard abbreviations of the 10–20 system and preferably by numbers also. For applications using entirely non-standard placements, such as electrocorticography or depth recording, a rectangular matrix of numbered sockets is more convenient than irrelevant letter codes on a head outline, and more inputs may be required than for routine EEGs.

One of the problems when preamplifiers are mounted in a headbox is that calibration is more complex, and when purchasing a new machine the buyer should ascertain whether the preamplifiers are tested by the internal calibration circuit.

Analogue filters Once amplified, the signal can be converted to a series of digital samples that can be processed conveniently in a computer using a variety of digital filters (see *Digital filters*, p. 318) to, for example, limit frequency contents to those relevant for the study at hand, suppress noise, or enhance certain signal contents. However, before this conversion can be done, analogue filtering needs to take place to ascertain that the bandwidth/frequency contents of the analogue signal to be converted are below a certain frequency. The usefulness of having a digital representation of an analogue signal stands or falls with the property that the former is completely representative of the latter. According to the Shannon theorem (also known as the 'sampling theorem') (Shannon 1949), a continuous time signal can be completely recovered from its sampled version if, and only if, the sampling frequency is greater than twice the highest frequency present in the signal

(the quantity '2 × signal bandwidth' is often referred to as the Nyquist frequency). If we use a sampling frequency, f_s, that is smaller than that, the digital signal will not be representative of the original signal and will suggest the presence of some slower frequency contents that are not really there. The effect is illustrated in *Fig. 8.3* in which a 20 Hz signal is sampled at rates of 100, 50 and 25 Hz.

In this figure, the Nyquist frequency is 40 Hz, so only the first two sampling rates will lead to a correct digital representation of the 20 Hz signal; the 25 Hz sampling rate results in a suggested signal of 5 Hz. Frequency contents in the analogue signal, f_{org}, that are higher than half the sampling rate will be 'folded back' over $f_s/2$ to form an 'aliased' frequency f_{alias}. Since the sampling frequency is fixed at the start of the measurement the only way to ensure that no aliasing occurs once the measurement has started is by removing all frequency contents higher than $f_s/2$ in the analogue signal. This is done by using an analogue low-pass filter – called an anti-aliasing filter.

Note that there is risk of aliasing whenever activity is present at more than half the sampling frequency, irrespective of whether or not that activity is of any interest. A 60 Hz mains interference will cause aliasing if sampled at 100 Hz, even if the wanted signal is an event-related slow potential, containing little information above 40 Hz (the 60 Hz mains interference would be present as an aliased frequency of 100/2 – (60 – 100/2) = 40 Hz in the reconstructed digital signal). Similar problems can occur when high-frequency EMG interference artefacts disturb EEG measurements.

Analogue filtering in this phase may comprise other filters than merely an anti-alias filter. Analogue filters may also be employed to minimize artefacts in this stage already. Examples include high-pass filters to remove low-frequency movement artefacts, or so-called 'notch filters' to selectively remove frequency contents in a small range (e.g. near the 50 Hz or 60 Hz mains frequency). Such uses will be further discussed in *Filtering methods* (p. 318). Details on analogue filters used in this context can be found in *Signal bandwidth and filters* (p. 35).

Analogue-to-digital conversion AD conversion is done using a so-called AD converter. Often it is implemented as a card that can be plugged into a standard PC. In digital EEG systems, analogue to digital conversion and possible multiplexing take place in the headbox.

Two steps constitute the AD conversion process: sampling and quantization.

If we know the highest frequency that can occur in the signal under study we can make an estimation of what the minimum sampling frequency needs to be in order to get a representative digitized signal. *Table 2.4* lists the approximate bandwidths of several neurophysiological signals. From this table one could conclude that a clinical EEG should be sampled at least 140 Hz, auditory brainstem EP recordings require a sampling frequency of at least 6 kHz, etc.

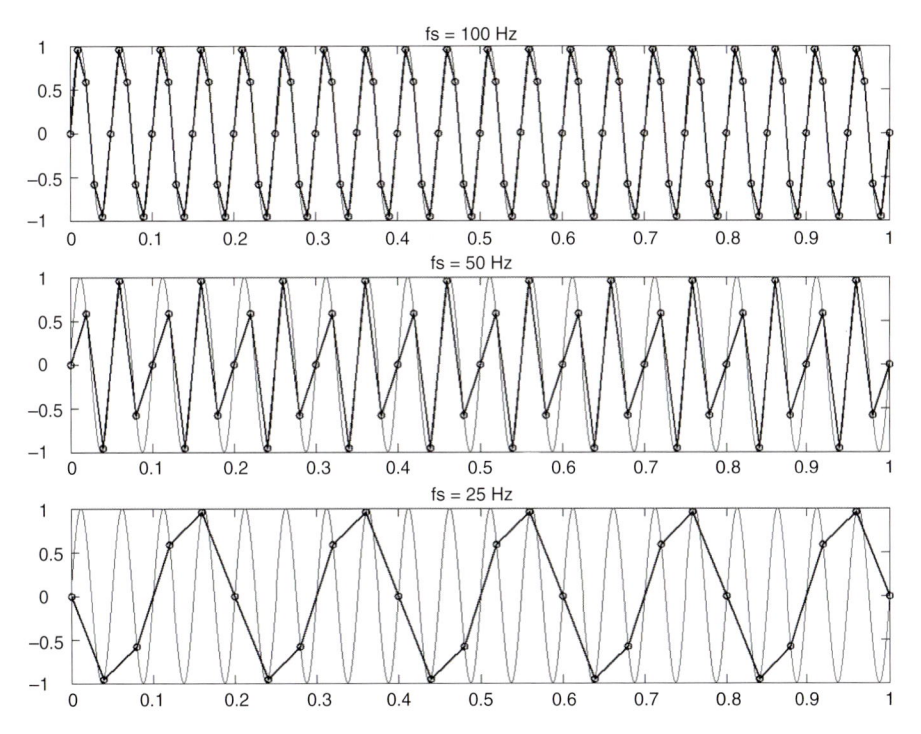

Fig. 8.3 Aliasing. Example of a segment of 1 second of a sine wave with a frequency of 20 Hz (f_{org}) sampled at different frequencies, f_s; f_s = 100 Hz, 50 Hz, and 25 Hz (the thin curved line is original analogue signal, the thick straight-segment line is digital signal as reconstructed by interpolating the samples taken at the marker positions). Only the figures in which the sampling frequency is higher than 2 times the original frequency ($2*f_{org}$ = 40 Hz) give digital signal curves that have a frequency of 20 Hz as well. For example, sampling at 25 Hz results in an 'aliased' frequency in the digital signal (f_{alias}) of 5 Hz as the analogue frequency of 20 Hz 'folds back' over $f_s/2$ (12.5 Hz); $f_{alias} = f_s/2 - (f_{org} - f_s/2) = 12.5 - (20 - 12.5) = 5$ Hz.

In practice one chooses higher sampling frequencies and uses a minimum of 5, and preferably 10, points per cycle to define a component, an EP wave or an EEG spike, for instance. (In *Fig. 8.3* one can see, for example, that although a sampling frequency of 50 Hz gives a correct result in the sense that it reproduces the 20 Hz frequency contents of the original signal it does not give a very accurate representation of the curve shape in itself. The 100 Hz sampling approximation is already a somewhat better representation.) Using a sampling frequency higher than the Nyquist frequency is referred to as oversampling. Another reason for using a higher frequency than the Nyquist frequency lies in the fact that the frequency response curve of an analogue low-pass filter does not have a strict 'brick wall' shape, passing all frequency contents below the cut-off frequency and abruptly totally blocking all frequencies directly above it, but rather has a graded 'roll-off'. This implies that some of the higher frequency contents near the filter's cut-off point will still remain in the filtered signal. For this reason it is better to 'err on the safe side' and use a higher sampling frequency.

An internal clock in the AD converter takes care of the timing of successively taking samples of the voltage at the analogue input(s). The result of this step is that the continuous signal is transformed into a discrete time series.

The next step concerns quantization of the signal; each obtained sample is coded into a binary number that approximates the original analogue value as closely as possible (e.g. using rounding or truncation). The length (in bits) of the binary number determines the number of different values that can be represented. For example, using 12-bit data gives the opportunity to use $2^{12} = 4096$ different digital values. Obviously, adding more bits potentially improves the quality of quantization. As Picton et al (1984) have shown, 12-bit registration is adequate for most neurophysiological applications, giving, for example, a resolution of approximately 1 µV over a range of ± 2 mV. It should be noted that good digital representation is only achieved when the amplitude range of the input signal covers the set input range of the AD converter as well as possible. If we set the AD converter's input range to ± 2 mV, but present an analogue signal to it with an amplitude in the range of only ± 0.5 mV, we would in practice only use one-fourth of the possible 4096 quantization levels, leaving us with a quantization resolution of the order of only 1000 steps to describe the different values of the signal. Thus, for this reason as well, proper analogue amplification is necessary.

Once obtained, digital values of the different inputs are interleaved into one stream of data (multiplexed) and transmitted sequentially by a single cable to an interface card in a standard computer that takes care of further processing, storage and display of the data.

In summary, the signals are recorded in referential format, as the potential difference between each electrode and a common reference, converted to digital form and input to a computer. Thereafter, all operations on the data are mathematical functions. The various display montages are set up by simple subtraction. For instance the potential difference between electrodes A and B, V_{A-B}, is given by:

$$V_{A-B} = V_{A-ref} - V_{B-ref}$$

where V_{A-ref} and V_{B-ref} are the potential differences recorded between A and B and the common reference. The remontaged signal values calculated in this way are plotted on a suitable display device, usually a VDU. The sensitivity of the display, the scale of trace deflection to signal amplitude (in µV/cm), the time scale (mm/s) and any frequency filtration, are all determined by computer software.

Some form of automatic calibration is usual. When the computer is switched on, or when the recording program is started, calibration signals are injected into all the preamplifiers. The resulting digital signals are compared with the expected values and corrections calculated for DC offset and sensitivity. If these are within tolerated limits, the necessary corrections are applied to signals subsequently recorded, either by adjusting the amplifier characteristics or by simple digital manipulation of the input values (subtracting the DC offset and multiplying by a sensitivity correction factor). If the errors are outside tolerated limits a message is issued indicating an amplifier fault. This approach is reliable and efficient but totally invisible to the user, who may regret the inability personally to check amplifier performance, as on a traditional machine. 'Calibration' now tells us only that the manufacturer's DC and sensitivity correction procedures are working adequately and that no amplifier has totally failed.

Storage and archiving The data are stored as they are recorded, either on a local hard disc or over a network to a file server. Ideally, even if the server option is generally used, local storage facilities should be available, so that recording is possible when the network is shut down or at sites without access to a network socket. As an hour of 30-channel digital EEG recorded with acceptable spatial and temporal resolution typically occupies about 40 megabytes, a modest hard drive (anno 2004) built into the recorder will have a storage capacity of upwards of 2000 h, and the server may store many months' output of a busy department. For long-term archiving, records are backed-up onto optical discs, DVDs or CD ROMs. It is prudent to take advantage of the reduced cost and volume of storage to keep duplicate archives on different sites in case of fire or other mishaps.

In order to be able to access the stored data for analysis in a later phase one needs software to read the data files created during the recording session. Neurophysiological data are typically stored in binary data files following specific, usually manufacturer-unique, formats so that they cannot be accessed using general-purpose software. In some cases software may exist to read, convert and save data in different formats, including manufacturer-independent formats such as the European Data Format (EDF) (Kemp et al 1992). The issue of storing the 'pure EEG' data in conjunction with other data, such as features derived from the EEG, manually entered annotations, general observations, indications of artefacts, occurrence of alarms, and maybe digital video, all properly time stamped, is much more complex altogether and requires a database-like approach. Several proposed standards for archiving and exchanging biomedical signals exist, ranging from the relatively straightforward EDF to the more complex formats such as ASTM-E-1467-94 (American Society for Testing Materials 1994) and the format based on the CEN/TC251/WGIV model (vital signs information representation (European Committee for Standardisation 2000)). As discussed by Värri et al (2001) each approach has its advantages and disadvantages.

Checking the performance of the recording set-up

Various checks of calibration and other aspects of technical performance, and of safety should be carried out on delivery of a new EEG and at regular intervals thereafter. Basic tests of performance should be carried out by the technologist at the start of every working day and briefly repeated before and after each investigation. Some tests, notably those for safety, require special equipment and the attendance of a qualified engineer.

Routine daily checks The routine checks performed before and after each recording establish that the machine is fit for use, allow faults requiring immediate correction to be identified, and, no less importantly, provide a permanent record of machine characteristics which may be needed during a subsequent interpretation of the EEG to determine whether an anomaly is real or artefactual. These records of machine checks are also of value to the engineer required to investigate an intermittent fault. Meticulously to measure and adjust before every recording all those features of machine performance tested by these checks is a policy of perfection, which may be impractical to follow. If recalibration is carried out comprehensively at suitable, regular intervals (weekly or monthly, depending on the reliability and stability of the machine) and in an abridged form at the start of each working day, it should be sufficient before each recording to measure sensitivity, check most other characteristics by eye and to correct or adjust obvious faults.

Environment It should go without saying that prior to any other preliminaries the fitness of the recording environment should be checked. Is the bed or couch correctly positioned and the linen or disposable sheeting clean? Is the area free of obstructions and litter from the previous investigation? Are fire extinguishers and resuscitation equipment in place? Are all necessary electrodes, other transducers and disposables present? In case of using sensors with an expiry date, check whether they are still 'valid'. Is all recording software starting up correctly?

Expendables   The amount of free hard-disc space on the computer should be checked beforehand to guarantee that the whole recording will fit on it. Network connections should be verified and the availability of material to make backups checked.

Recording system   For a digital recording system routine checks on the characteristics of the amplifier system are superfluous, as are those on filters and sensitivity controls as these are purely mathematical functions. However, the need to check sensitivity and other amplifier functions remains.

In a conventional system, the amplifiers and other analogue circuits are specific to the individual channels. A defect in an amplifier or a filter will consistently affect one channel only. In a digital machine, by contrast, the only analogue amplification is associated with the headbox sockets, and therefore with individual electrodes, not channels. The concept and term 'channel' are still applied to digital machines but relate to a software function, not a signal pathway. As interchannel differences can arise only by unsound programming or corruption of the software, there is little point in regularly checking for these. The headbox amplifiers are automatically checked when the program is initiated or when the computer is booted. Appropriate corrections are made to compensate for any minor errors but this is not evident and cannot be checked by the user, except possibly if a file containing the correction factors is accessible.

Calibration as such can be limited to recording a 100 μV square wave calibration signal on the reference montage used for data acquisition. It is likely that this will show impeccable equal deflections (after corrections that are transparent to the user), or possibly an error message or grossly abnormal response on a channel outside the tolerance limits. If calibration signals over a range of amplitudes are available, it may be useful to test linearity of the amplifiers. For research purposes or any other situation where precise calibration is essential, it may be necessary to check the sensitivities for signals injected at the headbox, using the montage intended for the subsequent recording. These automatic checks do not test noise levels that can be assessed as in an analogue machine.

The printers available on many machines are digital peripherals and it is unnecessary to check their write-out characteristics, after establishing on delivery that the sensitivities and time scales on the printout are correct. It probably is advisable to check the paper transport, however, as a defect could distort the tracing as with a pen recorder.

The battery of tests described above should not only be carried out meticulously at the start of each day's work, but also briefly at the beginning and end of each recording. Machines of different designs require different tests or procedures. For any particular machine it should be possible to devise an optimal sequence so that completion of one check as nearly as possible leaves the controls set up ready for the next. With ingenuity, practice and dexterity, it is possible to carry out the entire test battery suggested above within less than 1 minute on most machines (excluding time spent on measuring tracings).

Occasional checks   At intervals determined by intensity of use, but typically every 3 months, digital machines require some housekeeping:

- *Hard drives*. The hard drive of each workstation and also of any server should be checked for redundant files, for instance EEGs or reports that have been archived and should have been deleted, records that were copied back onto the system for temporary use such as a case conference on a patient investigated some months previously. Deleted files may have been transferred to a special 'Recycle' or 'Trash Bin' directory where they are still occupying disc space. The hard drive may need defragmenting and a disc surface integrity check; the utilities provided with the operating system will indicate whether this is necessary.
- *Montage*. Temporary montages may have been created for a special investigation with non-standard electrodes, or because a headbox socket was temporarily malfunctioning. These should be identified and deleted unless it is anticipated that they will be needed again.
- *Customization*. Many systems allow users to change the layout, colour scheme, toolbars, etc., of the acquisition and review programs. It may be possible to save specific combinations of settings as 'views'. Unwanted or idiosyncratic views or settings should be removed.
- *Analysis software*. If the review program includes analysis features, brain mapping, power spectra, etc., it is usually possible to change the parameters of these. For a particular purpose, settings may have been chosen that are unsuitable for routine use. For instance, the lower frequency limit of a spectral display may have been increased in order to view higher frequencies in an EEG containing large amounts of slow activity, or the amplitude resolution for isopotential mapping may have been altered to accommodate an EEG of unusually high or low amplitude. Such parameters should be reset to their standard default values.

Safety issues

A general description of safety issues associated with neurophysiological monitoring was given in Chapter 2. However, a summary and reiteration of some key items are given here (*Table 8.1*).

Table 8.1 A summary of safety issues (see also *Safety during neurophysiological recordings in the ICU and operating theatres*, p. 55)

Issues	Preventive actions
General considerations for electrical safety during biomedical measurements (International Electrotechnical Commission Standard IEC 60601-1, 1988) Types of fault that may give rise to injury: 1. An electrode is connected to relatively high voltage source (typically equipment fault) 2. The earth connection becomes disconnected within the apparatus, at the mains outlet or in the electrical wiring system	• Use of optical or high-frequency transformer coupling • Use of single-point earthing • Regular safety checks to check for leakage and correct equipment functioning • Documentation of events and performed safety and functional tests is essential

Special considerations for monitoring during intensive care and surgery	• Use:
	  – isolated power supply
	  – single point earthing
	  – electrical isolation of the patient from monitoring equipment
	• Regular calibration checks on monitors (where possible!)
	• Checks for component failure
	• Checks for computer memory faults
	• Precautions against power supply transients and failure
	• Precautions against blocking; use relatively low gain (DC), low impedance to ground
	• Care to prevent ingress of small metallic items or liquids into monitoring equipment
	• Ensuring adequacy of general training for monitoring and understanding of the implications of failure to interpret a trace or respond to warnings or alarms correctly
	• Care to avoid mechanical hazards: take into account basic limitations of space with a high concentration of apparatus around the patient that may lead to a danger of dislodging equipment such as ventilator connections or vascular lines; a preliminary induction to appropriate behaviour should be sought from the chief nurse in charge of the operating suite; and use clear placing and labelling of cables
	• Be aware of potential problems in 'electrically susceptible patients' who have direct lines to the heart or great vessels via pacemakers, cardiac catheters or central venous pressure lines
	• Action to reduce problems in monitoring associated with surgical diathermy, fibrillators and defibrillators and balloon pumps. Preliminary checks must be made that any neurophysiological monitoring equipment to be used has isolation amplifiers that are protected against potentials from these sources
	Surgical diathermy:
	  – Check what level of protection is provided in neurophysiological monitoring equipment and, indeed, whether disconnection during diathermy is advised
	  – Decrease risk of blocking using radiofrequency (RF) filtering and clamping RF signals before they reach the level where the input amplifier will rectify it
	  – Use of a deblocking circuit is highly recommended
	  – If such a circuit is not available technologists need to be taught to short the input leads when the amplifier is blocked
	• Flammable vapours may be present in the operating theatre; check AG/APG markings (International Electrotechnical Commission, IEC-60601-1, 1988)
	• Reduction of static charges on patients or staff are usually avoided by antistatic drapes and gowns and the use of a fairly humid atmosphere
	• Intraoperative EP recording:
	  – The safest arrangement is for monitoring to be undertaken by a dedicated team (not a surgeon or anaesthetist) consisting of at least two competent persons, one of whom checks each action as it is performed
	  – Intensities of stimulation should not be excessive, particularly in view of the need to stimulate continuously at high rates for long periods
	  – Deblocking (and stop averaging) facilities during diathermy are highly desirable
	  – If such a circuit is not available technologists need to be taught to short the input leads when the amplifier is blocked

Safety specifications for electromedical equipment are detailed in the International Electrotechnical Commission Standard IEC 60601-1 (1988) (*Medical Electrical Equipment. Part 1: General Requirements for Safety*, 2nd edition) (which is to be updated in 2005) and British Standard BS EN 60601-1 (1990). In the USA the corresponding regulation is UL2601-1, which specifies lower earth leakage currents than IEC60601-1, and superceded the previous standard, UL544, on 1 January 2005 (Marcus and Biersach 2003). Particular requirements for EEG machines were published as IEC 60601-2-26 (2003) and adopted as European Standard 60601-2-26 (2003). Particular requirements for the safety of high frequency surgical equipment are detailed in IEC 60601-2-2 (1998). More information concerning these standards can be obtained at www.iec.ch, www.bsi-global.com and www.ihs.com (for the USA). American standards are described in American Electro-encephalographic Society Guideline Nine (1994).

IEC 60601-1 states: 'Equipment shall be so designed that the electric shock in normal use and in single fault condition is obviated as far as practicable.' There are two types of faults that may give rise to injury:
1. An electrode may become connected to a relatively high voltage source (with respect to earth) so that there is current flow through the electrode on the body to the earth electrode. This could be caused by a fault in the equipment.
2. The earth connection becomes disconnected within the apparatus, at the mains outlet or in the electrical wiring system. This defect can easily escape detection, unless routine safety checks are performed. If a second fault then develops it is possible that the equipment casing or chassis or patient 'earth' connection could be at mains voltage without the fuse blowing. If the patient (or operator) then touches items of equipment or water pipes that are earthed, lethal current will flow.

General safety matters for neurophysiological monitoring are wide ranging in ICU and intraoperative work. It is essential to obtain specific expert advice on electrical safety from the appropriate hospital safety officer before conducting any form of monitoring work in ICUs or operating theatres.

Protection of the patient from the power supply by input isolation is important and is mandatory if non-medical equipment, such as a computer, is connected to a non-isolated medical device.

Special knowledge and technical advice from medical physicists about safety in the operating theatre is needed and recordings should be made by experienced neurophysiology technologists with expertise in operative monitoring. The potential problems facing the inexperienced are such that some workers, even in centres with a strong tradition of monitoring, have suggested that it is virtually impossible to obtain 'clean' EEG recordings in the operating theatre or ICU because of external electrical interference from other equipment.

In addition to the general safety measures for EEG monitoring there are some particular hazards that may be encountered during EEG monitoring in the operating theatre (Prior and Maynard 1986, Bruner and Leonard 1989). These include surgical diathermy (electrocautery), defibrillators and fibrillators, possible hazards from widely spaced electrodes, flammable vapours, static electricity, sparks, high temperatures and mechanical hazards. The general principles underlying electrical safety which have been detailed in Chapter 2 are also expounded with great clarity by Burke et al (1999).

IMPROVING SIGNAL QUALITY

Neurophysiological measurements (and in fact practically all measurements) suffer from that fact that what we record is a mix of a 'pure' signal (originating from the physiological system we are trying to study), and disturbing signal(s) that originate from other sources. These 'other sources' can be of biological as well as physical origin. Examples of the former are electro-ocular, electromyogenic or electrocardiographic activity recorded during EEG measurements; examples of the latter are effects of mains interference, machine fault, movements of cables or electrodes. These disturbing signals are usually referred to as 'artefacts' or 'noise'. Illustrations of typical artefacts occurring during neurophysiological monitoring are given in Chapter 2 (and in Binnie et al 2003, Section 4.2.8). Here we shall illustrate some of the basics of signal processing methods used to minimize noise and artefacts.

Before resorting to 'artefact rejection' or 'noise removal' methods however, it is good to remember that the 'contaminating' signals do contain information. Although most 'physical artefacts' are highly undesirable and should be prevented or eliminated by good recording techniques, the much commoner 'biological artefacts' arise from events within or relating directly to the patient and may be extremely informative and reveal much about the state of the patient. They may imply anxiety or tension, somnolence or biological reactivity to external stimuli in an apparently comatose individual. In many cases a biological artefact may change from a problem to a helpful facet of the recording. Examples of the value of biological artefacts are: the electro-oculogram (EOG), which can be employed as a built-in calibration signal as anomalies in its typical appearance in commonly used montages should alert the observer to the presence of either a technical problem or a biological abnormality; and artefacts in an unconscious patient in the ICU or operating theatre, where the onset of biological artefacts, whether spontaneous or in response to noxious stimuli, may provide an early indication of emergence from coma, of undue lightness of anaesthesia, or of breakthrough from muscle relaxants. For these reasons we can regard the biological artefacts with rather more enthusiasm than the physical ones and hope that the reader will enjoy the experience of using them in understanding more about the patient behind the EEG.

In order to have a 'clean', artefact-free record in which to examine both the electroencephalographic and other biological features of interest, we must deal first with the origin and management of physical artefacts before resorting to signal processing methods. It has to be acknowledged that the traditional classification into physical and biological artefacts is somewhat arbitrary since there is considerable overlap of mechanisms of causation: for instance, a loose electrode will dramatically indicate patient movement, as well as giving rise to spontaneous artefacts of its own. It must also be borne in mind that specific signal processing methods, whether by straightforward filtering, or by more advanced automatic rejection manoeuvres may introduce problems by modifying the appearance of features that one wishes to preserve.

In critical care work, in the operating theatre or an ICU or special care baby unit, physical artefacts are usually a great problem. This is attributable to high levels of interference, for example from surgical diathermy or other equipment in the vicinity of the patient. Evidence of interference may, of course, highlight possible faults in other apparatus or connections. It is essential to plan a strategy for all recording work at the neurophysiology department before actually undertaking it. This requires careful checking of both the venue and the mobile neurophysiological apparatus in collaboration with an experienced engineer who can advise on the electrical safety aspects and certify that the equipment is suitable for the purpose. A 'dummy run' with recording in, for example, the operating theatre, to check for unexpected interference or difficulties with cables and mains supply is essential. Practical problems commonly arise in this type of work because of the number of personnel actively involved in the care of the patient, and because of a possible lack of conveniently placed mains supply points. Safely placed and preferably short cable paths should be devised. Earth loops and current leakage from electrical equipment are other potential problems that must be rigorously addressed before recording in the operating theatre or units where patients may have indwelling cardiac lines, pacemakers, etc. Signal processing methods may be helpful in improving a signal's quality, but not even the fanciest algorithm can match the use of careful planning and common sense in effectiveness.

Filtering methods

Probably the most common method used to improve the signal quality is to use a filter that emphasizes certain parts of the frequency contents of the signal and suppresses others. Filters can be implemented to operate on analogue signals (analogue filters) or on digital signals (digital filters).

Analogue filters

An example of an analogue filter is the low-pass, anti-aliasing filter, as discussed in *Signal bandwidth and filters* (p. 35). The simplest analogue filters consist of a resistor and capacitor. The impedance of a capacitor decreases as the frequency of the applied signal increases; for a DC signal its impedance is 'infinite'. Thus, by using these components in different set-ups a circuit can be obtained that suppresses certain frequency contents of the input signal. An example of a very simple realization of a high-pass and a low-pass filter using this concept is given in *Fig. 8.4*.

The exact frequency response of the filter is defined by the capacitance and resistance values of the components. The examples given here are obviously very simple realizations of analogue filters. More complex configurations of components are used in practice to realize desired frequency responses. More details about the design and realization of analogue filters can be found in, for example, Van Valkenburg (1995).

Digital filters

Digital filters are implemented in computer hardware or software and employed once the data have been digitized. They thus operate on discrete samples. A digital filter operates by taking a number of samples (often the N 'most recently measured' ones), multiplying each of them with a specific weight factor (also called a 'filter coefficient'), summing all the results of the multiplications, and using that as filter output.

A big advantage of digital filters over analogue ones is their flexibility; whereas changing the behaviour of the latter requires redimensioning of the components used, the behaviour of digital filters can be done 'on the fly' without any cost associated with changing the filter coefficients using computer software instructions. For applications where real-time behaviour in high-speed measurements is an issue dedicated signal processing hardware boards can be used to implement digital filters (comparable to, e.g., high-speed graphics boards in a PC). For applications where this is less of an issue digital filters can be easily

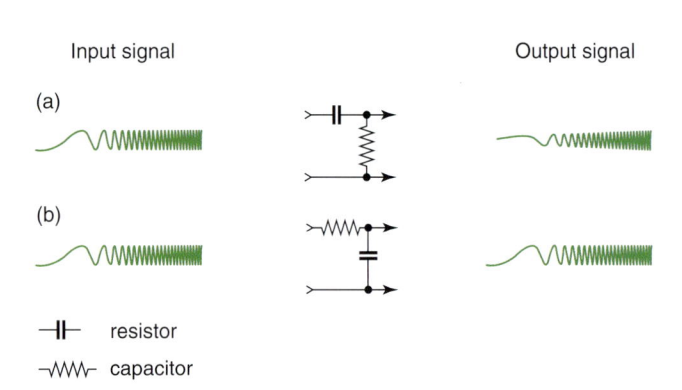

Fig. 8.4 Example of an analogue high-pass filter (a) and an analogue low-pass filter (b). The input signal is a sine wave of increasing frequency. In (a) the capacitor 'blocks' the DC signals, attenuates the low frequencies and passes the high frequencies. In (b) the high frequencies are attenuated because the impedance of the capacitor decreases as the frequency increases and 'shunts' the high frequencies. (modified from Binnie et al (2004), by permission.)

implemented in software as functions used as part of a bigger program.

As the flexibility of implementing digital filters is so high one can imagine that many different classes of filters exits. We will limit the discussion here to the most common and most useful filters for neurophysiological monitoring applications.

Linear time-invariant filters Linear time-invariant (LTI) filters are perhaps the most common type of digital filters. Time-invariant means that, once defined, the properties of the filter do not change ('the frequency response of the filter at this moment is the same as it was an hour ago'). Linear means that: (a) if one increases the amplitude of the input signal by a factor *a* then the output signal will increase by exactly the same factor *a*, and (b) if we consider the following two ways to process two signals with a linear filter:

1. we process two signals *x1* and *x2* separately by the filter, generating outputs *y1* and *y2* respectively, and then sum the filter outputs – we will get as the end result a summed signal *y1* + *y2*; or
2. we sum signals *x1* and *x2* and use the merged signal *x1* + *x2* as input to the filter, the filter's output will be the same *y1* + *y2* as we got in (1) above.

Linearity of a system makes calculations of its properties and evaluation of its behaviour considerably easier than would be the case in a non-linear situation. However, non-linear filters can have very powerful properties and outperform linear filters in some applications (see *Non-linear filters*, p. 321).

Finite impulse response (FIR) filters Without loss of generality, we assume here that our sampling rate is 1 Hz (the samples are taken at 1 second intervals). To calculate the filter output at time *t* we consider the $N + 1$ most recently recorded samples: i.e. $x(t)$, the 'current' sample; $x(t-1)$, the sample obtained 1 second ago; $x(t-2)$, the sample recorded 2 seconds ago; until $x(t-N)$, the sample recorded N seconds ago. Each sample value is multiplied with a dedicated weight factor (called a 'filter coefficient'), the most recent sample $x(t)$ is weighted with weight factor number 0, $x(t-1)$ with weight factor number 1, etc. The results of all $N + 1$ multiplications are then summed and the sum is presented as the filter output. In mathematical terms the calculation is given as:

$$y(t) = x(t) \cdot w_0 + x(t-1) \cdot w_1 + x(t-2) \cdot w_2 + \ldots$$
$$+ x(t-N) \cdot w_N$$
$$= \sum_{i=0}^{N} x(t-i) \cdot w_i$$

The output of the filter is thus fully dependent on the filter coefficients (and the input signal, of course). The number of samples to use, $N + 1$, is called the order of the filter. A higher order allows for a more refined filter behaviour (but obviously also increases computational demands). By defining the 'right' set of coefficients one can obtain the desired behaviour of the filter.

Figure 8.5 gives an example of how to calculate the output of a filter of order 5 with all filter coefficients equal to 0.2. Closer examination of the functionality reveals that this particular filter in effect does nothing more than calculate the average of the 5 most recent samples and use that as output (adding the 5 samples together and dividing the result by 5 gives the same result). One can intuitively expect that this 'moving average' filter has a smoothing effect that cancels out rapid variations in the signal and outputs the general underlying trend; it has low-pass behaviour.

The calculated output of the filter is given in *Fig. 8.6*. As another illustration, the effect of another filter, with filter coefficients alternating between 0.2 and –0.2 is given. This emphasizes differences between samples and thus suppresses low-frequency trends and emphasizes high-frequency components. The magnitudes of the frequency responses of the two filters are given in *Fig. 8.7*.

Apart from our confirmation on the assumption that one filter has a smoothing/low-pass behaviour and the other one has a high-pass behaviour, we can learn several more things from *Fig. 8.6*. We can see that the output signal is delayed with respect to the input signal; it can only start to produce sensible output once 5 data samples are available, and since the filter output is based on samples from the past, its current output will also reflect parts of the signal 'from the past', and there is thus a delay in the output. In the general case for a filter like this, if the filter uses the N most recent samples as input, the delay in output will be $(N-1)/2$ samples. This delay can easily be corrected for in software when presenting the filtered results to the user. Filters operating solely on the current sample and samples recorded in the past are called *causal* filters. We can also envisage filters that use samples 'to the right' of the current sample, i.e. samples that are to be measured 'later', such filters are called *non-causal* filters. It may be obvious that these filters cannot be implemented directly in real-time since that would require them to 'look into the future'. Instead, a block of recorded data is stored in a buffer first and subsequently the non-causal filter can be run on the buffered data.

Figure 8.7 shows the magnitude of the frequency response of the filters. A value of 1 indicates no suppression at all, a value of 0 indicates total suppression. From this figure one can conclude that the first filter passes DC signals (0 Hz) without any attenuation but suppresses high frequencies; the second filter, on the other hand, reduces a DC amplitude to about 20% of its original value. It should be noted that the frequency response of a filter is not only defined by the way it affects amplitude, but also by the way it affects the phase of the signal; certain frequencies in the input signal will be delayed more than others. This behaviour can be depicted in a phase response plot (not shown here). In practical applications the filter

Chapter 8 Further Signal Analysis

used will have a much better suppression of unwanted frequencies than the 20% that is shown in this example and we use a logarithmic scale on the *y*-axis to plot the behaviour.

When designing a filter the aim is to find a suitable set of filter coefficients that results in a certain frequency response. Many computational methods exist to estimate filter coefficients that match a user-defined frequency response. One starts, for example, with defining how the frequency response ideally should look by creating so-called pass- and stopbands in the frequency range and entering this information as input to a filter design algorithm. Together with additional information (e.g. the desired order of filter and some limits on how much the actual filter behaviour may deviate from an ideal one) the algorithm then calculates the filter coefficients. More information on digital filter design methods can be found in DeFatta et al (1988) and Proakis and Manolakis (1996). An example of a realized filter as produced by the Remez filter design algorithm is given in *Fig. 8.8*.

The particular type of filter discussed here is called a *finite impulse response* (FIR) filter. It is so called because its response to an impulse-like input, or any signal that 'ends' at a given moment, will eventually go to zero once the input signal has 'died out'. As an alternative, *infinite impulse response* (IIR) filters not only operate on recorded input

5 filter coefficients:
(w0 w1 w2 w3 w4) = (0.2 0.2 0.2 0.2 0.2)

Most recently recorded sample is multiplied by w0
2nd most recently recorded sample is multiplied by w1
3rd most recently recorded sample is multiplied by w2
4th most recently recorded sample is multiplied by w3
5th most recently recorded sample is multiplied by w4

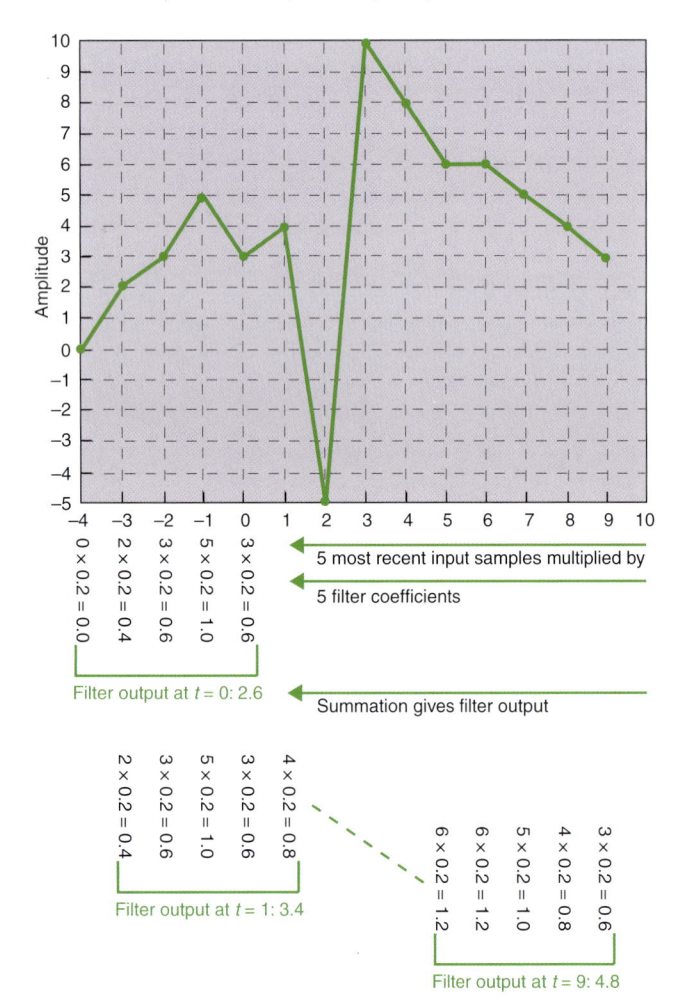

Fig. 8.5 Example of a digital filter implementing a low-pass behaviour.

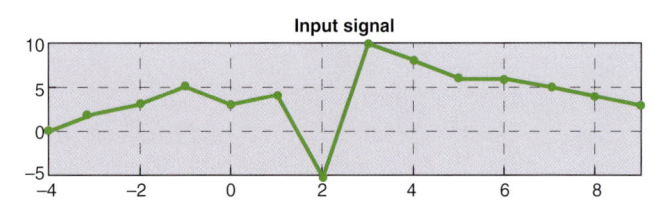

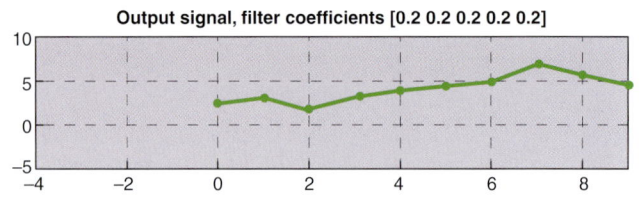

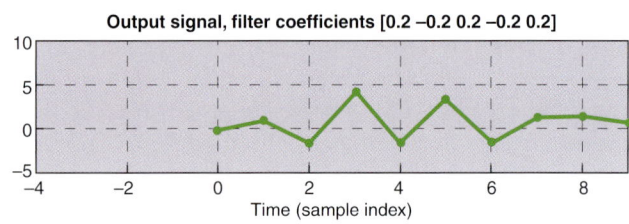

Fig. 8.6 Digital filtering example: the input signal (top) is the signal in *Fig. 8.5*, outputs as produced by a low-pass filter (middle); and a high-pass filter (bottom).

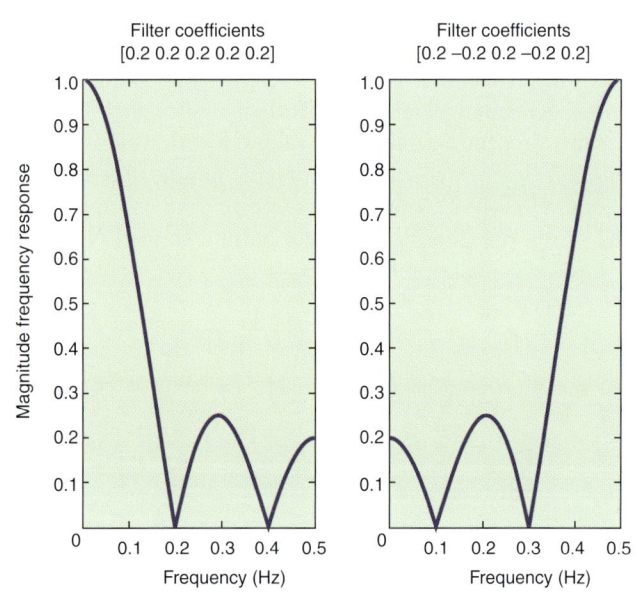

Fig. 8.7 Magnitude of the frequency responses of the filters presented in *Fig. 8.6* (assumed sampling frequency is 100 Hz).

signal samples, but also on previous filter output values (in the equation above an extra term would be added involving $y(t-1)$, $y(t-2)$, etc.). IIR filters are potentially more powerful than FIR filters (i.e. they can realize similar frequency response behaviour with a considerably smaller number of filter coefficients), but there are some disadvantages as well. The output of the filter does not necessarily go to zero once the input signal has stopped, and this can thus lead to instability in the output (if the filter has not been designed carefully). Also, IIR filters have a so-called non-linear phase shift which implies that different frequency components of the input signal appear with different time delays in the output signal. This may cause problems, for example when the aim is to measure exact latencies of complexes in a (filtered) signal.

Non-linear filters It may be obvious that the equation used to calculate a filter output as a linear combination of recorded samples is just one alternative out of many. One might expect that for some processing tasks higher order equations (using quadratic terms etc), or the use of functions more advanced than just multiplying and summing would be more effective to obtain a certain behaviour. For some applications this is true; a simple non-linear filter can outperform a linear version for certain tasks, but for the vast majority of cases the linear filter remains the best choice. Reasons to prefer a linear filter over a non-linear one are: its behaviour is well-defined for all possible input signals that can occur, it is very easy to implement in hardware or software, and well-established methods exist to design a linear filter behaving in a certain desired manner. Non-linear filters are more often designed on an empirical basis, their design process is more difficult to interpret, and some of the methods may be very sensitive to certain parameter settings yielding unexpected results in some situations.

Median filters One type of non-linear filter that is used in many biosignal processing schemes is the median filter. It is particularly effective in removing spike-like disturbances from data. A median filter is presented with a number of input samples $(2M + 1)$, which it puts in ascending order of sample magnitude and then outputs the value of the sample at position $M + 1$ (in the middle of the ordered set) as filter output. The effect is depicted graphically in *Fig. 8.9* in which a 3-point median filter is used to process a signal containing a spike disturbance and a step-change in the signal. For example, at $t = 5$ the input of the median filter consists of the sample values at $t = 4$, 5 and 6: [1, 1, 10]. The median of this series is 1, which is then used as the filter output at $t = 5$. At $t = 6$ the filter input is [1, 10, 1], again the median, and filter output, is 1. The reader can easily calculate the filter output for other samples and verify that the median filter does not affect the shape of the step change in the signal. As a comparison, the effect of using a simple low-pass FIR filter is shown as well. Although the FIR filter suppresses the amplitude of the spike it does not completely remove it. Also, since an FIR filter is linear, if the spike had an amplitude 2 times bigger, the output of the filter would also be 2 times bigger. For the median filter this does not hold; even if the spike had an amplitude of 1000, the median filter would still have removed it entirely. If the spike had

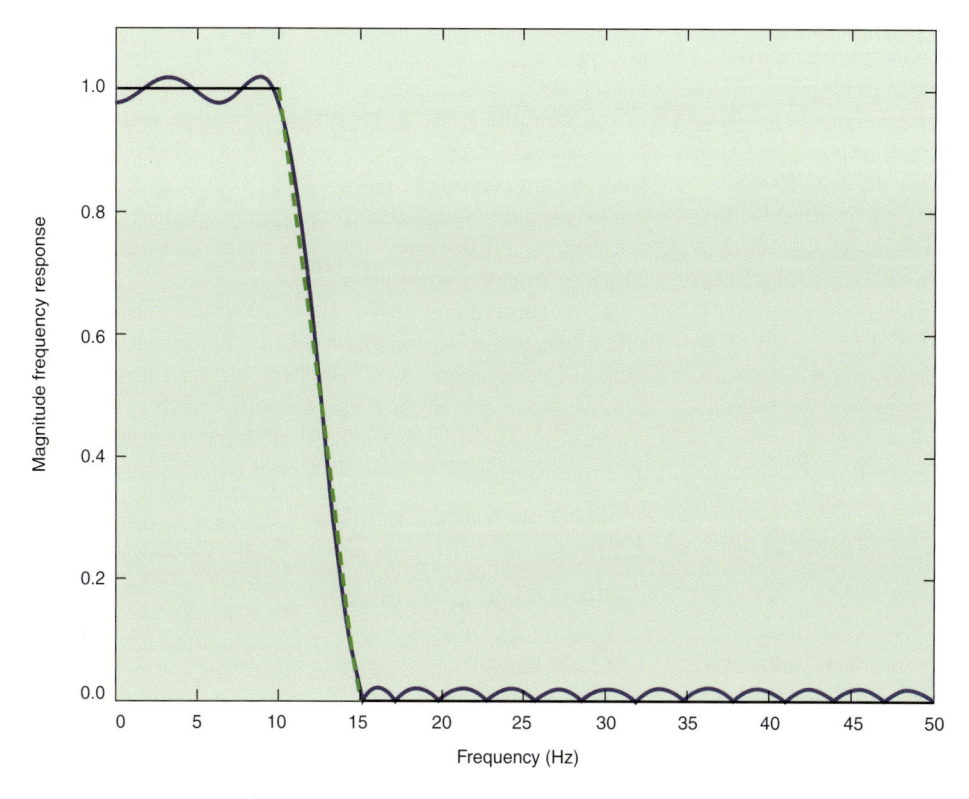

Fig. 8.8 Characteristics of a digital filter as created by a filter design algorithm (the Remez algorithm). The straight lines indicate the input to the algorithm (the 'ideal' filter); passband 0–10 Hz, stopband 15–50 Hz, and a 'don't care region' between 10 and 15 Hz (dashed line). The curved line is the frequency response of the actually realized filter (the desired filter order was set to be 33).

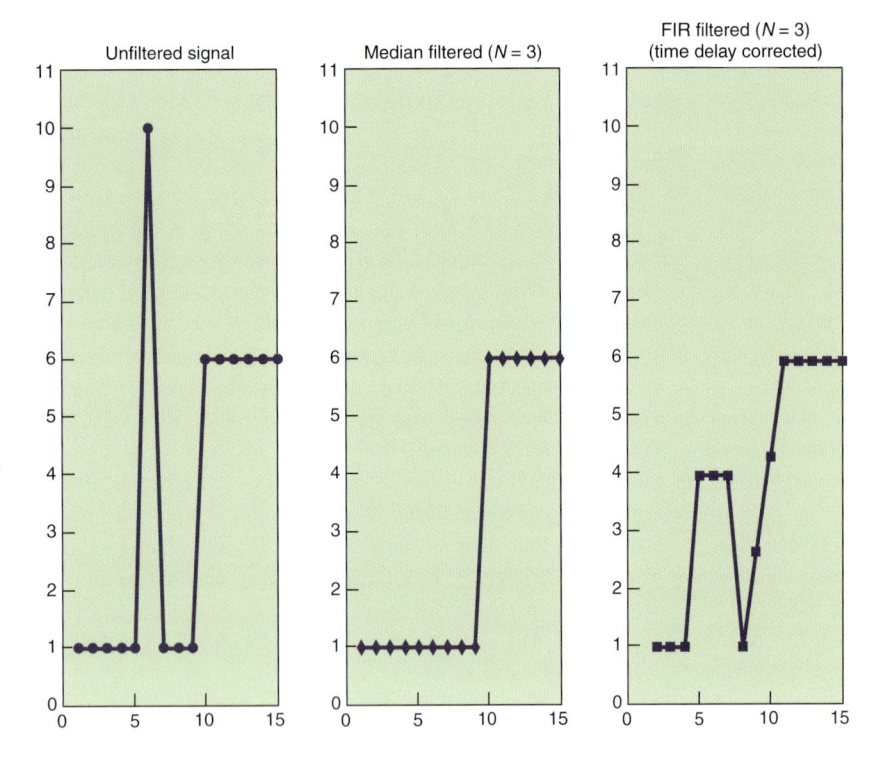

Fig. 8.9 A signal (left) containing a spike-like disturbance at t = 6 and a step change at t = 10 is filtered using a median filter (middle) and a low-pass FIR filter (right). The median filter removes the spike effectively and leaves the step change intact whereas the FIR filter suppresses, but does not completely remove, the spike and 'smears out' both the location of the spike as well as the step change.

been more samples wide we would have needed a median filter with more inputs in order to remove it completely. In general, if the spike (or any other similar local disturbance that can be considered an 'offset from a baseline') is S samples wide one needs a median filter with at least $2S + 1$ inputs to remove it. Note that the median filter as used here is a non-causal filter and we thus need buffering of data to use it.

As median filters are very effective in removing spikes and so-called 'shot-noise' they are found in many biosignal processing systems as a means to improve the signal quality. Also, as median filters leave step changes intact they are also useful in detecting changes in DC levels of signals, this is for example useful in burst suppression analysis tasks (Lipping et al 1994, Cerutti et al 1996).

A disadvantage of methods using median filters is that they do not have as elegant a theoretical background, as for example FIR filters have; one cannot simply draw a frequency response plot of a median filter as the response is different for different types of signals, and the design phase of such a filter is typically a more empirical process. Also, when the input size of a median filter grows, the number of computations needed to obtain the median will increase rapidly, thereby increasing processing load on the system. Therefore, in practice one cannot use too wide median filters. Details about the use of median filters in (neuro)physiological signal processing applications can be found in, for example, Lipping (2001).

Optimal filters In the examples in the previous sections we used filters for examples of cases where the noise was present in a certain frequency band and the underlying pure signal in another. For example, in the case in *Fig. 8.8*, we implicitly defined 'noise' in the signal as being anything above 15 Hz, and the pure signal as anything below 10 Hz. In such cases one can straightforwardly define pass- and stopbands and design an appropriate filter. This is applicable if, for example, the aim is to remove movement artefacts from an EEG recording and we define 'everything below 0.5 Hz' as non-EEG noise/artefacts, or if we want to remove high-frequency disturbances and we remove 'everything above 40 Hz' as being 'EMG and other noise'. It may be obvious that in many cases this is an overly simple approach. It may very well be that artefacts have higher frequency contents than 0.5 Hz, we may be interested in studying the EEG above 40 Hz as well, and it is quite likely that the EMG extends below that defined 40 Hz as well. So, in reality we often have overlap of noise and underlying 'pure' signal in the frequency domain and it is not that straightforward anymore to use pass- and stopbands to design a good filter.

The problem is depicted in *Fig. 8.10*; the right-hand side of the picture shows the realistic situation of overlapping noise and signal. With the depicted set-up we would remove some of the noise, but certainly not all. An alternative would be to change the filter characteristics so that all the noise would be removed (make the passband narrower), but in that case also some part of the underlying 'pure' signal would be removed, which is probably also not what we want. So, which of the two is better? Or, is there maybe some other alternative?

Finding the optimal filter for a given task is done using a so-called optimality criterion. For different applications we may have different goals and different priorities. Often used optimality criteria are *minimization of the mean squared error* and *maximization of the signal-to-noise ratio*.

The former aims to yield a filter output that is as close as possible to the pure signal. If we denote the recorded noisy signal as x, consisting of pure signal, s, and a noise component, n, then the realized filter output, y, should be such that at any given time the expected absolute value of $(y - s)$ is as small as possible. The quantity $y - s$ is called the error of the filter. For mathematical reasons, and since in practice we do not care whether the error has a positive or negative sign, we aim to find a filter that minimizes the squared error $(y - s)^2$. Performing this minimization using mathematical methods leads us to the definition of the frequency response of a filter that is optimal in the mean squared error sense; the *Wiener filter* (see Proakis and Manolakis 1996). A complicating factor is that to calculate a Wiener filter we need to know beforehand how the frequency contents (power spectra) of s and n look. In many research applications this is obviously not the case. Methods exist to make estimations of the power spectrum of s and n and then use these to come to a so-called *a posteriori Wiener filter* (De Weerd 1981).

The 'maximization of signal-to-noise ratio' criterion uses an entirely different philosophy. Here, the aim is not to reproduce a clean underlying signal as faithfully as possible but rather to detect whether a certain signal characteristic ('typical shape') occurs at a given moment or not. Applications are, for example, the detection of the R-waves in a noisy ECG signal, or detection of EEG periods in which typical eye-movement artefacts occur. An exact shape definition is not needed, a rough generalization will do. Using this shape definition a set of filter coefficients can be designed that constitute a filter that is very sensitive to input signals containing the defined shape. In other words, if the noisy input signal contains the defined shape (like a generalized R-wave, or a 'typical' eye-movement EEG) the output of the filter will be high, otherwise it will be low. Such a filter is called a *matched filter*.

Adaptive filters In many measurements not only do we have overlap of noise and underlying signal, but as an extra complication these components will probably not be stationary. Thus the frequency properties of the signal and noise may change, and what appeared to be an 'optimal filter' at the start of the recording may not be optimal at all anymore after some time. This type of problem calls for filters that can be adjusted (i.e. the filter coefficients can be changed) in response to changes in the properties of the recorded signal: *adaptive filters* (Widrow and Stearns 1985).

Adaptive filters 'learn' the signal and noise properties by processing samples, and adapt their behaviour in response to them. Such filters use a 'performance criterion' to evaluate how well they perform and an algorithm that makes a decision, on the basis of the filter's current performance and the recorded signal, on how to update the filter coefficients. Again, different performance criteria exist and a wide selection of adaptive filter types is available.

Adaptive noise cancelling with reference input In many cases where we have a recorded signal of interest contaminated by non-stationary noise we have the opportunity to record an extra input signal that consists mainly of noise. The assumption is that the noise in the extra channel is somehow correlated to the noise in the recorded signal that we want to study. We denote the recorded noisy signal as x, and we assume it is a summation of a pure signal, s, and a noise component, n (and we thus can write $x = s + n$). The aim is to use a second recording channel that can be thought of as 'reference noise', n_{ref}. The second channel has to be chosen in such a way that we can assume that n_{ref} is somehow related to n. The general outline of such a filter is depicted in *Fig. 8.11*.

The functionality of this approach hinges on the assumption that there exists some (linear) filter that, when presented with input n_{ref} will generate an output n_{approx} that is a good approximation of the noise n in the signal under

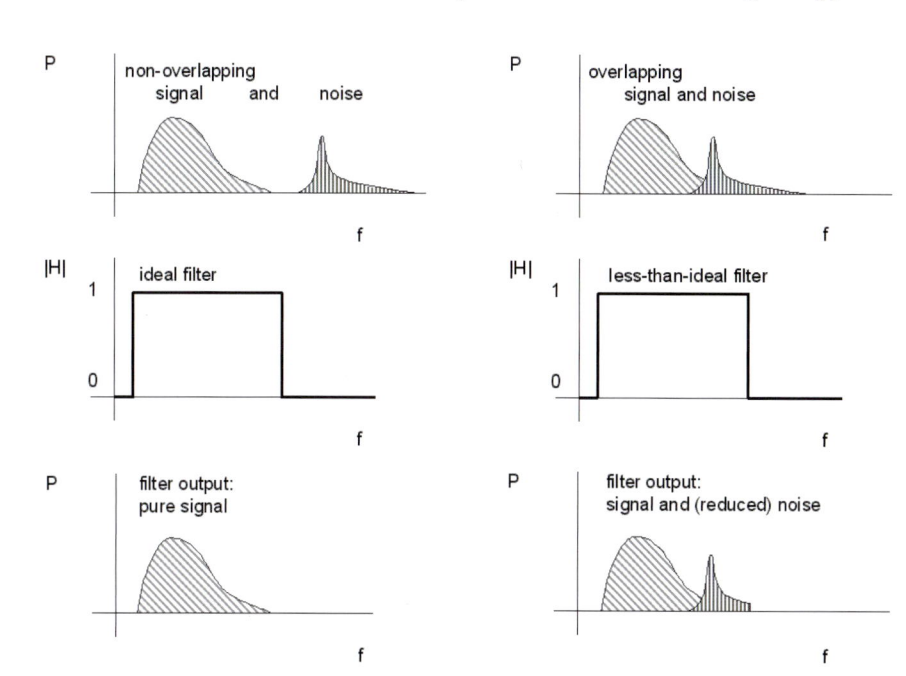

Fig. 8.10 Effects on the frequency contents of a recorded signal in the non-overlapping and overlapping signal-and-noise cases. f indicates frequency, P is an indication of the power of the signal at a given frequency, and |H| indicates the magnitude of the frequency response of the filter. (Adapted from Cohen (2000).)

study. As a result, if we subtract n_{approx} from x we will get a final output signal that is a good approximation of s. As 'goodness' of the filter we use a performance measure reflecting the error of the filter. In this particular case the error is defined as y, and to optimize performance we need to minimize the error/expected value of y. This works here because minimizing the expected y implies making n_{approx} equal to n as nearly as possible; we cannot influence the value of s. Thus by minimizing the error we will get a better filter. Again, for mathematical reasons in practice we choose to minimize the squared error. An algorithm receives information about how well the current filter is doing (in the form of the squared error) on the recently recorded samples and updates the filter coefficients in response. Intuitively, if the error is large a relatively big update of filter coefficients is needed, and if the error is small only minor updates may be needed. A simple (but effective) algorithm to calculate the needed updates of weights is called the least mean squares (LMS) algorithm. In the start-up phase such a filter performs very poorly (as the coefficients are initialized with random numbers), but after an adaptation phase the filter will perform satisfactorily – it will eventually converge to the 'optimal' Wiener filter for the signals at hand. Examples of application of adaptive filters can be found in cancellation of fluctuating 50/60 Hz interference, where we can obtain the reference noise easily from the mains, or other noise due to mains and equipment interference (Ferdjallah and Barr 1994). Other applications can be found in enhancement of evoked potential recordings (e.g. Yu et al 1994) and in the measurement of foetal ECGs. Here we can write x as the foetal ECG and n as the noise constituted by the maternal ECG from another location, for example the chest (Widrow et al 1975).

Common applications of filters in EEG recordings during surgery and intensive care The design principles for filters in EEG monitoring apparatus concern the balance between an adequate bandwidth for faithful recording of the signals of interest and one sufficiently restricted to minimize unwanted signals. This generally implies preprocessing with a frequency selective filter, which sharply reduces signals of less than 1–2 Hz to avoid contamination by the common low frequency artefacts encountered during ICU and surgical monitoring. The filter also sharply reduces signals above 20 Hz to reduce high frequency artefacts such as muscle potentials. There is usually an additional notch filter at mains frequency (50 or 60 Hz) to reduce line interference; this may be an adaptive type to take into account frequency variations in power-line signals (Ferdjallah and Barr 1994).

A filter may be used that weighs the EEG spectrum to counteract the normal tendency of slow components to be of larger amplitude than faster ones. To understand this, it is helpful to consider that, when a frequency analysis of the EEG is displayed on linear amplitude and frequency scales, it will appear to have an approximately exponential fall towards zero amplitude at high frequencies. If logarithmic amplitude and frequency scales are used, the underlying fall will appear approximately linear. Multiplying the logarithmically displayed spectrum by a function with the opposite slope will diminish this underlying trend and equalize the logarithmic frequency–amplitude spectrum to a horizontal line (Prior and Maynard 1986). This avoids inadequate representation of beta and alpha rhythms compared to that of delta and theta components during those signal processing procedures that assume an equal amplitude across the frequency spectrum. Sometimes described as 'pre-whitening', this process may be likened to the use of non-parametric methods as a means of dealing with biological factors that invalidate conventional parametric statistical handling of data.

Averaging techniques: recording evoked potentials

In *Optimal filters* (p. 322) the problem of overlap of the frequency contents of noise and signal was discussed and different methods were described to increase the signal quality as much as possible in such a situation.

Under certain circumstances a different technique can be used to improve the signal quality; repeating the measurement many times and averaging the results. The central assumption for such a technique to work is that the underlying signal, s, remains constant during all repeated measurements, and the noise, n, is a stationary random signal with average 0. Such averaging techniques are used in the measurement of EPs. In the following, x denotes the signal picked up at the electrodes, s is the underlying reaction to a stimulus (auditory, visual or somatosensory) we want to study, and n is the superimposed spontaneous/background EEG activity that we consider as noise and that we would like to reduce as much as possible. We can then use the averaging technique by applying a number of stimuli *if the following assumptions hold*:

1. the noise is a random, zero-mean signal whose statistical properties (mean, variance) do not change over time;
2. the response to stimuli, s, does not change over time (i.e. phase, form, latency and amplitude stay the same); and

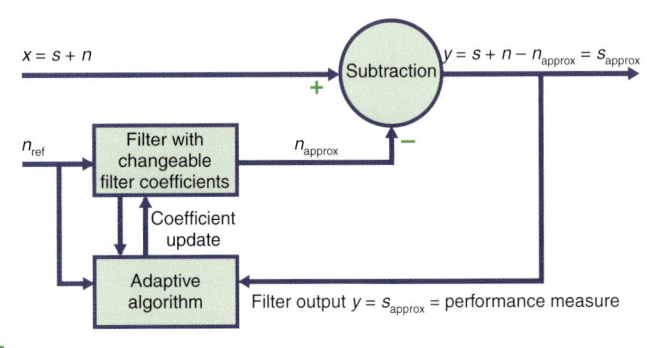

Fig. 8.11 An adaptive filter using a reference input to cancel noise.

3. signal, s, and noise, n, are not correlated, and we can write $x = s + n$.

If we present N stimuli, record after each stimulus i the signal x_i (consisting of the constant evoked response s, and a randomly varying n_i) and average the results we get as average of x

$$x_{average} = \frac{1}{N}(x_1 + x_2 + ... + x_N)$$
$$= \frac{1}{N}((s+n_1) + (s+n_2) + ... + (s+n_N))$$
$$= \frac{1}{N}(N \cdot s + n_1 + n_2 + ... + n_N)$$

or, rewrite it more compactly as:

$$x_{average} = s + \frac{1}{N}\sum_{i=1}^{N} n_i$$

The expected value of $x_{average}$ is

$$E\left[x_{average}\right] = E\left[s + \frac{1}{N}\sum_{i=1}^{N} n_i\right] = E[s] + E\left[\frac{1}{N}\sum_{i=1}^{N} n_i\right]$$
$$= s + 0 = s$$

The variance of $x_{average}$ is (assuming s is constant, and n_i is uncorrelated with n_j so that $E[n_i \cdot n_j] = 0$ for $i \neq j$):

$$\text{var}\left[x_{average}\right] = E\left[\left(\frac{1}{N}\sum_{i=1}^{N} n_i\right)^2\right] = \frac{1}{N}\text{var}(n)$$

Eventually the averaged x will converge to its expected value of s, and the variance of the averaged x will decrease with increasing N. To put it the other way around, and in amplitude terms (not in squared amplitudes that are used for the variance), this means that the signal-to-noise ratio (SNR) improves with $\sqrt{N}$. So, increasing the number of stimuli by a factor of 4 leads to an increase in SNR of 2, etc. The choice of the number of N in practice depends on the application. For some event related potential (ERP) studies the use of several tens of stimuli may suffice, whereas for brainstem auditory evoked potentials (BAEPs) several thousands of stimuli may be needed to 'lift' the small EP (of the order of less than a few microvolts) from the spontaneous EEG background noise (with an amplitude that is of the order of ten times bigger).

The essence of the averaging technique is that the responses remain constant and the intrinsic background noise activity is a random zero-mean process independent of the response activity throughout the collection of the trials. However, it is known that the amplitude and latency of some responses, particularly cognitive ERPs, depend upon the attitude of the subject to the task at hand and that changes can occur during a set of trials. In this situation the average will have an amplitude and latency close to the mean of the individual trials, but it is impossible to determine from the average whether it is composed of variable or stable EPs.

Averaging regarded as a filtering process The properties of the averaging process in the recording of evoked responses can be examined further if we realize that the process is in fact the application of a linear digital filter on corresponding samples in the sequence of sweeps. The process is depicted in *Fig. 8.12*.

For example, if we recorded 5 sweeps to calculate an EP, we could view the averaging as the application of a filter with coefficients [1/5 1/5 1/5 1/5 1/5] to an input sequence consisting of the samples with index number X in each successive sweep. The output is the average at index (time) X after stimulus presentation. Note that this is exactly the same filter as the 5-point averaging filter described in *Linear time-invariant filters* (p. 319). Consequently, it has the same frequency response as depicted on the left-hand side of *Fig. 8.7*. In this case though, the 'sampling frequency' to use is the rate at which the stimuli are applied, *not* the EEG sampling frequency. (In the example of the ERP recording in *Fig. 8.12* it is 1 Hz, whereas for an intraoperative BAEP recording it could be 11.1 Hz.) If we look at the frequency response in *Fig. 8.7* we can interpret this particular averaging process using 5 sweeps as follows: the filter passes frequencies lower than 1/5 of the sampling frequency (lower than 0.2 Hz) and suppresses frequencies above that; in other words, components that stay more or less constant over the sweeps are maintained and the variable components are suppressed. If we used more sweeps, the filter would get longer (higher order) and its frequency response representation would change accordingly. Using 100 sweeps we would get a 100th order filter with coefficients [1/100 ... 1/100], etc. Examples of different responses are given in *Fig. 8.13*.

The frequency response is characterized by a 'low-frequency passband' ranging from 0 to $1/N$ Hz, implying that the higher that N is chosen the narrower the definition of 'non-stationary' gets. Also it is clear from the figure that the amplitude suppression of anything with a frequency higher than $1/N$ gets more and more powerful with increasing N.

Some workers have designed 'optimized' bandpass filters to make them suitable for EP monitoring, while others have preferred to develop non-conventional, adaptive filtering strategies. Among the former, Fridman et al (1981), working with BAEPs, tried to estimate which frequencies were preferentially present in the EP signal and which in the background noise. On the basis of their estimates these authors proposed a relatively narrow digital bandpass (450–1300 Hz) for BAEP monitoring. This necessarily entails the loss of a substantial part of the amplitude and morphology information contained in the low frequencies, but allows satisfactory automatic latency calculation

(John et al 1982); use of this method should therefore be limited to instances where amplitude information is considered non-relevant.

A different technique for improving the SNR has been developed by Boston (1985), who used a second channel to estimate the noise content for each response in the average. A noise-cancellation function was derived for each individual response, and successive 'cancelled' responses were averaged. The major difficulty with this technique is that the 'noisy' channel should record only noise, for if a part of the signal is present, it will also tend to be cancelled. This implies spatially separated signal and noise derivations, from which only widespread noise affecting both channels simultaneously will be effectively cancelled. This is probably the reason why this technique results in good SNR improvements when noise is due to diffuse electrical interference, but not if it is mainly muscular or electroencephalographic in origin. Consequently, it is not of much value operationally.

As discussed in *Optimal filters* (p. 322), Wiener filters belong to the family of 'optimal' filters that minimize the squared error between the filtered response and the actual evoked signal, provided that the spectra of signal and noise are previously known. The performance of standard Wiener filters when applied to averaged EPs has been shown to be rather unsatisfactory in conditions with low SNRs (De Weerd 1981, Wastell 1981) such as those usually encountered in ICU recordings. A second, important drawback of Wiener filtering results from the variation of the frequency content of EPs with latency, from high frequencies in the earlier components to lower frequencies in the later ones. When short- and long-latency EPs are simultaneously monitored, as for example in the case of somatosensory evoked potentials (SEPs), different filters are optimal for different latency ranges. This leads to the solution proposed by De Weerd and Kap (1981) who devised a method, derived from Wiener's, that considers simultaneously the frequency and time domains, and thus entails changes of the filtering characteristics that are a function of latency ranges. This 'time-varying filter' has already been implemented in commercial systems, but not yet used for monitoring purposes.

Bertrand (1985) and Bertrand et al (1987) proposed a modification of Wiener filtering to be applied to BAEP monitoring in the ICU. Since conventional Wiener filters are not reliable when spectra are estimated under low SNR conditions, he proposed: (a) calculation of the spectra from small groups of averages of 200 stimuli instead of individual responses, thus ameliorating SNR of data from which the filter is derived; and (b) application of this modified Wiener filter to the small averages which stand for 'individual' responses rather than the grand-averaged EP. This greatly enhances SNR in small averages and allows reproducible BAEPs to be obtained after only 200–400 stimuli

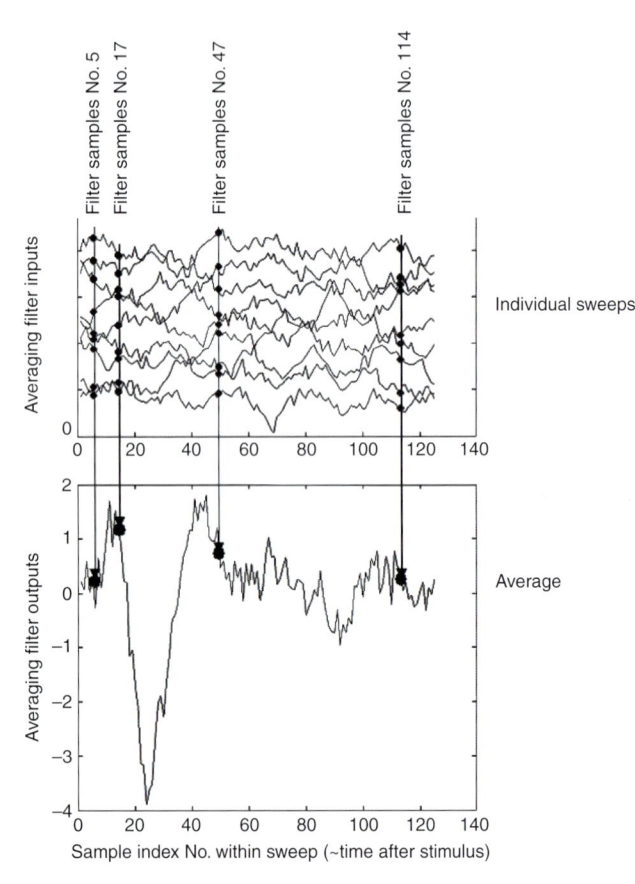

Fig. 8.12 Averaging process of sweeps of EEG recorded after application of auditory stimuli during an event-related potential (ERP) recording using the oddball protocol. The upper graph shows examples of some of the individual sweeps; nine sweeps each 125 samples long, the bottom graph shows the average (calculated over a total of 420 sweeps). (EEG sampling frequency is 250 Hz). The process can be viewed upon as repeated application of a low-pass (averaging) FIR filter to corresponding samples in the sweeps. The filter output is the average of the input samples. There are as many filter passes as there are samples in a sweep.

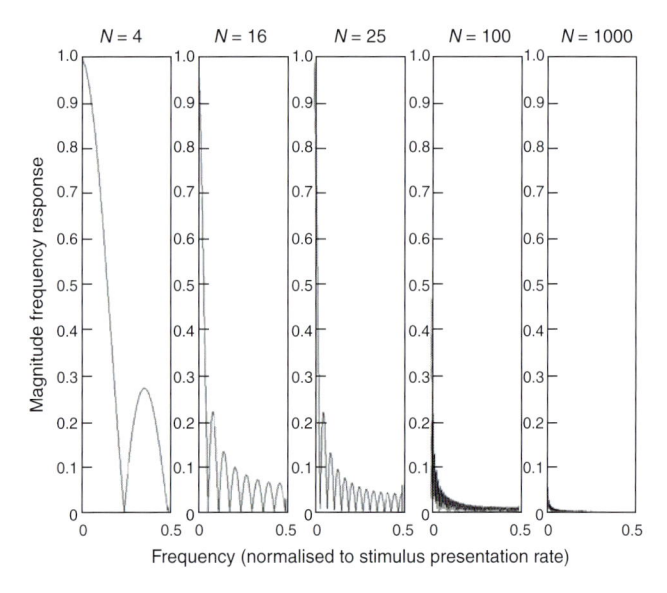

Fig. 8.13 Equivalent frequency response for averaging process for different numbers of sweeps, N.

(Bertrand et al 1987). In its application to monitoring this can be considered an adaptive filter, in the sense that the frequencies eliminated are not necessarily the same from one EP to the next, and thus 'adapt' to changes in the noise and signal content of successive small averages. Since the frequency content of BAEPs does not vary significantly from the beginning to the end of the response (within a 10 ms window), a time variation of the filter is not necessary in this context. This adaptive filtering technique has proved highly satisfactory for BAEP monitoring in the ICU and operating theatre. Although further improvements have been made to this technique, notably with the use of the wavelet transform (Bertrand et al 1990, 1994), these have not yet been applied to ICU monitoring in a large series of patients.

Many other proposed, but not yet widely applied, methods exist that use alternatives to the conventional averaging scheme to speed up data acquisition. Examples include extraction of EPs from single sweeps by using parametric models of the response (Spreckelsen and Bromm 1988), methods that use wavelet transforms (Bartnik et al 1992) and methods that use random stimulation application schemes (Cluitmans 1990, van de Velde et al 1993).

Artefact detection and rejection methods

The range of different artefacts that can occur during neurophysiologic measurements is extremely wide; each type of artefact has its own effects on the signal, often calling for application-specific signal processing approaches. We will limit ourselves here to an overview of typical artefacts and suggested techniques to deal with them. A summary overview is given in *Table 8.2*.

As can be inferred from *Table 8.2* considerable research efforts have been dedicated to development of procedures for the identification and handling of artefactual components in EEG data processing. Progress has been substantial for several specific forms of artefact such as mains frequency interference, ECG pickup, EOG signals and scalp EMG potentials, particularly in the context of automatic recognition of seizure discharges in long-term epilepsy monitoring and sleep monitoring systems (Ives and Schomer 1988, Panych et al 1989).

In computer processing, a similar approach can be automated at various levels, one of the simpler being a switching system which disconnects all channels during surgical diathermy (Barlow 1985). More complex systems may identify specific artefacts and delete contaminated segments (Eckert 1998). Unfortunately, simple subtraction methods using templates (e.g. for ECG contamination) may be successful in normal subjects, but fail when used in monitoring critically ill patients in the ICU or during surgery in whom the ECG complex may not only be abnormal but variable in morphology and rhythm.

Compared with sleep or epilepsy monitoring, there has been rather little attention to the complex problems of long-term ICU and intraoperative monitoring, where artefacts of both extrinsic and biological origin abound. Van de Velde et al (1999) studied the features of approximately 1000 artefacts occurring during seven 24 h recordings in a general ICU. Using time-varying autoregressive modelling and slope detection, and relatively short duration periods (20–40 s) for contextual information, the automated system found approximately 90% of the artefacts identified by two independent observers.

These observations make it clear that strategies for artefact identification and handling in automated monitoring systems must be sophisticated and developed in consultation with clinicians (Nakamura et al 1996, Celka et al 2001). A good policy is to retain but mark artefacts in recordings, so that the valuable evidence of arousal or cyclical changes in reactivity to external stimuli are available for diagnostic and prognostic purposes.

Monitoring work during major surgery has the additional problem of interference from other equipment near or attached to the patient. Particular problems arise due to surgical diathermy, fibrillators, defibrillators, pacemakers and other cardiac support systems (see *Dangers of misinterpretation due to unrecognized artefacts in EEG monitors displaying only processed signals*, p. 62). Most of these produce artefacts which have little clinical relevance and strenuous efforts are required to identify and eliminate or mark them.

As noted in *Dangers of misinterpretation due to unrecognized artefacts in EEG monitors displaying only processed signals* (p. 62) considerable care is needed when aiming to automate interpretation during critical care, in the course of which both well-recognized and unusual artefacts may be encountered. Display of the unprocessed EEG signal throughout monitoring and parallel quality checks by means of continuous display of electrode impedance, line frequency and other external potentials is essential. In addition, special attention is required in the design of algorithms for computerized monitoring systems to deal with the specific problem of artefacts. Approaches range from the use of simple amplitude limits, to linear digital filters, to non-linear adaptive systems, to complex expert systems ,and further. Much progress has been made, but for many types of artefacts there is still a lot of work to be done. The availability of cheaper and more powerful computing power facilitates the exploration of more advanced methods to attack the unsolved problems.

SIGNAL PROCESSING AND INTERPRETATION

As we proceed further into the processing chain the tasks at hand get less generic and more and more specialized. Whereas the data-acquisition methods are quite generic and, in the sense that the principles are valid for most neurophysiological applications, the issues of filtering and artefact treatment already showed that a choice of methods exists to deal with different types of signals and disturbances. Going further, we enter the stage of processing

Table 8.2 A summary of the most common artefacts and suggested approaches to deal with them*

Sources	Effect on signal	Proposed approaches
Physical artefacts		
Electrodes and input leads; electrochemical changes at the electrode–tissue interface due to movements	Ranges from gross deflections to occasional pops present in the EEG	Ensure proper application of electrodes/cables, make sure patient is relaxed
Hardware/software fault in data-acquisition equipment	Can be anything, ranging from obvious 'nonsense signal' to subtle loss of a few bits or blocks of data at irregular intervals that may be very hard to notice. Spikes due to interference from equipment working at performance limits	Use extensive test runs/pilot cases (including the whole processing and signal interpretation chain)
Electrical interference (electrostatic and electromagnetic induction)	Interference patterns, 50/60 Hz activity in the signal. In EP recordings mains interference can show up in the averaged waveform	Shielding, avoidance of cable loops, use of (adaptive) notch filters (Van der Weide and Pronk 1979, Ferdjallah and Barr 1994). In EP recordings use phase-synchronized triggering at alternating opposite phases of the sinusoidal signal (Emerson and Sgro 1985), otherwise use stimulus presentation rates unsynchronized to the mains period (Stecker and Patterson 1996)
Other types of interference	Spikes due to interference from equipment working at performance limits may occur (e.g. when working at high sampling rates). Mobile phones in vicinity of EEG equipment may cause interference	Guidelines to keep interfering equipment at a safe distance (Robinson et al 1997)
Biological artefacts		
Oculogenic potentials	Series of typical deflections in EEG curves due to eyeball movements	Ask subject to close eyes, use of eye pads. Systems for their automatic reduction are discussed by Cooper et al (1980) and Barlow (1985). Brunia et al (1989) and Jervis et al (1988) survey several methods of ocular artefact correction. van den Berg-Lenssen et al (1989) describe a method that uses an autoregressive model. Sadasivan and Dutt (1995) propose adaptive filters. More recently, ICA[†] methods have been proposed by Vigário (1997). Vigon et al (2000) quantitatively compare four different methods (including ICA)
Myogenic potentials	Intermittent or continuous high-frequency disturbances in EEG traces due to muscle activity	Help the patient to relax.[‡] Low-pass filtering (with risk of removing beta components from EEG). Systems for the on-line reduction of EMG contamination of up to 16 channels of EEG using a microprocessor have been described, for example by Gotman et al (1981) and Ives and Schomer (1988) Adaptive algorithms can demonstrate very efficient artefact filtering while leaving intact important EEG features (Panych et al 1989)
Electrocardiac potentials	ECG signal is picked up by EEG electrodes	Use average or source derivations; change patient's posture. Computer-based methods. Bickford et al (1971) described a method in which the ECG was averaged (using the R-wave as trigger) and a suitable proportion subtracted from the (single) channel of EEG. Barlow and Dubinsky (1980) extended the method to eliminate the residual ECG when using the balanced non-cephalic reference of Stephenson and Gibbs (1951).

Continued

Table 8.2   Continued

Sources	Effect on signal	Proposed approaches
		Nakamura et al (1990) proposed a method in which an ECG average from a concurrent ECG recording was subtracted from the EEG. Although this still resulted in an R-top artefact in the EEG, with exclusion of these remaining artefactual periods, the authors demonstrated that reliable short-latency SEPs could be recorded
Pulse artefacts and ballistocardiogram due to placement of electrode near artery	A saw-toothed waveform synchronous with the pulse is sometimes recorded due to disturbance of electrical double layer at the electrode/skin interface	Change of electrode position, bunching of leads. 'Straightforward' filtering is not possible without affecting the underlying EEG contents
Changes of skin potential or resistance	Long duration baseline sways in EEG in patients who are hot or worried	Good planning, use of comfortable rooms
Stimulus related artefacts contaminating SEP recordings	Stimulus-locked activity picked up at the scalp in EP measurements. Too loud stimuli in AEP recordings inducing muscle artefacts from the stapedius	Hardware blanking inputs during artefactual periods. Parsa et al (1998) propose adaptive filtering techniques
Movement and tremor	Movement of electrodes and cables (see *Electrodes and input leads* artefact, above) due to rocking, hiccups, head scratching, chewing, seizures, etc.	Careful observation of patient (use of accelerometers to detect movement if necessary)

*In all cases the first approach should be use of a well-thought out measurement set-up, the signal processing approaches listed here should be regarded as a last resort rather than as a first expedient.

†ICA (independent component analysis) is a method to decompose 'mixed' signals into separate independent signals representing activity from different sources (see e.g. Hyvärinen et al 2001).

‡We do not advocate the use of muscle relaxants to 'clean up' an EEG or EP recording. This may have disastrous consequences (Verma et al 1999).

---

to obtain a specific 'higher-level' goal. This can range from long-term trend monitoring of a patient in the ICU, to detection of ischaemia during an operation, detecting epilepsy seizures, to assessing adequacy of anaesthesia, and so on. Each of these goals can again be approached with different methods (signal processing paradigms). This section aims to give an overview of the most common processing tasks encountered in neurophysiological applications during intensive care and surgery (*Common processing tasks*, below), and give the reader an introduction to the underlying principles of the most relevant methods (*Processing methods*, p. 332, and *Integration of features into the clinical context – pattern classification*, p. 346). *Statistical process control* (p. 354) concentrates on methods to solve the problem of identifying significant changes in variables using statistical process control methods.

Common processing tasks

This section discusses some typical examples of the signal processing tasks carried out in surgical and intensive care monitoring environments.

Segmentation

An EEG is often described as a non-stationary time series because the pattern of fluctuation in one section of record changes either 'spontaneously' or because the person taking the record has changed the recording conditions, for example by giving a drug or simply by saying 'open your eyes'. Analytically, 'non-stationary' means that the statistical parameters that describe the signal (mean, variance and all higher order statistical descriptors) change during the recording. If all parameters remain the same, the series is said to be 'stationary' in the strict sense. In practice, often a looser definition of stationarity is used. For example, the signal is defined 'stationary' if the mean amplitude and variance remain 'relatively' constant. An EEG is a mixture of epochs when it is (more or less) stationary (e.g. resting with eyes closed) and when it is non-stationary (eyes opening, dozing, during an absence seizure, bursts of theta activity, sharp waves, and so on). Note that it is only the start of the new activity that is the non-stationary element – the EEG while the eyes remain open, or during a seizure, can be stationary for a period of time.

Many signal processing methods use the assumption that the signal they act upon is stationary. Perhaps the best known example is the calculation of the frequency contents of the signal using the Fourier transform (FT), which requires that the segment of data from which the signal's power contribution in different frequency bands is to be calculated is part of a stationary signal. A way to fulfil this requirement is to divide the signal into 'almost' stationary segments and use processing methods with those segments of data as input. The problem is then: how do we divide the signal so that we obtain a usable 'piecewise-stationary' signal? On the one hand it is favourable to use as long segments as possible since that means that we can present more data to the signal processing algorithm, usually leading to a more accurate estimate of calculated features. On the other hand, we do not want to make segments so long that they contain different states of the signal since that would lead to results that would not necessarily make sense.

Electroencephalographers scan the records looking for non-stationarities which separate stationary but different patterns of EEG activity. As they know only too well, the changes are rarely predictable and neither is the duration of (new) stationary activity. For this task they make use of their experience and human visual pattern recognition capabilities. To automate this task using computer algorithms is far from trivial.

Two approaches are in common use: *fixed-duration segmentation* and *adaptive segmentation*. The former is especially applied in situations were a signal is very highly non-stationary, the signal is divided into short, constant-duration segments. This is a very straightforward way of dealing with the problem, and its popularity lies exactly in that fact. It is also used in many EEG processing applications, for example in frequency representations that follow the frequency of the signal over time, such as short-time Fourier transform (STFT), compressed spectral array (CSA), colour density spectral array (CDSA) or practically any method that tracks features (e.g. amplitude, power, BIS, spectral entropy) for use in an on-line monitoring device that tracks short-term signal properties (e.g. during surgery). The optimal duration for the segments is to be defined beforehand, and is dependent on the type of application. For EEG processing a common choice of segment length to achieve 'wide-sense' stationarity is in the order of several seconds (van de Velde 2000).

In some applications, where stationary segments can be of long duration, the division of the signal into short segments is not the most effective one. Examples include the representation of large amounts of data by a summary of occurrences of 'typical patterns' either to provide an efficient temporal profile of the data or for data compression purposes. This usually concerns situations where longer term trends are processed, such as for example in EEG recordings in the ICU or during sleep studies. In such a situation one would like to make the segments as long as possible without violating stationarity assumptions. The length of the segments is adapted on the basis of the underlying signal properties. As long as the signal remains stationary in a predefined sense the segment can keep growing in size, but as soon as the properties of the signal change significantly the segment is 'closed', and a new segment is started. There are many different algorithms proposed, an overview can for example be found in (Keogh et al 2004). There are three main strategies to divide signals into stationary segments:

1. *Sliding window approach*: define a starting stationary segment as reference, compare data from a next data window with the reference; if the data are comparable to the data in the reference segment, increase the size of the stationary reference segment. If not, start a new segment.
2. *Top-down approach*: partition the data recursively into smaller segments until a stopping criterion is reached.
3. *Bottom-up approach*: start from a very fine segmentation, and subsequently merge similar segments until a stopping criterion is reached.

In applications where analysis it to be done on-line, the sliding-window approach is the only practically suitable one (the other two require the whole recording to be completed first). It therefore is also the most useful approach in medical signal monitoring applications and we will limit our discussion to this approach.

A fixed reference window is used to define an initial segment of the signal. The duration of the reference window is determined such that it is long enough to allow calculation of a descriptor of the segment (e.g. variance, of the signal or power in different frequency bands) yet short enough that the segment can be considered stationary. A second, sliding window is shifted along the signal. The descriptor of the segment defined by the sliding window is estimated at each window position. The two descriptors are then compared. As long their difference remains below a certain decision threshold, the reference segment and the sliding segment are considered close enough and are thus supposed to be related to the same stationary segment. Once the difference exceeds the decision threshold, a new segment is defined. The process continues by defining the last sliding window as the reference window of the new segment. Some of the segmentation algorithms use growing reference windows rather than fixed ones. In yet other methods, both windows slide.

Various segmentation methods differ in the descriptors they use. Bodenstein and Praetorius (1977) proposed the use of autoregressive (AR) models (see *AR modelling*, p. 335), Michael and Houchin (1979) used the autocorrelation function of the signal in 1 s windows. In the method used by Creutzfeldt et al (1985), the characterization of the various stationary patterns of activity within the EEG is carried out by clustering them according to the mean frequency and logarithm of the power in the segments. Barlow (1984) compared selective analogue filtering, matched inverse digital filtering and automatic adaptive segmentation during carotid clamping during which clinical and EEG changes were reported. Of the three methods, segmentation was best correlated with the changes of the EEG that had been reported by visual analysis. Agarwal et al (1998) applied the so-called non-linear energy operator (NLEO) approach to EEG, which requires minimal parameter adjustments. The NLEO operator calculates the sum of the values of $x(i-1) \cdot x(i-2) - x(i) \cdot x(i-3)$ from all samples i in a window at position N and a similar sum from samples in the window at $N+1$ and uses as the segmentation criterion the difference between the two sums. They further extended the method to reach segmentation on the basis of multiple channels (rather than the more usual single channel segmentation). As an application they used adaptive segmentation as a method to reduce the data acquired during long term monitoring in the ICU (Agarwal and Gotman 2001). They separated the EEG into stationary segments and extracted and classified the features and printed out representative samples of each group of patterns. An example of segmentation is given in *Fig. 8.14*. Särkelä et al (2002) used the NLEO operator to locate burst suppression periods and segments with artefacts in the EEG.

Detection of events, complexes and trends

Another common task is the detection of characteristic events in the signal. In the case of EEG monitoring, one can think, for example, of detection of epileptic seizures, burst suppression, arousal, ocular artefacts, K-complexes, sleep spindles or sudden loss of EEG amplitude. In the case of EPs one can think of the detection of characteristic peaks and troughs whose presence needs to be identified to obtain their latency and amplitude values. In the case of long-term measurements the 'characteristic' to be detected could be a trend in a signal; an increase or decrease in the signal that develops slowly over time and thus is difficult to notice by taking one look at a monitor.

In the case of *event detection* the main task is to track the signal, looking for characteristics in the signal that are associated with the event to be found. The problem then typically concerns an exact as possible determination of the time of occurrence of a specific waveform in a noisy signal. Approaches to solving the problem are again application specific. A commonly followed avenue uses the construction of templates forming a 'prototypical shape' of the event and comparing periods of the signal at hand with the template. This can be done by, for example, calculating their correlation or using other similarity measures. If there is high similarity between the signal and template there is a high indication that the sought event is present at that specific time.

Matched filters are a specific implementation of this. These are a type of optimal linear filter (see *Optimal filters*, p. 322) with their weight coefficients chosen so that the filter output becomes high if the prototypical shape is present in the noisy signal and low otherwise. In the special case where we can consider the noise as 'white noise' (i.e. noise with equal power at all frequencies) the values of the filter coefficients can be found very easily by choosing them to fit the shape of the template (sought signal shape), but inverted ('run backwards in time').

Methods can range from the rather straightforward to very complex. For example, for burst suppression detection there exist relatively simple (e.g. Prior et al 1978, Lipping et al 1995) as well as complex methods using extensive mathematical transformations and neural networks (Arnold et al 1996, Galicki et al 1997).

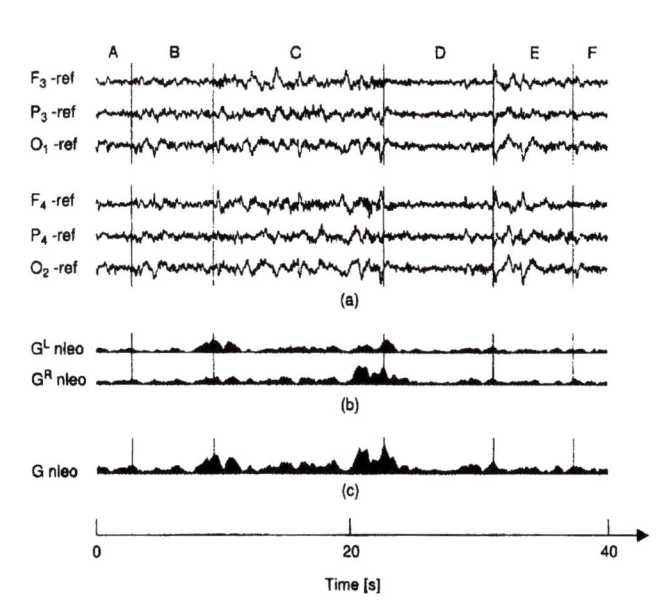

Fig. 8.14 Example of EEG segmentation techniques applied to a six-channel reference recording shown in (a) from left and right frontal, parietal and occipital electrodes. Vertical bars indicate segmentation boundaries. (b) Segmentation criterion for the left- and right-sided channels, respectively. (c) Overall segmentation criterion used for final simultaneous segmentation of the left and right sides. (From Agarwal R, Gotman J 2001 Long-term EEG compression for intensive-care settings. A method of summarizing many hours of EEG data for quick identification of problems/easier evaluation. IEEE Engineering in Medicine and Biology 20:23–29, by permission © 2001 IEEE.)

The same range of 'simple to complex' approaches can be found, for example, in methods that aim to *recognize characteristic peaks in AEPs*. These include digital filtering in combination with 'peak picking' algorithms (Fridman et al 1981), matched filtering techniques (Delgado and Özdamar 1994), syntactic pattern recognition techniques (Grönfors 1993), rule-based systems (Boston 1989) and use of artificial neural networks (ANNs) (van Gils 1995).

Detection of whether or not changes in observed signal values imply an actual *'significant' change* requires use of techniques that take into account the individual variability in the recording. The sections *Assessment of results* (p. 7) and *Statistical process control* (p. 354) discuss such methods. As demonstrated by Lipping et al (2000), a signal processing approach to *detect steps and trends* in the RMS amplitude of the EEG can be relatively simply realized using the median filtered signal (*Median filters*, p. 321), examine differences between means of sliding time windows to detect sudden changes, and use line fitting to detect slopes in the signal. A review of *trend-detection* methods can be found in Avent and Charlton (1990).

Calculation of patient condition indices

Many efforts have been undertaken to use data processing methods to simplify the view of the 'patient state' by merging different physiological monitored data into one index number. An illustration from the field of surgery and intensive care monitoring is the use of indices that reflect 'depth of anaesthesia' and 'depth of sedation'. The idea behind such an index is that it provides one, objective, easy interpretable, value displayed on a monitor instead of having to rely on a subjective opinion made by observing many different variables originating from different sources.

As 'anaesthesia' comprises various components (hypnosis, analgesia and muscle relaxation), covering the whole concept by one number is a tall order. The majority of available 'depth of anaesthesia' indices are based upon cortical measures like the EEG and AEPs, and as such only reflect the hypnotic component; therefore the term 'depth of hypnosis' would be more appropriate. Muscle relaxation is commonly assessed by measuring motor nerve stimulation responses. The development of a method to measure 'level of analgesia' is still a very open research area (van Gils et al 2002).

One of the main problems in being able to test the functionality of a developed index against, for example, depth of hypnosis is the choice of a physiological endpoint. For index development purposes, scales like the Observer's Assessment of Alertness/Sedation (OAA/S) Scale (Chernik et al 1990) modified for anaesthesia (Glass et al 1997) or the Ramsay scale (Ramsay et al 1974) are employed widely. The modified OAA/S scale is given in *Table 8.3*. Another commonly used 'reference scale' is the estimated effect site concentration of the administered anaesthetic drugs. The aim is then to develop an index that behaves such that it moves in the same direction as the OAA/S/Ramsay/drug concentration value (Smith et al 1996). That is, a decreasing index value is associated with a decrease in OAA/S value, and an increase is associated with an increasing OAA/S value. Moreover the absolute value of the index should indicate a working range, for example on a scale from 0 to 100 (100 representing a fully conscious state, 40–60 typically the target workable region for surgical levels of hypnosis, and 40 and below the area of unpractical deep anaesthesia prolonging recovery).

A myriad of signal processing methods has been applied for development of indices that correlate with the level of hypnosis. Methods based on the EEG include time-domain, frequency-domain, time–frequency, higher order spectra, and complexity descriptors approaches. Methods based on AEP recordings use latency and peak values or measures that describe the curve shape. Literature overviews concerning these approaches can be found in Chapter 3, van Gils et al (2002) and Senhadji et al (2002).

Processing methods

The most common and certainly least complex processing method is simply visualizing the raw traces on paper or a computer screen and using human visual pattern recognition techniques by browsing along to qualitatively assess the data. However, in many applications, especially where fast on-line interpretation of multiple signals is needed, a computerized approach would be helpful. The computer-based equivalent of visual browsing aims to extract certain features from the straightforward amplitude versus time traces. Such analysis is called *time-domain* analysis. Another popular way of looking at the properties of signals is by examining their frequency contents. After transforming the signal into the *frequency domain* (e.g. using the FT) features are extracted that say something about the frequency distribution. *Time–frequency methods* provide an alternative way of looking at the signal, providing us with information on both domains.

Most used methods are linear approaches, inherently assuming that the underlying system they describe is linear in nature; linear features perfectly describe second-order statistics of a system (mean, variance, autocorrelation function). In the case of non-linearities, higher order statistics are required, an example of this is the bispectrum. Since neurophysiologic signals are not generated by a linear system, non-linear methods are potentially useful for processing them.

When the features obtained using (a combination of) the above techniques are used as input to further analysis they can be considered as representing the EEG or EP in a condensed form that is particularly suited for the application at hand. In the case of EEG signals one often refers to such a feature set as quantitative EEG (qEEG).

Time-domain methods

Amplitude measures Amplitude information is the most obvious feature of EEGs or EPs, as it can be obtained directly by visual assessment. Several methods have been used to quantify it. Probably the most commonly used measure is the RMS value of the signal calculated over segments of data. It is calculated by squaring all the sample values, calculating the mean of those squared values

---

Table 8.3 Responsiveness scores of the modified Observer's Assessment of Alertness/Sedation Scale* as often used for development of depth of hypnosis monitoring indices

Response	OAA/S score
Responds readily to name spoken in normal tone	5 (Alert)
Lethargic response to name spoken in normal tone	4
Responds only after name is called loudly or repeatedly	3
Responds only after mild prodding or shaking	2
Does not respond to mild prodding or shaking	1
Does not respond to noxious stimulus	0

*From Glass PS, Bloom M, Kearse L, Rosow C, Sebel P, Manberg P 1997 Bispectral analysis measures sedation and memory effects of propofol, midazolam, isoflurane, and alfentanil in healthy volunteers. *Anesthesiology* 86:836–847, by permission.

(i.e. the variance or power of the signal) and subsequently calculate the square root of that mean. It provides an efficient measure of mean absolute amplitude. The RMS value is equivalent to the standard deviation in statistics to represent deviation from a mean. Other used measures include the median amplitude and amplitude envelope descriptors (see e.g. Binnie et al 2003).

Zero crossing computations   Simple amplitude measures of time series take no account of waveform or frequency, both of which have great significance in EEG interpretation. However, simple time-domain methods can be used to obtain some assessment of the frequency contents. The time between successive crossings of the baseline of a sine wave is half the period of the waveform; the reciprocal of the period is its frequency. Since zero crossings are much simpler to compute than the FT (*Frequency-domain methods*, p. 336) they have been used to measure the frequency of EEG components. However, in a complex wave some components might be too small to cross the baseline and various strategies have been used to get over these difficulties (Burch 1959). Pigeau et al (1981) compared zero crossings and Fourier analysis and found that they 'share similar types of information'. Despite its limitations, a form of period analysis has been used successfully in various special-purpose devices such as the first version of the Cerebral Function Analysing Monitor (RDM Consultants, Uckfield, Sussex, UK, http://www.cfams.com) (see *Historical development of EEG monitors*, p. 73).

Normalized slope descriptors   Hjorth (1970) developed a method of quantifying the EEG based on the measurement of three variables, two of which are obtained from the first and second time derivatives (slopes) of the amplitude fluctuations of the signal. The first derivative is the slope or rate of change of the amplitude of the signal. At peaks and troughs this is zero; at other points it is positive or negative depending on whether the amplitude is increasing or decreasing with time. The steeper the slope of the wave, the greater is the first derivative. The second derivative is the derivative (slope) of the first derivative. Peaks and troughs in the first derivative (corresponding to points of greatest slope in the original signal) result in zero in the second derivative.

The first measure derived by Hjorth is called 'activity' and is the variance of the amplitude fluctuations in the epoch. The second, 'mobility', is found by dividing the variance of the first derivative by the variance of the primary signal and taking the square root of the result. It gives a measure of the mean frequency of the signal. The third measure, 'complexity', is the ratio of the mobility of the first derivative to the mobility of the signal itself. A sine wave has a complexity of 1; more complicated waveforms have greater values. Thus for each analysis epoch three numbers are obtained that are descriptive of the pattern. Hjorth showed that these measures describe the signal wave shape and are mathematically related to the power spectrum of the signal. The measures are easily derived with less computation than Fourier analysis (with modern computing resources, however, this is no longer an issue).

Correlation analysis   Correlation analysis is a method of determining whether there is a relationship between two or more variables, and for measuring its degree. For example, it may be used to quantify any relationship between the height and weight of individuals or between the amount of alpha activity and scores obtained from psychological tests. Correlation may also be used to test how similar the signal in one channel of an EEG is to that occurring simultaneously in another. The usual correlation measure is the correlation coefficient, referred to as the product–moment correlation coefficient or 'Pearson's *r*', as described in statistics texts. It measures the degree of similarity between two variables; it is independent of their magnitudes; it can have any value between −1 and +1, zero indicating no common feature. The correlation of two time series such as two EEG channels can be determined from the amplitude values by calculating the correlation coefficient. The mean of the cross-products of these values, i.e. the sum of the cross-products divided by the number of data points in the epoch, is called the covariance. This number is dependent on the amplitude of the signals in the two traces and may be large, either because of small amplitude signals which are similar throughout the two epochs, or because there are signals which are larger but only partly similar. To make the covariance independent of the signal amplitudes it is normalized by dividing by the square root of the products of the variances of the two time series. It is this number, the normalized covariance, that is called the correlation coefficient. If the two signals are of identical pattern, the correlation coefficient is +1, even though they may be of different amplitudes. If the signals are of identical waveform but of opposite polarity, the correlation coefficient is −1. Any other degree of similarity will fall between +1 and −1, with 0 signifying complete dissimilarity.

Sometimes EEG signals recorded from different parts of the scalp may have similar patterns, but are not synchronous; i.e. there is a time delay between the signals from different electrodes. One way of measuring this is with the cross-correlation function. This is a plot of the correlation coefficient calculated for each of several consecutive displacements in time of one signal with respect to the other, each displacement being one sampling interval. The cross-correlation function will have a maximum at the displacement corresponding to the time difference between the two signals.

If there is a strong rhythmical component common to the two time series, the cross-correlation function will fluctuate rhythmically, as the increasing time displacements cause the peaks of one signal alternately to coincide with and oppose the peaks of the other. This is shown in *Fig. 8.15(a)* in which the correlation function is calculated for two channels of alpha activity. As the similarity and rhythmicity decrease from (a) to (c) the peak correlation coefficient decreases and the correlation function shows a more rapid fall with time displacement.

If a signal is multiplied with itself, the result is called an autocorrelation function. The peak correlation coefficient is then always maximum for zero time displacement because the two signals are exactly alike (*Fig. 8.15(d)*). When progressive time shifts are introduced the autocorrelation function is generated. The oscillations of the autocorrelation function can be used to measure the rhythmicity of the signal. For a sine wave it will oscillate with constant amplitude (±1) as the displacement increases. For less regular waveforms, the function decays over a few cycles with increasing displacement; thus autocorrelation can be used to detect the rhythmicity in a single channel and to determine its frequency. The FT of the autocorrelation function is the power spectral density (PSD) of the signal, which will be discussed in further detail in *Frequency-domain methods* (p. 336).

Sometimes it is sufficient to plot the covariance without normalizing each value to get the correlation coefficient. This is then called the cross- or auto-covariance

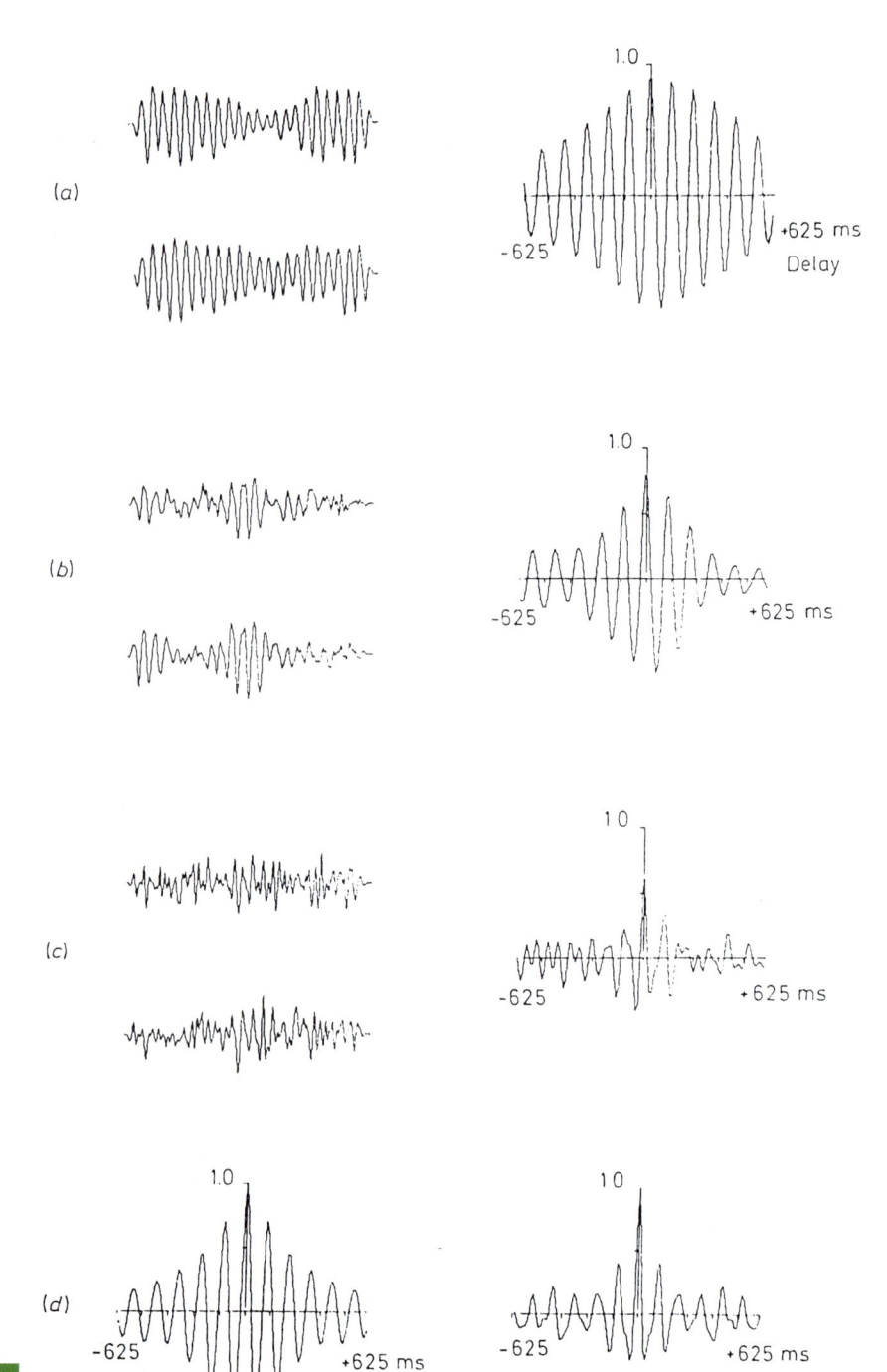

Fig. 8.15 Correlation functions on the right for the pairs of EEG signals on the left. The autocorrelation functions of the lower of the two signals in (b) and (c) are shown in (d). (From Cooper et al (1980), by permission.)

function and will show any rhythmical activity in the signals, but will be sensitive to amplitude.

Because of conduction delays, signals having similar patterns can occur in different parts of the nervous system at different times. The correlation coefficient between two such signals will be a maximum when one signal is displaced in time with respect to the other until they fluctuate together. The time difference can be found by measuring the displacement at which the maximum correlation coefficient occurs in the cross-correlation function. Correlation analysis is useful for determining the time delay (phase shift) between EEG channels because it provides a value for the average delay over the whole epoch. An extensive review of correlation methods and applications is given by Gevins (1987).

Phase and time delay analysis As already mentioned, EEG signals occurring over a large area of the head can show small time differences at different sites. The functional significance of these time delays is not known but it would seem reasonable that the waves that occur earlier are closer to the generator and thus provide information about the source of the activity. This is useful in corticography (see Binnie et al 2003, Section 8.3.5.4) and intracerebral recording (see Binnie et al 2003, Sections 4.12.12.4 and 8.3.5.1) in which significant delays of abnormal activity can occur. The time delays are usually too small to be measured by visually examining curves and analytical methods, such as the cross-correlation function (which shows a time offset of the peak coefficient), are required.

It should be noted that when the difference between two signals is expressed as phase, conversion to time must take into account the frequency of the waveform. Thus, a phase difference of 9° is equivalent to 9/360 of one cycle or a delay of 2.5 ms for a 10 Hz signal, 6.25 ms for 4 Hz, and 10 ms for 2.5 Hz. Care should be taken to ensure that multiplexing in an AD converter (in which channels are usually sampled sequentially) does not introduce significant systematic interchannel delays which can appear in such analyses. Conversely, for a fixed time lag, phase difference is proportional to frequency. If there is a genuine functional relationship between the activities in two channels, the phase spectrum should show a linear ramp over those frequencies for which there is a high coherence.

AR modelling Many physiological signals, like spontaneous EEGs, are unpredictable in that no matter how much is known about the fluctuations up to the present moment, the pattern of fluctuation in the future cannot be forecast. Thus, particular measures may be different from epoch to epoch even when the recording conditions, the state of the subject, etc., are apparently unchanged. As these features are a function of the generating process, it may be more appropriate to try to derive a model of a generator that could produce them rather than try to describe the signals themselves. The output of such a generator must have the same statistical properties as the signal, although the waveform generated may look considerably different from the signal from which the model is derived. It should be realized that approaching the problem from a statistical point of view does not necessarily imply that the brain actually generates random noise. All these activities may have a function in the brain or be the consequence of another function (even though we have no idea what this is) in which the apparently random element is not random but complex and unknown. Description of the EEG in terms of random processes presupposes only a mathematical and not a physical model.

A frequently used model is called the autoregressive (AR) model. In this the amplitude at any time instant has two components, one that is correlated with all past values while the other is an added random perturbation. The amplitude of the signal at a particular time instant is expressed as a weighted sum of amplitudes at preceding instants plus a random component. An extension of this is the autoregressive moving average (ARMA) model in which not only amplitudes at preceding instances but also the earlier estimates of the random perturbations are used. We will concentrate on the AR model here. Mathematically it is expressed as

$$y(k) = a_1 \cdot y(k-1) + a_2 \cdot y(k-2) + \ldots + a_p \cdot y(k-p) + e(k)$$
$$= \sum_{i=1}^{p} a_i \cdot y(k-i) + e(k)$$

where $y(k)$ is the current sample whose value is to be estimated as a weighted sum of the preceding p samples plus a random noise (i.e. error) term $e(k)$. The value p is called the model order. The values a are called model parameters, or AR coefficients. If we write down similar equations for all samples y in an epoch we get a series of linear equations. From these we can solve estimations of the optimal AR coefficients $a_1 \ldots a_p$. There are various methods to do this; examples are the Yule–Walker method, the Burg method and adaptive methods (see e.g. Akay 1994).

If (and only if) the model fits the data correctly, the AR coefficients provide an efficient means of describing the characteristics of the underlying signal. However, there are some caveats:

- One has to make sure that the model is valid – check whether the error terms, e, can really be considered to be random noise (this can and should be tested *a posteriori*) and that they are not correlated with amplitude values, y;
- The model order, p, needs to be defined in advance. A too low model order will lead to a poor signal description; a too high model order will lead to 'overfitting' and create increased variance in the model parameters. Some guiding aids exist to find an optimal order (e.g. Akaike's information criterion (Akaike 1974). In practice it involves some trial and error experimenting to see what order works best.

Studies on EEG signal processing using AR models report orders up to $p = 15$; relatively low orders of $p = 5$ or $p = 6$

have been reported in various types of EEG (e.g. Jansen et al 1981, Cerutti et al 1986, Pardey et al 1996).

Autoregressive models have been used in clinical neurophysiology to correct the EEG for ocular artefacts (van den Berg-Lenssen et al 1989) and to track the EEG during neurosurgery to measure reactions of the central nervous system (CNS) to hypotensive drugs (Cerutti et al 1986). Another application concerns use of tracking the AR parameters for non-stationarity detection/segmentation purposes (Lopes da Silva et al 1976, van de Velde 2000). AR parameters are also used to assess depth of anaesthesia in the Advanced Depth of Anaesthesia Monitor (ADAM) (Thomsen 1992) (see p. 83).

Frequency-domain methods

A typical EEG is composed of quasi-rhythmical components, such as alpha activity, and transients, for example a K-complex or an epileptiform spike and wave. Frequency analysis is the most common method of analysing the EEG and emphasizes the rhythmic components.

The result of frequency analysis of a particular length of time (an epoch) is a frequency spectrum which is often presented as a graph showing the amplitude of the components as a function of frequency. Instead of amplitude, the *y*-axis is often scaled in terms of the square of amplitude, traditionally called power, and the term 'power spectrum' is used. If individual harmonic components within a single or averaged power spectrum are divided by the bandwidth in Hz, units of power/Hz are obtained. This is called the power spectral density (PSD). Since power is proportional to the square of the voltage at the input of the EEG machine, spectral units are usually $\mu V^2/Hz$. This power is not necessarily related to energy, the usual concept of power.

There are various methods to estimate the power spectrum; they may be classified into *non-parametric* and *parametric* methods. An example of the former is given in *Power spectrum estimates using the discrete Fourier transform* (below), and an example of the latter in *Power spectrum estimates using AR models* (p. 339).

Power spectrum estimates using the discrete Fourier transform
The FT was developed by Jean Baptiste Fourier (1768–1830), the French mathematician, who first showed that it was possible to transform any repetitive (periodic) time series into a finite set of frequency components. Conversely, any periodic signal, however complex, can be formed from a series of harmonically related sine waves the fundamental frequency of which is the repetition frequency of the signal. An example of this is given in *Fig. 8.16* in which a sharp wave is formed from eight sine waves.

An assumption in the Fourier series analysis of a particular signal epoch is that the signal is repeated before and after the epoch, so that the analysis has a fundamental period equal to the duration of the epoch. For example, in *Fig. 8.16* there would have to be sharp waves occurring periodically before and after the section of signal in the illustration.

When an EEG sample of finite duration is Fourier analysed, it is implicitly assumed that the result is representative of a much longer sample and is a measure of the EEG generating process for that particular record. The signal is assumed to be 'stationary', i.e. its statistical properties do not change over time. Such an interpretation requires consideration of the statistical properties of signals (see *Segmentation*, p. 329). As already realized, the EEG is never exactly repetitive (stationary), but violation of this principle has not inhibited many people from applying Fourier series analysis to brain activity. A specific period of time (say 10 s) can be analysed, but this epoch will give only a statistical estimate of the (brain's) generating process, just as measuring the height of a group of 10 people will give only a statistical estimate of the height of the whole population. Taking a longer sample (say 2 min) would provide a better estimate, but this too will be invalid if the state of the patient changes, for instance by opening the eyes or falling asleep, or even by thinking! As we have only limited control of the state of the patient, even under general anaesthesia, any frequency analysis of the EEG will represent the frequency components only under the existing conditions. Data should be collected as far as possible in standardized circumstances and with appropriate epoch length. It is usual to analyse successive epochs of 4–10 s duration; this enables the variability of frequency spectra to be determined, though the use of significance tests on successive epochs does need a degree of stationarity. Frequency analysis is a useful tool in the hands of a person who understands its limitations.

The first stage in calculating the FT using classical mathematics is to multiply the data by a series of sine and cosine waves of frequencies corresponding to the harmonics of the fundamental frequency, and to calculate the average of these products. This gives the amplitudes of a pair of sine and cosine waves for each harmonic, representing the amount

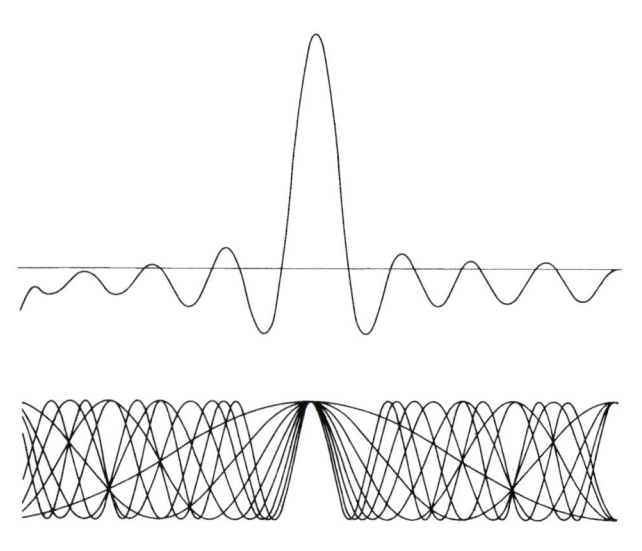

Fig. 8.16 Pulse wave (top) formed by the summation of eight harmonically related sine waves at the bottom. (From Binnie et al (2004), by permission.)

of activity of that frequency in the data. This is equivalent to correlating the original signal with sine and cosine waves at each frequency (strictly, finding the covariance; see *Correlation analysis*, p. 333). If the signal does not contain a component at a particular frequency, the correlation will be zero. This method is too slow for frequency analysis in real time. If there are 128 sample points in the original data, there are 64 sine amplitudes and 64 cosine amplitudes in the transform (Fig. 8.17). Each pair of sine and cosine values (which can be positive or negative) are then squared and summed to give the power of each frequency component (*Fig. 8.17*, right); amplitude values can be found by taking the square root of the power. This reduces the data by a factor of 2 (64 power or amplitude values in the example above), but involves a loss of phase information.

Because the pairs of sine and cosine components contain all the amplitude and phase information, the time series can be reconstituted from these values by the inverse transform. If sine and cosine components of particular frequencies are made zero before applying the inverse transform, the reconstituted epoch will not contain these frequencies and the process will have acted as a filter. The advantage of such a filter is that there is little or no phase distortion to cause the change of phase (or latency of EPs) that may accompany analogue filtering processes. If all sine and cosine components except one are eliminated before doing the inverse transform, the reconstituted waveform is of one frequency the amplitude of which is a measure of that frequency in the original signal. This can be useful in determining the amplitude of the fundamental and harmonic.

It is important to understand that the derivation of the frequency components by the FT is only a mathematical *transformation* of the signal. The resulting components do not necessarily exist physiologically but, when converted into spectra and frequency bands, indicate the proportion of the total signal occupying these bands. The result is a quantitative measure of the contribution of the observed EEG frequencies.

Before the advent of digital frequency analysis, a form of analysis was performed using analogue filters which gave outputs that approximated the Fourier spectrum. Modern methods invariably use digital data and a version of the FT called the discrete Fourier transform (DFT). The calculations are done very rapidly using the fast Fourier transform (FFT) algorithm developed by Cooley and Tukey in 1965. This can be done on a general-purpose computer or by special hardware. There are many computer algorithms for implementing the FFT, but it is convenient to make the number of digital samples in the time series a power of 2, for example 128, 256, 512, etc. Consequently, it is common to sample at a rate that is a power of 2, for example 256 points per second, and to use epochs of 1, 2, 4, 8 or 16 s (often the epochs are inconvenient durations of 1.024 s, 2.048 s, etc.).

The frequency resolution of a spectrum depends on the duration of the epoch that is transformed as well as the sampling frequency. If the duration of the epoch is N samples, sampled with a sampling rate of Fs Hz, then the spectrum will have components spaced by Fs/N Hz. In the case of the *discrete* FT there are an infinite number of components, but the series is periodic, repeating itself with a period of Fs; there is one component subseries centred around 0 Hz, another (identical) one around $1Fs$ Hz, a third one around $2Fs$ Hz, etc. The point in the middle, $Fs/2$, thus acts as the border between repeating subseries and in practice we cut off the frequency component series at that point to study the spectrum. If the analysed signal contains frequency components higher than $Fs/2$, the subseries will

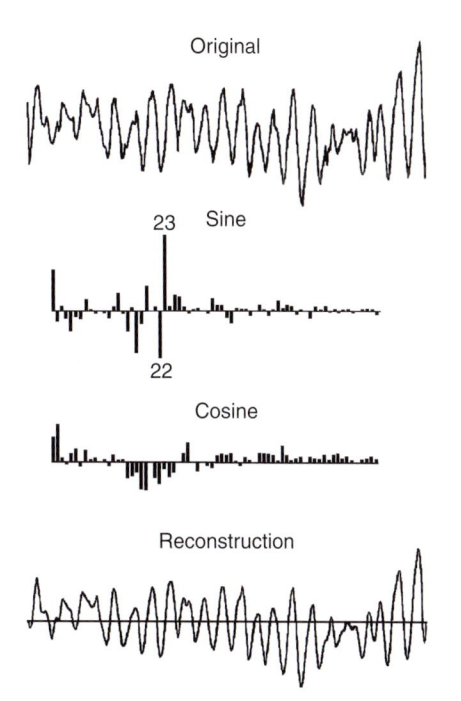

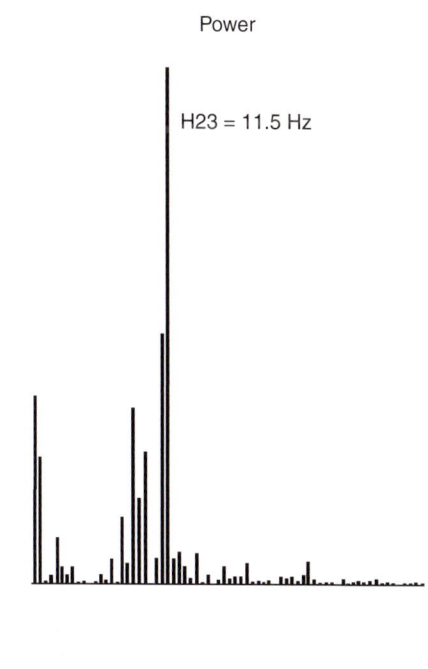

Fig. 8.17 Fourier transform of 2 s of alpha activity (upper left) into 64 sine (second left) and cosine (third left) frequency components. These components have positive and negative amplitudes. Squaring and adding them for each frequency generates the power spectrum (right). This has a peak at the 23rd harmonic of 11.5 Hz (harmonics are spaced at 0.5 Hz intervals). Application of the inverse transform to the sine and cosine components reconstructs the original waveform (lower left). (From Binnie et al (2003), by permission.)

be so wide that they cross the border and overlap each other, leading to the same problem we saw in *Analogue filters* (p. 318); aliasing. For example, if an epoch of 4 s contains 400 samples (sampling rate 100 Hz) we will use 200 components of power (or amplitude) at 100/400 = 0.25 Hz intervals, starting with a DC component at 0 Hz, a component at 0.25 Hz and subsequent components at 2/4 (0.5), 3/4, 4/4 (1), 5/4, ..., 199/4 (approximately) 50 Hz. It is usual to average groups of adjacent terms of the complete spectrum to reduce the number of components. Further averaging to give frequency ranges corresponding to the conventional EEG bands is also often done. The spectra can be displayed as bar graphs or as a smooth curve, the latter being used for displays in three dimensions (power, frequency and time) (*Fig. 8.18*).

It has already been stated that Fourier series analysis assumes that the signal in the epoch being analysed is repeated outside the epoch (as though the end is joined on to the beginning of an identical epoch). If the signal at the beginning and end is not zero (or of the same amplitude and slope) there will be an abrupt change or discontinuity, which introduces additional (small) components into the spectrum that are not in the original signal, this is called 'spectral leakage'. This abrupt discontinuity is in practice always introduced if we select an epoch for analysis using a straightforward time window. Spectral leakage effects are more important for short than for long epochs but can be decreased by 'tapering' the data points so that the signal increases from zero at the start of the epoch and decreases to zero at the end. Tapering the ends of an epoch in this way is equivalent to multiplying the signal by a function that is unity except for the tapered ends. The function is usually called a 'window function'. Examples of window functions include the Bartlett, Hamming, Hanning and Blackman windows – they are depicted in *Fig. 8.19*.

Examples of the effect of varying the epoch length and choosing a different type of window are presented in *Fig. 8.20*.

One of the advantages of the transformation from the time domain into the frequency domain is that the spectrum is a compact quantitative description of the pattern of the EEG and enables comparisons to be made with other EEG signals (in other channels or in a different record). Each value in the amplitude or power spectrum is representative of the whole epoch, whereas in the original signal each amplitude value describes only one instant of time. For example, if the signal contains a small but persistent component (alpha activity or flicker-following) that is masked by other frequencies, it will become clear in the spectrum. The disadvantages are that the power or amplitude spectra do not contain information about the time (phase) relationships of the harmonics (which are contained in the sine and cosine components), and that there is no information about the time distribution of the activity

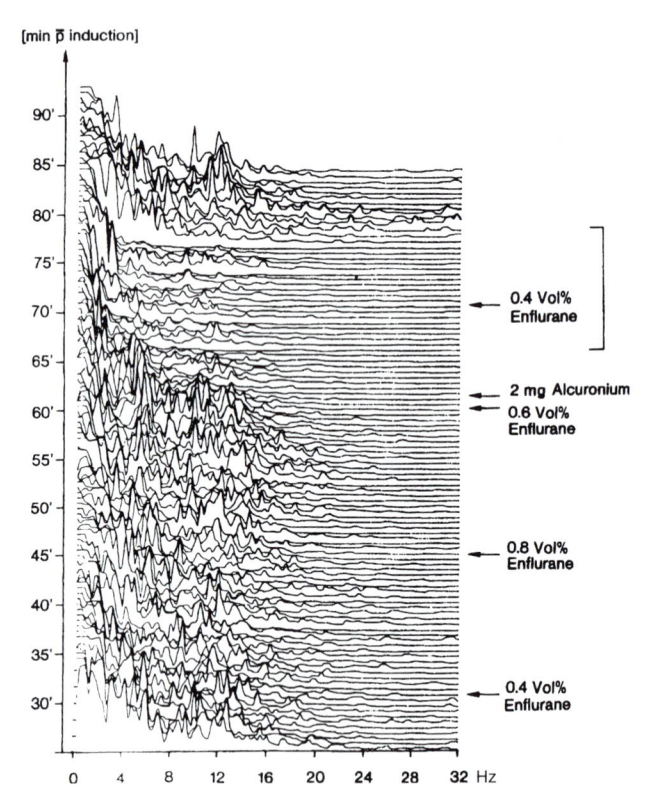

Fig. 8.18 Clinical example of a compressed spectral array (CSA) in a 72-year-old man during surgery under combined epidural and enflurane anaesthesia. Time runs from the bottom to top of chart and is given as minutes from induction of anaesthesia. EEG from Cz–A1; TC 0.3 s, HF 70 Hz. FFT averaged over 30 s epochs. Considerable EEG depression occurs with a 10 min decrease in blood pressure to 70/30 mmHg in the later part of recording. (From Pichlmayr I, Lips U, Künkel H 1984 *The Electroencephalogram in Anesthesia. Fundamentals, Practical Applications, Examples*, English edition (trans. E Bonatz, T Masyk-Iversen). © Springer-Verlag, Berlin, with kind permission of Springer Science and Business Media).

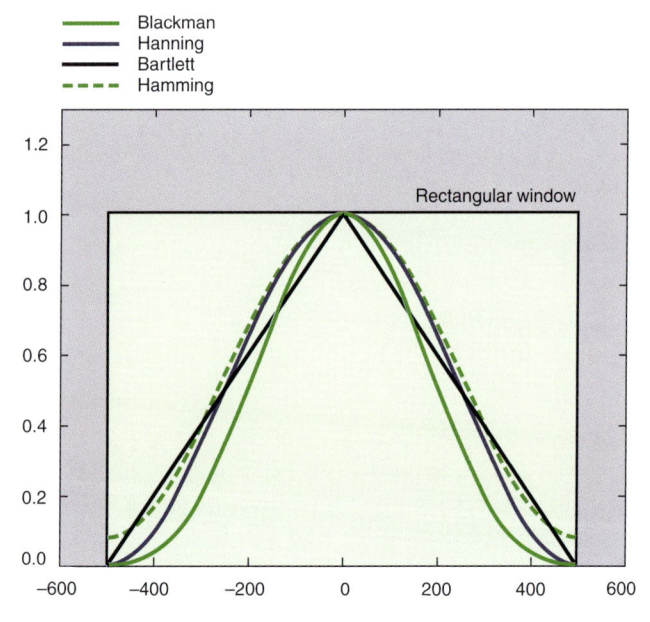

Fig. 8.19 Examples of different window functions used to reduce spectral leakage. In this example the window width (epoch length) is 1000 samples and is centred around time *t* = 0.

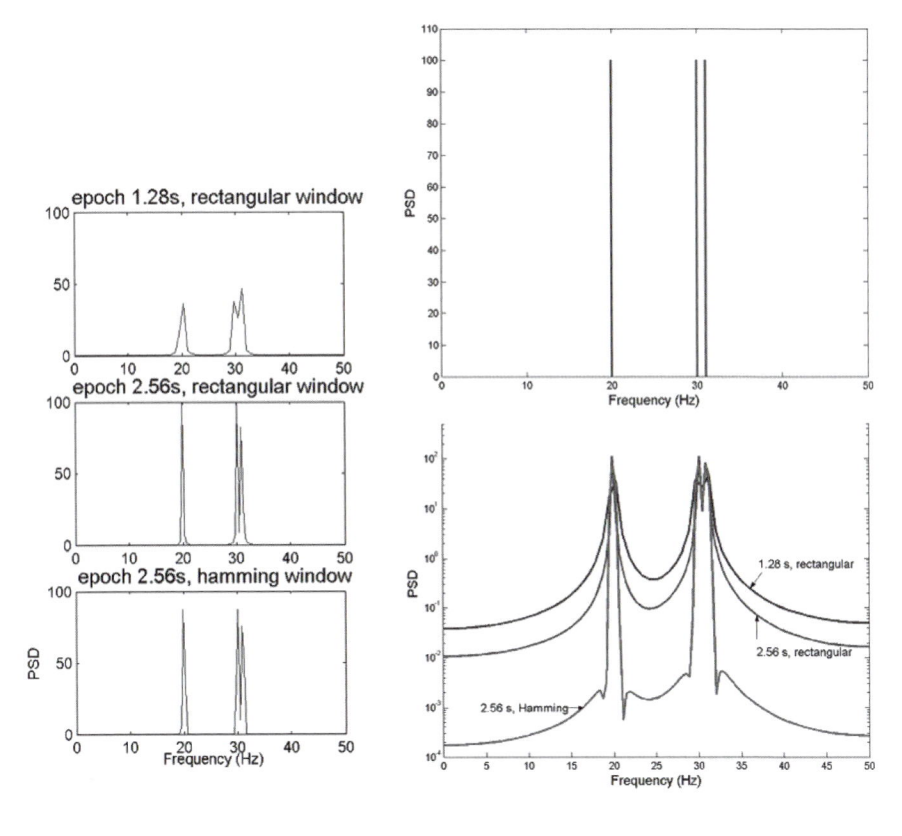

Fig. 8.20 The estimated PSD of a signal consisting of summation of three sine waves, of 20 Hz, 30 Hz and 31 Hz, using different windows and epoch lengths. The sampling rate is 100 Hz. Top right: The theoretically correct PSD. Left: The PSD as obtained using combinations of an epoch length of 1.28 s and 2.56 s and two different windowing functions. Bottom right: The same three curves in one graph but now the PSD axis is on a logarithmic scale to make the differences clearer. It can be seen that spectral leakage (and poor resolution) is especially prominent for small epoch sizes. Increasing the epoch length increases frequency resolution, and using a different window decreases spectral leakage.

within the epoch. This means that if we look at the spectrum only, there is no way of knowing whether a particular frequency component is derived from a short burst of high amplitude activity or from prolonged low amplitude activity, nor where in the epoch it occurs. This problem may be approached by using time–frequency approaches such as wavelet analysis (see *Time–frequency models – wavelet analysis*, p. 340).

A measure of the variability of spectra is necessary when assessing the significance of changes observed in different clinical or experimental conditions, such as changes caused by use of drugs, changes of mental state and level of arousal. Fluctuations of amplitude or power of individual spectral lines from one epoch to another can be caused by a number of factors other than a change of frequency. For example, fluctuations of amplitude of a single frequency component in the EEG (amplitude modulation) will generate components called 'side-bands' at frequencies dependent upon the frequency of the basic rhythm and of the modulation. For this reason fluctuations of individual spectral lines, especially in long epochs (say of 10 s when the spacing is 0.1 Hz), cannot sensibly be tested for significant differences. Spectral variability can be reduced by averaging the spectra of several epochs (which extends the analysis period), by averaging several adjacent spectral lines (which gives a spectrum of lower frequency resolution) or by forming a 'running average' spectrum over successive sometimes partly overlapping epochs. Averaging spectra or adjacent harmonics has the advantage of greatly reducing the amount of data to be handled, displayed and stored and is useful in monitoring.

In this section we have estimated the power spectrum or PSD using the squared values of the results of the DFT. However, the definition of the PSD is that it is the FT of the autocorrelation function (*Correlation analysis*, p. 333), and that very definition provides another avenue for calculating the same spectrum. That method (called the *Blackman–Tukey* method) uses a finite time interval to get a (biased) approximation of the autocorrelation function and employs the FT on that. Due to the bias in the approximated correlation function it is not without its problems. The method described here using the squared DFT is often referred to in the literature as the *periodogram* approach (the term *Welch's method* is used if partly overlapping segments are averaged to smooth the result).

Power spectrum estimates using AR models Besides their use as efficient signal descriptors, the AR coefficients as described in *AR modelling* (p. 335) can be used for another purpose as well. The PSD of the signal can be calculated directly as a mathematical function of the AR coefficients (see e.g. Bruce 2001). As the AR coefficients are also called model parameters, the AR modelling method to calculate the PSD is also referred to as a *parametric approach*. It gives us an attractive alternative to the DFT approach, because:
- unlike the DFT, it does not make drastic assumptions about the data outside the analysis window;
- the estimation of the PSD using AR modelling has a smoother, sometimes easier interpretable shape than that obtained with the DFT;
- the PSD does not suffer from spectral leakage as the DFT does;

- the frequency resolution is not as dependent on the number of data samples; and
- the PSD is more statistically consistent, especially when short segments of data are analysed.

The earlier mentioned caveats related to the AR method are still valid, i.e. the model has to be valid and the model order needs to be predefined. Another drawback is that the method is more complex than the DFT/FFT approach. These caveats may explain why in the vast majority of commercial systems the PSD calculation is done using the non-parametric DFT/FFT approach.

Commonly used features extracted from the power spectrum

Once reliable spectra have been obtained there are basically two ways that one can track changes in the spectrum. For each spectrum calculated over a segment either: (1) try to visualize the shape of the whole spectrum in a plotted curve and see how that changes over time; or (2) extract a set of parameters that describe the most important features of the spectra and follow those variables as a function of time. The first way leads to methods that plot the EEG spectra in a semi-three-dimensional graph, with frequency on the *x*-axis, time on the *y*-axis and the PSD on the *z*-axis (the so-called compressed spectral array (CSA) (see *Fig. 8.18*)); or plot a two-dimensional graph in which the PSD is represented by colour coding or, originally, the intensity of one colour (colour density spectral array (CDSA) and density-modulated spectral array (DSA) respectively). These methods are excellent when interpretation of the spectra is to be done via the human eye; however, in order to facilitate processing via a computer preference is often given to extracting a set of single variables that are thought to quantitatively represent the main features of the EEG spectra. They can be fed directly into any algorithm and can easily be plotted in a simple time graph. Parameters that are commonly derived from the frequency representation include:

- the peak power frequency (PPF) – the frequency at which the highest power exists;
- the median power frequency (MPF) – the frequency below which half of the total spectral power is present;
- the spectral edge frequency (SEF) – the frequency below which 90% or 95% (depending on the definition used) of the total spectral power can be found; and
- the power present in the different frequency bands (typically, delta (0–4 Hz), theta (4–8 Hz), alpha (8–12 Hz) and beta (> 12 Hz)) – this can be either the absolute power or the relative power (with respect to the total signal power).

Although these measures have been shown to be correlated with changes in the brain, one should bear in mind that they cannot be used to reduce the EEG to one single number without obvious loss of important information. The measures can be compared within the same record, from record to record in the same patient or across groups of patients in different conditions (drugs, clinical diagnosis, coma depth, etc.).

Many papers are published each year on the use of frequency analysis. In most of them spectral analysis is used as a means of quantifying the EEG (for an overview see Binnie et al 2003, Section 4.12), for example during anaesthesia, during carotid artery and cardiac surgery, and after cerebral ischaemia due to stroke.

Time–frequency methods – wavelet analysis

In *Power spectrum estimates using the discrete Fourier transform* (p. 336) a fundamental problem of the DFT was noted, namely detecting the time of occurrence of events in the signal that do not occupy the whole epoch being analysed is not possible. If a transient signal occupies say 2 s of a 10 s epoch, the DFT will contain spectral values of the transient throughout the whole 10 s. Differing times of occurrence of the transient within the epoch will make no significant difference in the spectrum. Similarly, a signal with a sine wave component present throughout a 10 s epoch will give an DFT identical to that of the same signal, but with double the amplitude, present for only half the epoch duration.

Analysis of a signal to measure the time of occurrence, as well as the frequency components of events in the signal is referred to as time–frequency analysis. One technique, (adaptive) segmentation, has already been described in *Segmentation* (p. 329) (Barlow 1984). There are a number of other methods; for a general review see Lin and Chen (1996).

One way of solving the problem is to analyse the signal using successive short epochs, sometimes overlapping, of say 1 or 2 s duration. This is done in the so-called short time Fourier transform (STFT). One problem with this method is that, because the frequency resolution is Fs/N, where N is the number of samples in the epoch, and Fs the sampling frequency, short epochs have poor frequency resolution (and very short epochs may/will violate statistical assumptions about the signal outside the epoch window). We can improve the frequency resolution by lengthening the epoch, but that would obviously give us poorer time resolution so we would not be able to track fast changes in the signal. Moreover, longer epochs increase the risk of having non-stationarities in the signal, degrading the value of the frequency analysis further. This problem is called the time–frequency trade-off problem. Wavelet analysis is one alternative method to STFT; it provides more flexibility with respect to the problem of time–frequency resolution trade-off, and moreover it can provide a more effective means to compactly describe signals that are not very well represented by sine wave shapes. By analogy with Fourier analysis, we can do the analysis using a continuous approach, the continuous wavelet transform (CWT), or we can employ a discrete version, the discrete wavelet transform (DWT).

Just as the sine and cosine components of FT are called the basis functions of the Fourier analysis, a wavelet is the basis function used in wavelet analysis. It is a signal having a *finite* duration, with a zero mean value, and often with a shape such that its FT occupies a certain band of frequencies. Examples of some wavelets are shown in *Fig. 8.21*.

In practice, a wavelet shape is chosen to suit a particular type of signal at hand. Just as in the FT a signal can be thought of as consisting of a summation of basis functions. However, there are two essential differences when considering the basis functions of the FT and WT:

1. The basis functions of the FT are infinitely continuing functions (sine and cosine functions), whereas the basis functions of the WT are distinctly localized functions (see the curves in *Fig. 8.21* that really have a start and an end point).
2. The basis functions of the FT are predefined; they are always sine and cosine functions, whereas the shape of the basis functions of WT can be chosen (subject to certain mathematical conditions). Therefore, results of wavelets analyses can vary depending on the choice of wavelet shapes, and studies reporting their results should always explicitly mention the type of wavelets used.

Whereas in FT we use sine and cosine functions with different frequencies to decompose (and recompose) the original signal, in WT we use stretched (dilated) and displaced (translated) versions of the original wavelet shape (the 'mother wavelet') to represent the original signal. Thanks to the ability to choose different wavelet shapes, the WT can be more effective than the FT in describing certain signals. For example, steeply varying or distinctly localized signals, like spikes or bursts or certain complexes, may be much better represented by a summation of a few appropriately chosen wavelets rather than a long summation of smooth sine waves that do not fit the character of the signal very well.

In a simple wavelet analysis, the wavelet function is in effect correlated (mathematically convolved) with the signal to be analysed, starting at the beginning of the epoch and at consecutive displacements throughout the epoch. Any part of the signal having a similar underlying shape as the wavelet will then give a positive value of covariance. The displacements are called 'translations' in wavelet terminology. The covariance is displayed as a function of time, so that the time of occurrence of any parts of the signal with the same frequency content as the wavelet are easily seen.

In order to carry out the analysis for another frequency band, the duration of the wavelet is changed, usually increased (dilated); i.e. the signal is stretched along the time axis. Performing the same correlation analysis now with the stretched version of the wavelet will lead to detection of slower components in the signal. This process can be repeated iteratively, every time detecting successively lower frequency components in the signal. Thus, with every step towards the direction of lower frequency we use a longer wavelet shape. This has the result that for lower frequencies we are using a longer time window (epoch length). As we have seen earlier, a longer epoch length

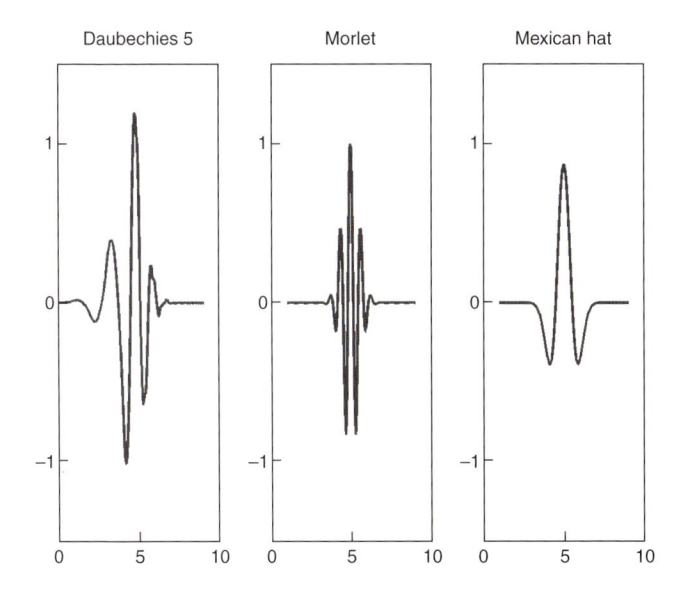

Fig. 8.21 Some examples of wavelet functions. Many more exist, some in 'families', like 'Daubechies' from which 'Daubechies 5' is one member.

gives us better frequency resolution. Thus, in WT the frequency resolution increases towards lower frequencies (i.e. it decreases towards higher frequencies). This is another essential difference from the FT where the frequency resolution is the same throughout the whole frequency range. Overall, the product of time resolution with frequency resolution is exactly the same for both WT and FT, but for WT we have an extra parameter to tune the trade-off balance at different frequencies. This property is highly useful in many biomedical signal analysis applications. If the aim is to detect 'bursts' or sharp complexes (high frequency components) we typically want to determine at exactly what time they occur, not so much what their exact frequency content is (we want good time resolution, and can accept poorer frequency resolution). Conversely, if we are interested in analysis of slow underlying trends we typically want to have a good frequency resolution in the first place, the exact phase information is of secondary importance.

When using FT we use *two* basis functions for (de)composition of signals; the sine and the cosine function (they form a so-called orthonormal basis). The same approach is common in wavelet analysis. For (de)composition, we use not one type of shape, but *two* to allow a more complete representation of the signal. Confusingly enough in wavelet terminology one of them is called 'wavelet function' and the other one 'scaling function'. An example of each is given in *Fig. 8.22*.

If we consider the values of the points in a wavelet function or a scaling function as a set of filter coefficients then we can calculate associated filter characteristics (see *Fig. 8.22* bottom graphs). Convolving the signal now with the different scaling and wavelet functions effectively results in filtering the signal with the associated filters (in this particular example they are low- and high-pass filters, often they are bandpass filters). Thus, the scaling func-

Chapter 8 Further Signal Analysis

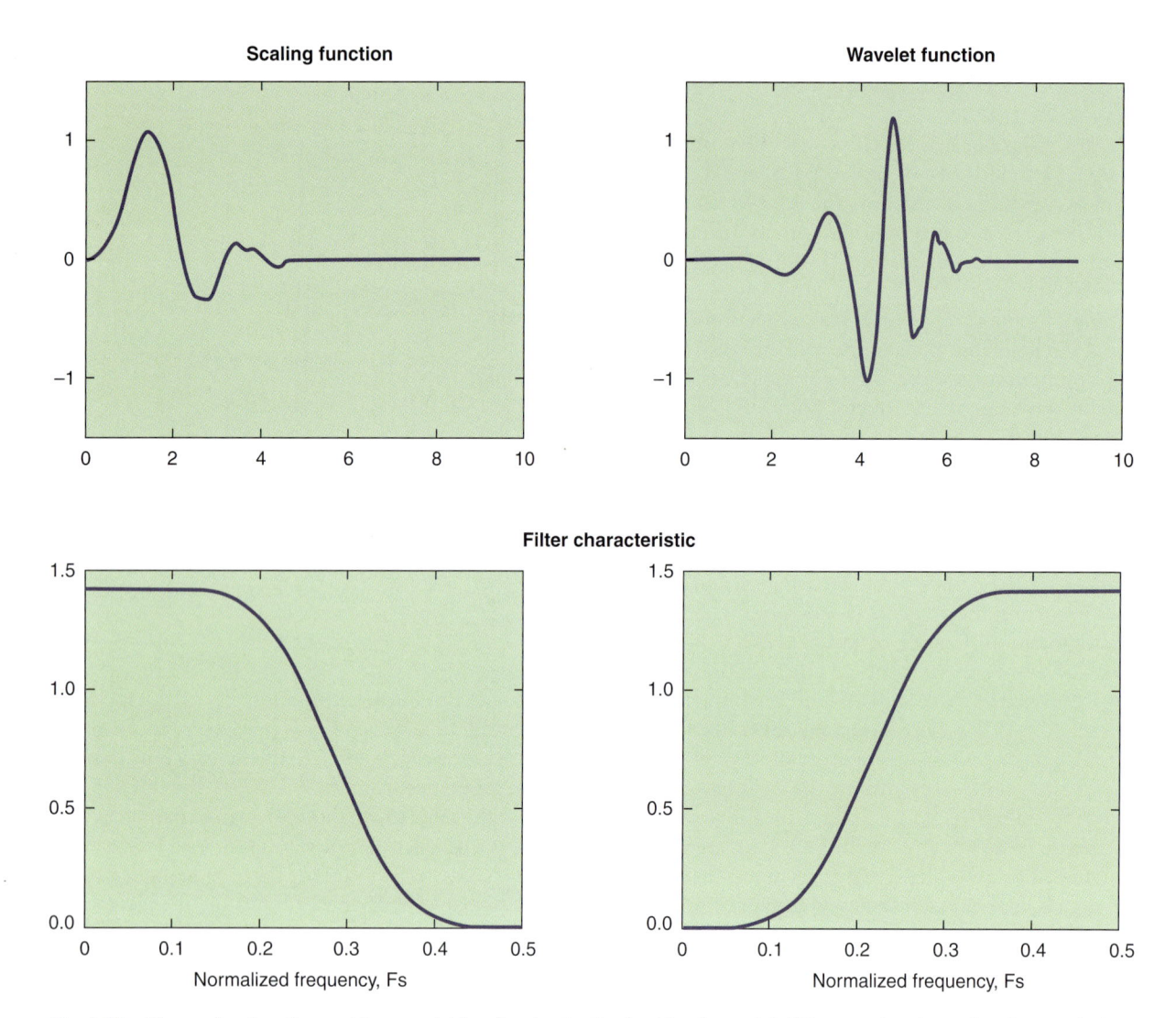

Fig. 8.22 The scaling function and the wavelet function for the Daubechies 5 wavelet. If the sample values of each curve in the upper graph were used as filter coefficients a filter would result with a frequency response depicted in the lower graphs. The scaling function is associated with a low-pass filter passing frequencies of < 0.25 Fs, the wavelet function with a high-pass filter that passes frequencies > 0.25 Fs (the high end border for the analysis is obviously at 0.5 Fs).

tion (<–> low-pass filter) can be thought of as being used to capture the slow ('coarse') components of the signal, whereas the wavelet function (<–> high-pass filter) captures fast ('detail') components. Instead of stretching or compressing the wavelet and scaling function to change the frequency response of the associated filters, in practice one varies Fs by subsampling the signal. For example, let us assume we have 1000 samples representing data recorded in 1 second; Fs = 1000 Hz. Low-pass and high-pass filtering will give us information about two frequency bands – one containing frequencies between 0 and 250 Hz, and the other one between 250 and 500 Hz. Next, we can subsample the signal by keeping only every other sample, in effect obtaining a new sampling rate Fs of 500 Hz. Applying the same filters now on the subsampled signal will lead to two other new signals; one representing 0–125 Hz and another one for 125–250 Hz. This process can be repeated, every time decreasing Fs by subsampling further and thus getting finer and finer frequency bands. This process is called wavelet multiresolution decomposition. An example of the process is depicted in *Fig. 8.23*.

Examining in which bands the most energy is contained gives the opportunity to create a compact representation of the signal. In the example in *Fig. 8.23* we can make a good approximation of the total signal by a weighted summation of the deepest level low frequency component (representing the trend curve) and the high-frequency component (representing the burst) at the second level of the tree. In order to be able to recompose a good approximation of the signal we thus need to store the fact that we need only the wavelets at those two specific levels plus their associated weights; a considerable data compression compared to the original 1000 samples.

The above is a simplified, non-mathematical, description of wavelet analysis. The theory has been considerably extended and is mathematically complex (Coifman and Wickerhauser 1996, Figliola and Serrano 1997, Hubbard 1998). Extensive tutorials which do not use difficult

Signal processing and interpretation

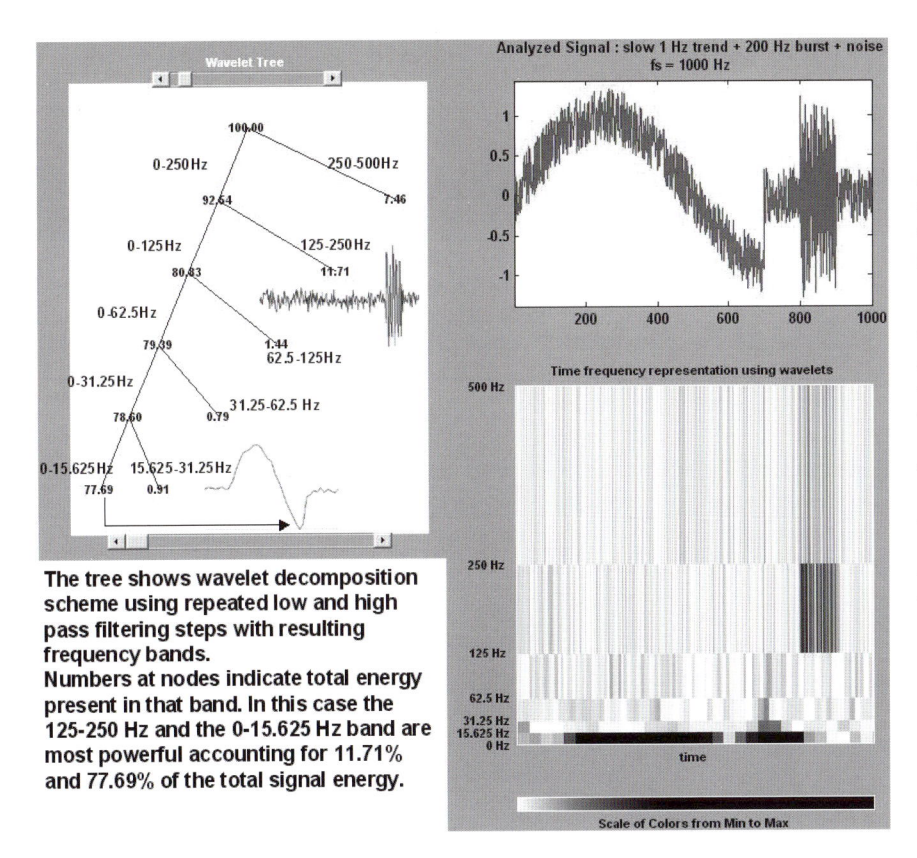

Fig. 8.23 Multiresolution decomposition of an artificial signal (top right) consisting of a slow 1 Hz wave, a 200 Hz burst and added noise (Fs = 1000 Hz). The signal is low- and high-pass filtered using Daubechies 5 wavelets (*Fig. 8.22*) in the process described in the text, leading to the tree structure of frequency bands as depicted in the top left figure. The bands with the highest energy are the lowest frequency band (< 15.625 Hz) and the 125–250 Hz band. (Extending the tree further would improve the frequency resolution at the low end.) The associated time–frequency representation of the signal is depicted at the bottom right. Note the poor frequency resolution but good time resolution at high frequencies and the opposite at low frequencies. The timing of the burst can be detected accurately (but its frequency contents less so) and the frequency of the slow trend can be detected accurately (but its exact location not). In most practical situations this division of priorities with respect to resolutions is highly useful.

The tree shows wavelet decomposition scheme using repeated low and high pass filtering steps with resulting frequency bands.
Numbers at nodes indicate total energy present in that band. In this case the 125-250 Hz and the 0-15.625 Hz band are most powerful accounting for 11.71% and 77.69% of the total signal energy.

mathematics have been published by Graps (1995) and Samar et al (1995, 1999). The main clinical application of wavelet analysis has so far been the detection and analysis of transients such as EPs and epileptiform spikes (Schiff et al 1994, Samar et al 1995, Senhadji et al 1995, Clarencon et al 1996, D'Attellis et al 1997, Goelz et al 2000). An application to spike detection is shown in *Fig. 8.24* and a multiresolution analysis of a spike and wave record is shown in *Fig. 8.25*.

Browne and Cutmore (2000) used a method based on wavelet analysis to extract ERPs from the on-going EEG in preference to trial averaging. Another clinical application is for assessment of the level of anaesthesia using wavelets on the EEG (Zhang and Roy 1999) as well as on SEPs (Hoppe et al 2002). Stockmanns et al (1997) as well as Roy and colleagues (Nayak and Roy 1998, Huang et al

Fig. 8.24 Detection of an epileptiform spike by wavelet analysis: (a) EEG; (b) spike detection. (From D'Attellis CE, Isaacson SI, Sirne RO 1997 Detection of epileptic events in electroencephalograms using wavelet analysis. *Ann Biomed Eng* 25:286–293. © Springer-Verlag, Berlin, with kind permission of Springer Science and Business Media.)

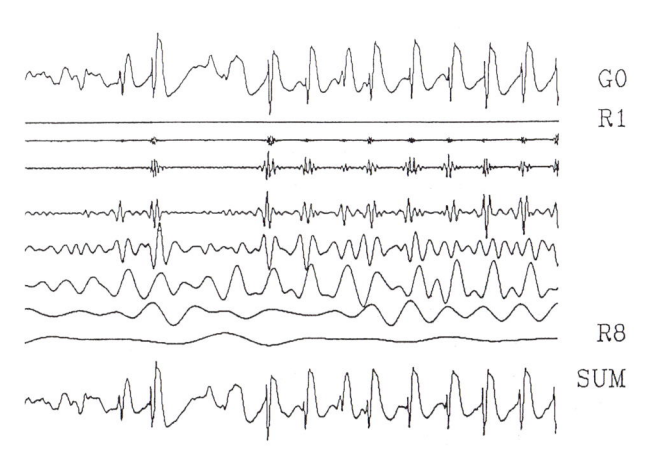

Fig. 8.25 Multiresolution decomposition wavelet analysis of a spike and wave episode. Traces R1 to R8 are the results of the wavelet analysis of the signal in the top trace (G0). Wavelet dilation increases from R1 to R8. SUM shows the reconstruction of the EEG by summing traces R1 to R8. (From D'Attellis CE, Isaacson SI, Sirne RO 1997 Detection of epileptic events in electroencephalograms using wavelet analysis. *Ann Biomed Eng* 25:286–293. © Springer-Verlag, Berlin, with kind permission of Springer Science and Business Media.)

Chapter 8 Further Signal Analysis

1999) employed the DWT to compact the MLAEP while still capturing its relevant time and frequency contents successfully. The latter used those wavelet components that were found to separate best between non-responders and as input to an ANN that would subsequently help control an infusion pump for propofol administration. Wavelets can be designed to act as a matched filter to detect particular evoked or ERPs, such as the P300, in single trials (Samar et al 1995, Effern et al 2000). A comprehensive review on the use of wavelets to detect EEG transients may be found in Senhadji and Wendling (2002).

It seems likely that wavelet analysis will be used increasingly for EEG analysis. A problem with using wavelets is that there is a subjective element in the choice of the wavelet for a particular analysis. This may deter its application, although it can be argued that most signal analysis techniques involve some subjectivity – for example the choice of epoch length and window shape in Fourier analysis.

Alternative methods

The interpretation of the relationship between the complex spatiotemporal neurophysiological activity recorded using sensors and the functioning of the nervous system, both in health and disease, presents a formidable problem. Assumptions that underlie the methods described in the previous sections are that the signal results from a linear process and is (weakly) stationary. However, the non-linear characteristics of brain functioning give rise to a signal that has very complex dynamics (Koch and Laurent 1999), this may be one reason why the linear methods do not perform satisfactorily in all applications. For this reason, the search for alternative methods, including non-linear methods, is receiving more and more interest.

Complex methods exist to study the relationships between the neurophsyiologic measurements and brain function, and attempt to discover the physiological basis of the recorded signals. Clinical studies are confined to those departments that can afford the considerable computing skills and hardware required. While some clinical studies using new analysis techniques show significant differences between diagnostic groups, as with so much of signal interpretation a specific diagnosis of an individual patient does not necessarily follow. Some of the newer methods also tend to concentrate on the time–frequency analysis of a single channel whereas the clinical use of the EEG depends heavily on multichannel recordings. The extra dimension of topography is a formidable hurdle for these new techniques to overcome (see Koenig et al 2001).

No attempt is made here to describe the rather advanced mathematical and computing techniques that some of the newer methods require, as they go beyond the remit of this book. However, as an example we summarize two examples of processing techniques as used in EEG monitoring during surgery: higher order spectral analysis and complexity analysis using spectral entropy. Examples of other noteworthy techniques include chaos theory, Nunez' wave dispersion model, Barlow's extrema slopes, and independent component analysis (ICA). Summaries of the first three methods can be found in Binnie et al (2003) and a description of the latter in Hyvärinen et al (2001). Novel methods of analysing the EEG are continually being developed and the literature should be searched by readers interested in this topic.

Higher-order spectra – the bispectrum We have seen in *Correlation analysis* (p. 333) that the covariance and the cross-correlation coefficient are measures of the similarity of two time series. If we are interested in exploring whether specific frequencies or frequency bands might be interrelated we can determine this by computing cross-correlations in the frequency domain. This is the coherence function or coherence spectrum. It is similar to a frequency spectrum, but the ordinate scale is not a measure of amplitude or power but a correlation coefficient showing the similarities at particular frequencies. Detailed descriptions of coherence are given in Shaw (1981, 1984), and Dumermuth and Molinari (1987). The first of these is an easy tutorial that includes a worked example using simple arithmetic.

However, there are several problems which make the coherence spectrum difficult to apply and interpret (Shaw and Simpson 1997). In practice, interpretation requires averaging over several components in the spectrum over several epochs. Selecting the components for averaging within a spectrum is a rather arbitrary process because it assumes that adjacent spectral values are functionally related; this may not be the case. For example, in averaging over six terms, three terms may be functionally related and three may not. The resultant coherence will then be lower than correct for the three related terms and higher than it should be for the three unrelated terms. A way of solving this problem is to determine the frequency components that are functionally related, or calculate correlations on functionally significant waveforms. One way to address this problem is to use models of signal generation. Another is to calculate so-called higher-order spectra.

In the description in *Correlation analysis* (p. 333) it was mentioned that the PSD of a signal is the FT of its autocorrelation function. The autocorrelation function (acf) was obtained correlating the signal with shifted versions of itself. In mathematical terms, if we have a signal x recorded as a function of time (or 'sample index') n:

$$\mathrm{acf}_x(i) = E\{x(n) \cdot x(n+i)\}$$

where the $E\{\ \}$ expression indicates the expectation of the outcome of the product of the signal x with a copy of itself shifted by i. The value $\mathrm{acf}_x(0)$ is calculated using a time shift 0 ($i = 0$), the other points of the function are calculated by varying i. Applying the FT to the thus obtained acf leads us to an estimation of the power spectrum of the signal. Another name for this 'usual spectrum' is the second-order spectrum of the signal. We can calculate higher-order spectra by extending the above defined acf function. For EEG processing, the third-order spectrum (more commonly known as the bispectrum) has received particular attention. For example, to calculate the third-order spectrum we would use a function C, dependent on two time shifts, defined as

$$C(i, j) = E\{x(n) \cdot x(n + i) \cdot x(n + j)\}$$

The function C is also called the third-order moment sequence (and acf the second-order moment sequence – the reader may infer that the first-order moment sequence is the mean of the signal). The bispectrum can now be viewed as being calculated from a two-dimensional FT of C (leading now to a three-dimensional graphical spectrum representation with two frequency axes, $f1$ and $f2$, and power on the third axis; see *Fig. 3.14*). In practice the calculation is not trivial and, as with the 'normal' power spectrum, again various different approaches exist to calculate it (see e.g. Petropulu 2000).

As we noted the power spectrum treats the signal under study as a superposition of uncorrelated harmonic components but does not provide information about phase relationships. The bispectrum has several useful properties:

1. the bispectrum of a Gaussian process is zero, so we can use the bispectrum to verify the validity of assumptions about the Gaussianity of the signal, or we can use it to suppress Gaussian noise with unknown mean and variance;
2. with it we can obtain information not only about the magnitude but also about the phase response of signals; and
3. we can use it to detect and characterize non-linearities in the data. When two signals with frequencies $f1$ and $f2$ are passed through a non-linear system, a new frequency $f1 + f2$ appears at the output, with its phase coupled with the phases of the input signals. This phase coupling is detected by a high value in the bispectrum at coordinate $(f1, f2)$. This can be used in the earlier discussed problem of finding out which frequencies are coupled and which are independent.

All three properties can play a role in EEG signal processing applications. The exact physiological meaning of the quantitative bispectral values of the EEG is unclear. One intuitive explanation could be that the more and the larger bispectral components there are in the signal, the stronger the phase coupling between the sources of the EEG, which would suggest that there is a smaller number of independent sources. The bispectrum has been used in research on vigilance states (Ning and Bronzino 1989), and it is being used as one component in a composite parameter used to assess the depth of anaesthesia, i.e. the Bispectral Index (BIS).

Bispectral Index (BIS) This composite parameter was developed by Aspect Medical Systems, Inc. (Newton, MA, USA, http://www.aspectmedical.com) using a large database of in the order of 2000 patient recordings. The empirically obtained index consists of a weighted sum of the following four parameters (for a more in-depth description see e.g. Rampil 1998):

- from the time-domain; burst suppression ratio (BSR), the fraction of time in an epoch when EEG is suppressed, and QUAZI suppression, which allows burst suppression detection in presence of a wandering voltage baseline;
- from the frequency domain; the relative beta ratio, the log ratio of power in the frequency bands 30–47 Hz and 11–20 Hz (the borders of these bands have been empirically obtained)
- from the bispectrum; the SynchFastSlow parameter, the log ratio of the sum of all bispectrum peaks in the range 0.5–47 Hz over the sum of the bispectrum in the area 40–47 Hz.

There is no reason why new parameters could not be added to the index if they are shown to improve performance. As the index thus may be subject to change with different release versions one should document the particular BIS version number used during a recording.

The parameters incorporated in BIS are combined in a non-linear fashion to obtain one single number which correlates to the level of hypnosis on a scale from 100 to 0 (100 representing a fully conscious state, 40–60 typically the target workable region for surgical levels of hypnosis, and 40 and below the area of unnecessarily deep anaesthesia prolonging recovery). The weights of the summation change within this range; during light sedation, most emphasis is on the relative beta ratio, during surgical levels of hypnosis the SynchFastSlow parameter is emphasized more, and during deep anesthetic levels the burst suppression and QUAZI suppression play a key role (Rampil 1998, Bruhn et al 2000). BIS does not work perfectly in all situations (Dahaba 2005): for example, it is unreliable during ketamine anesthesia and sedative concentrations of nitrous oxide do not appear to affect BIS (Barr et al 1999); also, in the absence of neuromuscular blockade the EMG may have a biasing effect on the BIS value (Renna et al 2002).

No physiological explanation has been found for this correlation, although some studies have found certain correlations with physiological processes. For example, studies using positron emission tomography (PET) during propofol and isoflurane anesthesia show that BIS correlates linearly with a reduction in the cerebral glucose metabolic rate (Alkire and Pomfrett, 1997). The method is popular thanks to the associated strong empirical evidence (Glass et al 1997).

Measures of complexity – entropy Another means of describing the signal properties is by characterization of the amount of 'disorder' present in the signal. A sometimes used thought behind this type of description applied to EEG is that different states of the brain result in different complexities of the EEG signal due to the variation of the number of independent signal generators. The complexity of the EEG can then be associated with the brain state (e.g. level of consciousness). However, it is probably safest to regard these quantities in the first place as pure signal descriptors only.

Examples of complexity measures applied to EEG are: entropy (Rezek and Roberts 1998); Lempel–Ziv complexity, characterizing chaotic temporal components in non-linear systems (Lempel and Ziv 1976); fractal spectrum, separating random fractal components from a time series in the frequency domain (Yamamoto and Hughson 1993); and non-linear correlation index (Widman et al 2000).

Entropy (S) reflects the amount of irregularity of the signal, or amount of information embedded in the signal. The higher the entropy, the more information is required to predict (or describe) the signal. As such, entropy can be used as one measure of complexity of the signal. High entropy (close to 1) may be associated with a large number of processes contributing to the signal, whereas low entropy (close to 0) may be due to a small number of dominating processes. The entropy can be estimated from a relatively small number of samples. The two most frequently used methods for entropy estimation are calculation of spectral entropy and of approximate entropy (ApEn).

- *Spectral entropy* quantifies the spectral complexity by employing the definition of Shannon's channel entropy (Shannon 1948) on the normalized power spectrum, P, with i frequencies, and $P(f_i)$ the power at frequency f. The spectral entropy, S, is calculated as:

$$S = \sum_i P(f_i) \cdot \ln\left(\frac{1}{P(f_i)}\right)$$

If the signal contains a wide range of almost equally contributing frequencies, the spectral entropy value will be high, if there are only a few dominant frequencies, the spectral entropy will have a low value.

- *Approximate entropy (ApEn)* was introduced as a quantification of regularity in sequences and time-series data, initially motivated by applications to relatively short, noisy data sets (Pincus 1991, Pincus et al 1991). ApEn assigns a non-negative number to a sequence or time-series, with larger values corresponding to greater apparent serial irregularity, and smaller values corresponding to instances of more recognizable features or patterns in the data. ApEn measures the logarithmic likelihood that runs of patterns that are 'close' for m contiguous observations remain close on next incremental comparisons. It is relatively insensitive to low level noise and robust to occasional artefacts, which makes it useful for application to EEG signals.

Viertiö-Oja et al (2001) compared the values of fractal spectrum, Lempel-Ziv complexity, approximate entropy, spectral entropy, and the BIS components SynchFastSlow and BetaRatio, during different levels of hypnosis (propofol-, sevoflurane- and/or N_2O-induced). Most of these measures show comparable and acceptable results with respect to prediction probabilities for conscious/unconscious prediction. They concluded that characteristics of disorder, such as entropy and complexity, indicate a patient's level of hypnosis with a prediction probability at least as high as that of the BIS. Furthermore, the methods calculating the descriptors have the advantage that they can be applied consistently from light levels of sedation down to total suppression (whereas in BIS adjustment of the algorithm depending on the level of hypnosis is required).

Entropy measures are used in a commercial monitoring system (M-ENTROPY Module, GE Healthcare, Helsinki, Finland, http://www.gehealthcare.com) to assess the level of hypnosis during anaesthesia. It uses entropy values calculated from the EEG and frontal EMG (called response entropy (RE)) and an entropy value calculated mainly from the EEG (called state entropy (SE)). The former has faster dynamics, allowing for faster reactions that are needed, for example, on impending arousal. A similar 100–0 scale as BIS' is used to indicate the level of hypnosis. RE, SE and BIS show comparable results, although RE tends to indicate emergence from anaesthesia earlier than SE or BIS (Vakkuri et al 2004). The algorithm used in the M-ENTROPY Module has been published (Viertiö-Oja et al 2004).

Summary

The preceding sections have shown that a very large variety of methods exist, often providing many different avenues to achieve similar goals. There is no such thing as a 'best' overall method (if there was, these sections could have been considerably shorter!). The choice of which method to use is in the first place governed by the purpose of the measurement and what exactly we want to get out of the measurements; secondary considerations may be the practical implementation differences between similar methods.

Table 8.4 summarizes the main characteristics of the discussed methods. In practice, research efforts will try different approaches for a given application and eventually use what works best in a particular case. Any non-trivial system will most likely use a combination of methods to extract features using different methods. These features are then combined and further processed in higher level signal interpretation system in order to integrate them into the clinical context. Such systems are the subject of the next section.

Integration of features into the clinical context – pattern classification

The methods described in the previous sections produce numbers as their output, usually as a function of time (or frequency), that say something about the signal characteristics. Often these numbers are called *features*. A common use of these is to plot them directly in a trend curve that can be displayed in real-time for monitoring purposes or redisplayed later, for example on a PC screen or printouts for off-line analyses. Visual examination of trends of separate features (together with other clinical data and the raw data) can be a very powerful tool for examining the patient state – provided the meaning of the features is clear to the person examining them. Often, however, it is advantageous to combine features using 'next-level' processing methods that generate outputs that are more readily interpretable in a clinical context. This is especially so in situations where experts are not always available. For example, in the EEG monitoring case, the International Federation for Clinical Neurophysiology (IFCN) recommendations for clinical practice of EEG and EPs in comatose and other unresponsive states (Guérit et al 1999) consider continuous neuro-monitoring as 'a neurophysiological tool to be used by non-neurophysiologists' and emphasize that neurophysiological information must be 'coded in language that is readily interpretable by the user'. To this end they suggest a three-step approach consisting of:

1. quality screening, followed by automatic feature extraction;
2. selecting and combining features to define patterns; and
3. integration of patterns into a clinical context.

The first step has been discussed in the preceding sections. This section will briefly discuss the second step and then proceed with the third step, discussing methods that use references to evidence-based data libraries showing clusters or patterns of high sensitivity or specificity.

From features to patterns

Examples of features include time-domain descriptors, frequency descriptors, wavelet coefficients, complexity quantities, etc. As one single feature describes only one aspect of the signal we usually use several features to provide a more useful summary of the signal. We thus create a *feature set*. An example of a feature set extracted from an EEG channel could be (RMS amplitude (μV), SEF95 (Hz), MedianFrequency (Hz), SpectralEntropy (–)) and from an MLAEP it could be for example (LatencyPeakV (ms), AmplitudePeakV (μV), LatencyPeakNb (ms), AmplitudePeakNb (μV), AreaUnderCurve (ms·μV)). Differences in the recorded signals will obviously lead to different values for the extracted features; for the first feature set one measurement may lead to the values (50, 45, 8, 0.88) whereas another may lead to the values (80, 39, 5, 0.61). We can call these different instances of the feature set values *patterns* and then devise a system that is able to relate different patterns to, for example, different patient states. It uses patterns as its input and generates an associated patient state as its output. The system thus performs a pattern classification (recognition) task. As we will see, it can do this by, for example, using the statistical properties of the features, trying to find clusters in the sets of patterns, or creating certain reasoning rules.

It may be evident that the performance of such a pattern classification system is highly dependent on the definition of the feature set as that is the only information about the underlying measurement it receives. *Feature selection*, selecting the most appropriate feature set for a given application, has elements of both a science and an art. Usually, background knowledge, results from earlier performed studies and other experience give some good first pointers to features that are most likely to be useful. Candidate other features may originate from speculations on what might be useful, or interesting, to try out. One might be tempted to include as many features as possible following the reasoning that more features convey more information, which is good for the classifier. This is not a good idea. Literally hundreds of features can be extracted from one single signal by simply using a wide range of methods and different parameter settings. Providing all these features to a classifier would mean that it has to create knowledge about the typical distributions these hundreds of features follow. The complexity of the task increases exponentially with the number of features used. The problem is sometimes referred to as the 'curse of dimensionality'. In practice, one wants to use a relatively small but effective feature set in order to create a classifier that is able to generalize well (i.e. that works both on data that is used to create (train) the classifier functionality as well as on unseen data that may be recorded in the future). A rough rule of thumb is that the ratio number of observations/number of features should at least be 10 (Jain et al 2000). The more complex the classification paradigm used, the higher this ratio should be. There exist formal algorithmic approaches to determine an optimal subset from a given set of candidate features for a given application and classification method (for an overview see e.g. Jain et al (2000)). A classifier using a limited feature set has other advantages; it simplifies implementation, allows for faster processing and requires less resources of the system it runs on (obviously there is a limit; a too small feature set will most likely not lead to any useful classification at all).

Other than selecting features from a candidate set one may create new ones by using combinations of original features and so create a smaller, but maybe more effective, set. Such combinations may be done by, for example, linearly combining features that are highly correlated into a smaller set of new features that cover as much as possible information present the original set, but are as dissimilar (orthogonal) as possible. Principal component analysis (PCA) is an example of such an approach. It should be noted that the resulting new features, principal components, are a mix of original features and as such do not necessarily have any physiological meaning (e.g. Principal Component 1 = 0.5 SEF95 + 0.1 SpectralEntropy + ...). If the aim is to maintain some form of interpretability of the features, principal factor analysis (PFA) with so-called rotation of factors is more appropriate. The methods are described in detail in many statistical textbooks, a good non-mathematical description can be found in, for example, Sharma (1996).

Chapter 8 Further Signal Analysis

Table 8.4 Summary of main characteristics of processing methods discussed in *Processing methods* (p. 332) (many more methods exist!)

	General description	Typical use	Strengths	Caveats
Time-domain methods	Methods that describe the time–amplitude characteristics of the signal	General signal descriptor	Can track fast changes in signal	Limited frequency information
Amplitude measures (RMS, median, …)	Simple descriptor based on amplitude values	General signal descriptor	Simple and effective	–
Zero-crossing computations	A simple way to estimate frequency contents	Frequency estimation (rough)	Fast	Does not always work (e.g. with complex waves)
Normalized slope descriptors	Three descriptors that describe wave shape and are related to the power spectrum	Fast signal characterization/ frequency description	Fast	With modern computing power the real spectrum can also be calculated quickly
Correlation analysis	Determination of relationship between variables	Comparison of signals measured from different locations	Simple	Results not always easy to interpret
Phase and time-delay	Measure time differences in signals analysis	Studies on generators of signals at different locations	Simple	Not always easy to measure, other factors may also influence delays
AR modelling	Model signal as a parametric time-series plus noise	Compact representation of signal	Can be highly effective summary of signal	Model has to be valid; model order has to be predefined; computationally complex
Frequency-domain methods	Methods that describe frequency characteristics of the signal	Frequency contents description of signal	Good frequency representation	Limited time resolution, stationarity assumptions
DFT/FFT	Spectrum estimation using discrete version of Fourier transform	General signal descriptor	Fast implementation, readily available	Makes assumptions about the data that may not be fulfilled; cannot (should not) be used on small segments of data
Periodogram	Estimate spectrum directly following its mathematical definition	General signal descriptor	Simple	Uses biased estimate of the acf

Method	Description	Advantages	Disadvantages
PSD estimation using AR modelling	AR model parameters can be used to directly calculate the PSD	Alternative to DFT/FFT in 'problem situations'	Statistically more consistent than DFT/FFT	Model has to be valid; model order has to be predefined; computationally complex
Spectrum features	Features describing the character of the PSD (SEF, power in frequency bands, etc.)	Compact representation of spectrum	Effective data reduction	Risk that some relevant information of the original spectrum is not covered in the descriptors
Time-frequency methods	*Methods that describe both time and frequency contents of signal in one output*		*Allows more complete view of signal*	*Not as easy to use as pure time or frequency methods*
STFT	DFT/FFT performed on subsequent segments of data	Plots depicting frequency content versus time	Fast implementation, readily available	Time-frequency resolution trade-off plus same considerations as that of DFT/FFT above
Wavelets	Describe the signal using localized functions as alternative to sine waves	Plots depicting frequency content versus time, data compression, feature extraction, detection of complexes	Versatile	Requires choice of wavelet type, may be less easy to interpret than STFT
Alternative methods		*Additional signal descriptors*	*Depends on method*	*Depends on method*
Higher-order spectra	Extension of normal spectrum using two frequency axes	Able to detect coupling between frequencies, BIS index	Adds information about frequency contents	Not straightforward to calculate and interpret
Complexity measures	Describe amount of disorder in signal	General signal descriptors, e.g. used in depth of anaesthesia indices	Provide additional means of describing signal	Do not necessarily have any physiological meaning

Finally, depending on the application, features can sometimes be processed to create higher level features, or features can be used that do not directly come from the data-acquisition system but are, for example, annotations or observations. Examples are 'occurrence of epileptiform discharge (yes/no)', 'burst suppression ratio', 'arousal (yes/no)' and 'OAA/S score'. These usually provide very powerful inputs for further classification.

Many techniques for pattern classification exist. A common division is into three main groups: statistical, syntactic, and artificial intelligence (AI) approaches. Statistical techniques base the classification upon the estimated probability distribution of measurements that are characteristic for certain clinical context classes. Syntactic techniques use a structural description to describe the shape of the patterns and interrelationship between features. AI techniques are usually divided into two main groups: symbolic AI techniques, which manipulate information by employing logical statements (e.g. if … then … else); and connectionist AI techniques, which learn to relate measurements (features) to clinical context classes using example cases. For symbolic AI the term 'expert systems' is often used, while for connectionist AI techniques the term 'artificial neural networks' is common. Many textbooks discussing these methods exist – overview books covering most of these methods are those by Weiss and Kulikowski (1991) and Schalkoff (1992).

Statistical methods

Once we have defined our set of, say, N features to use, we may consider the resulting patterns as vectors in an N-dimensional space. For example, each measurement from which the earlier suggested feature set (RMS amplitude (µV), SEF95 (Hz), MedianFrequency (Hz), SpectralEntropy (–)) would be calculated would lead to a four-dimensional vector. We commonly call the vectors *feature vectors*, and the N-dimensional space in which they occur *feature space*. The classification problem now becomes one of dividing the feature space into certain regions that each represent a class (e.g. patient state). Once a measurement is done, its feature vector is calculated, it is checked in which region in feature space it falls, and the associated patient state is given as output of the classification. Examples of the problem in a two-dimensional space (thus, using only two features) are depicted in *Fig. 8.26*.

To perform its task the classifier thus has to use the boundaries in feature space that fit the given classification problem. Constructing a classifier then means finding the optimal boundaries. It may now become clearer why it is not a good idea to use, for example, a hundred features in a feature set; it would lead to the task of solving such a problem in 100-dimensional space. To say that this is a difficult task would be an understatement – both for human and computer.

Different methods exist to find such decision boundaries. One approach is to define an optimal classifier that is able to convert the *a priori* class probability (the probability that the patient state is 'S' 'in general') into an *a posteriori* class probability (the probability that the patient state is 'S' given the measurement X we have just done). Another approach is to minimize the expected error (number of wrong classifications) the classifier will make. Methods using the first approach use *Bayes' theorem* (e.g. Russek 1983) for this, which states (applied to the 'patient state' example) that the probability of patient state S is proportional to the likelihood of measuring data X given that we have patient state S, times the prior probability of S. Estimating probabilities in practice usually requires considerable computing power. Bayesian methods have experienced a resurgence during the last few years due to the increased amount of computing power that is now available (e.g. Tan 2001).

Alternatively, one can try to find clusters in the data (that may, or may not, be associated with certain patient states) and construct boundaries that separate them (clustering techniques). Or, if it is know beforehand to which class each measurement belongs, one can try to construct borders that are optimal for implementing this specific division of measurements over classes. Examples of such approaches are linear discriminant methods and logistic regression classifiers.

The field of statistical pattern recognition techniques is very broad but the theory is well established. Consequently, there exists a large body of literature in the form of books and articles. For example, Schalkoff (1992) (also describes other methods), Sharma (1996) (general multivariate statistics) and Jain et al (2000) provide useful starting points for exploring the subject further.

Syntactic methods

Whereas statistical techniques aim at classifying feature vectors with the help of assumed distributions, syntactic techniques concentrate on the structure of the patterns. They are not very useful for random data (like EEG) but are applicable to deterministic curves like ECG, blood pressure curves, or EP curves that have a typical morphology. The basic tenet is that patterns belonging to the same class can be characterized by a common structural description (they may 'look similar'). Such descriptions are made using low-level features (called *primitives*), such as 'peak', 'trough', 'upgoing trend', 'downgoing trend' or

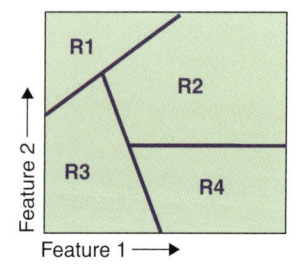

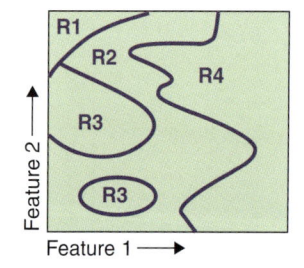

Fig. 8.26 Examples of divisions a two-dimensional feature space into regions, R. (a) A relatively simple situation where the regions can be separated by linear decision boundaries. (b) A problem with a complex grouping of regions with non-linear boundaries. (Based on Schalkoff (1992).)

'flat segment'. A recorded signal (e.g. an EP curve), thus can be deconstructed into a sequence of primitives. This is called a structural description. A mechanism that matches the extracted structural description against known structures, often referred to as a 'parser', recognizes the patterns. These techniques use grammars that reflect class structures, or they match so-called relational graphs.

The usefulness of syntactic techniques in practical situations is sometimes believed to be limited (Schalkoff 1992), mainly because of parsing difficulties when noisy data are presented. There are applications where syntactic techniques are powerful, for example in the recognition of hand-written characters. Also, studies have shown that a combination of syntactic and statistical techniques may provide powerful pattern representation capabilities (Fu 1986). Examples of how syntactic methods are applied to EPs can be found in Bruha and Madhavan (1988) and Grönfors (1993).

Artificial intelligence (AI) methods

A third group of methods, the AI methods, concentrate on two main concepts:
- *knowledge presentation* – the choice of a method to represent facts; and
- *learning* – the AI based system should be able to increase its efficiency and effectiveness when a similar task is processed repeatedly.

Two main approaches can be distinguished: *symbolic AI methods* and *connectionist methods*.

Symbolic AI methods – expert systems A symbolic AI system uses *symbols* to describe and identify structures and states. An example of a symbol can be the word 'square' to represent a square-shaped part of an image, or the words 'high blood pressure' to indicate a mean SAP higher than 100 mmHg, or the words 'high burst suppression ratio', 'medium RMS amplitude', 'low beta-band power', etc. Usually these symbols are then manipulated using logical statements (**if** 'high RMS amplitude' **and** 'low beta-band power' **then** ... **else** ...).

Expert systems are well-known implementations for manipulating symbolic information. They use knowledge of the problem domain to interpret data and to provide feedback to the user, i.e. they can 'explain' how they interpreted the data. The domain knowledge is put into the expert system as a result of the interaction between a domain expert and a knowledge engineer: a specialist skilled in analysing the reasoning behind the expert's problem-solving process and encoding it into a computer system. A schematic representation of an expert system is given in *Fig. 8.27*.

The *user interface* lets the expert system interact with the outside world, for example during input of knowledge, performing tests and explaining the reasoning process followed. The *data base* contains the current problem data. The *knowledge base* contains the domain-specific knowledge, it includes: *concepts* (basic terms of the problem domain, often obtained from textbooks); *rules* (empirical associations that link events, characteristics and relationships, typically obtained by consulting an expert); *models* (collections of interacting rules, they can be associated with a certain problem hypothesis or a diagnostic conclusion); and *strategies* (that can help to use the rest of the knowledge base by indicating which part of the knowledge base to use for a problem). The *inference engine* searches the most suitable part of the knowledge base during a problem-solving task and also evaluates the various rules in the knowledge base to add results to the data base. The inference engine can use different approaches in its search process, for example backward chaining (start with the goal of the problem and repeatedly infer subgoals on the way to the final solution of the problem), or forward chaining (start with inferring the immediate consequences of the data and use consequences to come to further consequences, repeat until the final goal of the problem has been reached). Ideally the inference engine is constructed in such a way that no software changes are needed to enable its use in different domains.

A strong point of expert systems is that they operate in a transparent manner, i.e. the path to their conclusions can be traced – this allows explanation or justification of their conclusions. They have been successfully applied in a variety of fields, but also have their limitations. One essential step in the construction of an expert system is the storage of knowledge in the knowledge base. To achieve this, the knowledge has to be present in a structured form. The transition from the domain expert's knowledge to rules cannot always be made; this is especially the case when this knowledge is partially based on experience. Another problem may occur when the domain expert's opinion or

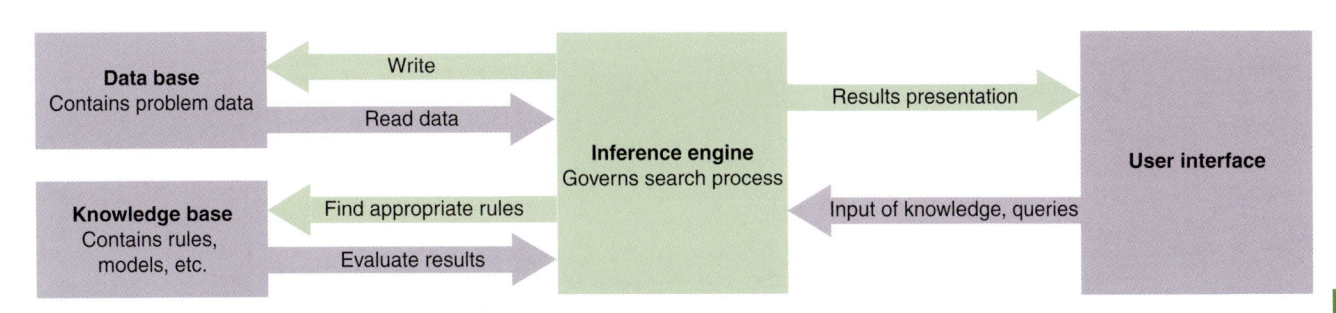

Fig. 8.27 A schematic representation of an expert system.

method of acting changes over time; it may then be difficult to change the knowledge base to reflect the altered situation. More on the subject can be found in, for example, Weiss and Kulikowski (1991). Examples of the practical use of expert systems can be found in *Summary and examples of usage* (p. 353).

Connectionist AI methods – artificial neural networks
Connectionist AI approaches in pattern classification mainly concern the use of *artificial neural networks* (*ANNs*). They are especially useful in situations where the domain expert's knowledge is not structured (e.g. largely based on experience) and rules for an expert system cannot be extracted. The classification functionality is then learned by providing many example cases and extracting input data–output class relation from that.

ANNs consist of a large number of rather simple and typically non-linear processing units (sometimes called *nodes* or *neurones*) that are densely interconnected. Typically, they are ordered in *layers*. A schematic representation of an ANN is given in *Fig. 8.28*.

The processing units in the *input layer* receive the values of recorded data (feature vector) to be processed (either to learn from, during a training phase, or to directly classify, during the 'normal use' of the ANN). The units in the *output layer* produce an estimation of the class that is most likely to be associated with the input data. Following our earlier example, the ingoing arrows on the left-hand side of *Fig. 8.28* could represent values of RMS amplitude, SEF95, etc., and the outgoing arrows on the right-hand side represent indications of patient states. In the figure there are three output units that could be used, for example, to distinguish between states A, B and C. Each unit's output value is associated with the occurrence of a particular class/state; the unit that has the highest output value (usually a value between 0 and 1) then indicates the class most likely associated with the presented input values. There are many other schemes thinkable to encode the output with a different number of output units. Units in layers between the input layer and the output layer are called hidden units, which are organized in *hidden layers*. Each connection between units has a *weight* associated with it (some example weights are indicated in *Fig. 8.28*). The information flow throughout the network is established by calculating the product of a processing unit's output value with the associated weight for the connection from that unit to a unit in the next layer. A unit in that next layer sums all the weighted outputs it receives from units in the previous layer and performs a non-linear function, *f*, on that sum. The value of *f* is typically high (close to 1) if the weighted summation exceeds a certain threshold, and low (close to 0) otherwise. The output of that function is used as the unit's output, and the weighted summing process is repeated towards the next layer until the output layer is reached. The outputs of the units in the output layer are then used as network output.

For a given network the main factor that influences what gets generated at the output in response to a certain input are the values of the connection weights. These determine how inputs (features) are processed to finally come to a classification. The problem then is to find the right weight configuration to solve a given problem. This is done by using a process called *training*. The weights are initialized with random values (and the network thus generates random output in response to inputs). Subsequently, a (large) number of example measurements with known classes (patient states) are presented to the network, for example so that we know that in response to measurement (50, 45, …) the desired output of the network should be 'class A' (1, 0, 0). We calculate the actual output of the network and compare it with the desired output. Initially, the difference between the actual and the desired output will be quite big and thus the network will most likely make wrong classifications. We can now update the weights in the network by using information about the difference between the actual and the desired output (its magnitude and whether it is positive or negative) so that the next time we present that (or a similar) measurement as input, the actual output will be (a little bit) closer to the desired output. At that next time the weights will probably still need some updating, but not as much as before. If we repeat this process many (hundreds or thousands) of times we may eventually converge to a weight configuration that classifies satisfactorily, fulfilling some predefined performance criterion. The more example measurements we have the better it is for creating a network that is able to generalize and implement the classification task as robustly as possible. In fact, having a large enough example set is a prerequisite for using an ANN approach; a practical minimum is of the order of a hundred example cases (this depending also on the complexity of the task, size of the feature set, and the type of ANN used). If the training phase was successful and we have created an ANN that is

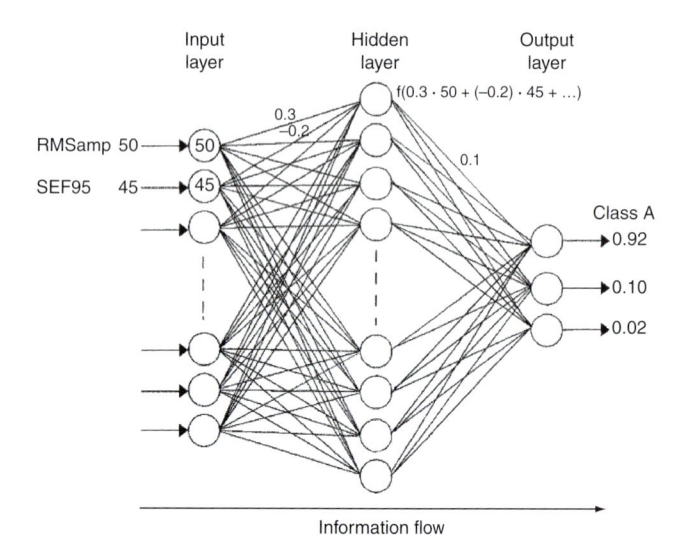

Fig. 8.28 A schematic representation of an ANN with one hidden layer.

able to generalize well (i.e. has not learnt to process only the example cases perfectly 'by heart' but is able to make a generally valid mapping from measurements to associated classes) we can use it as a fast classifier that calculates its output as a series of multiplications, additions and non-linear functions.

The above description of an ANN and its training is just one example of a possible realization. A large variety of ANN architectures exists. Architectures can be characterized by the characteristics of the processing units, the topology of the network (how the units are interconnected), and the scheme to train the ANN. In the training description above we assumed that for each given example measurement we can use a known 'desired output' (perhaps provided by a human expert who has labelled the data as belong to certain patient states). In such a case we speak of *supervised learning*. Sometimes we do not have any right outputs but we still can learn something by processing the example measurements (e.g. find clusters in the data or find information about the distribution of different features). In such cases we can use *unsupervised learning* approaches. Probably the most popular ANNs that use supervised learning are networks using the so-called back-propagation algorithm for training (Rumelhart and McClelland 1986). A well-known example of a network using unsupervised learning is Kohonen's self-organizing map (SOM) network (Kohonen 1988).

Neural networks may be successfully applied in situations where the knowledge required to perform a classification task is mainly based on experience and not on straightforward reasoning rules or mathematical models and equations. Neural networks can be trained to solve very complex classification tasks (e.g. like the one depicted in *Fig. 8.26*) thanks to their non-linearity and lack of assumptions on, for example, underlying data distributions. However, training an ANN correctly requires some knowledge about the problem, the underlying data and the functionality of the ANN, and calls for considerations on how to assess general performance, and often involves some trial and error. Having sufficient data to train an ANN sensibly is an absolute prerequisite. A disadvantage of ANNs compared to other methods is that their functionality may be experienced as a black-box giving the user no clue about how it comes to its classifications. Although this is not entirely true (e.g. one can examine weights of the connections to see how certain features are processed), an ANN is certainly less transparent than an expert system or most statistical approaches.

Summary and examples of usage

The classification methods discussed in the previous sections are summarized in *Table 8.5*.

In practical non-trivial problem solving situations we rarely see just one method being used. It is quite usual that, for example, part of the problem can be described with rules and some part is solved by experience. In such a case one would use an expert system based module and an ANN-based module in combination to solve the problem. Such systems are called *hybrid systems*. They provide the most versatile solution to a classification problem, combining the advantages and solving the disadvantages of individual methods. Obviously, this also involves more complexity during the classifier construction phase.

The literature gives many examples of the discussed methods applied to neurophysiological recordings in intensive care and surgery. Applications include warning and alarm systems, depth of anaesthesia or sedation assessment, servocontrol, detection of seizure discharges or arousal responses, and sleep staging.

Some selected examples of applications include the use of ANNs to *detect seizure discharges* (Webber et al 1994, 1996). The learning ability of the ANN can be 'tailor-made' to match the performance, in terms of specificity and sensitivity, of individual experts in visual analysis

Table 8.5 A summary of the main characteristics of classification methods discussed in *Integration of features into the clinical context – pattern classification* (p. 346)

Method	Principle	Advantages	Disadvantages
Statistical methods	Determine class probability based on given measurements and distribution estimation	Theoretically well-established, many different methods available	For very complex problems other approaches may be easier to apply
Syntactic methods	Classify on basis of underlying shape of signal	Sensible approach for deterministic signals	Not very useful for noisy signals
AI methods:			
Expert systems	Use reasoning rules to come to classification	Transparency of operation, works well if expert's reasoning process can be coded	Less suited if expert's knowledge is mainly based on experience, may be difficult to update
Artificial neural networks	Use example cases to learn to classify correctly	In principle can solve very complex problems that cannot be solved by reasoning or modelling approaches	'Black-box' functionality, needs a large set of example cases

(Pradhan et al 1996) or to construct indices, for example of anaesthetic depth (Ortolani et al 2002). Gabor (1998) has validated and compared a self-organizing neural network (Gabor et al 1996) with other detection strategies in 181 seizures in 200 EEGs recorded from 65 patients. The monitored seizures had frequency–amplitude features which were identified as distinct from other EEG patterns. Detection rates of 93% were achieved. However, when the signal was transformed into the audio frequency range (Oxford Medilog system), seizures could be detected aurally by trained technologists with greater sensitivity (98%) than with other strategies used!

An ANN system for assessing *depth of anaesthesia* and *recognition of arousal responses* produced by pain stimuli during surgery was trained with spectral information from more than 25,000 30 s epochs of EEG from 196 patients (Eckert et al 1997). These authors defined clusters of spectral patterns associated with the sequence of EEG changes from 5 min before induction until intubation, but had to add a 'suppression parameter' to overcome the problems associated with spectral measures when the EEG showed burst suppression.

The Advanced Depth of Anaesthesia Monitor (ADAM, see p. 83) (Thomsen et al 1989, 1991, Thomsen 1992) uses pattern classification techniques based on statistical techniques, unsupervised learning and hierarchical analysis, in combinations with rules to classify patterns characterizing various depths of anaesthesia. Results with the ADAM system improved when a burst suppression marker was included as a variable; it was then possible to demonstrate a more rational sequence of changes following the course of anaesthesia than with simpler univariate methods (Thomsen and Prior 1996). Somewhat similar approaches have been used with more recent versions of the Cerebral Function Analysing Monitor (CFAM3, see p. 83). As with the ADAM system, this monitor compares new data with its knowledge base on-line; the clinician can therefore view the position and statistical significance in relation to the limits of the training data. It has the additional advantage that non-typical patterns are detected in new patients so that the user can be warned that they are unsuitable for such comparative classification. An overview of ANN approaches on the subject is presented by Robert et al (2002). Zhang and Roy (2001) have shown the promise of a hybrid approach using Lempel and Ziv complexity, ApEn and Spectral Entropy as inputs to an adaptive network-based fuzzy inference system to discriminate between awake and asleep levels for propofol, isoflurane and halothane anaesthesia in 15 dogs. More suggestions for hybrid approaches are presented in Zhang et al (2002), who make a case not only for merging methods but also sources, such as EEG, MLAEP and ECG.

Trend analysis, warning and alarm systems: as with techniques for artefact detection (van de Velde et al 1999), the design of systems to interpret signals to produce intelligent alarms must be set in the particular clinical context of the data (Bloom 1993). Regardless of the signal processing technique used (Bloom discusses the problem as it affects cluster analysis, discriminant analysis and statistical predictors), changes in the individual patient's clinical state with time, or due to medical or surgical interventions, will alter the significance of the very trends that might be used to generate warnings or alarms. There is a considerable body of work on anaesthetic depth and indicators of potential for 'awareness', but as yet there are few systems in use for detection of brain ischaemia.

Development of a *warning system for paediatric ICU work* has utilized an expert system using fuzzy logic and neural networks with the aim of automating regular statements about the level of abnormality in the EEG. The method uses segmentation, feature extraction and classification prior to presentation (Agarwal et al 1998, Si et al 1998). The aim of this work was to warn the neurophysiologist when the EEG might be abnormal and require examination.

Servocontrol and other feedback systems: early attempts to use simple EEG frequency-based, servocontrolled systems for delivery of single agent anaesthetics have been described in Binnie et al (2003, Section 8.4.2.1). Following their fall into disuse with the advent of 'balanced anaesthesia', such systems have re-emerged for informed assistance in the control of intravenous anaesthesia and of intravenous sedation in the ICU (Schwilden et al 1987, Veselis et al 1991, Frenkel et al 1995, Yli-Hankala et al 1999).

Statistical process control

The term 'statistical process control' covers a number of techniques for identifying a significant change in a series of values. The need for such techniques arose when machine production became widespread, because corrections could no longer be made by hand in the course of manufacture, so variability in the output of mechanized processes had to be minimized. The first statistical approaches to manufacturing were made in the 1920s by BP Dudding in Britain and Walter Shewhart in the USA (Shewhart 1931). Since then the original techniques have been refined and extended, and new methods have been introduced. Modern introductions to the subject are provided by Woodall and Adams (1998) and NIST/SEMATECH (2004), while more detailed accounts are provided by Ryan (2000) and Montgomery (2004). The methods to be described, although developed independently, are examples of an approach arising from the concepts of likelihood and efficient score statistics (Fisher 1925), from which are derived the general class of Cuscore charts, of which the charts used in practice are particular examples (Box and Luceño 1997).

Fundamental to all these methods is the concept of a process being 'in control'. This implies that the output is predictable, so that we can confidently expect future values to lie within clearly defined limits. To be predictable in this sense the output must have a stable mean and standard deviation, with no unexpectedly high or low values (*Fig. 8.29*).

The 'process' was originally a manufacturing process of some sort, and the values were the sizes, weights, etc.,

Signal processing and interpretation

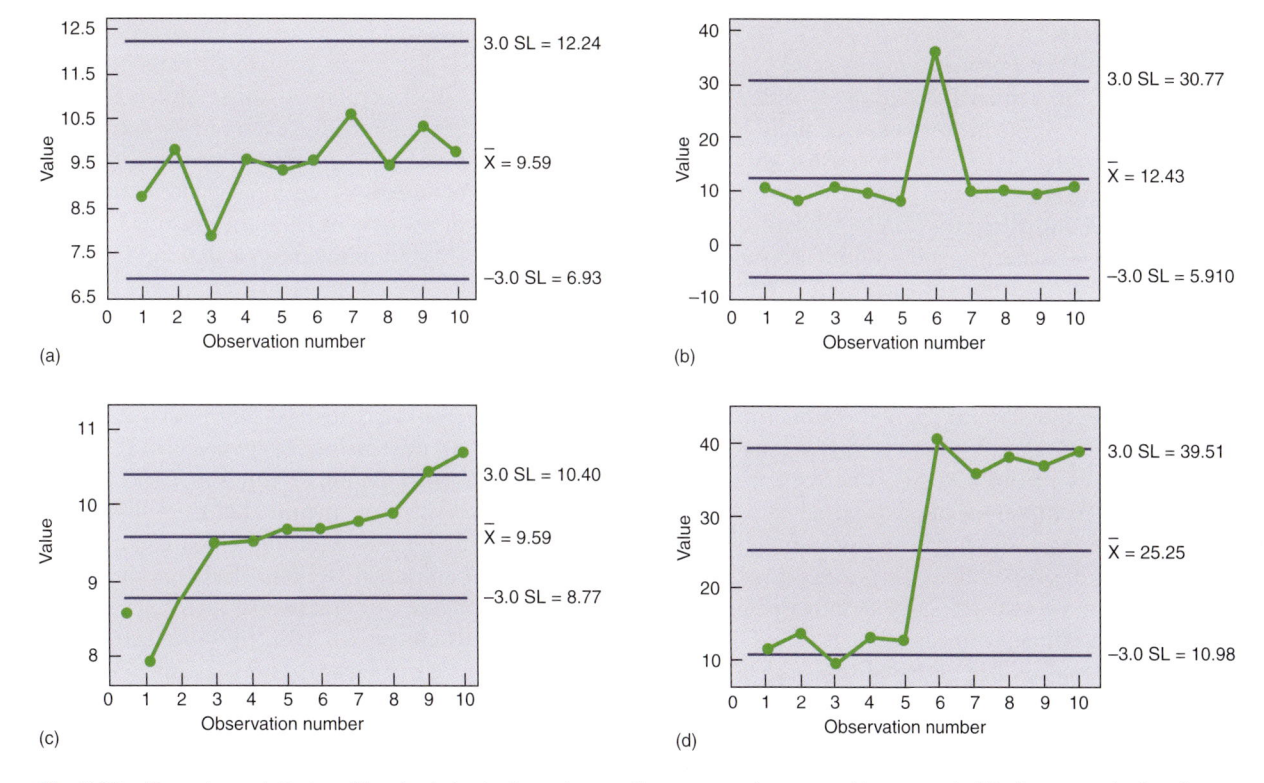

Fig. 8.29 Ten values plotted on Shewhart charts. In each case the mean and upper and lower control limits were calculated from all ten values. All points in (a) are within the control limits, although many more values would be required to be certain that the process was in control. A single exceptional value in (b) indicates that the process is out of control. In (c) there is a clear trend, while in (d) there is a sudden change in the mean; in each case some points are outside the control limits. SL denotes σ limit.

of the articles produced. However, it has long been accepted that measurement itself is subject to variation, so that repeated measurements of the same quantity will not produce identical values, but the values should be predictable within limits. Repeated values obtained by measurement can therefore be regarded as the output of a process, and the values must be 'in control', otherwise the measurement process is totally unreliable (Eisenhart 1963).

When a process is 'in control' variation is not completely eliminated, but the remaining variation is random and arises from a number of different causes, referred to as 'common causes'. If the number of sources of variation is large, and the contribution of each is relatively small and purely random, then the values will follow a normal or Gaussian distribution. When the process goes 'out of control' (departs from the previously established limits of variation: becomes unpredictable) a 'special' or 'assignable' cause is responsible. This may be a change in the raw material, a machine fault, an operator error, or, for measurements made on a patient, a change in the patient's condition. A process can go out of control in many different ways, for example there may be a single exceptional value, a sudden shift in the mean, a gradual drift, or a sudden increase in variation (see *Fig. 8.29*). If the expected manner of going out of control is known then a Cuscore chart can be designed to provide maximum sensitivity, but usually we do not know precisely how the process will go out of control, so whatever type of control chart is used it will be less than ideal (Box and Luceño 1997). The three charts described below (Shewhart, EWMA and Cusum), while each being the appropriate Cuscore chart to detect a particular type of out-of-control process, have been found to be applicable to a wide variety of situations.

The conventional method of using a control chart can be divided into two phases. In phase 1 we acquire sufficient values to give a reliable estimate of the process characteristics (mean and standard deviation), from which control limits are calculated. If any values are outside the limits we must look for an assignable cause that might account for the process being out of control, but if all the values lie within the limits then the process is in control, and we can use the limits to monitor the subsequent performance of the process in phase 2.

The performance of a control chart is assessed by calculating the probability of detecting an out-of-control point, but the result is usually expressed as the reciprocal of this probability, known as the average run length (ARL). This is the average number of points plotted for each out-of-control point. If the process is in control a large ARL is desirable to minimize false alarms, but when the process goes out of control a short ARL is required for rapid detection. A graph of ARL against the magnitude of the change to be detected is called the operating-characteristic curve, and has been widely used to compare different charts and detection criteria (Montgomery 2004).

Chapter 8 Further Signal Analysis

Shewhart charts

To determine whether a process is 'in control', Shewhart (1931) estimated the standard deviation (σ) and set limits at 3σ on either side of the mean (Wheeler and Chambers 1992). The multiple of 3σ was chosen for the simple empirical reason that experience showed it provided adequate detection of out-of-control values, while at the same time minimizing false alarms, and 3σ limits are now generally used in industry. Originally each value was the mean of a sample consisting of a number of determinations, the number being sufficient to ensure that each sample included a reasonable representation of the variability of the process, but it is quite possible to apply the same method to a series of individual values, and this is what is usually done with values obtained from patients. The standard deviation is estimated from the mean range of values in a sample (the *moving range*), or the mean difference between consecutive pairs of single values (the *mean consecutive difference*), divided by a constant to give an unbiased estimate of the standard deviation. For pairs of single values the constant is 1.128. The procedure for establishing the limits for individual values can therefore be summarized as follows:
1. Determine the mean value $\bar{X}$.
2. Determine the mean consecutive difference $\bar{R}$.
3. Estimate the standard deviation (σ) as $\bar{R}/1.128$.
4. Set the upper control limit at $\bar{X} + 3\sigma$, and the lower control limit at $\bar{X} - 3\sigma$.

For normally distributed values, the probability that any value will lie outside the 3σ limits is 0.0027, or 0.27%; this is the probability of a false alarm if the process is in control. The ARL for an in-control process is therefore 1/0.0027, or 370. As explained above, this is the average number of points plotted for each out-of-control point. The Shewhart chart is the appropriate Cuscore chart for the detection of a single exceptional value, so the ARL for a single value more than 3σ from the mean is 1. However, the ARL for a 2σ shift in the mean is 6.3, and for a 1σ shift in the mean the ARL is 43.96 (Montgomery 2004). The Shewhart chart is therefore relatively insensitive to small shifts in the process mean (*Fig. 8.30(a)*). Detection of shifts and trends can be improved by a number of additional criteria, called *runs rules*, such as 'six points in a row all increasing or all decreasing' or 'two out of three points more than 2σ from the mean' (Nelson 1984). However, these additional criteria assume that the values are normally distributed, an assumption which is not necessary for Shewhart's empirical 3σ limits. Even if values are normally distributed, a combination of runs rules leads to an in-control ARL less than 100, therefore considerably increasing the risk of false alarms (Champ and Woodall 1987). Many users therefore regard these criteria as warning signs only, rather than reliable indications that the process has gone out of control.

How many values are needed to establish that the process is in control? Shewhart (1931) originally suggested that if 25 samples, each of 4 values, did not indicate any lack of control then the process could be regarded as showing a reasonable degree of control. This implies that if we use

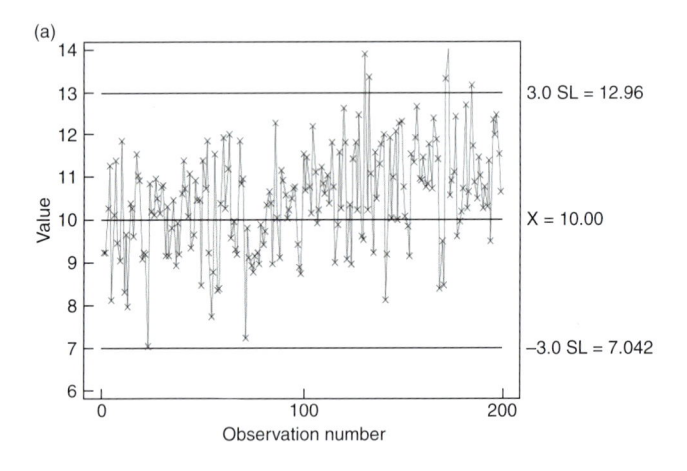

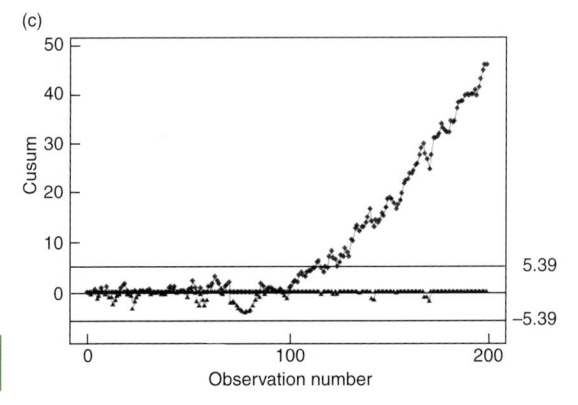

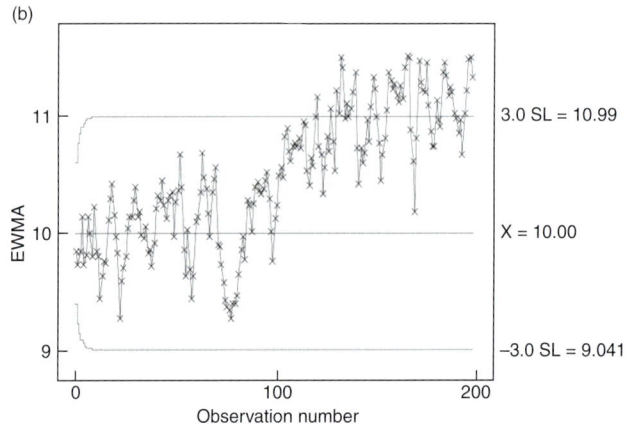

Fig. 8.30 Shewhart, EWMA and Cusum charts consisting of 200 values. The first 100 values constitute phase 1, and are used to estimate the mean and standard deviation, and to calculate the control limits, which are applied to the next 100 values (phase 2). The mean of the second 100 values is 1σ greater than the mean of the first 100. In the Shewhart chart (a) only a few points are above the upper control limit because of the relative lack of sensitivity of the Shewhart chart to small changes in the mean, whereas many points in the EWMA chart (b) are above the upper control limit. The Cusum chart (c) places nearly all points in phase 2 above the upper limit, illustrating that the Cusum is the optimal chart to detect small changes in the mean. SL denotes σ limit.

individual values rather than sample means, we need 100 consecutive values before we can regard the process as being in control. More recent work indicates that in general with subgroups of *n* values, 400/(*n* – 1) values are required, and for single values at least 300 would ideally be needed (Quesenberry 1993). However, we can never be absolutely certain that a process is in control; when all the points on a chart lie within the limits the chart merely fails to demonstrate that the process is out of control. If we need greater certainty that the process is really in control we must include more values, perhaps as many as 1000 (Shewhart 1931). On the other hand, the fact that a process is *not* in control is often apparent when far fewer values are available (Wheeler and Chambers 1992). In *Fig. 8.29(b)* one value is considerably higher than all the others, and lies above the upper limit; in *Fig. 8.29(c)* the values clearly follow a rising trend, and several values lie outside the 3σ limits, while in *Fig. 8.29(d)* there is a sudden change in the mean. In all these cases a large departure from being in control is identified by plotting a small number of points.

The exponentially weighted moving average (EWMA) chart

Exponential smoothing was used for economic forecasting in the 1950s, and the method was first applied to control charts by Roberts (1959), initially under the name of 'geometric moving average', but the term 'exponentially weighted moving average' is now commonly used. On a Shewhart chart each data point is plotted, but on a EWMA chart the values plotted are the weighted means of each value and all previous values, so that if Y_i is the *i*th value acquired, the *i*th EWMA, m_i, is given by:

$$m_i = m_{i-1}(1 - \lambda) + Y_i \lambda$$

The weighting constant λ is between 0 and 1; if λ = 1 the EWMA chart is identical to the Shewhart chart. The contribution of the *j*th value Y_j to the *i*th EWMA m_i ($j < i$) declines exponentially as *i* increases because Y_j is multiplied by $\lambda(1 - \lambda)^{i-j}$ (Roberts 1959).

A EWMA chart is usually constructed by setting the initial mean to the mean of values obtained in phase 1, or to a target value if there is one. When individual values are used, rather than the means of samples, the control limits are

$$3\sigma[\lambda/(2 - \lambda)]^{0.5}$$

above and below the mean. The standard deviation σ is estimated from the mean consecutive difference between individual values, as in the Shewhart chart.

The EWMA chart is the appropriate Cuscore chart to detect an exponential increase or decrease in the mean, where the ratio of consecutive values is (1 – λ) (Box and Luceño 1997), but the weighting factor λ can be chosen so that the EWMA chart provides satisfactory ARLs to detect a linear trend, or a step change. For example, a λ value of 0.25 gives an ARL of 493 for an in-control process, and detects a 1σ shift in the mean with an ARL of 11 (Ryan 2000) (*Fig 8.30(b)*). For the same shift in the mean, the Shewhart chart gives an ARL of 43.96, but for shifts in the mean greater than 2.75σ the Shewhart chart produces a shorter ARL than a EWMA chart with λ = 0.25 (Roberts 1959). For the detection of a linear trend the performance of the EWMA chart is similar to that of a Shewhart chart with additional 'runs rules' (Hunter 1986, Box and Luceño 1997). Because the two charts complement each other in this way, it is sometimes informative to plot the EWMA and Shewhart data and control limits on the same chart (Hunter 1986).

The Cusum chart

In the Cusum chart, introduced originally by Page (1954), the plotted points are the cumulative sums of the deviations from a target value, or from the mean of the process established in phase 1. The Cusum chart is the appropriate Cuscore chart to detect a step change in the process mean (Box and Luceño 1997). Provided the process mean remains constant, the chart will be a line of constant gradient, and any sustained change in the mean produces a change in the gradient. If the target value and the process mean are identical then the line is horizontal; a difference between the process mean and the target is indicated by a sloping line, the difference being proportional to the gradient. The use of Cusum charts in this simple or 'classical' form has been strongly advocated for medical purposes, and there is an obvious application to everyday tasks such as detecting a change in a patient's temperature (Chaput de Saintonge and Vere 1974).

While the simplicity of this 'classical' form of the Cusum chart is appealing, there are no control limits, and determining whether a change in gradient is significant is not straightforward. The method proposed originally was to use a V-mask, consisting of two lines forming a V pointing to the right (>), with the angle between the lines bisected by the horizontal (Barnard 1959, Lucas 1976). The V-mask is placed on the chart with its vertex at each point as it is acquired, and if previous values lie outside the arms of the V then the process is out of control. However, determining the appropriate angle between the arms of the V is not straightforward, and the V-mask is cumbersome to use, so it is not now recommended (Montgomery 2004).

The preferred Cusum method is to plot the sum of deviations above and below target separately as upper and lower Cusums. The two Cusums, C_i^+ and C_i^-, are:

$$C_i^+ = \begin{cases} x_i - (T+K) + C_{i-1}^+ & \text{if } x_i - (T+K) + C_{i-1}^+ > 0 \\ 0 & \text{if } x_i - (T+K) + C_{i-1}^+ \leq 0 \end{cases}$$

and

$$C_i^- = \begin{cases} (T-K) - x_i + C_{i-1}^- & \text{if } (T-K) - x_i + C_{i-1}^- < 0 \\ 0 & \text{if } (T-K) - x_i + C_{i-1}^- \geq 0 \end{cases}$$

where x_i is the ith value acquired, T is the target value, or mean from phase 1, and K is called the *slack* or *allowance*. The two Cusums can be plotted on the same chart without causing confusion because one is always positive or zero, and the other is always zero or negative. Control limits are set at $T + H$ and $T - H$, where H is the *decision interval*. Values of H and K are selected to give the required ARL for the change to be detected. For example, to detect a 1σ change in the mean, suitable values for H and K are 5σ and 0.5σ, which give an ARL of 465 for an in control process, and 10.4 for a 1σ shift in the mean (Montgomery 2004). These are similar to the ARLs obtained with the EWMA chart, and superior to those from the Shewhart chart without additional runs rules. However, like the EWMA chart, the Cusum gives longer ARLs than the Shewhart chart for large changes in the mean, so combined Shewhart and Cusum charts are sometimes useful (Westgard et al 1977).

The conventional method of using a control chart cannot be used where there is neither a clearly defined target, nor any possibility of establishing the mean and variability of the process in a lengthy phase 1. This applies to many manufacturing processes, and to clinical situations in which control charts might otherwise be useful (Del Castillo et al 1996). Several methods have been suggested to overcome this difficulty (e.g. Quesenberry 1991, Wasserman 1995); none is entirely satisfactory, although a method based on the Cusum chart (Hawkins 1987) seems more promising (Montgomery 2004).

USE OF THE TOOLS

Decision assistance

The output of the data processing and, possibly, classification steps as described in previous sections is typically used in systems to assist in decision-making tasks. Decision assistance systems are computer-based systems developed to support and help professionals in (clinical) decision-making tasks. There are many different roles such systems can play (Nykänen and Saranummi 2000): they can act as a system providing reference material; they can act as a 'watch-dog' for screen routines; they can be used to forecast outcomes of different alternatives; they may work as a tutoring or guidance system, representing a more experienced consultant for the user; and they may produce information from data by interpretation.

The first applications of decision assistance systems were in diagnosis and treatment planning, where they typically used Bayesian statistics. Such systems proved to work well, provided that reliable data were sufficiently available and variables could be considered independent. However, in many areas in medicine both demands cannot be met. Following the realization that the statistical data approach is not enough and extracted expert-knowledge representation was needed to emulate decision steps, research employing AI methods started in the 1960s. This led to the use of knowledge-based systems and expert systems. Research efforts mainly concentrated on how to represent expert knowledge, how to acquire it and how to use it in problem solving. Other issues for research and development emerged as well; human–computer interaction, interfaces to databases and database management systems, and integration of natural-language processing with knowledge-based systems. The increased availability of powerful and cheap hardware in the last decades has led to the situation that affordable systems can be used on site. Developments in computer graphics, graphical user interfaces and image processing have further promoted the development and use of such systems.

Nowadays the most common view is that knowledge-based systems are rarely meant to replace human expertise and that making them perform at 'expert level' is in practice extremely difficult to realize. Thus, the term expert system is becoming less used, and it is more common to speak of (clinical) decision assistance systems.

Medicine is a very complex application area for decision assistance systems. The complexity arises from the fact that it has a strong empirical basis (in comparison with other sciences), and there is limited understanding of many disease processes as well as the decision-making process itself. Much of the expertise is developed through experience using 'soft' knowledge, which is difficult to employ in formal reasoning systems. For this reason, hybrid approaches using modules with ANNs, expert systems, and statistical methods, etc., are considered to be the most promising.

In neurophysiological applications during intensive care and surgery it is essential to view all the neurophysiological findings in the context of the functional state of the patient's other physiological systems and the effects of any pharmacological or physical interventions. This means that we must understand not only the clinicopathological disturbances due to the presenting disease, but also the effects of drugs, exhaustion, pain, disturbed sleep–wake cycles, multisystem failure, and a continuous background of 'activity' (noises, bright lights, frightening procedures) which can affect the patient. Decision-making in such a setting is an extremely complex cognitive task demanding the application of different kinds of knowledge and a clear understanding of the purpose of activities and the system of possible actions one is able to initiate. It involves use of structures of possible actions, and consequences are based on empirical data (case records, observations, information related to hospital procedures) combined with the ability to process and interpret these data theoretically.

Relevant information is maximized by combining recordings from the CNS with those from other physiological systems into a single display. As we have seen in the previous sections signal processing techniques can assist in detection of significant trends and interrelationships. The IFCN recommendations for clinical practice of EEG and EPs in comatose and other unresponsive states (Guérit et al 1999) consider continuous neuromonitoring as 'a neurophysiological tool to be used by non-neurophysiologists' and emphasize that the neurophysiological information must be 'coded in a language that is readily interpretable by the user'. The suggested three-step approach to achieve this (quality screening, feature selection to form patterns and integration of patterns into the clinical context) was noted in *Integration of features into the clinical context – pattern classification* (p. 346). Guérit et al emphasize that 'because the neurophysiologist cannot be present throughout the entire period of neuromonitoring, it is essential that intensivists and the nursing team be trained in some basic principles of EEG interpretation and the way to react to the most commonly encountered technical problems'. They add that 'a trained neurophysiologist and an EEG technologist should be reachable for the remaining problems'.

Examples of systems that assist in decision-making have been presented in *Summary and examples of usage* (p. 353). These systems typically concentrate on a single problem (detection of epileptic seizures, or assessment of depth of anaesthesia, or dedicated warnings) rather than provide a fully integrated 'general' system that is used routinely.

Development and uptake of methods

There is a discrepancy between the number of research projects having resulted in potentially useful neurophysiological signal interpretation methods and the amount of cases where such methods are actually implemented in a commercially available system.

Before a new method can hope to achieve widespread adoption in routine clinical work, it must be clinically validated and evaluated at many different sites and then be implemented as a commercial product. However, commercial competitiveness will only be achieved once the new method is accepted for use amongst a large group of clinicians. Such acceptance can only be reached once implementation has been demonstrated in a real clinical environment on a clinically feasible platform. Such a dissemination cycle can lead to a propagation delay of 5 years or more between obtaining results from novel signal interpretation methods and their emergence in products designed for the critical care marketplace (Saranummi et al 1997). To shorten this delay, the development of new methods needs to be accompanied by full consideration of the issues of implementability and usability, and requires strong interactions between method developers, potential users and members of the patient monitor equipment industry. Goethe and Bronzino (2000) give a list of recommendations to be used in the design process, especially targeted to foster user-acceptance of proposed methods in decision support and monitoring systems. The most critical issues include:

- The establishment of a project team with at least one member of clinical staff as a representative of a clinical task force and at least one outside consultant from the industry.
- A clinical task force representing all clinicians involved (possibly from different institutions). The task force's activities begin at the very start of the project and well before any hardware or software decisions come into play so that requirements and suggestions from the clinical staff can be taken into account. During the development, implementation and testing phases issues will be raised by the end-users – time for such issues to be resolved must be included in the schedule.
- Evaluation of the system on site prior to implementation in a 'real device'. This allows an effective assessment of performance (*Performance assessment*, below), usability (*Display of results*, p. 360) and practical issues (that may be related to institutional needs and processes that were not evident in the system design phase), gathering of feedback from the end-users and improvement of the system. This ideally requires some type of flexible 'test-platform' providing the possibility to use software plug-in in the devices/monitors to be tested.

Following such an approach the development and uptake cycle can be kept as short as possible. An example of a project in which the above three points were specifically addressed to demonstrate the feasibility of such an approach was the EU project IBIS (Carson et al 2000) in which signal processing methods for detecting brain dysfunction during intensive care and surgery were developed, tested and implemented using an interdisciplinary team of participants, a data library for early evaluation, and a dedicated test platform for algorithms.

Performance assessment

One of the main issues that will be considered when assessing the potential usefulness of a method is its performance. However, the performance of a method is a very loosely defined concept. For one user it may be related to measures, such as the percentage of correct classifications a certain method makes (using measures like sensitivity, specificity, ROC curves, etc., as described in *Determining the optimal cut-off point*, p. 11) and its clinical significance, for another user it may be the correlation of a method with an existing assessment scale, such as the Ramsay or OAA/S scale. For a third user it may be more related to the cost savings the proposed method potentially brings thanks to the fact that patients need a shorter time to recover from operations or can be released sooner from intensive care. And for a fourth user it may be related to a potential increase in the quality of life of the patient in the years after his or her stay in hospital.

Methods of assessing clinical significance of detected changes in values presented by a method implemented in a monitor are described in *Assessments of results* (p. 7). Typically such assessment calculations are done once the development process of the proposed method has reached a phase where it is implemented in a monitor and running on-line. Before this phase is reached, however, many preliminary performance assessments of candidate methods will have been done during research and development.

The approach in such a phase is typically that a set of data cases is used to develop and test a method off-line under laboratory conditions. The aim is to have data cases (consisting of recorded signals, annotations made by nurses, etc., recorded during patient stays in the ICU or during operations) that are as representative as possible for the 'real-life' situation in which the monitor algorithm to be developed will have to work. Such a set of data cases is often referred to as a *data library*. A good quality data library is an extremely useful tool for research and development purposes as it can speed up the process of introducing new methods (Carson et al 2000). Examples of physiological databases are the IMPROVE database (Nieminen et al 1997, Korhonen et al 1997, Thomsen et al 1997) (physiological signals during stays in the ICU, with several EEG cases), the IBIS database (Carson et al 2000) (physiological signals during stays in the ICU and surgery, with EEG, EPs and ERPs), and the Siesta database (Klösch et al 2001). For an overview of biosignal data libraries and aspects involved in creating them, see Cohen and Korhonen (2001).

An often occurring problem associated with data sets concerns the division of the set of cases into subsets. Typically, one wants to use as many cases as possible for developing the method (e.g. training the neural network, developing statistical classifiers) in order to make the method perform as well as possible. But, on the other hand, one wants also to use as many cases as possible as test cases to get an as accurate as possible assessment of the performance. Using the same cases for both training/development purposes as well as testing purposes is in general not acceptable as this will lead to a highly positively biased estimation of the performance. A classifier that has been trained to classify a specific set of cases perfectly may have been optimized ('overtrained') so that it has learned the processing of these particular training cases but is not able to make any generalization to other cases. It then has memorized how to deal with the training set instead of having created a generally applicable classification of functionality. The aim is to develop a system that is able to deal with new data in a real-life situations, and thus it is the generalization ability that we want to assess, not the performance on a specific training set. Using training cases for testing/performance assessment purposes should thus be avoided, as it will lead to an unrealistic and overoptimistic view of the method's potential usefulness. A certain part of the data library should be reserved for performance assessment purposes and be kept away from the method development process. The collection of data libraries containing physiological data is often costly, and the number of cases limited. Therefore, the division of the available data into training and testing subsets has to be thought out carefully. Different schemes exist to estimate performance in situations were the available data set is limited, these include k-fold cross-validation, leave-one-out and bootstrapping paradigms (Weiss and Kulikowski 1991). It should be noted that even if we use a test set that contains different cases from the training set the performance assessment may be positively biased. This can happen if we develop a method, test its performance, redevelop (e.g. fine-tune some parameters), test again, redevelop, and test again, etc. In that case we are tuning our method so that it will perform as well as possible on the test cases. Even though the test cases are not formally part of the training subset they are implicitly used to steer the development process and we are thus developing a method specialized in performing well on these specific test cases. This phenomenon is also known as 'training on the test set' and will also lead to a positively biased assessment of performance. The only way to avoid this is to reserve a part of the data library at the very start of the project and keep it locked away throughout the whole research and development phase. Only at the very end, once a stable prototype has been developed and no changes to the methods are to be made, can it be made available to make a final assessment of the performance of the method. It should be kept in mind though that the only 'real' performance assessment can be done once the method is being used on-line on new cases.

Display of results

Outputs from monitors provide the crucial interface between the user and the signals of interest. A clear display for communication with the user is essential, although the psychology of this process appears to have received scant attention in the design of clinical monitoring equipment. Fortunately, with the increasing use of PCs and the Internet, most hospital staff have considerable experience of modern interactive interfaces and with training readily attain confidence and competence in the use of clinical monitors. Several key points that play a role in the use of a neurophysiological monitoring system are listed in *Table 8.6* (see also Chapter 2).

It has been noted earlier that universal inclusion of facilities to view the unprocessed EEG signal throughout monitoring and to have parallel quality checks by means of continuous display of electrode impedance, line frequency and other external potentials is essential when interpreting recordings.

Facilities for synchronous display of variables from, for example, the cardiovascular and respiratory systems, with neurophysiological signals have led to a significant increase in understanding. This has proved one of the most useful aspects of the introduction of modern clinical monitoring, in that polygraphic displays encourage appraisal of the whole patient and not just individual systems (*Fig. 8.31*).

Superimposition on the EEG–polygraphic recording of a display of some form of EEG analysis is also helpful (*Fig. 8.32*) (Thomsen et al 2000).

Equipment which allows simultaneous display of EEG and EP also helps to give an overall view of the peripheral and central neurophysiological pathways, together with cortical function (*Fig. 8.33*).

Table 8.6 Key features in the interaction between a neurophysiological monitor and users

Issues	Requirements
Quality control	(a) Electrode contact impedance monitored continuously
	(b) Unprocessed ('raw') data available
	(c) Minimize external interference, detect and mark biological artefacts
	(d) Frequency spectra: display in full, not just derived measures
Annotation facilities	Easy on-line annotation and event marker tools
Advanced analysis and display of clinically important features	(a) Steady state and changing features
	(b) Brief events: arousals, seizure discharges, short depressions
	(c) Trends: over days, hours, minutes
	(d) EEG variability: e.g. 2–4/min changes, sleep features, burst suppression, etc.
	(e) Isoelectric EEG: indicated immediately
	(f) Averaged EP curves with possibility to manually and automatically indicate characteristic peaks, stability of amplitude and peak estimations
Output	(a) Measurable display/tracing
	(b) Digital output available for further analysis or signal processing
	(c) Facility to switch between primary signals, digital data and trend plots
	(d) Statistical comparison with appropriate, validated, database available on-line
	(e) Permanent record with text report meeting medico-legal requirements
Feedback or alarms?	Follow local policy or establish needs with clinicians

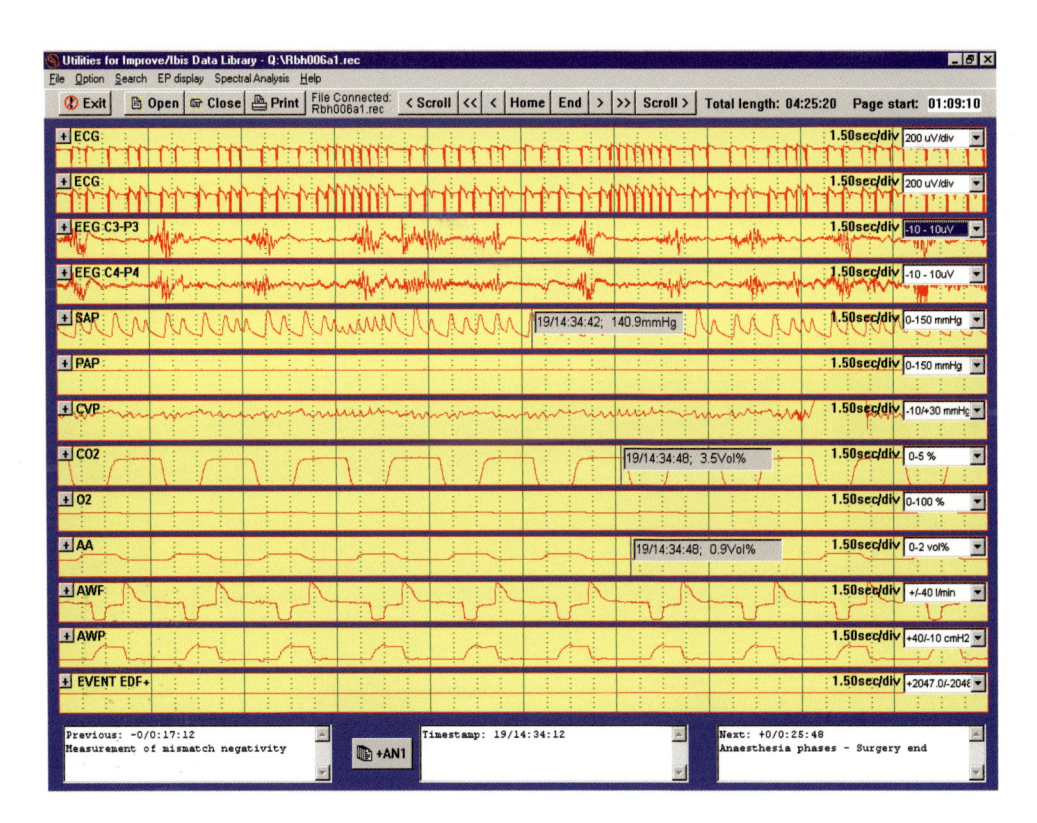

Fig. 8.31 Retrospective EEG–polygraphic display, using the WinDisp data library browser, of compressed data from 1 min of recording towards the completion of surgery in a 71-year-old woman who had undergone sequential right carotid endarterectomy and coronary artery grafting. SAP, systemic arterial blood pressure; PAP, pulmonary artery pressure; CVP, central venous pressure; CO_2, airway carbon dioxide concentration; O_2, airway oxygen concentration; AA, airway concentration of isoflurane; AWF, airway flow; AWP, airway pressure; event EDF, marker channel. Note the burst suppression pattern in the two EEG channels. (From Thomsen et al (2000), by permission.)

Chapter 8 Further Signal Analysis

The EEG can be displayed during recording, scrolling across the screen in much the same way as the conventional chart rolls across the table of the recorder. It can also be recalled from disc for subsequent off-line inspection and interpretation, or for computer analysis, brain mapping, etc. Sections of particular interest can be plotted on a high resolution printer/plotter, or transcribed onto a chart recorder, if available. Further issues related to the display of digital EEG can be found in *Displaying digital EEG* (p. 43).

Accurate on-line interpretation of EP changes depends on the quality of feature extraction. Given the enormous number of EPs collected after several hours of recording, there is a need for some sort of data compression; this is usually obtained by the extraction of selected parameters from raw tracings. Algorithms for automatic detection of peaks are can be implemented, provided the SNR is acceptable (see *Detection of events, complexes and trends*, p. 331), but reduction of EP data to latency and amplitude values is not enough for monitoring purposes. Derived parameters, such as the stability (reciprocal of variance) of latency and amplitude over time, or the slope (first derivative) of their changes, which measures the 'aggressiveness' of the causative process, are more specifically relevant to monitoring. Such techniques permit chronobiological correlation of changes in monitored EPs with those in other clinical data. This requires the construction and display of 'trend curves' where information is presented sequentially over a time axis, making its evolution over hours or days readily accessible.

Since the *raison d'être* of EP monitoring is to provide rapid warning in the event of CNS homeostatic changes, systems should be equipped with alarms to call for the clinician's attention if any one of the monitored parameters is no longer detected, or if a recorded value goes beyond accepted normal limits.

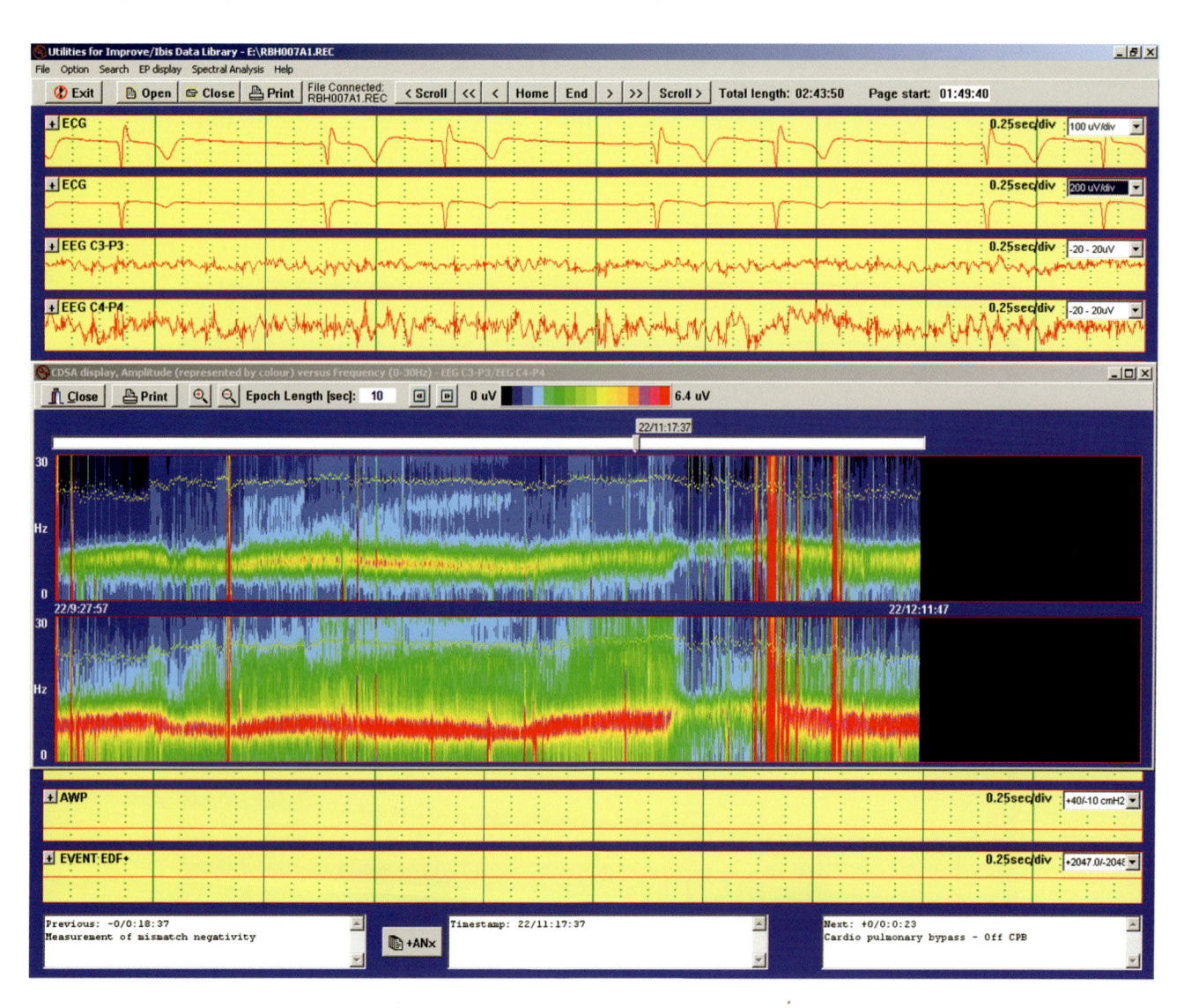

Fig. 8.32 Retrospective EEG–polygraphic display, similar to that in *Fig. 8.31* from a 65-year-old man in whom coronary artery grafting under cardiopulmonary bypass (CPB) had just been completed and the patient was about to come off CPB. Note the superimposed colour density spectral array (CDSA) display using 2 s epoch analysis, with a superimposed (yellow) spectral edge frequency marker, summarizing the whole 2 h 44 min procedure; the grey marker (22/11.17.37) shows the timing of the 10 s, left-sided, EEG sample channels. (Modified from Thomsen et al (2000), by permission.)

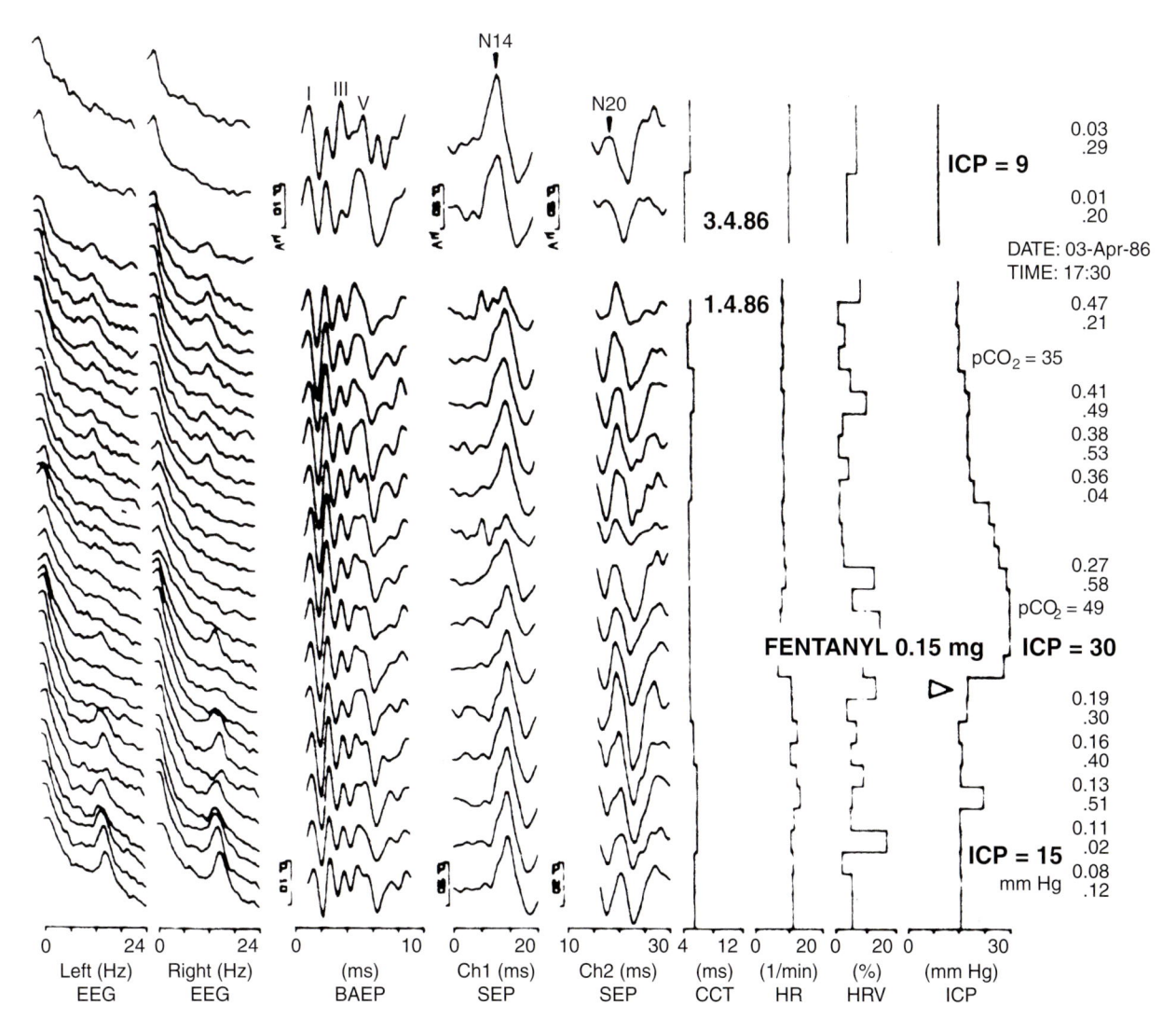

Fig. 8.33 Example of a combined display of a processed EEG and EPs recorded during intensive care from a 19-year-old man in post-traumatic coma following a severe head injury. The display, from left to right, shows compressed spectral array (CSA, log scale), brainstem auditory (BAEP) and somatosensory (SEP) evoked potentials, central somatosensory conduction time (CCT), heart rate (HR), heart rate variability (HRV) and intracranial pressure (ICP) values. Time runs from bottom to top of the recording. Note the transient rise in ICP with associated loss of faster EEG components, but little change in the short-latency EPs, after fentanyl administration. (From Pfurtscheller G, Schwarz G, Schroettner O, Litscher G, Maresch H, Auer L, List W 1987 Continuous and simultaneous monitoring of EEG spectra and brainstem auditory and somatosensory evoked potentials in the ICU and the operating room. *Journal of Clinical Neurophysiology*, 4:389–396, by permission. Lippincott, Williams and Wilkins Inc. Philadelphia © 2001 American Clinical Neurophysiology Society.)

Use of IT facilities

The use of information technology (IT) to perform neurophysiological recording and processing functions creates the possibility for digital communication between functions, thus leading to an integrated environment. Compatibility between equipment facilitates communication considerably. Translation routines between different types of systems are usually available but they are a less than ideal solution. The use of the facilities offered by the digital environment varies depending on factors such as: the size of the laboratory (number of recording systems), homogeneity in level of digitalization (mix of analogue and digital or fully digital), computer literacy of users, availability of technical knowledge and imagination. Certainly, this requires a considerable initial investment, but it is to be expected that costs will be recovered through more cost-effective practice. Some features of IT in the context of neurophysiology include the following (see also Binnie et al 2003, Chapter 7.8).

Analysis Algorithms for processing, analysis, classification, alarming, statistical postprocessing, or comparison to reference values can be implemented in algorithms in software modules that can be installed and updated flexibly.

Display As discussed in *Display of results* (p. 360) the use of modern computer display hardware provides great flexibility with respect to results presentation.

Remote access to data In the busy laboratory, it is necessary to have separate 'review stations' for offline analysis, for rounds with colleagues and for education. The recording stations are then only occupied during the actual investigations. The review stations have access to data from all recording stations (usually from a server via a network). In the case of monitoring a remote recording (e.g. long term EEG, intraoperative recording), the office module should be an on-line slave-monitor or on-line active unit. In some instances a small video camera on the recording equipment may be helpful or even necessary for accurate understanding of the ongoing recording when observed at a distance.

Reporting The integration of signals, analysis with comparisons to reference values, display of results in graphical or numerical form, and a written report help to convey the neurophysiological results to the non-expert neurophysiologist. Reporting is a very important part of the entire study and an effective use of resources in a digitized laboratory may improve general report quality considerably.

Storage An important feature is the ease with which data can be stored and retrieved. All biosignals, reports, related data (pathology reports, photos) can be stored in data bases. If data-handling is efficient, all previous information about a patient can be obtained immediately, a useful feature at follow-up. For research, such databases are of unique value. One may now also fulfil legal requirements of the storage of raw data much more easily than before. The stored data are, of course, so vital to the operation of the department that rigorous backup procedures are essential. Regular backup every 24 h, or even every 6 h, is reasonable. The backup data must be physically separated from the server in case of disaster. Some backup systems are very sophisticated but also rather opaque, saving separately all changes since the previous backup, rather than single complete records. Make sure that you can in fact retrieve data from your backup discs.

Server In the simplest case with a stand-alone recording system, the recording workstation runs the analysis software and stores results, and a connected printer is used for reporting. If two or more recorders and other digital functions are part of the routine set-up, a common server is recommended. This may be local within the laboratory or be connected to external nets. The advantages of a server are: (1) the same program is used for all users; (2) it is easy to update, so the same data are available to all users, even when files have been revised by different people at different times; (3) it is easy to back-up; and (4) it has integrity. A disadvantage of this approach is increased vulnerability – if the server (or network access to it) becomes dysfunctional there may be considerable repercussions on the functionality of all the workstations connected to it.

Networks As soon as more units or functions of various kinds are routinely used, a network is typically used. A decision then has to be made as to whether this should be an independent local area network (LAN) or whether all equipment should be directly connected to a hospital-wide system. The storage and transmission of large volumes of EEG, and particularly video data, on the main Hospital Information System is not usually welcomed by IT managers. It is therefore usually better to have a departmental LAN that can communicate with the main system, as required, to export reports and import patient administration material, referrals or clinical records. Within the laboratory, users of recording equipment or review equipment and administrators may share common resources such as a server, backup system, laser printer, colour printer and scanner. With increasing degree of integration, wider connections may be required. This gives the possibility to connect to systems within the same hospital or even to other hospitals connected on a wide area network (WAN). Finally, for general communication, for example to reach various databases, a connection to the Internet may be installed. Effective firewalls, up-to-date virus protection software and other precautions are of utmost importance in this case.

Risk management An action plan has to be available for maintenance of security and for use if the system's integrity is compromised due to hacking attacks via the network or in case of a vital computerized administrative or medical function breakdown due to network failure, robbery, hard disc crash, fire, flooding, etc. Complete hazard elimination is almost impossible to achieve and would be far too expensive. Therefore acceptance limits for data loss, time of breakdown and so on have to be defined and measures to meet corresponding criteria be implemented. The hazard analysis may result in a number of safety precautions, for example: automatic backup of all data every night for total system reconstruction; backup tapes and original discs are stored in a fireproof safe on a separate site; mirror hard discs on the server for instant backup; star network instead of a circular chain; recording equipment shall be able to run locally without connection to the server; data acquisition workstations need sufficient local storage to run for, say, 24 h without access to the server; servers in secure rooms; servers installed above the floor in case of flooding; uninterruptible power supplies for all critical equipment; up-to-date firewalls and virus protection software and appropriate security settings for Internet access.

Administration Referrals and reports can be made in digital form and treated in paperless mode. The patient data should only be inserted once, preferably through the central patient administration system, where it can be validated and updated as necessary, and then automatically transferred to appointment and recording systems, report headers, bills and statistical summaries.

Telemedicine Telemedicine comprises various types of remote communications. In neurophysiology it usually

comprises transmission of biosignals and of video communication.
- *Biosignals taking up small volumes of data* (e.g. EMG or EPs) recorded at one site can be transferred to specialists at another site (for analysis, to obtain second opinions, for collection of reference values, or for scientific collaboration). This can be done using a secure Internet connection giving access to certain segments in the servers of the remote hospitals (IP address giving access only to neurophysiological data), or as properly encrypted e-mail attachment.
- Transfer of *data taking up large volumes of data* (e.g. EEG) is more difficult, unless a dedicated fibre optic network is available. However, the problems have been addressed in various ingenious ways over many years, because the potential benefits are clearly considerable, for several reasons: (1) In most countries, EEG facilities are widely dispersed. (2) Patients may require investigation outside the laboratory, for instance in the operating theatre or ICU, or in the home, school, or workplace to monitor seizures in a natural environment. (3) Expertise in some specialized applications such as neonatology is not widely available. (4) Immediate interpretation may be required, for instance in status epilepticus, during electrocorticography and intensive care. (5) EEG is a small discipline and its practitioners often work alone; audit, clinical governance, research, and continuing medical education therefore require collaboration and multicentre review of records. Both routine and urgent reporting would be easily accomplished, with a minimum cost in time or expense, by transmission over telephone lines. This is not trivial, as the volumes of data are considerable: typically some 40 MB/h. Off-line transmission of EEGs as complete files is surprisingly difficult without dedicated wide bandwidth facilities. An irrecoverable interruption of transmission is highly probable whilst sending a 40 MB file over standard telephone lines. On-line synchronous live EEG transmission would be preferable but is beyond the capacity of the public telephone system without data compression. The ideal is even more demanding: off-line EEG transmission at the data rates employed by experienced electroencephalographers reading EEGs, one or two 10 s pages per second, requiring the EEG display to be updated at 10–20 times recording speed. All three approaches will doubtlessly be feasible when wide-band networks for telephony, video and other applications become the norm. Data compression algorithms are available to allow review of EEGs even over the public telephone network in approximately 20% of real time, i.e. at a presentation rate acceptable to the experienced reviewer (Holder et al 2003). This is a facility particularly useful for obtaining expert opinion out-of-hours on emergency EEGs, for instance in neonates or intensive care patients.
- *Video-conferencing* facilities can be arranged either within the hospital department or at generally accessible locations. Typical use is to allow colleagues outside the hospital to participate in rounds, consultations, meetings and training sessions. The use of video-conferences and lectures within medicine is increasing and will develop further with faster communication links, and smaller and less expensive equipment and traffic cost.

Successful implementation and acceptance of IT features in an environment is not self-evident and requires taking into account several issues. Factors for a successful integration of IT in clinical neurophysiology include:
- the integration is driven by medical needs and by available technology;
- ideas for solutions have been defined;
- consensus among staff and partners; and
- defined plans for development and implementation.

REFERENCES

Agarwal R, Gotman J 2001 Long-term EEG compression for intensive-care settings. IEEE Eng Med Biol 20:23–29.

Agarwal R, Gotman J, Flanagan D, et al 1998 Automatic EEG analysis during long-term monitoring in the ICU. Electroencephalogr Clin Neurophysiol 107:44–58.

Akaike H 1974 A new look at the statistical model identification. IEEE Trans Autom Control 19:716–723.

Akay M 1994 *Biomedical Signal Processing*. Academic Press, San Diego, CA.

Alkire MT, Pomfrett CJD 1997 Toward a monitor of depth: Bispectral Index (BIS) and respiratory sinus arrhythmia (RSA) both monitor cerebral metabolic reduction during isoflurane anaesthesia. Anesthesiology 87:A421.

American Electroencephalographic Society 1994 Guideline eleven: guidelines for intraoperative monitoring of sensory evoked potentials. J Clin Neurophysiol 11:77–87.

American Society for Testing and Materials ASTM 1994 *Standard Specification for Transferring Digital Neurophysiological Data Between Independent Computer Systems*. ASTM E1467–94 (www.astm.org), Philadelphia, PA.

Arnold M, Doering A, Witte H, et al 1996 Use of adaptive Hilbert transformation for EEG segmentation and calculation of instantaneous respiration rates in neonates. J Clin Monit 12:43–60.

Avent RK, Charlton JD 1990 A critical review of trend-detection methodologies for biomedical monitoring systems. Crit Rev Biomed Eng 17:621–659.

Barlow JS 1984 Analysis of EEG changes with carotid clamping by selective analog filtering, matched inverse digital filtering and automatic adaptive segmentation: a comparative study. Electroencephalogr Clin Neurophysiol 58:193–204.

Barlow JS 1985 A general-purpose automatic multichannel electronic switch for EEG artifact elimination. Electroencephalogr Clin Neurophysiol 60:174–176.

Barlow JS, Dubinsky J 1980 EKG-artifact minimization in referential EEG recordings by computer subtraction. Electroencephalogr Clin Neurophysiol 48:470–472.

Chapter 8 Further Signal Analysis

Barnard GA 1959 Control charts and stochastic processes. J R Stat Soc B 21:239–271.

Barr G, Jakobsson JG, Owall A, et al 1999 Nitrous oxide does not alter Bispectral Index: study with nitrous oxide as sole agent and as adjunct to IV anesthesia. Br J Anaesth 82:827–830.

Bartnik EA, Blinowska KJ, Durka PJ 1992 Single evoked potential reconstruction by means of wavelet transform. Biol Cybern 67:175–181.

Bertrand O 1985 Système informatisé d' enregistrement sequentiel des potentiels évoqués auditifs précoces adapté à la surveillance des malades comateux. PhD Thesis, University of Lyon.

Bertrand O, García-Larrea L, Artru F, et al 1987 Brain-stem monitoring. I: A system for high-rate sequential BAEP recording and feature extraction. Electroencephalogr Clin Neurophysiol 68:433–445.

Bertrand O, Bohórquez J, Pernier J 1990 Technical requirements for evoked potentials monitoring in the intensive care unit. Electroencephalogr Clin Neurophysiol 41:51–69.

Bertrand O, Bohórquez J, Pernier J 1994 Time–frequency digital filtering based on an invertible wavelet transform: an application to evoked potentials. IEEE Trans Biomed Eng 41:77–88.

Bickford RG, Sims JK, Billinger TW, et al 1971 Problems in EEG estimation of brain death and use of computer techniques for their solution. Trauma 12:61–95.

Binnie CD, Cooper R, Mauguière F, et al (eds) 2003 *Clinical Neurophysiology*. Volume 2: *EEG, Paediatric Neurophysiology Special Techniques and Applications*. Elsevier, Amsterdam.

Binnie CD, Cooper R, Mauguière F, et al (eds) 2004 *Clinical Neurophysiology*. Volume 1: *EMG, Nerve Conduction and Evoked Potentials*, revised and enlarged edition. Elsevier, Amsterdam.

Bloom MJ 1993 Techniques to identify clinical contexts during automated data analysis. Int J Clin Monit Comput 10:17–22.

Bodenstein G, Praetorius HM 1977 Feature extraction from the electroencephalogram by adaptive segmentation. Proc IEEE 65:642–652.

Boston JR 1985 Noise cancellation for brainstem auditory evoked potentials. IEEE Trans Biomed Eng 32:106–1070.

Boston JR 1989 Automated interpretation of brainstem auditory evoked potentials: a prototype system. IEEE Trans Biomed Eng 36:528–532.

Box G, Luceño A 1997 *Statistical Control by Monitoring and Feedback Adjustment*. Wiley, New York.

Bronzino JD (ed.) 2000 *The Biomedical Engineering Handbook*, 2nd edition. CRC Press/IEEE Press, Boca Raton, FL.

Browne M, Cutmore TR 2000 Adaptive wavelet filtering for analysis of event-related potentials from the electroencephalogram. Med Biol Eng Comp 38:645–652.

Bruce EN 2001 *Biomedical Signal Processing and Signal Modeling*. Wiley, New York.

Bruha I, Madhavan GP 1988 Use of attributed grammars for pattern recognition of evoked potentials. IEEE Trans. Systems Man Cybernetic 18:1046–1049.

Bruhn J, Bouillon TW, Shafer SL 2000 Bispectral Index (BIS) and burst suppression: revealing a part of the BIS algorithm. J Clin Monit Comput 16:593–596.

Bruner JMR, Leonard PF 1989 *Electricity Safety and the Patient*. Year Book Medical Publishers, Chicago, IL.

Brunia CHM, Möcks J, van den Berg-Lenssen MMC, et al 1989 Correcting ocular artifacts in the EEG: a comparison of several methods. J Psychophysiol 3:1–50.

Burch NR 1959 Automatic analysis of the EEG. Electroencephalogr Clin Neurophysiol 11:827–834.

Burke D, Nuwer MR, Daube J, et al 1999 Intraoperative monitoring. In: *Recommendations for the Practice of Clinical Neurophysiology. Guidelines of the International Federation of Clinical Neurophysiology.* EEG Suppl 52 (eds G Deuschl, A Eisen). Elsevier Science, Amsterdam, pp. 133–148.

Carson ER, van Gils M, Saranummi N (eds) 2000 IBIS: improved monitoring for brain dysfunction in intensive care and surgery. Comput Method Program Biomed 63:157–235.

Celka P, Boashash B, Colditz P 2001 Preprocessing and time–frequency analysis of newborn EEG signals. IEEE Eng Med Biol Mag 20:30–39.

Cerutti S, Liberati D, Avanzini G, et al 1986 Classification of the EEG during neurosurgery. Parametric identification and Kalman filtering compared. J Biomed Eng 8:244–254.

Cerutti S, Carrault G, Cluitmans PJM, et al 1996 Non-linear algorithms for processing biological signals. Comp Meth Prog Biomed 51:51–73.

Champ CM, Woodall WH 1987 Exact results for Shewhart control charts with supplementary runs rules. Technometrics 29:393–399.

Chaput de Saintonge M, Vere DW 1974 Why don't doctors use Cusums? Lancet i:120–121.

Chernik DA, Gillings D, Laine H, et al 1990 Validity and reliability of the Observer's Assessment of Alertness/Sedation Scale: study with intravenous midazolam. J Clin Psychopharmacol 10:244–251.

Clarencon D, Renaudin M, Gourmelon P, et al 1996 Real-time spike detection in EEG signals using the wavelet transform and a dedicated digital signal processor card. J Neurosci Method 70:5–14.

Cluitmans PJM 1990 Neurophysiological monitoring of anesthetic depth. PhD Thesis, Eindhoven University of Technology.

Cobbold 1988 *Transducers for Biomedical Measurements*. Wiley, New York.

Cohen A 2000 Biomedical signals: origin and dynamic characteristics; frequency-domain analysis. In: Bronzino JD (ed) *The Biomedical Engineering Handbook*, 2nd edn. CRC PRess/IEEE Press, Boca raton, FL, pp. 52-1 to 52-24.

Cohen A, Korhonen I 2001 From the guest editors: biomedical signals databases. IEEE EMB Mag 20:23–24.

Coifman RR, Wickerhauser MV 1996 Wavelets, adapted waveforms and de-noising. Electroencephalogr Clin Neurophysiol 45(Suppl):57–78.

Cooley JW, Tukey JW 1965 An algorithm for the machine calculation of complex Fourier series. Math Comput 19:297–301.

Cooper R, Osselton JW, Shaw JC 1980 *EEG Technology*, 3rd edition. Butterworth, London.

Cooper R, Binnie CD, Billings RJ (eds) 2005 *Techniques in Clinical Neurophysiology – A Practical Manual*. Churchill Livingstone, Oxford.

Creutzfeldt O-D, Bodenstein G, Barlow JS 1985 Computerized EEG pattern classification. Clinical evaluation. Electroencephalogr Clin Neurophysiol 60:373–393.

Dahaba AA 2005 Different conditions that could result in the Bispectral Index indicating an incorrect hypnotic state. Anesthesiology 101:765–773.

D'Attellis CE, Isaacson SI, Sirne RO 1997 Detection of epileptic events in electroencephalograms using wavelet analysis. Ann Biomed Eng 25:286–293.

DeFatta DJ, Lucas JG, Hodgkiss WS 1988 *Digital Signal Processing: A System Design Approach*. Wiley, New York.

Del Castillo E, Grayson J, Runger G, et al 1996 A review of SPC methods for short runs. Commun Stat Theory 25:2723–2737.

References

Delgado RE, Özdamar Ö 1994 Automated auditory brainstem response interpretation. IEEE Eng Med Biol Mag 13:227–237.

Devasahayam SR 2000 *Signals and Systems in Biomedical Engineering: Signal Processing and Physiological System Modeling*. Kluwer Academic/Plenum, New York.

De Weerd JPC 1981 Facts and fancies about '*a posteriori* Wiener' filtering. IEEE Trans Biomed Eng BME-28:252–257.

De Weerd JPC, Kap JI 1981 *A posteriori* time-varying filtering of averaged evoked potentials (Parts I, II). Biol Cybern 41:211–134.

Dumermuth G, Molinari L 1987 Spectral analysis of EEG background activity. In: *Methods of Analysis of Brain Electrical and Magnetic Signals* (eds AS Gevins, A Rémond). *Handbook of Electroencephalography and Clinical Neurophysiology*, revised series, volume 1. Elsevier, Amsterdam, pp. 85–130.

Eckert O 1998 Automatic artefact detection in intraoperative EEG monitoring [in German]. Biomed Tech 43:236–242.

Eckert O, Werry C, Neulinger A, et al 1997 Intraoperative EEG monitoring using a neural network [in German]. Biomed Tech 42:78–84.

Effern A, Lehnertz K, Fernandez G, et al 2000 Single trial analysis of event related potentials: non-linear de-noising with wavelets. Clin Neurophysiol 111:2255–2263.

Eisenhart C 1963 Realistic evaluation of the precision and accuracy of instrument calibration systems. J Res Nat Bur Stand-C Eng Instrum 67C:161–187.

Emerson RG, Sgro JS 1985 Phase synchronized triggering: a method for coherent noise elimination in evoked potential recording. Electroencephalogr Clin Neurophysiol 61:17P.

European Committee for Standardisation (CEN) 2000 *Vital Signs Information Representation*, ENV 13734. CEN, Brussels.

Ferdjallah M, Barr RE 1994 Adaptive digital notch filter design on the unit circle for the removal of powerline noise from biomedical signals. IEEE Trans Biomed Eng 41:529–536.

Figliola A, Serrano E 1997 Analysis of physiological time series using wavelet transforms. IEEE Eng Med Biol Mag 16:74–79.

Fisher RA 1925 Theory of statistical estimation. Proc Camb Phil Soc 22:700–725.

Frenkel C, Schuttler J, Ihmsen H, et al 1995 Pharmacokinetics and pharmacodynamics of propofol/ alfentanil infusions for sedation in ICU patients. Intensive Care Med 21:981–988.

Fridman J, John ER, Bergelson M, et al 1981 Application of digital filtering and automatic peak detection to brain-stem auditory evoked potential. Electroencephalogr Clin Neurophysiol 53:405–416.

Fu KS 1986 A step towards unification of syntactic and statistical pattern recognition. IEEE Trans Pattern Analysis and Machine Intelligence PAMI-8:398–404.

Gabor AJ 1998 Seizure detection using a self-organizing neural network: validation and comparison with other detection strategies. Electroencephalogr Clin Neurophysiol 107:27–32.

Gabor AJ, Leach RR, Dowla FU 1996 Automated seizure detection using a self-organizing neural network. Electroencephalogr Clin Neurophysiol 99:257–266.

Galicki M, Witte H, Dörschel J, et al 1997 Common optimization of adaptive preprocessing units and a neural network during the learning period. Application in EEG pattern recognition. Neural Networks 10:1153–1163.

Gevins AS 1987 Correlation analysis. In: *Methods of Analysis of Brain Electrical and Magnetic Signals* (eds AS Gevins, A Rémond). *Handbook of Electroencephalography and Clinical Neurophysiology*, revised series, volume 1. Elsevier, Amsterdam, pp. 171–194.

Glass PS, Bloom M, Kearse L, et al 1997 Bispectral analysis measures sedation and memory effects of propofol, midazolam, isoflurane, and alfentanil in healthy volunteers. Anesthesiology 86:836–847.

Goelz H, Jones RD, Bones PJ 2000 Wavelet analysis of transient biomedical signals and its application to detection of epileptiform activity in the EEG. Clin Electroencephalogr 31:181–191.

Goethe JW, Bronzino D 2000 Design issues in developing clinical decision support and monitoring systems. In: *The Biomedical Engineering Handbook* (ed JD Bronzino), 2nd edition. CRC Press/IEEE Press, Boca Raton, FL.

Gotman J, Ives JR, Gloor P 1981 Frequency content of EEG, EMG at seizure onset: problems of removal of EMG artefact by digital filters. Electroencephalogr Clin Neurophysiol 52:626–639.

Graps A 1995 An introduction to wavelets. IEEE Computat Sci Eng 2:50–61.

Groen GJ, Patel VL 1988 The relationship between comprehension and reasoning in medical expertise. In: *The Nature of Expertise* (eds M Chi, R Glaser, M Farr). Lawrence Erlbaum, Hillsdale, NJ.

Grönfors T 1993 Peak identification of auditory brainstem responses with multifilters and attributed automaton. Comp Meth Prog Biomed 40:83–87.

Guérit J-M, Fischer C, Facco E, et al 1999 Standards of clinical practice of EEG, EPs in comatose and other responsive states. In: *Recommendations for the Practice of Clinical Neurophysiology: Guidelines of the International Federation of Clinical Neurophysiology* (EEG Suppl 52) (eds G Deuschl, A Eisen). Elsevier, Amsterdam, pp. 117–131.

Hawkins DM 1987 Self-starting Cusum charts for location and scale. Statistician 36:299–315.

Hjorth B 1970 EEG analysis based on time domain properties. Electroencephalogr Clin Neurophysiol 29:306–310.

Holder D, Cameron J, Binnie CD 2003 Tele-EEG in epilepsy: review and initial experience with software to enable EEG review over a telephone link. Seizure 12:85–91.

Hoppe U, Schnabel K, Weiss S, et al 2002 Representation of somatosensory evoked potentials using discrete wavelet transform. J Clin Monit Comput 17:227–233.

Huang JW, Lu Y-Y, Nayak A, et al 1999 Depth of anesthesia estimation and control. IEEE Trans Biomed Eng 46:71–81.

Hubbard BH 1998 *The World According to Wavelets: the Story of a Mathematical Technique in the Making*, 2nd edition. AK Peters, Wellesley, MA.

Hunter JS 1986 The exponentially weighted moving average. J Qual Technol 18:203–210.

Hyvärinen A, Karhunen J, Oja E 2001 *Independent Component Analysis*. Wiley, New York.

International Electrotechnical Commission (IEC) 1988 *Medical Electrical Equipment Part 1: General Requirements for Safety*, 2nd edition. Publication 60601-1. IEC Secretariat, Geneva.

International Electrotechnical Commission (IEC) 1998 *Medical electrical equipment Part 2-2: Particular Requirements for the Safety of High Frequency Surgical Equipment*, 3rd edition. Publication 60601-2-2. IEC Secretariat, Geneva.

International Electrotechnical Commission (IEC) 2003 *Medical electrical equipment Part 2-26: Particular Requirements for the Safety Of Electroencephalographs*, 2nd edition. Publication 60601-2-26. IEC Secretariat, Geneva.

Ives JR, Schomer DL 1988 A 6-pole filter for improving the readability of muscle contaminated EEGs. Electroencephalogr Clin Neurophysiol 69:486–490.

Jain AK, Duin RW, Mao J 2000 Statistical pattern recognition: A review. IEEE Trans Pattern Anal Machine Intell 22:4–37.

Jansen BH, Bourne JR, Ward JW 1981 Autoregressive estimation of short segment spectra for computerized EEG analysis. IEEE Trans Biomed Eng 28:630–638.

Jervis BW, Ifeacher EC, Allen EM 1988 The removal of ocular artefacts from the electroencephalogram: a review. Med Biol Eng Comp 26:2–12.

John ER, Baird H, Fridman J, et al 1982 Normative values for brain stem auditory evoked potential obtained by digital filtering and automatic peak detection. Electroencephalogr Clin Neurophysiol 54:153–160.

Kemp B, Värri A, Rosa AC, et al 1992 A simple format for exchange of digitized polygraphic recordings. Electroencephalogr Clin Neurophysiol 82:391–393.

Keogh E, Chu S, Hart D, et al 2004 Segmenting time series: a survey and novel approach. In: *Data Mining in Time Series Databases* (eds P Last, A Kandel, H Bunke). World Scientific, Singapore.

Klösch G, Kemp B, Penzl T, et al 2001 The Siesta project. Polygraphic and clinical database. IEEE EMB Mag 20:51–57.

Koch C, Laurent G 1999 Complexity and the nervous system. Science 284:96–98.

Koenig T, Marti-Lopez F, Valdes-Sosa P 2001 Topographic time–frequency decomposition of the EEG. Neuroimage 14:383–390.

Kohonen T 1988 *Self-organization and Associative Memory*, 2nd edition. Springer-Verlag, New York.

Korhonen I, Ojaniemi J, Nieminen K, et al 1997 Building the IMPROVE data library. IEEE EMB Mag 16:25–32.

Lempel A, Ziv J 1976 On the complexity of finite sequences. IEEE Trans Inf Theory 22:75–81.

Lin Z, Chen J De Z 1996 Advances in time frequency analysis of biomedical signals. Crit Rev Biomed Eng 24:1–72.

Lipping T, Jäntti V, Yli-Hankala A, et al 1995 Adaptive segmentation of burst suppression pattern in isoflurane and enflurane anesthesia. Int J Clin Monit Comput 12:161–167.

Lipping T, Loula P, Jäntti V, et al 1994 DC-level detection of burst-suppression EEG. Meth Inf Med 33:35–38.

Lipping T 2001 Processing EEG during anaesthesia and cardiac surgery with non-linear order statistics based methods. PhD Thesis, Tampere University of Technology, Publication 316.

Lipping T, Mandersloot G, Prior P 2000 Monitoring trends in cardiac surgery. Proc 22nd Ann Int Conf IEEE EMBS, Chicago 4:2821–2824.

Lopes da Silva FH, ten Broeke W, van Hulten K, et al 1976 EEG non-stationarities detected by inverse filtering in scalp and cortical recordings of epileptics: statistical analysis and spatial display. In: *Quantitative Analytic Studies in Epilepsy* (eds P Kellaway, I Petersén). Raven Press, New York, pp. 375–388.

Lucas JM 1976 The design and use of V-mask control schemes. J Qual Technol 8:1–12.

Marcus ML, Biersach BR 2003 Regulatory requirements for medical equipment. IEEE Instrum Meas Mag 6(4):23–29.

Michael D, Houchin J 1979 Automatic EEG analysis: a segmentation procedure based on the autocorrelation function. Electroencephalogr Clin Neurophysiol 46:232.

Montgomery DC 2004 *Introduction to Statistical Quality Control*, 5th edition. Wiley, New York.

Nakamura M, Sugi T, Ikeda A, et al 1996 Clinical application of automatic integrative interpretation of awake background EEG: quantitative interpretation, report making, and detection of artifacts and reduced vigilance level. Electroencephalogr Clin Neurophysiol 98:103–112.

Nakamura M, Shibasaki H, Nishida S 1990 Method for recording short latency evoked-potentials using an ECG artifact elimination procedure. J Biomed Eng 12:51–56.

Nayak A, Roy RJ 1998 Anesthesia control using midlatency auditory evoked potentials. IEEE Trans Biomed Eng 45:409–421.

Nelson LS 1984 The Shewhart control chart: tests for special causes. J Qual Technol 16:237–239.

Nieminen K, Langford RM, Morgan CJ, et al 1997 A clinical description of the IMPROVE Data Library. IEEE EMB Mag 16:21–24:40.

Ning T, Bronzino JD 1989 Bispectral analysis of the rat EEG during various vigilance states. IEEE Trans Biomed Eng 36:497–499.

NIST/SEMATECH 2004 *e-Handbook of Statistical Methods*. Available at: http://www.itl.nist.gov/div898/handbook (accessed 20/11/2004).

Nykänen P, Saranummi N 2000 Clinical decision systems. In: *The Biomedical Engineering Handbook* (ed JD Bronzino), 2nd edition. CRC Press/IEEE Press, Boca Raton, FL.

Ortolani O, Conti A, Di Filippo A, et al 2002 EEG signal processing in anaesthesia. Use of a neural network technique for monitoring depth of anaesthesia. Br J Anaesth 88:644–648.

Page ES 1954 Continuous inspection schemes. Biometrika 41:100–115.

Panych LP, Wada JA, Beddoes MP 1989 Practical digital filters for reducing EMG artefact in EEG seizure recordings. Electroencephalogr Clin Neurophysiol 72:268–276.

Pardey J, Roberts S, Tarassenko L 1996 A review of parametric modeling techniques for EEG analysis. Med Eng Physics 18:2–11.

Parsa V, Parker PA, Scott RN 1998 Adaptive stimulus artifact reduction in noncortical somatosensory evoked potential studies. IEEE Trans Biomed Eng 45:165–178.

Petropulu AP 2000 Higher-order spectral analysis. In: *The Biomedical Engineering Handbook* (ed JD Bronzino), 2nd edition. CRC Press/IEEE Press, Boca Raton, FL.

Pfurtscheller G, Schwarz G, Schroettner O, et al 1987 Continuous and simultaneous monitoring of EEG spectra and brainstem auditory and somatosensory evoked potentials in the intensive care unit and the operating room. J Clin Neurophysiol 4:389–396.

Picton TW, Hink RF, Perez Abalo M, et al 1984 Evoked potentials: how now? J Electrophysiol Tech 10:177–221.

Pichlmayr I, Lips U, Künkel H 1984 *The Electroencephalogram in Anaesthesia. Fundamentals, Practical Applications, Examples*, English edition (trans E Bonatz, T Masyk-Iversen). Springer-Verlag, Heidelberg.

Pigeau RA, Hoffmann RF, Moffitt AR 1981 A multivariate comparison between two EEG analysis techniques: period analysis and fast Fourier transform. Electroencephalogr Clin Neurophysiol 52:656–658.

Pincus SM 1991 Approximate entropy as a measure of system complexity. Proc Natl Acad Sci USA 88:2297–2301.

Pincus SM, Gladstone IM, Ehrenkranz RA 1991 A regularity statistic for medical data analysis. J Clin Monit 7:335–345.

Pradhan N, Sadasivan PK, Arunodaya GR 1996 Detection of seizure activity in EEG by an artificial neural network: a preliminary study. Comput Biomed Res 29:303–313.

References

Prior PF, Maynard DE 1986 *Monitoring Cerebral Function. Long-term Monitoring of EEG, Evoked Potentials*. Elsevier, Amsterdam.

Prior PF, Maynard DE, Brierley JB 1978 EEG monitoring for control of anaesthesia produced by Althesin infusion in primates. Br J Anaesth 50:993–1001.

Proakis JG, Manolakis D 1996 *Digital Signal Processing: Principles Algorithms and Applications*, 3rd edition. Prentice Hall, Englewood Cliffs, NJ.

Quesenberry CP 1991 SPC-Q-charts for start-up processes and short or long runs. J Qual Technol 23:213–224.

Quesenberry CP 1993 The effect of sample size on estimated limits for X and X-control charts. J Qual Technol 25:237–247.

Rampil IJ 1998 A primer for EEG signal processing in anesthesia. Anesthesiology 89:980–1002.

Ramsay MAE, Savege TM, Simpson BRJ, Goodwin R 1974 Controlled sedation with alphaxalone–alphadolone. BMJ 2:656–659.

Rangayyan RM 2002 *Biomedical Signal Analysis: a Case-study Approach*. IEEE Press/Wiley Inter-Science, Piscataway, NJ.

Renna M, Wigmore T, Mofeez A, Gilbe C 2002 Biasing effect of electromyogram on BIS: a controlled study during high-dose fentanyl induction. J Clin Monit Comput 17:377–381.

Rezek IA, Roberts SJ 1998 Stochastic complexity measures for physiological signal analysis. IEEE Trans Biomed Eng 45:1186–1191.

Robert C, Karasinski P, Arreto CD, Gaudy JF 2002 Monitoring anesthesia using neural networks: a survey. J Clin Monit Comput 17:259–267.

Roberts SW 1959 Control charts based on geometric moving averages. Technometrics 1:239–250.

Robinson MP, Flintoft ID, Marvin AC 1997 Interference to medical equipment from mobile phones. J Med Eng Technol 21:141–146.

Rumelhart DE, McClelland JL (eds) 1986 *Parallel Distributed Processing: Explorations in the Microstructure of Cognition*, volumes I and II. MIT Press, Cambridge, MA.

Ryan TP 2000 *Statistical Methods for Quality Improvement*, 2nd edition. Wiley-Interscience, New York.

Sadasivan PK, Dutt DN 1995 ANC Schemes for the enhancement of EEG signals in the presence of EOG Artifacts. Comp Biomed Research 29:27–40.

Samar VJ, Swartz KP, Raghuveer MR 1995 Multiresolution analysis of event-related potentials by wavelet decomposition. Brain Cognition 27:398–438.

Samar VJ, Bopardikar A, Raghuveer MR, et al 1999 Wavelet analysis of neuroelectric waveforms: a conceptual tutorial. Brain Language 66:7–60.

Saranummi N, Korhonen I, van Gils M, et al 1997 Framework for biosignal interpretation in intensive care and anaesthesia. Meth Inf Med 36:340–344.

Särkelä M, Mustola S, Seppänen T, et al 2002 Automatic analysis and monitoring of burst suppression in anesthesia. J Clin Monit Comp 17:125–134.

Schalkoff R 1992 *Pattern Recognition: Statistical, Structural, and Neural Approaches*. Wiley, New York.

Schiff SJ, Aldroubi A, Unser M, et al 1994 Fast wavelet transform of EEG Electroencephalogr Clin Neurophysiol 91:442–455.

Schwilden H, Schuttler J, Stoekel H 1987 Closed-loop feedback control of methohexital anesthesia by quantitative EEG analysis in humans. Anesthesiology 67:341–347.

Senhadji L, Wendling F 2002 Epileptic transient detection: wavelets and time–frequency approaches. Neurophysiol Clin 32:175–192.

Senhadji L, Dillenseger JL, Wendling F, et al 1995 Wavelet analysis of EEG for three-dimensional mapping of epileptic events. Ann Biomed Eng 23:543–552.

Senhadji L, Wodey E, Ecoffey C 2002 Monitoring approaches in general anesthesia: a survey. Crit Rev Biomed Eng 30:85–97.

Shannon C 1948 Mathematical theory of communication. Bell System Technol J 27:379–423, 623–656.

Shannon CE 1949 Communication in the presence of noise. Proc Inst Radio Engrs 37:1:10–21.

Sharma S 1996 *Applied Multivariate Techniques*. Wiley, New York.

Shaw JC 1981 An introduction to the coherence function and its use in EEG signal analysis. J Med Eng Technol 5:279–288.

Shaw JC 1984 Correlation and coherence analysis of the EEG: a selective tutorial review. Int J Psychophysiol 1:255–266.

Shaw JC, Simpson D 1997 EEG coherence – caution and cognition. Bull Br Psychophysiol Soc 30/31:7–9.

Shewhart WA 1931 *Economic Control of Quality of Manufactured Product*. Macmillan, New York.

Si Y, Gotman J, Pasupathy A, et al 1998 An expert system for EEG monitoring in the pediatric intensive care unit. Electroencephalogr Clin Neurophysiol 106:488–500.

Smith WD, Dutton RC, Smith T 1996 Measuring the performance of anesthetic depth indicators. Anesthesiology 84:38–51.

Spreckelsen MV, Bromm B 1988 Estimation of single-evoke cerebral potentials by means of parametric modelling and Kalman filtering. IEEE Trans Biomed Eng 35:691–700.

Stecker MM, Patterson T 1996 Strategics for minimizing 60 Hz pickup during evoked-potential recording. Electroenceph Clin Neurophysiol 100:370–373.

Stephenson WA, Gibbs FA 1951 A balanced non-cephalic reference electrode. Electroencephalogr Clin Neurophysiol 3:237–240.

Stockmanns G, Nahm W, Thornton C, et al 1997 Wavelet-analysis of middle latency auditory evoked potentials during repetitve propofol sedation. Anesthesiology 87:A465.

Tan SB 2001 Introduction to Bayesian methods for medical research. Ann Acad Med 30:444–446.

Thomsen CE 1992 Hierarchical cluster analysis and pattern recognition applied to the electroencephalogram – development of ADAM (Advanced Depth of Anaesthesia Monitor). PhD thesis, University of Aalborg.

Thomsen CE, Prior PF 1996 Quantitative EEG in assessment of anaesthetic depth: comparative study of methodology. Br J Anaesth 77:172–178.

Thomsen CE, Christensen KN, Rosenfalck A 1989 Computerized monitoring of depth of anaesthesia with isoflurane. Br J Anaesth 63:36–43.

Thomsen CE, Rosenfalck A, Nørregaard Christensen K 1991 Assessment of anaesthetic depth by clustering analysis and autoregressive modelling of electroencephalograms. Comput Meth Prog Biomed 34:125–138.

Thomsen CE, Gade J, Nieminen K, et al 1997 Collecting EEG signals in the IMPROVE data library. IEEE EMB Mag 16:33–40.

Thomsen CE, Cluitmans L, Lipping T 2000 Exploring the IBIS data library contents – tools for data visualisation, (pre-) processing and screening. Comp Meth Prog Biomed 63:187–201.

Vakkuri A, Yli-Hankala A, Talja P, et al 2004 Time–frequency balanced spectral entropy as a measure of anesthetic drug effect in central nervous system during sevoflurane, propofol an thiopental anesthesia. Acta Anesthesiol Scand 48:145–153.

van de Velde M 2000 Signal validation in electroencephalography research. PhD Thesis, Technical University Eindhoven.

van de Velde M, Cluitmans PJM, Declerck AC 1993 Optimization of stimulation frequency in middle latency auditory evoked potentials in humans. Technol Health Care 1:340–341.

van de Velde M, Ghosh IR, Cluitmans PJM 1999 Context related artefact detection in prolonged EEG recordings. Comput Methods Prog Biomed 60:183–196.

van den Berg-Lenssen MMC, Brunia CHM, Blom JA 1989 Correction of ocular artefacts in EEGs using an autoregressive model to describe the EEG; a pilot study. Electroencephalogr Clin Neurophysiol 73:72–83.

Van der Weide II, Pronk RAF 1979 Interference suppression for EEG recording during open heart surgery. Electroencephalogr Clin Neurophysiol 46:609–612.

van Gils MJ 1995 Peak identification in auditory evoked potentials using artificial neural networks. PhD Thesis, Eindhoven University of Technology.

van Gils M, Korhonen I, Yli-Hankala A 2002 Methods for assessing adequacy of anesthesia. Crit Rev Biomed Eng 30:99–130.

Van Valkenburg ME 1995 *Analog filter design*. Oxford University Press, Cary, NC.

Värri A, Kemp B, Penzel T, et al 2001 Standards for biomedical signal databases. IEEE Eng Med Biol Mag May/June:33–37.

Verma A, Bedlack RS, Radtke RA, et al 1999 Succinylcholine induced hyperkalemia and cardiac arrest: death related to an EEG study. J Clin Neurophysiol 16:46–50.

Veselis RA, Reinsel R, Sommer S, et al 1991 Use of neural network analysis to classify electroencephalographic patterns against depth of midazolam sedation in intensive care unit patients. J Clin Monitoring 7:259–267.

Viertiö-Oja H, Meriläinen P, Särkelä M, et al 2001 Spectral entropy, approximate entropy, complexity, fractal spectrum, and bispectrum of EEG during anesthesia. In: *Proceedings of the 5th International Conference on Memory Awareness, and Consciousness*. Memorial Sloan-Kettering Cancer Center, New York, pp. 49–50.

Viertiö-Oja H, Maja V, Särkelä M, et al 2004 Description of the Entropy algorithm as applied in the Datex-Ohmeda S/5 Entropy module. Acta Anaesthesiol Scand 48:154–161.

Vigário RN 1997 Extraction of ocular artefacts from EEG using independent component analysis. Electroencephalogr Clin Neurophysiol 103:395–404.

Vigon L, Saatchi MR, Mayhew JEW, et al 2000 Quantitative evaluation of techniques for ocular artefact filtering of EEG waveforms. IEE Proc Sci Meas Technol 147:219–228.

Wasserman GS 1995 An adaptation of the EWMA control chart for short run SPC. Int J Prod Res 33:2821–2833.

Wastell DG 1981 When Wiener filtering is less than optimal. An illustrative application to the brain stem evoked potential. Electroencephalogr Clin Neurophysiol 51:678–682.

Webber WR, Litt B, Wilson K, et al 1994 Practical detection of epileptiform discharges (EDs) in the EEG using an artificial neural network: a comparison of raw and parameterized EEG data. Electroencephalogr Clin Neurophysiol 91:194–204.

Webber WR, Lesser RP, Richardson RT, et al 1996 An approach to seizure detection using an artificial neural network (ANN). Electroencephalogr Clin Neurophysiol 98:250–272.

Webster JG (ed) 1992 *Medical Instrumentation*, 2nd edition. Houghton-Mufflin, Boston, MA.

Weiss S, Kulikowski C 1991 *Computer Systems That Learn: Classification and Prediction Methods from Statistics, Neural Nets, Machine Learning, and Expert Systems*. Morgan Kauffman Publishers, San Mateo, CA.

Westgard JO, Groth T, Aronsson T, et al 1977 Combined Shewhart–Cusum control chart for improved quality control in clinical chemistry. Clin Chem 23:1881–1887.

Wheeler DJ, Chambers DS 1992 *Understanding Statistical Process Control*, 2nd edition. SPC Press, Knoxville, TN.

Widman G, Schreiber T Rehberg B, et al 2000 Quantification of depth of anesthesia by nonlinear time series analysis of brain electrical activity. Phys Rev E 62:4898–4903.

Widrow B, Stearns SD 1985 *Adaptive Signal Processing*. Prentice-Hall, Englewood Cliffs, NJ.

Widrow B, Glover J, MacCool J, et al 1975 Adaptive noise cancelling: principles and applications. Proc IEEE 63:1692–1716.

Woodall WH, Adams BM 1998 Statistical process control. In: *Handbook of Statistical Methods for Engineers and Scientists* (ed HM Wadsworth), 2nd edition. McGraw Hill, New York, pp. 7.1–7.28.

Yamamoto Y, Hughson RL 1993 Extracting fractal components from time series. Physica D 68D:250–264.

Yli-Hankala A, Vakkuri A, Annila P, et al 1999 EEG Bispectral Index monitoring in sevoflurane or propofol anaesthesia: analysis of direct costs and immediate recovery. Acta Anaesthesiol Scand 43:545–549.

Yu X-H, He Z-Y, Zhang Y-S 1994 Time-varying adaptive filters for evoked potential estimation. IEEE Trans Biomed Eng 41:1062–1071.

Zhang X-S, Roy RJ 1999 Predicting movement during anaesthesia by complexity analysis of electroencephalograms. Med Biol Eng Comp 37:327–334.

Zhang X-S, Roy RJ 2001 Derived fuzzy knowledge model for estimating the depth of anesthesia. IEEE Trans Biomed Eng 48:312–322.

Zhang X-S, Huang JW, Roy RJ 2002 Modeling for neuromonitoring depth of anesthesia. Crit Rev Biomed Eng 30:131–173.

Chapter 9

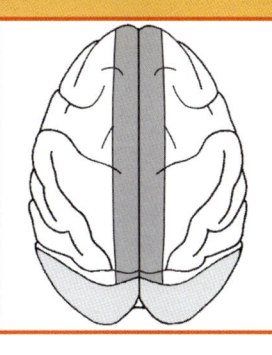

Legal Implications of Neurophysiological Monitoring

In addition to their clinical and scientific responsibilities, clinical neurophysiologists undertaking diagnostic and/or monitoring work in the intensive care unit (ICU) and the operating theatre need to be concerned about a range of medico-legal issues in their professional lives. These may be divided into:
1. matters affecting safety of patients and staff;
2. responsibilities for training and supervising staff to ensure a high quality performance of techniques;
3. action to ensure proper reporting of results and that storage of material relating to investigations complies with legislation about confidentiality and data protection rights of the individual;
4. specific medico-legal applications of neurophysiological techniques; and
5. broader areas where neurophysiology may assist indirectly in matters with possible medico-legal implications.

The last two of these are discussed in Chapter 1, which should be read in conjunction with this section.

In discussing legal questions it is important to recognize that different laws apply in different countries, and that the procedure for dealing with a mishap or settling a dispute depends on the legal system adopted by the country in which the problem arises. The situation is becoming more complicated in Europe, where directives from the European Union are mandatory in member states, and are usually incorporated into national law, but applied by very different legal systems. The following paragraphs therefore merely cover certain general principles, rather than giving a detailed account of the law in any particular country, which might be misleading to readers living elsewhere.

Before proceeding further it would be wise to clarify the legal status of laws, regulations and guidelines (Petch 2002).

Laws are made by the national executive: in an autocracy by royal proclamation or dictatorial edict, in a (democratic) republic by resolution passed in a representative legislative assembly, and in a totalitarian state by edict which purports to express the will of the people. Laws are binding, in the sense that the law-maker has the power to enforce compliance, and penalties for non-compliance. In the modern world, as a result of wars or revolutions, some countries find themselves, for a time at least, in a state of anarchy or chaos, in which activities that should be controlled by law are not controlled at all. Medicine must be practised, and neurophysiological monitoring may be required, under autocratic, democratic, totalitarian or anarchic regimens, where the legal implications of neurophysiological monitoring will be very different

Regulations are of two types. The first type ensure the detailed application of the law, and are either included in the law, or they are made by government departments or government-sponsored organizations under powers granted by law. Some regulations of this type are made by international bodies such as the International Organization for Standardization (ISO), and adopted as national standards in many countries. The second type are made by non-government organizations to regulate the activities of their members, clients or employees. While regulations of this second type lack the force of law, there are still penalties for non-compliance.

Clearly everyone should be aware of, and comply with, the laws and regulations in force in the country in which they live and work, and any regulations made by their employer.

Guidelines, in contrast to laws and regulations, are merely recommendations about good practice. Clinical guidelines began as an attempt to contain costs, but are now usually intended to ensure that patients receive the best, or at least adequate, treatment (Andrews and Redmond 2004). They are made by non-government organizations, such as the International Federation of Clinical Neurophysiology or the American Electroencephalographic Society, and have no legal status. In most countries the fact that guidelines have been followed provides no immunity from legal action, or from an unfavourable verdict, in the event of a mishap (Damen et al 2003), although this may change as bodies such as the National Institute for Clinical Excellence (NICE) in the UK become more influential (Samanta et al 2003). While supposedly based upon evidence, guidelines are never perfect, usually being drawn up by committees, and therefore represent the view that is the least unacceptable to the majority, having optimal effect when applied to the average patient. Rigid adherence to guidelines when the circumstances are exceptional could have disastrous results for the patient, and lead to the doctor being severely criticized.

SAFETY OF PATIENTS AND STAFF

The work of a department of clinical neurophysiology brings staff into contact with many aspects of legislation about safe practice with regard to electro-medical equipment, drugs and dangerous substances, and control of infection, as well as the more general features of the safe and healthy environment for patients and staff, such as reduction of hazards from fire, slippery floors, excessive noise and extremes of temperature and humidity.

A hospital has a duty to ensure the safety of both patients and staff, and this is a legal requirement in many countries, although in some cases hospitals are exempt from full compliance with the measures applicable to other places of work such as factories. In many countries, therefore, a hospital would be held negligent in law if it did not provide satisfactory and safe working conditions for its staff. The clinical neurophysiologist in charge of a department would be expected to maintain a safe working environment for all the staff, and to be concerned about their health. However, managers are increasingly taking over responsibility for running departments, so the neurophysiologist's direct responsibility for staff safety is diminishing, but there is still a moral obligation, if not a legal duty, to draw deficiencies to the attention of management. The prevailing health and safety at work regulations may require that a specific individual is designated Safety Officer. The Safety Officer will be required to keep regular dated records pertaining to checking of electro-medical equipment, drugs and lotions, fire fighting and escape equipment, infection control, first aid, personal security and risk avoidance. The Safety Officer will generally be responsible for informing staff of national and hospital safety rules and procedures and for drawing up a local set of rules and ensuring that all staff receive, understand and comply with them.

The welfare of patients who use the department is the responsibility of both the clinical neurophysiologist and the hospital managers, who would be expected to maintain the safety of the department and be satisfied that all the furniture, fittings, lighting, heating, safety and electromedical equipment are in good order. Clear sign-posting of exits, fire apparatus and departmental facilities should be provided in accordance with local and national regulations. The situation with respect to neurophysiological monitoring is complicated by the fact that monitoring work is carried out in operating departments or ICUs, which have their own management and often their own safety regulations. Neurophysiology staff working away from their own department would be well advised to identify themselves to the person in charge on arrival in the operating department or ICU, and to make sure that they are aware of any safety regulations specific to the department in which they are working for the time being. Staff working in other departments should be identifiable both by name and as belonging to the Clinical Neurophysiology Department.

A risk assessment should be carried out, to identify, before they cause actual harm, all possible risks within the department, and in other departments where monitoring may be required, whether electrical, chemical, microbiological, or related to fire hazards, possibility of accidental injury, etc. This may be backed up by procedures for reporting all actually or potentially adverse incidents, from a defective mains lead to prescribing errors or damaged floor coverings. Attention to risk assessment does not only protect those responsible from possible litigation but, more practically, avoids harm to staff and patients and motivates managers to fund necessary improvements in facilities, if for no better reason than avoidance of personal blame.

Electromedical equipment

Specific regulations exist about the safety of electromedical equipment and are outlined in *Practical aspects of intraoperative and ICU neurophysiological recording* (p. 55). The physician in charge of clinical neurophysiology is expected, when buying equipment for the hospital, to recommend those machines which conform to the standards laid down by the government or appropriate national agencies, for example the international standard for electrical safety, IEC 60601-1. In the USA the corresponding regulation is UL2601-1, which specifies lower earth leakage currents than IEC 60601-1, and superseded the previous standard (UL544) on 1 January 2005 (Marcus and Biersach 2003). In addition, manufacturers are increasingly required to certify that equipment is fit for its purpose; in countries belonging to the European Union, for example, the safety of all new medical equipment is demonstrated by the CE (Conformity European) marking, and the safety of equipment not so marked should be carefully assessed prior to clinical use (Spencer et al 2003). If substandard machinery were to be installed, and an accident was to occur, then although the managers would be held to be negligent, the clinical neurophysiologist nevertheless would have to carry some share of the responsibility. Proper documentation of safety checks, by appropriately trained personnel, on installation and at regular intervals thereafter are essential. In the event of inadequacies of equipment which have a bearing on the provision of the clinical service, the neuro-physiologist should ensure that the hospital authorities are informed in writing of his or her professional opinion that there is a need for improved maintenance facilities or replacement of equipment. Risk of an erroneous result or equipment failure during a procedure, for example intraoperative monitoring, could well have medico-legal implications. Any report based on results where equipment failure or artefacts make a recording inadequate should clearly state this fact.

A particular difficulty arises where equipment is needed for a purpose, a research project for example, for which no equipment has been certified as safe. Specially designed equipment may be produced locally, or a manufacturer's newly developed device may be under test. Alternatively, a machine certified as safe for one purpose may be modified or used in circumstances which invalidate its safety certificate (e.g. CE mark). This is analogous to using a drug to treat a condition for which it was not licensed. In all these cases the possible risks must be identified, minimized, and shown to be outweighed by the benefits of using the equipment (Spencer et al 2003).

Infection control
General guidelines about control of infection are given in *Practical aspects of intraoperative and ICU neurophysiological recording* (p. 55). It is incumbent on the clinical neurophysiologist to keep abreast of current patterns of infection and of regulations designed to avoid risk to patients or staff. This may mean that changes in practice are required, for example in cleaning and sterilization methods for electrodes and ancillary apparatus and in changing to exclusive use of disposable electrodes, where there is any form of invasive technique such as needle EMG, intracranial recording or epidural spinal monitoring. Regular review should be made of all procedures from an infection control point of view and changes documented in departmental safety manuals issued to all staff. The emphasis should always be on the risk of the procedure and not on trying to predict the risk that any individual patient might be carrying any specific infection. Frequent hand-washing and use of disinfectants is now regarded as vital in preventing spread of infection, and is mandatory in some hospitals before performing any procedure on a patient, particularly in the ICU.

Regular and specific advice should be given to staff about reduction of personal risks by proper use of techniques and on the advisability of immunization against transmissible infections, such as hepatitis B, which might be considered an occupational hazard.

Drugs
Prior to monitoring during surgical operations, patients may undergo baseline investigations in the clinical neurophysiology department. A secure supply of drugs for resuscitation and other medical emergencies is always required in the department in addition to any drugs used in conjunction with clinical neurophysiological examinations. The physician-in-charge will be expected to satisfy him or herself that the department complies with regulations concerning drugs liable to cause dependence or misuse (Controlled Drugs) for which there are legal requirements regarding prescription and safe keeping. National and local regulations should be checked regularly and the clinical neurophysiologist should ensure that all staff are familiar with them. In the UK, it is required that certain drugs be kept in a locked cabinet with a double-locked inner compartment, whilst others can be kept in the outer locked cupboard. The drug cupboard must be firmly fixed to the wall so that it cannot be removed, and when the main lock is undone, a warning light must show above the cabinet. The keys for the drug cupboard may only be held by specifically designated and qualified people, for example the doctor, nurse or chief technologist. No drug should be prescribed or administered for a test without checking with the patient about previous allergies or adverse reactions to drugs. Any such drugs must be recorded in the records of the patient. There must be an established procedure whereby any drugs given to patients are recorded correctly in the departmental drug book, together with the signatures of two persons, who both check the drug and witness its administration.

During intensive care monitoring, and occasionally during surgery, there are sometimes requirements to test the effect of a medication (e.g. to see whether an intravenous drug will control seizure discharges in a patient with subclinical status epilepticus). Such decisions should be taken jointly with the medical staff in clinical charge of the patient; authorization, delivery and recording of the details of appropriate medication is then the responsibility of the ICU staff.

Identification of patients and consent to procedures
Clearly, it is essential to ensure that all investigations and monitoring procedures are carried out on the correct patient. Whilst inpatients are generally issued with identification bracelets, some outpatients may be confused as to person and place and careful checks are necessary before undertaking a test (e.g. a presurgical evaluation). It should hardly be necessary to point out that a patient who resists examination, appears to be mentally disturbed, or becomes violent should be handled with caution and wherever possible in the presence of more than one member of staff. Patients known to have expressed suicidal thoughts should not be left unaccompanied. Chaperoning is prudent when investigations (e.g. those of the pelvic floor or sphincters) concern intimate examinations by a person of either sex. A written record of any untoward events or incidents should be kept and the hospital administration informed. This applies as much to incidents or accidents affecting staff as to patients.

Statute law and common practice in respect of obtaining consent will vary greatly between countries and is subject to continual review. Only general guidelines can be given here. Potentially hazardous procedures such as surgical operations may require a two-stage process, of first providing information, then obtaining consent after the patient has had time to reflect and ask questions. A legally competent adult may be presumed to be consenting if he or she submits to a simple procedure such as application of EEG electrodes after adequate explanation. More invasive techniques and those that carry a risk of inducing seizures or involve administration of drugs may require formal information and consent, and this also applies to some types of innovative procedures carried out as part of the development of new techniques. When the patient, after being informed, needs to weigh up the possible risks and benefits of what is proposed, it is not good practice to confront him or her with this only a few minutes before the investigation. Consent for neurophysiological monitoring during surgical operations will usually be included in the consent obtained for the proposed operation, but it is advisable to confirm with the surgeon that this is the case, particularly if the monitoring involves invasive techniques such as intracranial or intraspinal electrodes, or transcranial stimulation, for which consent should be obtained specifically because the patient may give consent for operation, but in spite of full explanation, withhold consent for monitoring. Patients requiring monitoring in ICUs are sometimes conscious, and may then give consent if mentally competent, and perhaps more importantly, they may refuse to give consent. Usually, however, intensive care patients are unconscious, or at least not mentally competent, when the need for monitoring becomes apparent. In this situation the responsible physician must certify that a proposed procedure is in the patient's best interest, and be prepared to defend that decision. In some countries, the UK in particular, no-one can give consent on behalf of an adult, but it is prudent to obtain the informal agreement of the next of kin of an unconscious patient before undertaking any procedure. Some hospitals have a clearly defined protocol for this eventuality. Special considerations apply to children. For the very young parental consent will suffice, but the age at which older children are regarded as capable of consenting varies in different countries. Obviously, the use of neurophysiological techniques during research studies will require the usual ethical evaluation and formal consent by the subjects involved.

An important aspect of obtaining consent is informing the patient about any risks associated with the proposed procedure. Clearly the patient must be told about significant risks, but it is questionable whether knowing about a number of low risks (e.g. ~$1:10^6$) will help the patient to make a decision. The level of risk that must be communicated is different in different countries.

It may occasionally be necessary to collect normative data for a newly developed neurophysiological monitoring technique. Any research on human subjects, whether healthy subjects or patients, should be approved by the appropriate research ethics committee (Working Party Report 1986, 1990), who will usually also consider the information to be given to subjects or patients. It is wise to be prepared for unexpected abnormal findings, by including a section in the consent form so that all 'healthy' volunteers can indicate their preference as to action to be taken if investigations reveal abnormality. Options might include no action unless the abnormality is serious, discussion with the clinical neurophysiologist, notification of their personal physician, or referral to an appropriate specialist.

Video monitoring may be required (e.g. in patients with severe seizure disorders) to investigate precise temporal relationships between neurophysiological and clinical events by means of split-screen technology. It may also be used during other types of neurophysiological recording for training purposes, including in simulators. If it is to be used for presentation at meetings or as teaching material it should be made with due sensitivity to the patient's privacy and considerations of confidentiality. Formal consent procedures are mandatory when videos which were made for investigation of specific clinical problems for the benefit of the patient are then used for other purposes. Hospital policy may dictate written informed consent for all video recording. UK guidelines have been given by the General Medical Council (1997).

Inherent risk of procedures

Certain investigations carry an implicit, albeit small, risk. For example any provocative procedure during an EEG, even the widely used routine hyperventilation or stroboscopic stimulation, may provoke an epileptic seizure in a susceptible patient. It would be irresponsible to continue with the provocative technique once warning signs were evident in the EEG or clinically, unless the investigation was specifically aimed at recording a seizure, for example during consideration of surgical treatment for intractable epilepsy when precautions to avoid damage or complications must be available. In any patient, risk of physical injury during a seizure, risk of impairment of the airway or oxygenation, or inhalation of vomit, must be the concern of the physician and technologists in charge. Appropriate resuscitation equipment and training must be available. Similarly, the use of needle electrodes should be avoided in a patient with a bleeding diathesis. Most pitfalls can be avoided by a combination of well-understood codes of practice for each procedure and careful discussion and planning of the investigation with the referring physician or surgeon.

During surgical monitoring electrical equipment is likely to be used in the presence of anaesthetic gases, some of which are inflammable. Although the risk is low with modern equipment, and any released gases are scavenged, some manufacturers specifically state that certain machines should not be used in the presence of inflammable vapours. Fire risks associated with anaesthetic gases are covered in Annex E of a standard applicable to the USA produced by the National Fire Protection Association (NFPA 99 2002).

Two neurophysiological monitoring procedures give rise to particular risks. The first is the introduction of intracranial or intraspinal electrodes, where the electrode may damage the brain or spinal cord, and may be the source of infection. The second is the risk of epilepsy and cortical damage arising from transcranial electrical stimulation (MacDonald 2002). Both these risks are low, but different legal systems do not agree about the need to inform the patient about a low risk as part of the process of obtaining informed consent.

RESPONSIBILITIES IN TRAINING AND SUPERVISING STAFF

The neurophysiologist in charge of a neurophysiology department will be responsible for the standard of medical and technological practice within the department. In many countries, it is not yet a legal requirement for either technologists or medical personnel to attend postgraduate education in neurophysiology. However, regular (e.g. weekly) seminars within the department as part of post-basic training and for audit of the quality of technological and medical work is good practice. Many departments also undertake or participate in training programmes for junior medical staff and student technologists. Although not a legal requirement in all countries, every clinical neurophysiology technologist is expected to take the appropriate qualifying examinations, where these exist. No technologist should work unsupervised before passing the relevant examinations and having demonstrated adequate understanding and practical competence to undertake the various types of neurophysiological recordings. By the same token, no trainee doctor should undertake investigations (e.g. EMG techniques) unsupervised, until such time as observation by an experienced clinical neurophysiologist has confirmed his or her competence.

RESPONSIBILITIES IN RESPECT OF REPORTS ON INVESTIGATIONS

The department will be expected to have an efficient system to ensure proper reporting of results and that storage of material relating to investigations complies with legislation about confidentiality and the data protection rights of the individual. There must be a day-book in which details of every investigation is recorded, whether carried out in the department or in the ICU or operating suite.

In the case of neurophysiological monitoring procedures, reporting is often in the form of verbal discussion between the clinical neurophysiologist and the clinician, backed up by a contemporaneous written documentation in the patient's notes. A full report may be generated by use of facilities within the monitoring equipment which typically includes a summary display of key waveforms and the clinical neurophysiologist's report. This should be available in the patient's medical notes shortly/promptly after completion of the procedure and copies filed in departmental records in the usual way.

The medical officer in charge of the department is entirely responsible for reports sent to other clinicians. Although it is usual practice for the monitoring technologist in the department to write a factual description of the recording procedure undertaken, nevertheless, the departmental head remains responsible for the entire report. Good medical practice dictates that reports are concise, clear and attempt to answer the questions asked by referring physicians. Information stored about patients must be treated as strictly confidential and held in compliance with legislation about the data protection rights of the individual. Since in many countries this includes the right of the patient to view information held, care must be taken to avoid irrelevant personal remarks in documentation. Difficulties may arise when unexpected information is revealed by a patient, for example about driving when disallowed because of epilepsy. Discussion between the clinical neurophysiologist and the referring physician may be the best approach.

REFERENCES

Andrews EJ, Redmond HP 2004 A review of clinical guidelines. Br J Surg 91:956–964.

Damen J, van Diejen D, Bakker J, et al 2003 Legal implications of clinical practice guidelines. Intensive Care Med 29:3–7.

General Medical Council 1997 *Making and Using Visual and Audio Recordings of Patients.* Guidelines available from: Standards Section General Medical Council, 178–202 Great Portland Street, London WIN 6JE.

MacDonald DB 2002 Safety of intraoperative transcranial electrical stimulation motor evoked potential monitoring. J Clin Neurophysiol 19:416–429.

Marcus ML, Biersach BR 2003 Regulatory requirements for medical equipment. IEEE Instrum Meas Mag 6(4):23–29.

NFPA 99 2002 *Standard for Healthcare Facilities.* National Fire Protection Association, Quincy, MA.

Petch MC 2002 Heart disease, guidelines, regulations and the law. Heart 87:472–479.

Samanta A, Samanta J, Gunn M 2003 Legal considerations of clinical guidelines: will NICE make a difference? J R Soc Med 96:133–138.

Spencer SA, Nicklin SE, Wickramasinghe YA, et al 2003 An essential 'health check' for all medical devices. Clin Med 3:543–545.

Working Party Report 1986 *Research on Healthy Volunteers.* Royal College of Physicians, London.

Working Party Report 1990 *Research Involving Patients.* Royal College of Physicians, London.

Index

Page numbers in **bold** are main entries. Page numbers followed by F denote figures. Page numbers followed by T denote tables.

A

Accidents, and
    patients or staff, 371
    safety procedures, 55, 315
Acetylcholine, and
    botulinum poisoning, 242
    end-plate noise, 4.4T
    neuromuscular junction, 146
Acoustic neuroma, 260–268, 7.1T, 7.8F
Action potentials, **127**, 4.20F
    compound muscle (CMAP), 52, 53, 2.6T, 2.8T, 175, **231**, 266
    eighth cranial nerve (8NAP), 133, 4.18F, 261, **263**
    muscle relaxants and, 175
    sensory (SAP), 51, 2.7T, 4.17F, 138, 140, 175, 256, 270, 278
    temperature effects on SAPs, 147, 4.38F
Acquired immune deficiency syndrome, 58
Acute inflammatory demyelinating polyneuropathy (AIDP), 234–237
Acute motor axonal neuropathy (AMAN), 234, 237
Acute motor sensory axonal neuropathy (AMSAN), 234, 237
Acute onset neuropathies, 234
Adaptive filters, 95, 97, **323**, 8.11F, 3.25F, 327, 8.2T
Adaptive noise cancelling, 323
Advanced Depth of Anaesthesia Monitor (ADAM), **83**, 3.1T, 3.13F, 3.24F, 5.13F, 7.30F, 336, 354
AEP Monitor/2, 98
Age, and
    anaesthetic and sedative drug changes, 83, 85, 163, **170**
    consent for investigations, 373
    EEG, 3.12dF, 109
    EPs 134, **144**, 256, 258
AIDS, Acquired immune deficiency syndrome, 58
Alcohol, and
    acute porphyria, 242
    channelopathies, 243, 6.6T
    risk factors in cardiac surgery, 281
    severity of EEG abnormality, 194
    skin preparation, 24, 2.3T
Aliasing, 38, 2.23F, 42, 47, 48, 312, 8.3F, 318, 338
Alignment, of pens or jets, 41
Alpha activity, rhythm, 3.1T, 3.1F, 3.2F, **110**, 4.1F, 4.2F, 114, 191, 6.6F
Alpha coma, 118, 193, 196, **198**

American Academy of Neurology reports
    brain death in adults, 229
    intraoperative neurophysiology, 4, 6
    persistent vegetative state, 228
American Academy of Pediatrics guidelines
    brain death in children, 229
American Association of Electrodiagnostic Medicine (AAEM, formerly AAEE), 145
American Electroencephalographic Society recommendations on
    digital EEG recording, 47
    electrode placement nomenclature, 29
    evoked potentials, 56
    infectious diseases, 58
    intraoperative EP monitoring, 7, 65
    safety, 56, 58, 65, 317, 371
    suspected cerebral death, 124
American Society for Testing and Materials (ASTM), 314
Amplifiers, **32**, 2.14F, 2.15F, **311**
    blocking and interference, 19, 34, 60, 62, 66
    connecting electrodes to amplifiers and the recording convention, 29
    DC, 19
    differential (balanced), 29
    discrimination (common mode rejection), 33
    early development, 32
    EEG, 39, 311
    EMG, 54, 311
    EPs, 48, 311
    frequency characteristics, 35, 2.19F
    isolation, 62, 8.1T
    noise, 34, 2.17F
    safety, 62, 8.1T
Amplitude measures, 3.16F, **332**, 8.4T
Anaesthesia, **161**
    dose–response curves, 3.10, 159, 5.7F, 5.10F
    effect on CBF, 163
    nomenclature, 162
    see also Anaesthesia, effects on EEG, **161**
    see also Anaesthesia, effects on EPs, **172**
    see also Awareness during anaesthesia, **176**
    see also Combined EEG and EP measures, **176**
    see also Depth of anaesthesia, assessment methods, **173**

Anaesthesia, effects on EEG, **Chapter 5, 161**
    antagonists for central and respiratory depression, 169
    'balanced' anaesthesia, 163
    depth of anaesthesia, EEG assessment, 169
        influence of arterial hypotension, 170
        influence of hypothermia and hyperthermia, 171
        influence of maturation and ageing, 170
    epileptogenicity and EEG activation, 163
    intravenous anaesthetics, 163
        barbiturates, 163
        thiopental sodium, 164
        etomidate, 164
        ketamine (phencyclidine), 165
        propofol, 165
    inhalational anaesthetics, 165
        nitrous oxide, 166
    local anaesthetics, 169
    muscle relaxants (neuromuscular blocking drugs), 169
    sedative and analgesic perioperative drugs, 167
    'neuroleptic anaesthesia', 169
Anaesthesia, effects on EPs, **Chapter 5, 172**
    depth of anaesthesia, EP assessment, 175
    influence of temperature, 176
    influence of arterial hypotension, 176
    drug effects
        barbiturates, 172
        diazepam, 175
        etomidate, 172, 175
        fentanyl, 175
        halogenated anaesthetics, 172
        halothane, 5.18F
        ketamine, 172, 175
        lignocaine, 175
        morphine, 175
        muscle relaxants, 175
        nitrous oxide, 175
        propofol, 172, 175
Anaesthesia, methods to assess depth of, see Depth of anaesthesia, assessment methods, **73**
Analogue filters, **36**, 2.16F, 289–290, **312**, 318, 330, 337–338
Analogue to digital (AD) conversion, 38, 39, 312, 7.1T
Aneurysm

Index

aortic, surgical monitoring, 274, **277**
intracranial, surgical monitoring, 225, **253**, 7.1T
   artefact from intracranial clips, 52
   postoperative EEG abnormalities, 118
   EP monitoring during surgery, 7.6F
   vasospasm and prognostic qEEG signs, 225
Anoxia, cerebral, see Cerebral hypoxia–ischaemia
Anterior horn cell disease, 146
Anteroinferior cerebellar artery (AICA) occlusion and BAEPs, 208
Anti-aliasing filters, 42, 47, 312
Arm length, effect on SEPs, 136, 137, **144**
Arousal, paradoxical, 117, 118, 4.10F, 155, 156, 177, **191**, 6.2F, 195, 253
Arousal phenomena, 114, 118, 155, 167, 177, 5.23F, 5.25F, 181, 6.1T, 6.1F, **191**, 192, 6.2F, 6.3F, 6.5F, 331, 354
Arc distortion, 41, 2.26F
Archiving 42, 314
Artefact detection and rejection methods, 65, 75, 3.5F, **90**, 3.20F, 159, 165, **317**, **327**
   dangers of misinterpretation due to unrecognised artefacts in EEG monitoring, 62
   defibrillators, fibrillators, pacemakers and balloon pumps, 62, 2.34F
   interference due to surgical diathermy, 60, 2.32F, 2.33F
   summary table, **8.2T**
Artefacts, biological and physical, **90**, **327**, 8.2T
   cardiac massage, 2.34F
   diathermy, 60, 2.32F, 2.33F
   electrocardiogram (ECG), 4.16F
   electrodes, electrode, lead or cable movement, 19–24, 31, 34, 39, 312, 8.2T
   eye movement and blinking, 2.4F, 2.30F, 2.3T, 4.1F
   digital EEG, 46, 47
   EP recording, 48, 49, 66, 95, 212
      stimulus induced, 48, 51, 68, 262, 266, 278
   ICU patients, 3.18F, 3.19F
   movement, 19
   muscle, 2.3T, 38, 56, 85, 5.12F, 5.25, 6.4F, 196
   pacemaker, 62, 2.34F, 8.2T
   patient disturbance and care, 59, 3.18F, 3.19F, 163
   patient movement, 19, 3.18F, 3.19F, 155
   respiratory, 3.18F, 4.7F, 4.14F
   surgical appliances and instruments, 57, 62, 2.34F
   value in patient state assessment, 1, 2
Artificial intelligence (AI) methods, 350, **351**
   connectionist AI, 350, 351
   symbolic AI, 350, 351
Artificial neural networks (ANNs), see Neural networks
Assessment of results of monitoring, **7**
   clinical significance of changes, 8

comparing two methods of monitoring, 11
confidence intervals, 10
determining the optimal cut-off point, 11
effect of corrective action, 12
is the risk eliminated by monitoring?, 13
statistical significance of changes, 7
Asymptomatic carotid artheroscIerosis study group (ACAS), 286
Audit, 286, 365
Auditory evoked potentials (AEPs)
   brainstem auditory EP monitoring
   electrocochleography (ECochG)
   long latency auditory EPs (LLAEPs)
   middle latency auditory EPs (MLAEPs)
   see also Brainstem auditory EPs (BAEPs)
Auditory evoked potentials, normal findings, 132
   latency and amplitude values of normal adult BAEPs, 133
   middle latency auditory evoked potentials, 133
   waveform and origins of normal adult BAEPs, 132
Auditory evoked potential index (AAI), 5.20F, 5.21F, 5.22F
Auditory monitoring, practical aspects, 262
Auditory nerve, **126**, 205, **261**, 7.8F, 264, 266
Auditory stimulation, 50
   click, 50
   tone, 51
Autocorrelation function (Acf), 334, 8.15F, 339, 344
Autoregressive (AR) models, 83, 98, 327, 8.2T, 330, 335, 336
Averaging, **48**, 55, 73, 79, **125**, 4.31F, 142, 177, 289, **324**, 344
   (±) average, 66
   moving block averaging, 67, 2.35F
   noise estimation in average, 66
   pseudorecursive averaging, 67
   signal-to-noise ratio (SNR), 326
   viewed as low-pass filtering, 324–325, 8.12F
Axons
   damage in vegetative states, 226
   neuropathies, 232–237, 6.34F, 642, 6.6T
   propagation, 126, 127, 4.19F, 4.20F
Awareness during anaesthesia, **176**
   medico-legal aspects of awareness during, 180
   neurophysiological features useful in prediction of possible awareness during anaesthesia, 177

B

Bandwidth of neurophysiological signals, **35**, 2.4T
Barbiturates, and
   acute intermittent porphyria, 242
   EEG, 3.2F, 3.7F, 115, 123, 156, 5.3F, 163, 224
   EPs, 158, 172, 175, 194, 202
Basic rest and activity cycle (BRAC), **113**, 3.2T, **192**
Basis functions, 340, 341
Berger, Hans, 73, 3.1F, 3.2F

Beta activity, rhythm, 110
   see also EEG waveforms and interpretation
Biomedical engineering, definition of, 309
Biosignal interpretation (BSI), 309
   use of tools, **358**
      decision assistance, 358
      development and uptake of methods, 359
      display of results, 360
Bipolar recording, 24, 26, 31, 32, 2.13F, 4.3, 113, 139, 5.19F, 5.18F, 260, 262
Bispectral (bispectrum) analysis, **84**
Bispectral Index (BIS) Monitor, **85**, 3.14bF, 101, 3.17F, 3.25F, 3.26F, 160, 5.6F, 5.8F, 5.20F, 5.21F, 5.22F, 180, **345**
Bispectrum, **84**, 3.14aF, 332, **344**
Bispectrum (bispectral) analysis, **84**
Blackman–Tukey method, 339
Blood pressure, arterial
   EEG and, 113, 170, 5.15F
   EPs and, 176, 7.18F
Botulism, 242
Brachial plexus
   injuries, 278
   SEPs, **136**, 142, 6.7F
Brain protection, 80, 158, 160, 164, **255**, 284
Brainstem auditory evoked potentials (BAEPs) and
   anaesthetics, 172
   brainstem death, 209, 230
   clinical effectiveness, 3, 4
   comatose patients, 200, 204, 6.4T
   drugs effects, 3, 207
      anaesthetics, 172
      barbiturates, 158, 172
      lidocaine, 159
      phenytoin, 159
      sedatives in ICU, 157
   hypothermia, 205, 214, 229
   intraoperative use of BAEPs in
      brainstem function, 266
      posterior fossa surgery, 260
      preservation of hearing, 265
   normal values for latency and amplitude, **132**, 4.2T, 4.26F, 5.16F
   prognostic value in coma, 6.4T
      absent BAEPs, 205
      late normal BAEPs, 206
      normal BAEPs, 206
      present but abnormal BAEPs, 205
   recording technique, 50
   traumatic coma, 204, 206–209
   waveforms and origin, 132
Brainstem auditory evoked potential monitoring, 67, 3.1T, 3.22F, **94**, **208**, 6.16F, 7.1T, 260, 262, 265
   displaying BAEPs with SEPs and EEG, 3.23F, 8.33F
   ICU monitoring, 209
Brainstem death (brain death), **228**
   combined EEG and EP monitoring, 6.33F, 231
   EEG and brainstem death, 229, 230, 231, 6.34F

Index

EPs and brainstem death, 133, 207, 208, 209, 211, 212, 6.15F, 213, 215, 230
historical introduction, 228
preconditions and rationale for neurophysiological testing, 229, 230
present status of neurophysiological testing in brainstem death, 230
Brainstem lesions
alpha coma, EEG in, 118, 198
BAEP monitoring in, 209, 211–215, 6.16F
multimodality EPs in, 208
spindle coma, EEG in, 199
theta coma, EEG in, 198
Breach rhythm, 113
British Standards Institute (BSI) requirements for electromedical equipment, 317
Burst suppression (EEG pattern), 75, **123**, 4.15F
anaesthetics and sedatives producing, 1, 75, 156, 163, 5.2F, 5.4F, 5.8F, 5.9F, 165–166, 5.11F, 170, 172, 175
brain protection and, 160, 256, 287
burst suppression ratio (BSR,) 85, 3.17F, 345, 350, 351
$CMRO_2$ and, 224, 5.7F
comatose patients, 6.3F, 6.5F, 198
definition of, **123**, 4.15F
detection methods for monitoring, 77–80, 3.6F, 83, 85, 3.2T, 3.17F, 100, 3.3T, 6.1T, 331, 354, 8.6T
drug overdose and, 3.7F
encephalopathy, encephalitis, producing, 218
hypothermia, 166, 172
ischaemia producing, 254, 7.24F, 288, 7.3T
interburst intervals (IBI),123, 160, 170
mechanisms and modelling, 123, 166
variance of interburst intervals in, 114
Burst suppression, detection methods, 77–80, 3.6F, 83, 85, 3.2T, 3.17F, 100, 3.3T, 6.1T, 331, 354, 8.6T
'BUN' (Black Up Negative) convention 30, 41

C

Calibration, of
analogue EEG machines, 18, 37, **41**, 125
digital EEG machines, **46**, 47
EMG and nerve conduction, **54**
EP recorders, **49**
monitoring equipment, 309, 312, 314, 315, 8.1T
Capacitance (C), **20**, 37, 49, 66, 318
Cardiac arrest, **216**, and
EEG, 121, 4.14F, 4.15F, 123, 4.16F, 6.2F, 6.6F, 198, **216**, 6.5T, 6.22F, 6.23F, 6.24F
EPs, 201, 6.3T, 207, 216
neocortical damage, 225
outcome prediction, 3, 216, 6.5T, 6.24F
Cardiac surgery, **279**
abnormalities of preoperative EEG, 281
characteristics of intraoperative EEG changes, 282
historical background, 279
incidence of neurological and neuropsychiatric sequelae, 280
intraoperative factors affecting the EEG and incidence of brain damage, 283
arterial hypotension, 283
arterial carbon dioxide concentration, 284
hypothermia, with or without addition of high-dose barbiturates, 284
SEPs in cardiac surgery, 284
Cardiopulmonary bypass (CPB), and artefacts due to pulsation, 59
EEG, 3.9F, 170, 7.2T, 7.25F, 8.32F
SEPs, 176, 284
Carotid endarterectomy, **285**
assessment of significant changes in the EEG, 287
choice of methods for qEEG monitoring during carotid artery surgery, 298
historical background, 285
intraoperative factors affecting the EEG and incidence of stroke, 286
SEPs in carotid endarterectomy, 291
surgical approach and place of monitoring, 285
value of EEG monitoring, 286
Cavitron ultrasonic surgical aspirator (CUSA), 5.1T
Central conduction time (CCT), **141**
methods
lower limb stimulation, 142
upper limb stimulation, 141
normative data in adults, 4.3T
Central conduction time (CCT), and abnormalities, 202, 6.2T
arterial hypotension, 176
barbiturates, 158
carotid endarterectomy, 291
cerebral blood flow, 7.5F
cerebral ischaemia during intracranial surgery, 258
coma, 191, 200, 6.2T, 202–204, 6.3T, 208, 8.33F
hypothermia, 176
phenytoin, 159
Central motor conduction time
spinal surgery and, 270, 7.14F
Central core disease, 243, 6.6T
Cerebellopontine tumours, BAEPs in, 260, 265, 266
Cerebral Function Monitor (CFM), **76**, 3.1T, 3.6F, 79, 3.7F, 5.2F, 5.15F, 6.4F, 223, 6.26, 6.27F, 6.28F, 225, 6.30F, 6.31F, 280, 282, 7.25F
Cerebral Function Analysing Monitor (CFAM), 3.1T, 78, 79, **83**,3.12F, 3.24F, 5.1F, 5.11F, 5.12F, 5.15F, 5.25F, 6.3F, 6.26F, 282, 7.26F, 333, 354
Cerebral hypoxia and ischaemia, and BAEP changes, 214
brain protection regimens, 80, 158, 160, 164, **255**, 284
cardiac arrest, 216
cardiac surgery, 279, 284
carotid endarterectomy, 285, 7.3T, 7.29F
CCT during intracranial surgery, 258, 7.5F
cerebral arterial boundary zone changes, 2.31F, 4.7F, 279, 291
neuropathology, 279–280
comparisons between EEG methods for monitoring, 100, 225
controlled/induced hypotension, 170–171, 5.15F, 253
hypothermia and brain protection, 284
intracranial aneurysm surgery, 279
intracranial surgery for vascular malformations, 253
profound hypotension and localised cerebral arterial boundary zone ischaemia, 2.31F, 116, 4.7F, 6.2F
SEP monitoring for ischaemia, 258
subarachnoid haemorrhage and vasospasm, 80, 225, 253
thresholds for ischaemia, 7.4F
Cerebral ischaemia, see Cerebral hypoxia and ischaemia
Cerebral metabolic rate for oxygen ($CMRO_2$) and EEG, 5.7F
Cerebral oedema, 260
Cerebral perfusion pressure (CPP), 116, 176, 192, 209, 216, 221, 223, 225, 6.30F, 280, 284, 285
Cerebrospinal fluid (CSF) pressure and BAEPs, 213, 263
Cerebral vascular accident (CVA, stroke)
alpha coma and, 198
brainstem stroke and locked-in syndrome, 197
complicating anaesthesia, 6
complicating cardiac surgery, 280–281, 284, 7.3T
complicating carotid surgery, 11, 102, 285, **286**, 287, 288, 290, 291, 7.3T
complicating subarachnoid haemorrhage, 80, 81, 225
EEG asymmetries and asynchronies in, 2.27F, 113, 114, 340
EPs in, 6.3T, 260
periodic lateralised epileptiform discharges (PLEDs) in, 4.12F
Cervical cord
compression of, 7.19F
CCT to cortex, 144
lesions of, 6.8F, 243
SEPs, 136
spinal shock, 243
Channelopathies, **242**, 6.6T
Chaos theory applied to EEG, 123, 344
Classification methods, **346**, 8.5T
artificial intelligence (AI) methods, 351
artificial neural networks (ANNs), 352
expert systems, 351
statistical methods, 350
syntactical methods, 350
Clinical conditions affecting the central nervous system encountered in the ICU, **189**
brainstem death (brain death), **228**

379

Index

cardiac arrest and hypoxic–ischaemic encephalopathies, 216
coma and related states, **189**
    EEG recording and interpretation in coma and related states, **189**
    EEG features in comatose patients, **193**
    EPs in comatose patients, **199**
    EP monitoring, **208**
    EPs combined with other variables, 215
encephalitis, **218**
epileptiform discharges and status epilepticus, **218**
head injury and other neurosurgical applications, **221**
metabolic and toxic encephalopathies and multiple organ failure, **218**
vegetative states, **225**
Clinical neurophysiology monitoring service
    setting up and running, 3.4T
    checking performance of recording set-up, 65, **314**
    see also Safety
    good practice, 5, 7, 19, 190, 371, 373, 375
    medico-legal considerations and responsibilities, 1, 5, 6, 7, 13, 56, 58, 65, 3.2T, 93, 3.4T, 176, 180, 226, 228, 278, 8.6T, 364, **Chapter 9, 371**
    quality control and audit, 1, 49, 73, 81, 83, 3.2T, **93**, 102, 3.4T, 8.6T, 365
    safety issues, **55, 315, 371**
    training, 3, 4, 5, 6, 18, 47, 56, 59, 65, 73, 3.2T, 93, 102, 3.4T, 181, 220, 274, 8.1T, 360, 365, 371, 374
Coarction of aorta, surgical monitoring, 7.1T, 268, 277
Cochlear nerve, **126**, 127, 132, 133, 204, 209, **261**, 7.10F
Coherence analysis, 60, 81, 198
    bi-coherence 84, 3.14aF
Coherence function, coherence spectrum, 344
Collodion, 22, 2.3T, 23, 2.5F, 2.6F, 59
Colour density spectral array, colour modulated density spectral array (CDSA), 3.1T, 79, 3.13F, 5.13F, 309, 330, 340, 8.32F
Coma and related states, **189**
    alpha coma, 118, 193, 196, **198**
    brain protection regimens, 158, 160, 164, 255
    brainstem death (brain death), **228**
    cardiac arrest and hypoxic–ischaemic encephalopathies, 121, 4.14F, 4.15F, 4.16F, **216**, 6.2F, 6.5T, 6.24F
    clinical scoring systems and prognostication, 3, 191, 224
    encephalitis, **218**
    epileptiform discharges and status epilepticus, **218**
    head injury and other neurosurgical applications, **221**
    locked-in syndrome, **196**
    metabolic and toxic encephalopathies, **218**
    hepatic, 114, 4.5F, 4.13F, 121, 190, 194, 218
    renal, 194, 218
    sepsis-related, 218
    multiple organ failure, 218
    spindle coma, 199
    subarachnoid haemorrhage and vasospasm, 80, **225**, 253
    theta coma, 198
    toxic encephalopathies, **218**
    vegetative states, **225**
Coma, EEG features, **189, 193**
    arousal patterns in stuporous and comatose patients, 3.3T, 156, 6.1T, **191**, 6.2F
    procedure, 191
    nature of responses, 191
    paradoxical slow wave arousal responses, 191
    periodicity of arousal patterns in comatose patients, 192
    burst suppression pattern, 123
    EEG scoring systems and prognostication, 3, 78, 191, 196, 216, 6.5T, 6.24F, 224
    electrocerebral inactivity–electrocerebral silence (ECS), the isoelectric EEG, 124
    episodic low amplitude events (ELAEs), 115, 193
    triphasic waves (complexes) 120, 4.13F, 193, 218
Coma, EP features, **199**
    auditory evoked potentials, 204
    central somatosensory conduction time, 200
    clinical use of evoked potentials during coma, 200
    EP scoring systems and prognostication, 3, 191, 200, 6.2T, 6.3T, 6.4T, 208, 224
    multimodality EP assessment, 208
    EPs in comatose or pseudocomatose states caused by primary brainstem lesions, 208
    non-pathological causes of EP abnormalities in the ICU, 199
    somatosensory evoked potentials, 200
Coma, EP monitoring, **208**
    BAEP monitoring in the ICU, 209
    clinically relevant abnormal features, 209
    irreversible changes, 209
    transient BAEP changes, 212
    EPs combined with other variables, 215
    why evoked potential monitoring?, 208
Combined EEG and EP measures for depth of anaesthesia, 98, 176
Common average reference derivation, 43
    comparison with other methods, 2.13F
Common mode rejection, 66
Common-mode rejection ratio (CMRR), 33, 48, 54
Common peroneal nerve, 235
Common reference derivation, 2.13F, 31, 39, 311, 314
    in digital systems, 37–38, 42, 43, 314
Complexity, measures of, 345, 8.4 T
    see also Entropy
Compound action potential (CAP), 120, 4.20F, 127, 4.18F, 132, 134, 139, 142, 148
Compound muscle action potential (CMAP), 2.8T, 231–233, 6.34F, 235, 6.36F, 237, 6.37F, 6.6T, 242–243
Compound nerve action potentials, recording of, 2.6T, 2.7T
Compressed Spectral Array (CSA), 3.1T, **78**, 3.8F, 3.9F, 3.23F, 5.24F, 216, 6.22F, 225, 253, 282, 7.28F, 7.29F, 330, 8.18F, 340, 8.33F
Concentric needle electrode (CNE), 29, **53**, 2.28F, 145
Condensation vs rarefaction clicks, 50–51, 262–263
Conduction block
    AIDP form of GBS, 235
    EPs
        BAEPs, 261, 7.8F, 263
        SEPs (spinal cord), 176, 268
Conduction of action potentials, 127, 4.20F
    neuromuscular blockade, 271
    slowed, 231
    temperature effects, 147, 4.37F, 4.38F
Conduction studies, **53**
    equipment, 53
    stimulation, 55
    temperature, effects on, 147, 4.37F, 4.38F
    practical aspects in ICU and surgery, **68**
    see also individual nerves
Conduction velocity, 51, 141, 233
    temperature effects, 147, 4.37F, 176, 272
    see also Motor conduction velocity
Confidential Enquiry into Perioperative Deaths (CEPOD), 6
Consent to investigations, 373
Contingent negative variation (CNV), 48
Continuous wavelet transform (CWT), 340
Correlation analysis, **333**, 335, 337, 341, 8.4T
Cortical lesions
    hypoxic–ischaemic (arterial boundary zone, 'watershed'), 60, 2.31F, 4.7F, 279, 283, 291
Cortical localisation
    intraoperative SEPs for, 7.1T, **260**
Cortical stimulation
    transcutaneous stimulation of the motor cortex, 52, 271
    electrical stimulation, 52
    magnetic stimulation, 52
Covariance, 333–334, 337, 341
Creutzfeldt–Jakob disease, (CJD)
    EEG features, 121, 4.14F, 218
    infection control, 58
Critical damping, 41
Critical illness, neuromuscular syndromes of, 231
Cross-correlation function, 333, 335
Cup electrodes, 22, 2.3T
Cyclical alternating pattern (CAP), 113, 192

Index

D

Damping, critical, 41
Data acquisition, 42, 47, 48, 67, 3.2T, **311**, 8.2F, 8.2T
   checking the performance of the recording set-up, 314
      environment, 314
      expendables, 315
      occasional checks, 315
      recording system, 315
      routine daily checks, 314
   recording process, 311
      amplifiers, 311
      analogue filters, 312
      analogue-to-digital conversion, 312
      sensors and transducers, 311
      storage and archiving, 314
   safety issues, 315
Data formats, 48, 314
Data libraries, 3.3T, 6.1T, 347, 359, 360, 8.31F
Decibels (dB), 36, 50
Decision assistance, **358**
Decision boundaries, 350, 8.26F
Development and uptake of methods, **359**
Delta activity, 110
   see also EEG waveforms and interpretation
Demyelination, 141
   demyelinating neuropathies, 231, 234, 235
Dendrites, 4.19F
Denervation, 68, 146, 4.4T, 232–234, 237
Density spectral array, density-modulated spectral array (DSA), 79, 3.1T, 3.13F, 5.13F, 309, 330
Depth of anaesthesia, assessment methods, **Chapter 3**, 73
   see also Advanced Depth of Anaesthesia Monitor (ADAM)
   see also AEP Monitor/2
   see also Auditory evoked potential index (AAI)
   see also Bispectral Index (BIS) Monitor
   see also Burst suppression detection methods
   see also Cerebral Function Monitor (CFM)
   see also Cerebral Function Analysing Monitor (CFAM)
   see also Colour density spectral array, colour modulated density spectral array (CDSA)
   see also Combined EEG and EP measures for depth of anaesthesia
   see also Compressed spectral array (CSA)
   see also Density spectral array, density-modulated spectral array (DSA)
   see also Entropy EEG Monitor
   see also Narcotrend Monitor
   see also Patient State Index (PSI) Monitor
Derivation, effect on display of background activity, 111
   electrocerebral silence (isoelectric EEG), 124–125

qEEG monitoring, 288
   spatiotemporal patterns, 131
Derivation, methods of, **31**, 2.13F
   bipolar derivation, 31
   choice of method, 32, 43
   common average reference derivation, 31
   common reference derivation, 31
   digital EEG and reformatting signals, 42, 43
   source derivation, 31
Diathermy (surgical), 57, 58, **60**, 2.32F, 66
   detachment detection circuits, 62, 2.33F
   disconnection of patient during, 67
   heating effect on facial nerve, 267
   notch filters to reduce artefact, 90
   safety, 8.1T, 317, 318
Digital filters, 36, 47, 66–67, 90, 209, 263, 312, 318, 8.5F, 8.6F, 8.8F, 331
   digital filtering for noise reduction, 66, 96, 97, 263, 312, 8.10F
Digital recording systems, 37
Dipoles
   modelling in EEG, 43
   localisation of EP sources, 128–131, 4.21F, 4.22F, 141
Disc electrodes, 2.3F, 2.5F, 23–24, 58–59
Discrete Fourier Transform (DFT), **336**, 340, 344
   frequency resolution, 337
   spectral leakage, 338, 8.19F, 8.20F, 339
   window functions, 338, 8.19F
Discrete wavelet transform (DWT), 340
Display
   conventions, 30
   EEG
      analogue systems, 40–41
      digital systems, 38, 43, 2.27F, 47
   EEG and EP monitoring, 3.2T, 93
   EPs, 49, 66–67
   EMG and nerve conduction studies, 55
Display of results, 360
Distal motor latency, 4.37F, 6.35F, 6.36F
Dose–response curves, 3.10, 159, 5.7F, 5.10F
Dynamic range of recorders
   Cerebral Function Monitor (CFM), 78
   Cerebral Function Analysing Monitor (CFAM), 3.12F
   EEG, analogue systems, 41
   EEG, digital systems, 47
   requirements for EEG monitors, 70

E

Earth leakage current, 317, 372
Earth loops, 57, 318
EEG features in comatose patients, **193**
   alpha, theta and spindle coma, 198
   locked-in syndrome, 196
EEG recording and interpretation in coma and related states, **189**
   arousal patterns in stuporous and comatose patients, **191**
      procedure, 191
      nature of responses, 191
      paradoxical slow wave arousal responses, 191
      periodicity of arousal patterns, **192**

EEG recording equipment, **38**
   analogue EEG systems, **40**
      calibration of analogue EEG systems, 41
      EEG write-out systems, 40
      paper transport and time marking, 41
   digital EEG systems, **42**, 2.27F
      calibration of digital EEG systems, 46
      data storage for digital EEG, 42
      displaying digital EEG, 43
      general features of EEG recording workstations, 42
      inputs for digital EEG recording, 42
      limitations, problems and opportunities of digital EEG, 46
   general features of EEG machines, 38
      amplifiers for EEG recording, 39
      filters for EEG recording, 40
      input circuits for EEG recording, 39
EEG recording set-up
   performance checks, **314**
   safety, 55, 315, 8.1T, 317
EEG waveforms and interpretation, **109**
   describing EEG phenomena, 109
      amplitude, 110
      frequency of repetition, 110
      inherent variability of the EEG and other biological rhythms, 113
      reactivity, 114
      rhythmicity, 110
      spatial distribution, 113
      spatiotemporal patterns, 113
      symmetry and synchrony, 113
      transients, 112
      wave shape (morphology), 110
   general categories of EEG abnormality, 114
      altered reactivity, 118
      amplitude reduction, 115
      burst suppression pattern, 123
      change in frequency content, 114
      electrocerebral inactivity – electrocerebral silence (ECS) – the isoelectric EEG, 124
      epileptiform activity, 119
      excessive beta activity, 115
      localised and lateralised abnormalities, 116
      periodicity, 120
         periodic lateralised epileptiform discharges (PLEDs), 121
         triphasic complexes (or waves), 120
         other repetitive patterns, 121
      rhythms at a distance (projected rhythms), 117
      slowing, 114
Eighth (vestibulocochlear) cranial nerve, 261
   auditory, **126**, 205, **261**, 7.8F, 264, 266
   cochlear, **126**, 127, 132, 133, 204, 209, **261**, 7.10F
Electrical safety, **56**, 65, 66, **316**, 8.1T, 372
Electrocerebral inactivity (electrocerebral silence, ECS, isoelectric EEG), 68, 3.1T, 3.2T, **124**, 4.16F, 125
   anaesthetic and CNS depressant drugs, 125, 161, 5.7F, 163, 175

Index

brain death, 124, 230, 6.33F
cardiac arrest, 6.24F,
hypothermia, 176
vegetative states, 226
Electrocerebral silence (ECS), *see* Electrocerebral inactivity
Electrocochleogram, electrocochleography (ECochG), **126**, 4.18F, 205
   in surgical monitoring, 7.1T, 260–266, 7.8F, 7.9F
Electrodes, **18**
   desirable characteristics of electrodes, 21
   electrodes for EEG recording
      self-retaining electrodes, 22
      disc and cup electrodes, 22
      needle electrodes, 24
   electrodes for EMG and nerve conduction studies, **53**
      concentric needle electrodes, 53
      single fibre needle electrodes, 53
      surface electrodes, 53
   electrodes for EP recording, 48
   electrodes for use in MRI unit, 24
   equivalent circuit of the electrode–tissue junction, 19
   general features of electrodes, 18
Electrodes, preparation, application, and care
   application, 2.3T, 2.5F, 2.6F, 23–24, 59
   chloriding, 21
   collodion, application and removal, 23
   conductive, adhesive pastes, 24
   contact jelly, 23
   impedance, 19, 20, 2.3F, **21**, **34**, 59, 65, 83, 124, 327, 360
   removal of, 23
   safety, 22, **58**, 3.2T
   skin preparation, 23
   sterilisation, 58
Electrode placement systems, **24**
   general considerations in the design of placement systems, 24
   International 10–20 system of electrode placement, 26
      anatomical studies, 28
      designation of electrode positions, 27
      method of measurement, 26
      reduced placement for monitoring, 59
Electrode types, 2.2T, 2.3F, 2.4F, 2.3T
   chlorided silver, 19, 20, 2.5F
   for DC recording, 21
   for ECochG recording, 263, 7.9F
   for EMG and nerve conduction studies, 20, **53**, 2.2.8F
      concentric needle electrodes, 53
      single fibre EMG needle electrodes, 53
      surface electrodes, 53
   for EP recording, **48**
   for ICU and intraoperative EEG recording, 2.3T, 59
   MRI, for use in, 24
   needle (disposable), **24**, 2.5F
   pad, 23
   self-retaining, **22**
      cup, **22**, 2.5F
      disc, **22**, 2.5F
   silver/silver chloride, 19, 20, 2.5F
   sphenoidal, 21
   stick-on, 2.6F
Electroencephalogram, electroencephalograph, electroencephalography, *see* EEG
Electromagnetic interference, 24, 32–33, 48, 54, 57, 2,32F, 2.33F, 310, 8.2T
Electromyography, *see* EMG
Electro-oculogram (EOG), 34, 317
Electroretinogram (ERG), 49, **126**, 256
EMG and nerve conduction studies (NCS), **53**
   amplifiers for EMG and nerve conduction, 54
   displaying EMG and nerve conduction data, 55
   electrodes for EMG and nerve conduction studies, **53**
      concentric needle electrodes, 53
      single fibre needle electrodes, 53
      surface electrodes, 53
   stimulation for nerve conduction studies (NCS), 55
EMG findings and interpretation, **145**
   effects of limb temperature on nerve conduction, 147
   features of motor units recorded by needle electrodes, 145
   other normal EMG phenomena, 145
      insertion activity, 145
      end plate noise, 145
      fibrillations at single sites, 146
      fasciculations, 146
   nerve conduction findings and interpretation, **147**
EMG, single fibre recording (SFEMG), 55, 2.1T, 2.4T, 233, 242
   analysis of data, 55
   electrode, 53, 2.28F, 2.29F
   fibre density, 55
   jitter, 55, 233, 242
   turns/amplitude analysis, 55
Encephalitis, encephalitides, **218**
   'brainstem encephalitis', 237
   Creutzfeldt–Jakob disease (CJD), 58, 121, 218
   EEG features, 121, 190
   herpes simplex, 121, 5.4F, 190, 218
   herpes zoster, 3.11F, 6.5F
   postinfluenzal, 6.4F
   rubella, 6.27F
   SEP predictive value, 6.3T
   subacute sclerosing panencephalitis (SSPE), 121, 218
   viral encephalitis, 121
Encephalopathies, 76, 118, 121, 6.1T, 190, 229
   hepatic, 114, 4.5F, 4.13F, 121
   hypoxic–ischaemic, 121, 123, 4.15F, **216**, 6.22F
   multiple organ failure, 218
   other metabolic and toxic, 218
   renal, 218
   sepsis-related, 218
End-plate
   activity, 4.36F
   noise, 145, 4.4T
Endoneurium, 261
Entropy, 85, **345**
   approximate entropy (ApEn), 346
   response entropy (RE), 85, 3.17F, 3.26F, 101, 346
   spectral entropy, 309, 330, 344, 346, 347, 350, 354
   state entropy (SE), 85, 3.26F, 101, 346
Entropy EEG Monitor, 3.1T, **85**, 3.17F, 3,26F, 101, 346
Epilepsy
   cortical localisation of foci, 260
   EEG mapping of discharges, 2.27F, 46
   EEG patterns, **119**, 4.11F, 4.12F, 121, 4.14F, 170
   non-convulsive status epilepticus, 121, 4.14F, 190, 193
   risk with magnetic stimulation, 52
   status epilepticus, 2, 3.12cF, **218**
Epileptic seizures, and
   EEG monitoring, 2, 81, 3.2T, 3.3T, 6.1T
   training in recognition of EEG seizure patterns, 94, 113, 220
Epileptiform discharges, and
   anaesthetics, 163, 165, 166, 167, 5.14F, 169
   carotid endarterectomy, 287
   definition of, 119
   detection of short-term events, 79, 310, 331, 336, 8.24F, 8.25F
   non-convulsive status epilepticus, 121, 4.14F, 190, 193
   status epilepticus in ICU patients, **218**, 6.25F, 6.27F, 6.28F
Episodic low amplitude events (ELAEs), 115, 193
EPs, *see* Evoked potentials
Equivalent circuit of electrodes, 19, 20, 2.2F
Erb's point, 134, 142, 4.3T
Ergonomy, ergonomic problems, 46
Etomidate, effects on EEG and EPs, 3.10F, 159, 164–165, 172, 175
European Carotid Surgery Trialists' Collaborative Group (ECST), 286
European data format (EDF), 314
Event detection, 331
Event related potentials, 112, 178
   contingent negative variation (CNV), 48
   mismatch negativity, 101, 207
Evoked potentials in comatose patients, **199**
   auditory evoked potentials, 204
      absent AEPs, 205
      'late' normal BAEPs, 206
      middle and long latency AEPs, 207
      normal BAEPs during acute coma, 206
      present, but abnormal, AEPs, 205
      short latency AEPs, 204
   clinical use of evoked potentials, 200
   combination of several EP modalities, 208
   EPs in comatose or pseudocomatose states caused by primary brainstem lesions, 208
   non-pathological causes of EP abnormalities in the ICU, 199

Index

somatosensory evoked potentials, 200
  absence of cortical SEPs, 200
  abnormal central conduction time, 202
  central somatosensory conduction time, 200
  middle latency components, 200
  normal SEPs, 204
Evoked potential monitoring in comatose patients, **208**
  BAEP monitoring in the ICU, 209
  clinically relevant abnormal features, 209
    irreversible changes, 209
    transient BAEP changes, 212
  EPs combined with other variables, 215
  why evoked potential monitoring?, 208
Evoked potential recording systems, **48**
  amplifiers for EP recording, 48
  averaging, 48
  calibration of EP recorders, 49
  electrodes for EP recording, 48
  EP display, 49
  EP stimulators, 49
    auditory stimulation, 50
      click, 50
      tone, 51
    somatosensory stimulation, 51
      electrical, 51
    transcutaneous stimulation of the motor cortex, 52
      electrical stimulation, 52
      magnetic stimulation, 52
      safety considerations with magnetic stimulation, 52
    visual stimulation, 49
      flash, 49
      LED goggles, 49
  filters for EP recording, 48
Evoked potential waveforms and interpretation, **125**
  action potentials, 127
  electroretinogram and electrocochleogram, 126
  general definition – limitations – clinical utility of evoked potentials, 125
  how to localise EP sources from surface recordings, 130
  near-field versus far-field EPs, 128
  postsynaptic potentials, 128
  relation between neuronal responses and surface EPs, 127
  responses and EP components, 126
Evoked potentials, normal findings by modality, **Chapter 4, 131**
  see also Auditory evoked potentials, normal findings, **132**
  see also Flash VEPs, normal findings, **131**
  see also Somatosensory evoked potentials, normal findings, **134**
Excitatory and inhibitory postsynaptic potentials (EPSPs and IPSPs), 128
Expert systems, 81, 89, 350, **351**, 8.27F, 353, 8.5T, 354, 358
Explosion risks with flammable vapours, 58, 165

Eye movement and blink potentials, 2.4F, 48, 59, 2.30F, 90, 4.1F, 4.4F, 113, 4.16F, 5.14F, 196, 197, 227, 323

F

F waves, 52, 55, 236–237
  recording of, 2.9T
Facial nerve, 262, 267–268, 7.12F,
Facial palsy, complicating removal of cerebellopontine angle mass lesions, 262, 268
False negatives (FN), false positives (FP), 6, 8–12, 1.1T, 1.2T, 65, 6.3T, 209, 230–231, 257, 259, 265, 268, **273**, 289, 291, 355–356
Far-field potentials, **128**, 4.17F, 4.21F, 132, 133, 4.27F, 4.28F, 137, 4.29F, 140–142, 4.32F, 4.33F, 144–145, 4.34F, 6.7F, 230
  see also Near-field potentials
Fast Fourier transform (FFT), see Fourier analysis
Faults and fault finding
  computer, 8.1T
  daily checks, 314
  electrical (injurious), 56–57, 60, 62, 8.1T, 317–318
Features (in signal processing), 3, 81, 3.2T, 89, 92, 95, 98, 3.3T, 163, 169, 177, 179, 6.1T, 253, 280, 7.3T, 309, 346, 8.4T, 8.5T
  feature selection, 347, 359
  feature set, 347, 352
  feature space, 350, 8.26F
  feature vector, 350
Femoral nerve
  stimulation, 141, 275
Fibre characteristics
  muscle, 53–54, 145–146, 4.4T, 4.17F
  neural, 51, 128, 132–134, 137–139, 144, 176
Fibre density, on single fibre EMG recording, 55
Fibrillation
  in EMG recordings, 68, 146, 4.4T, 232, 233, 6.6T, 242
  cardiac, ventricular, 56, 58, 7.26F
Filters
  adaptive filters, 95, 97, **323**, 8.11F, 3.25F, 327, 8.2T
  analogue filters, **36**, 2.16F, 289–290, **312**, 318, 330, 337–338
  anti-aliasing filters, 42, 47, 312
  averaging as a filtering process, 325, 8.12F
  causal filters, 319, 322
  digital filters, 36, 47, 66–67, 90, 209, 263, 312, **318**, 8.5F, 8.6F, 8.8F, 331
  digital filtering for noise reduction, 66, 96, 97, 263, 312, 8.10F
  finite impulse response (FIR) filters, **319**, 8.9F, 8.12F
  filter coefficients, **318**, 8.5F, 8.6F, 8.7F, 8.11F, 331, 341, 8.22F
  high pass filters, 18, 34, 2.20F, **40**, 47, 48, 55, 312, 8.4F, 8.6F, 341–343, 8.22F, 8.23F

infinite impulse response (IIR) filters, **320**
linear time-invariant (LTI) filters, 319
  linearity of, 41, 319
low-pass filters, 35, 2.20F, 36, 40–42, 66, 313, 318, 8.4F, 8.6F, 8.2T, 8.22F, 342
matched filters, 323, 331, 344,
matched inverse filters, 289
median filters, 321–322, 8.9F
moving average filter, 319
non-causal filters, 319, 322
non-linear filters, 321
  median filters, 321–322, 8.9F
  optimal filters, 96, 322, 323, 326, 331
notch filters, 34, 40, 90, 256, 312, 324, 8.2T
optimal filters, 96, 322, 323, 326, 331
phase response of a filter, 319
phase shift of a filter, 37, 321
pre-whitening filters, 3.6F, 89, 90, 324
time-varying filters, 96, 326
Wiener filters, 96, 323, 326
Filters for EEG and EP monitoring, 3.3F, 3.5F, 3.6, 89, 3.14F, 3.16F, 95, 314
Flash, stroboscopic, 49
  EEG, 49, 374
  ERG, 126
  VEPs, 131, 4.24F, 4.25F, 131, 4.1T, 166, 256
Flash VEPs, normal findings, **131**
  latency and amplitude values of the normal flash VEP, 131
  normal waveforms of in adults, 131
Fourier analysis, 74, 3.1T, 78, 333, 340, 344
  compressed spectral array (CSA) 3.1T, **78**, 3.8F, 3.9F, 3.23F, 5.24F, 216, 6.22F, 225, 253, 282, 7.28F, 7.29F, 330, 8.18F, 340, 8.33F
  colour density spectral array (CDSA), 3.1T, 79, 3.13F, 5.13F, 309, 330, 340, 8.32F
  density spectral array, density-modulated spectral array (DSA), 3.1T, 79, 340
  Dietsch's manual calculation, 74, 3.2F
  spectral measures, **79**
Fourier, Jean Baptiste, 35
Fourier transform (FT), 329, 8.17F, 8.4T
  discrete Fourier transform (DFT), 336, 337,340
  fast Fourier transform (FFT), 3.8F, 3.16F, 309, 337, 8.18F, 340, 8.4T
  short-term Fourier transform (STFT), 330, 340
Fracture
  skull, 3.1F, 204
  spinal, surgical monitoring, 260, 264, 268, **275**, 7.18F, 7.19F
Frequency-domain analysis, 332, **336**, 344, 345, 8.4T
  use in clinical monitoring, 60, 76, **78**, 289, 322
Frequency range (signal bandwidth, filters), **35**, 2.4T, 2.19F, 313
  EEG recordings, 2.4T

Index

EP recordings, 2.4T
EMG, nerve conduction recordings, 2.4T
frequency response curve, 35, 313
Frequency resolution, 36, 337, 8.20F, 339–341, 8.23F, 8.4T
Friedreich's ataxia
   SEPs in, 275
Frontal intermittent rhythmic delta activity (FIRDA), 60, 117, 4.9F, 191, 253
Functional MRI (fMRI), EEG electrodes for use during, 24

G

G_1/G_2 convention EEG ('BUN'), 30, 41
Galvanometer, 40, 2.25dF, 3.1F
Gamma activity, rhythm, 110
   see also EEG waveforms and interpretation
Genetics, and
   neuromuscular disorders (channelopathies), 242, 6.6T
Giant SEPs, 175
Glasgow Coma Scale (GCS), 3, 160, 190. 191, 207, 224, 229–230
Glasgow Outcome Score, 190, 6.3T
Guillain–Barré syndrome (GBS), 234, 6.36F, 6.37F, 6.38F
   acute inflammatory demyelinating polyneuropathy (AIDP), 235
   acute motor axonal neuropathy (AMAN), 237
   acute motor sensory axonal neuropathy (AMSAN), 237
   Miller–Fisher syndrome (MFS), 237

H

H reflex, recording of, 2.10T
Headbox, 26, 39, 42, 62, 312, 315
Head injury, **221**
   brain protection regimes, 80, 5.2F, 160, 224, 6.21F
   EEG, 113, 118, 4.10F, 197, 199, 225–226
   EEG and EP monitoring, 6.21F, 6.29F, 6.30F, 8.33F
   electrode problems, 24
   evoked potentials, 201, 202, 6.2T, 6.3T, 204, 205, 207, 209, 219, 6.21F
   stimulus-induced arousal, 191
Hearing loss, and
   BAEPs, 133, 204, 6.9F, 7.8F
   hearing level (HL), 50
   hearing preservation in posterior fossa surgery, 260, 7.8F, **265**
   threshold level, 50
Heart rate, and
   anaesthetic depth, 166, 175
   artefact in recording, 181
   burst suppression EEG, 166
   comatose patients, 4.10F, 191, 6.2F, 227
   cyclical changes, 113
   polygraphic monitoring displays 3.1T, 3.13F, 216, 226, 7.30F, 8.33F
Heavy-metal-induced neuropathies, 232
Height, effect on SEPs, 139–142, **144**
Hemiplegia, SEPs in, 201

Hepatic encephalopathy, 114, 121, 4.5F, 4.13F, 121, 190
Herpes simplex encephalitis, 121, 5.4F, 190, 218
Herpes zoster encephalitis, 3.11F, 6.5F
Higher order spectra – the bispectrum, 332, **344**, 8.4T
Historical development of EEG monitors, 73, 81
   Berger and early workers 1920s to 1960, 73
   early time- and frequency-domain monitors, 76
      Cerebral Function Monitor (CFM), 76
      Fourier analysis with compressed spectral arrays (CSA), 78
      spectral measures, 79
         univariate spectral measures, 80
   more recent approaches to the development of
      Advanced Depth of Anaesthesia Monitor (ADAM), 83
      Bispectral Index Monitor, 84
      Cerebral Function Analysing Monitor (CFAM), 83
      Entropy Monitor, 85
      Narcotrend Monitor, 85
      Patient State Index (PSI), 85
HIV/AIDS, 58
   see also Acquired immune deficiency syndrome
Hjorth descriptors (normalised slope descriptors), 333, 8.4T
Homeostasis, cerebral, 209
Human immunodeficiency virus (HIV), 58
   see also AIDS
Hyperkalaemic periodic paralysis, 243, 6.6T
Hyperthermia, 172, 219, 6.6T
Hyperventilation
   activation of seizures, 374
   effect on EEG, 117
Hypokalaemic periodic paralysis, 6.6T
Hypotension, arterial, and
   comatose patients, 221, 6.29F
   controlled, 163, **170**, 5.15F, 253, 254
   during cardiopulmonary bypass, 283
   EPs and, 176, 6.13F, 214, 263, 265, 269, 7.18F, 278
   localised cerebral arterial boundary zone ischaemia, 2.31F, 116, 4.7F, 6.2F
Hypothermia, and
   BAEPs, 205, 214
   brain protection, 255
   EEG, 79, 115, 123, 125, 166, **171**, 229, 280, 282, 7.2T, 7.25F, 284–285
   SEPs, 145, 4.34F, 176, 229
   VEPs, 256
Hypoxia–ischaemia, cerebral, see Cerebral hypoxia–ischaemia

I

Improving signal quality, **317**
   artefact detection and rejection methods, 327
   averaging techniques: recording evoked potentials, 324
   averaging regarded as a filtering process, 325

filtering methods, 318
   adaptive filters, 323
   adaptive noise cancelling with reference input, 323
   analogue filters, 318
   digital filters, 318
   linear time-invariant filters, 319
   finite impulse response (FIR) filters, 319
   non-linear filters, 321
      median filters, 321
      optimal filters, 322
   filters, applications in EEG recordings during surgery and intensive care, 324
Inhibitory postsynaptic potential (IPSP), 128
Instrumentation, connection with patients and recording methods, **Chapter 2**, **17**
   see also Amplifiers, **32**
   see also Connecting electrodes to amplifiers and the recording convention, **29**
   see also Digital systems, **37**
   see also EEG recording equipment, **38**
   see also EMG and nerve conduction studies (NCS), **53**
   see also Electrodes for neurophysiological recording, **18**
   see also Electrode placement systems, **24**
   see also Evoked potential recording systems, **48**
   see also Practical aspects of intraoperative and ICU neurophysiological recording, **55**
   see also Recording the electrical activities of the nervous system, **18**
   see also Signal bandwidth and filters, **35**
Integration of features into the clinical context – pattern classification, **346**
   artificial intelligence (AI) methods, 351
   connectionist AI methods – artificial neural networks, 352
   from features to patterns, 347
   symbolic AI methods – expert systems, 351
   statistical methods, 350
   summary and examples of usage, 353
   syntactic methods, 350
Intensive care unit, applications of clinical neurophysiology in, 1
   advantages of co-ordinated EEG, EP and EMG assessment, 2
   common clinical questions, 1
   comparisons with coma scores and other outcome predictors, 3
   diagnostic versus monitoring applications, 2
Intensive care unit, neurophysiological work in, **Chapter 6**, **189**
   neurophysiological parameters to be monitored and procedures, 189
   sleep in the ICU, 189
   see also Clinical conditions affecting the central nervous system encountered in ICU, **189**

Index

see also Neuropathies, myopathies and related disorders encountered in ICU, **231**
Interburst intervals (IBI) in burst suppression pattern, 123, 160, 170
Interference
   electrical, 21, **32**, 39, 54, 57, 68, 8.2T
      electromagnetic, 32, 48, 54
      electrostatic, 32, 54
      EP stimulus rates to reduce, 68, 262
      fibrillators, defibrillators, balloon pumps, 62, 2.34F
      pacemakers, 62
      recording, 3.12F
      reduction, 75, 3.5F, 3.6F, 96, 263, 270, **317**
      surgical diathermy, 60, 2.32F, 2.33F
   physical
      surgical apparatus, 57
      reduction of, 88
International Electrotechnical Commission (IEC) standards, 49, 50, 53, 56–58, 62, 317–318, 8.1T, 372
International Federation for Medical and Biological Engineering (IFMBE), 309
International Federation for Clinical Neurophysiology (IFCN), standards, **7**
   brainstem death, 229
   comatose patients, 1, 125, 198, 347, 358
   digital EEG, 47
   electrocerebral inactivity (isoelectric) EEG, 124
   impaired cerebral perfusion, 288, 7.3T
   intensive care and intraoperative monitoring, 1, 3, 7
   intraoperative monitoring, 7, 65, 288
   terminology, 109, 120, 124
International Federation of Societies for Electroencephalography and Clinical Neurophysiology (IFSECN), report on methods
   electrode placement (10–20) system, **24**
International Organization for Standardization (ISO), 7, 371
International Organisation of Societies for Electrophysiological Technology (OSET), 7, 65
Intraoperative and ICU neurophysiological recording, practical aspects, **55**
   dangers of misinterpretation due to unrecognised artefacts in EEG monitors displaying only processed signals, 62
   electrodes for intraoperative EEG recording, 59
Intraoperative and ICU EEG recording, **59**
   interference due to surgical diathermy, 60
   interference due to defibrillators, fibrillators, pacemakers and balloon pumps, 62
Intraoperative and ICU EP recording, **65**
Intraoperative and ICU EMG and NCS, **68**
Intracranial pressure (ICP)
   brain protection regimens, 160
   brainstem death, 207, 209
   EEG and, 114, 6.30F
   EPs and, 6.11F, 6.12F, 211–215, 6.14F, 6.16F, 6.17F, 6.18F, 6.20F
   inherent variations, 113, 195
   in head injury, 115, 116, 160, 221, 8.33F
   in ICU, 67, 3.1T, 3.2T, 221
   in intracranial surgery, 253, 7.1F
   stimulus-induced changes, 190, 191
Intracranial surgery, **253**
   EEG monitoring in intracranial surgery, 253
      brain protection regimens in neurosurgery, 255
      historical background, 253
      intracranial aneurysms and vascular malformations, 253
      procedures, 253
      subarachnoid haemorrhage and vasospasm, 253
   Intraoperative SEPs for cortical localisation, 260
   posterior fossa surgery, **260**
      BAEPs, 263
      cochlear nucleus, 263
      contribution of monitoring to preservation of hearing, 265
      definition of a significant change in auditory monitoring, 264
      ECochG, 263
      experimental studies, 261
      factors affecting auditory responses during operations, 263
      interruption of cochlear blood supply, 261
      intracranial 8th nerve action potential, 263
      manipulation of the eighth cranial nerve, 261
      methods of monitoring, 262
      monitoring brainstem function, 266
      monitoring cranial nerve motor function, 266
      practical aspects of auditory monitoring, 262
      recording, 263
      stimulation, 262
   SEP monitoring in intracranial surgery, **257**
      experimental studies, 257
   SEP monitoring for ischaemia, 258
      aneurysm surgery, 258
   visual evoked potential monitoring, **256**
      methods of VEP monitoring, 256
      recording, 256
      practical aspects of VEP monitoring, 256
      stimulation, 256
Ischaemia, cerebral, *see* Cerebral hypoxia–ischaemia
Isoelectric EEG, *see* Electrocerebral inactivity
IT facilities, applications of, 363
   administration, 364
   analysis, 363
   display, 363
   networks, 364
   remote access to data, 364
   reporting, 364
   risk management, 364
   server, 364
   storage, 364
   telemedicine, 364

J

Jelly, electrode, 19, 20–23, 2.3T, 2.5F, 58, 3.7F
Jitter, on single fibre EMG recording, 55, 233, 242

K

K-complex, 4.4F, 113, 120, 155, 191, 6.2F, 195, 331, 336
Ketamine
   effects on BIS, 345
   effects on EEG, 164, 165
   effects on EPs, 159, 172, 175
Kyphoscoliosis, surgical monitoring, 268, **274**

L

Lateral lemniscus (LL) ad BAEPs, 133
Least mean squares (LMS) algorithm, 324
Legal implications of neurophysiological monitoring, **Chapter 9**, **371**
   responsibilities in respect of reports on investigations, 375
   responsibilities in training and supervising staff, 374
   safety of patients and staff, 371
      drugs, 373
      electromedical equipment, 372
      identification of patients and consent to procedures, 373
      infection control, 372
      inherent risk of procedures, 374
Light intensity, 49, 311
Light-emitting diodes (LED) for visual stimulation, 43, 49, 166, 256
Lilly waveform, 51
Limitations, of
   coma scoring systems, 3
   digital EEG, 46
   EEG monitoring in carotid endarterectomy, 286
   EEG frequency analysis, 336
   evidence-based medicine, 5
   evoked potentials, **125**, 142, 206, 6.10F, 209, 214
   expert systems, 351
   period analysis, 333
   recording channels in EEG monitors, 59, 78
   spectral representation, 79, 309
   statistical concepts of 'normality', 206
   zero-crossing analysis, 79, 333
Line thickness, digital EEG trace, 47
Linear time-invariant (LTI) filters, 319
Linearity, of
   amplifiers, 315
   data distribution and neural networks, 353
   EEG filters, 41, 319
Local area network (LAN), 47, 364
Locked-in syndrome, 2, 190, **196**, 6.6F, 208, 226

Index

Long latency auditory evoked potentials (LLAEPs), 199, 207–208
Lumbosacral plexus, 4.28F, 140
Luminance, 126

M

Magnetic resonance imaging (MRI)
    confirmation of localised ischaemic damage, 282
    electrodes for use in, 24
    electrode–cortex relationships, 26
Magnetic stimulation, **52**
    of motor cortex, 271
    safety considerations, 52
    seizure activation 52
Malaria, cerebral, 3.12F, 219, 6.25F
Malignant hyperpyrexia, 172
    see also Neuroleptic malignant syndrome
Mapping, topographic, of
    EEG features, 42, 43, 2.27F, 287, 315, 353, 362
    EPs, 130, 4.21F, 4.23F, 141, 160
Masking, with white noise, 262
Maturation and ageing
    EEG and anaesthesia, 170
    EPs 144, 127
Median nerve
    CCT, 144
    CMAPs, 2.6T, 2.8T
    NAPs, 7.23F
    SAPs, 2.7T, 4.38, 233, 235, 236, 6.35F, 6.37F
    SEPs, 4.21F, 130, 4.23F, 133, 134, 4.3T, 4.27, 137, 4.30F, 140, 142, 4.32F, 5.16F, 5.17F, 5.18F, 6.7F, 237, 258, 259, 260, 7.7F, 7.21F, 284, 7.31F, 291
Median nerve conduction studies in ICU, 235, 6.35F, 6.36F, 6.37F
Median power frequency (MPF) 159, 165, 340
Medico-legal issues
    awareness during anaesthesia, **176**, 180
        neurophysiological predictors, 177
    brainstem death (brain death), 228
    consent to procedures, 373
    equipment faults or failure, 372
    good practice, 5, 7, 3.2T, 3.4T, 8.6T, 371, 375
        quality control, 1, 81, 3.2T, 93, 8.6T
        performance assessment, 359
    intraoperative monitoring work, 5, **371**
        international guidelines to good practice, 7
        medico-legal organisation reports, 6
        national enquiries into perioperative mortality, 6
        nerve damage, intraoperative, 278
        objective studies of effectiveness, 6
        risk assessment and reduction, 372
        staffing implications, 6
    responsibilities in respect of reports on investigations, 375
    responsibilities in training and supervising staff, 374
    safety of patients and staff, 371
    drugs, 373

electromedical equipment, 372
    identification of patients and consent to procedures, 373
    infection control, 373
    inherent risk of procedures, 374
Metabolic and toxic encephalopathies and multiple organ failure, 218
Methicillin-resistant *Staphylococcus aureus* (MRSA), **58**
Methods for continuous EEG and evoked potential monitoring, **Chapter 3**, **73**
Methods for EEG monitoring, **73**
    historical development of EEG monitors, 73
    quality control and validation, 93
    technical requirements for EEG monitoring in ICU and during surgical operations, 85
Methods for EP monitoring, **94**
    technical requirements for EP monitoring, 94
Methods of derivation, see Derivation
Middle latency auditory evoked potentials (MLAEPs), analysis and classification, 98, 344, 347, 354
    comatose patients, 199, 6.4T, 207
    depth of sedation and anaesthesia, 101, 159, 172, 175, 5.17F, 176, 5.21F
    normal waveforms, **133**, 5.15F
    prognostication, 6.4T, 207, 209
    recording parameters and procedures, 94, **133**
Miller–Fisher syndrome (MFS), 235
Monitoring, assessment of results, **7**
    clinical significance of changes, 8
    comparing two methods of monitoring, 11
    confidence intervals, 10
    determining the optimal cut-off point, 11
    effect of corrective action, 12
    is the risk eliminated by monitoring?, 13
    statistical significance of changes, 7
Motor conduction velocity (MCV), temperature and, 147, 4.37F
Motor pathway monitoring during spinal surgery, 270, 7.15F
    transcutaneous stimulation of motor cortex, 52, 271
        electrical stimulation, 52
        magnetic stimulation, 52
        safety, 52
Motor neuropathy
    acute intermittent porphyria, 242
Motor unit, motor unit potentials
    analysis, 68
    features of, 68, **145**, 4.4T
        end-plate noise, 145
        fasciculations, 146
        fibrillations, 146
        insertion activity, 245
    in ICU patients, 68, **231**, 6.6T, 243
    neuromuscular junction structure, 146, 4.36F, 4.4T
    diseases of, 232–233
    recording, 53, 2.28F
        needle electrode, 2.1T, 53, 2.28F, 2.29F, **145**, 4.4T

single fibre (SFEMG) electrode, 54, 2.28F, 2.29F
MRSA (methicillin resistant *Staphylococcus aureus*), **58**
Multimodal evoked potentials, in coma, 2, 42, 198–199, **208**
Multiple sclerosis, EP findings, 125, 200, 6.8F, 205
Multiresolution decomposition, 342, 8.23F, 8.25F
    see also Wavelet analysis
Muscle fibre, see 'Fibre' entries
Muscle relaxant drugs
    adverse effect during EEG, 56
Musculocutaneous nerve, 2.7T, 7.23F
Myasthenia gravis, 231, **242**
Myelinated nerve fibres, 51, 134, 137, 148
    demyelination, 141
    demyelinating neuropathies, 231, 234, 235
Myopathy, 232–234, 6.6T
Myopathies and related disorders encountered in ICU, see Neuropathies, myopathies and related disorders encountered in the ICU
Myotonia congenita, 243, 6.6T
Myotonic discharges, 55
Myotonic dystrophy, 6.6T

N

Narcotrend Monitor, 3.1T, **85**, 3.16F, 101, 3.25F, 5.10F
National Confidential Enquiry into Perioperative Deaths (NCEPOD), 6
National Institute for Clinical Excellence (NICE), 371
Near-field potentials, **128**, 137
    see also Far-field potentials
Needle electrodes, 20, **24**, 2.3T, 2.3F, 2.5F, 2.6F, 34
    care of reusable electrodes, 24, **58**, 2.29F
    concentric for EMG, 20, 29–30, **53**, 2.28F
    dangers, **24**
        in bleeding disorders, 24, 59
        in SEP stimulation, 51
        transmission of viral infections, 24, 58, 59
    single fibre EMG, **53**, 2.28F, 2.29F
Nerve action potential (NAP), 30, 53, 2.6T, 126, 256, 261, 263, 7.10F, 271, 272, 278
Nerve conduction studies (NCS), **53**
Nerve conduction, clinical measurement of, 2.1T, **53**, 55, 68
    ICU patients, 236–237
Nerve conduction
    fibre diameter and conduction velocity, 51, 144
    maturation, 144
    temperature effects, 144, 147, 4.37F, 176, 272
    velocity, 144
Nerve fibre
    see Fibre
    see Nerve conduction

Index

Nerves and plexuses, *see* individual nerves and plexuses
   see Brachial plexus
   see Eighth (vestibulocochlear) cranial nerve
   see Facial nerve
   see Femoral nerve
   see Lumbosacral plexus
   see Median nerve
   see Musculocutaneous nerve
   see Peroneal nerve
   see Radial nerve
   see Saphenous nerve
   see Sciatic nerve
   see Sural nerve
   see Tibial nerve
   see Ulnar nerve
Neural networks, 159, 196, 331, 344, **352**, 8.28F, 8.5T
Neuroleptic malignant syndrome, 172
   see also Malignant hyperpyrexia
Neuropathology, of
   brain(stem) death, 229
   critical illness neuropathy, 233
   Guillain–Barré syndrome, 234
   head injury, 226
   hypoxic-ischaemic brain damage, 116, 4.7F, 199, 229, 279, 282
   neocortical damage, 25
   persistent vegetative states, 226, 228
Neuropathies, myopathies and related disorders encountered in ICU, **231**
   acute intermittent porphyria, 237
   acute onset neuropathies, 234
   acute weakness due to disorders of neuromuscular transmission, 242
   botulism, 242
   familial periodic paralysis and the channelopathies, 242
   Guillain–Barré syndrome (GBS), 234
   myasthenia gravis, 242
   neuromuscular syndromes of critical illness (critical illness neuropathy), 231
   non-peripheral causes of acute onset generalised weakness, 243
   other causes of acute or subacute peripheral neuropathy, 242
Neurophysiological monitoring
   assessment of results, **Chapter 1**, **7**
      clinical significance of changes, 8
      comparing two methods of monitoring, 11
      confidence intervals, 10
      determining the optimal cut-off point, 11
      effect of corrective action, 12
      is the risk eliminated by monitoring?, 13
      statistical significance of changes, 7
Neurophysiological monitoring during surgical operations, **Chapter 7**, **253**
   cerebral ischaemia during non-intracranial surgery, 279
   intracranial surgery, 253
   spinal cord function monitoring, 268
   spinal root surgery and peripheral nerve surgery, 278
Neurophysiological work in the intensive care unit, **Chapter 6**, **189**
   clinical conditions affecting the central nervous system encountered in ICU, 189
   clinical conditions affecting the peripheral nervous system and muscles encountered in ICU, 231
Noise
   see also Signal-to-noise ratio
Noise, electrical, 34, 2.17F, 39, 47, 49, 272, 310, 8.1F, 315, 317, 322, 324, 331
Noise reduction
   adaptive noise cancellation, 96, 323, 8.11F
   digital filtering, 66, 96, 97, 263, 312, 8.10F
Noise (sound), in ICU and surgery, 11, 50, 52, 113, 114, 155, 167, 170, 178, 180, 189, 6.2F, 195, 197, 358, 372
Noise, thermal, 34, 35
Noise, white or masking, 51, 262
Non-linear energy operator (NLEO), 330
Normal and pathological phenomena in EEG, evoked potential, EMG and nerve conduction studies, **Chapter 4**, **109**
   see also EEG waveforms and interpretation, **109**
   see also EMG findings and interpretation, **145**
   see also Evoked potential waveforms and interpretation, **125**
Normal hearing level (nHL), 50
Normalised slope descriptors (Hjorth descriptors), 333, 8.4T
North American Symptomatic Carotid Endarterectomy Trial Collaborators (NASCET), 286
Notch filters, 34, 40, 90, 256, 312, 324, 8.2T
Nyquist frequency, 38, 48, 312, 313

O

Observer's Assessment of Alertness/Sedation Scale (OAA/S,), 332, 8.3T
Occipital intermittent rhythmic delta activity (OIRDA), 118
Optimality criteria/criterion, 322
OSET, International Organisation of Societies for Electrophysiological Technology, 7, 65
Overshoot, 59

P

Pad electrodes, 21, 23, 26, 58
Paper (chart) speed, 41, 73
Paramyotonia congenita, 6.6T, 243
Paroxysm, paroxysmal activity, 120
Patient State Index (PSI) Monitor, 3.1T, **85**, 3.15F, 160, 5.5F
Peak power frequency (PPF), 338
Peak recognition in EP monitoring, 67
Pen
   alignment, 41
   arc distortion, 2.26F
   damping effects, 41, 2.25F
   deflection, direction of, 30
   limitations of mechanical response, 36
   linearity, 41
   motors (galvanometers), 40
   sensitivity, 41
   stiction (friction), 41
Performance assessment (methods), **359**
Performance checks (EEG recording set-up), 314
Periodic paralysis, **242**, 6.6T
Periodic lateralised epileptiform discharges (PLEDs), **121**, 191, 193, 4.12F, 287
Periodicity of EEG features
   arousal patterns in coma, 92, 3.3T, 103, 109, 6.1T, **192**, 6.5F, 327
   basic rest and activity cycles (BRACs), 113
   cyclical alternating pattern (CAP), 113
   periodic complexes, **120**
      generalised periodic complexes, 121, 4.12F, 4.14F, 196
      periodic lateralised epileptiform complexes (PLEDs), 121, 4.12F
      triphasic complexes or waves, 120, 4.13F
Periodogram, 339, 8.4T
Peroneal nerve
   common, 235, 2.6T
   deep, 2.8T
Persistent vegetative state, permanent vegetative states (PVS), **225**
   clinical picture, 226
   EEG patterns and recording techniques, 197, 199, 226–228, 6.31F, 6,32F
   EP assessment, 200, 6.2T, 206, 228
   legal and ethical aspects, 225–226, 228
   neuropathology, 226, 228
Phenytoin
   effects on EPs, 159
   causing acute intermittent porphyria, 242
Pituitary tumours, VEPs in, 256
Polymorphic delta activity (PDA), 116, 5.4F
Postsynaptic potential (PSP), **128**, 263
Power spectrum
   EEG, 2.27F, 336, 8.4T
      estimates using parametric methods (AR models), 339
      estimates using non-parametric methods (DFT/FFT), 336, 8.17F
      features extracted from, 340
      filtering, 96, 323
   SFEMG, 55
Power spectral density (PSD), 7.25F, 334, 336, 8.20F, 339, 340, 344
Pre-whitening filters, 3.6F, 89, 90, 324
Proximal weakness, 6.6T
Pseudo-comatose states
   EEG in, 190
   EPs in, 208
Pulse artefact, 8.2T

Index

Q

Quality control, 1
   EP recording, 49
   EEG monitoring, 81, 83, 3.12F, 3.13F, 3.2T, **93**, 102, 5.13F
   neurophysiological monitoring services, 3.4T, 8.6T
   neurophysiological signals, 317
Quality of patient outcome, 56, 94, 359
Quantitative EEG (qEEG; QEEG), and,
   coma depth assessment, 196
   control of anaesthetic depth, 163
   frequency bands, 110
   induced hypotension, 171, 5.15F
   trend analysis, 93, 6.28F, 253, 354
   warning of impending cerebral ischaemia, 225, 253, 287
Quantization of signals, 313

R

Radial nerve, 2.8T
   injuries to, 278
   superficial radial, 2.7T
Radiofrequency (RF) interference, 32, 60, 61, 66, 2.32F, 8.1T
Ramsay scale, 160, 5.5F, 332, 359
Receiver operating characteristic (ROC) (curve) 11, 1.2F, 7.27F
Recommended international non-proprietary name (of drugs) (rINN), 163
Recording the electrical activities of the nervous system, **18**
   amplifiers, 32
   connecting electrodes to amplifiers and the recording convention, 29
   digital systems, 37
   electrode placement systems, 24
      general considerations in the design of placement systems, 24
      International 10–20 system of electrode placement, 26
         anatomical studies, 28
         designation of electrode positions, 27
         method of measurement, 26
   electrodes for neurophysiological recording, 18
      general features of electrodes, 18
      equivalent circuit of the electrode tissue junction, 19
      desirable characteristics of electrodes, 21
      self-retaining electrodes, 22
      disc and cup electrodes, 22
      needle electrodes, 24
      electrodes for use in MRI unit, 24
   signal bandwidth and filters, 35
Regional cerebral blood flow (rCBF) and EEG, 285, 287
Reliability, inter-rater, EEG, 94
Renal failure, 4.9F, 194, **218**, 232, 6.6T, 283
Requirements, for
   digital EEG equipment, 47
   drug prescription and storage, 373
   EEG monitoring in ICU and surgery, 85, 3.2T, 8.6T
   electrical safety, 56
   electrocerebral silence, recording of, 124
   electromedical equipment safety, 56, 317
   EP monitoring in ICU and surgery, 66, 94
   infection control, 58
   monitor displays, 93, 356
   patient identification and consent, 373
   quality control, 3.2T
   safety of EEG equipment, 56, 317
   storage of data, 364
Responsibilities in respect of reports on investigations, 375
Responsibilities in training and supervising staff, 374
Rhabdomyolysis, 6.6T
Rhizotomy, dorsal, surgical monitoring, 278, 7.1T
Richmond Agitation–Sedation Scale (RASS) 160
Root mean square (RMS) amplitude measure, 6.5F, 309, 331, 332, 333, 347, 8.4T, 350, 351, 352

S

Safety during neurophysiological recording in ICU and operating theatres, **55**, **315**, **371**
   electrical safety, 56
      diathermy and diathermy safety circuits, 60
      defibrillators, fibrillators, pacemakers and balloon pumps, 62
      earth leakage currents, 57
      earth loops, 57
      faults in equipment, 57
   electrode care and sterilisation, 59
   electromedical equipment, 372
   explosion risks, 58
   flammable vapours, static electricity, sparks and high temperatures, 58
   general care of the patient, 56
   infection control, 58, 373
   input isolation from power supply, 57
   magnetic stimulation, 52
   misinterpretation of processed EEG contaminated by artefact, 62
   pacemakers, 57
   particular physical hazards and safety considerations, 58
Safety of patients and staff, **371**
   drugs, 373
   electromedical equipment, 55, 315, 372
   identification of patients and consent to procedures, 373
   infection control, 58, 373
   inherent risk of procedures, 374
Sampling
   aliasing, 38, 2.23F, 42, 312, 8.3F
   digital EEG systems, 47
   EEG monitors, 3.2T, 340
   EPs, 137
   number of muscle sites in EMG, 4.4T
   Nyquist frequency, 38, 48, 312
   oversampling, 38
   sampling frequency, 36, 38, 312, 325, 8.12F, 337
   sampling theorem (Shannon theorem), 312
Sampling rate, 38, 48, 137, 310, 312, 319, 8.2T, 337, 338, 8.20F, 342
Saphenous nerve, 2.7T
Sciatic nerve, injuries to, 278
Schwann cell, 234
Sedation, assessment with EEG and evoked potentials, **155**
   assessment of sedation, 159
   barbiturates, 158
   effects of sedative drugs on the EEG, 155
   effects of sedative drugs on EPs, 157
   other intravenous CNS depressants, 159
Sedation–Agitation Scale (SAS), 160
Self-organising map (SOM), 353, 354
Segmentation, 289, **329**, 330, 331, 336, 8.14F
   adaptive, 92, 289, 290, 330, 331, 340
   fixed duration, 330
Sensation level (SL), auditory, 50
Sensitivity
   amplifiers, 39
   displays, 314
   EEG machines, 2.22F, 38, 40–49, 2.25F, 125, 314, 315
   EMG machines, 145
   filters, 36
Sensitivity and specificity of investigations, 4, 8, 55
   anaesthesia and sedation monitoring, 100, 181
   assessment of results, **7**, 1.1T, 1.2T, 1.2F
   cardiac arrest, 230
   carotid surgery, 286, 287, 290, 291
   comatose patients, 201, 6.3T, 204,
   EEG monitoring, 3.2T, 92, 93, 3.3T, 3.20F
   neurophysiological monitoring services, 3.3T, 3.4T, 6.1T
   performance of methods, 353, 359
   seizure discharge detection, 220, 354
   spinal monitoring, 272, 274
   statistical control methods, 354
   surgical monitoring, 8–11, 1.1T, 6.1T, 260, 269, 277
Sensors and transducers, **311**
Sensory action potential (SAP)
   abnormalities, 68, 231–236, 237
   recording with surface electrodes, 2.7T, 68
   temperature effects, 147, 4.38F
Servo-anaesthesia, 75, 3.4F, 169–170
Shannon theorem, 312
Shannon's channel entropy, 85, 346
Short-term Fourier transform (STFT), 330, 340, 8.4T
Sigma activity (sleep spindles), 113, 4.4F
Signal analysis, **Chapter 8**, **309**
   artefact detection and rejection methods, 327
   data acquisition, 311
   improving signal quality, 317

Index

integration of features into the clinical context – pattern classification, 346
safety issues, 315
signal processing and interpretation, 327
    statistical process control, 354
technology, 311
use of the tools, 358
Signal bandwidth and filters, 35
Signal to noise ratio (SNR), and averaging, 325
    EPs, 66, 271, 325
    filtering, 322, 323
Signal processing and interpretation, **327**
    common processing tasks, 329
        segmentation, 329
        detection of events, complexes and trends, 331
        calculation of patient state indices, 331
Signal processing methods, **332**
    alternative methods, 344
        higher-order spectra, 344
            Bispectral Index (BIS), 345
        measures of complexity – entropy, 345
    frequency-domain methods, 336
        commonly used features extracted from the power spectrum, 340
        power spectrum estimates using the discrete Fourier Transform, 336
        power spectrum estimates using AR models, 339
    time-domain methods, 332
        amplitude measures, 332
        AR modelling, 335
        correlation analysis, 333
        normalised slope descriptors, 333
        phase and time delay analysis, 335
        zero crossing computations, 333
    time–frequency methods
        short-term Fourier transform (STFT), 340
        wavelet analysis, 340
Single fibre EMG (SFEMG), 55, 2.1T, 2.4T, 233, 242
    analysis of data, 55
    electrode, 53, 2.28F, 2.29F
    fibre density, 55
    jitter, 55, 233, 242
    turns/amplitude, 55
Sleep, and
    arousal patterns, 117, 118, 155, 4.10F, 191
    BAEPs, 209
    BRAC cycles, 192
    comatose patients, 92, 192, 6.4F, 195–196, 6.5F
    cycles, 3, 80, 81, 3.2T, 94, 109, 195, 198, 330, 353, 358
    cyclical alternating pattern (CAP), 113, 192
    deprivation, 155, 189, 6.1F
    ICU patients, 3, 155, 189, 6.1F, 191
    head injury, 3.9F

K-complex, 4.4F, 113, 155, 191, 331
locked-in syndrome 196–197
monitoring system capabilities, 3.2T, 109, 327, 8.6T
neonates, 3.12dF
neuroleptic anaesthesia, 169
persistent vegetative states, 225–228
spindle coma, 198–199
spindles, 112, 4.4F, 113, 331
vertex sharp waves, 4.4F
Sleep–wake cycles, 3, 81, 6.4F, 195, 198
Sodium channels, 6.6T, 243
Somatosensory evoked potentials (SEP)
    anaesthetics and sedatives, 159, 172, 5.16F, 5.17F, 5.18F, 5.19F
    central conduction time, 141, 200
    CNS depressants, 159
    comatose patients, 125, 200, 6.2T, 6.3T
    cortical generators, 4.22F
    maturation, 144
    normal findings, 134, 4.3T, 5.16F
        upper limb stimulation, 4.21F, 134, 4.27F, 6.7F
        lower limb stimulation, 139, 4.31F
    monitoring during surgery, 7.1T
        cardiac, 284
        carotid endarterectomy, 291, 7.31F
        cortical localisation, 260
        intracranial, 257
        ischaemia, 258, 279
        spinal cord function, 268, 7.14F, 272
        spinal root and peripheral nerve, 278
    non-pathological causes of variation
        body height and arm length, 144
        skin and core temperature, 144, 4.34F
    stimulation, 51
Somatosensory evoked potentials, normal findings, **134**
    central conduction time in the dorsal column system, **141**
    lower limb CCT, **142**
    upper limb CCT, **141**
    fibre tracts involved in the genesis of SEPs, **134**
        body height and arm length in adults, 144
        maturation, 144
        non-pathological sources of SEP variation, **144**
        skin and core temperature, 144
    short latency SEPs to electrical stimulation of the upper limb, 134
        early cortical components, 138
        peripheral components, 134
        spinal components, 136
    short latency SEPs to lower limb stimulation, **139**
        cortical potentials, 140
        far-field scalp positivities, 140
        peripheral components, 139
        spinal potentials, 139
Sound pressure level (SPL), 50
Specificity of investigations, *see* Sensitivity and specificity of investigations
Spectral measures, in EEG monitors, **79**, 8.4T

spectral edge frequency (SEF), 3.1T, **80**, 3.13F, 100, 159, 5.13F, 6.5F, 6.22F, 340, 8.32F
    SEF90, 3.11F, 100, 101, 179, 5.21F, 340
    SEF95, 340, 347, 350, 352
    univariate spectral measures, 80
Sphenoidal electrode
    impedances 20, 2.3F, 21
Spinal cord function monitoring, **268**
    criteria for identification of a significant change, 273
    experimental studies and mechanisms of spinal cord damage, 268
    false positives, false negatives and the predictive value of spinal monitoring, 273
    methods of monitoring cord function, 269
        sensory monitoring, 270
        motor monitoring, 270
        comparison of methods, 271
    practical aspects of spinal cord monitoring, 272
    recording, 272
    stimulation, 272
Spinal cord monitoring, clinical applications, **274**
    aortic surgery, 277
    cervical spine surgery, 275
    spinal fractures, 275
    spinal tumours and arteriovenous malformations, 276
    surgery for spinal deformity (kyphosis and scoliosis), 274
Spinal root surgery and peripheral nerve surgery, **278**
Spinocerebellar tract, 272
Spinothalamic tract, 134
Standard deviation
    EEG measures 112, 329, 333
    EPs, 206, 258, 333
    statistical process control, 8, 355, 356, 8.30F, 857
State entropy (SE), 85, 3.26F, 101, 346
Stationarity, 60, 79, 329, 330, 336, 8.4T
Statistical methods (for pattern classification), 350, 8.5T, 358
Statistical process control, **354**
    Cusum chart, 357
    exponentially weighted moving average (EWMA) chart, 357
    Shewhart charts, 356
Steady potentials, 18, 19, 22, 34, 37
Steady-state EPs
    AEPs, 175
    VEPs, 38
Sterilisation of electrodes, 58, 373
Stick-on electrodes, 26, 2.6F
Stiction of pen recorder, 41
Stroboscopic light stimulation, **49**, 256, 374
Storage, of
    digital EEG data, **42**, 47, 314
    EP data, 67
    neurophysiological and patient data, **364**, 375
Subacute sclerosing panencephalitis (SSPE), 120, 218

389

Index

Subarachnoid haemorrhage (SAH), 80, 225, 226, **253**
Summating potential
   BAEP, 261, 263
   ECochG, 127, 4.18F, 261, 263
Superimposition display, in
   EEG–polygraphic monitoring, 360, 8.32F
   F wave studies, 2.9T
Supplementary motor area (SMA), 139
Sural nerve, 2.7T
   EMG, 233, 235, 236
   SEP, 134, 139, 141
Surgical operations and neurophysiological monitoring, **3**
   assessment of results, 7
   cost-effectiveness and staffing implications, 5
   medico-legal issues in intra-operative monitoring, 5
   specific clinical reasons for intraoperative neurophysiological monitoring, 4
Synapses
   effect on EPs 125, 128, 4.17F, 176
Synaptic cleft, effect on postsynaptic potentials, 128
Syntactic methods, **350**, 8.5T

T

Technical requirements for EEG monitoring in ICU and during surgical operations, **85**
   artefact identification and handling, 90
   feature extraction and basic signal processing and interpretation, 92
   pre-processing, 89
   requirements for monitor displays, 93
   quality control and validation, 93
Technical requirements for EP monitoring, **94**
   automation, 94
   feature extraction and display, 98
   filtering, 95
   comparisons between currently available methods, 100
Telemedicine, 364
Temperature, and
   amplifiers – effect on bandwidth, 35
   brain protection, 284, 285
   EEG, hypothermia and hyperthermia, 3.21F, **171**, 221, 7.25F
   EPs, hypothermia and hyperthermia, **176**, 199, 215
   BAEPs, 176, 212, 214, 263
   SEPs, 144, 145, 4.34F, 176, 258, 285
   nerve conduction, 68, 144, **147**, 4.37F, 4.38F
   safety in high temperatures, 57, 58, 317, 372
Ten–twenty (10–20) system, **26**, 2.9F, 2.10F, 2.11F, 109
Theta activity, rhythm, 110
   see also EEG waveforms and interpretation
Thermal noise, 34, 35
Tibial nerve, and
   CAP, 148
   CCT, 142

CMAPs, 2.8T, 6.36, 6.37
H reflex, 2.10T
SEPs 4.3T, 139–141, 4.31F, 144, 148, 175, 176, 260, 272, 275, 7.19F, 7.21F
spinal EPs, 275, 7.16F
Time constant (TC), 18–21, 2.16F, 2.20F, 2.21F, 2.22F, 2.24F, 34–41, 53, 2.5T, 112, 311
Time-domain analysis, 66, 326, **332**, 345, 347, 8.4T
   use in clinical monitoring, 76, 79, 96, 218
Time (phase) delay analysis, 335
Time marker, 41, 43
Timing errors due to arc distortion, 41, 2.26F
Tönnies, Jan Friedrich, 32
Transducers, 39, 262, **311**, 315
Transient ischaemic attack (carotid artery surgery for prevention), 102
Traumatic plexus lesions
   brachial: 278, 7.1T
Trend detection, 65, 289, 331
Trigeminal neuralgia, 7.11F
Triphasic waves (complexes) 121, **120**, 4.13F, 156, 163, 190, 193, 218
True negatives (TN), true positives (TP), 6, 8–12, 1.1T, 1.2T, 273
Turnover frequency, 37, 2.5T, 40
Turns/amplitude analysis of EMG, 55
Tympanography, 204

U

Ulnar nerve, 2.6T, 2.7T, 2.8T
   CCT, 142
   CMAPs, 2.6T, 2.8T
   damage during surgery, 278
   dorsal ulnar, 2.7T
   sensory action potentials, 2.7T
   SEPs, 134, 137, 7.19F
   train of four stimulation, 175, 267
Unmyelinated nerve fibres, 51
Usefulness of monitoring, 4
   see also Sensitivity and specificity of investigations

V

Variability of biological rhythms, **113**
Variability, EEG, **113**
   anaesthetic and sedative drug effects on, 166, 167, 170, 181, 5.1F, 5.11F, 224
   arousal in comatose patients, 101, 6.4F, 195, 6.31F
   burst suppression activity, 77, 79
   prognostic implications, 6.4F, 195, 196, 224, 6.31F, 253
   signal processing considerations, 3.1T, 3.2T, 79, 80, 5.1F, 5.11F, 170, 195, 218, 225, 253, 336, 339, 8.6T
Variability of electrode placement to anatomy, 26, 28
Variability of EPs, 200
   in anaesthetised and comatose patients, 200, 256, 258, 270, 271, 273, 275,

inter- and intra-individual 48, 125, 131, 132, 133, 137, 141, 144
in EP monitoring, 73, 176
test–retest, 48, 49, 4.25F
Variability of heart rate, 8.33F
Variability, statistical, 7, 8, 331, 354, 356, 358
Variance, in EEG and EP measurements, 66, 98, 3.26F, 112, 114, 123, 324, 325, 329, 330, 332, 333, 335, 345, 362
Vegetative states, 225
VEPs to flash stimulation, normal findings, 131
   latency and amplitude values of the normal flash VEP, 131
   normal waveforms of in adults, 131
Viral infections, control of, **58**, 373
Video display unit; visual display unit (VDU), 36, 38, 39, 40, 42, 43, 46, 49, 58, 110, 112, 314
Visual evoked potential (VEP), and
   anaesthesia, 167, 172, 175–176
   arterial hypotension, 176
   bandwidth, 2.4T
   brainstem death, 230
   clinical use, 125, 128
   in multimodality EP monitoring in ICU, 199
   in surgical monitoring, **256**, 7.1T, 7.3F
   normal findings, **131**, 4.24F, 4.25F, 4.1T
   stimulation, 49, 256
   flash, 49
   LED goggles, 49
Vulnerability, cerebral, 192, 209, 282

W

Walter, W. Grey, 30, 161, 253, 7.1F
Wide area network (WAN), 364
Wavelet analysis, 97, 176, 327, 339, **340**, 8.4T
   continuous wavelet transform (CWT), 340
   discrete wavelet transform (DWT), 340, 344
   frequency resolution, 340, 341
   multiresolution decomposition, 342, 8.23F, 8.25F
   scaling function, 341, 342, 8.22F
   wavelets, 340,.21F
   wavelet function, 341, 342
   wavelet transform, 97, 176, 327
Welch's method, 339
White noise, 331
Wiener filtering of evoked potentials, 96, 97, 323, 324, 326
WinDisp (display system), 8.31F
Window functions 338, 8.19F

Z

Zero-crossing computations, 79, 3.1T, 3.21F, 177, 333, 8.4T
Zeta waves, 116, 287